Plasma, Electron and Laser Beam Technology

Development and Use in Materials Processing

Yoshiaki Arata
Director General of the
   Welding Research Institute
Director of Research Center for
   Ultra High Energy Density Heat Sources
Prof. Dr. Osaka University

American Society for Metals
Metals Park, Ohio 44073

Production coordination by
Carnes Publication Services, Inc.

Library of Congress Catalog Card Number: 85-73740
ISBN: 0-87170-254-1
SAN: 204-7586

PRINTED IN THE UNITED STATES OF AMERICA

Preface

Modern industry depends heavily on heat sources for heat processing techniques. These heat sources include electromagnetic energy sources (for example, arc heat, joule heat and sunlight), mechanical energy sources (for example, the heat of friction), and chemical energy sources (for example, the heat of combustion). Among these, the arc heat source has been widely used for heat processing since its first application to welding in 1885 by Benardos and Olszewski. Indeed, it should be remarked that arc heat sources have played a fundamental role in the development of welding science and technology.

Conventional arc heat sources, however, have not kept pace with recent rapid advances in other technologies. Presently, advanced technical fields require heat sources with higher energy densities and with superior controllability, precision and quality, yet with lower power consumption, than can be obtained with arc heat sources. It has thus become imperative to develop new heat sources and related technologies for application in heat processing.

After finishing my doctoral thesis on the heat treatment and phase transformation of various steels, I entered a new field of study on the development and application of high energy density heat sources. For the past thirty years, I have researched plasma beams, charged particle beams (electron and ion), laser beams and neutral particle energy beams. These beams are characterized by an ultra high energy density (from 10 to 1000 times as dense as arc heat sources) and by high output. They can be precisely controlled over a wide range of conditions. I have devoted myself to the development of various types of new heat sources and to the study of their characteristics and the various features of thermal processing phenomena. I have systematized these results to develop a new academic field — Heat Processing Engineering — and have tried to apply this knowledge to actual materials processing.

Today, these new heat sources play a central role in the rapid development in technology. New heat sources are expected to propel industry into a newly competitive economic environment, even if the initial investment seems prohibitive. In fact, they will undoubtedly be the processing energy sources in the development and manufacture of very large scale integration (VLSI) devices and new materials in the surface modification, melting, refining and heat processing of composite materials; and in high precision welding of nuclear fusion reactors, vessels for deep sea exploration (down to 6000 m), space vehicles, airplanes and rockets.

My study of ultra high energy density heat sources began with Prof. M. Okada in 1956 with research on nuclear fusion reactions in which an ultra high temperature state was produced using extremely high current pulsed discharge "pinch plasma." It was the first experimental study of nuclear fusion in Japan. To obtain an ultra high temperature plasma other than the pinch plasma, I proposed a new method to obtain a large output pulsed electron beam: the Transtron accelerator. Research on the Transtron accelerator later became the basis for the development of the world's first high output electron gun for heat source in 1972, which subsequently led to the full-scale practical ultra high power EB heat source. At approximately the same time that we were studying the development and application of the high output electron beam, Dr. Maiman succeeded in the oscillation of the ruby laser (1960) and Dr. Patel in the oscillation of the CO_2 laser (1964). I was greatly impressed by these results, and focused my attention on the development of a high output CO_2 laser heat source and the "Arata Laser Focus System." This research opened the way for the new heat processing by high power CO_2 laser.

On the basis of these results, I was the first to propose and identify the general source and processing characteristics of ultra high energy density beam heat sources, indicating the common natures of powerful energy beams such as electron beams, laser beams, plasma beams, ion beams and so on. Recently, I safely rounded a first orbit of 60 years in my life, named "Kanreki" in the Oriental thought, and I was quite moved that I could establish the "first stage" of my life work in this period. My works have been recognized by other researchers around the world and rewarded with a number of international prizes. In June 1985, I was awarded the Japan Academy Prize, the highest award given to a scientist in Japan. I was deeply honored by this recognition of my contribution to the development and application of new heat sources. It also encouraged me to continue and expand these studies as the "second stage" of my life work.

*   *   *   *   *   *

To commemorate receiving the prize, I privately published "Ultra High Energy Density Heat Source and Its Application to Heat Processing," a collection of over 50 papers from a total of 550 I had published over the years. That volume was brought to the attention of the American Society for Metals, who expressed a desire to publish the present revised book of my work based on my own summary.

This book is, as the title indicates, a review of the development and application for heat processing of the three major high energy density beams — plasma beams, electron beams and laser beams. Following an introductory chapter, chapters 2 and 3 contain results of research on the development of high energy density beam heat sources and related beam characteristics, and industrial applications of these beams in welding, cutting, heat treatment, surface modification and other heat processing. I hope this book is of interest to researchers and materials processing specialists and also to graduate-level students as a guideline for research, and that it will provide a stimulus for further research and development.

The introductory chapter of this volume gives a general review of the characteristics of plasma, electron and laser beams. This chapter outlines the scope of the book; more-detailed research results are given in subsequent chapters. Various examples of the applications of these beams to heat processing are described, emphasizing the processing characteristics inherent to high energy density beam technology. Future trends in welding using ultra high energy density beam technology are discussed, with various applications to new materials technology.

In Chapter 2, studies on obtaining high power beams are described separately for plasma, electron and laser beams. This chapter gives a brief historical survey on the development of these beams.

Fundamental beam characteristics are studied experimentally and theoretically in Chapter 3. Practical methods for obtaining strong focusing of high energy density beams are emphasized.

Chapter 4 enumerates the physical characteristics of the weld zone in high energy density beam welding processes. These include hardness, fracture toughness and weldability. Dynamic behavior of the beam and the beam hole during welding and cutting is also described.

In Chapter 5, research results on the application of ultra high energy density beams to various heating processes are reported. Results on several welding methods using the electron beam heat source are described. Laser beam welding, cutting and heat treatment processes are discussed for various materials. Plasma spraying of ceramics and underwater welding are studied as examples of application of plasma beams.

*   *   *   *   *   *

I would like to express my sincere appreciation and thanks to my colleagues and many people who have assisted me over the years. A special word of thanks is due to Professor Emeritus M. Okada, the 8th President of Osaka University, who has given me invaluable support. Thanks are also due to Professor Emeritus Y. Imai, member of The Japan Academy and the Director of the Metals Museum in Japan, for encouraging the publication of this volume. Both of these professors have offered sharp insights into my work and have given me the courage to expand the range of my research. I would also like to thank Ms. S. Refsnes, Acquisitions Editor of ASM, and Dr. T. Eagar of MIT for their recommendations in publishing this volume.

Finally, I would like to express my sincere appreciation and thanks to my wife, to whom this book is dedicated, for her unfailing devotion, assistance and encouragement over the years.

YOSHIAKI ARATA
October, 1985

Contents

CHAPTER 1

INTRODUCTION

Development of Ultra High Energy Density Heat Sources and Their Application to Heat Processing
Y. Arata
[Trans. JWRI **13** (1984), 121.]

High Technology for Materials Processing Based on Welding
Y. Arata
[Paper presented at International Welding Congress (ASM Joining Division) during Materials Week 12-17 October 1985, Toronto, Ontario.]

Development of Ultra High Energy Density Heat Sources and Their Application to Heat Processing

Abstract

   This paper presents studies on the ultra high energy density heat sources, their characteristics and applications to heat processing.

   After some historical review of the various types of electron and laser beam equipment, the important problem of focusing large output lasers and electron beams is discussed, together with their characteristics as heat sources.

   Moreover, the basic weld zone characteristics produced by ultra high energy density beams are also examined, paying particular attention to hardness and the "Fracture Path Transition Temperature".

   Finally, the author refers to the experimental results of horizontal electron beam welding, Tandem Electron Beam Welding and laser beam welding, and describes the similarities and differences in processing characteristics of the laser beam and electron beam.

KEY WORDS: (High Energy Density Beam) (Heat Source) (Laser Welding) (Electron Beam Welding) (Heat Processing)

1. Introduction

Any currently known form of energy, including electromagnetic energy, mechanical energy and chemical reactive energy can be employed as a heat source for practical use such as for heat processing.

A wide variety of manufacturing heat sources are available through various technologies, but from the practical viewpoint they can be studied basically by generalizing the problem to two major points of their power and density.

In conventional studies on energy and heat source problems, considerable advances have been made in research on grades of energy, power magnitudes and their applicability, but it seems that not as much research has been devoted to the question of energy density and control.

Recently, however, electromagnetically accelerated particle beams such as electron, ion, neutral and special type plasma beams with large outputs have been developed. These beams have inherently high energy density and have begun to be used as high energy density heat sources. Heat sources of such large output and high density are capable of opening up new applications in the future, particularly if employed for processing and will not only contribute greatly to industry but will also help to develop new areas of science and technology.

High energy density beams can be produced in either a pulsed or continuous mode. However, to utilize them more widely as a popular heat sources such as the arc, they should have outstanding characteristics, particularly with regard to continuous output. From this standpoint

we have been endeavoring for over twenty years to develop equipments with a large continuous output and to clarify their characteristics.

We also believe that when considering the laser beam as a processing heat source, it should always be compared with an electron beam or plasma heat source. Thus, we believe that the basic task in utilizing high power lasers is to clarify the characteristics of each heat source and confirm the similarities and differences in processing characteristics. That is why I have selected the electron beam as well as the laser beam in this report, and the study results described later in this paper reflect this orientation.

2. Development of Ultra High Energy Density Beams

Among those who have contributed to the development of electron beam heat sources are Pierce[1], Steigerwald[2], Stole[3], and many other famous researchers. The researchers who helped to develop the laser beam heat source include such celebrated people as Maiman[4], known for the ruby laser; Patel[5], for the CO_2 laser; and Brides[6], for the Ar-ion laser, and other excellent researchers.

I have paid primary attention to the continuously oscillating CO_2 laser beam as a high power heat source for processing. First of all my interest in large output heat sources was triggered in 1957, when we created a pinch plasma using an electric discharge of 1.6 MA for research into nuclear fusion[7]. This was the origin of experimental studies of nuclear fusion in Japan. To obtain an ultra high temperature plasma other than pinch plasma, we planned to create a large output pulse electron beam, and proposed the "Transtron" accelerator[8], also to be called a linear betatron, and developed the first stage accelerating section[9], as shown in **Fig. 1**. Using this accelerator,

Fig. 1   Transtron Accelerator.

we obtained an energetic beam with a voltage of 1 MV and a current of 260 A using a hot cathode and 300 kV and several thousand amperes using a plasma cathode.

Those are pulse devices and their industrial applications are limited. Thus we found it urgently necessary to develop some large output equipment capable of a continuous output. In 1972 we succeeded in developing the world's first 100 kV, 100 kW electron beam welding device[10] as shown in **Fig. 2**. In 1975 we developed 300 kV, 100 kW

Fig. 2   100 kW 100 kV EB Welder.

device[11] as shown in **Fig. 3 (a)**. In 1980 we produced, by way of experiment, 600 kV, 300 kW electron beam device as shown in **Fig. 3 (b)** for welding etc.[12], having presently the largest output in the world. These devices in **Fig. 3** could be realized by the development of a new

(a) 100 kW 300 kV EB Welder.

Fig. 3   (b) 300 kW 600 kV EB Welder.

acceleration system of "electromagnetic acceleration unit"[13] shown in **Fig. 4**. Using these high power devices, we have established "a new welding method on ultra thick metal plates" of over 30 cm in thickness. Several years later, researches for practical use of such a source began in various countries[14].

A large output CO_2 laser was realized through research carried out by Whitehouse[15], Tifferny[16], Lock[17], Banas[18] and many others. In 1966 we introduced a 1 kW CO_2 laser device shown in **Fig. 5** and developed the world's first large output continuous CO_2 laser welding-cutting method. In particular, the name "laser gas cutting"[19] which we gave to the cutting method is now widely used. At about the same time, we developed a different cutting method as shown in **Fig. 6** and named it "electron beam gas cutting"[20]. In terms of its cutting property, the laser beam method seems superior but further detailed study on it will be required in the future. Although in the range of up to 3 kW (1970), we used trial equipment made by ourselves, we are now using 5 kW system made by Spectra-Physics and 15 kW one made by AVCO to carry out research on welding (of metallic or non-metallic

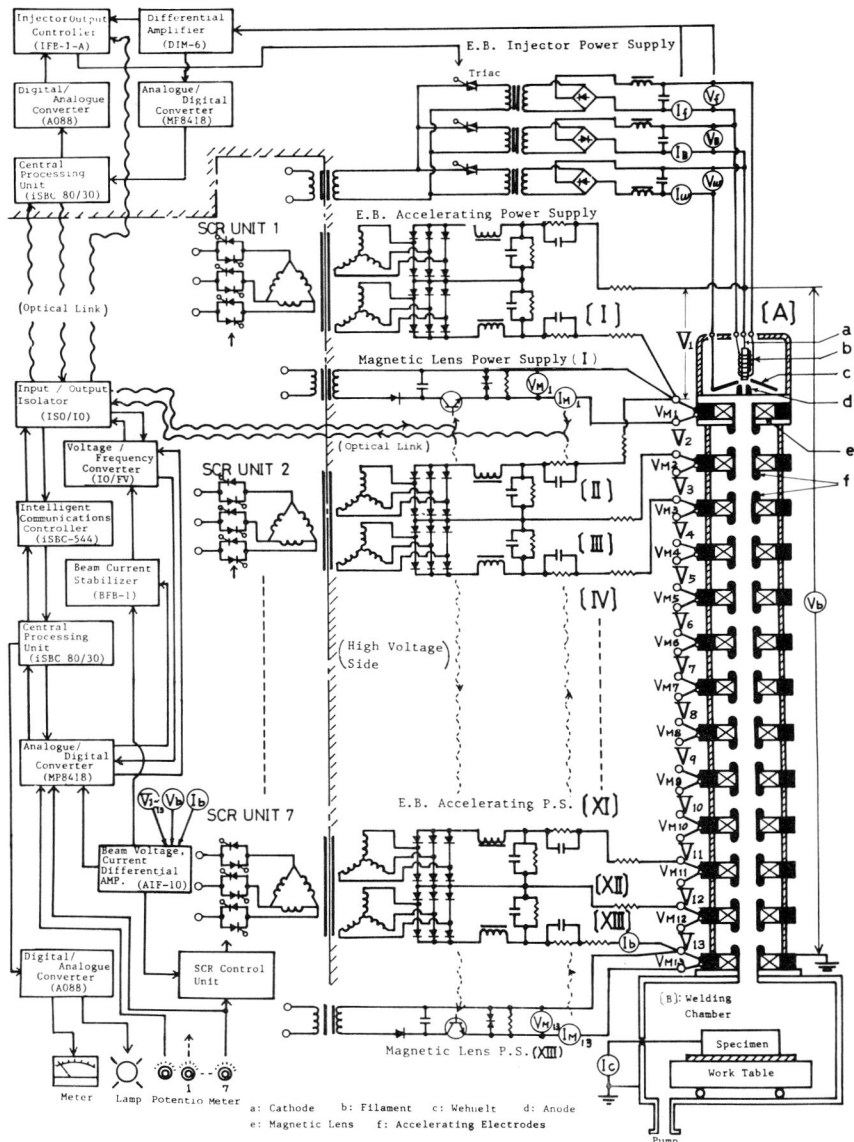

Fig. 4 600 kV–300 kW strong focusing type 13-stage accelerating electron beam heat source. [A]: EB gun with 13 electromagnetic acceleration unit.

materials (ceramics)[21], cutting, heat treatment (such as, laser surface hardening[22], laser gas hardening[23]) and alloying etc.

Such large output laser and electron beam equipments

have been installed in "The Research Center for Ultra High Energy Density Heat Source" of our institute established in 1980. Other new equipments for high power and high energy density heat sources are also being

Fig. 5 1 kW CO_2 Laser Welder (and applicable also to cutting). The first CO_2 laser for material processing.

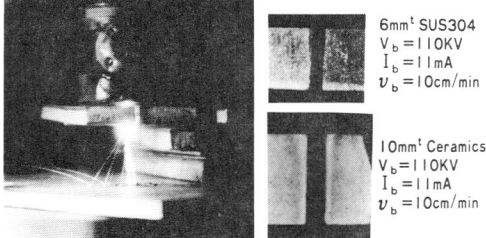

6mm^t SUS304
$V_b = 110$KV
$I_b = 11$mA
$v_b = 10$cm/min

10mm^t Ceramics
$V_b = 110$KV
$I_b = 11$mA
$v_b = 10$cm/min

Fig. 6 NV-EB gas cutting with two samples of the cutting cross section.

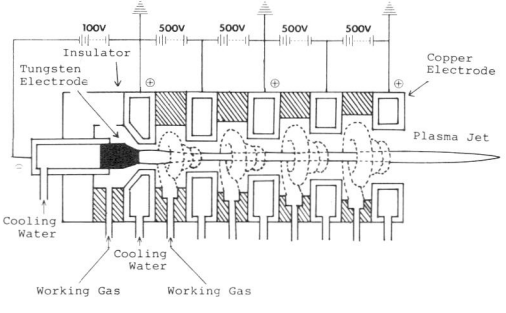

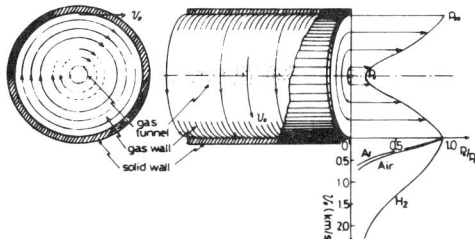

Fig. 7 Principle of the "Gas Tunnel" and its Special High Power Plasma Jet Device.

constructed and researched in this research center. One is a new type of special high power plasma jet device shown in **Fig.** 7. This device can produce a "gas tunnel" of low pressure by using a special strong vortex gas flow[24]. Inside this gas tunnel the pressure reaches a level below 20 Torr. Therefore, a very stable high power arc plasma beam can be created along its axis. By using this device we produced an Argon plasma beam of about 30000 K at 800 A with input power of 150 kW. Another device being developed is the ECR Plasma device shown in **Fig.** 8. Using a 60 GHz,

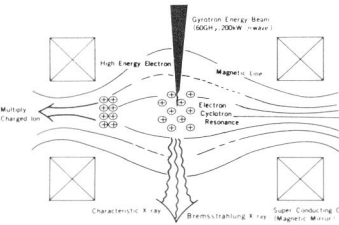

Fig. 8 ECR Plasma Device and its explanation drawing of ECR heating.

200 kW Gyrotron, a high temperature and high density plasma is produced in the magnetic mirror of super conducting magnets by electron cyclotron resonance heating.

3. Fundamental Characteristics of Ultra High Energy Density Heat Sources

Since the degree of energy density obtained from electron beams and CO_2 laser beams is almost identical, their characteristics as a heat source are similar in many respects. However, their processing characteristics differ in considerable cases because their spatial propagation characteristics and their interaction with materials are different.

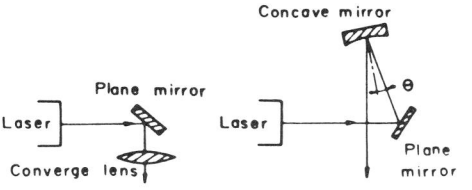

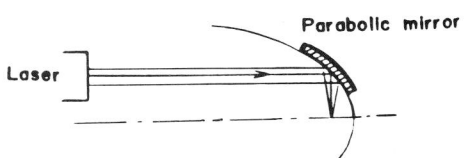

a) Converging lens focus system. (conventional system)

b) Arata laser focus system.

c) Parabolic mirror type focuss system.

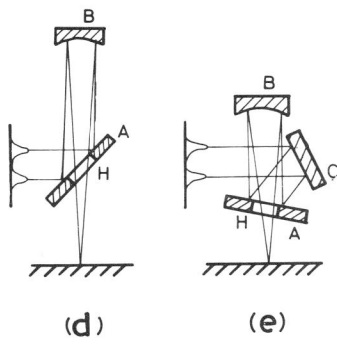

(d)   (e)

Fig. 9 Beam focusing methods. The principle of (d) and (e) is the same as system (b). In order to converge the beam downward, optical system using conversing lens, spherical mirror which are combined by a plane mirror can be used.

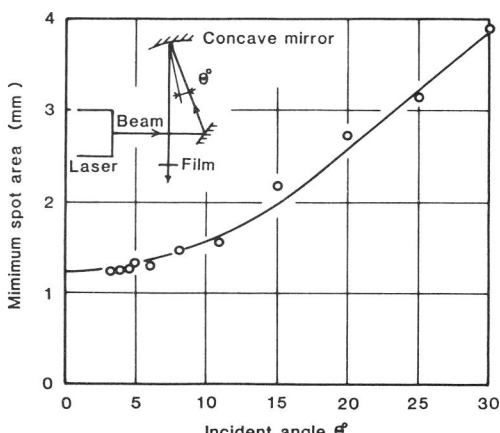

Fig. 10 Minimum spot area vs. incident angle (focal length: 151 mm, beam diameter: 15 mm).

3.1 Focusing

To utilize a laser beam in a state of high energy density, the method of beam focusing is extremely important. **Figure 9** shows five kinds of basic beam focusing systems. Traditionally used are (a) and (c), whereas (b) is the method proposed by the author about 15 years ago as a large output laser focusing method[25],[26]. It is called the Arata laser focus system[27] and its features are shown in **Fig. 10**. The preferred convergence angle should be smaller than a few degrees. Compared with other systems, in this system the greater the laser output the more effective the function becomes. The system of Fig. 9 (b) is suitable for a columnar beam such as a Gaussian beam. For hollow beams, systems (d) and (e) are suitable, whose principal is the same as system (b). Now we use system (e) for 15 kW CO_2 laser device.

The beam focusing, irrespective of whether it's a laser or an electron beam, can be performed as shown in **Fig. 11**. As an important parameter, the author defined the "a_b value" as follows and named it the "beam active

$$a_b = D_O/D_F$$
(D_F: focal distance, D_O: object distance)   (1)

parameter"[28],[29]:
With the incident angle represented as θ and the standard deviation at the focusing point as σ, we also set the beam active zone length ℓ_b as follows[30],

$$\ell_b = 4.46\sigma/\tan\theta \qquad (2)$$

which is essential for knowing the beam's characteristics for heat processing.

Any paper in which the a_b value is not clearly written can only be half evaluated or even not evaluated at all as a research paper in some cases. For example, as shown in **Fig. 12**, the beam penetration depth h_p depends greatly on the a_b value in the case of electron beams, and it varies much more violently in the case of laser beams. Moreover, since the laser heat source causes peculiar changes not only to the penetration depth but to the bead shape as shown in **Fig. 13**, we named this specific phe-

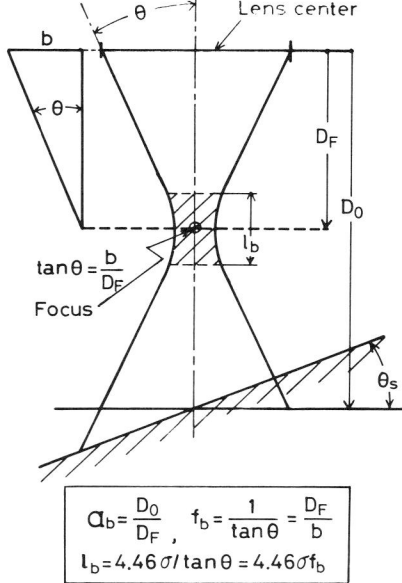

$$a_b = \frac{D_0}{D_F}, \quad f_b = \frac{1}{\tan\theta} = \frac{D_F}{b}$$
$$l_b = 4.46\sigma/\tan\theta = 4.46\sigma f_b$$

Fig. 11 General view of high energy density beam and its parameters.

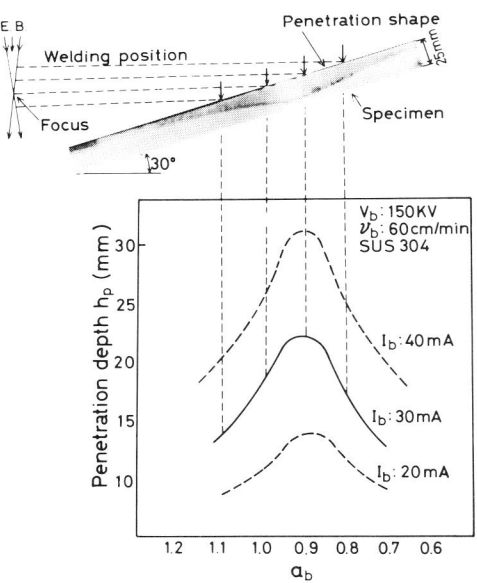

Fig. 12 Relation between active parameter a_b and penetration depth h_p for various current I_b.

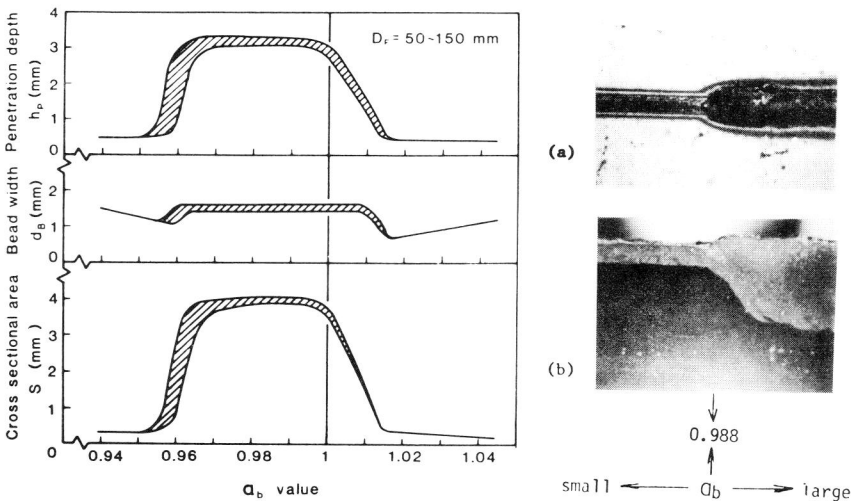

Fig. 13 Relation between a_b value and penetration depth h_p, bead with d_B and cross section S.

nomenon "bead transition"[31]. This is a phenomenon not seen in the case of electron beams. Thus, in utilizing a laser as a heat source at high energy density, the allowable range of a_b values ($\pm\Delta a_b$) is extremely narrow as compared with the case of electron beam (also see **Fig. 14**).

This fact necessitates far higher control in laser welding than in electron beam welding. We have already reported in detail the mechanism of how bead transition is generated[32], so I will not discuss it in this paper.

So far a number of researchers have proposed shapes

8

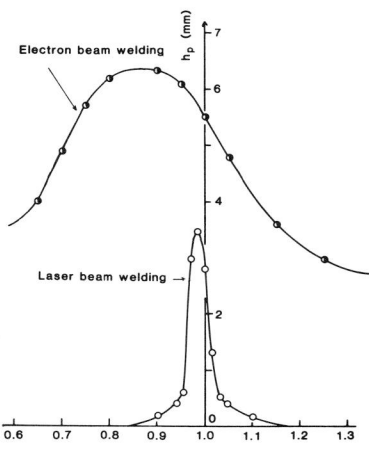

Fig. 14 Comparison of laser welding with electron beam welding for $h_p - a_b$ relation.

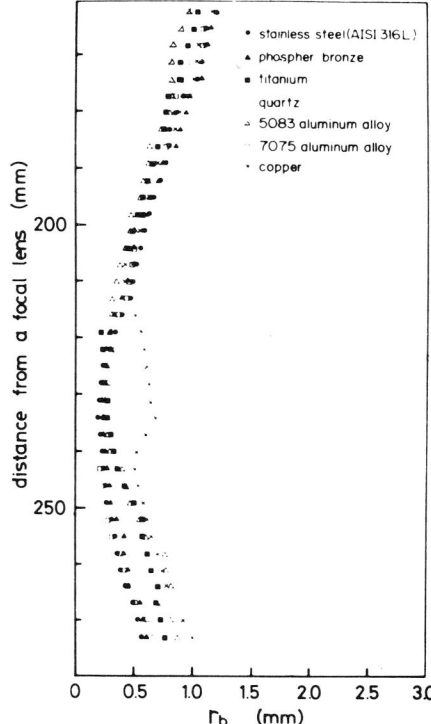

Fig. 16 Summarization of proper beam shape for each type of material.

of high energy density convergent beams and how to measure their energy densities[33)−35)]. However, their measuring methods are generally complicated and short of accuracy in many cases. Measurement is particularly difficult in the case of large outputs, and even impossible during practical heat processing. The author has proposed a method of measuring the beam shape and mean energy density under such conditions[36),37)]. This method of measurement shown in **Fig. 15**, is now widely used as the

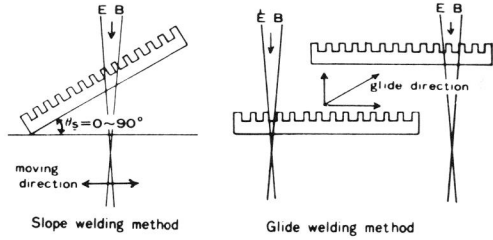

Fig. 15 Schematic drawing of AB test method.

"AB test" (Arata beam test) method[35),37)]. It utilizes the sharp edge effect of a specimen as illustrated, and can be applied to other materials according to output. Ceramics can be used for lasers of not so large an output. **Figure 16** shows an example of an electron beam which is often applied to conventional stainless steel. The beam spot strength $w_b(r)$ in this active zone is normally distributed in a Gaussian form. For example, in the case of a CO_2 laser, **Fig. 17** shows the actual measurements at focus, which are approximated by the Gaussian curve[38)]:

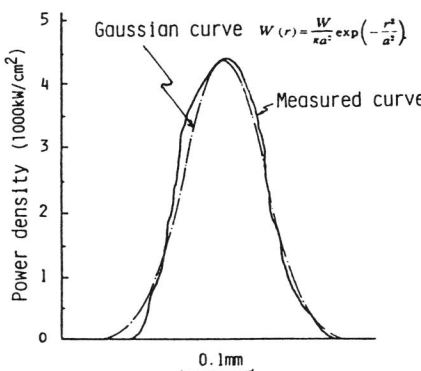

Fig. 17 Power density distribution at focal point.

$$w_b(r) = \frac{W_b}{\pi a^2} \exp\left(-\frac{r^2}{a^2}\right) \tag{3}$$

where W_b is the beam output, "a" is the beam spot radius and "r" is the distance from the beam axis. In Fig. 17 W_b = 1 kW, $d_b = 2a = 0.17$ mm and its central energy density is as high as $w_b = 5000 \, kW/cm^2$: the focal depth, however, is so small that the energy density decreases by half when separated 1 mm from the focal position ($D_F = 64$ mm). In the case of a multiplex mode of $W_b = 5$ kW, d_b became 0.4 mm and the central density was almost identical to that in the single mode of 1 kW.

The greatest difference between an electron beam and laser beam in terms of heat source characteristics is found in their absorptivity to metal materials. The absorptivity of an electron beam is extremely high, whereas that of a laser is extremely low. This greatly affects the phenomena, and is the most probable cause of the above-mentioned "bead transition". When a CO_2 laser beam was applied to a polished metal surface with a conductivity σ ($= 1/\eta$, η: specific resistance), the absorptivity A_b was well compatible with our actual measurements[39] as shown in **Fig. 18** using the Hagen-Rubens equation[40].

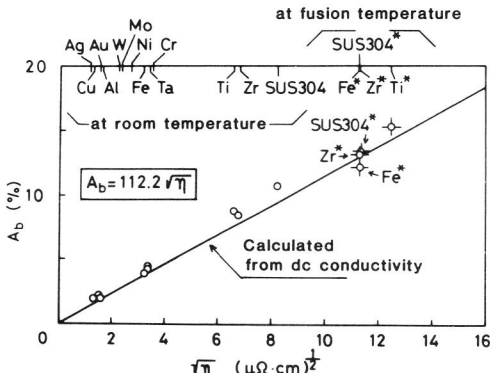

Fig. 18 Absorptivity of metals at room temperature and fusion temperature marked with symbol*. SUS-304 corresponds to AISI 304 stainless steel.

$$A_b = 112.2\sqrt{\eta} \qquad (4)$$

Thus, using these results, we attempted to find the minimum beam power W_{bM} required to heat the metal surface up to the melting point T_M.

$$
\begin{aligned}
W_{bM} &= 1.58 \times 10^{-2}\kappa d_b T_M\sqrt{\sigma_M} & \text{(a)} \\
&= 8.92 \times 10^{-3}\sqrt{\sigma_M}\, W_{bM}^* & \text{(b)}
\end{aligned} \qquad (5)
$$

This is shown in **Fig. 19**. Here W_{bM}^* ($= \sqrt{\pi}\,\kappa d_b T_M$) corresponds to W_{bM} at $A_b = 100\%$. As can be understood from Figs. 18 and 19, the absorptivities are less than 15% even in Ti, Zr, Fe, and SUS 304 in the molten state of the highest absorption. This indicates the need for surface treatment to raise the A_b values when metal is to be welded using a laser of several hundred watts. However, if

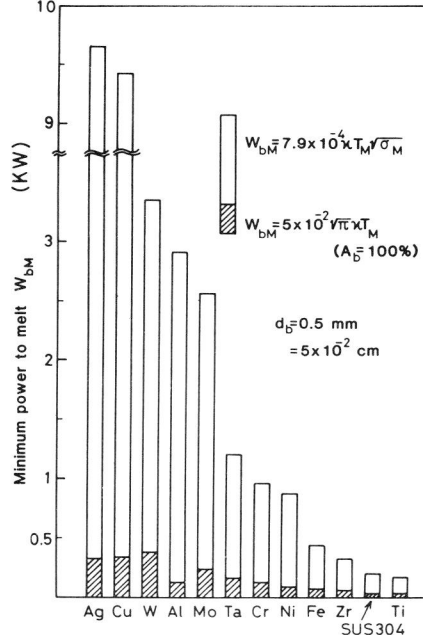

Fig. 19 Minimum laser power to weld various metals with infinite thickness. In this figure σ_M represents dc conductivity at fusion temperature (constant power, Gaussian distribution).

the incident laser beam output W_b increases and the energy density reaches several hundred kW/cm^2, then, as in the case of the electron beam, beam holes will be formed and the beam will make multiple reflections on the wall surface (number of times $N = \pi/tan^{-1}$ (s/h)). By this phenomenon, laser energy concentrates on the wall and on the bottom as well. We called this phenomenon the "wall-focusing effect"[41),42]. The laser's effective absorptivity $\tilde{A}_b$ ($= W_{bA}/W_b$) or absorption beam power W_{bA} rapidly rises due to this effect along with the beam hole depth h_p as shown in the following formula and the illustration in **Fig. 20**, so these parameters become independent of the state of the material surface.

$$\frac{W_{bA}}{W_b} \; (\equiv \tilde{A}_b) = 1 - (1 - A_b)^{\pi/tan^{-1}(s/h)} \qquad (6)$$

Here it is assumed that the beam hole is wedge-shaped as shown in **Fig. 20**, that is, s, h and A_b are the beam inlet width, depth and wall surface absorptivity, respectively, and there is no effect from the plasma, etc. Usually we may assume that $s = d_b$ and $h = h_p$. On the other hand, on the assumption that the beam is a line heat source travelling at speed v_b and that the weld bead width d_B and penetration depth h_p are formed in a material with melt-

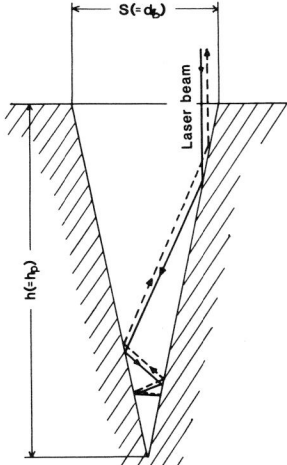

Fig. 20 Reflection path of laser beam in wedge shape cavity.

ing point T_M, thermal conductivity κ and thermal diffusivity κ_D, the beam power W_{bC} required becomes[43]:

$$W_{bC} = 8\kappa T_M \left(0.2 + \frac{v_b d_B}{4\kappa_D}\right) h_p \qquad (7)$$

As shown in **Fig. 21**, the above W_{bC} increases in proportion to h_p and intersects the W_{bA} curve ($W_{bC} = W_{bA}$). Thus, the beam hole can grow deep enough to reach the intersection. This is a heat source characteristic of laser beams, which have a large reflection loss, and is one of the main characteristics which differ substantially from electron beams, whose loss is smaller.

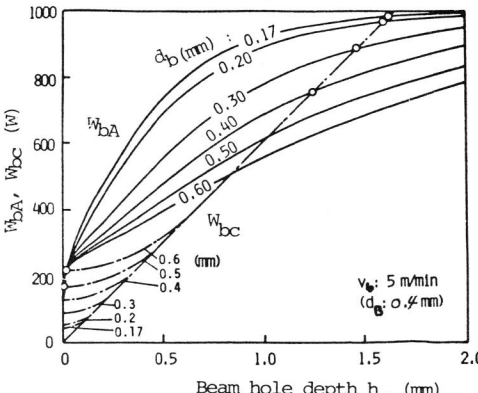

Fig. 21 Relation between h_p and W_{bA}, W_{bC}.

3.2. Penetrated heat sources

In many cases, so far analysis on physical properties of

the weld zone was based on thermal conduction theory, the "Point" or "Line" heat source theory. However, these theories were not sufficient for establishing the characteristic properties of high energy density beam heat processing. Therefore, the author proposed following two new theories: The "$\alpha\beta$-Distributed Heat Source[44]" and the "Band Heat Source[45]".

3.2.1. $\alpha\beta$-distributed heat source

Figure 22 shows various types of heat sources. The $\alpha\beta$-distributed heat source is a generalized type of heat source, (Fig. 22 (c)), which includes the point (Fig. 22 (a)) and line heat sources (Fig. 22(b)) as its special cases.

The dimensionless input energy $Q^*(\beta)$ is expressed as follows:

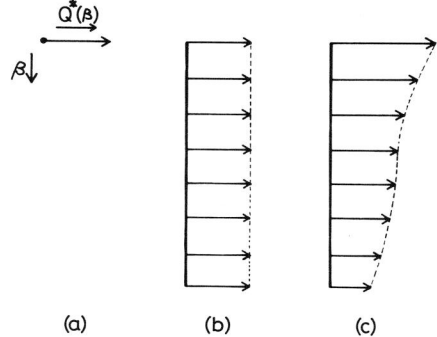

Fig. 22 Various types of heat source.

$$Q^*(\beta) = (1 - \beta)^\alpha \qquad (8)$$

where β is a dimensionless distance and α is the index to

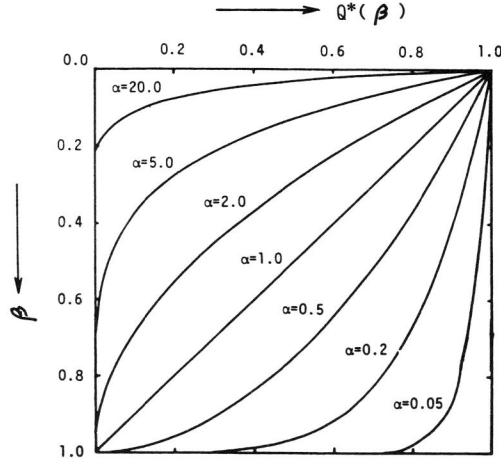

Fig. 23 Input energy distribution function $Q^*(\beta)$.

the input energy distribution.

When $\alpha = 0$,

$$Q^*(\beta) = 1 \qquad (9)$$

and it corresponds to a line heat source of uniform input energy distribution.

When $\alpha = \infty$,

$$\begin{aligned} Q^*(\beta) &= 1 \quad (\beta = 0) \\ &= 0 \quad (\beta \neq 0) \end{aligned} \qquad (10)$$

and it corresponds to a point heat source. The dependence of $Q^*(\beta)$ on β is shown in **Fig. 23**.

3.2.2. Band heat source

As shown in **Fig. 24**, a rectangle is considered to be equal to the weld zone bead cross section S_B ($= \int_{hp}^{hp} d_B(z)\,dz$), and $\widetilde{d}_B$ is given so that $S_B = h_p \widetilde{d}_B$. This is to be called the "effective bead width" as compared with the actual surface bead width d_B. This kind of rectangular heat source can be made similar to a band heat source if $h_p \gg \widetilde{d}_B$ within an infinite solid body. We obtain the following formula[46]:

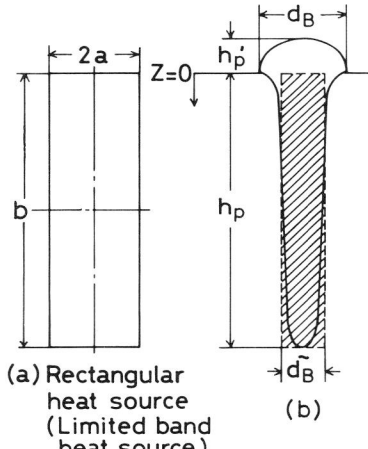

(a) Rectangular heat source (Limited band heat source)

(b)

Fig. 24 Relation between rectangular heat source and actual weld bead.

$$T_M^* = \frac{1}{\pi v_b^*} \left(\frac{\pi}{2} - \int_0^{\pi/2} \exp\left(-\frac{2v_b^*}{\cos\varphi}\right) d\varphi \right) \quad (a)$$

$$= \frac{1}{2v_b^*} \quad (\text{where } v_b^* \gtrsim 1) \qquad (b)$$

$$(11)$$

where T_M^* is the dimensionless quantity of melting temperature T_M, and v_b^* corresponds to the dimensionless quantity of welding speed v_b. With κ and κ_D as the material's thermal conductivity and thermal diffusivity, respectively:

$$T_M^* = \frac{T_M}{\left(\dfrac{W_b}{4\kappa b} \right)} = \frac{1}{\left(\dfrac{W_b}{4\kappa T_M b} \right)} = \frac{1}{W_b^*} \quad (a)$$

$$= \left(\frac{4\kappa T_M}{W_b} \right) h_p \quad (\text{where } b = h_p) \quad (b)$$

$$(12)$$

$$v_b^* = \frac{a v_b}{2\kappa_D} \qquad (c)$$

$$= \frac{v_b \widetilde{d}_B}{4\kappa_D} \quad (\text{where } 2a = \widetilde{d}_B) \qquad (d)$$

$$(13)$$

From formulae (10)(b), (11)(b) and (12)(b), the penetration depth can theoretically be induced as

$$h_p = 0.5 \left(\frac{W_b}{\kappa T_M} \right) \left(\frac{\kappa_D}{v_b \widetilde{d}_B} \right) \qquad (14)$$

where, however, it is assumed that 100% of the beam power W_b is absorbed into the material, and such problems as slight differences between actual bead shape and rectangular bead shape, actual molten pool temperature, beam hole formation and vapor pressure in the beam hole, etc. are ignored, in order to process purely by the theory of heat conduction alone. For that reason, with the parameters kept unchanged, the theoretical formula (14) can be written as the following empirical formula:

$$h_p = K \left(\frac{W_b}{\kappa T_M} \right)^n \left(\frac{\kappa_D}{v_b \widetilde{d}_B} \right)^m \qquad (15)$$

where K, n and m are experimental constants. As shown in **Fig. 25**, if K = 0.35, n = 1 and m = 0.83, you can see

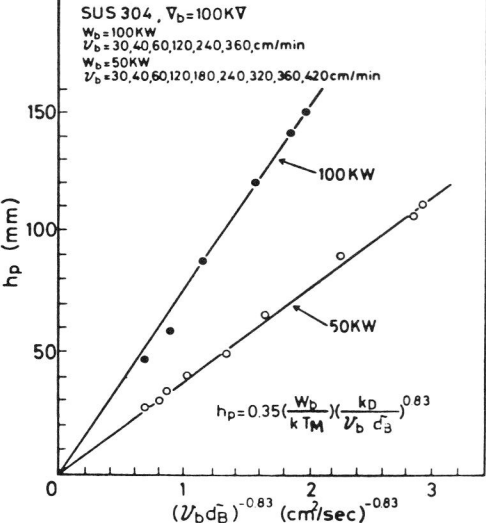

Fig. 25 Relation between h_p and $(v_b \widetilde{d}_B)^{-0.83}$.

how the calculated value according to formula (15) meets the value of the actual measurement.

However, in other tests, there are cases where one parameter is chosen to express an experimental formula such as $h_p \propto 1/\sqrt{v_b}$ or $h_p \propto W_b^{1.7}$ for example. This means that compared with formula (15), there seems to be a considerable difference in the extent of the factors' contributions. That is because the phenomenon of penetration is affected by many closely interrelated factors. Their degree of contribution appears different due merely to the selection of a specific factor as a parameter or constant to simplify the relative formulae. Also, care should be taken as to the applicable limits of the formulae if only one parameter is chosen to make an experimental formula. When formulae (14) and (15) are compared, no great difference is found between the theoretical formula and the experimental one except that the K-value changes from 0.5 to 0.35 and the m-value from 1 to 0.83. These differences are considered to have been caused by the combination of the various conditions mentioned just after formula (14), but it can be seen that their effects are not so conspicious.

Formulae (14) and (15) may further be changed and rewritten as formulae (16) and (17) as follows:
From formulae (11)(b) and (12)(a), the relation between speed and the incoming heat of a dimensionless quantity can be found by:

$$v_b^* = 0.5W_b^* \quad \text{(provided that } v_b^* \gtrsim 1) \qquad (16)$$

This is nothing but formula (14), and W_b^* is the dimensionless quantity of incoming heat W_b. An experimental formula corresponding to formula (16) is then set up:

$$v_b^* = K_1 W_b^{*\ell} \qquad (17)\,(a)$$

And if the experimental parameters of formula (15) (K = 0.35, n = 1, m = 0.83) are employed, we get $K_1 = 0.375$ and $\ell = 1.2$ from the above formula, giving us:

$$v_b^* = 0.375 W_b^{*1.2} \qquad (17)\,(b)$$

As shown in **Fig. 26**, formulae (16) and (17)(b) conform well in a very wide range to the linear equation

$$v_b^* = 0.59W_b^* - 0.34 \qquad (18)$$

which was found by the method of least squares, based not only on the actual measurements obtained by the

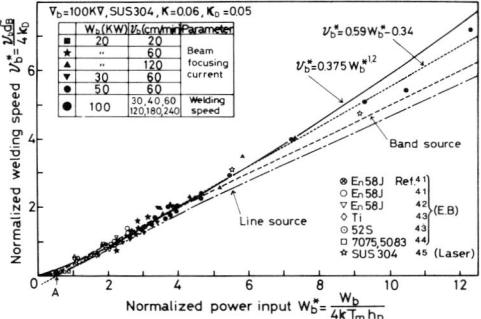

Fig. 26 Comparison of the experimental values with calculated values on v_b^* and W_b^*.

Table 1 Experimental conditions of other researchers.

Reference	41		42	43		44		45
Material	En58J		En58J	52S	Ti	7075	5083	SUS 304
K(cal/cm.sec.°C)	0.07		0.07	0.4	0.065	0.31	0.28	0.06
K_D(cm²/sec)	0.05		0.05	0.76	0.08	0.48	0.46	0.05
Beam	Electron Beam							Laser
Beam	130KV 40mA	27-33KV,150mA	30KV, 100mA	150KV, 30mA		150KV, 7mA		2.0KW
v_b(cm/min)	120-300	66	66	50		150		126-430
Parameter	Welding speed	Beam focusing current	Focusing current and focus Position	Work distance				Welding speed

Table 2 Chemical compositions of materials used. W_t (%)⟵→PPm

	C	Si	Mn	Cu	Ni	Cr	Al	Mg	Mo	Ti	Zn	Fe	P	S	N	O
En58J	0.06	—	—	—	10.89	—	17.65	—	2.85	—	—	Re	—	—	—	—
52S	—	0.1	—	0.04	—	0.17	Re	2.4	—	—	—	0.18	—	—	—	—
7075	—	0.11	0.03	1.6	—	0.22	Re	2.4	—	—	5.56	0.27	—	—	—	—
5083	—	<0.4	0.3~0.1	<0.1	—	—	Re	38~48	—	<0.2	<0.1	<0.4	—	—	—	—
SUS 304	0.05	0.74	1.74	0.12	10.9	19.5	0.015	—	0.16	—	—	Re	0.030	0.010	365	70
SM 41	0.18	0.47	0.71	—	—	—	0.050	—	—	—	—	Re	0.015	0.010	68	36

Re:remainder

author (shown in black symbols), but also on many actual measurements obtained by other researchers[47]−[51] (indicated in white symbols). **Table 1** shows the experimental conditions and **Table 2** the chemical compositions of the materials used. Formula (18) is found in a wide range of W_b^*, from low power to 100 kW class. However, in a low power range, $v_b^* = 0$ is obtained at $W_b^* \approx 0.57$ (point **A** on the horizontal axis in **Fig. 26**). Hence it is obvious that any value below this value is inapplicable. In the case of electron beam welding, the applicable range of the most popular conventional line heat source is, as seen from **Fig. 26**, $W_b^* \approx 1 \sim 4$ at most, so both the theoretical and experimental formulae based on the band heat source agree well with the linear formula obtained by using the method of least squares on a fairly wide range of actual measurements.

It seems that the melting efficiency of electron beam welding has not been fully clarified. The melting efficiency may be defined as follows:

$$\eta_M = \frac{\text{Quantity of heat required to raise the temperature of bead area to the melting temperature (cal/sec)}}{\text{Input heat to base metal (cal/sec)}}$$

$$= \frac{h_p \tilde{d}_B v_b (\rho C T_M + Q_L)}{\eta_b W_b} \qquad (19)$$

where, ρ, C and Q_L are density of the base metal, specific heat and latent heat, respectively, and the other symbols are all those already mentioned. If the beam's melting efficiency of the base metal is represented by η_b and $Q_L \rightarrow 0$, $\eta_b \rightarrow 1$, then formula (19) becomes

$$\eta_M = \frac{v_b^*}{W_b^*} \times 100 \ (\%) \qquad (20)$$

where if theoretical formula (16) is employed, $\eta_M = 50\%$ is obtained. This is inexplicable because normally about 60% is obtained in actual measurements as shown in **Fig. 27**. Actually, however, $Q_L = 0$ and the molten pool temperature is greater than T_M, so from formula (19), it is obvious that $\eta_M > 50\%$. Thus it may be quantitatively expressed using formula (18).

$$\eta_M = \frac{0.59}{1 + \dfrac{0.34}{v_b^*}} \qquad (21)$$

From this we can get $\eta_M \lesssim 60\%$ and see that it almost fully conforms to the actual measurement. Even in this case, a line heat source can be used only in a narrow range where the value of v_b^* is not more than 1.5 or so.

3.3 Surface heat source

Here we describe the characteristics of a "surface heat source" employed large output high energy density beams by which the material surface is not or only slightly melted.

This applied mainly to such processes as quenching and

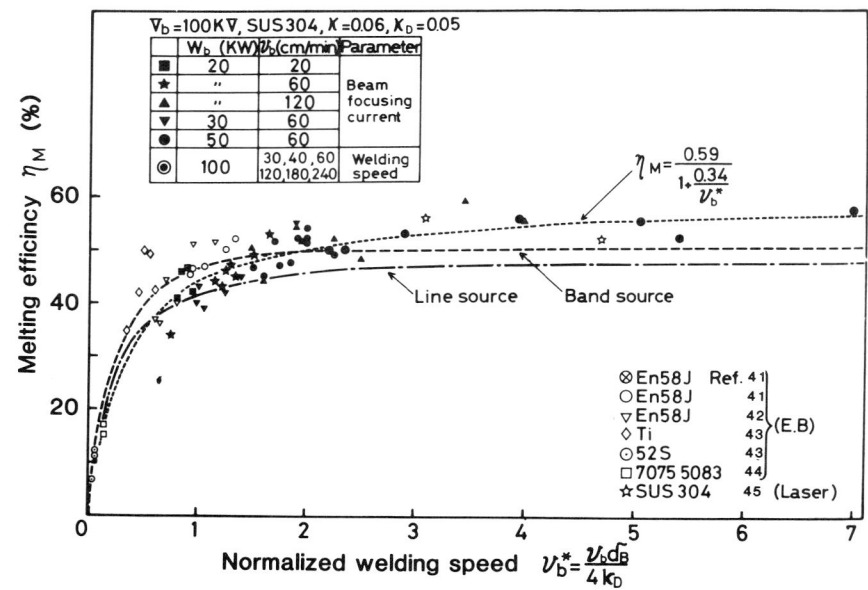

Fig. 27 Comparison of experimental melting efficiency and calculated ones.

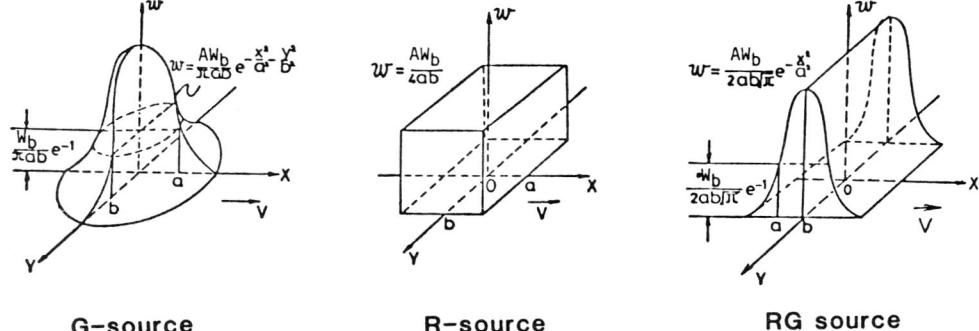

G-source   **R-source**   **RG source**

Fig. 28 Intensity distribution of surface heat source.

other surface treatments, surface alloying and so on. In most cases, as shown in **Fig. 28**, the temperature distribution of surface heat sources can be considered to have three types: a Gaussian distribution (its cross section is generally elliptical, but in special cases circular; G-heat source), a rectangular distribution (uniform temperature; R-heat source), and a rectangular-Gaussian distribution (the above two types are overlapped; RG-heat source)[52]. Such a heat source distribution can be obtained by properly modifying the multiple mode of the beam, by using a special optical system with a segment mirror, or by using a beam oscillator.

As mentioned earlier, the beam's reflection loss against metal is small in the case of an electron beam, but extremely great in the case of a laser. For example, with a CO_2 laser, even on a molten surface with the highest absorptivity, you can only achieve $A_b \approx 15\%$ at best. Thus, to quench steel material with low output CO_2 lasers, we tried to increase A_b by lightly coating the surface with a phosphite film for example. This enabled us to attain A_b = 50 to 90% under conditions where the surface was not melted, as shown in **Fig. 29**. In this case, the

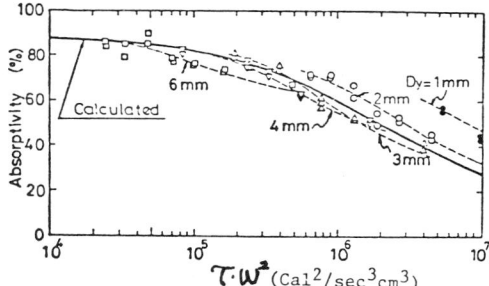

Fig. 29 Absorptivity plotted as a function of τw^2 (τ = interaction time, w = power density). The absorptivity can be predicted from this figure independently of beam spot shape, laser power and travelling velocity.

important parameter is τw^2 where the laser strength is $w(r)$ and its dwell time τ. Thus we can find the A_b value of the surface irrespective of the beam spot, beam output and beam travelling speed. Under such surface condition, if scanned by large output high density beams, the surface is rapidly heated and its heat affected zone appears only in the vicinity of the surface. Thus, in this case, the temperature distribution in the stationary state can be obtained from the following formula (dimensionlessly displayed) providing that the said distributed heat source travels on a semi-infinite plate at a speed of v_b.

$$T_G^* = \frac{16}{\pi} \int_0^\infty \frac{1}{\sqrt{(a^{*2}+t^{*2})(b^{*2}+t^{*2})}}$$
$$\times \exp[-\frac{(2x^*+v^*t^{*2})^2}{4(a^{*2}+t^{*2})} - \frac{y^{*2}}{b^{*2}+t^{*2}} - \frac{z^{*2}}{t^{*2}}]dt^*$$
(G-source) .... (22)

$$T_R^* = \int_0^\infty (\text{erf}\, \frac{2x^*+v^*t^{*2}+2a^*}{2t^*} - \text{erf}\, \frac{2x^*+v^*t^{*2}-2a^*}{2t^*})$$
$$\times (\text{erf}\, \frac{y^*+b^*}{t^*} - \text{erf}\, \frac{y^*-b^*}{t^*})\exp(-\frac{z^{*2}}{t^{*2}})\, dt^*$$
(R-source) .... (23)

$$T_{RG}^* = \frac{4}{\sqrt{\pi}} \int_0^\infty \frac{1}{\sqrt{a^{*2}+t^{*2}}} \exp[-\frac{(2x^*+v^{*2})^2}{4(a^{*2}+t^{*2})} - \frac{z^{*2}}{t^{*2}}]$$
$$\times (\text{erf}\, \frac{y^*+b^*}{t^*} - \text{erf}\, \frac{y^*-b^*}{t^*})\frac{dt^*}{b^*}$$
(RG-source) .... (24)

providing that $T^* = 16\sqrt{\pi}\, \kappa r T/A_b W_b$, $v_b^* = r v_b/2\kappa_D$,

$x^* = \frac{x}{r}$, $y^* = \frac{y}{r}$, $z^* = \frac{z}{r}$, $a^* = \frac{a}{r}$, $b^* = \frac{b}{r}$, $r^2 = ab$

and $\text{erf}\xi = \frac{2}{\sqrt{\pi}}\int_0^\xi \exp(-u^2)du$ with κ = thermal conductivity, κ_D = thermal diffusivity, T = temperature, W_b = laser power, A_b = beam absorptivity and v_b = travell-

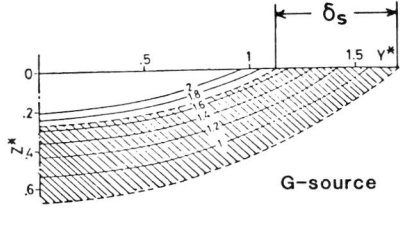

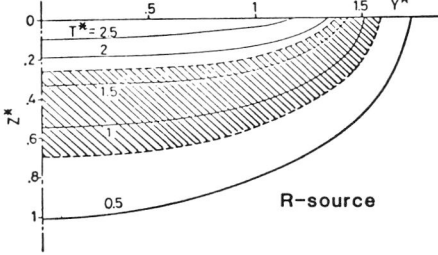

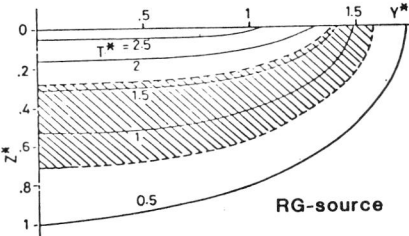

Fig. 30 Isothermal lines ($v_b^* = 4$). The hatched region corresponds to the softened zone. The value of T in this region varies from $(4/15)T_{max}^*$ and $(8.5/15)$ T_{max}^*. T_{max}^* is defined to be the temperature at the origin (Z = 0, Y = 0).

ing speed.

Figure 30 shows the isothermal lines drawn using the

above formulae (22) to (24) for each heat source. The hatched area in the figure shows the cross sectional configuration of the temperature range from $\frac{4}{15} T_{max}^*$ to $\frac{8.5}{15} T_{max}^*$ and corresponds to the softening zone affected by heat. The magnitude of the width of this softening zone, "softening width" δ_s, on the metal surface is important from a practical viewpoint. This indicates that a G-source is not practical because it enlarges the softening width, whereas R- and RG-sources are more useful because their softening widths are narrower.

Figure 31 shows the maximum hardening depth (hardening temperature: 850°C, surface temperature: 1500°C and a* = 0.7) in the case of RG-source. This is an example calculated with beam output W_b = 1.5 kW and W_b = 5 kW. To find out the extent of the effect according to the type of heat source, the result for R-source is also shown. This indicates that there is no great difference between R and RG heat sources.

In the same figure we also plotted the softening width δ_s when steel is quenched by a laser. The quenching efficiency η_Q is the ratio of the thermal capacity required (when the softened zone of the metal is adiabatically heated to 850°C) to that of the input laser. η_Q increases proportionally to v_b^* as shown in **Fig. 31**.

Figure 32 shows the hardened zone of tool steel SK-5 quenched by a 1 kW laser of RG distribution, and **Fig. 33**

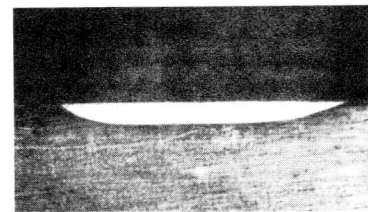

Fig. 32 Cross section of laser hardened SK-5 (Case depth = 0.25 mm).

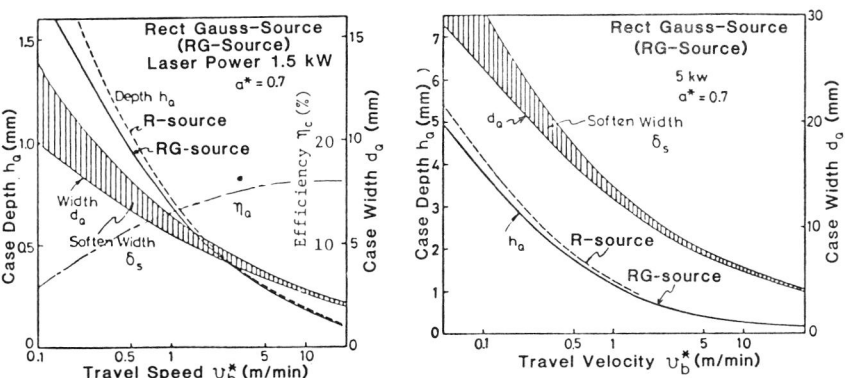

Fig. 31 Hardened depth h_c, hardened width d_c and softened with δ_s for two laser powers as RG-source.

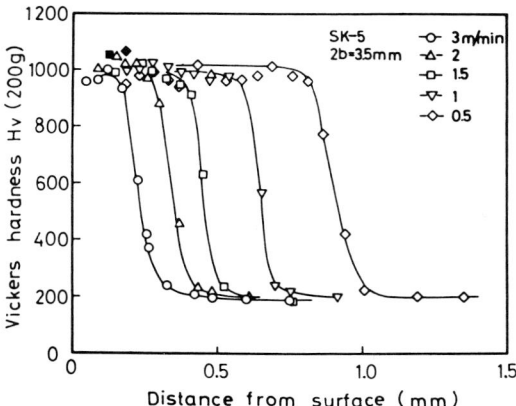

Fig. 33 Hardness distribution of SK-5.

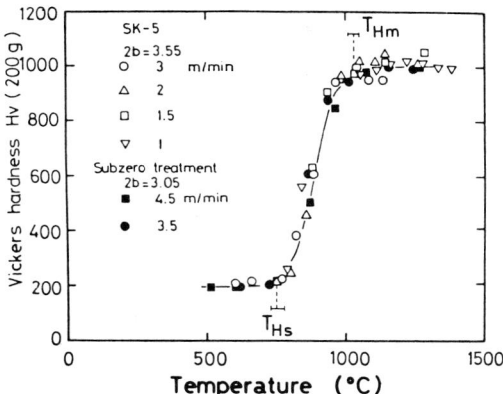

Fig. 35 Relation between heated temperature and hardness.

shows the hardness distribution in this case. This hardness is about $H_v = 100$ higher than in the case of water quenching, which indicates that it is difficult for austenite to remain. In this case, the cooling time $\tau_{800 \to 400}$(sec) is considerably short as shown in **Fig. 34**. In normal quenching, as further shown in **Fig. 35**, there is a relationship between the temperature at which the maximum hardness T_{Hm} is attained and the temperature at which to start hardening T_{Hs} (determined by C%). As shown in **Fig. 36**, T_{Hm} is usually considerably higher than the A_{c3} transformation. Up to about 0.4%C it rapidly drops from 1200°C to 900°C as the amount of carbon increases. After that, even if the amount of carbon increases, it stays constant at 900°C or so. T_{Hs} is not so affected by the amount of carbon and stays constant at about 750°C.

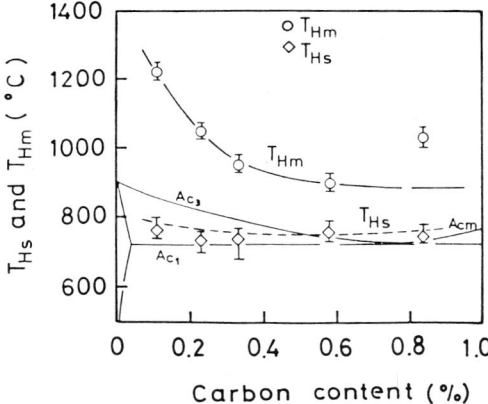

Fig. 36 Effect of C% on T_{Hm} and T_{Hs}.

4. Basic Physical Characteristics of the Ultra High Energy Density Beam Weld Zone

A weld zone by high energy density beam has a specially peculiar shape in the bead cross section with narrow band. Moreover rapid heating and cooling processes are dominant in this zone during welding. These features inherently give rise to also peculiar physical properties of the weld. Here we describe about hardness and some mechanical properties.

4.1. Hardness

We have examined the hardness of the weld in about 100 kinds of steel. Generally speaking, the hardness is the function of both the element term f (E) and cooling term

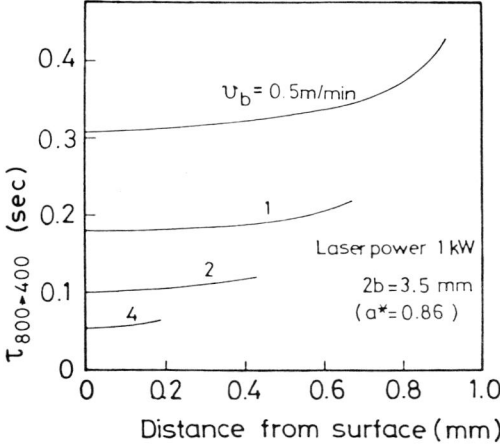

Fig. 34 Cooling time from 800°C to 400°C ($\tau_{800 \to 400}$).

$$H_V = F\left\{f(\tau), f(E)\right\} + F_0 , \qquad\qquad ----- \text{[A]}$$

$$H_V = F\left\{f(\tau_{T_1 \to T_2}), C_{eq}\right\} + B , \qquad\qquad ----- (1)$$

$$F = \frac{A}{\tau_{T_1 \to T_2}^k} C_{eq'} , \qquad\qquad ----- (2)$$

$$H_V = \left\{\frac{840}{\tau_{800 \to 500}^{0.22}} - C_{eq} + 58\right\} \pm 66 , \qquad\qquad ----- (3)$$

$$[C_{eq}] = [C] + \frac{[Mn]}{2.4} + \frac{[Si]}{24} + \frac{[Ni]}{14} + \frac{[Cr]}{16} + \frac{[Mo]}{60} ,$$

$$\dot{\tau}_{800 \to 500} \fallingdotseq 3.8 \times 10^{-2} \left[\frac{0.8 I_b V_b}{v_b h_p}\right] \left[\frac{1}{(500 - T_0)^2} - \frac{1}{(800 - T_0)^2}\right] .$$

Fig. 37 Arata electron beam weldability.

f (τ) as expressed in equation (A) of Fig. 37. However, conventional hardness equations ever proposed have been based on the element term only by using carbon equivalent and the term f (τ) was completely neglected. So, we developed[53] the hardness prediction equation which includes both terms, as shown in equation (3) of Fig. 37. This relation was named as "Arata Electron Beam Weldability"[54]. **Figure 38** shows that the predicted hardness

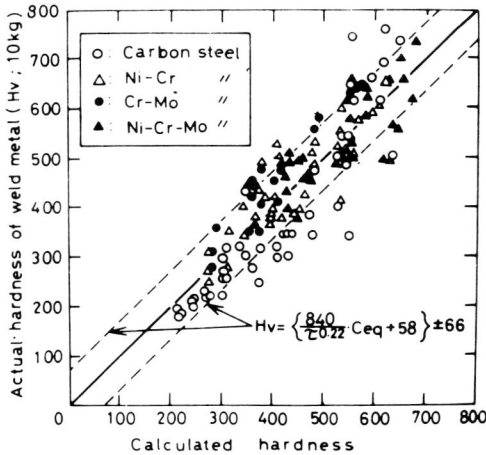

Fig. 38 Comparison of actual hardness with calculated hardness.

by this equation lies within a deviation of 66 for data published on EB welds of various types of steels conducted at a number of different research laboratories. We further developed the new hardness prediction equation by regression analysis, in order to increase the accuracy and to be applicable not only to EB and laser welds but also to arc welds[55].

Figure 39 shows the basic concept of introducing the hardness prediction equation. As is well known, the hardness changes with the cooling rate, as the curve shown in the figure shows. This characteristic curve has a close correlation with the amount of Martensite structure. Points A and B represent the points where the amount of Martensite is 100% and 0%, respectively, in the microstructure. The cooling rate and hardness at each point are formulated by regression analysis as shown at the bottom of the figure and thus the hardness at an arbitrary cooling rate is expressed by the equation in the rectangular box. In **Fig. 40** the accuracy of several hardness equations[56], [57] is compared for the same data for many kinds of steel. It is clear that the new Arata's equation can predict the hardness with much higher accuracy.

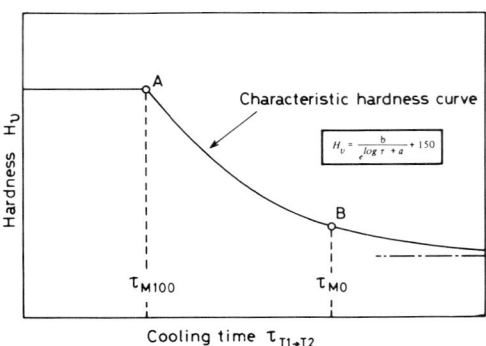

A: $H_v(\tau_{M100}) = 835[C] + 287$, $\log \tau_{M100} = 2.55([C] + \frac{1}{6.3}[Mn] + \frac{1}{3.6}[Si]) - 0.92$.

B: $H_v(\tau_{M0}) = 273([C] + \frac{1}{13}[Mn] + \frac{1}{9.7}[Si]) + 133$.

   $\log \tau_{M0} = -0.37([C] - \frac{1}{1.1}[Mn] - \frac{1}{0.44}[Si]) + 1.02$.

Fig. 39 Relation between cooling time and hardness.

4.2. Fracture path transition temperature: $_{\mathrm{I}}T_F$

The other problem of mechanical properties that is inherently caused by the narrow bead shape is the problem of "Fracture Path Transition Temperature, $_{\mathrm{I}}T_F$".[58] **Figure 41** shows some examples of Charpy, deep notch and COD test. In every test, the behavior of the fracture path is almost the same and is dependent on the temperature. This is a significant problem, particularly when we evaluate the impact test values.

In order to evaluate the change in fracture path quantitatively, we have defined the following parameters as shown in **Fig. 42**. δH_v is the relative hardness difference between the weld and mother plate, and F_p is the fracture path parameter which represents the shape factor of the

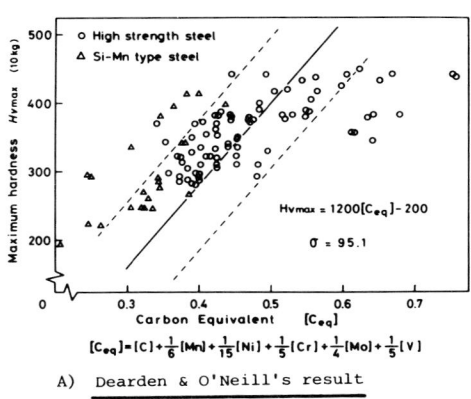

$$[C_{eq}] = [C] + \tfrac{1}{6}[Mn] + \tfrac{1}{15}[Ni] + \tfrac{1}{5}[Cr] + \tfrac{1}{4}[Mo] + \tfrac{1}{5}[V]$$

A) Dearden & O'Neill's result

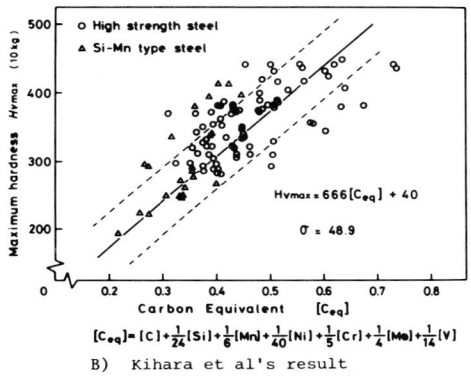

$$[C_{eq}] = [C] + \tfrac{1}{24}[Si] + \tfrac{1}{6}[Mn] + \tfrac{1}{40}[Ni] + \tfrac{1}{5}[Cr] + \tfrac{1}{4}[Mo] + \tfrac{1}{14}[V]$$

B) Kihara et al's result

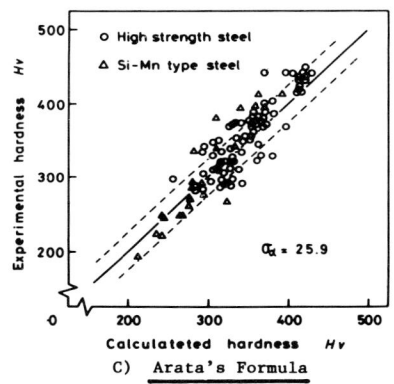

C) Arata's Formula

Fig. 40 Comparison between traditional equations and Arata's formula.

Heat input 10 kJ/cm	Testing temperature	
	0°C	−60°C
Charpy test (HT50)	10 mm	10 mm
COD test (HT50)	10 mm	10 mm
Deep notch test (HT80)	10 mm	10 mm

Fig. 41 Some examples of COD, deep notch and charpy tests.

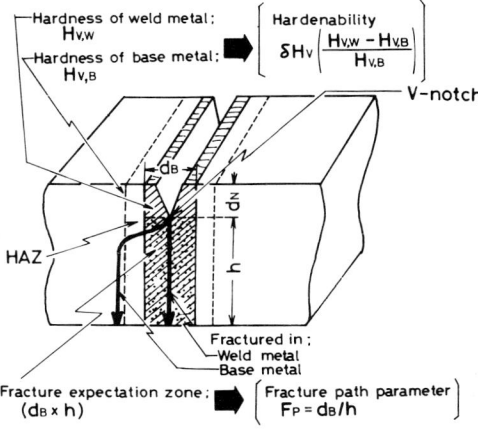

Fig. 42 Schematic illustration for two types of fracture path observed in charpy test etc.

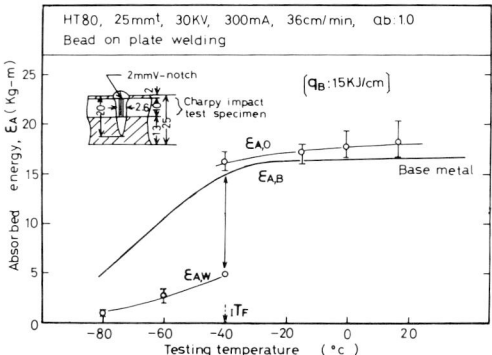

Fig. 43 Transition temperature curve.

bead. **Figure 43** shows an example of impact test, in which one can clearly see the existence of fracture path transition temperature $_IT_F$ when Charpy test is conducted on a welded joint. ϵ_{AO} shows when fracture path turns aside from the weld to the base metal, and ϵ_{AW} shows when the path remains in the weld metal. This figure also includes the test results of the base metal shown by curve ϵ_{AB}. In this way, the impact value for EB welded joint differs considerably depending on the fracture path. It is, therefore, very important to consider this phenomenon when one examines the impact values for the welded joint. While **Fig**. 44 shows an example how the transition

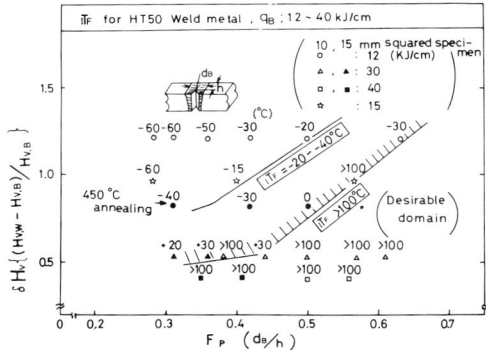

Fig. 44 Settlement for $_IT_F$ by using δH_v and F_p parameters.

temperature changes also with the difference in hardness and the fracture path parameter. These parameters are a function of the welding conditions and material, so one has to choose the optimum conditions using this kind of diagram. In the case of laser welding, similar results are obtained.

5. Application of Ultra High Energy Density Beams

5.1. Electron beam welding

5.1.1. *Horizontal electron beam welding*

In electron beam welding, it is essential that the electron gun should operate efficiently, and since the gun is considerably heavy and a high degree of accuracy is required, it must in general be fixed in position or supported by a machine as a robot. Therefore, it was said to be too difficult to perform electron beam welding in a variety of positions. It has been made possible by developing a beam deflector[59], as shown in **Fig. 45**. Figure 45 shows both

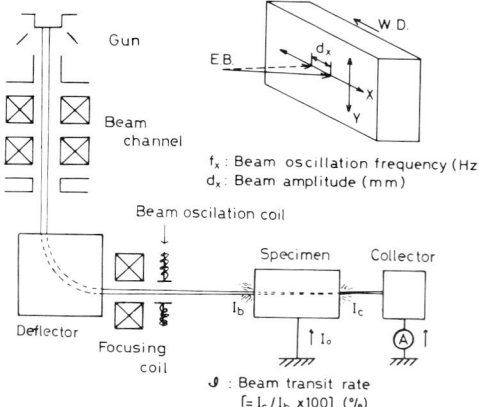

Fig. 45 Schematic diagram of beam oscillation at horizontal beam.

the experimental method and conditions. Namely, we used the 90° deflected horizontal beam. Some important beam parameters are explained here. We defined the welding direction as the X axis and the perpendicular direction as the Y axis. In general, X-oscillation is effective for materials with low heat conductivity such as steel, and Y-oscillation is good for highly conductive materials such as Aluminum. Another important parameter is the beam transit rate, which defined by I_c over I_b. The materials used in this experiment were Cr-Mo steel (2¼Cr-1Mo), stainless steel (SUS 304), high tensile strength steel (HT 50, 80), centrifugally cast steel pipe for welded structures (SMK 50) and thick plates of a thickness of 100 mm or more. In the case of these thick plates, however, a number of defects such as porosity appear if suitable beam conditions are not chosen. One means of inhibiting such phenomena is to produce beam oscillation, i.e., to cause the beam to oscillate along the transverse line of welding by applying X-oscillation, or to cause it oscillate perpendicularly through Y-oscillation by means of a suitable magnetic field, as shown in **Fig**. 45. The oscillations are indicated by frequency f_X, f_Y (Hz) and amplitude d_X, d_Y (mm).

The oscillation which should be chosen and the question of whether a compound oscillation (circular or elliptical in shape) is suitable depends on the material. **Figure 46** shows the effect of X-oscillation (f_X) when 100

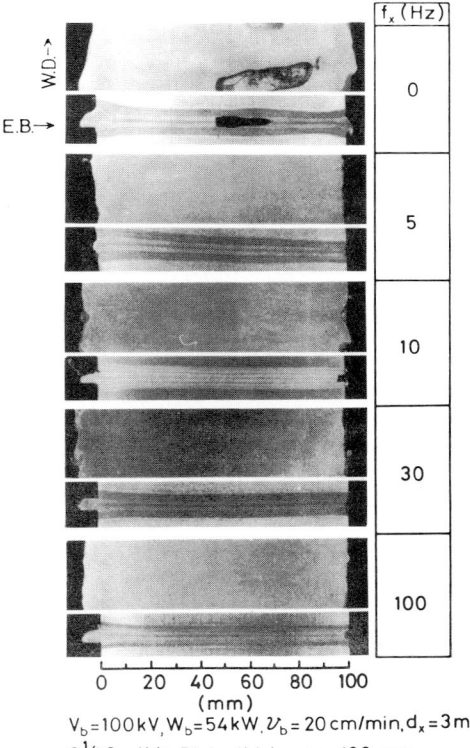

V_b=100kV, W_b=54kW, $\mathcal{V}_b$ = 20 cm/min, d_x =3mm
2¼Cr-1Mo. Plate thickness= 100 mm

Fig. 46 Fully penetrated bead sections of horizontal position welding with various beam oscillation frequency.

mm thick 2¼Cr-1Mo steel was welded at a welding speed of $\mathcal{v}_b$ = 20 cm/min[60]. **Figure 47** further illustrates these bead conditions. The results obtained are indicated by the symbols in the figure, which show the porosity rate R_P (= S/S_o) and under-fill rate R_u (= S_u/S_o). Conditions are optimum when $f_x \simeq 10$ and $d_x \simeq 2 \sim 5$. R_p and R_u being limited to almost 0. Furthermore, a condition called "parallel bead" can be obtained in the welded part, where the bead width is almost uniform throughout.

Figure 48 shows the effect of the oscillation amplitude. By selecting the optimum frequency and amplitude, it is possible to obtain a sound weld with a narrow band-shaped bead.

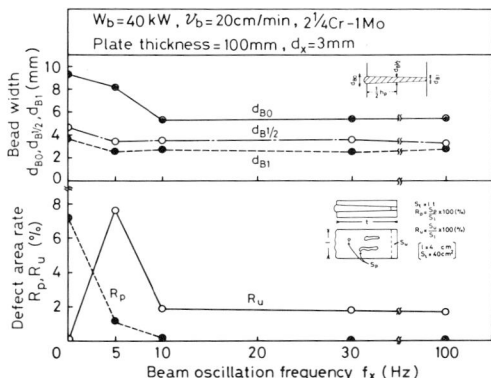

Fig. 47 Relation between bead width, defect rate and beam oscillation frequency.

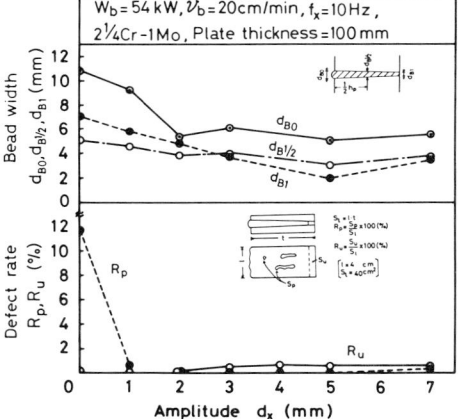

Fig. 48 Relation between bead width, defect rate and beam amplitude.

In full penetration welding, the important factor is the value of $\mathcal{I}$, the beam pass rate of the beam current. $\mathcal{I} = I_c/I_b$: I_b indicates the incident beam current, I_c indicates the collected beam current [shown in **Fig. 45**]. The influence of the $\mathcal{I}$ values is shown in **Fig. 49**, and the range of efficiency is $\mathcal{I} \simeq 10 \sim 50$. When $\mathcal{I} \simeq 10$ and $f_x \simeq 10$, porosity completely disappeared although some under-fill still remained. When $f_x \simeq 30 \sim 100$, the under-fill also almost totally disappeared (except for $1 \sim 2\%$). The relationship of f_x, d_x and $\mathcal{I}$ is most important.

Figure 50 shows the examples of good results obtained by beam control for various kinds of steel. It is clear that the appropriate selection of beam conditions is very important to obtain a sound weld.

Figure 51 shows the influence of dissolved oxygen and nitrogen contents in steel on defect formation. As seen in

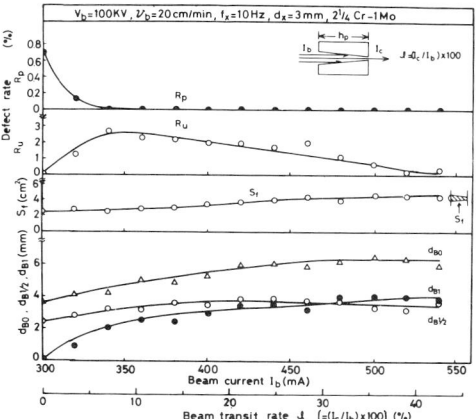

Fig. 49 Relation between bead width, defect rate and I_b, $\mathcal{J}$.

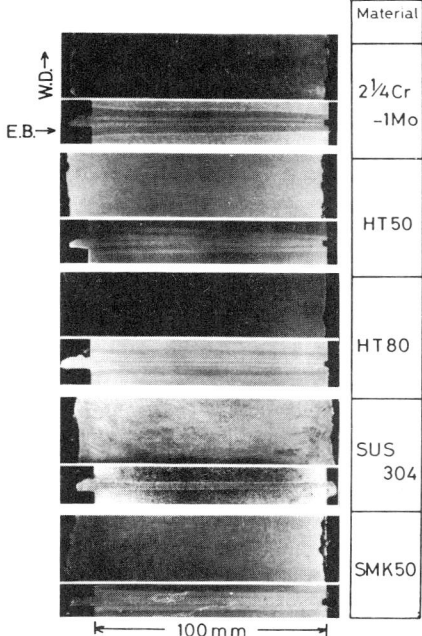

$V_b=100kV$, $W_b=40kW$, $\mathcal{V}_b=20\,cm/min$
Plate thickness= 100 mm, $f_x=10\,Hz$, $d_x=3mm$

Fig. 50 Fully penetrated bead sections of horizontal position welding with various metals.

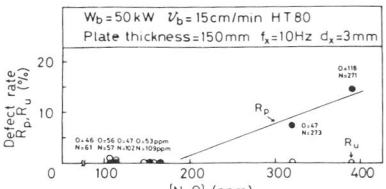

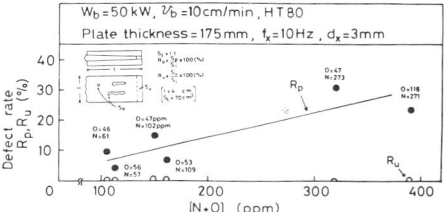

Fig. 51 Relation between defect rate and contents of [N + O].

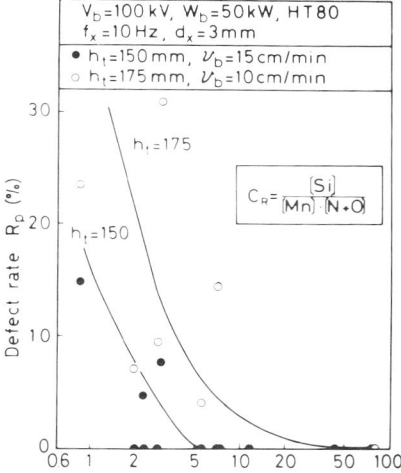

Fig. 52 Relation between defects rate and C_R.

this picture, a slight increase in plate thickness —for example, from 15 to 17.5 cm— bring about considerable effect on porosity formation. For conventional steels, the defect rate is well evaluated by the total amount of oxygen and nitrogen.

Figure 52 shows the effect of volatile elements in steel on the porosity appearance in the case of full penetration welding. In order to evaluate the contribution of these elements, we have introduced a new parameter C_R defined in the box of the figure.

In the vertical upward position welding by electron beam, the penetration depth does not saturate even at very low speed and the beam hole looks stable. It is, therefore, possible to weld very thick plates, even plates as thick as 30 cm, in this position. However, as in **Fig. 53** the grain size becomes coarse as in electro-slag welding. This lowers the impact value and make it undersirable for practical use. Therefore, it is necessary to choose an opti-

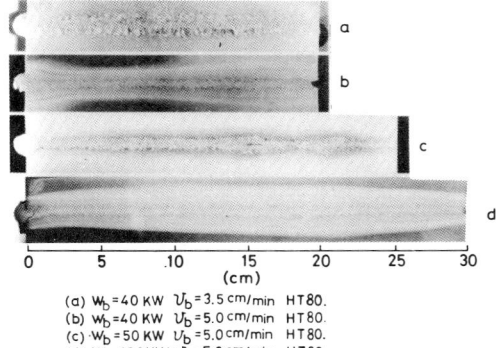

(a) W_b =40 KW v_b =3.5 cm/min HT80.
(b) W_b =40 KW v_b =5.0 cm/min HT80.
(c) W_b =50 KW v_b =5.0 cm/min HT80.
(d) W_b =100KW v_b =5.0 cm/min HT80.

Fig. 53 Fully penetrated bead cross sections of upward vertical position welding.

Fig. 54 Appearance of welded HT 60 steel pipe (diameter 1.4 m, thickness 50 mm).

mum welding speed to yield the appropriate cooling rate for the material used, in order to establish the practical EB welding of 20 to 30 cm thick plate. For this sake, it became clear that a conventional 100 kW EB welder still lacks the necessary power.

Figure 54 shows[61] the example of HT 60 steel of 50 mm thickness welded by local vacuum horizontal electron beam welding. Such local vacuum EB was also applied to a vacuum chamber of SUS 304L for nuclear fusion research.

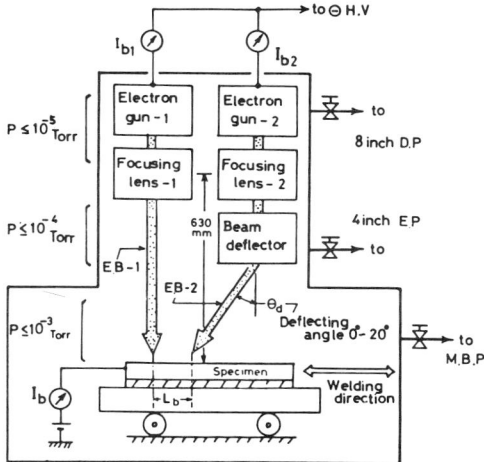

Fig. 55 Schematic diagram of Tandem Electron Beam welder.

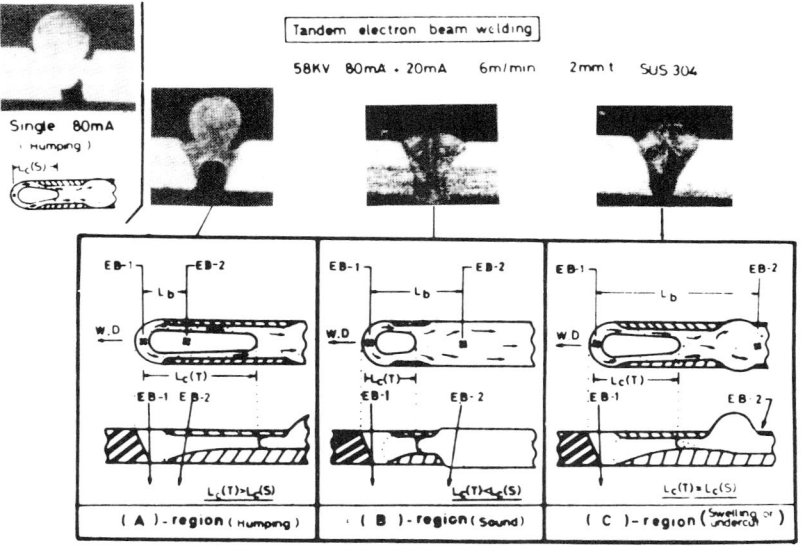

Fig. 56 Humping phenomenon and its suppression mechanism by TEB welding.

5.1.2. *Tandem electron beam welding*

Electron beam welding has many advantages compared with ordinary arc welding. However, it also has disadvantages caused by its high energy density, such as humping, spiking and root porosity. In order to overcome these problems, the author developed "Tandem Electron Beam Welding (TEB Welding)" as shown in **Fig. 55**[62]. This method utilizes two electron beams at the same time, one beam as a conventional single electron beam and the other as sub-beam for the repairment of welding defects.

In the case of high speed welding humping frequently occurs, as shown in **Fig. 56**. By impinging the second beam in the proper position, and separating it from the first beam by a "Tandem Gap", the humping phenomenon is suppressed as shown in **Fig. 56**. Suppression mechanism of humping is explained as shown in this figure. The second beam impinging in the proper position of molten pool changing the flow of molten metal so that it flows smoothly backward.

TEB welding can also suppress the welding defects such as spiking and root porosity which occur in deep penetration welding as shown in **Fig. 57**[63]. The suppression mechanism can be explained as follows: In the case of deep penetration welding, the second beam impinges onto the beam hole of the first beam. When the energy density of the second beam is properly low, the beam can stabilize strongly the beam hole and reheats the root zone, making any spiking and root porosity disappear. This process was revealed by high speed observa-

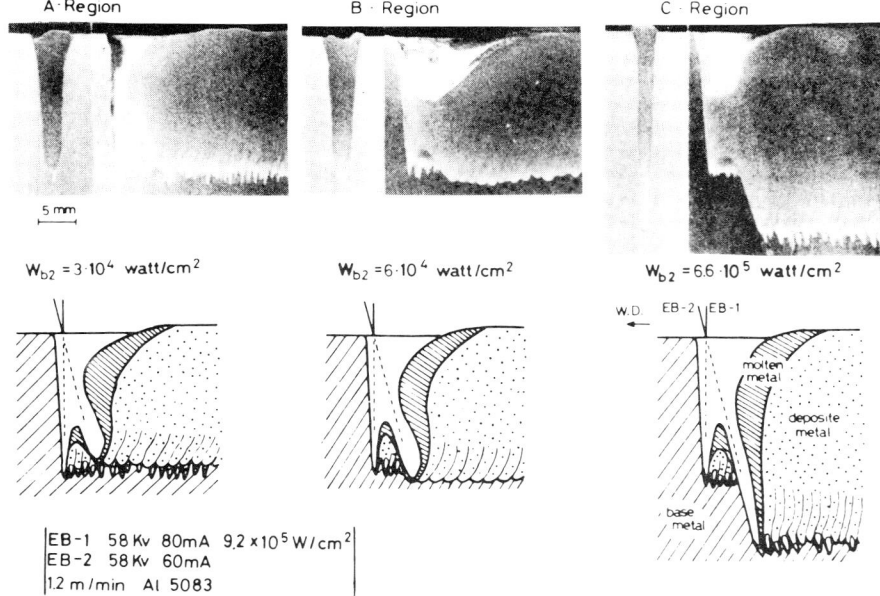

Fig. 57 Spiking phenomenon and its suppression mechanism by TEB welding.

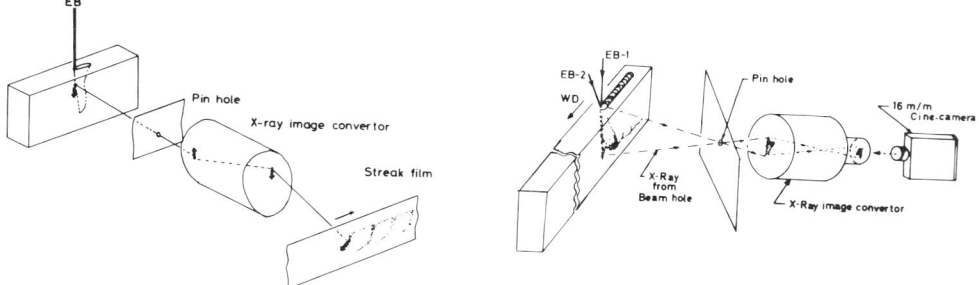

Fig. 58 High speed observation method of beam hole during welding.

24

tion of the beam hole X-ray[64] and transmission X-ray method[65] shown in **Fig. 58**.

The optimum power ratio between the first and second beam was determined to be 10% for aluminum alloy of 20 mm penetration[65]. For carbon steel of 30 mm penetration, it was also determined to be about 15%.

5.2. Laser welding

5.2.1. *Atmospheric laser welding*

Laser welding using a high power CO_2 laser can easily weld the specimen under atmospheric condition. However, in the case of atmospheric laser welding, so called "laser plasma" occurs and affects strongly the welded bead. This laser plasma can be suppressed by using inert gas as an assist gas, as shown in **Fig. 59**. The pressure of such assist

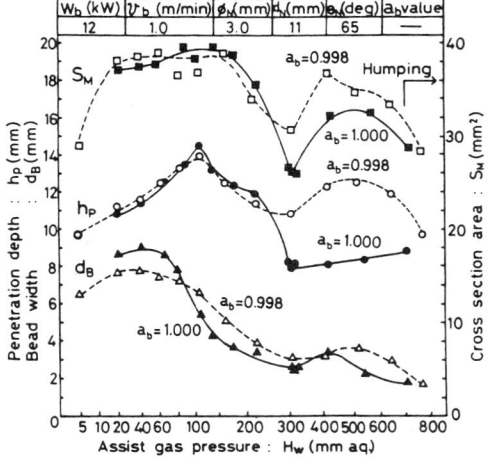

Fig. 59 Assist gas nossle for high power CO_2 laser welding.

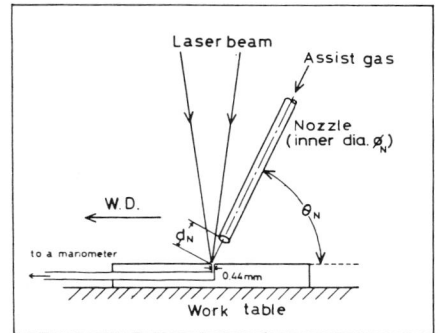

Fig. 60 Influence of assist gas on penetration depth and bead width.

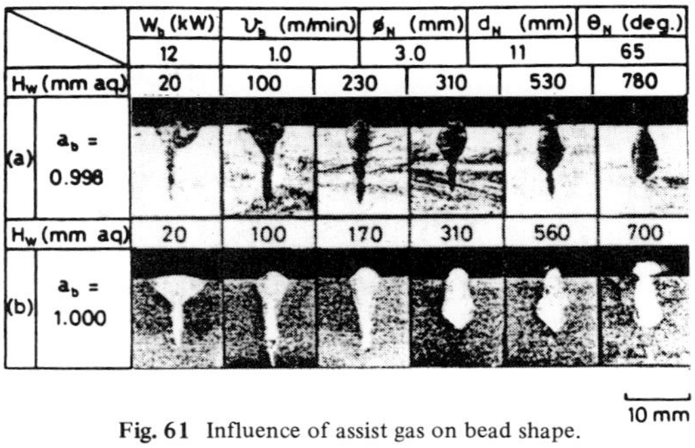

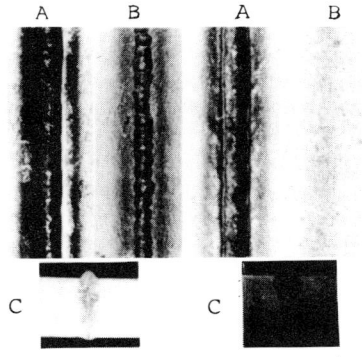

Fig. 61 Influence of assist gas on bead shape.

gas has a strong influence on the penetration depth, bead width and bead shape, as shown in **Fig. 60** and **Fig. 61**[66].

As the pressure of the assist gas increases, the amount of laser plasma decreases and the shape of the bead cross section changes from a wine-cup shape to an egg shape passing through a nail-head shape. A particular assist gas pressure should be selected for maximizing the penetration depth.

5.2.2. *Laser spike seam welding*

One of the most effective solutions for dealing with this laser plasma problem is the "Laser Spike Seam Welding" (LSSW) method invented by the author. In this process, the laser beam is oscillated so that it follows the movement of the specimen. The laser beam stops relative to the specimen for a certain period, it drills the specimen as a pulsed beam, then it quickly returns to its original point to keep away the laser plasma. **Figure 62** shows a

a) LSSW   b) Conventional

A; Front Bead, B; Back Bead
C; Cross Section

Fig. 62 Comparison of bead appearance and cross section between LSSW and conventional laser welding under the same welding conditions.

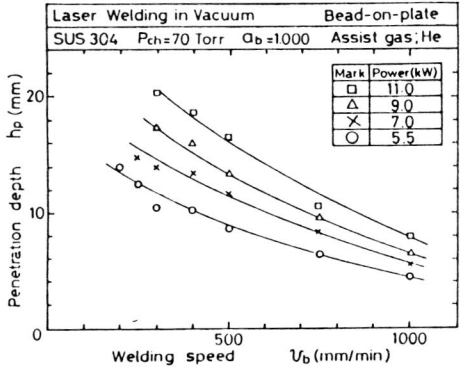 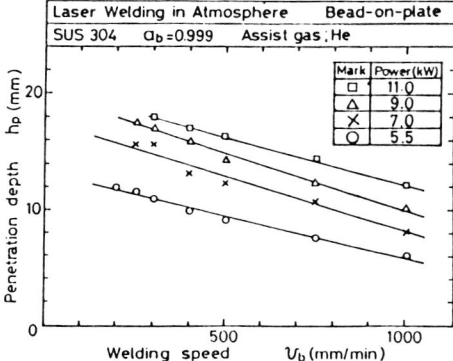

Fig. 63 Comparison of penetration depth between atmospheric and vacuum laser welding.

comparison of the bead appearance and cross section between LSSW and conventional laser welding under the same welding conditions: power, welding speed, gas flow rate etc. It can be seen that the LSSW process is superior in both the penetration depth and bead appearance[67].

5.2.3. *Vacuum Laser welding*

Vacuum laser welding is another solution for the problem of laser plasma. In a low vacuum of a few tens of Torr, less laser plasma is produced and the penetration depth increases with decreasing welding speed. This tendency is more evident than in atmospheric laser welding, as shown in **Fig. 63**[68].

5.3. Differences and similarities in processing characteristics of laser beam and electron beam

(1) The electron beam is easy to propagate and control in a vacuum, but its propagation range in air is extremely short. Particularly, its working distance for use as a heat source is too short to control, and so its working function is limited. On the other hand, the CO_2 laser beam is easy to propagate and control in air, and has good workability. However, during passage through hot gas or plasma, its energy absorption is so high that its working function is restricted. Thus, the larger the laser output the harder the plasma removal in air. However, in a vacuum, since the laser plasma disappears, the workability of these two beams (EB and LB) becomes similar. These phenomena were revealed by dynamic observation of the beam hole during actual welding[64],[69].

(2) "Bead transition" occurs in laser beams, but never in electron beams. Thus, extremely high precision is required for controlling the working distance of laser

beams.

(3) Laser beams have a particularly high reflection coefficient to metallic materials. Although the energy absorption rate of laser beams is not so high, it can be high to surface oxidation, non-metallic coating or non-metallic material. An electron beam causes a charge-up to non-metallic materials.

(4) Electrical power efficiency of beam generation is high for the electron beam but low for the laser beam. However, the latter generally has a better working performance in air.

(5) In electron beam welding, the thicker the plate the more effective the welding characteristics, but in the case of laser beam, the weldable plate thickness is limited practically, on the contrary, the laser beam is more suitable for cutting.

(6) Concerning surface treatment, each of the two beams has respective advantages and disadvantages.

References

1) J.R. Pierce; J. App. Phys., 11 (1940).

2) K.H. Steigerwald; Verhandl. deut. Physik., Gas, 4, 123 (1953).

3) J.A. Stohr; Fuel Elements Conference, Paris, November, 18-23 (1957).

4) T.H. Maiman; Nature, Vol. 187 (1960).

5) C.K.N. Patel; Phys. Rev., 136-5A (1964).

6) W.B. Bridges; Appl. Phys. Lett., 4-128 (1964).

7) M. Okada, Y. Arata, K. Nishiguchi and H. Maruo; Reports in the First Symposium of Atomic Energy in Japan, 1, 409 (1957).

8) Y. Arata; Chokoon Kenkyu, 1-6 (1957).

9) M. Okada and Y. Arata; Tech. Report of Osaka Univ., 13-567 (1963).

10) Y. Arata and M. Tomie; Trans. of JWRI, 2-1 (1973).

11) Y. Arata and M. Tomie; Journal of JWS, 46-7 (1977).

12) Demonstration of the Research Center for Ultra High Energy Density Heat Source in JWRI, October (1980).

13) Y. Arata and M. Tomie; Journal of High Temperature Society, 10-3 (1984).

14) Y. Arata and M. Tomie; Trans. of JWRI, 2-1 (1973), 4-1 (1975), 5-1 (1976); IIW. Doc VI-112-73 (1973); 2nd International Symposium of JWS, (1975); 7th International Conference of Electron and Ion Beam Science and Technology (at Washington, D.C., USA) (1976); Proc. of Inter. Conf. on Welding Research in 1980's, Poster Session, October (1980) (at JWRI).
A. Sanderson; Metal Construction BWJ, 6-1 (1974).
G. Sayagh, P. Dumonte and T. Nakamura; 2nd International Symposium of JWS, (1975).
K.S. Akop' yant, et al.; Automatic Welding, 28-4 (1975).

15) D.R. Whitehouse; Contract No. DA-01-021-AMC-12427(z).

16) W.B. Tiffany, R. Targ and J.D. Foster; App. Phys. Lett. 15-3 (1969).

17) E.V. Locke; E.D. Hoag and R.A. Hella; IEEE Quantum Electronics, QE-8-2 (1972).

18) C.M. Banas; Final report under Naval Research Laboratories Contract N00173-76-M-0107, UTRC. Report, No. R76-912260-1, August (1976).

19) Y. Arata and I. Miyamoto; Tech. Report of Osaka Univ., 17-285 (1967).

20) Y. Arata and M. Tomie; Trans. of JWS, 1-2 (1970).

21) H. Maruo, I. Miyamoto and Y. Arata; Proc. 1st Int. Laser Processing Conf. (1981); Proc. Annual Meeting of JWS (in Japanese) (1978).

22) H. Maruo, I. Miyamoto, T. Ishida and Y. Arata; Proc. 1st Int. Laser Processing Conf. (1981).

23) S. Katayama, A. Matsunawa, A. Morimoto, S. Ishimoto and Y. Arata; Proc. 3rd CISFFEL (1983).

24) Y. Arata; J. Phys. Japan 43-3 (1977).

25) Y. Arata and I. Miyamoto; Tech. Report of Osaka Univ., 19-887 (1967).; Y. Arata, I. Miyamoto and M. Kubota; IIW Doc. IV-4-69 (1969).

26) Y. Arata and I. Miyamoto; Trans. of JWS, 3-1 (1972).

27) AVCO HPL Lasers Tech. Note, No. 6 (1977).; R.M. Feinberg; Proc. Int. Conf. on Welding Research in the 1980's (1980).

28) Y. Arata; Journal of JWS, 41-11 (1972).

29) Y. Arata, M. Tomie and Y. Kato; Trans. of JWRI, 2-1 (1973).

30) Y. Arata, T. Ishimura and I. Miyamoto; Trans. of JWRI, 2-1 (1973).

31) Y. Arata, K. Inoue, H. Maruo and I. Miyamoto; Lectures of the International "Beam Technologegy" Conference in Essen on the 7th and 8th, May (1980) DVS-BERICHTE.

32) Y. Arata, H. Maruo, I. Miyamoto and F. Kawabata; Journal of JWS, 49-10 (1980).

33) A. Sanderson; British Welding Journal, 15-10 (1968).

34) P. Bwmonte et al.; IIW Doc. IV-131-73 (1973).

35) G. Sayagh; IIW Doc. IV-276-79 (1979).

36) Y. Arata, M. Tomie, H. Nagai and T. Hattori; Trans. of JWRI, 2-2 (1973).

37) H. Irie, T. Hashimoto and M. Inagaki; Journal of JWS, 46-9 (1977).

38) Y. Arata and I. Miyamoto; Journal of JWS, 39-12 (1970).

39) Y. Arata and I. Miyamoto; Trans. JWS, 3-1 Report 1 (1972).

40) E. Hagen and H. Rubens; Ann. Physik, 11 (1903).

41) Y. Arata and I. Miyamoto; Trans. JWS, 3-1, Report 2 (1972); Trans. of JWRI, 2-2 (1973).

42) Y. Arata and I. Miyamoto; Second Int. Symp. of JWS, (1975).

43) A.A. Wells; Weld. J., 22-5 (1952).

44) Y. Arata and K. Inoue; Trans. of JWRI, 2-1 (1973).

45) Y. Arata and I. Miyamoto; Trans. of JWRI, 1-1 (1972).

46) Y. Arata and M. Tomie; Journal of JWS, 46-8 (1977).

47) M.J. Adams; Brit. Weld. J., 15-3 (1968).

48) A. Sanderson; Brit. Weld. J., 15-10 (1968).

49) Kawasaki Heavy Industries Ltd., Private communication.

50) Y. Arata, M. Ohsumi and Y. Hayakawa; Trans. of JWRI, 5-1 (1976).

51) E.L. Locke, E.D. Hoag and R.A. Hella; IEEE Quantum Electronics QE-8-2 (1972).

52) Y. Arata, H. Maruo and I. Miyamoto; IIW. Doc. IV-241-78 (1978); 212-436-78 (1978).

53) Y. Arata, F. Matsuda and K. Nakata; Trans. of JWRI, 1-1 (1972); 2-1 (1973).

54) M.J. Bibby, J.A. Goldak and G. Burbidge; Weld. J., 54-8 (1975).

55) Y. Arata, N. Ohji, N. Kohsai and K. Nishiguchi; 2nd Int. Colloq. for EBW and Melting (1978); IIW. Doc. IV-263-79 (1979); Trans. of JWRI, 8-1 (1979).

56) J. Dearden and H.O. Neill; Trans. Inst. Weld. (U. K.), 1-3 (1940).

57) H. Kihara, H. Suzuki and H. Tamura; IIW Doc. No. 1, IX-288-61 (1961).

58) Y. Arata, Y. Shibata and S. Fujihira; Trans. of JWRI, 3-2 (1974); 4-2 (1975).; Y. Arata; Journal of SMS, 27-301 (1978).

59) Y. Arata and M. Tomie; 2nd International Symposium of JWS, (1975).

60) Y. Arata and M. Tomie; Trans. of JWRI, 9-2 (1980).

61) K. Shinada, Y. Kondo, S. Satoh, T. Shimoyama, G. Takano, M. Minami, T. Tanaka and Y. Arata; Proc. Int. Conf. on Welding Research in the 1980's (1980).

62) Y. Arata and E. Nabegata; Trans. of JWRI, 7-1 (1978).

63) Y. Arata, E. Nabegata and N. Iwamoto; Trans. of JWRI, 7-2 (1978).

64) Y. Arata, N. Abe and S. Yamamoto; Trans. of JWRI, 9-1 (1980).

65) Y. Arata, N. Abe, H. Wang and E. Abe; Trans. of JWRI, 11-2 (1982).

66) Y. Arata, T. Oda and R. Nishio; Trans. of JWRI, 12-2 (1983).

67) Y. Arata, N. Abe and T. Oda; Proc. ICALEO'83 (1983).

68) Y. Arata and T. Oda; Journal of High Temperature Society, 10-1 (1984).

69) Y. Arata; What Happens in High Energy Density Welding and Cutting?, 1980).

High Technology for Materials Processing Based on Welding

Abstract

When viewing the recent progress in welding science and technology, it can be seen that such progress is not only due to the development and expansion of existing technology, but also to the integration of individual processes to produce higher level technologies. A variety of new technologies have also come into being based on welding and related fields. In this paper will be reviewed the latest highly intergrated welding processes, and also some new technologies based on welding engineering. Some typical examples of advanced welding technologies which will be examined include: 1) High energy density welding, such as electron beam and laser welding 2) Welding by robots. Several new technologies will also be discussed to show how integrated welding knowledge can greatly contribute to those areas. These include: 3) Surface modification or treatment of materials by the use of high energy density heat sources 4) Methods of producing composite materials by modifying conventional welding methods 5) Making substances in different states, such as in amorphous form, ultra fine particles, and so on.

1. Introduction

Welding technology has served as the basis for the development of today's advanced manufacturing technologies. Welding is one of the most important processes used in the production of manufactured parts and goods, and also plays a vital role in most structures. Advanced technical and economic considerations have created a strong demand for the development of high precision, high performance welding techniques and systems in fields ranging from major architectural superstructures to minute electronic components.

Indeed welding engineering consist of many fields of science and technology, and has progressed as a typical inter-disciplinary new field, materialized by the import and accumulation of knowledge and experience from different fields such as shown in **Fig. 1**.

I believe this welding field has recently matured well and been playing an important role in the development of other research fields. We may say it has grown up from the conventional welding body established on the "imported" knowledge, to the new body to be able to "export" various fruitful results to other fields.

Arc heat sources developed approximately a century ago have played a central role in the development of modern welding technology and provided the foundation for today's welding engineering and advanced welding technology, without which the world would never have seen the development of today's shipbuidling, architecture, mechanical, electrical, chemical, metallics, and electronics industries. As the heat sources for welding as well as other heat processing, arc heat sources provide many advantages over other heat sources, including high temperature and high energy density which is ten times that of a gas flame. However, it has become extremely difficult for arc heat sources to keep pace with the technological demands for higher precision, performance, and quality. Material degradation at weld joints, residual stress and strain, and excessive heat input into the material and so on are problems originating from the intrinsic properties of arc heat sources. In other words, the energy density of arc heat sources is still too low for many modern applications, and this has required the development of new heat sources.

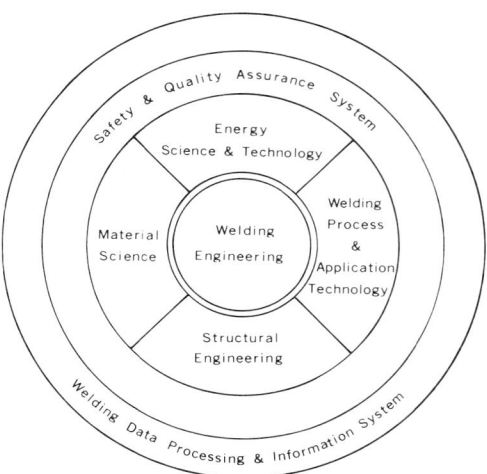

Fig. 1 Research fields related to welding engineering

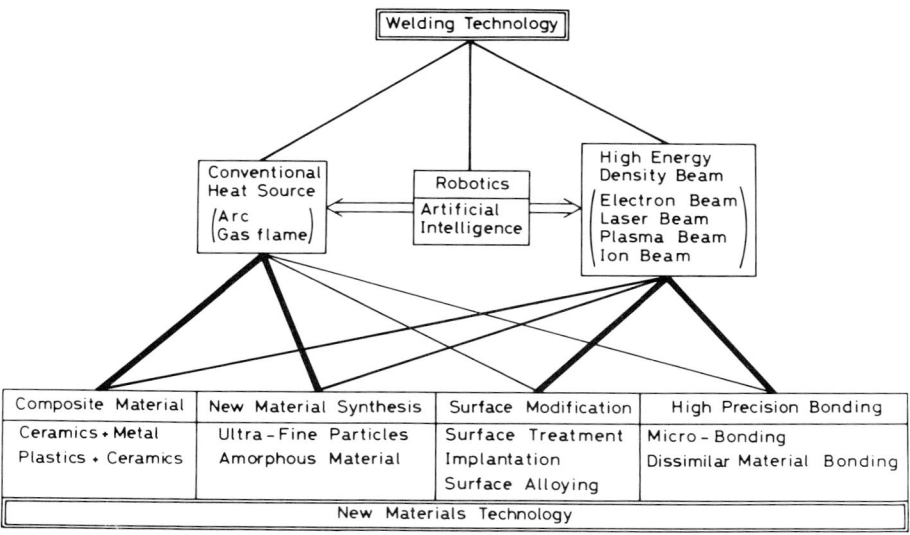

Fig. 2 New Technology based on welding

To overcome the drawbacks of arc heat sources, the author and his associates initiated research into the development and application of high energy density heat sources exhibiting characteristics quite different from arc heat sources. This research was begun approximately thirty years ago, with the greatest success coming in the last ten years[1]. These new heat sources include electron beams, laser beams and plasma beams.

Many recent advances in welding processes have come both from the above stated arc heat sources and these new high energy density beam sources. Furthermore, these technologies have been applied not only to welding and cutting techniques, but to the development of other new technologies for heat processing applications.

Figure 2 illustrates typical applications of these technologies. With two types of heat sources sustaining the development of welding technologies, consistently high precision welding quality has been achieved and welding efficiency maximized through the application of robotics. While, there are a number of new materials technologies as shown in the lower part of the figure gradually reaching a stage of development which will enable practical application in the 21st century. The latest welding engineering can bear a key to contribute to new materials technology with the development of fine control systems for processing. I believe this new trend in welding engineering will cultivate the growth and creation of other new fields. These include, for example, composite materials, and synthesis and qualitative improvement of new materials.

The relative importance of the current relationship between new materials technologies and the heat sources which play an important role in welding engineering is indicated by the thickness of the line. For example, improvement in methods utilizing conventional heat sources will enable bonding of ceramics to metals in the composite materials field. New arc control techniques are also being studied for the manufacture of ultrafine powders in the field of new material synthesis. Furthermore, high energy density beams are being examined with regard to high precision control of heat source for applications in the modification and treatment of other materials.

New materials technologies using high energy density beams are believed essential to the future of advanced control technologies. In this paper will be briefed the technical topics in welding processes using high energy density beams, and will also look at robotics and trends in integrated welding processes. Furthermore, trends in new research topics addressing the development of next-generation materials technologies based on the existing welding technologies will be considered in the context of composite materials and new material synthesis.

2. High Energy Density Beam Welding and Cutting

Advances in welding technologies refered to as the application of thermal energy to material processing paralleled the history of heat source development. Advances in

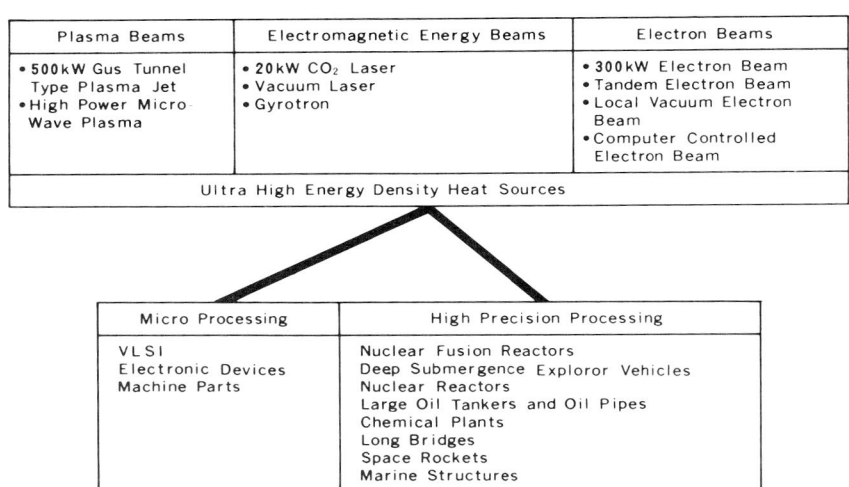

Plasma Beams	Electromagnetic Energy Beams	Electron Beams
• 500 kW Gus Tunnel Type Plasma Jet • High Power Micro Wave Plasma	• 20 kW CO_2 Laser • Vacuum Laser • Gyrotron	• 300 kW Electron Beam • Tandem Electron Beam • Local Vacuum Electron Beam • Computer Controlled Electron Beam

Ultra High Energy Density Heat Sources

Micro Processing	High Precision Processing
VLSI Electronic Devices Machine Parts	Nuclear Fusion Reactors Deep Submergence Exploror Vehicles Nuclear Reactors Large Oil Tankers and Oil Pipes Chemical Plants Long Bridges Space Rockets Marine Structures

Fig. 3 Ultra high energy density heat sources and their application

welding heat sources, the first of which utilized the heat of oxidation reaction, have grown steadily in both power output and energy density through the development and application of electric resistance Joule heat and arc heat sources. Requirements for increasing the power output and energy density of these heat sources still continue to develop and apply ultra high energy density beam heat sources, typified by electron beams. laser beams as electrmagnetic energy beams and plasma beams which are illustrated in **Fig. 3**. This section will examine current trends in welding processes utilizing these three high energy density heat sources.

2.1 Electron beam welding

Of the three ultra high energy density beams under consideration, the electron beam offers relatively higher output and the greatest ease of generation, and was therefore the first high energy density beam to be commercially applied.

The development of high power output, one major advantage of electron beam heat sources, has grown from the relatively low, several kilowatt power of early electron beam heat source to medium and high output beams generating over 100kW. Furthermore, low acceleration voltages of less than 100kV are sufficient for medium output beams, but demand for high output beams requires higher acceleration energy, too. Today 100kW class electron beam welding machines are available for commercial application.

Figure 4 shows output of strong focusing type "EBW"

apparatuses developed in recent 20 years from 1966 to 1985[2]. Circles show EBW guns for industrial use and squares show those for laboratory works. The solid line gives the maximum power for industrial use at each year. The broken line gives the maximum power in laboratory machine. As shown in this fugure, in 1967, 1972, 1974 and 1980, JWRI (Arata Lab.) was the first who developed the highest output machines at each period, and this result later led to the full-scale industrial application of high power, high density electron beams for ultra thick plate welding.

The largest electron beam welding machine currently under development is the 300kW class (acceleration voltage: 600kV, maximum beam current: 500mA) shown in **Fig. 5** and located at the Research Center for Ultra High Energy Density Heat Sources, JWRI[3]. To achieve this high voltage as high as 600kV, an unique acceleration unit, "multi-stage electromagnetic acceleration unit"[4] developed by the author is used. Single pass welding of ultra thick materials such as steel, aluminum alloys and so on is the developmental objective of this welding equipment. Realization of this objective will increase the thickness of weldable plate, enable super high speed welding, special large-scale castings beyond the reach of conventional welding techniques.

The "Electron Beam Welding Method for Ultra Thick Plates" was realized and established in 1972 for the first in the world at JWRI (Arata Lab.). **Figure 6** shows typical examples obtained by this method[5].

When these high power electron beams are applied to deep penetration welding, various welding positions are

30

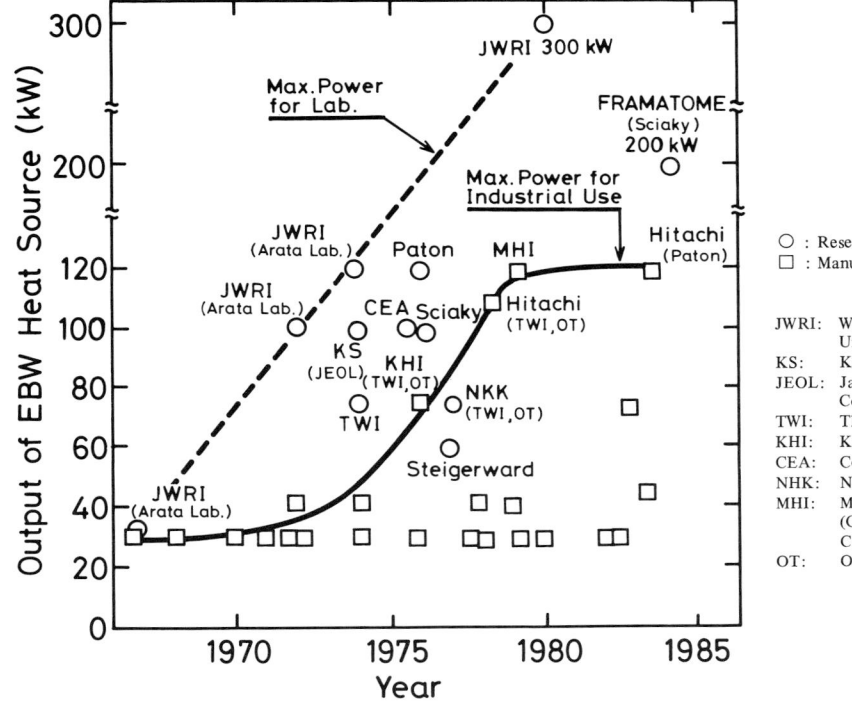

○ : Research
□ : Manufacturing

JWRI: Welding Research Institute, Osaka
        Univ.
KS: Kwasaki Steel Corporation
JEOL: Japan Electron Optics Laboratory
        Co., Ltd.
TWI: The Welding Institute, U.K.
KHI: Kawasaki Heavy. Industries, Ltd.
CEA: Commissariat a l'Energie Atomique
NHK: Nippon Kokan K. K.
MHI: Mitsubishi Heavy Industries, Ltd.
        (Cooperate with Mitsubishi Electric
        Co., Ltd.)
OT: Osaka Transformer Co., Ltd.

Fig. 4 Development of high power electron beams

Fig. 5 300kW electron beam welder

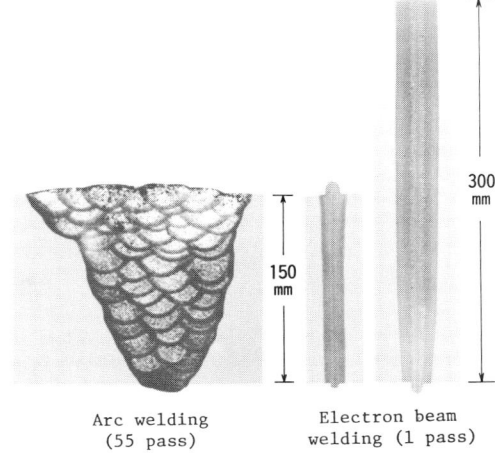

Fig. 6 Comparison of bead cross section between conventional welding and electron beam welding

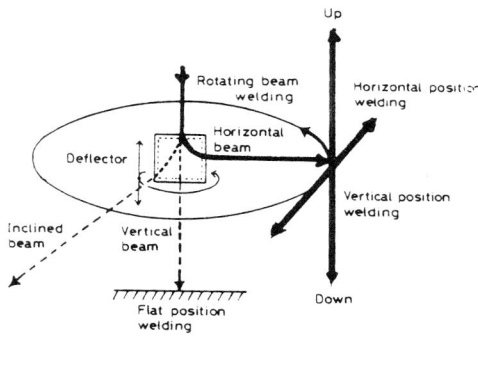

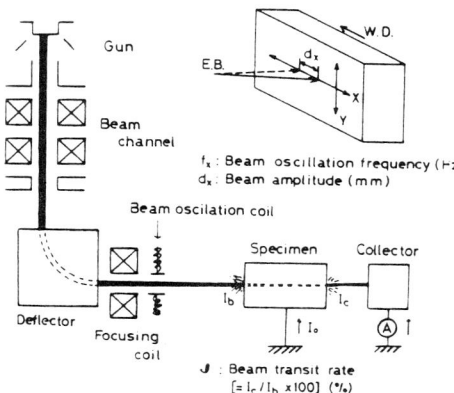

Fig. 7 All position electron beam welding method

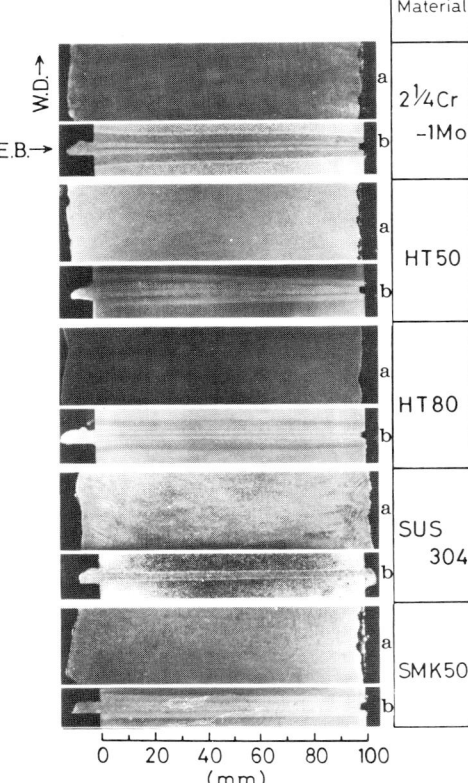

$V_b = 100 \, kV$, $W_b = 40 \, kW$, $\mathcal{V}_b = 20 \, cm/min$
Plate thickness = 100 mm, $f_x = 10 \, Hz$, $d_x = 3 \, mm$

Fig. 8 Bead cross sections of various kinds of material welded by electron beam
a. Longitudinal   b. Transverse

required. In the upper part of **Fig. 7**, a schematic diagram of all position electron beam welding method developed at JWRI is shown[6]. In the lower part, horizontal welding position is shown as an example. Deep penetration welding of ultra thick plates of various materials became possible by this method.

Figure 8 shows the bead cross sections of 10cm thick plates in various weld materials obtained by this system[6]. A narrow heat affected zone (HAZ) is obtained, thus minimizing stress and maximizing weld precision. This result led to the practical industrial application of the system.

Certain weld defects characteristics of high energy density beams, including spiking, porosity, and cold shut, can be somewhat improved by oscillating the electron beam, but one attempt to find a more fundamental solution is the Tandem Electron Beam welding method[7] illustrated in **Fig. 9**. Various kinds of weld defects are likely to be created in high energy density beam welding by the reason of high energy density itself. This Tandem

Electron Beam welding method combines two beams very skilfully as shown in Fig. 9 (a); the first beam offers the advantages characteristic of high energy density beams while the second beam works to correct or suppress the defects by the low energy density beam. Fig. 9 (b) and (c) show, examples of suppression of humping bead by this method. As the effectiveness of this welding machine has been confirmed at low output levels, the practicality of the machine shown in **Fig. 10** at medium output levels is now under study.

As the electron gun always works in a vacuum, electron beam welding is also characterized by an absence of atmospheric influences. This characteristic, however, also brings many kinds of weld defects which are inherent to high energy density beams, and restricts the workability

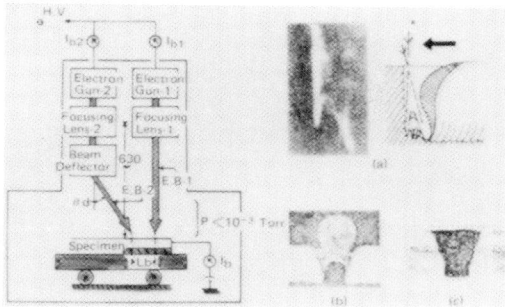

Fig. 9 Tandem electron beam welding

(a) Pinhole photograph and its explanation of beam hole X-ray
(b) Humping bead occurred in conventional high speed electron beam welding
(c) Sound bead obtain by tandem electron beam welding

Fig. 10 30kW Tandem electron beam welder

of process. Current research is attempting to resolve these shortcomings.

The vacuum welding chamber required in electron beam welding processes was thought during early developmental stages to offer significant advantages for welding reactive materials due to the clean atmosphere inherent in the vacuum chamber. But as attempts were made to utilize the deep penetration characteristic of this process in the manufacture of large structures, this apparent advantage became a major disadvantage. One attempt to solve this problem is the development of an electron beam welding machine in a low vacuum or a local vacuum environment[8]. **Figure 11** illustrates a local, low vacuum electron beam welding machine developed for welding 3.5m O.D. SUS304L toroidal vacuum chamber for nuclear fusion research. A local vacuum chamber composed of an inflation seal and lip seal maintaining a pressure of 10^{-2} Torr is moved for circumferential welding of the equatorial joint. This system produces good welding results.

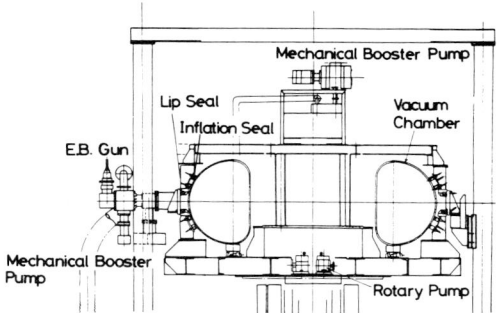

Fig. 11 Local vacuum electron beam welder

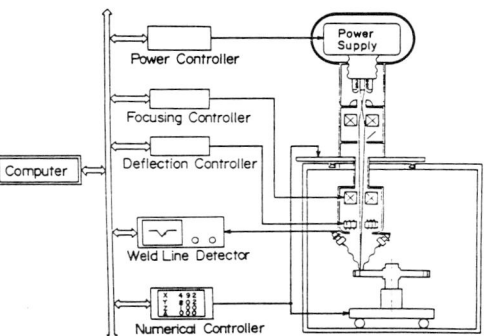

Fig. 12 Computer-controlled electron beam welder

Electron beams can be finely focused in principle, and be easily controlled. In fact, low energy electron beams for diagnostic use have been applied under the highly regulated conditions. Recent advances in electronics and computer technologies have enabled even higher precision control of high output electron beams for welding. In addition, the combination of computers and sensors has also enabled automatic weld line detection and automatic work positioning[9], as well as automatic focus detection[10]. **Figure 12** shows an on-line seam tracking system consisting of a computer and X-ray sensors; connecting this system with an NC table enables automatic welding of complicated welding lines.

2.2 Laser welding

Although the development of laser beams came later than electron beams, their relative ease of working and absence of vacuum working chamber requirements lead to rapid industrial application. At the same time, however, laser beams with high output and high quality assurance are difficult to produce. Recently numerous high output lasers in the 5kW to 20kW class have been developed and their welding applications are currently being investigated.

Practical high power CO_2 laser appratus for industrial use was first provided by AVCO in 1970's as an commercially available 15kW laser heat source, one of which was installed at JWRI as shown in **Fig. 13**. At present high output apparatuses over 10kW are produced in many countries. In the laboratory use a 25kW-calss apparatus has been developed[11] for a long time operation and a 100kW-class one for a very short time operation. It should be emphasized, however, that a practical appratus for the present day industrial use with a high quality assurance is still within 10kW-class.

The greatest problem faced in welding with such high output CO_2 lasers is laser plasma. When high energy density beam welding is conducted in a normal atmosphere, high density high temperature metallic vapour is greatly produced. The interaction of this metallic vapour with the laser beam creates so called laser plasma. This strongly absorbs or scatters the laser energy and prevents to obtain satisfactory weld beads.

Although an assist gas can be used to blow away the laser plasma, this gas also blows the molten metal. Strong jet of assist gas increases the size of the beam hole and prevents realization of a sufficient Wall Focusing Effect.[49] Therefore, even when an assist gas is used, it is not so easy to obtain the penetration depth of more than 20mm[12].

On the other hand, reducing the environmental pressure results in a reduction in plasma generation. **Figure 14** shows a quantitative comparison of penetration depth both in EB and laser beam welding. In both processes the penetration depth changes similarly with the ambient pressure. However, in case of EB welding the process is mainly governed by collisions with neutral and plasma particles and the total of these particle density in the beam pass directly affects the beam propagation and

attenuates the beam energy before reaching to the materials and/or beam hole wall. While in case of laser welding only charged particle density, that is, plasma density dominantly affects beam decay. So that in EB welding

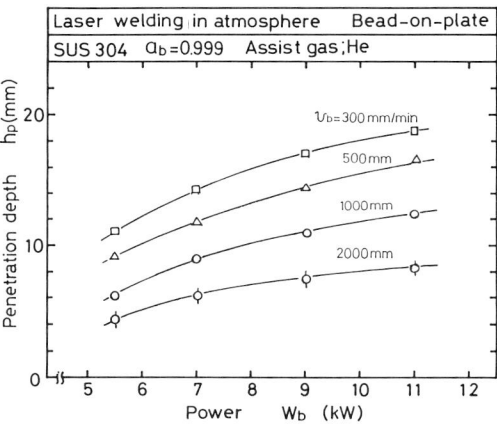

Fig. 13 15kW CO_2 laser and its power characteristics

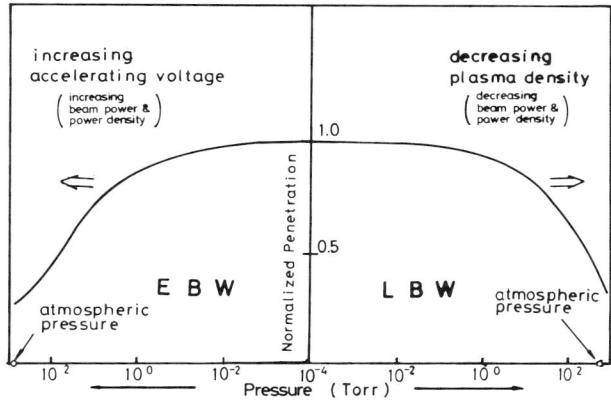

Fig. 14 Comparison of pressure dependence of penetration depth between electron and laser beam welding

34

the curve shifts to left side as the acceleration voltage is increased. While in laser process, it shifts to right side, if the plasma density is reduced, as we can find in the figure.

As the amount of plasma generated rapidly decreases at a pressure lower than several Torr, development and application of low vacuum laser welding method has been extensively studied by the author. This research effort seeks to improve laser welding performance without losing conventional laser functionability in low vacuum environ-

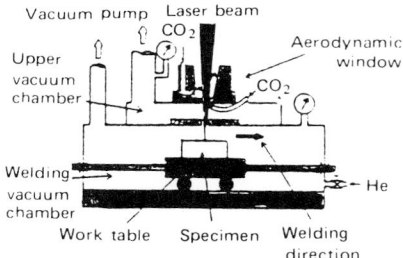

A) Partial vacuum type with aerodynamic window.

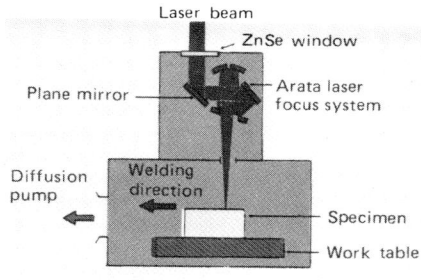

B) High vacuum type

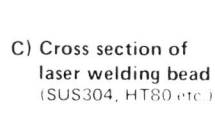 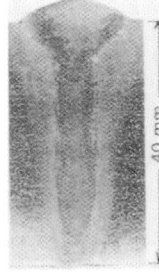

C) Cross section of laser welding bead (SUS304, HT80 etc.)

40 mm

Fig. 15 Experimental apparatus of vacuum laser welding

ment.

Figure 15 illustrates systems of a partial and a high vacuum laser welding[13]. The author named these new welding methods as "Vacuum Laser Welding". This high vacuum laser system can achieve a penetration depth of over 40mm even in a 10kW-class laser as shown in Fig. 15 (c), which is similar to that of electron beam welding.

In the partial vacuum welding system, a dynamic window replaces the conventional ZnSe or KCl lens, thus enables one to introduce a high output laser beam into the vacuum. This system achieves a penetration depth of over 25mm even in a low vacuum of 50 Torr.

As use of a low vacuum or local vacuum reduces laser operability, other investigations have southt to improve welding performance in a normal atmospheric environment. One of these techniques is called LSSW (Laser Spike Seam Welding)[14]. This method focuses on the delay between laser emission and laser plasma generation. The laser is concentrated on the same weld point and shifted forward the moment before production of laser plasma increases, by which the laser energy is used very effectively to attain a deep penetration. As shown in **Fig. 16**, this method offers better weld penetration than conventional techniques at the same power and welding speed.

In another attempt to improve weldability, a gap at the weld point is opened before welding begins, and the gap is filled with a filler wire to enable welding of thick plates[15]. This system is illustrated in **Fig. 17**. This technique also increases the allowable butt gap error, and is being considered as one means of compensating for the lack of high precision characteristics in laser welding system.

As beam output increases so does the effect of the laser beam focusing system on beam characteristics and welding characteristics. Three conventional system are shown in **Fig. 18** (a), (b), (c); in (a) the heat input to lens is incompatible with high output, and large aberrations in (b) and (c) result in poor focusing efficiency. In contrast, however, the effectiveness of "Arata Laser Focusing System"[16] shown in (d), (e), and (f) increases with the laser output. Though these three systems are designed with the same principle, system (d) is used for cylindrical beam, and (e) & (f) for coaxial one. System (f) is particularly effective with high output annular beams widely employed in multi-kilowatt lasers.

2.3 High energy density beam gas cutting

High energy density beams like EB and laser can be used also for a precise cutting as well as welding process. In regard to electron beam, in the early time before the development of EB welding Steigerwald et al. have studied

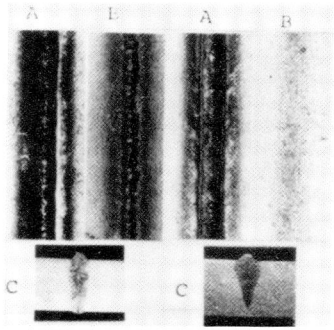

a) LSSW   b) Conventional
A; Front Bead, B; Back Bead
C; Cross Section

Fig. 16 Comparison of bead cross section between laser spike seam welding method and conventional method

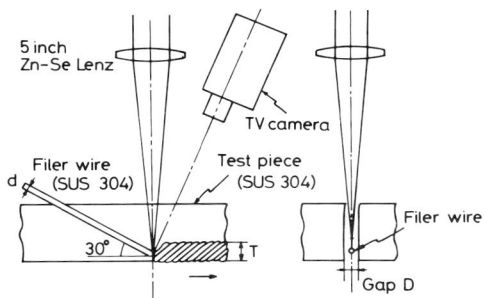

Fig. 17 Laser welding with filler wire

around 1950 piercing, drilling and cutting in a vacuum environment, though the output was very low. While the author has considered to apply EB and laser to cutting at an atmospheric pressure and in 1966 succeeded[54] in making a cooperative use of gas dynamic pressure and/or gas reaction energy with these beams and named them as "EB Gas Cutting" and "Laser Gas Cutting", respectively. By these methods the author has obtained good results in the cutting of various metals and ceramics[17, 18]. As is well

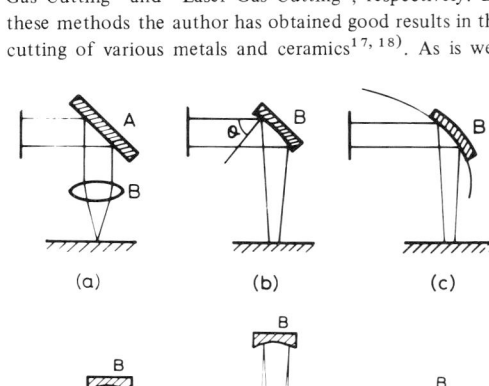

(a)   (b)   (c)

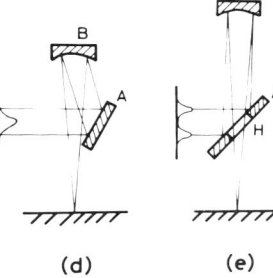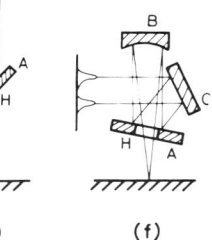

(d)   (e)   (f)

Fig. 18 Arata Laser Focus System

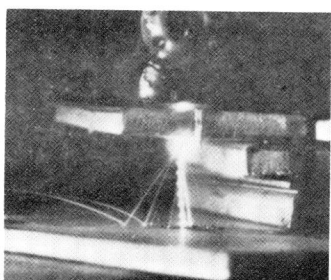

EB Gas Cutting

Laser Gas Cutting

Cross Section
(Stainless Steel)

10 mm

Cross Section
(Stainless Steel)

Fig. 19 High energy density beam gas cutting

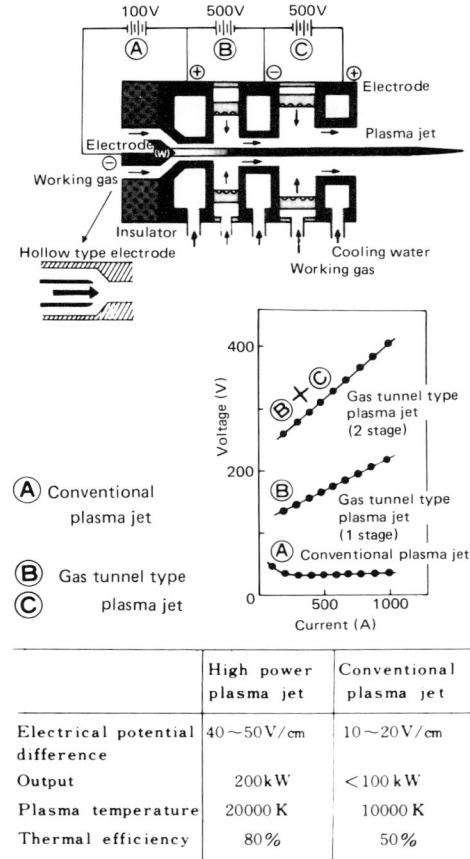

Fig. 20 Gas tunnel type plasma jet and comparison with conventional type plasma jet

	High power plasma jet	Conventional plasma jet
Electrical potential difference	40~50V/cm	10~20V/cm
Output	200kW	<100kW
Plasma temperature	20000 K	10000 K
Thermal efficiency	80%	50%

known the laser gas cutting are more frequently applied presently and the EB gas cutting has been studied for the cutting of rocks[50,51] rather than metals. In the laser gas cutting superior cutting results of a thick plate has been obtained recently by the development and use of plural nozzles[52,53].

Figure 19 shows typical example of high energy density beam gas cutting by EB and laser heat sources in open air, which has been developed by the author for the first in the world. On the right is shown laser gas cutting and on the left EB gas cutting. They give a similar cutting results as is shown in the lower part of the figure. These high quality cutting are obtained for various metal and non-metal plates.

2.4 Plasma beams

Plasma beams, one of high energy density beams, were

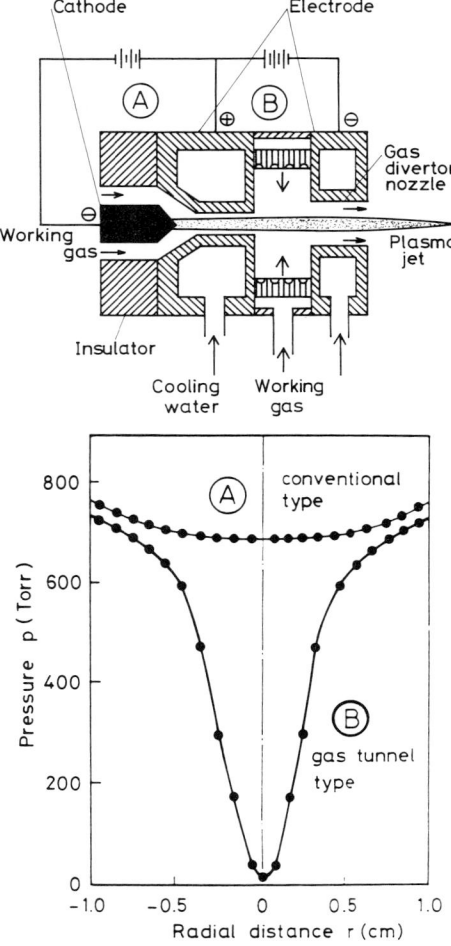

Fig. 21 Gas tunnel type plasma jet and its radial pressure distribution

developed as plasma jets. Today plasma beams are used with a variety of engineering applications, including melting of metallic materials, processing (cutting, welding, thermal spraying), and high temperature chemical reactions. The outstanding operability, low heat generation costs, high output, high temperature, and high heat efficiency of plasma jets indicates an expanding range of applications in the future. However, a conventional plasma jet has a drooping voltage-current characteristics which leads to the necessity of considerably large current to obtain a high output as shown in **Fig. 20**. This causes the erosion of electrode and other problems inhibit high output designs, and most plasma jets in use today feature output levels below 100kW.

To overcome these problems, a gas tunnel type high power plasma jet system in which the plasma beam is generated through a gas tunnel applying a powerful thermal pinch effect was developed for generation of high temperature, high energy density, high output plasma jets[19].

Figure 21 shows a simple connection of the conventional and the gas tunnel type plasma jet, and also the difference in the pressure distribution[20] between conventional type and gas tunnel type plasma jets. Output levels over 400kW have been obtained with the experimental device shown in Fig. 20. In the figure is also shown a comparison of their characteristics. Use of a special high speed vortex protects the electrodes, and the thermal pinch effect constricts the plasma jet, and suppresses the heat load on the tube walls. This system easily achieves high output levels with a stable plasma jet for high performance fusion of high melting point alloys, fusion of high melting point insulators such as ceramics, and metal refining.

Applications for materials processing, surface treatment, and thermal spraying with high melting point metallic materials, non-metallic materials and ceramics should bring future improvements in quality and efficiency. Other potential applications utilizing the high temperature characteristics of plasma jets include high temperature reaction production of compounds and crystals, production of high purity, spherical alumina powder, and destruction of toxic substance.

2.5 Industrial application of high energy density beams to manufacturing

Figure 22 shows a vacuum chamber manufactured by electron beam welding method for nuclear fusion research[2]. This structure has a very complex helical configuration and needs precise proccessing. It is impossible to produce such a chamber by arc welding method. The major radius of the chamber is 2.2m. The chamber has two parts of thickness with 20 and 33mm.

Figure 23 shows pressure vessels of 6000m class deep submergence exploror vehicle also produced by electron beam welding[2]. Development are progressing for these two kinds of materials shown as examples.

Figure 24 shows an example of nuclear reactor pressure vessel of stainless steel[2]. The diameter is 3m and the height is 5m. Thickness is 40mm.

3. Robotics and AI in Welding and Cutting

High performance and high precision process control are required to assure consistent quality in weld joints and welded structures produced with arc and/or high energy density beam processes. Robots have rapidly developed and been applied in a wide range of manufacturing fields in recent years to meet these requirements.

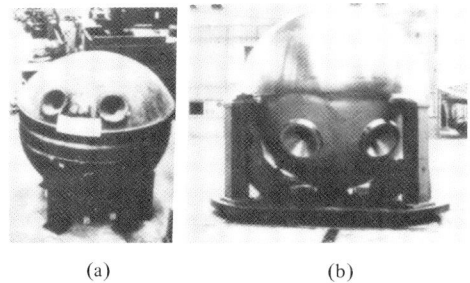

(a)   (b)

Fig. 23   Deep submergence exploror vehicle
(a) 10Ni-8Co, Diameter:2100mm, Thickness:80mm
(b) Ti-6Al-4V, Diameter:2000mm, Thickness:80mm

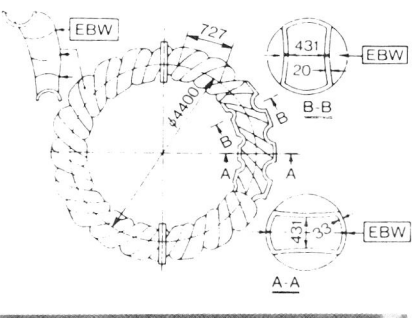

Fig. 22   Vacuum chamber for nuclear fusion research

Fig. 24   Pressure Vessel for Nuclear Reactor
(AISI 304, Diameter:3000mm, Highness:5000mm, Thickness:40mm)

The arc welding robot is the most common industrial robot today, and in this section will be examined the current state of welding robot production and use. **Figure 25** shows welding robot manufacturing trends in Japan[21]. As the chart indicates, arc robot manufacture increased dramatically through 1982, aided by severe price competition and reaching market saturation in 1983. Furthermore, while most arc welding robots are self-standing devices, they are increasingly used as part of larger welding and manufacturing system[22], whose example is illustrated in **Fig. 26** as the automobile chassis and frame welding line.

Teaching an arc welding robot is a complicated process, and the time required for robot teaching is one of the biggest roadblocks to greater use. The development of advanced sensors is expected to reduce robot teaching time requirements, simplify the procedure, and expand the range of robot applications.

Functions required of arc welding robot sensors include the ability to; (i) recognize the workpiece and those spots requiring welding; (ii) properly and accurately trace the welding seam even under a variety of poor welding conditions; (iii) detect the shape and gap of the joint, determine and compensate for any variance with the specifications; (iv) detect and prevent any interference and obstruction of foreign objects with the torch or robot arm; (v) detect weld quality during the welding process and to adjust the welding conditions when necessary; (vi) automatically conduct a nondestructive examination of the welding joint upon completion of the welding procedure.

Of these various functions, there currently exist only sensors to detect the beginning and end of the workpiece in (i), and arc sensors and optoelectronic sensors in (ii). All others are still in the experimental stages of development. New sensor types, high performance processors for sensor output processing, and cost reductions are current research and development objectives. Other research topics include the application of arc welding robots in the manufacture of such large structures as ships and ferroconcrete bridges. New actuator designs and the development of light-weight, portable robots are being examined[23] to solve these problems whose example is shown in **Fig. 27**.

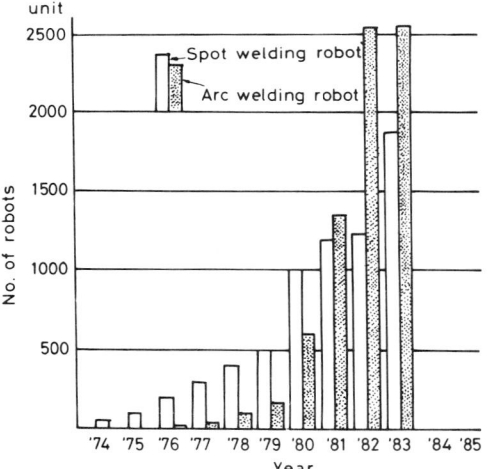

Fig. 25 Trend of production of welding robots in Japan

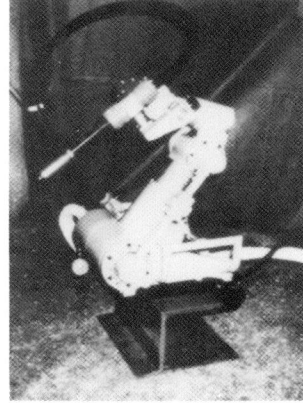

Fig. 27 Portable robot for shipbuilding

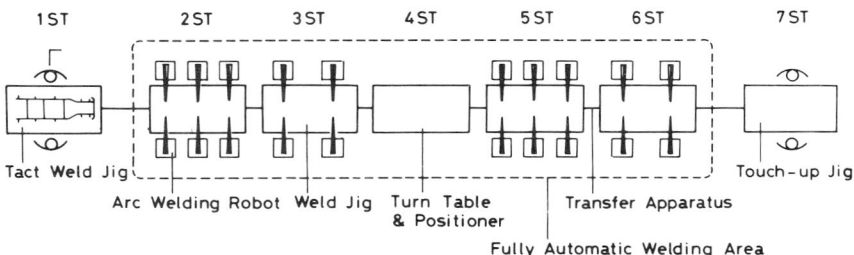

Fig. 26 Welding line for automobile chassis and frame

  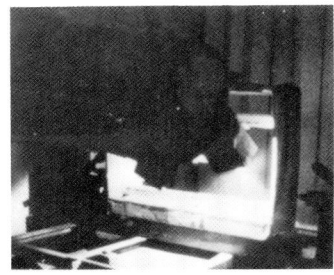

Fig. 28 Typical welding robots for arc welding, cutting and flame spraying

With the development of CAD systems, the desire to input CAD system output directly into the welding robot has naturally risen. While this requires equipping the robot with a language processing capability and standardizing the programming language. It also requires standardization of both robot software and CAD system disk operating systems (DOS). The diversification of microprocessors and the development of CAD systems for microcomputers increases the possibility such common operating systems as CP/M, MS-DOS, and UNIX will be enhanced with a real-time function and employed in robot operating system. Low software productivity is a problem not only for industrial robots, but for all computer-related technologies and, therefore, all industrial technologies.

Figure 28 shows typical examples of robots used today for, respectively, arc welding, cutting, and flame spraying. **Figure 29** shows a robot[24] developed very recent for automation of welding line of steel plate factories. **Figure 30** shows a three-dimensional laser welding and cutting robot[25] developed to automate high energy density laser beam welding and cutting processes. This system uses a 500W CO_2 laser and robot functions include five-axis,

fully integrated drive system. For instance, specification of the plate thickness and cutting speed enables the control system to automatically set optimum laser cutting conditions.

It is most likely that future robots will gradually grow away from teaching-playback control systems to intelligent robots offering automatic control and detection capabilities. Reasons behind these probable developments are a lack of skilled technicians and extreme diversity of welding processes and environments. Increased use of intelligent robots will promote factory automation and thereby effect other changes in the factory.

An Artificial Intelligence (AI) system consists of knowledge base and inference mechanism or inference engine as shown in **Fig. 31**. What should be remarked in Artificial Intelligence is its inference mechanism. A computer infers based on pieces of causalities by communicating with the user. By the conventional method shown below we have to formulate or develop a particular program for a particular condition. While by AI technique, we do not have to prearrange the piece of causalities in order to solve the problem, because the computer infers and picks up the appropriate pieces of causalities to solve the problem.

What is more important is that we can know what is lacking in the knowledge base by communicating with the

Fig. 29 Robot for automation of welding line of steel plate factories

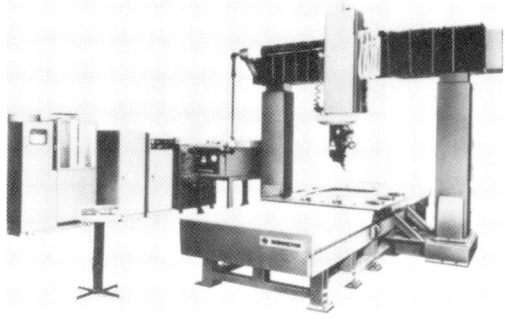

Fig. 30 Three-dimentional laser welding and cutting robot

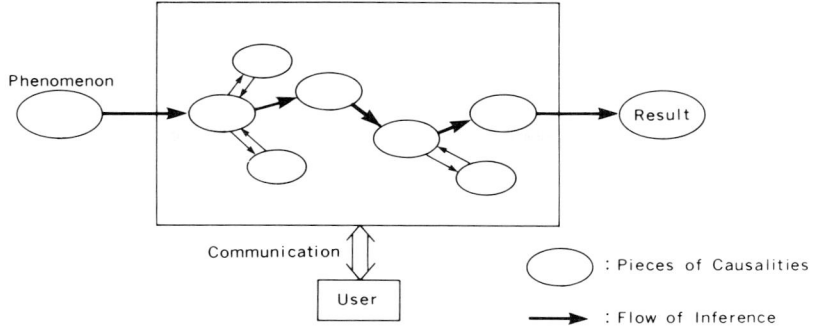

Processing by AI Technique

Phenomenon

Communication

User

◯ : Pieces of Causalities

⟶ : Flow of Inference

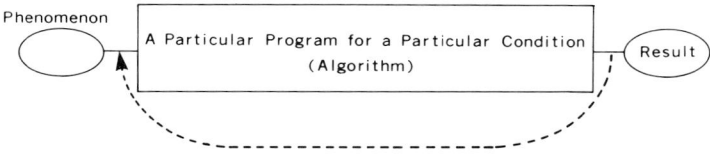

Processing by Conventional Methods

Phenomenon

A Particular Program for a Particular Condition (Algorithm)

Result

Fig. 31 Artificial intelligence and conventional algorithm

computer and we can add new pieces of information to the knowledge base.

On the basis of these considerations, an **EXPERT** system using **PROLOG** and capable of determining welding conditions for pressure vessels has been developed[26] at Osaka University. This system uses experience to compensate even for deviations in conditions requiring manual input by conventional algorithmic techniques.

The ability for tomorrow's intelligent robots to handle standard conditions alone will not be enough. They will need to be able to work under an extremely wide range of conditions. Accordingly, it is equally important to develop intelligent robots utilizing artificial intelligence technologies.

4. Surface Modification by High Energy Density Beams

As already mentioned high energy density beams offer good operability enabling precision control of heat input and its density in wide ranges. Furthermore, these beams can apply a large heat input in a short time, and are used as the heat source not only for welding, but in a number of different surface modification processes. This section will consider primarily laser beam applications.

4.1 Laser gas hardening

A surface hardening technique in which titanium is irradiated with a laser in a nitrogen atmosphere to form a TiN layer or Ti layer containing a high concentration of nitrogen at or directly below the fusion surface is being studied[27]. This technique employs the gas-liquid reaction occurring at high temperatures and is characterized by the direct formation of a ceramic coating on the base material.

Figure 32 shows a comparison of hardness distribution between laser gas hardening and conventional ion nitriding method. Laser gas hardening can harden titanium surface more rapidly and deeply than conventional method[28].

Figure 33 shows the relationship between the pressure of the nitrogen atmosphere and Vickers hardness, and between nitrogen pressure and TiN layer thickness. As these indicate, the higher the nitrogen atmosphere pressure, the thicker and harder the TiN layer.

4.2 Laser alloying

In subsequent studies, the surface of several materials (mild steel, SUS304, nickel, copper, and aluminum) were coated with an acrylic solution containing titanium powder. The surface was then irradiated in a nitrogen atmosphere with a pulse Nd:YAG laser (35J/pulse, 30mm defocus distance) to form a TiN alloy and harden the surface[29]. As the results shown in **Table 1** indicate, the irradiated surface was harder in each case than the base material, clearly demonstrating the ability to form a TiN alloy and harden the surface of the base material by coating the surface with titanium powder and irradiating this with a laser.

Figure 34 shows the results of hardness and titanium content tests in which an alloy of SUS304 and titanium was irradiated by a laser in a nitrogen atmosphere. At

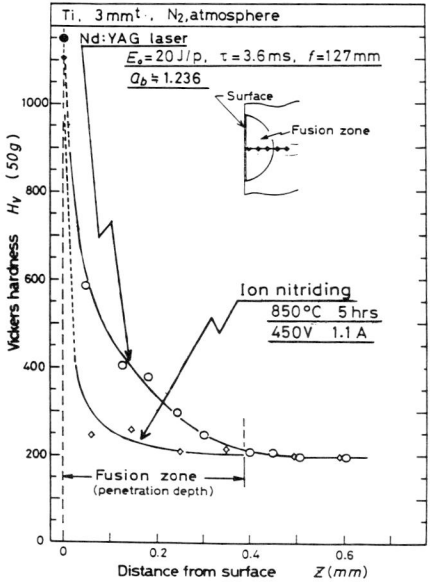

Fig. 32 Comparison between laser gas hardening and ion nitriding method

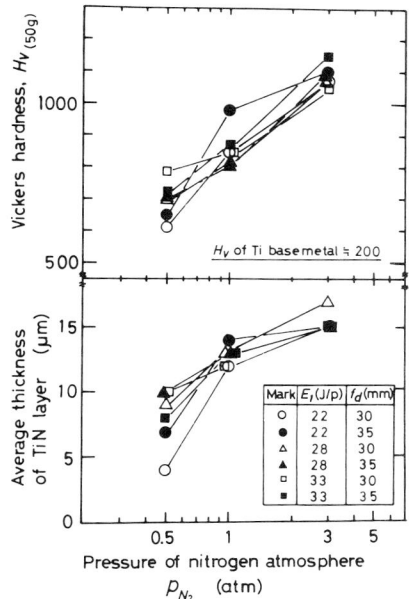

Fig. 33 Average thickness of TiN layer and Vickers hardness

Table 1 Increment of surface hardness for several materials

Materials	Vickers hardness, Hv (50g)		
	$(Hv)_S^*$	$(Hv)_b^{**}$	$(Hv)_S/(Hv)_b$
Mild steel	800 650–930	130	6.2
AISI 304	830 800–870	200	4.2
Nickel	650 580–760	110	5.9
Copper	740 630–810	85	8.7
Aluminum	580 500–650	45	13

* : Hardness of treated surface
**: Hardness of base metal

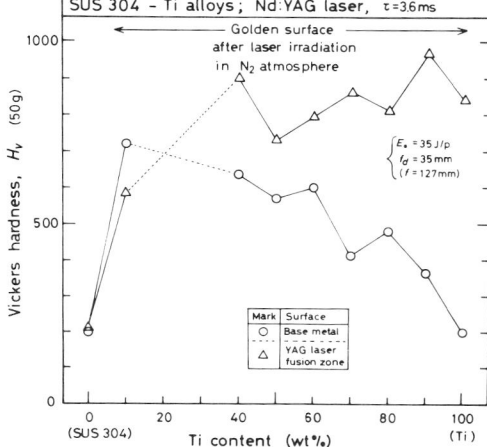

Fig. 34 Vickers hardness of SUS304-Ti alloys

titanium concentrations over 10% the surface appeared gold and at over 30% a hardness of Hv = 730 − 970 was obtained. In identical tests conducted with an aluminum-titanium alloy, however, virtually no change was observed until the titanium content exceeded 50%. While laser irradiation produced a TiN layer with increased hardness on the surfaces of each test sample, the effectiveness of the process varied with different materials.

4.3 Laser solution treatment

HAZ sensitization in SUS304 arising from the thermal history during extended use and from the welding heat cycle is the cause of intergranular stress corrosion cracking (IGSCC) and intergranular corrosion. Conventional preventive techniques include the use of low carbon steel or stable steels containing TiNb, stabilization heat treatment during the low weld heat gain stages or at 800° to 900°C after welding, or solution heat treatment at 1050° to 1100°C.

While solution heat treatment is extremely effective, it is not generally suitable with large structures due to problems with heat deformation and cooling speed arising from the high temperatures used. On the other hand, anticorrosion considerations do not require heating of the entire structure, and by using local processes which achieve sufficiently high cooling speeds by heating only the surface, solution heat treatment processes can also be used with large structures. Techniques to improve the corrosion resistance of sensitized stainless steel using local solution heat treatment processes and a laser beam heat source are also being considered[30]. **Figure 35** shows a cross section perpendicular to the bead following laser irradiation (1kW beam output, 7mm beam diameter, 1m/sec) of the HAZ sensitized area of S-2 steel, TIG-Welded with a 10kJ/cm heat input. Laser irradiation changed a ditch structure of 0.3mm deep in the HAZ sensitized area into a step structure, indicating desensitization of the area. Furthermore, as shown in **Table 2**, laser irradiation completely desensitized other HAZ areas 0.2 to 0.4mm from the surface.

4.4 Laser surface fusion treatment

Other studies have examined techniques to improve the oxidation resistance by modifying the composition and structure of alloy layers by surface fusion with a laser. **Figure 36** shows the results[31] of oxidation resistance tests conducted after a 1.7 to 2kW CO_2 laser (1 to 4mm beam diameter) scanned the surface of SUS430 and SUS304 either once or multiple times at $0.3 - 0.5$/min. Oxidation resistance was determined by the amount of weight gain and spalling of scale on SUS430 and SUS304, respectively. The oxidation resistance of both SUS430 and SUS304 was determined to have improved from laser fusion. XMA analysis and X-ray diffraction showed that the oxide layer of laser-irradiated materials has an extremely high chromium and silicon concentration, and that there are large amounts of Cr_2O_3 and SiO_2 in the film. It is believed that for the microstructural changes caused by laser irradiation to improve oxidation resistance, chromium and silicon are first dispersed over the surface, thus forming a protective layer.

5. Production of Composite Materials by Modified Conventional Menthods

Rapid development in the space, atomic power, and electronics industries in recent years have created new and demanding materials requirements. Furthermore, conventional metallic and organic materials are frequently incapable of withstanding these new conditions. On the other hand, new ceramics (nonmetallic inorganic materials) exhibit superior heat resistance, corrosion resistance, high temperature oxidation resistance, and wear resistance, and are thought to offer the performance required in many new industrial applications.

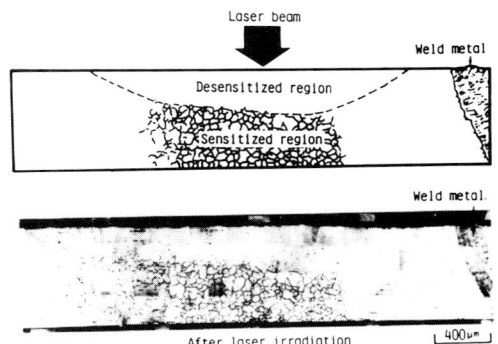

Fig. 35 Structure of bead after laser irradiation

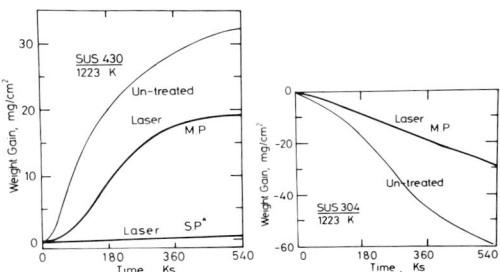

Fig. 36 Weight increment of SUS430 and SUS304 treated by laser

Table 2 Laser irradiation conditions of desensitization treatment

Heat input (kJ/cm)	Laser power (kW)	Beam traveling velocity (m/min)	Beam diameter (mm)	Result
10	1.0	1.0	7	Desensitized
	1.5	2.0	7	Desensitized
20	1.0	1.0	7	Desensitized
	1.5	2.0	7	Desensitized

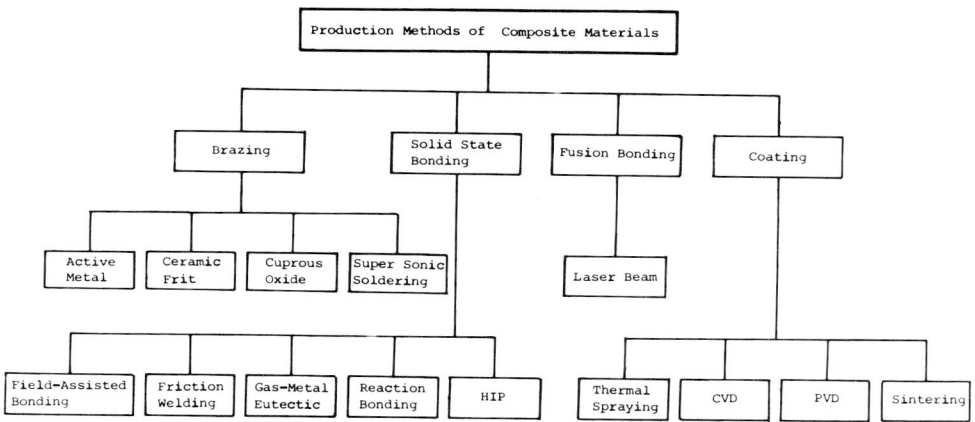

Fig. 37 Classification of composite material production techniques

Current ceramic compounds, however, suffer from some shortcomings (e.g. poor machinability, low resistance to thermal shock, manufacturing problems faced in the production of large and complicated structures), and their use is therefore limited. Developments in the composite materials field have come from efforts to compensate for deficiencies in ceramics, metallic materials, and plastics by composing two or of these more materials into a new material offering improved performance and new functions.

The bonds between the component materials are particularly important when manufacturing new composites. Conventional welding techniques, including brazing and diffusion welding, have been modified and effectively utilized in composite material manufacture. New heat sources, e.g. plasma beams and laser beams, are also being considered.

Figure 37 is a general classification of composite material production techniques[32] based on welding technologies, and this section will consider various composite manufacturing techniques and specific applications in producing ceramic-metal composites.

5.1 Composite technology as modifications of conventional methods

In brazing technique selection of the filler metal best suited to the base metals is difficult for bonding ceramics to metals. The importance of wetting the filler (metal, alloy, or ceramic) to the ceramic has led to adoption of techniques to improve wetting.

When a pure metal or alloy filler is used, brazing is performed in an inert, vacuum atmosphere with a filler metal containing an active, high melt point metal (e.g.

titanium, zirconium, niobium). New brazing techniques use amorphous alloy fillers[33] to bond such oxide ceramics as Al_2O_3 and non-oxide ceramics such as Si_3N_4 with various metals.

Figure 38 shows chemical compositions of amorphous alloy fillers used. The figure on the upper-left part, shows the result of alumina Al_2O_3 brazing, using these amorphous alloy filler. This filler is superior than the conventional powder as shown on the right.

Characteristics of amorphous fillers include (i) easy processing into thin films, rods, and other non-powder shapes, (ii) a constant filler structure which facilitates uniform application of the filler layer.

Use of oxide powders increases wetting of ceramic; Al_2O_3 brazing results[34] using cuprous oxide (Cu_2O) powder are shown in **Fig. 39**. The bond was created by the formation of $CuAlO_2$ through a reaction at the interface of Cu_2O and Al_2O_3. **Figure 40** shows cross sections of the brazed part[35] in a vacuum (5×10^{-5} Torr) at 1200°C of different ceramics and of a ceramic and SUS304 with cuprous oxide insert material. These brazing can also be achieved in air.

Solid state reaction bonding techniques apply metal-metal diffusion bonding methods to the direct bonding of ceramics and metals. Chemical reactions induced by externally applied pressure and heat at the ceramic-metal interface are utilized to create the bond. Cross sections of pressurized solid state reaction bonds[36] formed at 500°C in a vacuum between aluminum and various ceramics are shown in **Fig. 41**. This system also enables bonding in air, and the bond temperature is a major factor in the final bond. **Figure 42** shows composite bonds[37] of ZrO_2, aluminum, and titanium alloy. Although solid state

44

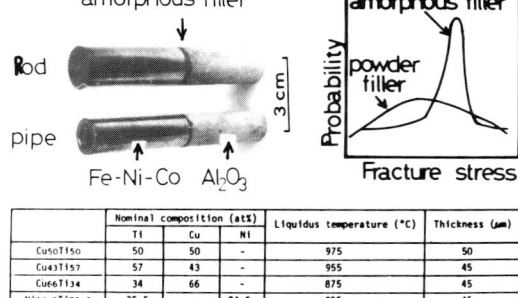

	Nominal composition (at%)			Liquidus temperature (°C)	Thickness (μm)
	Ti	Cu	Ni		
Cu50Ti50	50	50	-	975	50
Cu43Ti57	57	43	-	955	45
Cu66Ti34	34	66	-	875	45
Ni24.5Ti75.5	75.5	-	24.5	955	45

Brazing ceramics with amorphous filler metal in vacuum
(Excellent reliability, Easy brazing method)

Fig. 38 Brazing using amorphous alloy filler

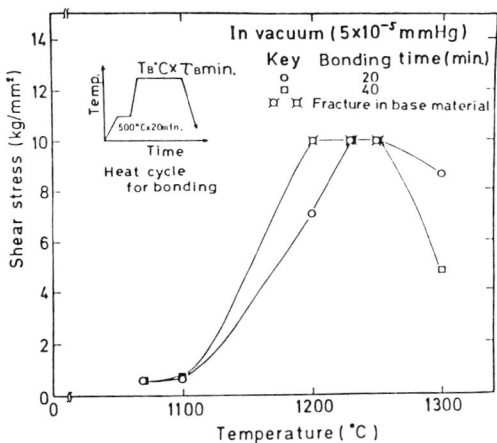

Fig. 39 Temperature dependence of shear strength of Al_2O_3 joint bonded by cuprous oxide method

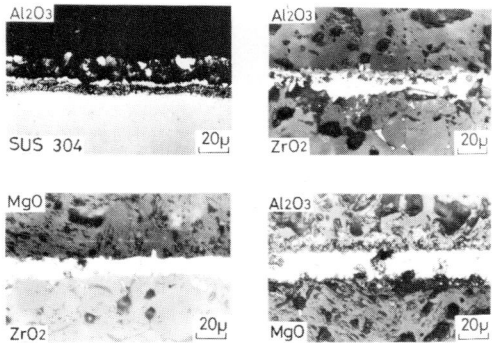

Fig. 40 Cross sections of brazed part of different ceramics and of a ceramic and SUS304 bonded by cuprous oxide method

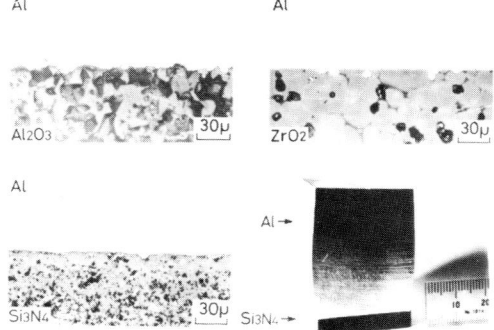

Fig. 41 Cross sections of bonded part of Al and various ceramics

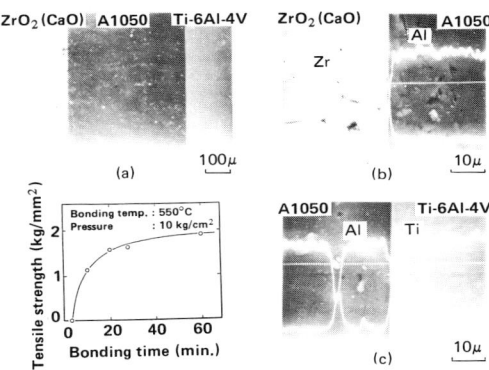

Fig. 42 Bond results of composite ZrO_2/Al/Ti alloy

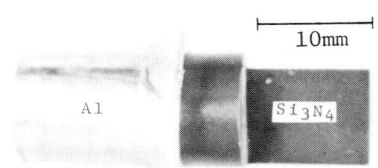

Fig. 43 Example of Al-Si_3N_4 joint (tensile strength; 15.9kg/mm²) bonded by friction welding

reaction bonding techniques are being studied for bonding various combinations of ceramics and metals, they are limited by the bonding atmosphere. Friction welding[38] in air of ceramics and metals is also being studied as shown in **Fig. 43**. Field assisted bonding[39] in which a DC voltage is applied to shift ions in the ceramic and form a ceramic-metal bond using the solid electrolyte characteristic of ceramics is also being studied. **Figure 44** shows some examples[40] of pressurized field assisted bonding of glass and aluminium. This technique can be used for

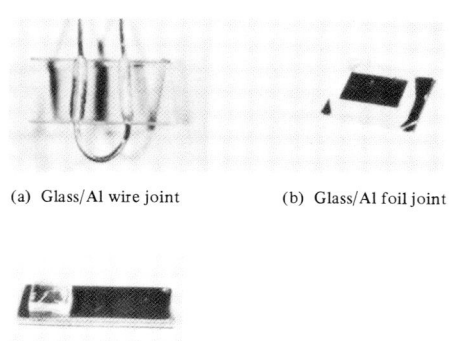

(a) Glass/Al wire joint   (b) Glass/Al foil joint

(c) Glass/Al plate joint
(bonded)

Fig. 44 Appearances of joints made by pressurized field assisted bonding of glass on Al

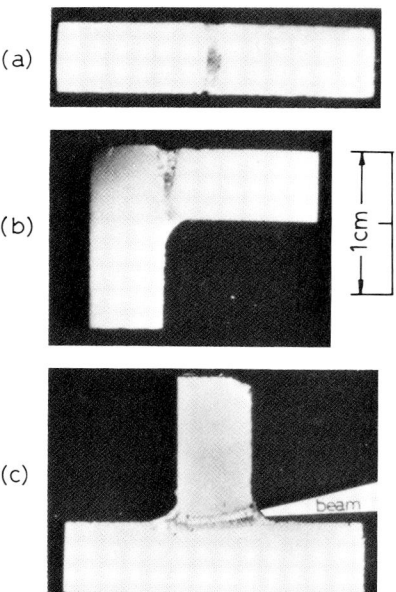

Fig. 46 Bond results of Al$_2$O$_3$ with laser beam

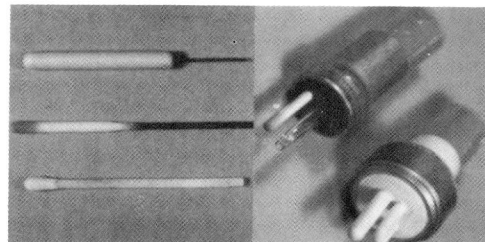

Fig. 45 Nozzles for injection molding made by HIP method

glass-semiconductor silicon, and β-Al$_2$O$_3$-metal bonding[41].

As to the bonding of dissimilar materials a HIP process is increasing its importance in many industrial fields, and the method is expected to be potential and promissing in production and processing of new materials including ceramics and composite materials. **Figure 45** shows an example[42] of nozzles for injection molding made by HIP method.

5.2 Composite technologies using high energy density beams

Research into ceramic fusion bonding applications of new high energy density heat sources are being conducted with laser beams. Results of tests on I-, L-, and T-Joints of Al$_2$O$_3$ formed with this technique[43] are shown in **Fig. 46**. Similar techniques using electron beams, and applications for bonding identical ceramics, dissimilar ceramics, and ceramic-metal composites are also being studied. The importance of these techniques is expected to grow for

Fig. 47 Needle sensor produced by thermal spraying (center) and tubular type sensor by sintering (top). On the right is shown the housed sensor.

applications in composite materials production.

Thermal spraying is a composite materials production technology employing plasma and high energy gas flame, and surface coating techniques to bond ceramics and metals, ceramics and plastics, and other materials to manufacture composite materials offering new surface characteristics.

The functions provided by coating are widely accepted as the actual applications in many industries. **Figure 47** shows sample applications[42] for sensor functions produced by ceramic plasma spraying. These sensors are used for measuring oxygen concentration in molten steel. The conventional sensor shown in the top of the left figure

Fig. 48 Ceramic-plastic composite produced by ceramic thermal spraying

uses sintered ZrO_2. The needle sensors manufactured by ceramic spraying consist of a Cr-Cr_2O_3/ZrO_2 composite on a molybdenum rod. These new needle sensors offer significant cost reductions and reduced response times.

The sign shown in **Fig. 48** was manufactured by ceramic plasma spraying of a ceramic-plastic composite[45]. Direct formation of a ceramic coat on plastic provides the heat resistance, wear resistance, and corrosion resistance of ceramics with the lower weight characteristics of ceramic-plastic composite.

While research into thermal spraying techniques capable of providing the surface of a base material with the characterisitcs of new ceramics is a relatively new field, the ability to manufacture composite structures featuring composite functions by simply coating the surface has been demonstrated, and the future importance of this field is expected to continue growing.

Other composite materials production techniques not covered here include FRP, FRM, and other ways to manufacture integrated composites, as well as coating techniques (e.g. CVD and PVD) which give new functions to the base material surface. The latter of these techniques is expected to become particularly important. While composite materials are being manufactured today using modifications of conventional welding techniques, composite material production techniques using laser beams and other high energy density beams will become increasingly important. Furthermore, these promise to be applied both as independent processing techniques and as composite processing technologies.

6. Other Applications of High Energy Density Beams

New heat source applications include the coating of conventional metallic materials with amorphous alloys and the manufacture of ultrafine ceramic particles.

Certain amorphous alloys are noted for extremely high corrosion resistance. Although conventional amorphous alloys can be quenched by ejecting the molten alloy on a high speed spinning roll, the shapes are limited to small

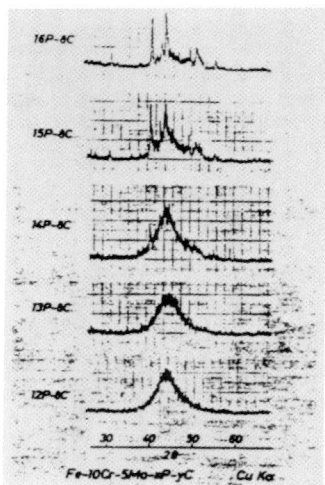

Fig. 49 X-ray diffraction pattern of laser irradiated Fe-10Cr-5Mo-xP-yC alloy

test pieces.

Irradiation of certain alloys for a short time with a high energy density laser beam can convert the surface of the alloy into an amorphous structure. **Figure 49** shows the X-ray diffraction pattern of an Fe-10Cr-5Mo-P-8C alloy, the surface of which has been converted to an amorphous structure by laser irradiation[46]. In alloys containing higher phosphorous content more than 14 at % sharp diffraction peaks resulting from a crystalline structure are observed in the broad peak resulting from a noncrystalline structure. However, in alloys containing 13 at % P and 12 at % P, however, only broad spectrum resulting from the noncrystalline structure were found, indicating that these alloys is a noncrystalline alloy containing no crystalline structures.

Arc heat sources are also being used in the trial manufacture of ultrafine particles of metals and ceramics[47,48]. High concentrations of metal atoms are evaporated from the constricted arc root operated in molecular gases such as hydrogen or nitrogen which have high cooling effect on arc, and they recombine to form ultrafine particles composed of many thousand atoms. The diameter of these particles is less than $0.1\mu m$.

This principle was used to design and manufacture the ultrafine particle production system shown in **Fig. 50**. Oxygen plasma can also be used for the plasma gas. **Table 3** shows some products manufactured with an active plasma-liquid phase reaction process. Nitrides, oxides (Al_2O_3, MgO), carbides (SiC, TiC), and metallic ultrafine particles are represented. Mixtures of nitrides (TiN, ZrN) and metallic ultrafine particles can also be manufactured

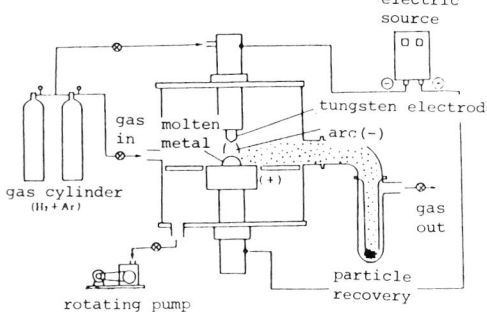

Fig. 50 Apparatus for preparation of ultrafine particles

Table 3 Composition of ultrafine ceramic particle produced by nitrogen, hydrogen and oxygen plasma-liquid phase reaction method.

Plasma Gas	Starting Material	Product (Ultrafine Particles)
Nitrogen	Ti , TiN	TiN (NaCl type)
	Zr	ZrN (NaCl type)
	Al , AlN	Al + AlN (Wurzite type)
	Si , Si₃N₄	Si (Diamond type)
Hydrogen	CaO	CaO (Lime), [Ca(OH)₂]
	MgO	MgO (Periclase)
	Al₂O₃	α – Al₂O₃ (Corundum)
	TiO₂	TiO₂ (Rutile)
	ZrO₂	ZrO₂ (Tetragonal > Cubic > Baddeleyite)
	SiC	β – SiC (Cubic) >> α – SiC (Hexagonal)
	Ti + C	TiC (NaCl type)
	W + C	WC (Hexagonal) > β – WC (Cubic) > α – W₂C
	WO₃ + C	WC > α – W₂C (Orthorhombic)
	WC	WC > β – WC
Oxygen	W	WO₃ (Monoclinic)
	Mo	MoO₃ (Orthorhombic)
	Nb	Nb₂O₅ (Monoclinic) or NbO₂ (Monoclinic)

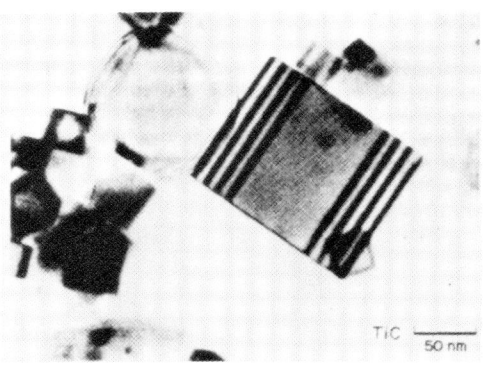

Fig. 51 Electromicrograph of TiC ultrafine particle produced by activated plasma-liquid phase reaction method.

depending on the metals used, thus increasing expectations for production of new materials with refined functions. **Figure 51** is a photo taken with an electron microscope, of ultrafine particle of TiC exhibiting a crystalline NaCl structure. Characteristic crystal growth is recognizable.

7. SUMMARY

The paper reviewed current trends in the development of welding and its allied processes using electron beams, laser beams and plasma beams and contrasted the application of these with conventional welding heat sources, most particularly arc heat sources. In addition discussion was made on the relationship between these technologies and new materials development technologies. It has been also reviewed the current status of and future trends in robotics, a field indispensable to increased manufacturing flexibility, productivity, and assuring high precision and quality in welding joints and cutting. The discussion can be summarized in the following five points.

(i) Future research in welding processes using high energy density beams can be expected to focus on the high performance and precision requirements. Examples of research into high precision and high performance welding include tandem electron beam welding and welding techniques combining computers and various sensors.

(ii) Research into welding robots incorporating artificial intelligence was examined with reference to specific examples, and the future necessity for robots with artificial intelligence was demonstrated.

(iii) Techniques using laser beams for surface modification like as nitriding and structural improvement of various metallic materials were introduced. The importance of electron beams, lasers and other high energy density beams in this field was also emphasized.

(iv) In reviewing research trends in composite materials technologies, the production of composites through the bonding of ceramics and metals was given as an example of the importance of integrated composites and coating techniques. The role of high energy density beams will also be more important in these areas.

(v) Amorphous alloys and ultrafine particles of ceramics and metals were offered as examples of new materials synthesis. The importance of reactive plasma and liquid phase reaction in ultrafine granule production were emphasized.

ACKNOWLEDGEMENT

The author wishes his gratitude to Prof. K. Inoue, Prof. A. Ohmori, Prof. M. Naka and Dr. N. Abe of Welding Research Institute (JWRI) at Osaka University for their individual discussions and arrangements of data and materials cited in this paper. He also appreciates very much for valuable comments made by Prof. K. Nishiguchi of Department of Welding Engineering of Osaka University, Prof. S. Miyake and Prof. A. Matsunawa and Dr. A. Kobayashi of JWRI.

References

1) Y. Arata, Rivista Italiano Della Saldatura, **39** (6), 1982, p.343.

2) Y. Arata, S. Sato, S. Shono and G. Takano, J. High Temp. Soc. **11** (5), 1985, p.165.

3) Y. Arata, Trans. JWRI **13** (1), 1984, p.121.

4) Y. Arata and M. Tomie, J. of High Temperature Society, **10** (3), 1984, p.110.

5) Y. Arata and M. Tomie, Trans. JWRI **2** (1), 1973, p. 17, J. of JWS, 46 (9), 1977, p.78.

6) Y. Arata and M. Tomie, Trans. JWRI **9**, 1980, p.157.

7) Y. Arata and E. Nabegata, IIW Doc. IV-221-77, Trans. JWRI 7 (2), 1977, p.101., Y. Arata, N. Abe, H. Wang and E. Abe, Trans. JWRI, **11** (2), 1982, p.1.

8) Y. Arata and M. Tomie, Trans. JWS 1 (2), 1970, p.176., Y. Arata, S. Satoh, T. Shimoyama, G. Takano, Trans. JWRI **11** (1) 1982, p.25.

9) H. Murakami, S. Sasaki, T. Iwami and M. Yasunaga, Report of EBW Committee of JWS, EBW-342-84, 1984.

10) Y. Sakamoto, M. Hiramoto and M. Ohmine Report of EBW Committee of JWS, EBW-358-85, 1985.

11) N. Tabata, H. Nagai, H. Yoshida, M. Hishii, M. Tanaka, Y. Myoi and T. Akiba, Proc. ICALEO '84, 1984, p.238.

12) Y. Arata, N. Abe and T. Oda, IIW Doc. IV-374-84, 1984.

13) Y. Arata and T. Oda, J. of High Temperature Society, **10** (1), 1984, p.24., Y. Arata, N. Abe, T. Oda and N. Tsujii, Proc. ICALEO '84, p.1.

14) Y. Arata, N. Abe, T. Oda and N. Tsujii, Proc. ICALEO '84, 1984, P.2.

15) Y. Arata, Trans. JWRI **2**, 1973, p.119., H. Maruo, I. Miyamoto, T. Ohya and Y. Arata, Proc. of Annual Meeting of JWS, No. 36, 1985, p.20.

16) Y. Arata, N. Abe and T. Oda, IIW Doc. IV-374-84, 1984.

17) Y. Arata and I. Miyamoto, Tech. Rept. of Osaka Univ. **17**, 1967, p.285, **19**, 1969, p.379.

18) Y. Arata and M. Tomie, Tech. Rept. of Osaka Univ. **17**, 1967, p. 303, Trans. JWS 1 (2), 1970, p.176.

19) Y. Arata and A. Kobayashi, J. of High Temperature Society, **11** (3), 1985, p.124

20) Y. Arata and A. Kobayashi, Trans JWRI **13** (2), 1984, p.173.

21) N. Noda, J. of JWS, **54** (1), 1985, p.35.

22) ibid, p.38.

23) M. Ohsawa, Private communication.

24) S. Satoh, Private communication.

25) K. Inoue, Private communication.

26) S. Fukuda, Private communication.

27) S. Katayama, A. Matsunawa, A. Morimoto, S. Ishimoto and Y. Arata, Proc. ICALEO '83, 1983, p.127

28) S. Katayama, A. Matsunawa, A. Morimito, S. Ishimoto and Y. Arata, The Metallurgical Soc. of AIME, 1984, p.159.

29) S. Katayama, A. Matsunawa and Y. Arata, Proc. of Annual Meeting of JWS, No. 35, 1984, p.80

30) Y. Nakao and K. Nishimoto, Report of Welding Metallurgy Committee of JWS, WM-10001-84, 1984.

31) N. Wade, Y. Koshihama and Y. Hosoi, Proc. of Annual Meeting of JWS, 1983, p.230.

32) Y. Arata and A. Ohmori, J. Japan Weld. Soc. **52**, 1983, p.24.

33) M. Naka, K. Asami, I. Okamoto and Y. Arata, Trans. JWRI, **12**, 1983, p.145.

34) I. Okamoto and A. Ohmori, Proceedings of Inter. Conference on Welding Research in the 1980's.: I. Okamoto, A. Ohmori, M. Kubo and Y. Arata, J. High Temp. Soc., 8, 1982, p.230.

35) Y. Arata, A. Ohmori and I. Okamoto, Private Communication.

36) Y. Arata and A. Ohmori, Trans. JWRI, **13**, 1984, p.41.

37) Y. Arata, A. Ohmori and S. Sano, Private Communication.

38) Y. Arata, A. Ohmori, A. Suzumura and et al., Private Communication.

39) G. Wallis and D.I. Domerantz, J. Appl. Phys., **40**, 1969, p.3946.

40) Y. Arata, A. Ohmori, S. Sano and I. Okamoto, Trans. JWRI **13**, 1984, p.35-40.

41) B. Dunn, J. Amer. Ceram. Soc., **62**, 1979, p.545-7.

42) H. Nishihara and et al., Private Communication.

43) H. Maruo, I. Miyamoto and Y. Arata, Proceedings of Inter. Laser Processing Conference, 1981.

44) Y. Arata, A. Ohmori. M. Matsuoka, K. Urata and T. Ogura, Annual Meeting of Japan Institute of Metals (Autumn, 1985).

45) Y. Arata and et al., Private Communication.

46) K. Hashimoto, Boshoku Gijitsu (Corrosion Engineering), **33**, 1984, p.335.

47) S. Ohno and M. Uda, J. Japan Inst. Metals, **48**, 1984, p.640.

48) M. Uda, Kinzoku (Metals), **52**, 1982, p.9.

49) Y. Arata and I. Miyamoto, Trans. JWS 3, 1972, p.143, Proc. 2nd Int. Symp. of JWS 1975.

50) B. W. Schumacher, Proc. 3rd Int. Conf. on Electron and Ion Beam Sci. and Tech. 1968, p.447.

51) M. Hamasaki, M. Katsumura and H. Utsumi, J. of High Temperature Society **11** (6), 1985, p.226.

52) Y. Arata, H. Maruo and S. Takeuchi, 1st Int. Laser Proc. Conf., 1981.

53) G. Sepold and R. Rothe, ICALEO '83, p. 156.

54) Y. Arata, Patent 438525 (1962), 531587 (1962), 764135 (1968).

CHAPTER 2

DEVELOPMENT OF ULTRA HIGH ENERGY DENSITY HEAT SOURCE

Electron Beam

Laser Beam

Plasma Beam

Electron Beam

The Transtron Accelerator S-1
M. Okada and Y. Arata
[Tech. Report Osaka Univ. **13** (1963), 289.]

Plasma Electron Beam of Very High Current
Y. Arata, M. Tomie and Y. Katayama
[Tech. Report Osaka Univ. **16** (1966), 485.]

Some Fundamental Properties of Nonvacuum Electron Beam
Y. Arata and M. Tomie
[Trans. JWS **1** (1970), 40. IIW IV−28−70.]

100-KW Class Electron Beam Welding Technology—Welding Apparatus and Some Aspects as a Heat Source
Y. Arata and M. Tomie
[Trans. JWRI **2** (1973), 17. IIW IV Doc. 112-73].

Development and Application of a Strong Focusing Electron Beam Gun With Multi-Stage Electromagnetic Accelerating Units
Y. Arata and M. Tomie
[J. High Temp. Soc. **10** (1984), 110.]

The Transtron Accelerator S-1

Abstract

The power source and the accelerating tube in the first acceleration step of Transtron accelerator were designed and produced so as to generate in the duration of 10 μsec. 1 MeV energetic electron beam of 260 amperes in the maximum rated electric power.

The apparatus was operated under acceleration voltage 160 to 600 KV. The effects of strong magnetic fields on electron beam diameter and radiant X-ray dose were investigated.

1. Introduction

Beam current of charged particles obtained by conventional accelerators are very small, usually the order of μA to mA. Transtron accelerator[6] was investigated in order to accelerate energetic charged particles of very large densities. 1MeV. beam current 10^2 to 10^4 A and duration 10 to 100 μsec.)

In 1957, the authors[7] commenced the design and fundamental experiments of the Transtron, and in 1960, produced Transtron S-1 which consists of the power source and the accelerating tube in the first acceleration step of the Transtron. Since then, the authors have investigated the characteristics of the apparatus and behavior of the large electron beam under the operation voltage 160 to 600 KV.

2. Experimental Apparatus and Procedure

Photo. 1 and Fig. 1 are panoramic photographs and schematic diagrams of the Transtron S-1. Each (a) and (b) shows the power source and the accelerating part respectively. The power source consists of a charging device, eleven discharging gaps, various resistors, and twenty condensers which were developed particularly for this source. (Characteristics of condenser; parastic inductance 0.009μH, capacitance 0.75μF, working voltage 50KV).

The accelerating part is composed of an accelerating tube, a D.C. generator (heating power source for the cathode and regulating source of Whenelt potential 0 to 1000 V.), a copper drift tube (30mm$\phi \times$ 200mm) with beam focusing coils (0 to 3000 gauss), a tungsten target, and a shunt resistor which gives time constant of accelerating duration.

(a)

(b)

Photo. 1 Photographs of the Transtron accelerator S-1
(a) Power Source   (b)   Accelerating part

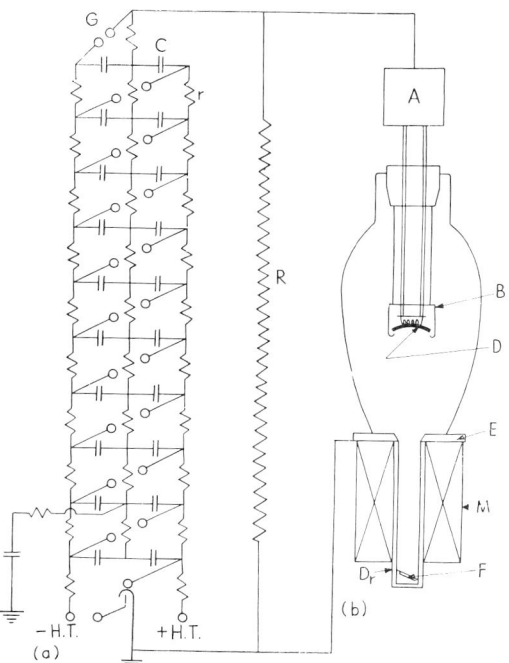

Fig. 1   Schematic diagrams of the
Transtron Accelerator S-1
(a)   Power Source
C: Condenser
r: Resistor
G: Spark gap
(b)   Accelerating division
A: D.C. generator,
B: Whenelt cylinder
D: Cathode
E: Anode
M: Focusing magnetic coil
F: Tungsten target
R: Shunt resistor
D_r: Drift tube

54

The main parts of the accelerating tube are the oxide coated Pierce type cathode 60mm in dia. and Whenelt cylinder whose electric potential is variable.

As all the condensers can be connected in a series with dischaging gaps after each has been charged up to 50KV, maximum voltage 1000KV is applied to the accelerating tube and stored energy of the condenser bank is supplied to the electron beam of 260 amperes in maximum within short duration of 10 μsec.

Electron beam currents are given by

$$I = GV^{\frac{3}{2}} = G_0 G* V^{\frac{3}{2}} \tag{1}$$

Where V=acceleration voltage

G=perveance of the accelerating tube

$G*$=relativistic factor[7]

G_0=perveance which is obtained by geometrical shape of the acc lerating tube, in above mentioned tube

$$G_0 = 0.34 \times 10^{-6} \ [A/V^{\frac{3}{2}}] \tag{2}$$

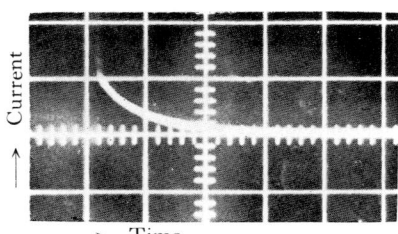

Photo. 2   The oscillogram of beam current.

Acceleration voltage: 300kv
Beam current: 50A
Time scale: 2μsec/div.

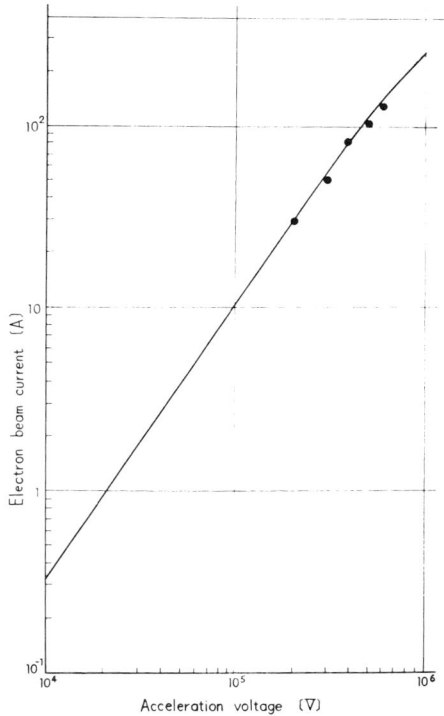

Fig. 2   Relation between acceleration voltage and beam current (perveance curve).

Fig. 2 is I-V characteristic of the apparatus and shows that the caluculated value by equation (1) is nearly equal to the experimental results. Photo. 2 is a typical oscillogram of an electron beam current obtained with this apparatus (circuit constant: $C=0.0375\mu F$, $R=250\Omega$, $L=25\mu H$).

Laminar electron flow originating from magnetically shielded cathode is focused and changed into completely non-laminar flow by a uniformly strong magnetic field.[8,9] Fig. 3 shows that such non-laminar electron flow appeared in the electrostatically shielded copper drift tube with a uniformly strong magnetic field. Notable perturbation of the beam diameter have been observed almost periodically with the increase of magnetic field intensity. While as shown in Fig. 4 and Photo. 3, such beam diameter in the drift tube is indirectly introduced by measuring, through many small holes or narrow

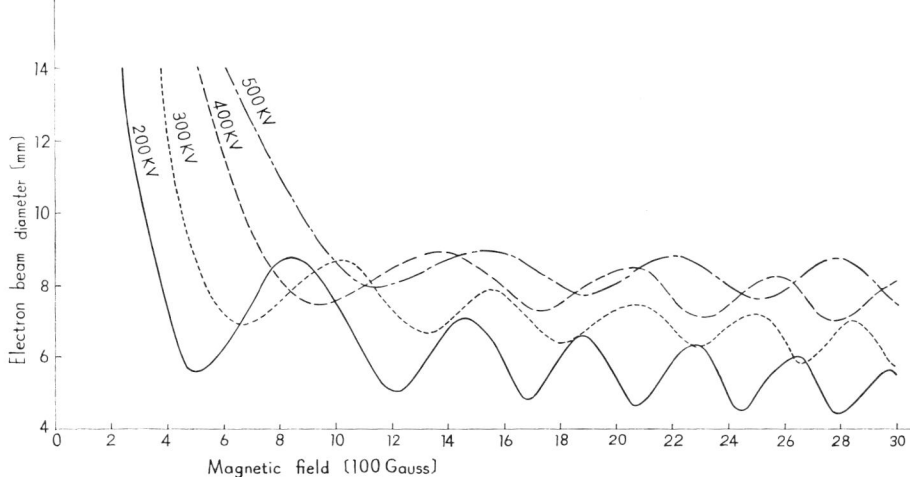

Fig. 3   Relation between magnetic field and electron beam diameter

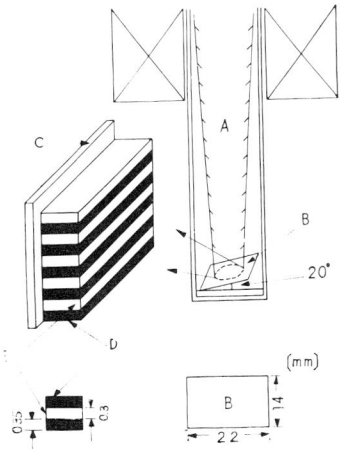

Fig. 4   Schematic illustration of instrument used to obtain electron beam diameter mediately

A: Electron beam,
B: Tungsten target,
C: X-ray film,
D: Lead plates,
E: Plastic plates.

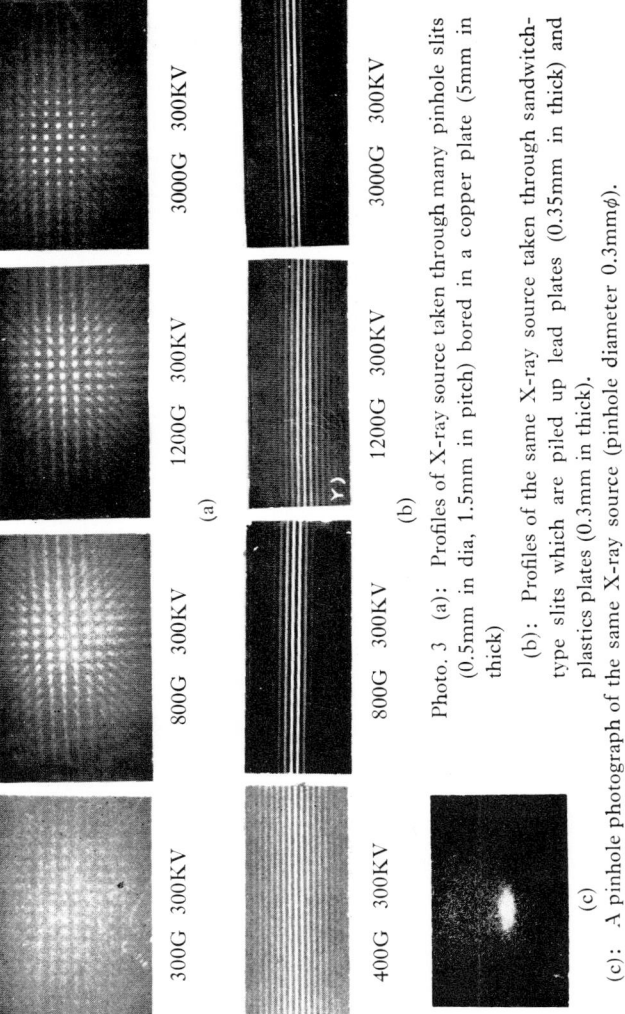

Photo. 3 (a): Profiles of X-ray source taken through many pinhole slits (0.5mm in dia, 1.5mm in pitch) bored in a copper plate (5mm in thick)

(b): Profiles of the same X-ray source taken through sandwich-type slits which are piled up lead plates (0.35mm in thick) and plastics plates (0.3mm in thick).

(c): A pinhole photograph of the same X-ray source (pinhole diameter 0.3mmφ).

slits, a shape and a size of X-ray source which originated owing to the electron beam bombardment on the target. Fig. 5 shows the dose of pulse X-ray radiated when the electron beam is accelerated by one pulse discharge and collides with the target. It seems that the ratio of the spread area of focused electron beam to the tungsten target area gives rise to such a characteristic of X-ray dose correlated to the focusing magnetic field. The tungsten target is plugged into a copper plate in such a manner their surfaces coinside.

The pulse X-ray dose which is produced by the energetic electron beam (600KV,

120A, 10μsec.) which was accelerated with the apparatus, comes up to fifty röntgen or so per pulse at 1cm. away from the center of the target. In this case the dose rate is five million röntgen per second.

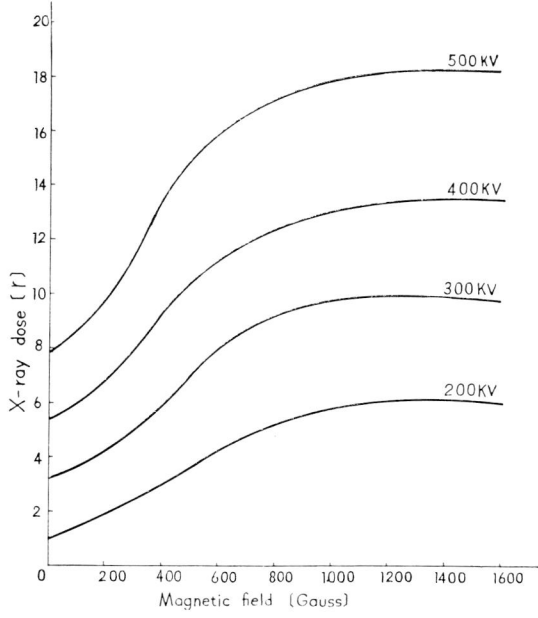

Fig. 5   Relation between magnetic field and X-ray dose.   Per one pulse discharge

Because of the characteristic of the accelerating tube, the electron beam is accelerated for the duration of only 10μsec. There is, however, no doubt that longer accelerating duration, larger accelerating voltage and beam current may be obtained in future. Then it is expected that the dose and the dose rate will increase much more than what obtained with this apparatus.

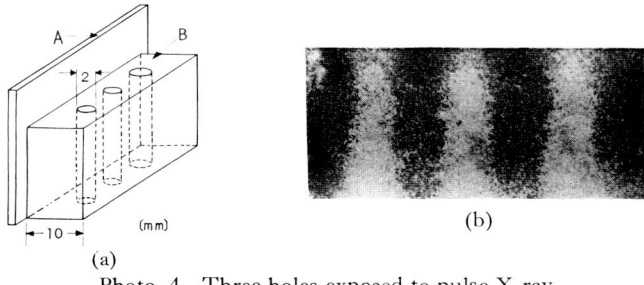

Photo. 4   Three holes exposed to pulse X-ray.
(a):   Schematic diagram
        A: X-ray film   B: Specimen
(b):   Photograph

Photo. 4 shows profile of several holes (2mmϕ), which were bored in a copper plate (10mm in thick), exposed to pulse X-ray generated with this apparatus.

3. Results and Discussion

a) A high energetic electron beam (600KV, 120A.) was obtained easily with Transtron accelerator S-1. The experimental results suggest that higher energetic electron beam will be obtained by scaling up this apparatus.

b) Energetic electron beam in the electrostatically shielded drift tube is focused by strong magnetic fields remarkably, and the beam diameter on the target fluctuates almost periodically the increase of magnetic field intensity.

These facts show that such electron beam is not Brillouin flow, but notable non-laminar flow.

The well-known equation, for the laminar beam without space charge penetrated into the uniformly strong magnetic field, is given below:

$$ r = r_0 \cos\left(\frac{\pi}{\lambda_c} z \right) \tag{3} $$

Where z is a distance from the entrance of the drift tube, r_0 and r are the beam radii at z=0 and z=z respectively, and λ_c is the cyclotron wave length defined by the next equation

$$ \lambda_c = \frac{2\pi\sqrt{2e^* V}}{e^* B} \quad \text{(non-relativistic)} $$

$$ \lambda_c = \frac{2\pi c \sqrt{1 - 1/m^{*2}}}{e^* B} \quad \text{(relativistic)} \tag{4} $$

Where e* equals the ratio of electron charge to mass (=e/m), B is the longitudinal magnetic field and V is an acceleration potential. c and m* are the velocity of light and the relativistic factor of electron mass (=m/m_0) respectively. It may be introduced by (3), (4) and Fig. 6[7] that r equal to r_0 appears periodically with respect to z in the drift tube even if magnetic field B is increased, and maximum amplitude of the electron beam diameter perturbation is independent of magnetic field B.

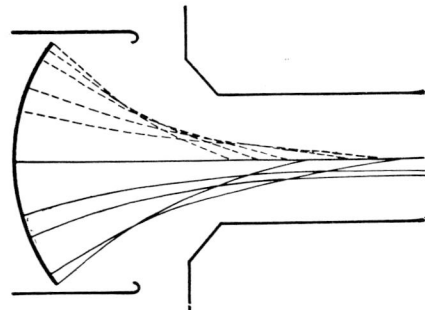

Fig. 6 Electron trajectories with and
without space charge.
········ Charge free
——— In presence of space charge

In case space charges are present, as shown in Fig. 6, accordingly the density rises; the electrons moving parallel to the axis of the drift tube increase in number, and thus the amplitude of the beam diameter perturbation will decrease as suggested by Fig. 3.

c) The energetic electron beam focused by strong magnetic fields shows interesting characteristics and has useful properties. For example, it changes into the source of intensive X-ray with very large dose and dose rate. One example of application to nondestructive inspection, as shown in Photo. 4, is illustrated.

Acknowledgement

This work was performed by the money in trust for the study on the Use of Atomic Energy for Peaceful Purposes. The authors are grateful to the cooperation of Mr. M. Tomie and Mr.H. Miyahara.

References

1) J.D. Cockcroft and E.I.S. Walton: *Proc. Roy. Soc.* (A) **136** 619 (1932)
2) R.G. Van de Graaff: *Phys. Rev.* **38** 1991 (1931): **43** 149 (1933)
3) E.O. Lawrence and M.S. Livingston: *Phys. Rev.* **37** 1707 (1931)
4) D.H. Sloan and Lawrence: *Phys. Rev.* **38** 2021 (1931)
5) D.W. Kerst: *Phys. Rev.* **58** 841 (1940): **60** 47 (1941)
6) Y. Arata: *Chokoon Shiryo (Japanease ultra high temp. review)* **1** No. 6 (1957) **2** No. 1 (1958)
7) M. Okada and Y. Arata: Physical Society of Japan, annual meeting (1963)
8) M. Chodorow, E.L. Ginzton, I.R. Neilsen and S. Sonkin: *Proc. I. R. E.* **41** 1584 (1953)
9) K.J. Harker: *J.App. phys.* **28** No. 6 (1959)
10) T.W. Johnston: *I.E.E.* Paper No. 2694R, Feb. (1959)
11) M.R. Barber and K.F. Sander: *Ibid* No. 2684R, Feb. (1959)

Plasma Electron Beam of Very High Current

Abstract

The authors designed and built the linear electric discharge accelerator which accelerates runaway electrons in plasma, and it produced the very high current pulse electron beam ($100 \sim 500$ [kV], $10^3 \sim 5 \times 10^3$[A], $< 1[\mu$ sec]).

The interaction of the beam with some kinds of magnetic fields such as flat, bell, mirror and cusp types, and the characteristics of the beam-generated X-ray with very high dose were investigated. It was proved that such X-ray source was useful for the study of a property of matter and non-destructive inspection etc.

Introduction

The authors commenced the design and fundamental experiments of Transtron Accelerator S-1 so as to generate 1 [MeV] energetic electron beam of 260 [Amp.] in the duration of 10 $[\mu$ sec][1),2)]. The accelerating tube with oxide-coated thermal emission type cathode is in the state of high vacuum, so it is difficult to inject such beam into other low pressure vessels (1 [atm.]$\sim 10^{-3}$ [mmHg]).

The authors[3)] developed new discharge accelerating tube which accelerates runaway electrons[4)] in plasma. The pulse plasma electron beam more than 1,000 [Amp.] could be obtained easily by this apparatus. The behaviour of such a high current beam, especially its interaction with magnetic fields and its relation of X-ray radiation, has not bean studied yet. Research in it has been made in detail and experiment on the application has been performed.

Experimental Apparatus and Procedure

Figure 1 shows the schematic representation of the apparatus, which consists of power source and discharge accelerating apparatus. The former is composed of condenser bank, and discharging gaps. When each condenser, which is charged in parallel, is jointed in series by the discharge of gaps, the terminal voltage of condenser bank rises up to 1 [MeV] at its maximum. The latter consists of the discharge accelerating tube (7 [cm] i.d.; $30 \sim 100$ [cm] in length), electrodes with ring anode and tungsten pole cathode, the drift tube (7 [cm] i.d.; $10 \sim 100$ [cm] in length), the magnetic vessel

61

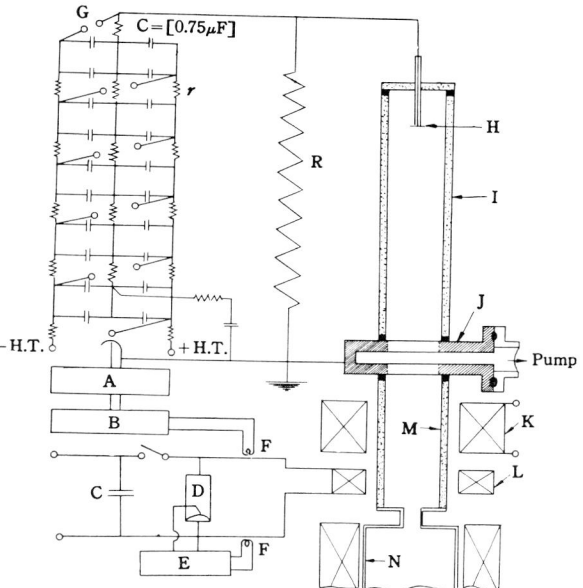

Fig. 1. Schematic diagram of the accelerating apparatu of plasma electron beam.

A. Starting pulse circuit
B. Delay circuit
C. Condenser
D. Ignitron
E. Delay circuit
F. Search coil
G. Gap
H. Cathode
I. Discharge accelerating tube
J. Ring anode
K. D.C. magnetic coil
L. Pulse magnetic coil
M. Drift tube
N. Beam Plasma experimental chamber
r. Resistor
R. Shunt resistor

(0~14,000 [gauss]) and the vacuum system. Runaway electrons are generated and accelerated immediately after the generation of plasma in the accelerating tube. They pass through the ring anode and enter into the drift tube or the magnetic vessel. Control magnetic vessel can generate D.C. field (0~3,700 [gauss]), and strong pulse field (14,000 [gauss] at the maximum). The diameters of the beam were decided by measuring the intensity distribution of X-ray on the backside of a thin target with X-ray film and by using pinhole camera. The energy of the beam was determined by means of mass spectrometer and the absorption characteristics of X-ray through materials. The X-ray dose was determined by using several dosemeters.

Experimental Results and Consideration

The number of runaway electrons, N_{er}, in the plasma in the state of thermal equilibrium is under the electric field E

$$N_{er} = N_{e0} \exp\left(-E_c/E\right) \tag{1}$$
$$E_c = 1.7 \times 10^{-13} n_{e0} z \ln \Lambda / T_e \quad [\text{V/m}]^{4),5)},$$

where N_{e0} and n_{e0} are the total number and the density of electrons in plasma respectively, T_e is the electron temperature, z is the charge number of each ion, Λ is the plasma parameter and E_c is the minimum electric field to generate runaway electrons in plasma.

Letting, L is the length of discharge accelerating tube, $L*$ its effective length, $S*$ its effective cross section, $V*$ its effective terminal voltage, P_0 and P_0^* are initial and effective initial gas pressure in it respectively.

$$N_{e0} = n_{e0} L* S* \propto P_0^* L* S* , \quad E = V*/L* , \quad L* = L-a , \quad P_0^* = P_0+b$$

where a and b are constants depending on the discharge accelerating tube and are given experimentally such as $a = 0.28$ [m] $b = 0.15 \times 10^{-3}$ [mmHg] in this case. Thus Eq. (1) becomes

$$N_{er} = K_0 P_0^* L* S* \exp\left(-k P_0^* L*/V*\right), \tag{2}$$

where K_0 and k are constants, and electron temperature in the accelerating tube is assumed to be nearly constant. Therefore, the beam current I, which can be taken into drift tube is

$$I = N_{er} e/\delta t \tag{3}$$

where δt is pulse time of the beam. Substituting Eq. (2) and (3) into Kramer's equation, $P_r \propto IV^2 Z$, we obtain

$$P_r = K P_0^* L* S* V*^2 Z* \exp\left(-k P_0^* L*/V*\right), \tag{4}$$

where the dosage of X-ray, P_r, which was generated by the bombardment of the beam with accelerating voltage V, $Z* = Z/Z_w$ (Z_w; atomic number of tungsten) and δt is assumed to be constant. So, K is constant in the case of a tungsten target, and the measured values of K and k are shown in Table 1[5]. Solid lines in Fig. 2 and Fig. 3 are obtained by substituting the values of K and k in Table 1 into Eq. (4). In these figures, each point is the measured value and coincides well with the calculated value

Table 1.   Values of K and k

L [m]	K ($\times 10^{-6}$)	k ($\times 10^8$)
0.3	2.4	16
0.4	2.8	7.2
0.5	3.1	6.4
0.8	2.9	3.2
1.0	2.6	2.6
1.1	1.1	1.8

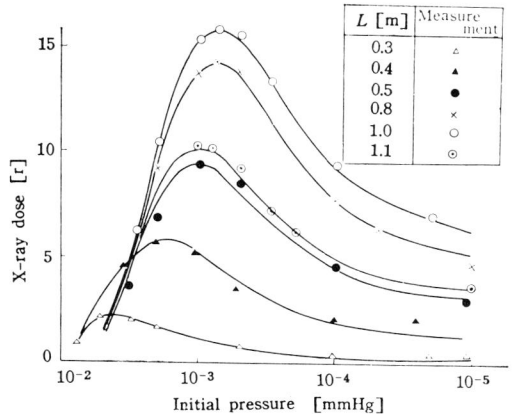

Fig. 2.　Relation among X-ray dose, the initial gas pressure and the length of discharge accelerating tube (curves are obtained by the equation (4) and $V = 160$ kV).

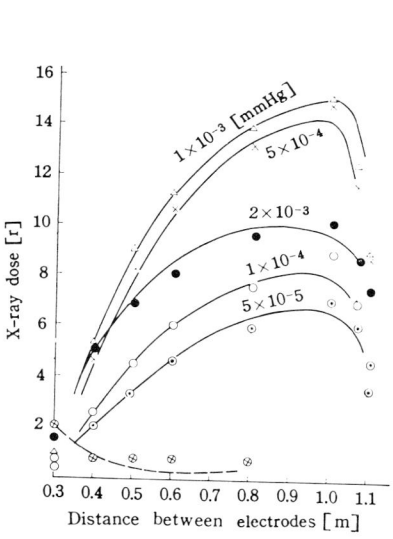

Fig. 3.　Relation between X-ray dose and the gap length of the electrodes.

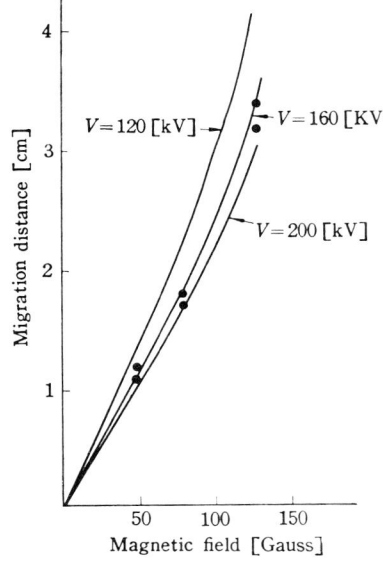

Fig. 4.　Relation between electron beam migration and magnetic field in mass spectrometer.

from Eq. (4).

When the migration distance of the beam on the film vertical to it is measured as a function of the energy of the beam which move in the magnetic field of this apparatus, characteristics of the mass spectrometer used is shown as solid lines in Fig. 4.　In this figure, each point is the plot of the measured value. Therefore, it can be seen that the beam energy gained in the discharge accelerating tube is 160~180 [KeV], when 300 [kV] as the terminal voltage of the condenser bank is impressed. The mean energy of the beam-generated X-ray was 80~90 [keV] in tungsten target. The electron distribution on the cross-section of the beam is not uniform in the r-direction in the drift tube. It is affected by the initial gas pressure in the discharge accelerating tube as shown in Fig. 5. More than 10^{-4} [mmHg], however, it seems to be uniform on the whole. Sampling out into the analizing vessel shown in the Fig. 6 some part of the beam with such uniform distribution and investigating the beam expanse due to space charge of it, partial beam

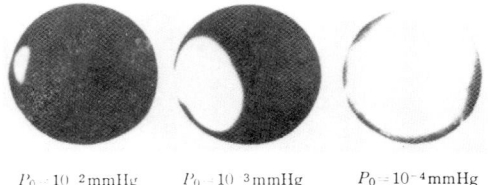

$P_0 = 10^{-2}$ mmHg $P_0 = 10^{-3}$ mmHg $P_0 = 10^{-4}$ mmHg

Fig. 5.   Radial distribution of plasma electron beam in the drift tube.

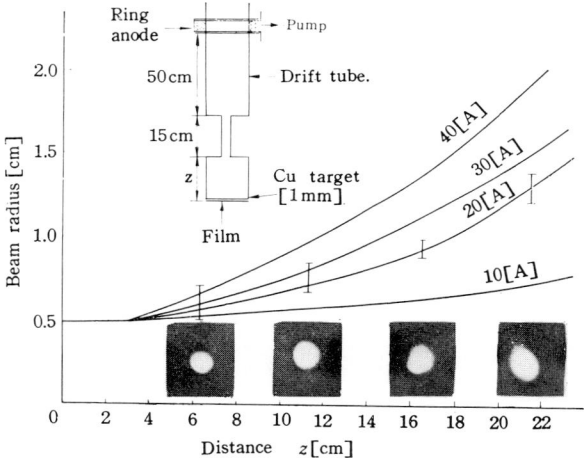

Fig. 6.   Beam expanse due to space charge of it.

current I_0 can be estimated using the next Eq. 5[6)]

$$\sqrt{I_0} = \frac{r_0}{z} \sqrt{\frac{\pi}{\mu_0} \cdot \frac{c_0}{e^*} \cdot \frac{\beta^3}{1-\beta^2}} \int_1^{r^*} \frac{dr^*}{\sqrt{\ln r^*}} \tag{5}$$

where $\beta = v/c_0$ (v; electron velocity, c_0; light velocity), z is distance from nozzle, $r^* = r/r_0$ (r; beam radius at z, r_0; r at $z=0$), $e^* = e/m$ (e; electron charge, m; electron mass), μ_0 is permeability of free space. Solid lines in Fig. 6 are plot of Eq. (5), and the bars

65

indicate the beam radius measured shows its fluctuating characteristics. The beam current I_0 flowing out of the nozzle (1 [cm] dia. 15 [cm] in length) is about 20 [Amp.] from Fig. 6. Considering the diameter of the drift tube (7 [cm]) and the beam expanse along nozzle (15 [cm] in length 1.8 [cm] dia.), it can be estimated that the actual beam current I is equal to $I_0 \cdot (1.8/1)^2 (7/1)^2$. Therefore, $I = 3,000$ [Amp.] for $I_0 = 20$ [Amp.].

When the above described beam is injected in the magnetic vessel with the magnetic flux density distribution shown in Fig. 7, the beam converges and its diamter varies as shown in Fig. 8. The beam diameter was measured on the vertical section

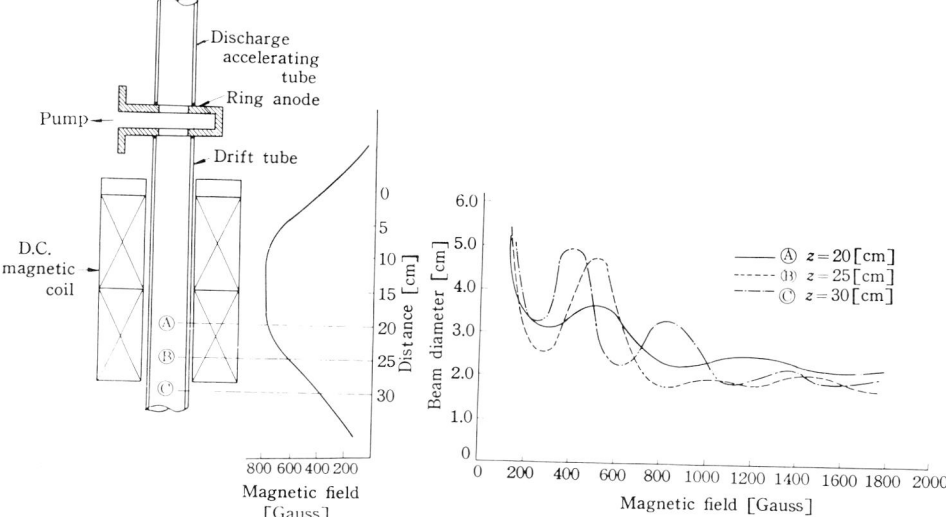

Fig. 7.   Magnetic vessel to determine beam diameter.

Fig. 8.   Relation between beam diameter and magnetic field.

(i.e. target) to the axis of the magnetic field at each point along it.   The beam diameter at each point and its perturbation become smaller, and the beam approaches to the laminor flow narrowed extremely, as the magnetic field becomes stronger.   In this case, the period of the perturbation of beam near the axis varies with the magnetic field intensity B_n.

$$B_n = \frac{4\pi\sqrt{V}}{\sqrt{e^* z}} \cdot n , \quad \left(= 3 \times 10^{-5} \cdot \frac{\sqrt{V}}{z} \cdot n \right)\left[\frac{Wb}{m^2}\right] \qquad (6)$$

n; positive integer

where beam radius $r = r_B (1+\alpha)$ $(\alpha \ll 1, r_B$; the beam radius of Brillouin flow[7]).   Eq. (6) is derived by the periodicity of α introduced by the equation for an electron path close to the axis.

In this experiment, the tantalum thin target (12 [μ] in thickness) was placed

66

vertically to the magnetic field at each point of Ⓐ ($z = 20$ [cm]), Ⓑ ($z = 25$ [cm]) and Ⓒ ($z = 30$ [cm]), and the perturbation with increase of the field at each point was examined. Beam diameters are determined accurately by measuring the intensity distribution of X-ray penetrated the thin target with X-ray film on its backside. B_n at each point Ⓐ, Ⓑ and Ⓒ is nearly equal to 600 n [gauss], 500 n [gauss] and 400 n [gauss] respectively from Eq. (6). These values are almost coincide with the experimental results in Fig. 8. Figure 9 shows the behaviors of the high current pulse plasma electron

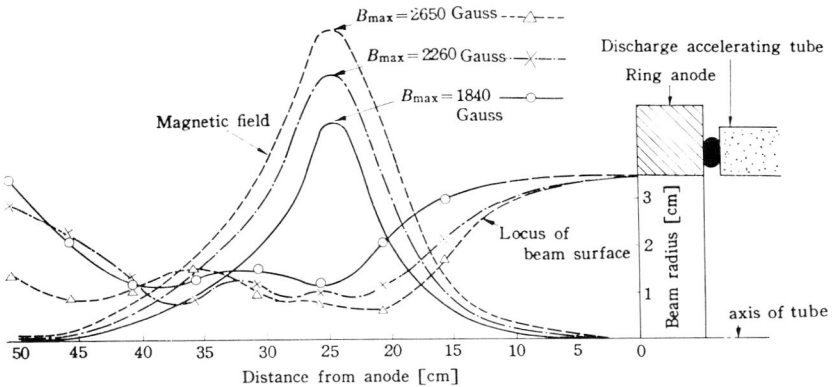

Fig. 9. Behaviour of the plasma electron beam focused by bell type magnetic field in the drift tube.

beam in the bell type magnetic field. It was found that the first focussing point of the beam shifted toward the anode and the beam converged with the increase of the field. The change of the beam diameter with the increase of the field measured by means of the method described above at the point of the strongest magnetic field is shown in Fig. 10. The curve of the beam perturbation in Fig. 10 varies almost periodically in

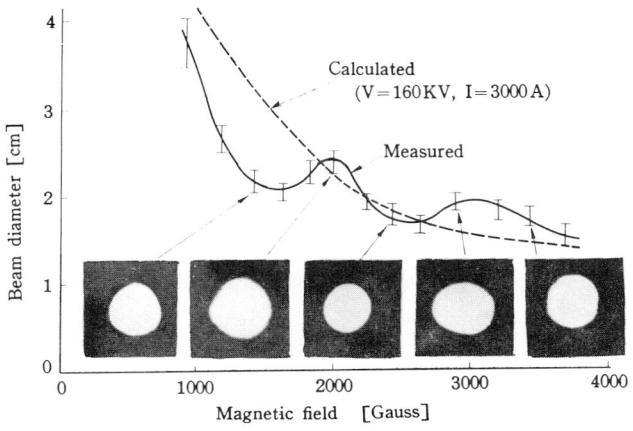

Fig. 10. Focussing effect of D.C. magnetic field on the beam diameter.

the neighbourhood of the dotted line given by the following equation, if $V = 160$ [kV] $I = 3,000$ [A]

$$r_B = \sqrt{\frac{2c_0\mu_0}{e^*\pi} \cdot \frac{1-\beta^2}{\beta} \cdot I} \cdot \frac{1}{B} \quad \text{[m]} \qquad (7)$$

This equation is obtained by equating the repulsive force of space charge and the magnetic forcusing force[6],[8]. Figure 11 shows the focusing characteristics of the plasma electron beam by the pulse strong magnetic field (14,000 [gauss] max.) which is obtained to pass the high current into a focusing magnetic coil connected with a crowbar circuit as shown in Fig. 1. Figure 12 shows the change of the dose of X-ray penetrated the

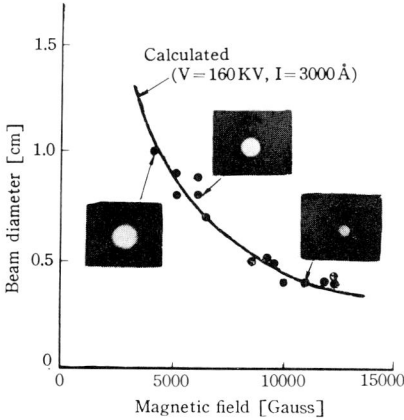

Fig. 11. Focusing effect of pulse magnetic field for beam diameter.

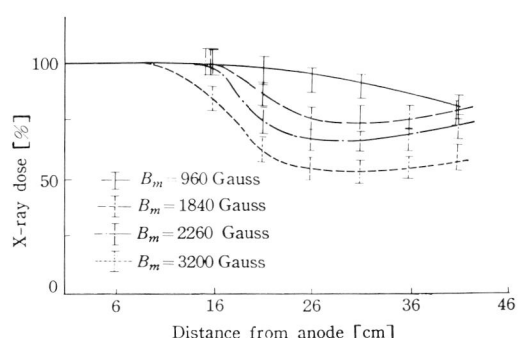

Fig. 12. Magnetic mirror reflection effect (bell type magnetic field).

tantalum thin target (12 [μ] in thickness) placed at each point along the axis of magnetic field. Assuming that this X-ray dose is proportional to the beam current, it proves that the beam current decreases during it passes through magnetic field. It may be considered that the electrons left from the beam appear because of the repulsive force, opposite to the beam direction, due to both the mirror reflection of the magnetic vessel and the space charge of the beam focused initially near the axis. According to this consideration, the repulsive force increases with the intensity of magnetic field, and thus the electrons bombarded on the wall of the drift tube near anode increase, and therefore X-ray radiated from the wall is expected to increase. Figure 13 shows this fact. Mirror and cusp field can be formed with combination of D.C. magnetic field (2650 [gauss] const.) and pulse magnetic field shown in Fig. 14. The radiated X-ray dose was more in mirror field. Figure 15 shows the fact that the hole (4 [mm] dia.) was made in the tantalum foil by the strong focused beam. This fact shows that the pulse plasma electron beam can be applies as a heat source. From the results of the penetrating tests shown in Fig. 16, it is found that beam focusing magnetic field is effective to improve resolving

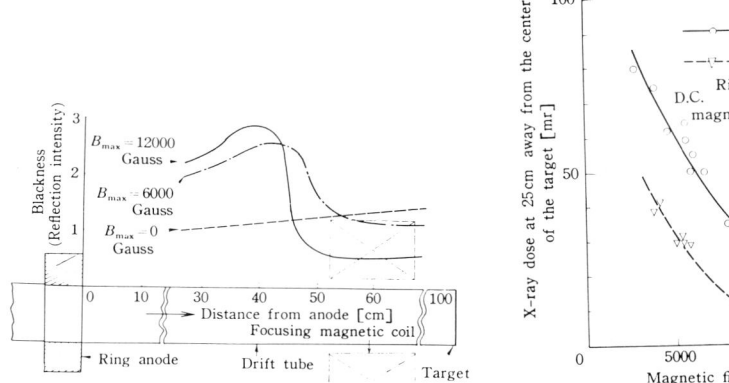

Fig. 13. Reflection effect of strong magnetic mirror on high current beam.

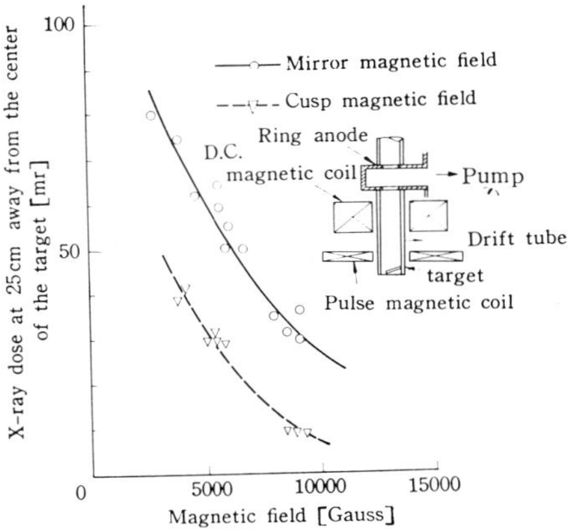

Fig. 14. The change of X-ray dose in cases of mirror and cusp type magnetic fields (pulse stronge magnetic field).

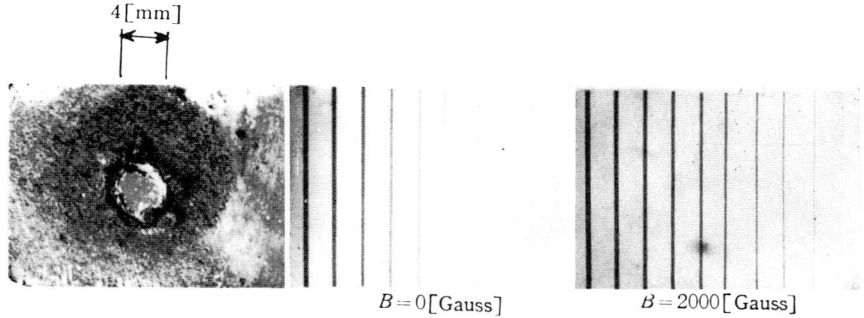

Fig. 15. Tantalum electrode (12μ in thickness).

$B \doteqdot 0 \lbrack \text{Gauss} \rbrack$ $B \doteqdot 2000 \lbrack \text{Gauss} \rbrack$

Fig. 16. Penetrameter tests.

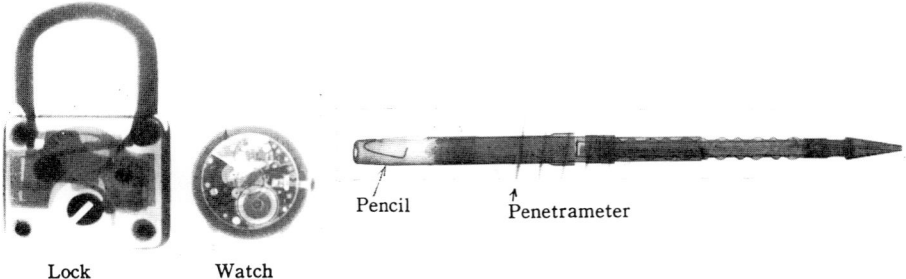

Lock   Watch   Pencil   Penetrameter

Fig. 17. Examples of non-destructive inspection (one pulse beam).

power. Figure 17 shows some examples of photographs of nondestructive inspections and the discharge accelerating tube can be applied to a high speed X-ray camera as its photographs can be taken within 1 [μ sec].

Conclusion

The authors designed and built the accelerator with the new discharge accelerating tube, and showed that the high current and high energy plasma electron beam could be accelerated easily. For example, the pulse plasma electron beam of 160~180 [kV] and about 3,000 [A] generated when the terminal voltage of condenser bank was 300 [kV]. The highest beam current was generated in the initial gas pressure of about 2×10^{-3}~8×10^{-4} [mmHg] in the discharge accelerating tube and the beam pulse time was less than 1 [μ sec]. It was shown that X-ray dose radiated was given by Eq. (4). In the strong focusing magnetic field, the beam is in the state of extreme focusing and reflecting. Particulary, in the much strong field the beam became more like narrow focused laminar flow.

The beam diameter at each point along the axis of magnetic field was ascertained besides its approximate formula was obtained. Mirror field was more effective for a radiant X-ray than cusp field. It was described that the pulse high current plasma electron beam could be applied to a heat source, non-destructive inspection and high speed X-ray camera etc.

References

1) Y. Arata: *Chokoon Shiryo (Japanese Ultra High Temp. Review)*, Vol. **1**, No. 6 (1957); Vol. **2**, No. 1 (1958). (in Japanese)
2) M. Okada and Y. Arata: *Technol. Repts. Osaka Univ.*, **4**, 567 (1963).
3) Y. Arata, M. Tomie and H. Miyahara: PrePrint of Spring Annual Meeting of the Appl. Phys. of Jap. (1965); *Non-Destructive Inspection*, Vol. **14**, No. 2 (1965). (in Japanese).
4) H. Dreicer: 2nd U.N.Conf. on Peaceful Uses of Atomic Energy, Vol. **32**, (1958).
5) M. Okada and Y. Arata: "Plasma Engineering" (1965) (in Japanese).
6) E. Sugata and K. Inaba: 132 Conf. on the beam of charged particles in Japan 1964.
7) L. Brillouin: *Phys. Rev.*, **67**, (1945).
8) J.R. Pierce: "Theory and Design of Electron Beam", (1954).

Some Fundamental Properties of Nonvacuum Electron Beam

Abstract

A fine laminar flow of high energetic electron beam, converged by the combined actions of a magnetic focussing effect as forming the brillouin flow and an plasma focussing effect, has been successfully introduced into the open atmosphere, through a long narrow path having a high resistance to the gas flow, as a high power nonvacuum electron beam (NV-EB).

This article describes the investigations on the behaviors of the beam in several kinds of gases under various pressures (10^{-4} to 10^3 mmHg), especially on the generation of a beam plasma, and a scattering and a range of the beam.

Several applications of the NV-EB as a heat source are also described.

1. Introduction

In the past decade around 1960, theoretical and technological progress in producing high energy-high power electron beam had been achieved in the field of plasma physics or thermo-nuclear fusion research.[1]~[11]

The interesting methods to radiate powerful electron beam into high pressure gases such as one atmospheric pressure or so, have been rapidly developed[12]~[22] in some countries since around 1960, and some of them are now utilized and researched in greater depth with vigor.

One of the authors[1] proposed the Transtron-accelerator which is able to generate an ultra-high power laminar flow of charged particles in 1957. And a new accelerator based on this was developed to generate a pulsive fine laminar flow electron beam (1 Mev, 260 Amp, 10^{-5} Sec; 500 KV, 1000~10000 Amp, 10^{-6}~10^{-9} Sec). This principle gives foundation to making the powerful fine laminar flow type NV-EB generator.

It has been thought[13] that such a powerful electron beam might be introduced smoothly into the open atmosphere as the NV-EB with aids of cooperative action of both magnetic focussing effect and plasma focussing effect, and this was proved in 1967.[20],[21]

Behaviors of the electron beam, produced by this NV-EB generator, in several kinds of gases under various pressures and some fundamental properties of the NV-EB as a heat source at the atmospheric pressure were studied.

2. Foundation of a Design for the NV-EB Generator

From a view point of industrial utility, the NV-EB generator should provide a new mechanism differing from the ordinary vacuum electron beam (V-EB) generator as follows:

1. Network of vacuum system
2. Beam-inducement to open atmosphere (or "beam channel")
3. Powerful energetic electron gun having a long life.

The NV-EB generator can be considered as a network of vacuum system. Air gas circulates through open atmospher, vacuum accelerating chamber and evacuation pump.

We call now the path of the beam between the open atmosphere and the vacuum accelerating chamber, the "beam channel". This beam channel is most important in the vacuum system and in the whole path of electron beam.

This beam channel should be so designed that it may have a very high flow resistance to a gas flow to ensure a very steep gradient in gas pressure, and the powerful electron beam must be aligned along the axis of the beam channel so as not to collide with the wall of the channel.

From this consideration, the beam channel will take the form of a long narrow pipe as shown in Fig. 1 (b) and (c) or Fig. 2 which was proposed by one of the authors.[13],[20]

On the contrary, the other researcher's apparatuses, as presented in Fig. 1 (a), consist in combination of slits having small diameters.

Preliminary experiments as shown in Fig. 3 were performed to investigate the characteristics of beam channel, and it was verified that the beam channel didn't disturb the passing of a fine laminar flow beam and also plasma focussing effect diminished the beam space charge as shown in Fig. 4.[21],[23]

On the converging electron beam along the axis, the plasma focussing has a simillar effect to the magnetic focussing effect, and by cooperation of both effects, the beam does not expand so much and becomes practically a laminar flow instead.

This co-operative action can only be utilized in (b) and (c) in Fig. 1 and this is impossible in the ordinary type (a) in Fig. 1.

The NV-EB generator was designed and constructed on the basis of these results as shown in Fig. 5.

The capacity of the necessary evacuation pump is not large compared with the oridnary type.

The main part of the NV-EB generator consists of an electron gun and a beam channel. Development of the electron-generating source having a long life is most necessary for industrial usage.

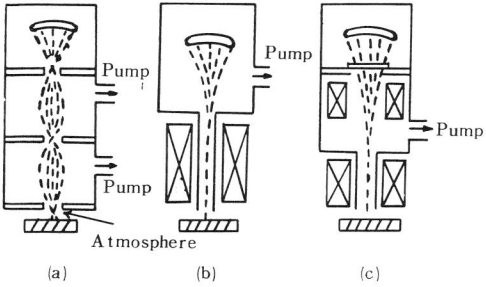

Fig. 1. Principle of the NV-EB apparatus
(a) Ordinary type
(b) and (c) New types proposed

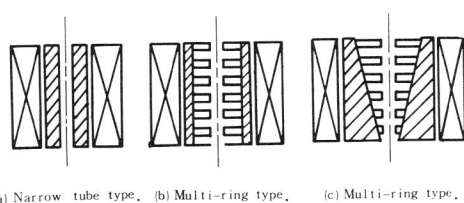

(a) Narrow tube type. (b) Multi-ring type. (c) Multi-ring type.
(same size) (not same size)

Fig. 2. Schematic illustration of main part of beam channel

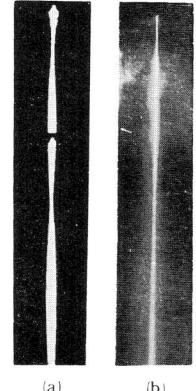

Fig. 4. Locus of electron beam indicating a plasma focussing effect.
(a) 320 volt beam (after D.A. Dunn etal[22])
(b) 10kvolt beam bunched along the axis

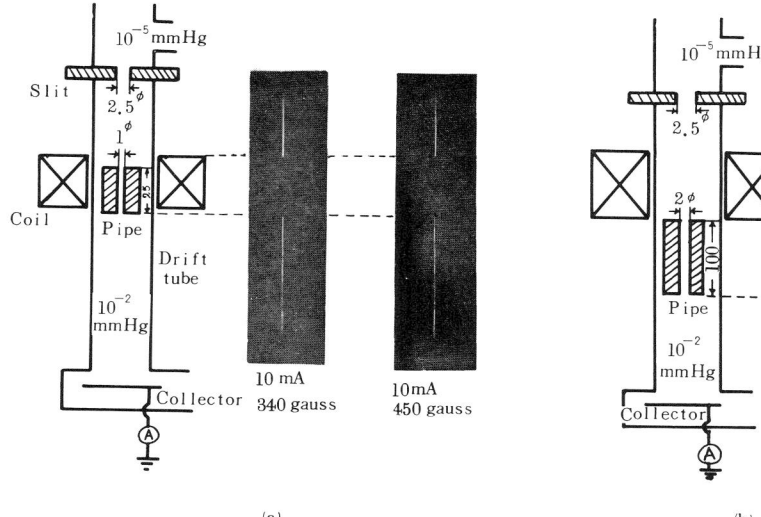

Fig. 3. Locus of fine laminar flow beam passed almost 100% through two kinds of long narrow pipes set in the low vacuum drift tube. (Accelerating voltage; 10 [KV])

72

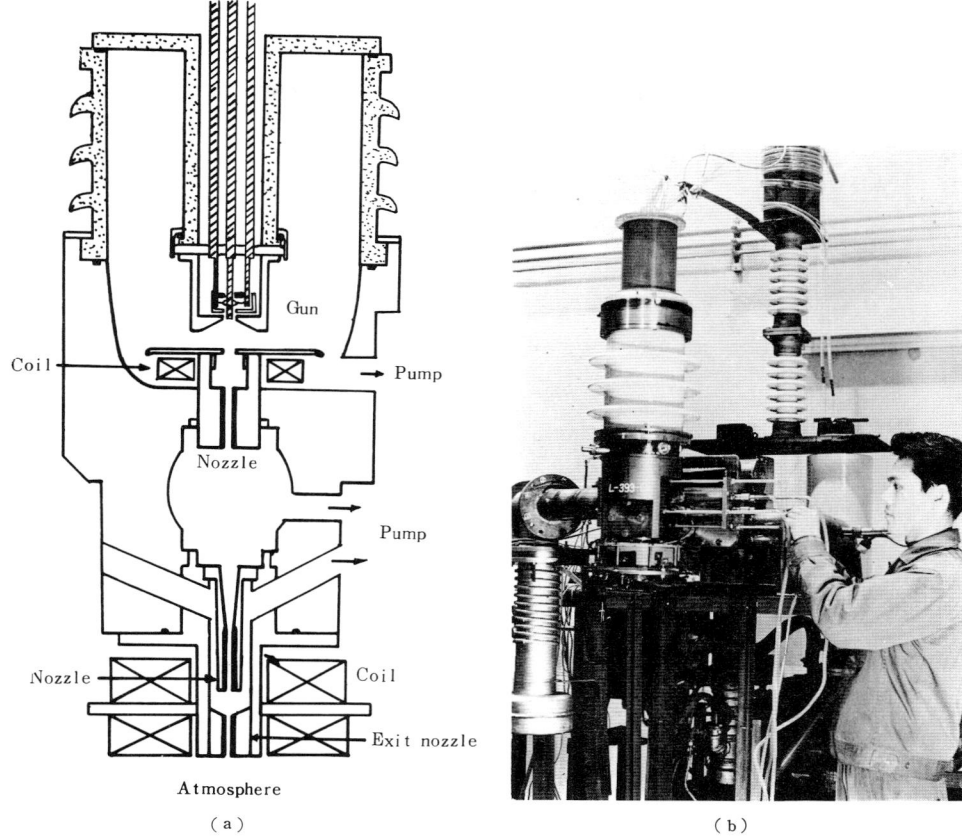

Fig. 5. (a) NV-EB generator containing electron gun and beam channel
(b) General view of NV-EB prototype welding apparatus

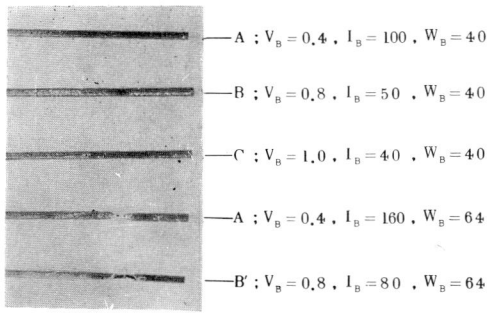

A ; $V_B = 0.4$, $I_B = 100$, $W_B = 40$

B ; $V_B = 0.8$, $I_B = 50$, $W_B = 40$

C ; $V_B = 1.0$, $I_B = 40$, $W_B = 40$

A ; $V_B = 0.4$, $I_B = 160$, $W_B = 64$

B' ; $V_B = 0.8$, $I_B = 80$, $W_B = 64$

V_B ; Bombardment voltage [kV]
I_B ; Bombardment current [mA]
W_B ; Bombardment power [W]

Fig. 6. Examples of consumption of tungsten rod cathodes

One example of the electron-generating source is an end of tungsten-rod containing ThO_2 which is heated homogeneously by electron-bombardment from the surrounding filament. Consumption of the tungstenrod cathode depends almost on bombardment energy W_B rather than bombardment current, I_B or bombardment voltage, V_B as shown in Fig. 6. The maximum energy of W_B was about 50~55 [Watts] in case of 1 mmϕ tungsten-rod containing 2% ThO_2 for the working time, $\tau_W = 10$ [hours].

Figure 7 shows the relation between W_B and τ_W to obtain a constant electron beam ($I_b = 30$ [mA]; $V_b = 40$ [KV]) and the relation between I_b and τ_W under constant conditions ($W_B = 50$ [Watts]; $V_b = 40$ [KV]), where V_b is the beam accelerating voltage and I_b is the beam current.

In general, both of curves $W_B - \tau_W$ and $I_b - \tau_W$ consist of A, B and C region. For the welding application, B flat-region is most convenient.

A forming mechanism of A, B and C in Fig. 7 will be discussed briefly in Appendix.

73

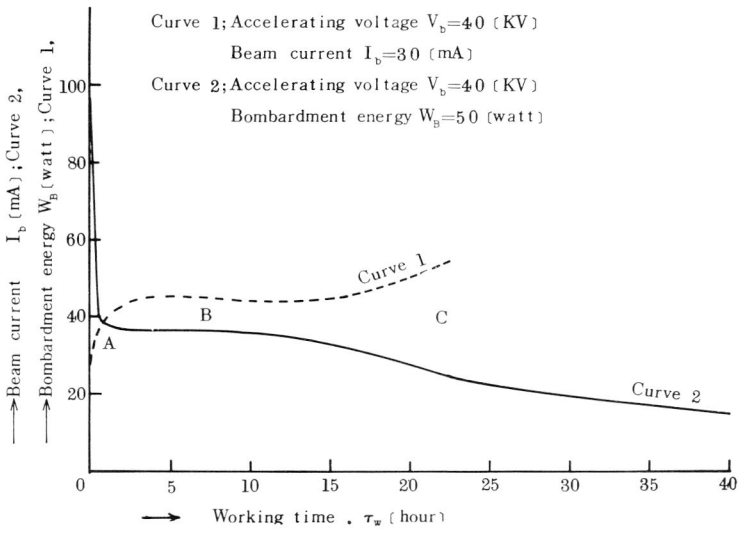

Curve 1; Accelerating voltage $V_b = 40$ [KV]

Beam current $I_b = 30$ [mA]

Curve 2; Accelerating voltage $V_b = 40$ [KV]

Bombardment energy $W_B = 50$ [watt]

Fig. 7. Relations between W_B and τ_W, I_b and τ_W

3. Experiments

Fig. 8 shows typical profiles of the beam and of the beam plasma obtained by this apparatus under various gas pressures. And Fig. 9 shows the profiles of the NV-EB plasma photographed with various exposures, from which the distribution of the beam intensity is roughly estimated. It is possible to observe roughly that the range of the beam in He is longer several times compared with air. This is explained in detail later. This phenomenon suggests that the function of NV-EB as a heat source is more intensified when the beam is shielded by such lighter gases as He and H_2.

Characteristics of the beam-generated plasma luminous ball tinged with blue in the air atmosphere were studied by methods as shown in Fig. 10, and the methods such as (a), (b) and (c) were used to search the apparent-thermal, electric and X-ray properties respectively.

The results obtained are as shown in Figs. 11, 12 and 13.

Fig. 11 shows the apparent-thermal property which is indicated by the pyroelectric potential occurring in the thermocouples placed on each point of the beam axis and has utility in practice, though it is the pseudo-properties, when this curve is represented by $T_0 e^{-\alpha x + \beta}$ ($\alpha = 0.6$, $\beta = 0.3$). This phemenon

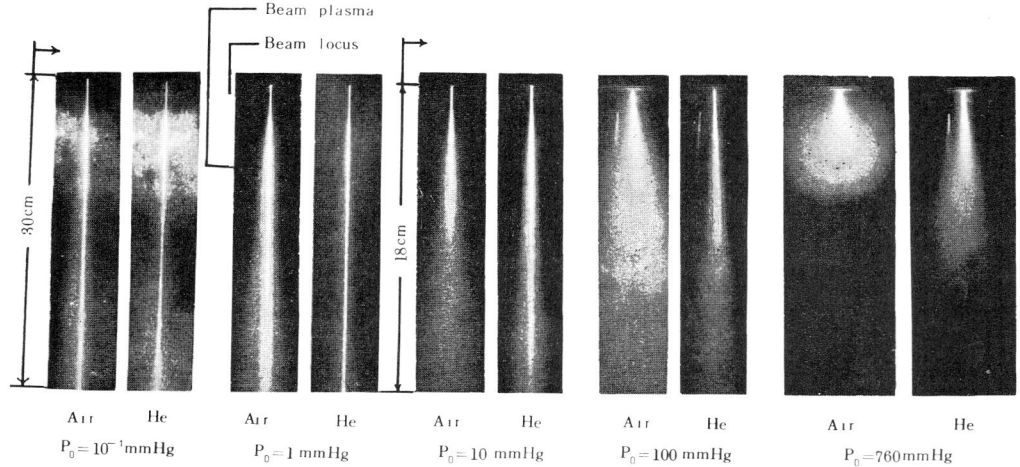

Fig. 8. Typical profiles of electron beam and beam plasma under various gas pressures (Accelerating voltage; 60 [KV])

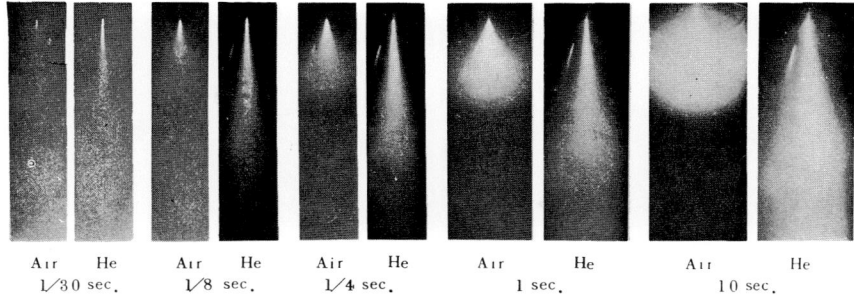

| Air | He | Air | He | Air | He | Air | He | Air | He |
| 1/30 sec. | | 1/8 sec. | | 1/4 sec. | | 1 sec. | | 10 sec. | |

Fig. 9.   Influence of the exposure on the profile of the NV-EB (Accelerating voltage; 60 [KV])

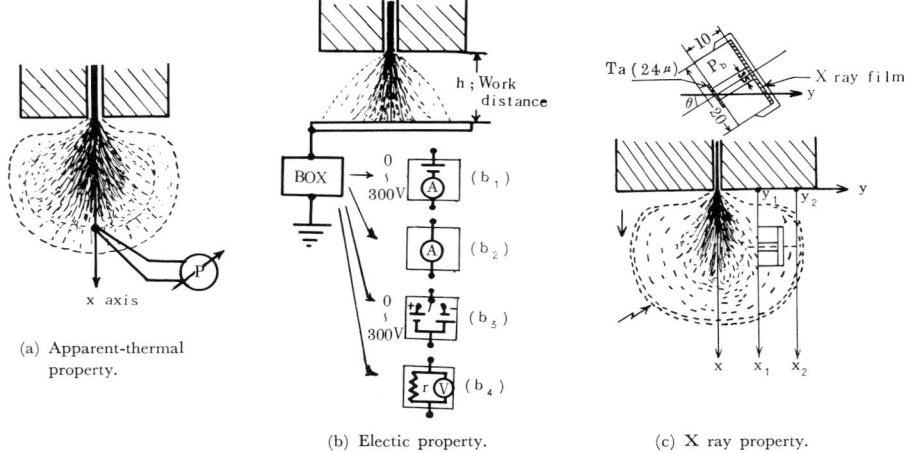

(a) Apparent-thermal property.

(b) Electic property.

(c) X ray property.

Fig. 10.   Measuring methods used to study various properties of plasma luminous ball in the atmosphere
(Accelerating Voltage; 60 [kV])

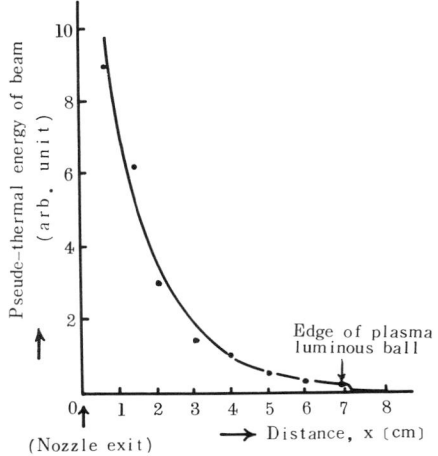

Fig. 11.   Pyroelectric property of beam-plasma ball on the beam axis (Accelerating voltage; 60 [KV])

occurs depending on the beam Power densities decreasing with distance, x rather than the absolute values of it.

Fig. 12 (a) shows the influence of the external electric field for the collector current in the atmospheric plasma luminous ball, utilizing the method of Fig. 10 (b) b_1, in which the curves 2 and 3 indicate that the electric conductivity of the luminous ball is very high near the nozzle exit as compared with the curve 1 in the case of the nonexternal electric field. This means that the high density of the plasma exists there.

Precise measurement of the NV-EB current to the collector or samples is very difficult in general.   The collector current corresponding to the apparent-beam current depends on the face area of nozzle end-plane as shown in Fig. 12 (b) which is obtained from Fig. 10 (b) b_2.   The collector current indicated in curve 1 seems to be close upon the real beam current compared with curves 2, 3 and 4 indicating the existence of the counter-beam current; especially in curve 4 indicating

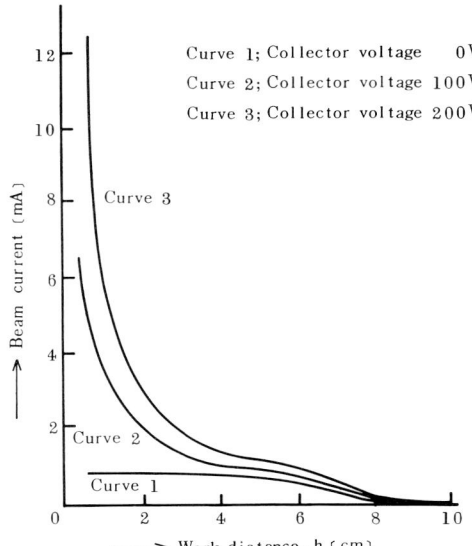

Curve 1; Collector voltage   0 V
Curve 2; Collector voltage 100 V
Curve 3; Collector voltage 200 V

Fig. 12 (a). Effects of the external electric fields on the plasma luminous ball in the atmosphere (Accelerating voltage; 60 [KV])

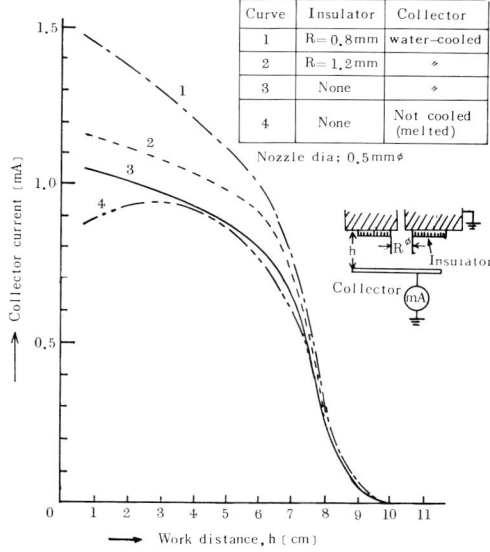

Curve	Insulator	Collector
1	R = 0.8mm	water-cooled
2	R = 1.2mm	〃
3	None	〃
4	None	Not cooled (melted)

Nozzle dia; 0.5mm φ

Fig. 12 (b). Effects of nozzle insulation on the collector current indicating the existence of counter beam current

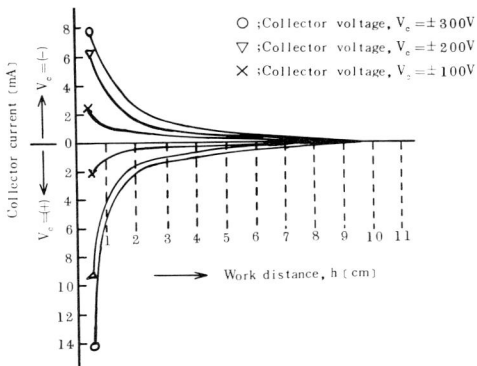

○ ;Collector voltage, $V_c = \pm 300$V
▽ ;Collector voltage, $V_c = \pm 200$V
× ;Collector voltage, $V_c = \pm 100$V

Fig. 12 (c). Effects of the external electric field polarity on collector current in the plasma luminous ball in the atmosphere. Beam current is not included in collector current. (Accelerating voltage; 60 [KV])

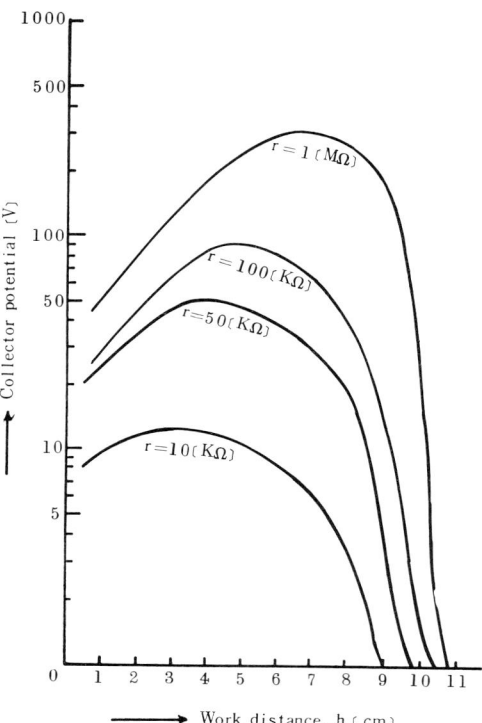

Fig. 12 (d). Charging effect on collector potential (Accelerating voltage; 60 [KV])

the decrease in the collector current even near the nozzle exit. Counter-beam current arises from a large amount of electrons emitted thermionically at the melting zone of the collector, water-cooled Cu, made by the NV-EB.

It was discovered that the atmospheric plasma luminous ball has rectifying action near the nozzle exit especially as shown in Fig. 12 (c) obtained from Fig. 10 (b) b_3. This ball showed 60 [KΩ] in resistance in case of collector potential, $V_c = \pm 300$ [V]

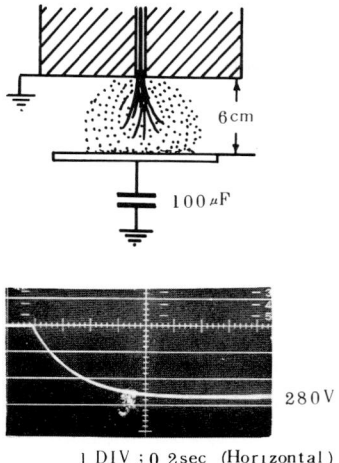

6 cm

100 μF

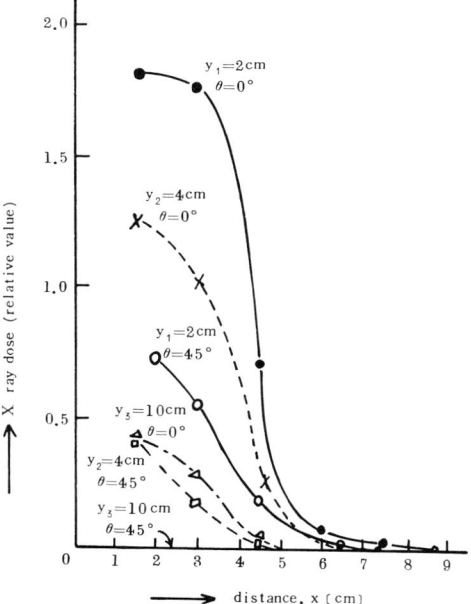

280 V

1 DIV : 0.2 sec (Horizontal)

100 volt (Vertical)

Fig. 12 (e). Charging to condenser by NV-EB

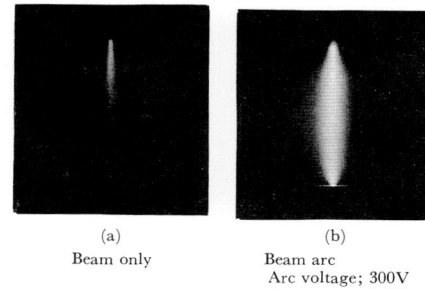

(a)
Beam only

(b)
Beam arc
Arc voltage; 300V
Arc current; 0.45A

Fig. 14. The typical appearance of beam arc (Accelerating voltage; 60 [KV], Work distance; 15 [mm], Beam current; 0.8 [mA])

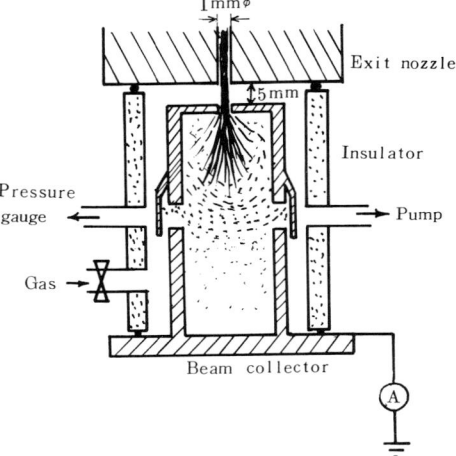

1 mm φ

Exit nozzle

5 mm

Insulator

Pressure gauge

Pump

Gas

Beam collector

Ⓐ

Fig. 15. Measuring method of the beam current

Fig. 13. X-ray dose radiated from Ta-film of X-ray probe set inside the plasma luminous ball

and the $h=1$ [cm], and the continuous discharge is generated in case of $V_c = \pm 350$ [V] and $h=0.5$ [cm].

Fig. 12 (d) shows the basic properties of the charge-transportation inside this ball. The beam-charged-collector potentials which vary with collector resistance, r, reach a maximum which is considerably higher than potentials near the nozzle exit.

Because of the counter-beam current as explained in Fig. 12 (b) and (c), and of high conductivity in

beam plasma generated near the nozzle, the potential cannot rise any more there. Provided that the condenser bank is set, it is possible to charge up them by the NV-EB most effectively at each maximum point, for instance, 100 μF condenser set at 6 cm apart from the nozzle was charged up to 280 volts in 0.3 sec using 60 [KV], 1.5 [mA] NV-EB as indicated in Fig. 12 (e).

NV-EB luminous ball does not irradiate X-ray detectable with X-ray probe indicated in Fig. 10 (c). But putting this X-ray probe inside the luminous ball, much dose of X-ray is observed as shown in Fig. 13.

This phenomenon occures because NV-EB collide with Ta window of the X-ray probe, therefore very high hard X-ray as same as in Fig. 13 is irradiated in case of NV-EB processing ordinary material.

Plasma arc through which the NV-EB flows at its central axis is called the "Beam Arc" in this article. In this beam arc, the range of electron beam is prolonged as the result of arc heating of beam path, as shown in Figs. 21 and 23 and also it may be possible

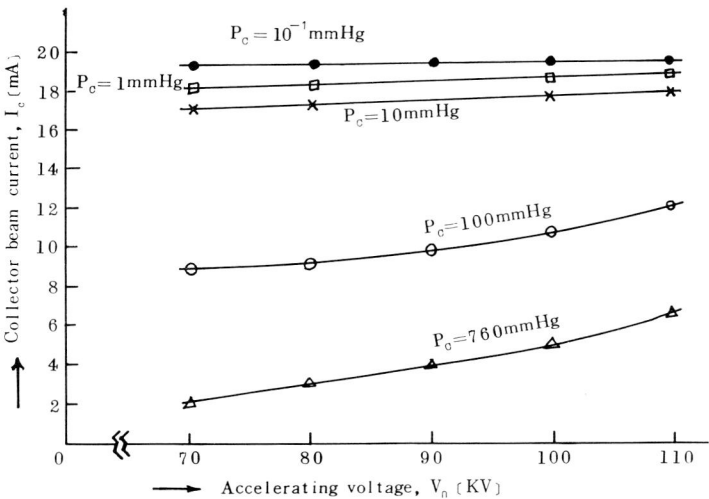

Fig. 16 (a). Relation between collector beam current and accelerating at voltage at various pressures

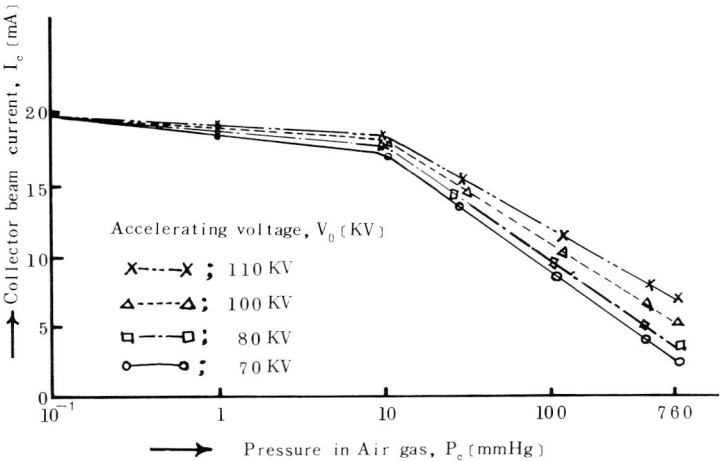

Fig. 16. (b) Relation between beam current and pressure in Air gas

to give some heat treatment actions to work pieces. Fig. 14 shows the typical appearance of a beam arc. This beam arc is fairly stable even at the arc length of $h=15$ mm and a white core, about 1 mm in diameter, can be observed in its axis. Plasma arc, however, disappeares whenever the NV-EB is interrupted.

It is important to know the total actual beam current out of the nozzle exit under various gas pressures and accelerating voltages in practice. This beam current is detectable using a special beam collector 5 mm in collector distance which is chosen considering the actual welding process and 1 mmϕ small hole drilled by the beam itself as indicated in Fig. 15. And then as shown in Fig. 16, (a) and (b),

the collector beam current, I_c was measured under several pressures, P_c and the accelerating voltages, V_b after it was adjusted to get a constant current, $I_c=20$ [mA], at $P_c=10^{-1}$ [mmHg] and $V_b=70$ [KV] because I_c was constant under $P_c \lesssim 10^{-1}$ [mmHg] and $V_b \gtrsim 70$ [KV]. The total beam current in the NV-EB or so is considerably affected by V_b, but it is not so influenced in the V-EB under $P_c \lesssim 10$ [mmHg] as shown in Fig. 16 (a).

All of the curves appear to bend around $P_c \approx 10$ [mmHg] as indicated in Fig. 16 (b). This seems presumable from the profiles of the beam plasma having the properties such that the plasma focussing action seems to have influence on the beam so as to keep the laminar flow under $P_c \lesssim 10$ [mmHg], but a

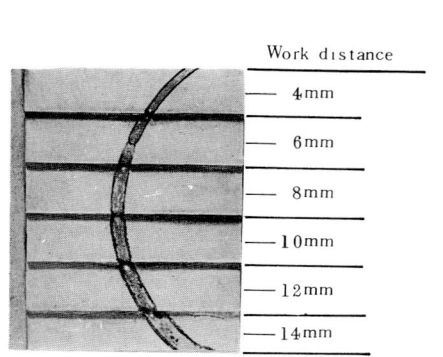

Work distance

—— 4mm

—— 6mm

—— 8mm

——10mm

——12mm

——14mm

(a) NV–EB weld bead on stepped
Specimen (Stainless steel)

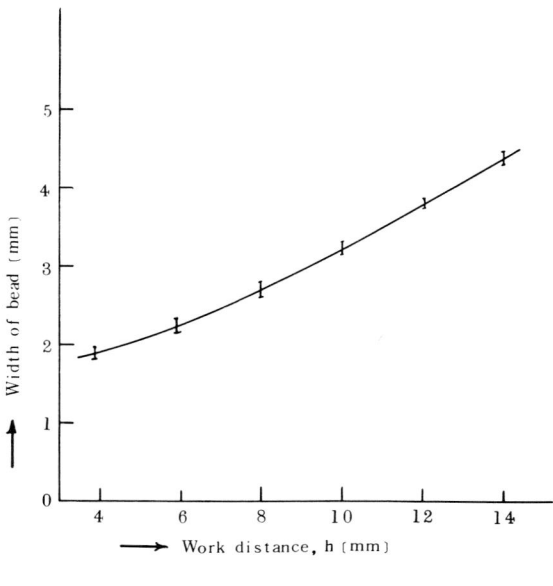

(b)

Fig. 17.

(cm)

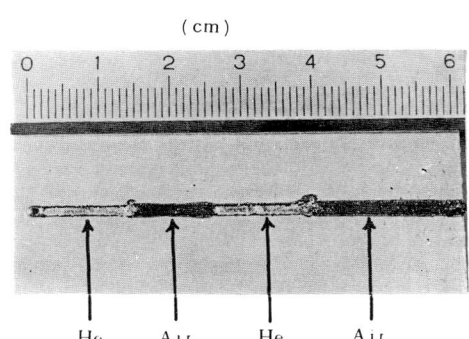

He   Air   He   Air

Fig. 18. (a)   Welding bead view of NV-EB; V_0=100 [KV]
(Accelerating voltage) v_W = 10 [cm/min]
(Welding speed)   shield gases; He, Air
(Stainless steel)

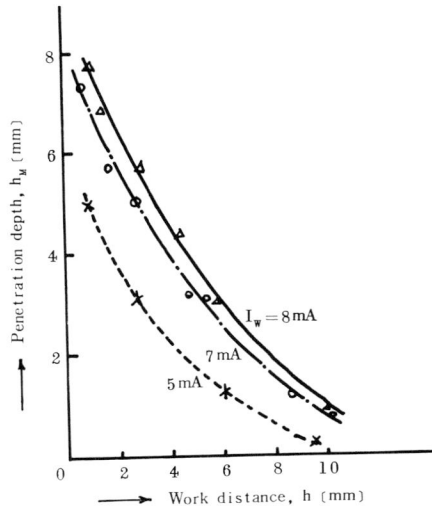

Fig. 18 (b).   Penetration depth in NV-EB Welding of
stainless steel V_0 100 [KV] v_W=10 [cm/min],
I_W: welding beam current

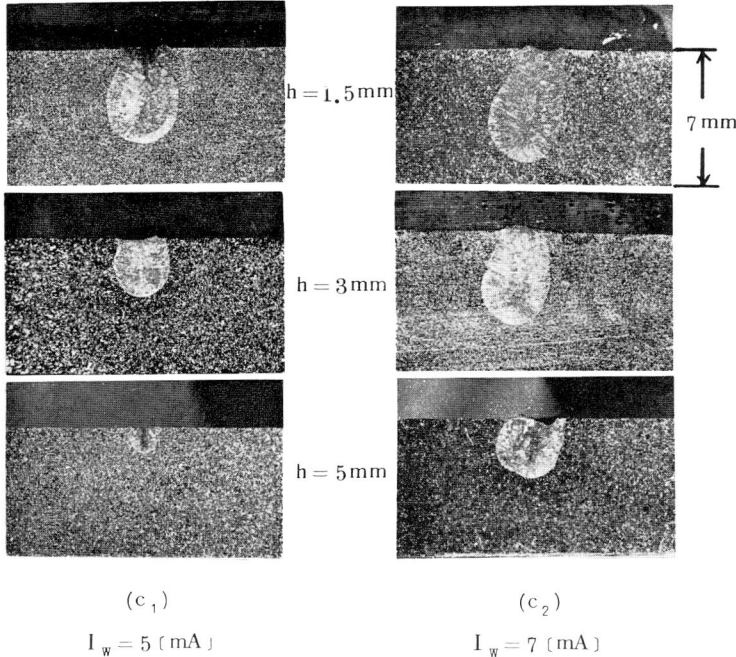

(c$_1$)   (c$_2$)

I $_w$ = 5 〔mA〕   I $_w$ = 7 〔mA〕

Fig. 18 (c).   Cross-sectional view of NV-EB weld bead.   Stainless steel, without shielding gas,
V_0=100 [KV], v_W=10 [cm/min]

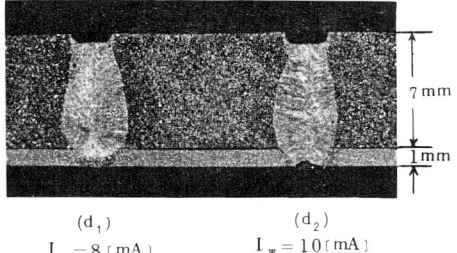

(d$_1$)   (d$_2$)
I $_w$ = 8 〔mA〕   I $_w$ = 1 0 〔mA〕

Fig. 18 (d).   Cross-sectional view of NV-EB lap weld bead,
7 mm and 1 mm stainless steel, V_0=100 [KV],
v_W=10 [cm/min], shielding gas; He

Fig. 19 (a).   NV-EB gas cutting

strong beam scattering action occurs rather than plasma focussing one under $P_c \gtrsim 10$ [mmHg], as seen from the photographs as shown in Fig. 8.

A work distance, h, defined as the distance between the processing material and the exit nozzle, is one of the most important parameters in utilizing the NV-EB as a heat source. It was appropriate to select several mm in h under $V_b \approx 100 \sim 150$ [KV], considering some problems such as the bead formation, as indicated in Fig. 17 and 18, the kerf as shown in Fig. 19, the exit nozzle and so on.

Fig. 20 shows the relation between the apparent NV-EB welding currents passed through various environment gases, I_W and one of the air, I_W (air) which was obtained from the experimental results such as

Fig. 20 (b) observed under the following conditions: V_b=100 [KV] h=5 mm and processing material used: 20 mm-Cupper circular plate rotated under 24 [r.p.m], radius = 5 [cm] i.e. working speed = 754 [cm/min]

Fig. 21 shows the relation of the apparent beam current, I_W and the theoretical range of the beam as illustrated in 5–Section (Discussion), R_e versus molecular weights of environment gases, where I_W was observed in various environment gases respectively under the same conditions of measuring I_W (air), which was variable by adjusting the bombardment power.

80

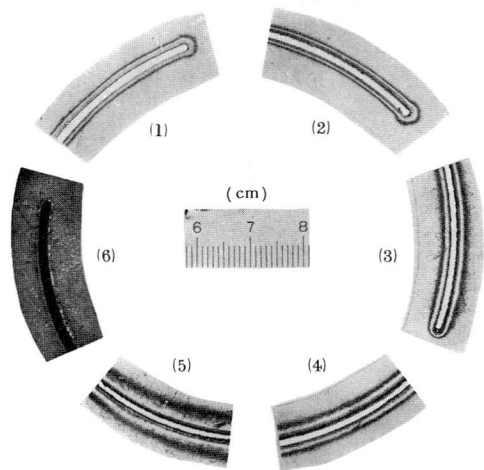

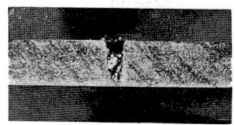

3mm Stainless steel
$V_0 = 85\,KV$
$I_W = 5\,mA$
$v_{cutt} = 20\,cm/min$

3mm Stainless steel
$V_0 = 90\,KV$
$I_W = 5\,mA$
$v_{cutt} = 20\,cm/min$

3mm Stainless steel
$V_0 = 100\,KV$
$I_W = 5\,mA$
$v_{cutt} = 20\,cm/min$

6mm Stainless steel
$V_0 = 110\,KV$
$I_W = 11\,mA$
$v_{cutt} = 10\,cm/min$

Fig. 19 (b). Kerf view of NV-EB gas cutting (Working gas; He)
(1) 1 mm Stainless, $V_0 = 100$ [KV], $I_W = 8$ [mA], $v_{cutt} = 20$ [cm/min]
(2) 2 mm Stainless, $V_0 = 100$ [KV], $I_W = 12$ [mA], $v_{cutt} = 20$ [cm/min]
(3) 4 mm Stainless, $V_0 = 105$ [KV], $I_W = 12$ [mA], $v_{cutt} = 15$ [cm/min]
(4) 6 mm Stainless, $V_0 = 110$ [KV], $I_W = 12$ [mA], $v_{cutt} = 10$ [cm/min]
(5) 8 mm Stainless, $V_0 = 110$ [KV], $I_W = 13$ [mA], $v_{cutt} = 8$ [cm/min]
(6) 10 mm Stainless, $V_0 = 110$ [KV], $I_W = 11$ [mA], $v_{cutt} = 10$ [cm/min]

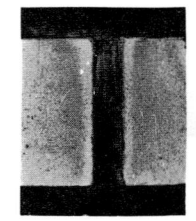

10mm Ceramics
$V_0 = 110\,KV$
$I_W = 11\,mA$
$v_{cutt} = 10\,cm/min$

Fig. 19 (d). Cross sections of NV-EB gas cutting specimen (Working gas; He)

1mm Stainless steel
$V_0 = 100\,KV$
$I_W = 8\,mA$
$v_{cutt} = 20\,cm/min$

2mm Stainless steel
$V_0 = 100\,KV$
$I_W = 12\,mA$
$v_{cutt} = 20\,cm/min$

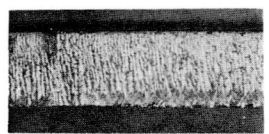

3mm Stainless steel
$V_0 = 105\,KV$
$I_W = 12\,mA$
$v_{cutt} = 15\,cm/min$

Fig. 19 (c). Appearance of surface cut by NV-EB gas cutting method

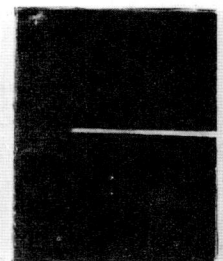

1 mm Pb
Cutting width 0.5 mm
$v_{cutt} = 15$ cm/min

Fig. 19 (e). Small hole on the ceramics which was instantly drilled by NV-EB

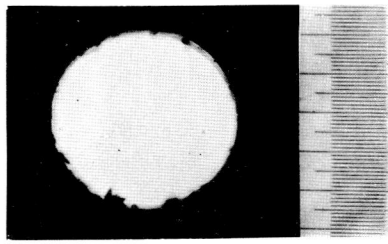

1.5 mm Ceramics
holedia 4.5 mm

Fig. 19 (f). Example of the NV-EB gas cutting of Soft material, Pb

4. Discussions

The range, R_e of the NV-EB travelling through the environment gas or the vaporized gas of processing material out of the nozzle exit is the most important parameter in practice.

It was solved as Eq. (2) obtained from formula (1) which was derived from the revised Bethe stopping power formula[24] under the following assumptions that the number density of environment gas or material, n, is constant along the beam axis (x-coodinate) and Q (V, V_{ex}) is equated to Q (V_0, V_{ex}) to simplify the calculation of the formula (1), where V is electron energy at distance x, [KV], V_0 is an initial electron energy ($V_0 = V$ at $x = 0$), [KV] and V_{ex} is mean excitation energy of target gas or material in volts, [V].

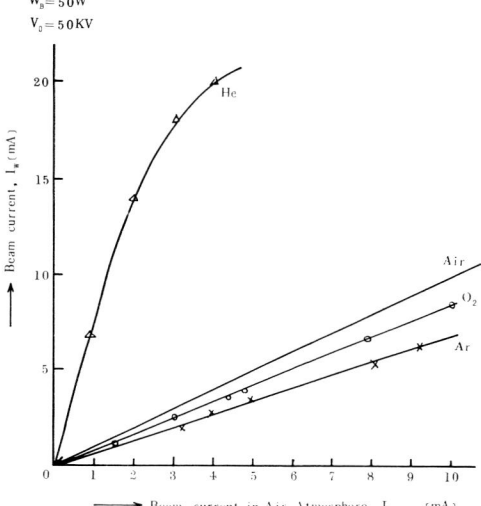

Fig. 20 (a). Comparison of beam current I_W in various environment gas atmospheres versus one in Air, I_W (air); $V_0 = 100$ [KV] $h = 5$ [mm]

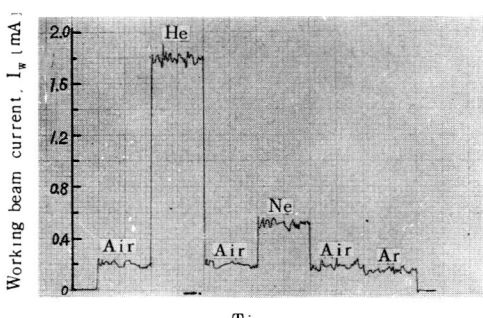

Fig. 20 (b). Working beam current in NV-EB welding with various environment gases under the same conditions; $V_0 = 100$ [KV] $h = 5$ [mm], $v_W = 754$ [cm/min]

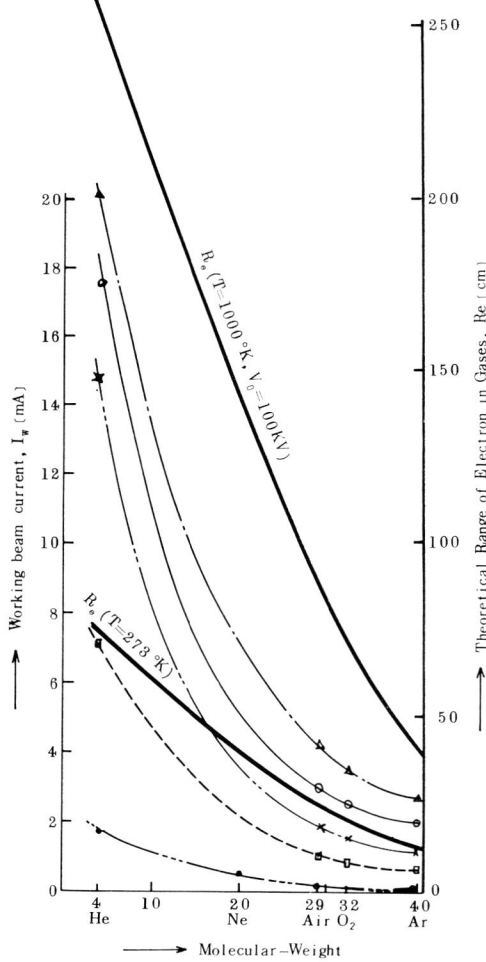

Fig. 21. Relation of I_W and R_e versus molecular-wieght of environment gas; same conditions as Fig. 20

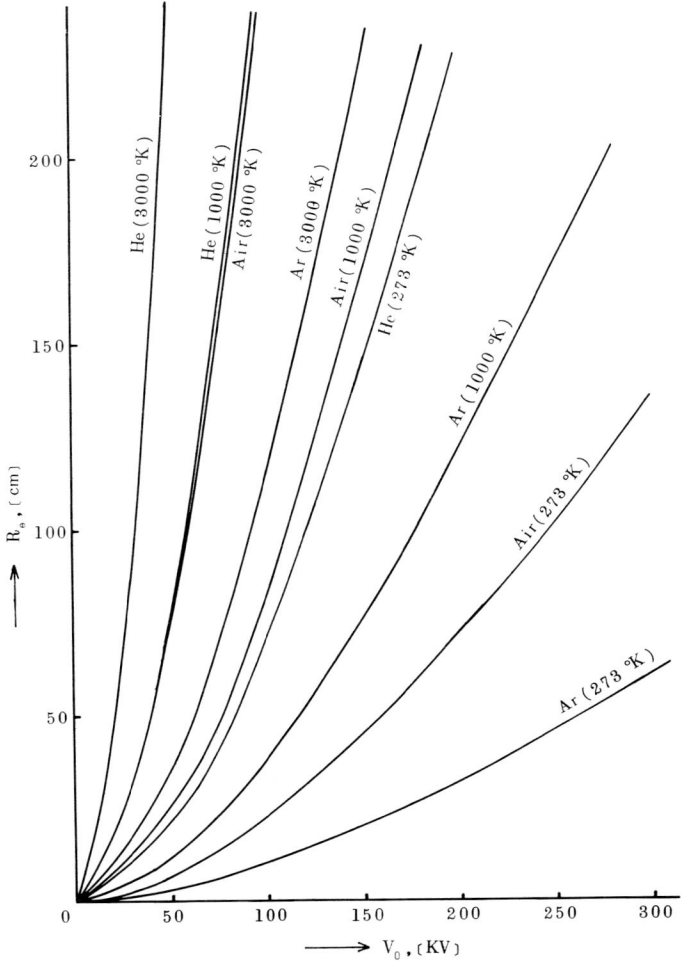

Fig. 22. Relation between the range, R_e and the accelerating voltage, V_0 in kinds and temperatures of environment gases

$$-\left(\frac{dV}{dx}\right)_{coll} = \left(\frac{2\pi r_e^2 m_{e0}c^2}{e}\right) \cdot \left(\frac{K_2^2}{K_1 K_3}\right)(nZ) \cdot Q(V, V_{ex}) \quad (1)$$

where the function $Q(V, V_{ex})$ is as follows:

$$Q(V, V_{ex}) = ln\,K_1 K_3 V - 2\,ln\,V_{ex} + \frac{2 - K_3^2}{K_2^2}\,ln\,2$$

$$+ \frac{K_1^2 + 8}{8K_2^2} + 13.82$$

where: $K_1 = 1.957 \times 10^{-3} V$
$K_2 = K_1 + 1$
$K_3 = K_2 + 1 = K_1 + 2$

In formula (1), r_e ($= 2.82 \times 10^{-13}$, [cm]) is the classical radius of an electron, Z is an atomic number of target gas or material.

$$R_e = \frac{1}{2\pi r_e^2} \cdot \frac{K_1^2}{K_2} \cdot \frac{1}{F(V_0 V_{ex})} \cdot \frac{1}{nZ}, \qquad [\text{cm}] \cdots \text{(a)}$$

$$= 2.08 \times 10^5 \frac{K_1^2}{K_1} \cdot \frac{1}{F(V_0, V_{ex})} \cdot \frac{T}{pZ}, \quad [\text{cm}] \cdots \text{(b)} \quad \Bigg\} \ (2)$$

where the function $F(V_0, V_{ex})$ is as follows:

$$F(V_0, V_{ex}) = 2\,ln\,\frac{V_0}{V_{ex}} + ln\,K_③ + \frac{2 - K_③^2}{K_②^2}\,ln\,2$$

$$+ \frac{K_①^2 + 8}{8K_②^2} + 13.82$$

where: $K_①, K_②, K_③ = K_1, K_2, K_3$ at $V = V_0$, respectively. T is temperature [°K] and p is environment gas pressure, [Torr]. R_e can be calculated easily from Eq. (2) using Tables 1 and 2,[27] and Fig. 22 shows

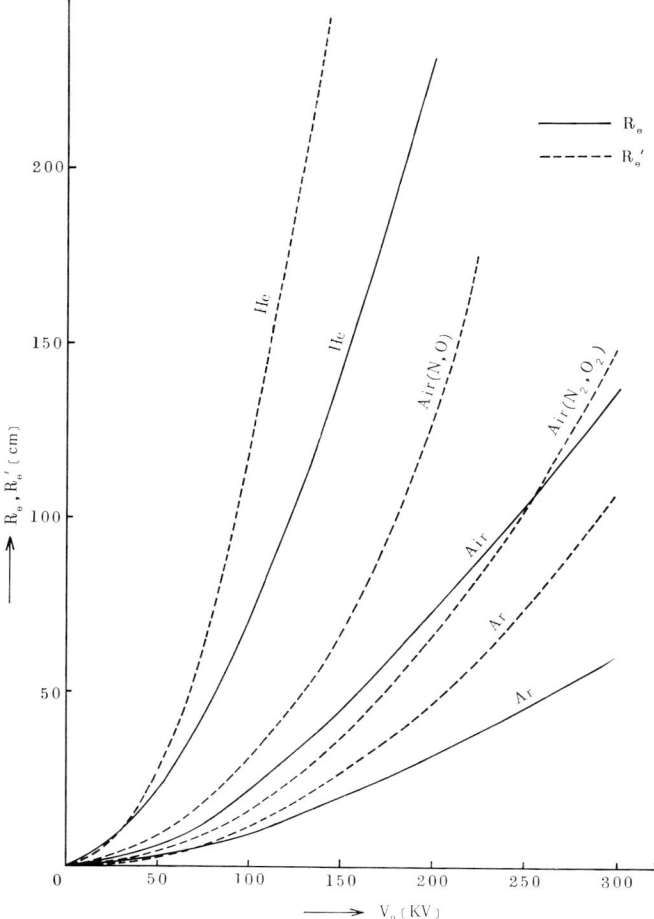

Fig. 23 (a). Relation between the range R_e, R_e' and the accelerating voltage V_0 in various environment gases at 273°K

Table 1.

$V_0[KV]$	10	50	100	150	200	300
$K_①$	0.020	0.098	0.196	0.294	0.392	0.587

Table 2.

	He	Air	N	O	Ar	Be	Al	Fe	Cu
$V_{ex}[V]$	26.9	80.5	87.7	98.0	192.0	60.4	150.0	241.0	276.0

it for various initial electron beam energies and gases.

Fig. 23 shows the difference between R_e and most simplified range, R_e', which had been usually used under $V_0 \approx 5 \sim 100$ [KV], $R_e' = 2.1 \times 10^{-6} \ V_0^2/\rho$, [cm]. Where ρ is density of target gas or material, [g/cm³]

Electron beam whose initial energy is V_0 [KV] approximately decreases the energy according to the following Eq. (3) while travelling distance R_e in the target gas or material.[28]

$$V = V_0\left(1 - \frac{h}{R_e}\right)^{1/2}, [KV] \qquad (3)$$

by using (3)

$$V^* = (1 - h^*)^{1/2}, \qquad (4)$$

where h is the distance from nozzle exit along x-axis, which corresponds to the work distance. ($V^* \equiv V/V_0$ and $h^* \equiv h/R_e$.)

Provided that I (h) is the beam current at h, the decreasing rate of beam current there, is as follows:

$$-\frac{dI}{I} = P \cdot dh \qquad (5)$$

where $P \cdot dh$ is the probability of electron loss from the beam axis generated while travelling the distance, dh, in environment gas.

In order to obtain this probability the following well-known Eq. (6) is used.

84

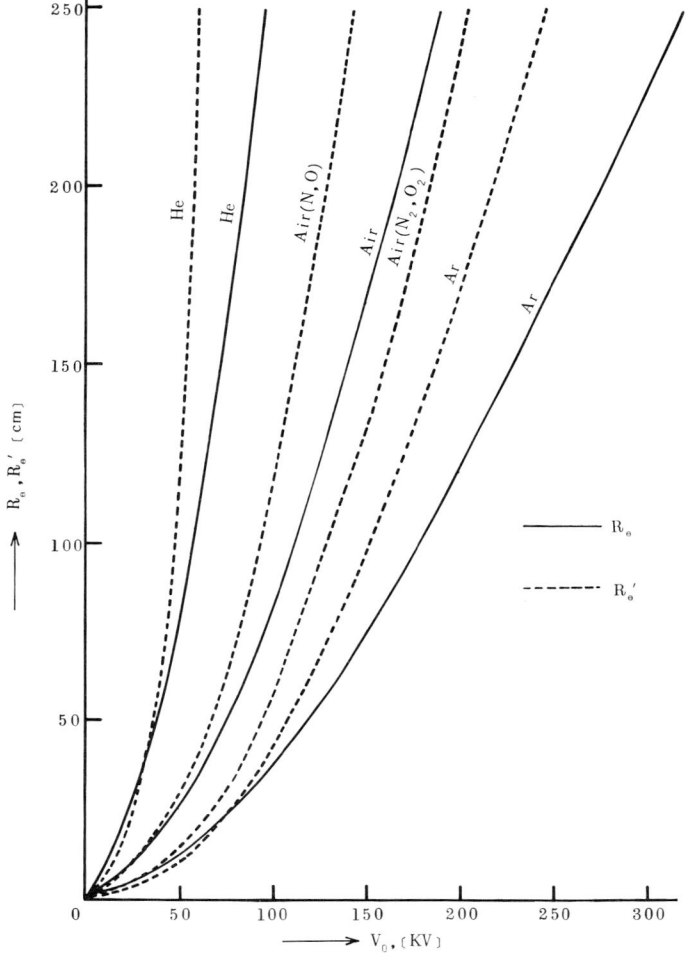

Fig. 23 (b). Relation between the range R_e, R_e' and the accelerating voltage V_0 in various environment gases at 1000°K

$$n(\theta) = n_e \cdot v_r \cdot N_G \cdot \sigma(\theta, K_2) \qquad (6)$$

where n_e is number density of electron [1/cm³], v_r is relative velocity of versus gas particle [cm/sec] ($\approx v_e$; electron velocity), N_G is total number of gas particles in the beam path ($N_G = n_G Sh$: provided that n_G is number density of target gas and S is the beam cross section) and next $\sigma(\theta, K_2)$ is multiple differential scattering cross section for the elastic scattering of electrons in light elements, and it is equivalent to Mott solution[25] simplified[26] with Born approximation.

$$\sigma(\theta, K_2) = \frac{1}{4}\left(\frac{r_e Z}{\beta^2 K_2}\right)^2 \cdot \frac{1}{\sin^4\frac{\theta}{2}}\left\{1 - \beta^2 \sin^2\frac{\theta}{2}\right.$$
$$\left. + \alpha\beta\pi Z \sin\frac{\theta}{2}\left(1 - \sin\frac{\theta}{2}\right)\right\} \qquad (7)$$

and $\alpha = 1/137$

Normalized beam current, I^* is obtained solving Eq. (5) using Eq. (6) as indicated in Eq. (8).

$$I^* = (1 - h^*)^{r \cdot \frac{K_{①}^2}{K_{②} V_0^2} \cdot \frac{Z}{F(V_0, V_{ex})}} \qquad (8)$$

where $I^* \equiv I/I_0$; I_0 is initial beam current ($I_0 = I$ at $h = 0$) and r is a constant value which is derived from calculation in Eq. (5) and (6) or results of experiment (curve 1) such as Fig. 12 (b).

It seems that r-value determined experimentally is more appropriate than the calculated one, because considerable inhomogeneity in temperature and density of environment gases along the beam axis takes place in practice.

Considering the curve 1 indicating a real beam

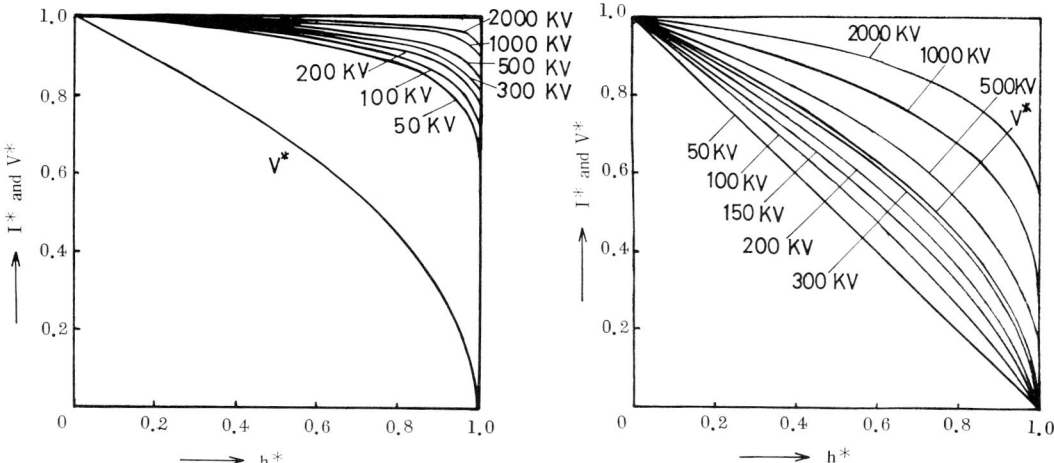

Fig. 24 (a). I^* and V^* versus h^* curves in He; Normalized beam current $I^*=I/I_0$, Normalized accelerating voltage $V^*=V/V_0$, Normalized work distance $h^*=h/R_e$

Fig. 24 (c). I^* versus curves in Ar

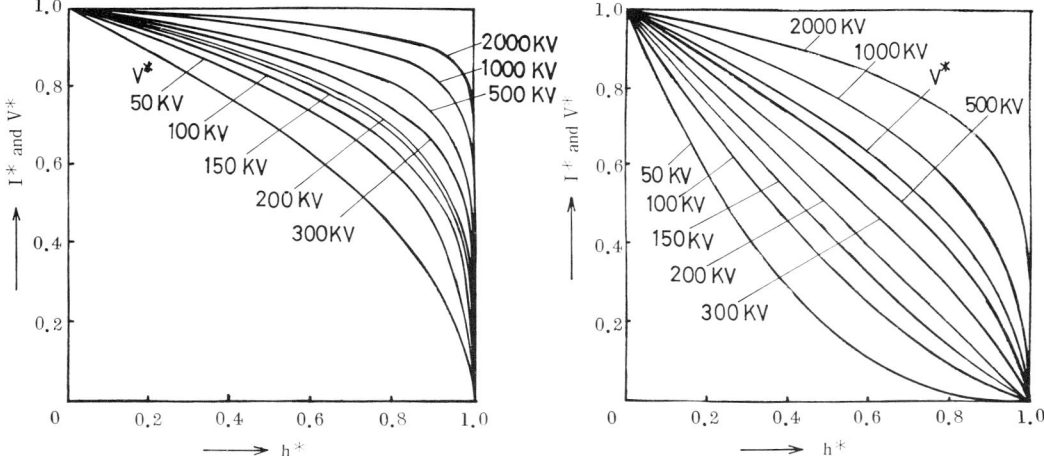

Fig. 24 (b). I^* and V^* versus h^* in Air

Fig. 24 (d). I^*, V^* versus h^* curves in Iron vapor

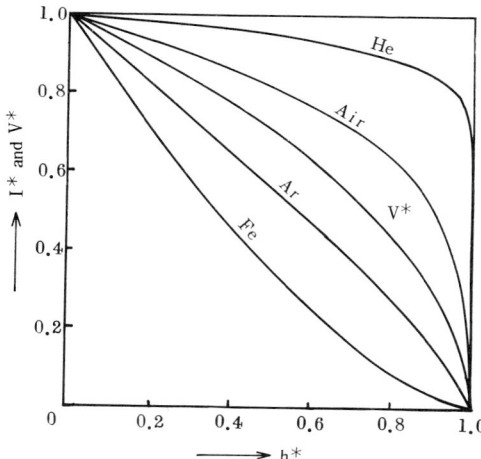

Fig. 25 (a). I^* and V^* versus h^* curves in various environment gases; $V_0=100$ [KV]

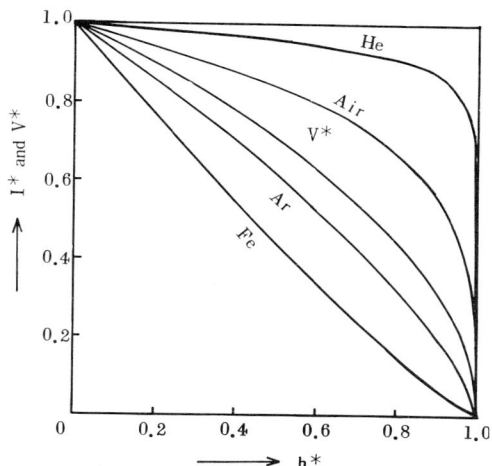

Fig. 25 (b). I^* and V^* versus h^* curves in various environment gases; $V_0=150$ [KV]

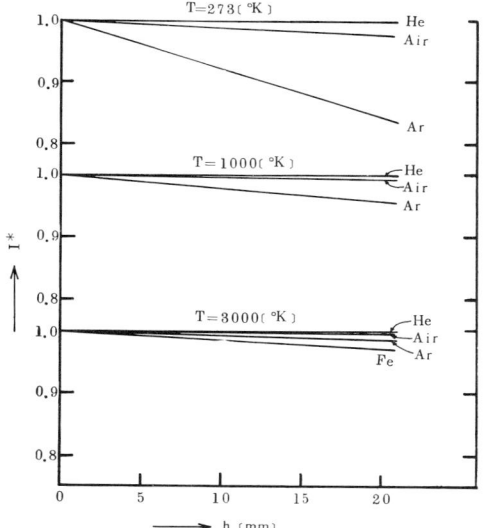

Fig. 26 (a). Effects of temperature and environment gas on the I^* at various work distances h; $V_0=100$ [KV]

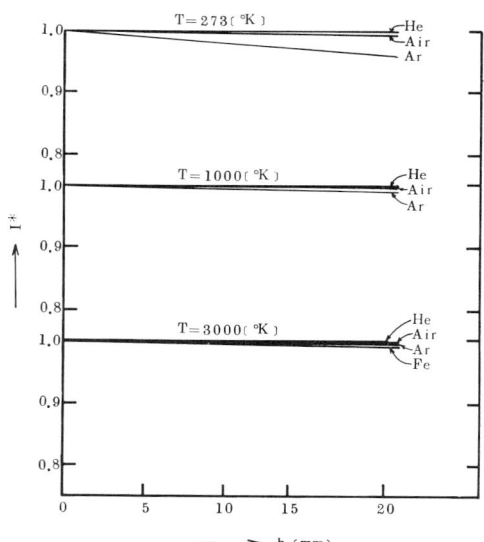

Fig. 26 (b). Effects of temperature and environment gas on the I^* at various work distances h; $V_0=200$ [KV]

current as shown in Fig. 12 (b), the r-value was 1.768×10^5. Fig. 24 shows the relation of I^* and V^* versus h^* under several kinds of gases (He, Air, Ar and Fe).

Fig. 25 shows the same relation as above under some initial beam energy ($V_0 = 100$, 150 [KV]).

Fig. 26 shows I^* versus h which is work distance, at several temperatures ($T = 273°K$, $1000°K$, $3000°K$, Fe alone; $3000°K$) and in different gases (He, Air, Ar, Fe).

The cosine of multiple scattering angle, $\langle \cos \theta \rangle_{AV}$ averaged over all electrons penetrating into material or gas, is derived as follows:

$$\langle \cos \theta \rangle_{AV} = \exp\left(-\int_{K_2}^{K_{②}} \frac{m(K_2)}{\frac{dK_2}{dh}} \cdot dK_2\right) \quad (9)$$

where the function $m(K_2)$ using Eq. (7) is as follows:

$$m(K_2) = 2\pi n_G \int_0^\pi \sigma(\theta, K_2)(1 - \cos \theta) \sin \theta \, d\theta \quad (10)$$

$\langle \cos \theta \rangle_{AV}$ is obtained as Eq. (11) using an approximation of the integrand in Eq. (9), in which the constants, a and b, are tabulated as 0.305 and 0.034 respectively in Al.[24)27]

$$\langle \cos \theta \rangle_{AV} = \frac{G(K_{②})}{G(K_2)} \quad (11)$$

where: $G(K_2) = \left(\frac{K_3}{K_1}\right)^{az} \cdot e^{bz/\beta}$

Fig. 27 shows the relation between average multiple scattering angle, $\langle \theta \rangle_{AV}$ versus h^* in Al under $V_0 = 100$ [KV] and 1000 [KV]. Accordingly, the beam power densities, w^*, at each point on the x-axis may be obtained as shown in Fig. 28, provided that the beam distribution is homogeneous in the beam mean section (πr^2) at distance, h, using the radius, r, as follows:

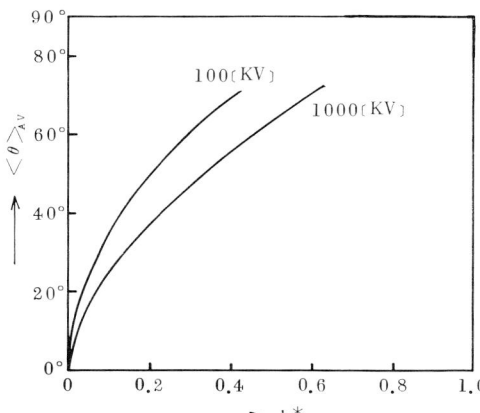

Fig. 27. Relation between average multiple scattering angle, $\langle \theta \rangle_{AV}$ versus h^* in Al; $V_0 = 100$ [KV] and 1000 [KV]

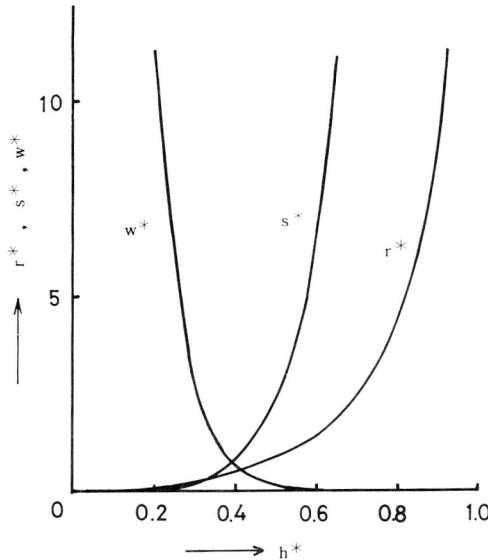

Fig. 28. Relation of r^*, S^*, and w^* versus h^* in Al; $V_0 = 100$ [KV] where: $r^* = r/R$, $S^* = \pi r^*$, $w^* = W^*/S^*$, $h^* = h/R_e$

$$r = \int_0^h \tan \langle \theta \rangle_{AV} dh \quad (12)$$

where it is assumed that $w^* = W^*/S^*$, $W^* = I^*V^*$, $S^* = r^{*2}$, $r^* = r/R_e$, $h^* = h/R_e$.

It seems that this was in good accordance with the experimental result as shown in Fig. 11 (excepting the vicinity of nozzle exit), which was converted to the case of $V_0 = 100$ [KV] and Al. These are considered to be useful for the estimation of the beam profiles and its penetration into the material when the beam is injected along the x-axis.

References

1) Y. Arata: Chōkōon Kenkyu Shiryo (Ultra high temperature Review; in Japanese) vol. 1, No. 6 (1957); vol. 2, No. 1 (1958).
2) N.C. Christofilos: Proc. 2nd U.N. Conf. on Peaceful Uses of Atomic Energy. vol. 32 (1958).
3) H. Dreicer: Ibid, vol. 32 (1958).
4) N.C. Christoflos: UCRL–5951–T (1960)
5) L.H. Lidsky, etal: J. Appl. Phys. (1962).
6) M. Okada and Y. Arata: Technol. Repts. Osaka University, 4, 567 (1963).
7) J.R. Morley: Proc. 5th Annual Electron Beam Symp. (1963); Trans. Vacuum Met. Conf. (1963).
8) K.A. Saudners and R.L. Sweell: UCRL–7363 (1963).
9) S. Graybill, etal: Proc. Electron and Laser Beam Symp. (1965)
10) Y. Arata, etal: Technol. Repts. Osaka University, 16, 724 (1966).
11) E. Abramyan and V. Gaponov: Atomnaya Energiya, vol. 20, No. 5 (May, 1966).
12) B.W. Schumacher: Optik 10, (1953) 116, 2nd Int. Vac. Cong. (1961) 1192.
13) Y. Arata: Private report. (1962) (Patent; 531587).
14) R.E. Kutschera: Trans. Vacuum Met. Conf. (1962).
15) L.H. Leonard, etal: Final Rept on Contract AF 33 (657)–7237 (1963) Interim Rept. AF 7–926 (1962).

16) R.R. Irving: Iron Age. March 29 (1962).
17) L.H. Leonard: Proc. 5th Annual Electron Beam Symp. (1963).
18) J.W. Meier: Welding journal, 42, 12 (1963).
19) J. Lempert, etal: Proc. Electron and Laser Beam Symp. p. 393 (1965)
20) Y. Arata: Chōkōon Kenkyu (Ultra-high temp. Review; in Japanese) vol. 4, No. 1 (1967); Electronics Magazine, vol. 12, No. 6 (1967).
21) Y. Arata and M. Tomie: Technol. Repts. Osaka University vol. 17, No. 777 (1967).
22) B.W. Schumacher: Electron and Ion Beam Science and Technology (Third International Conf.) p. 236 (1968).
23) D.A. Dunn and A.S. Halsted: Proc. of the Electron and Laser Beam Symp. p. 243 (1965).
24) F. Rohrlich and B.C. Carlson: Phys. Rev. 93, 38 (1954).
25) N.F. Mott: Proc. Roy. Soc. (London) A 124, 426 (1929); A 135, 429 (1932).
26) W.A. Mckinley and H. Feshbach: Phys. Rev. 74, 1759 (1948); R.H. Dalitz: Proc. Roy. Soc. (London) A 206, 509 (1951).
27) H.W. Lewis: Phys. Rev. 78, 526 (1950).
28) D.C. Schubert and B.W. Schumacher; Electron and Ion Beam Science and Technology, (3rd Inter. Conf.) p. 269 (1968).
29) G.J. Hine and G.L. Brownell; Radiation Dosimerty (1956).

Appendix

As shown in Fig. 7, electron beam emissibility of thoriated tungsten cathode varies in there stages. In region A, beam current decreases rapidly within 30~60 minutes from the beginning of E.B. generation.

Reversible property and irreversible property in beam emissibility of cathode will be considered as regards the reasons of this phenomenon.

The former is, as shown in Fig. 29, due to the cooling effect of rod cathode end-surface by the emission of electrons, and time constant of this process is fairly short in the order of 10~30 seconds.

On the other hand, the latter irreversible property depends upon the atomic diffusion in the cathode

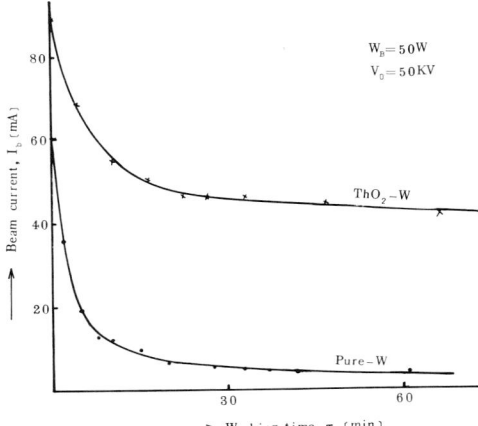

Fig. 30.   Relation between I_b and τ_W in Pure-W and ThO$_2$-W

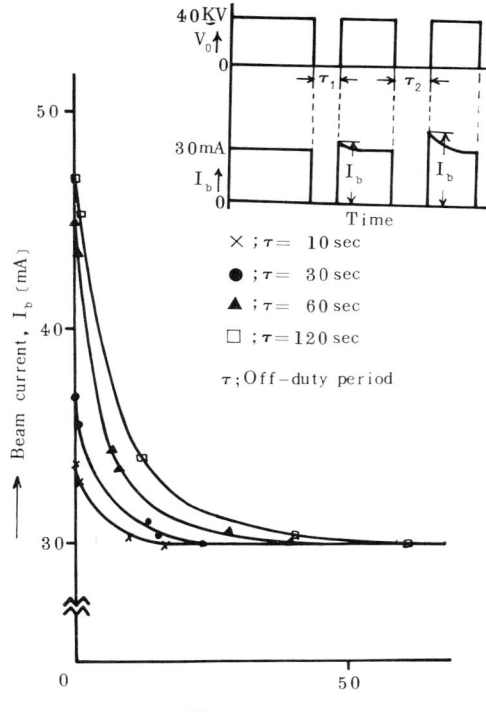

Fig. 29.   Relation between I_b and τ_W versus τ

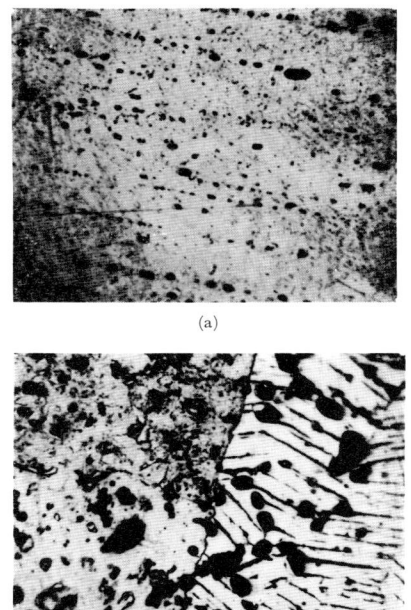

Fig. 31.   Recrystallization and growth of cathode material (a; initial state, b; after ten hours in use)

material, and its time constant is, therefore, such longer than in the former case.

As for the changes in region A, both characteristics of sintered pure tungsten cathode of thoriated tungsten cathode were very similar as shown in Fig. 30.

These results indicate that the activity of thorium oxide does not account for the existence of region A, nevertheless it affects the absolute value of e-lectron current.

The reasons for the decrease of current in region A seem to be the decrease of effective surface area contributing to the thermal emission of electron (actual area is believed to be wider than the apparent area, in general) and the changes in properties of sintered materials.

In Region B, the electron currnt is very stable and profitable for the practical use.

Recrystallization and growth of cathode material become very progressive in the stage of C in Fig. 7 as shown in Fig. 31, and the amount of thorium oxide decreases gradually through vaporization.

As the result of these phenomena, electron emissibility of cathode becomes inferior.

For the practical NV-EB apparatus, it is necessary to research and develop a profitable cathode material in which ThO_2 or other elements do not cohere and crack the cathode material.

Studies must also be conducted on the cathode structure capable of emitting enough beam current even with low bombardment power.

100-KW Class Electron Beam Welding Technology
—Welding Apparatus and Some Aspects as A Heat Source—

Abstract

A 100 KW-class electron beam welding apparatus was designed and assembled. Using this apparatus, character-istics of a powerful electron beam up to 120 KW as a heat source and of its bead penetration were both investigated. The possibility of both one-pass welding and cutting thick materials up to 10~20 cm was proposed.

1. Introduction

The heat source of the electron beam gives the deepest weld bead penetration compared with any others as it is well known. Characteristics of the electron beam heat source are that both the power and density become extremely high, moreover, these are controlled easily.

Such characteristics efficiently affect the thick materials as its power becomes much higher. The powerful electron beam of 100 KW-class or above, has the advantage over any other heat source, for the welding ultra high thick materials, 10~20 cm and/or over. From such a view point, the 100 KW-class EB-welder was produced and its capability was tested; moreover, some properties of the poweful electron beam as a heat source were studied. It was then applied to welding and cutting.

2. Characteristics of 100 KW-EBwelder

Fig. 1 and **Photo. 1** show the schematic dia-gram of the installation and its appearance respectively. Provided that V_b(KV), I_b(mA), and W_b(KW) are taken as beam accelerating voltage, beam current and beam power in the maximum value of a continuous respectively, its capability is as follows.

$$V_b=100, \quad I_b=1000, \quad W_b=100$$

And it is also possible to utilize it as follows:

$$V_b=200, \quad I_b=500, \quad W_b=100$$

And this welder was employed up to 120 KW to obtain experimental data although it was for a short interval. The volume of the vacuum work chamber is 1.3 m^3 and the capacity of the exhaust system is 2,000l / min rotary pump, 10,000 l/min mechanical booster pump, 360,000 l/min diffusion and ejection booster pump respectively. The maximum welding speed is 5 m/min and the cathode material is 3 mmϕ circular type Tungsten.

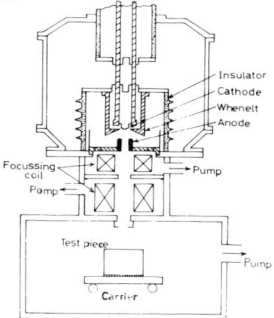

Fig. 1. Schematic diagram of EB-welder.

Photo 1. General view of 100 KW-EB welder which was developed by authors.

3. Some Results and Discussion

The welding bead test was performed using stainless steel SUS304, correspond to AISI304, whoes chemical composition is shown in **Table 1.** The results obtained are shown in **Figs. 2~4** and **Photos 2~5. Photo 2** indicates the typical transverse section of the bead of 50 KW and 100 KW electron beam weldment. **Figure 2** and **Photo 3** show one example of distribution of the vickers hardness and the micro-structure of the bead cross-section respectively.

Figure 3 shows the effect of welding speed on penetration depth at 100 KW, and using this figure, the relation of h_p against $1/\sqrt{v_b}$ (h_p: penetration depth, v_b: welding speed) is obtained as shown in **Fig. 4.** This is in agreement with formula, $h_p \propto I_b V_b^{1.3}/\sqrt{v_b}$, which is given by authors in another report of this JWRI[1].

Table 1. Chemical composition (W_t %) of stainless steel used (SUS304 correspond to SUS27 or AISI304).

Composition / Material	C	Mn	Si	Ni	Cr	Cu	Mo	Al	S	P
SUS 304	0.050	1.74	0.74	10.9	19.5	0.12	0.16	0.015	0.010	0.030

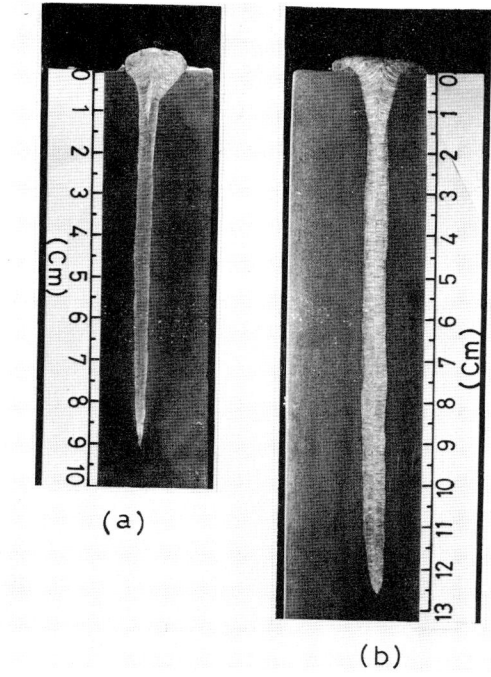

Photo 2. Cross-sectional view of the bead, welded under conditions of v_b=60 cm/min for SUS304,
  (a) : V_b=100 KV, I_b=500 mA, W_b=50 KW
  (b) : V_b=100 KV, I_b=1000 mA, W_b=100 KW

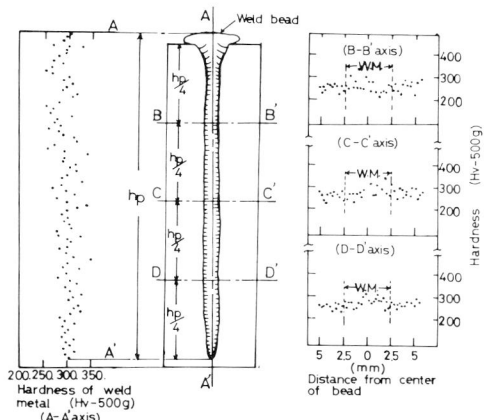

Fig. 2. Hardness distribution in cross-section of welds, (V_b=100 KV, I_b=1000 mA, W_b=100 KW, v_b=60 cm/min, SUS304).

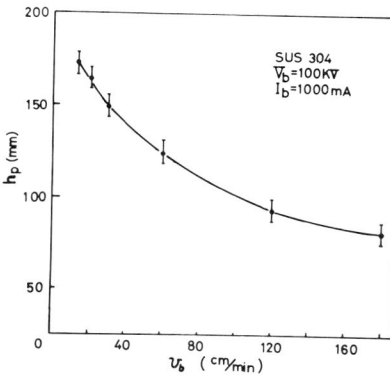

Fig. 3. Relations between v_b and h_p.

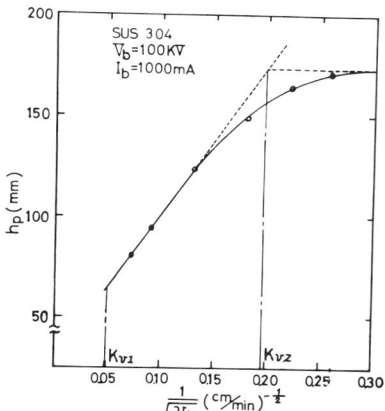

Fig. 4. Relations between h_p and $1/\sqrt{v_b}$.

92

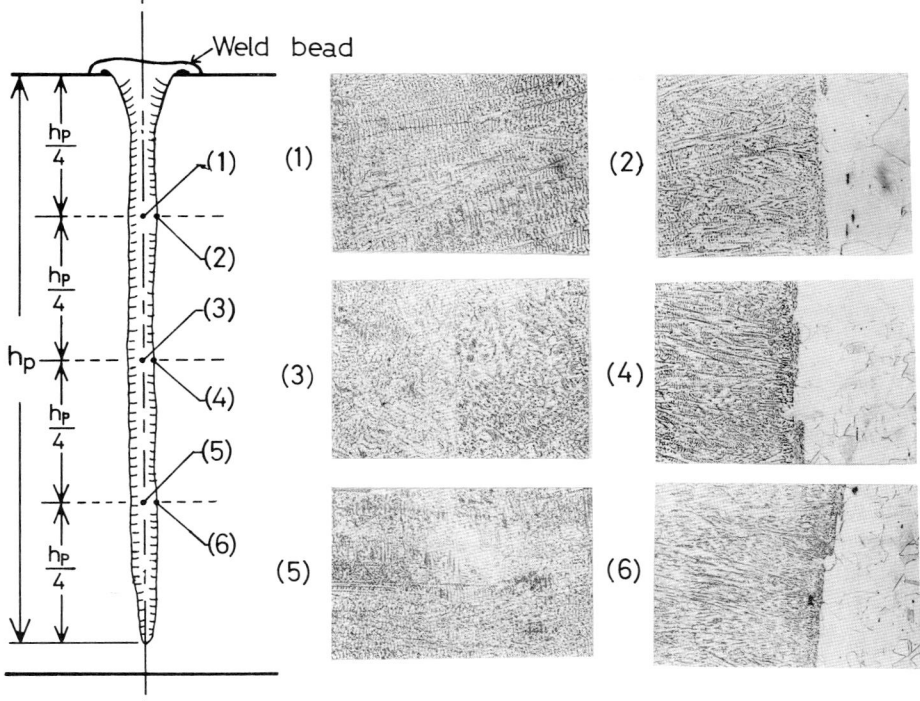

Photo 3. Microstructure at various positions in weld bead, (V_b=100 KV, I_b=1000 mA, W_b=100 KW, v_b=60 cm/min, SUS304).

Phot 4 indicates one example of 100 KW electron beam cutting, which demonstrates the usefulness of the powerful beam for cutting ultra thick materials. **Photo 5** shows the profile of the very long molten pool over 10 cm waved strongly during 100 KW-class beam welding, and its aspect was that of molten metal poured into the long narrow beam cutting zone which was similar to a deep narrow gap. Such a powerful beam looks like it causes severe gaps compared with a relatively low power beam such as the 50 KW-class beam or below, especially the evaporation due to violent boiling during welding.

Such "welding vapor" emitted violently from the molten pool near the beam, heavily contaminates both space and wall surface of the beam channel and the beam accelerating chamber, and it leads to fluent arcing in the electron gun.

Consideration for its prevention is as follows.

1) To make each slit or nozzle of the beam channel as small or long as possible.

2) To make the beam axis shift with each other inside accelerating chamber, the beam channel and the work chamber.

3) To make a relative high pressure gas layer in a

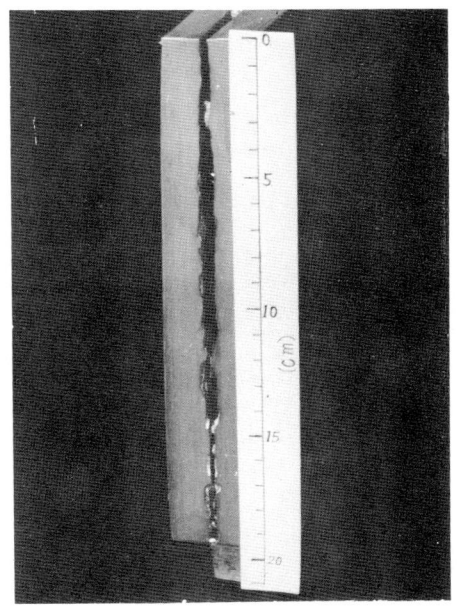

Photo 4. One example of beam cutting (V_b=100 KV, I_b=1000 mA, W_b=100 KW, v_b=20 cm/min, SUS304).

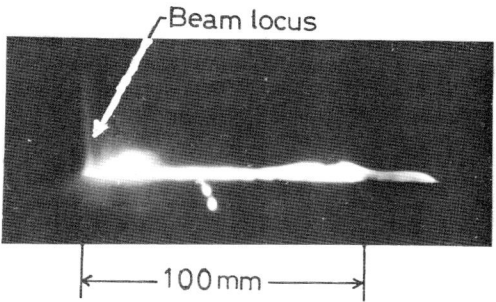

Photo 5. Typical appearance of molten pool in 100 KW beam welding (v_b=30 cm/min, SUS304).

region between the beam channel and the work chamber, and have its pressure much higher than that of the either.

4) It seems that the beam of higher voltage and lower current is better than the diametrically opposed beam to decrease welding vapor, in the case of the high power welder with the same power.

4. Conclusion

The 100 KW-class EB-welder (120 KW in a short time) was designed and installed. The possibility of one-pass welding and cutting for the ultra thick material was proved. A scheme to decrease the damage due to the violent welding vapor was given.

Acknowledgment

The authors thank Dr. Matsuda and electron beam group of JWRI for their discussions. Thanks are also expressed to the Education Ministry of Japanese Goverment assisted us.

Reference

1) Y. Arata, M. Tomie and Y. Kato: "Some Properties of 30 KW-class Electron Beam for Welding", Trans. JWRI Vol. 1, No. 2 (1973).

Development and Application of a Strong Focusing Electron Beam Gun With Multi-Stage Electromagnetic Accelerating Units

Abstract

New type steady-state electron beam guns were developed having strong focusing characteristics by "Multi-Stage Electromagnetic Accelerating Unit". One has 5-stage accelerating units up to 300kV with an output of 100kW and the other has 13-stage accelerating units up to 600kV with 300kW output. By these electron beams, "Low Pressure Electron Beam Welding" was found to be successful as well as the welding in a high and/or low vacuum environment. Furthermore it was certified that they also made fruitful results in the "Ultra-Thick Plate Welding Process".

1. Introduction

Development of a new technique for welding of ultra-thick plates with a thickness exceeding 20 or 30 centimeters in one pass or at super-high speed is considered an important step in the advance of welding technology. The authors thought that an urgent task would be to develop a strong-focusing electron beam heat source with ultra-high power in the range of several hundred kilowatts. We paved the way for full-scale practical use of high power electron beam welding in 1972 by developing a high power electron beam source capable of a maximum output of 120 kW, the world's first of its kind, and at the same time by applying it to the welding of ultra-thick plates[1]. Within a few years, it turned out to be the start of the application of a 100 kW-class electron beam heat source to practical welding on a worldwide scale[2]. Next we devised a new "electromagnetic accelerating unit" along with a special high power electron beam gun for welding. By connecting these units into multi-stages we realized an ultra-high voltage, high power electron beam heat sources, which had thus far been considered almost impossible to develop[3]. Some examples are the 300kV/100kW (in 1975) and 600kV/300kW (in 1980) gigantic electron beam heat sources. For use with these sources, we proposed a new welding process[4], and established the "ultra-thick plate electron beam welding method" whereby full-penetration welding was performed on metals with a thickness of more than 30cm.

This paper reports the details of the electromagnetic accelerating unit and the characteristics of the ultra-high voltage strong focusing electron beam heat source.

2. Outline of the Ultra-high Voltage Electron Beam Source

Usually an energy density of over 500 to 600 kW/cm² is required to obtain of weld beads deep penetration by an electron beam. The energy density of the beam W_b is given theoretically by an equation[5] with the effect of space charge taken into account,

$$W_b = K_s (V_b^{5.1}/I_b^{1.7}) \ldots \ldots \ldots \ldots \ldots \ldots \ldots \ldots \ldots (1)$$

where V_b = beam voltage, I_b = beam current, K_s = constant. This shows that the energy density is dependent stronger on beam voltage than on beam current. In fact, however, the "beam-plasma" inevitably occurs during the welding by a high energy density beam and the effect of the space

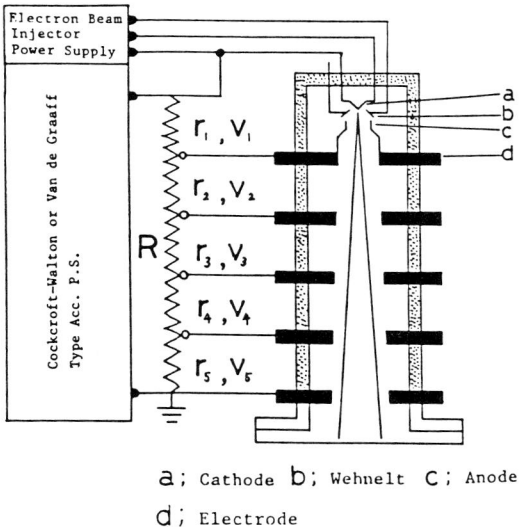

a; Cathode b; Wehnelt c; Anode

d; Electrode

Fig. 1 Schematic diagram of conventional high voltage E. B. accelerator.

charge on the energy density is markedly reduced[6], lowering the strong dependence on the beam voltage in equation (1). While dependence of the penetration depth h_p on V_b and I_b becomes much smaller, as was obtained by us[7] separately with the following equation:

$$h_p = K_m (I_b \cdot V_b^{1.3}/V_b^{0.5}), \dots\dots\dots\dots\dots\dots\dots\dots\dots\dots\dots\dots (2)$$

where V_b = welding speed and K_m = constant.

In obtaining eq. (2), the experimental values of V_b = 50 – 120 kV, I_b = 20 – 1,000 mA, and V_b = 40 – 380 cm/min were used. This equation was made clear to hold in the range of several kW to the order of 100 kW with the penetration depth sealing as $V_b^{1.3}$.

This result suggests that the ultra-high voltage type high power electron beam is better for welding. In virtually all practical electron beam welding equipment available so far both at home and abroad, the beam voltage was limited to the range of 30 to 175 kV. The greatest reason for the maximum of 175 kV was that in all such equipments, the one-stage accelerating method was used to energize the beam from the electron gun. In this method, high power electron guns inevitably experience the frequent breakdown phenomenon, particularly arcing, at an ultra-high voltage. Thus the beam voltage as well as the power output cannot be raised beyond certain limit. In order to overcome this limitation and obtain a practical and very high power beam, we developed a new multi-stage, ultra-high voltage electron gun. Conventionally multi-stage accelerators are of either the Van de Graaff or Cockcroft-Walton type, as shown in Fig. 1. As is well known, they supply voltage across each pair of accelerating electrodes through the resistive voltage divider R. However this method of voltage supply is incapable of accelerating high power beams. In that case each electrode experiences a large fraction of stray electrons from the high-power beam and from inevitably occurring beam-plasma flowing into it, causing the electrodes to be charged to irregular levels. This causes abnormal voltages and instant electrical breakdown across some pairs of electrodes, resulting in the frequent discharge in the accelerator. For this reason, these types of an accelerator is limited in the application to a high power electron guns for welding etc. It follows that new ideas and functions need to be added to the conventional multi-stage acceleration system. Some examples are:

1) Absorption of irregular electrode charge-up current to prevent abnormal voltage occurring across electrodes,
2) High stability of the accelerating voltages in each and all stages,
3) Optimum construction of cathodes and accelerating electrodes for high power, high-density electron beam generation,
4) Establishment of strong focusing capability for the extraction and transportation of a high-density electron beam.

3. Strong Focusing Electron Beam Heat Source with Five-stage Accelerating Unit (300 kV/100 kW-class)

We incorporated the above requirements for electron guns as a heat source into a prototype strong focusing electron gun with five-stage accelerating unit as shown in Fig. 2.

It includes five electromagnetic accelerating units connected together. The uppermost unit contains the electron generating assembly. The electrodes in each unit are fitted with electro-magnetic lenses for varying the diameter and the density of the beam to the desired values in each area of the units. These lenses also control electrons flowing into the accelerating electrodes.

Fig. 2 5-stage electromagnetic accelerating unit type E. B. gun.

97

Fig. 3 General view of 300kV, 100kW E. B. heat source.

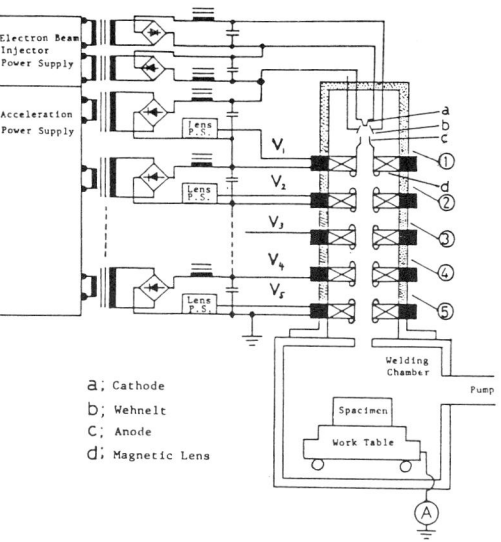

a; Cathode
b; Wehnelt
c; Anode
d; Magnetic Lens

Fig. 4 Schematic diagram of 100kW E. B. heat source which contains electric power supply circuits.
①~⑤ : electromagnetic accelerating unit.

The maximum beam voltage available by this five-stage accelerating electron gun is 300kV, but the potential across each pair of electrodes is lower than that for the typical one-stage accelerating gun, which acts favorably to suppress the breakdown. Figure 3 shows a general view of the prototype 100 kW-class electron beam heat source. Figure 4 shows its operating principle together with the power supply circuit. Let the limiting beam voltage, beam current and output of this equipment be V_{bm} (kV), I_{bm} (mA), and W_{bm} (kW), respectively at the continuous operation. Then, the performance of the source is represented by: $V_{bm} = 300$, $I_{bm} = 350$, and $W_{bm} = 100$ (120 at a short time).

The welding chamber has a volume of approximately 2 m³. The beam channel between the chamber and the electron gun performs the function of beam guidance and multi-stage acceleration. The chamber is designed for welding at all pressures from a high vacuum of 5×10^{-5} Torr to the atmospheric pressure of 760 Torr. It is equipped with a work-stage capables of scanning in the X, Y, and Z directions, and with a beam focusing deflector. These features allow a downward beam to be deflected to any desired angle up to 90° for all position welding. They also suppress the electrical breakdown due to a lot of metal vapor during welding. As stated earlier, for high power beams the electrodes of each unit are heavily charged and the tendency for abnormal voltage across each pair of electrodes becomes high. To prevent this tendency, each stage is provided with a constant-voltage power supply so that it can be controlled independently, as shown schematically in Fig. 4. This arrangement provides five-stage acceleration of stable, high power electron beams by absorbing irregular charge-up current at a fixed voltage across each pair of the electrodes.

4. Strong Focusing Electron Beam Heat Source With 12-Stage Accelerating Unit (600kW/300kW-class)

After the research on the 100kW-class heat source, we set out to develop another high power electron beam heat source in which $V_{bm} = 600kV$ and $W_{bm} = 300kW$. Figures 5 and 6 show, respectively, a general view of the prototype and the electron gun with 13-stage electromagnetic accelerating units contained in its insulating pressure vessel. These units are an improved version of the 5-stage 100kW-class heat source mentiomed in the former section and thirteen units are connected together as shown in the figure. Because the voltage at the electron gun of this new prototype reaches an ultra-high valve of 600 kV (1st stage accelerating voltage = 100 kV and that for each stage 2 through 13 = approximately 41.7 kV), it is installed in a pressure vessel kept at 2 kg/cm^2 in dry air. The welding chamber has a capacity of 23 m^3. The base pressure is 1 x 10^{-5} Torr. Inside three is a work-stage capable of scanning in the X, Y, and Z directions and a

Fig. 5 General view of 600kV, 300kW E. B. heat source.

Fig. 6 13-stage electromagnetic accelerating unit type E. B. gun.

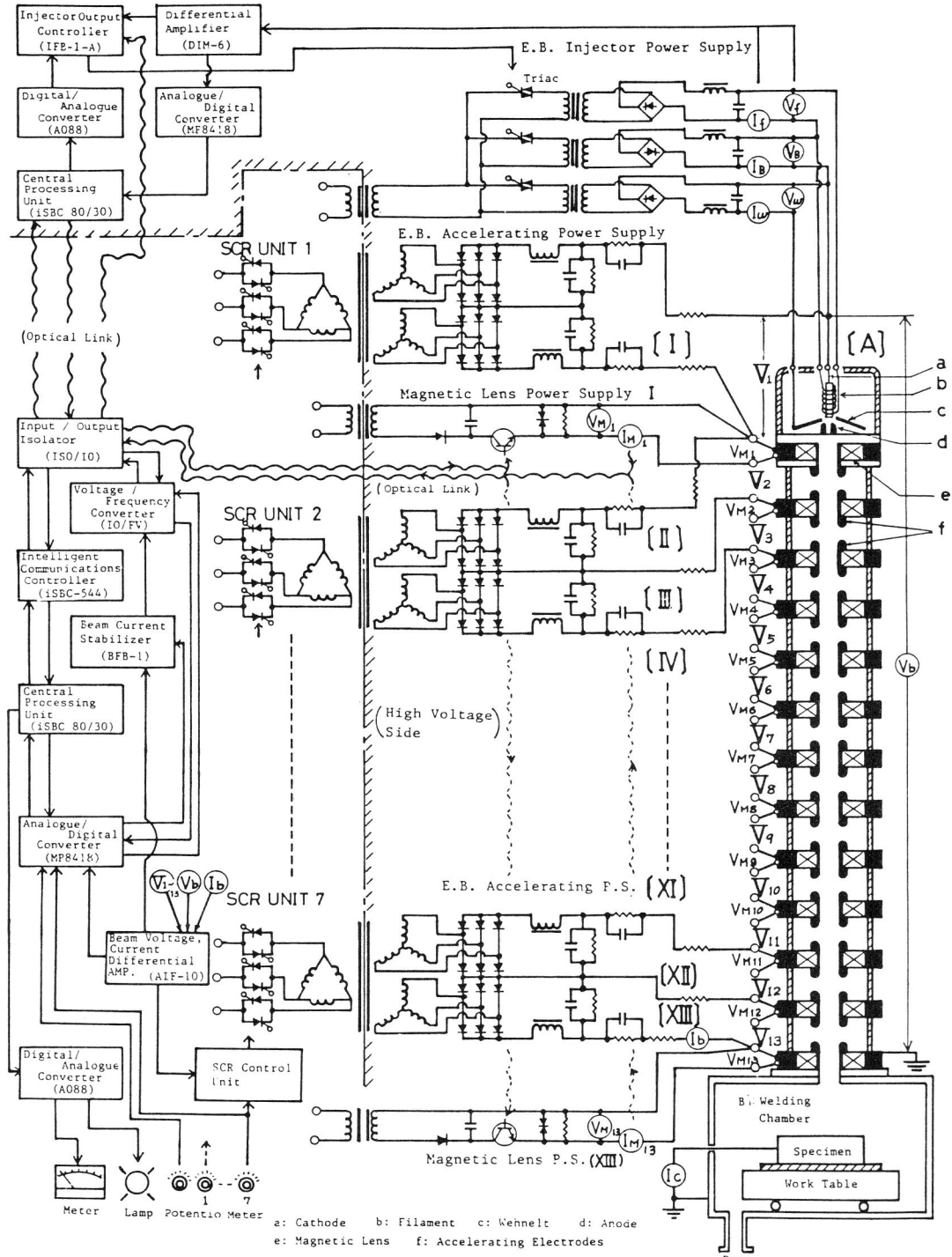

Fig. 7 Schematic diagram of 300kW E. B. heat source which contains electric power supply circuits.
[A] : E. B. gun with 13-stage electromagnetic accelerating unit.

100

special beam deflector for the prevention of the brenkdown phenomenon. Figure 7 shows block diagram of the heat source power supply circuit and the operating power supply circuit, with a drawing showing the accelerating principle. The operating power supply circuit contains that of the beam accelerater with independent control at each accelerating unit, the electromagnetic lens, the electron beam injector and the control system for these power supplies. The cathode is usually the indirect-heating type. For easy and precise control of the high power beam behavior, a micro-computer-based, optical control system is adopted. This system allows various informations for control to be transmitted to the ultra-high voltage side of the control circuit. Also, it enables monitoring and precisely controlling the operating condition of each accelerator and lens power supplies. It serves to suppress variations in the voltage and the current of the beam, which were measured to be within 0.5%.

The difficulty we met in the design of this heat source and in the preliminary experiment was the damage of the electron gun and the power supply due to the discharge phenomenon at an ultra-high voltage. For instance, in the gun at an accelerating voltage of 500kV in total, the initial stage accalerating unit sometimes experienced the discharge from the outer surface of the insulating tube. This resulted in the lowering of insulation along the discharge path due to its carbonization and failured the power circuit for the unit in question. In this case the electrodes at the initial stage accelerating unit was first exposed by the voltage several times higher than the original 80 kV during internal discharge and in turn it caused a discharge from the outer surface of the insulating tube.

According to the manner of the discharge this phenomenon was at one time the one in some parts of the 13-stage accelerating electrodes and at another time the whole discharge through all electrodes. In the partial discharge, the power supply of independent control type of the electro-magnetic accelerating unit absorbed all the discharge energy and restored the voltage to the original one in several seconds. Because of this, there was no adverse effects such as the occurrence of abnormal voltages on the electrodes of other units. On the other hand, in case of the whole discharge the discharge was initiated at the final accelerating unit, for example, and the cascade of the discharge went in sequence toward the initial stage. The initial stage electrode sometimes saw the abnormal voltage several times as high as the initially applied one, although of very short duration.

For the latter type of the discharge, causing the heavy damage to the system, we found the following solutions: provision of a discharge gap between each unit, use of an appropriate insulating gas mixture, good selection of the circuit components and raising of the withstand voltage of the power supplies by the adoption of protective circuits. Of cource there remain many problems to be soloved further, but we should remember that this is the world's first case of a high power strong focusing beam generation by the multi-stage acceleration method mentioned above. We are now investigating the properties of the beam under the steady-state and at an ultra-high voltage condition of 600 kV.

5. Characteristics of Ultra-High Voltage Electron Beam

The functional characteristics of ultra-dense beams extracted from the electron gun with electromagnetic accelerating units (such as the position of the focal point, the density distribution

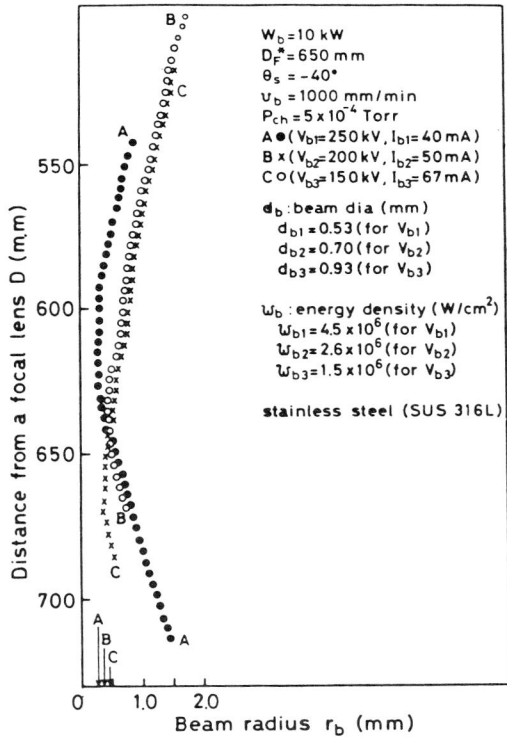

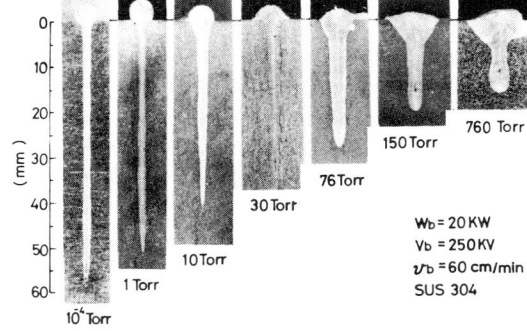

Fig. 8 Measurement results of beam shape by AB test method.

Fig. 9 Cross sectional view of weld beads for various gas pressures in welding chamber.

and the shape of the beams) is very important to evaluate the characteristics during welding. Below we describe some characteristics of the five-stage accelerating beam. Figure 8 shows an example of the results obtained from the AB test method[8]. The experiments were conducted at a pressure of 5×10^{-4} Torr. The beam power was fixed at 10kW. Three fixed values of the beam voltage were used with 250kV (I_b = 40mA), 200kV (50mA) and 150kV (67mA). The distance of the focal point of the beam at 250kV was found to be approximately 610 mm from the center of the final beam focusing lens, and the minimum diameter of beam d_b was 0.53 mm. The average beam energy density obtained was 4.4×10^6 W/cm^2, which was a high value of the density that couldn't be obtained by the conventional one-stage acceleration type electron gun. Using this ultra-high energy density beam, welding experiments were carried out at various gas pressures (10^{-4} Torr to atmospheric). Photographs of typical beads obtained are shown in Fig. 9. Detailed experiments showed that at the pressure of 10^{-4} Torr, the relationship between beam voltage, beam current and penetration depth agreed well with our empirical equation (2). Figure 10 shows photographs of typical beads on SUS304 stainless steel at a beam power of 20kW. Figure 11 shows the relation between various bead parameters (penetration depth h_p, bead width d_B and effective penetration parameter $\widetilde{P}_p$) and the gas pressure P_{ch}. At a high vacuum below 10^{-2} Torr, effective penetration parameters $\widetilde{P}_p$ (= h_p/d_b) is over 30 and an extremely narrow and sound bead is obtained. In the low pressure range of 1–10 Torr, h_p decreases by only 20–30% compared with

102

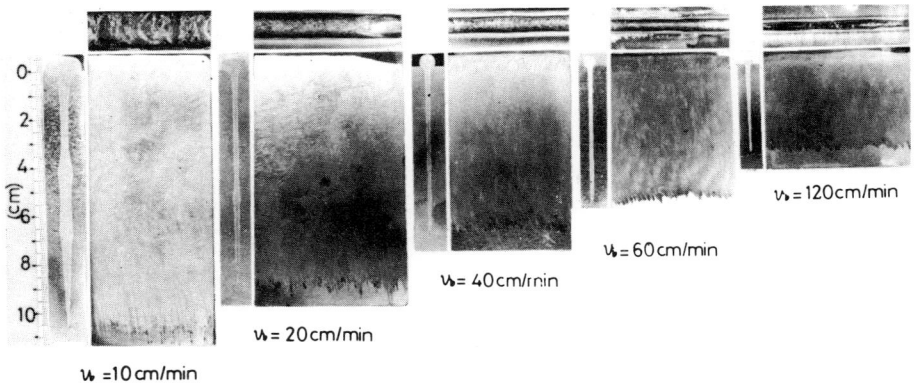

W = 20KW   V = 250KV   I = 80mA   SUS304

Fig. 10 Bead sections and bead appearances for various welding speeds at gas pressure 10^{-4} Torr in welding chamber.

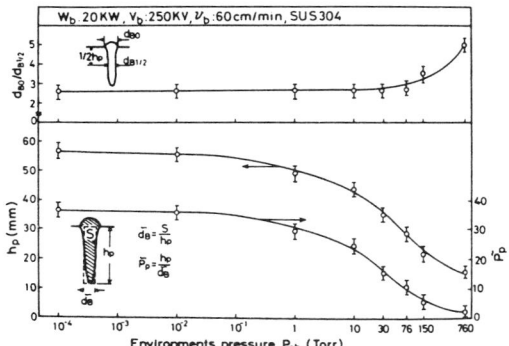

Fig. 11 Relations of bead width ratio $d_{B0}/d_{B1/2}$, penetration depth h_p, effective penetro-paramete. $\tilde{P}_p$ and gas pressures P_{ch} in welding chamber.

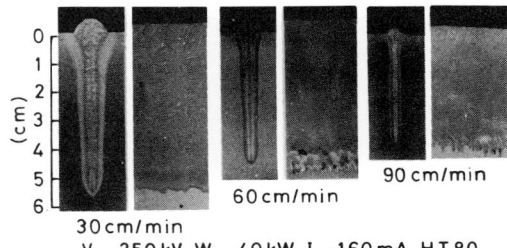

Fig. 12 Bead sections at various welding speeds for gas pressure 30 Torr in welding chamber.

that at 10^{-4} Torr. But at atmospheric pressure, scattering of the beam with the surrounding gas is violent and the penetration depth is lowered to $1/3 - 1/4$ of the value at 10^{-4} Torr with a broad bead width. By these results, we can say an ultra-high voltage type electron beam enables us to perform the "Low Pressure Electron Beam Welding" which has been considered to be hard to realize in the past. Figure 12 shows a typical example of bead configurations for a high-tensile steel (HT-80) obtained at 40kW of the beam power in a low pressure environment of 30 Torr. Even when the welding speed was varied, we could obtain a good beam configuration with a high practicality.

6. Conclusions

For precise, high quality welding of ultra-thick plates, it is desirable to adopt an ultra-high voltage, high-power electron beam welding technique. To realize such a technique, the authors developed a new accelerating unit and the "strong focusing electron beam gun with electro-magnetic accelerating units" by combining it in multi-stage configuration. This led to the production of prototype 300kV/100kW and 600kV/300kW ultra-high voltage, high power strong focusing beam heat sources, one after another. No similar heat sources with such a strong and high energy beam have yet to be seen in the world. This source showed that they had excellent beam characteristics giving high quality weld beads. Furthermore, this beam allowed the development of the "Low Pressure Electron Beam Welding" technique, which had been regarded difficult to pursue.

References

1) Y. Arata and M. Tomie: Trans. JWRI, 2-2 (1973) 17., IIW Doc. IV-111-73 (1973)., 2nd Inter. Symp. JWS, (1975) 45., 7th Inter. Conf. on Electron and Ion Beam Sci. and Tech. (at Washington D.C. USA), (1976). Y. Arata M. Tomie and Y. Kato: Trans. JWRI, 4-2 (1975) 1.

2) A. Sanderson: Metal Const, 6-1 (1974)., K. H. Steigerwald: 2nd Inter. Symp. JWS, (1975) 3. G. Sayegh, P. Damonte and T. Nakamura: 2nd Inter. Symp. JWS, (1975) 15. K. S. Akop'yant et al, Automatic Welding, 28-4 (1975) 57.

3) Y. Arata, M. Tomie: JWS 46 (1977) 429.

4) Y. Arata, M. Tomie: JWS 46 (1977) 686., Y. Arata and M. Tomie: Trans. JWRI, 9 (1980) 157., IIW Doc. IV-308-81 (1981)., Proc. of Inter Conf. on Welding Research in the 1980's (1980).

5) H. Schwarz: J. Appl. Phys. 33 (1962) 3464, Rev. Sci. Inter. 33 (1962) 688.

6) D. A. Dunn and A. S. Halsted: Proc. Electron and Laser Beam Symp. (1965) 243.

7) Y. Arata, M. Tomie: JWS 46 (1977) 514.

8) Y. Arata, M. Tomie, K. Terai, H. Nagai and T. Hattori: Trans. JWRI, 2-2 (1973) 1. IIW Doc. IV-114-73 (1973)., Y. Arata: IIW Doc. IV-340-83 (1983).

Laser Beam

Some Fundamental Properties of High Power CW Laser Beam as a Heat Source
Y. Arata and I. Miyamoto
[Tech. Report Osaka Univ. **19** (1969), 379.]

High Power Electric Discharge CO₂ Lasers
Y. Arata, I. Miyamoto, N. Kawanishi, T. Komori and A. Takayasu
[Trans. JWRI **2** (1973), 264.]

SOME FUNDAMENTAL PROPERTIES of HIGH POWER CW LASER BEAM as a HEAT SOURCE

Abstract

High power CW CO_2 gas laser apparatuses have been constructed and their fundamental properties as a heat source have been discussed. An optical system to focus the laser beam has been established and a diameter of beam focused has been measured. An absorption coefficinet of metals with various surface conditions is obtained.

New laser processing techniques (cutting, welding etc.) have been developed.

I. Introduction

A great deal of interest has been shown in the laser as a heat source due to the potential capability for extremely high density in energy. Lasers may be classified according to their mode of operation as continuous lasers or pulsed lasers. Until recently, lasers capable of melting and cutting materials were of the solid-state type with a pulsed output while the continuous lasers did not have sufficinet power output for processing the materials. However, the advent of the high power continuous wave (CW) CO_2 gas lasers has made it possible to apply the laser beam as the more general heat source.

We have constructed CW CO_2 gas laser apparatuses with high power output and high efficency, and some fundamental properties of CO_2 laser as a heat source are discussed. New techniques for metal processing, laser welding, laser cutting, etc. have been developed. They promise excellent fusion welding and cutting for materials.

II. CO_2 gas laser apparatus

High power CW laser action was recently reported in the rotational transitions of the $00°1$-$10°0$ vibrational band of CO_2 at wavelength near 10.6μ [1]. The mechanism proposed was collisions of the second kind between ground state CO_2 molecules and excited N_2 molecules;

$$CO_2 \; (00°0) \; + \; N_2 \; (v=1) \; = \; CO_2 \; (00°1) \; + \; N_2 \; (v=0) \; -\triangle E(=18cm^{-1})$$

Fig. 1 shows pertinent parts of the vibrational energy level diagram of CO_2 and N_2. Shortly afterwards, progress toward high power and high efficiency was achived with

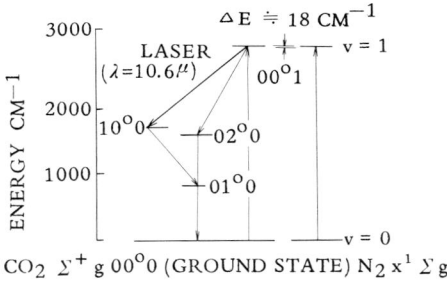

Fig. 1   Energy diagram of CO_2 and N_2.

the addition of He [2], H_2O [3] and H_2 [4] . The highest power is presently achieved by running a discharge in a flowing gas mixture of CO_2-N_2-He.

We have constructed the CO_2 laser apparatus with CW power output of 600W shown in Fig. 2 which consists of a 76mm diameter water cooled discharge tube, 12m long, placed between gold coated mirrors of 29.2m radius curvature spaced 13m apart. There are six sections to the tube, each section (2m long) having its own cathode with a series ballast resistor.

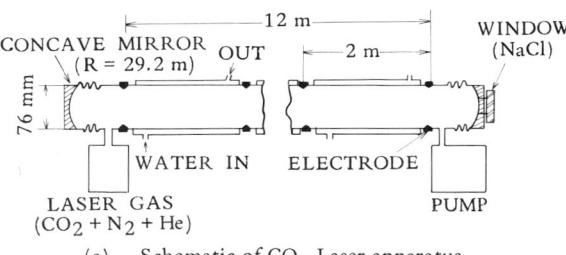

(a)   Schematic of CO_2 Laser apparatus.

(b)   Laser head.   (c)   Back side.

Fig. 2 CO_2 laser apparatus (12m long).

A cold cathode discharge, either ac or dc, is run in the flowing gas on each section with typical value 100mA and 8KV. These gases (CO_2, N_2 and He) are mixed previously in the tank with an optimum mixing ratio of $CO_2 : N_2 : He = 1 : 5 : 40$ and fed into one end of the tube at a pressure of 8~10 torr and are pumped out the far end with a mechanical pump. The present laser apparatus is operated from 60 cycles per second for simplicity, which gives output of 120 pulses per second. We have constructed another CO_2 gas laser apparatus illustrated in Fig. 3 consists of a 8m laser and a 16m laser amplifiers, folded with 8m length, with average power output of more than 1KW. The former apparatus has been used in this experiment mainly.

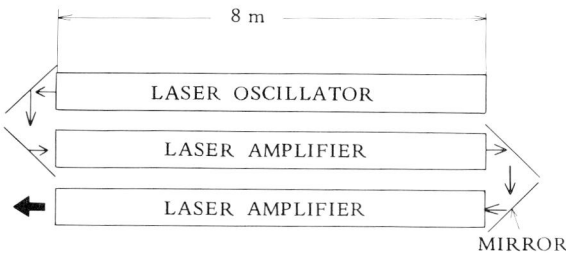

Fig. 3 CO_2 laser apparatus (24m long).

III. Power measurement

It is necessary to measure the power of the laser accurately with reasonably fast response in any laser work. Therefore it must be described how the power level are measured.

A flow calorimeter shown in Fig. 4 has been used to measure the high power level. Water is flowed through the calorimeter at a constant rate monitored by a flow meter. Thermitsters are used to measure the temperature difference of the input and output water accurately. To give a more accurate measurement of the power, a heater is incorporated to calibrate a detecting voltage of bridge V_b by means of a known heater input. An absorption coefficient of the calorimeters are calibrated by means of the calori-

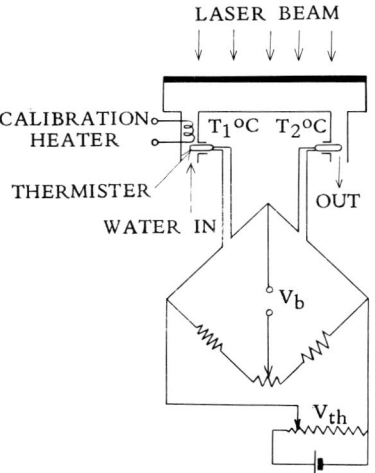

Fig. 4   Schematic of flow calorimeter.

meter shown in Fig. 5 ; although the absorption coefficient of this calorimeter is 100%, a response time is relatively long.

A surface of the absorption plate of the calorimeter shown in Fig. 4 consists of thin Cu plate etched with solved NaOH and $K_2S_2O_2$(100°C) for few hours. Thus excellent uniform surface can be obtained with the absorption coefficient of 65%, and a measurable range of power level of the calorimeter is from 0.1W to several KW and a minimum response time faster than a second can be obtained.

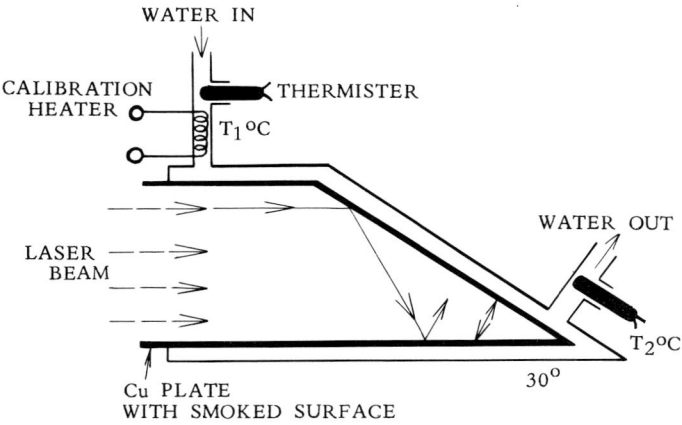

Fig. 5   Schematic of flow calorimeter. As a result of multiple reflections 100% of the incident beam is trapped, but response time is relatively long.

IV. Optical system

The laser beam can be focused by means of a converging lens or a concave mirror when the laser is used as a heat source. Fig. 6 shows how the beam is deflected downwards in both cases. The materials which can transmit 10.6μ laser are listed in Table 1.

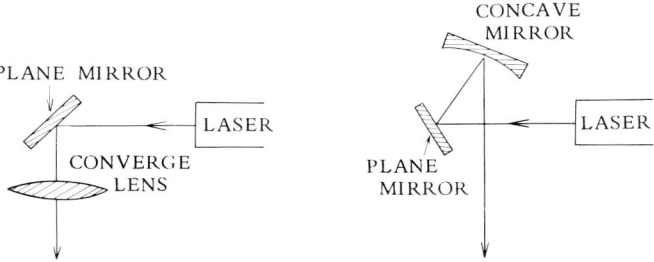

Fig. 6　Beam focusing methods.

(a)　As a converging lens KRS-5 is used.

(b)　The mirror surfaces are of vacuum deposited gold, silver or alminium.

Table. 1 Comparison of window and lens materials for CO_2 lasers[7].

Material	Index of Refraction	Fresnel Reflectivety	Fabry-Perot Reflectivity	Absorptivity	Hygroscopic
NaCl crystal	1.49	0.039	0.145	$0.01cm^{-1}$	Yes
KCL crystal	1.454	0.034	0.127	0.01	Yes
BaF_2 crystal	1.42	0.030	0.113	0.077	Slighty
KRS-5	2.38	0.167	0.490	0.02	Slighty
Irtran-2	2.19	0.139	0.429	0.55	No
Irtran-4	2.4	0.170	0.497	0.1	No
Germanium	4.0	0.360	0.779	0.07	No

In the case of the converging lens, it should be chosed from Table 1, but the materials are limited due to hygroscopic or unsuitable hardness and reflectivity. In addition, the inevitable optical absorption, no matter how it may be small, eventually leads to destruction in the case of high power laser. In the caseof concave mirror on the other hand, mirror coated with Au, Ag or Al with the reflectivity of near 100% at 10.6μ respectively can be used. Furthermore, a metal mirror can be cooled by water. Thus the optical system shown in Fig. 6 (b) is recommended.

To estimate the minimum spot size of focused laser beam, one should consider from the both view points of physical and geometrical optics.

From the view point of physical optics CO_2 laser beam may be focused theoretically to a diffraction limited spot by a nonaberration lens, since it is a plane paralell beam of coherent light in principle. But beam divergency of the practical laser is larger than choherent one because of an incompletion of the resonator, spectral broadening, concaved NaCl window etc., therefore it is more practical to think of focusing the actual beam down to 100 times the diffraction limit.

In the optical system of the practice, the aberrations of ray, i. e. its derivation from the path prescribed by the Gauss formulas are present, which are known as the five monochromatic aberrations. In this theory sin θ is replaced by $\theta - \theta^3/3!$. Especially aberrations become important for the optical system shown in Fig. 6 (b), and coma and astigmation are predomenant among the others.

A simple technique has been used for rough estimation of the sport size; laser beam has been irradiated on a film located at each position on the beam axis for a short time determined by a shutter, and sizes of a scorched area are measured; such as shown in Fig. 7. The minimum spot is considerably larger than diffraction limit because the divergency of the beam from the laser apparatus is large. We now are constructing a new laser with a small divergency to obtain a small spot size.

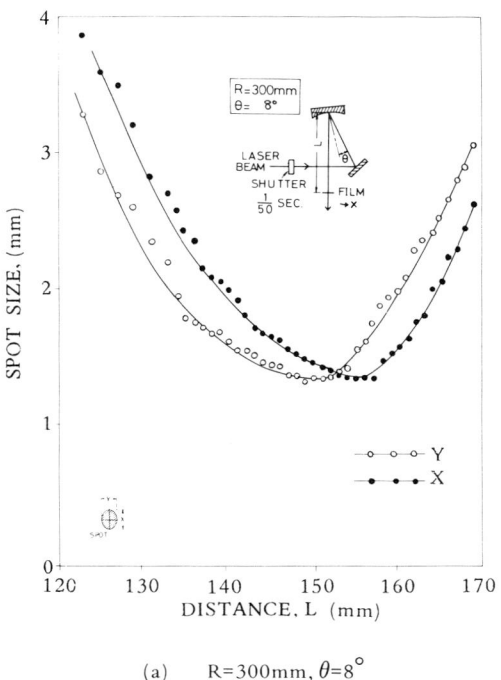

(a)   R=300mm, $\theta=8°$

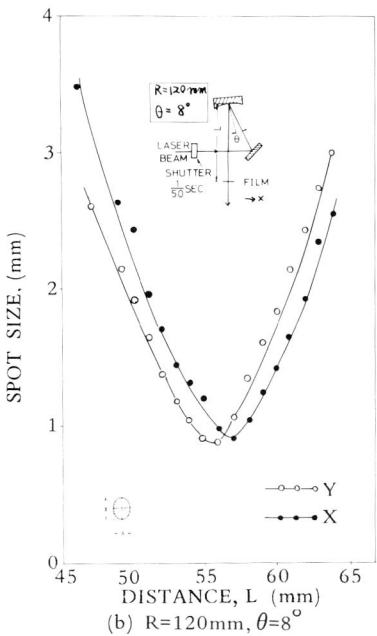

Fig. 7   Spot size of focused laser beam at each film position (Lmm).
Note; R is a radius of curvature of concave spherical mirror.

Fig. 8 shows astigmatic images of an off-axis object point at infinity, as formed by a concave spherical mirror. If a screen is placed at E and moved toward the mirror, the image will become a vertical line at S (Focal line), a circular disk at L (circle of least confusion) and horizontal line at T (focal line). The amount of astigmatism increases with increasing the angle between axis and chief ray, and two focal lines which are

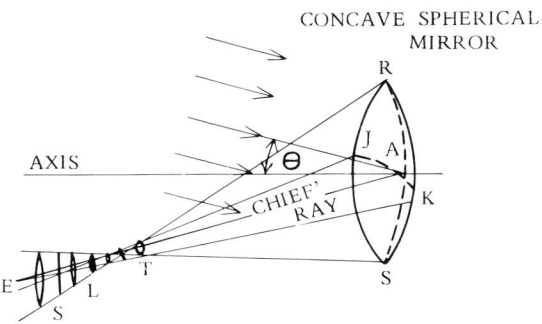

Fig. 8   Astigmatic images of an off axis object point at infinity, as formed by a concave spherical mirror. The focal lines T and S are perpendicular to each other. L is a circule of least confusion of astigmatism.

112

perpendicular each other become long. Fig. 9 shows the effects of the distance measured from the mirror to the film along the chief ray on the spot area and axial ratio. The astigmatism becomes predominant at higher angle. When the laser beam

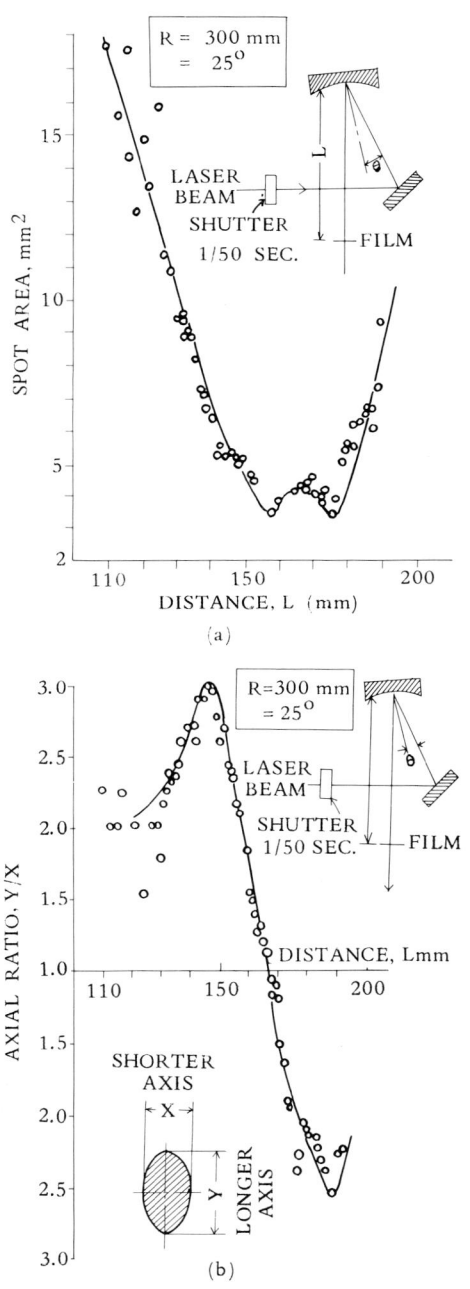

(a)

(b)

are applied as a heat source by means of the optical system shown in Fig. 6 (b), one should choose either a circule of least confusion of astigmatism with the small or a focal line with comparably large   according to its object.

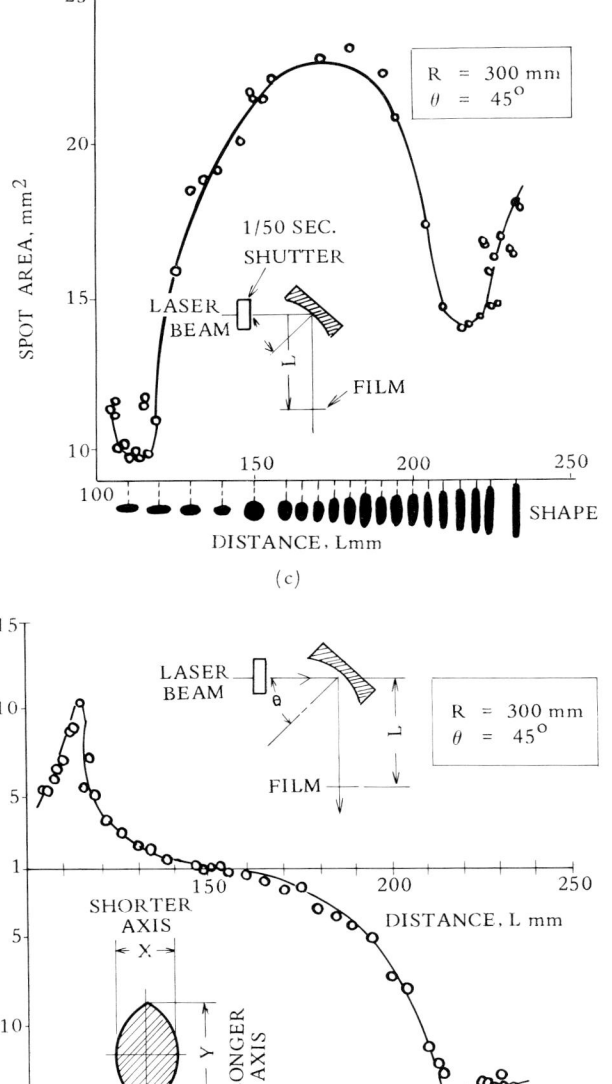

(c)

(d)

Fig. 9   Spot area and shape at each position when astigmatism is dominant.
(a) , (b) R=300 mm, θ=25° (c) , (d) R=300mm, θ = 45°

114

V. Reflectivity of metals

In general, highly polished metallic surfaces have a very high reflectance at wavelength 10.6 μ. It has been discussed that this optical property can be merit in the case of the reflecting mirror, but oppositly it becomes difficulty for the laser processing of the metals. The reflectivity of metals, however, varies with their surface conditions; for this example, Fig. 10 indicates a tendency of the temperature rise of the sheet steels

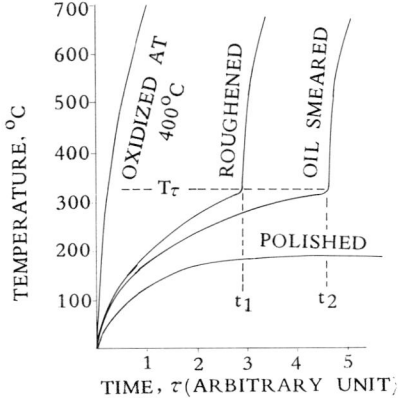

Fig. 10   A tendency of the temperature rise of th sheet steel with time after the laser beam irradiated on it.

with time after the laser beam irradiated on it. The temperature rise is usually gradual under a certain temperature T_τ, and on arriving at T_τ the rate of temperature rise becomes suddenly large to give an inflection of the curve. Though a timé until the absorption coefficient suddenly changes, τ, varies with surface conditions, temperature T_τ is almost constant indipendent of surface conditions. One can expect that an absorption coefficient of the steel becomes large suddenly at this temperature T_τ.

To confirm this phenomenon, steel specimens oxidized in a furnace at various temperature have been irradiated by the laser beam.   Fig. 11 indicates that time τ

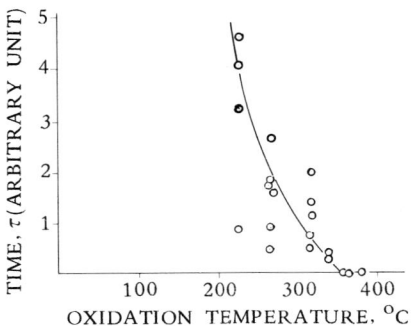

Fig. 11   Time until the absorption coefficient varies suddenly as a function of oxidation temperature.

decreases with increasing the oxidation temperature, and falls to zero at the temperature T_τ. This means that at emperature T_τ a certain oxide is formed, and we now think this is α-Fe_2O_3.

An absorption coefficient of metals has been measured under various conditions by a method shown in Fig. 12. Fig. 13 shows the absorption coefficient of polished and roughened metals at normal incidence. A solid line in Fig. 13 indicates theoretical value calculated from the dc conductivity at 20°C. Absorption coefficient of the polished metals is somewhat higher than theoretical one because polishing is comparably rough.

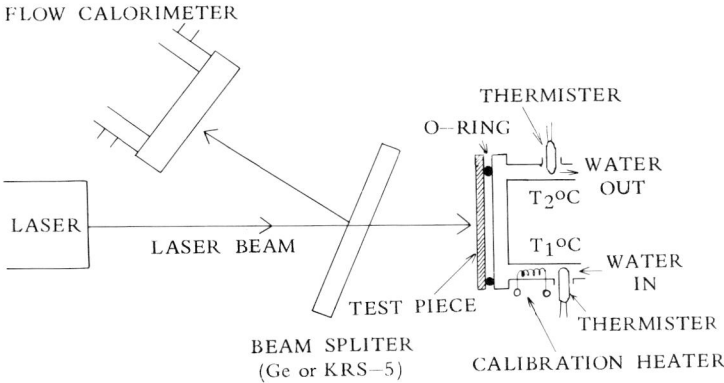

Fig. 12   Optical system for measurement of an absorption coefficient of metals and its oxide.

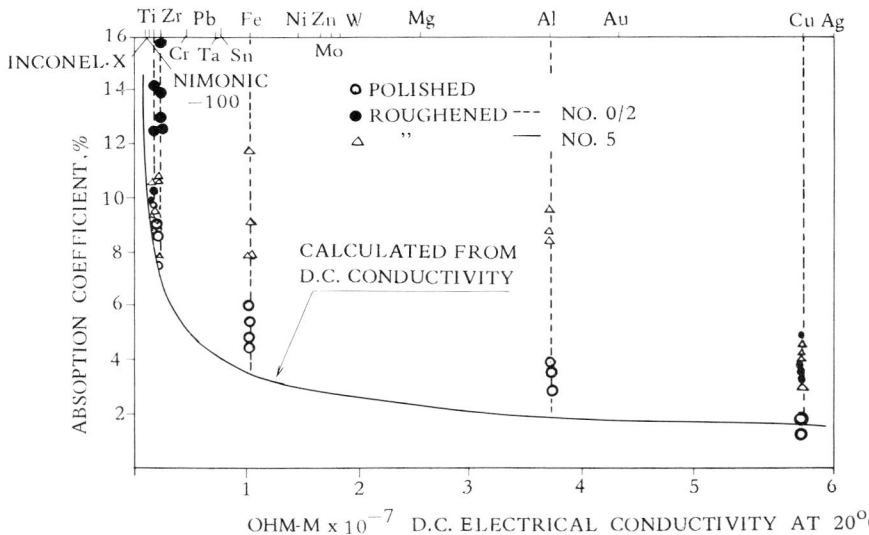

Fig. 13   Absorption coefficient of roughened and polished metals. A solid line indicates the theoretical value calculated from the dc conductivity.

116

Roughened metals have considerably higher absorption coefficient than polished because as a result of multiple reflections a large part of the incident beam is trapped. Table 2 shows the absorption coefficient of oxides of metals. Oxidation has been made at 600°C for two hours in a furnace. The absorption coefficient of oxydes is considerably larger than metals though these data scatter because oxidation films are not stable.

Table. 2 Absorption coefficient of oxidized metals.

METAL	Polished	Oxidized at 600°C for two hours*)
Al	3.4%	28 ∿ 50%
Fe	5.0%	38 ∿ 74%
Zr	8.3%	45 ∿ 56%
Ti	9.4%	18 ∿ 25%
Cu	1.4%	— **)

*) Oxidized test pieces have been roughened previously.
**) Oxidation films have peeled off naturally.

A chemical treatment is also effective to increase the absorption coefficient. For example provided that a polished cuppor plate immerses in solved $NaOH + K_2S_2O_8$ (100°C), the absorption coefficient becomes 65%, and in solved $NH_4CL + K_2S$ (room temperature) it becomes about 90%. As a simple method to increase the absorption coefficient, a thin smear of fine powder is recommended.

VI. Application to the metal processing

Because most dielectric materials absorb at wavelength of 10.6μ, considerable interest has been shown in CO_2 lasers for cutting and drilling applications. These applications have been well-covered in other literature[5] and will not be reviewed here. Excellent welding and cutting methods for relatively thin metals have been developed with CO_2 laser.

(1)　Laser cutting and laser gas cutting

High speed fine cutting of sheet steel with narrow kerf wide and heat affected zone has been accomplished which employs a focused laser beam for preheating in combination with an oxygen jet. This new cutting method, whose energy in combined laser energy, chemical reaction energy such as an oxidation energy, and dynamic energy of gas jet, is named as "laser gas cutting method" in distinction from laser cutting method which employs laser alone. Fig. 14 shows one example of the laser gas cutting system which consists of CO_2 laser apparatus, optical system and a O_2 gas nozzle. The work piece to be cut is set horizontally at the position of one of the focal lines of the astigmatism as shown in Fig. 15 and are moved horizontally to allow the cut. The distance between the nozzle and work piece is 1.5 ∼ 2mm. Fig. 16 shows the laser gas cutting. As

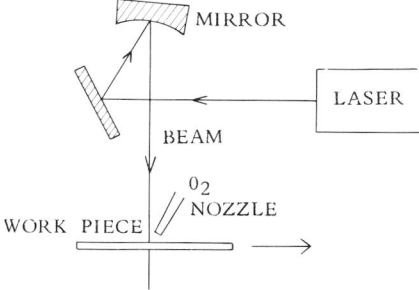

Fig. 14   Laser gas cutting system.

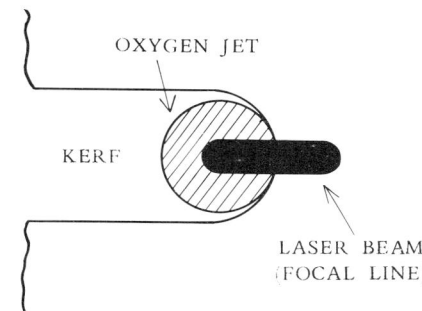

Fig. 15   Schematic of laser gas cutting with the focal line.

Fig. 16   Laser gas cutting.

a oxide of which absorption coefficient is large is formed on the surface of steel at elevated temperature, a sheet steel can be cut by laser beam alone. But molton metal tends of freeze out on the edge giving a rough cut and very low cutting speed accompanies with wide heat affected zone. An example of cut with laser gas cutting is shown in comparison with laser cutting in Fig. 17.

The quality of cuts with laser gas cutting method have been estimated by measureing the roughness of cutting edge h indicated in Fig.18: for $h < 0.05$mm the cut is

118

classified as a cut of smooth surface, and for h > 0.05mm as a cut of rough surface. DIN is unsuitable to classify the quality of the cut for very thin plate.

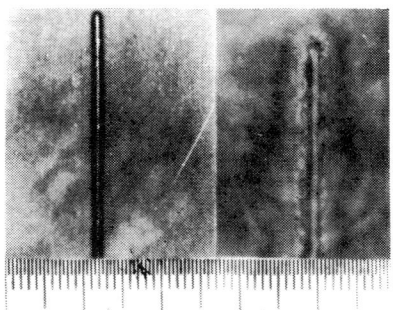

Fig. 17   Comparison between laser gas cutting and laser cutting (laser alone).
(a)   Laser gas cutting (carbon steel, 0.6mm thickness)
(b)   Laser cutting (carbon steel, 0.3mm thickness)

h < 0.05 mm :   SMOOTH SURFACE
h > 0.05 mm :   ROUGH SURFACE

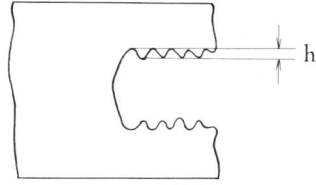

Fig. 18   Classification of the quality of cuts.

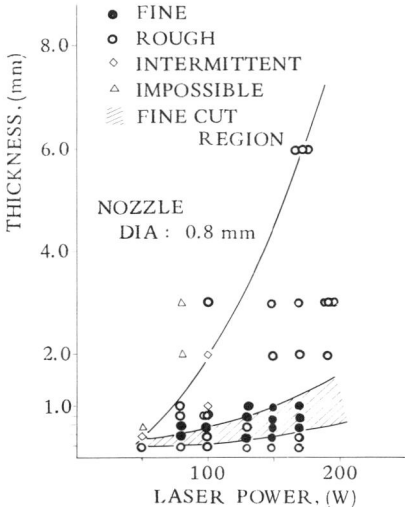

Fig. 19   Laser power vs. thickness possible to be cut.

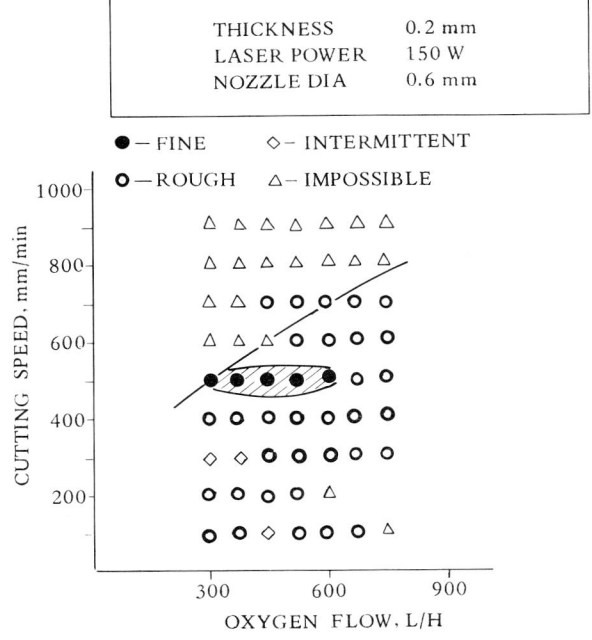

Fig. 20 (a)

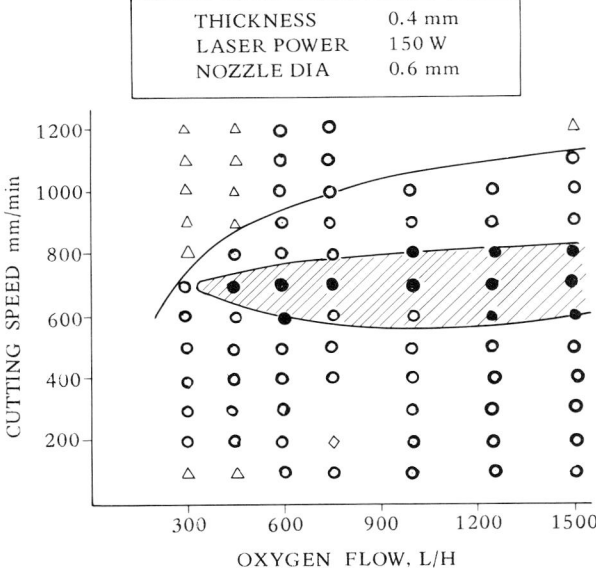

Fig. 20 (b)

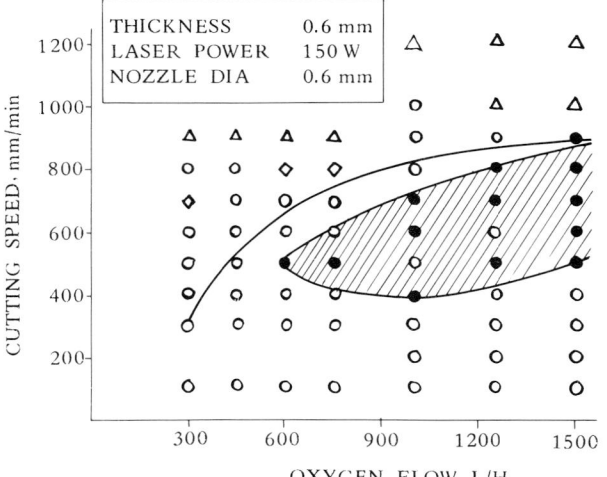

Fig. 20 (c)

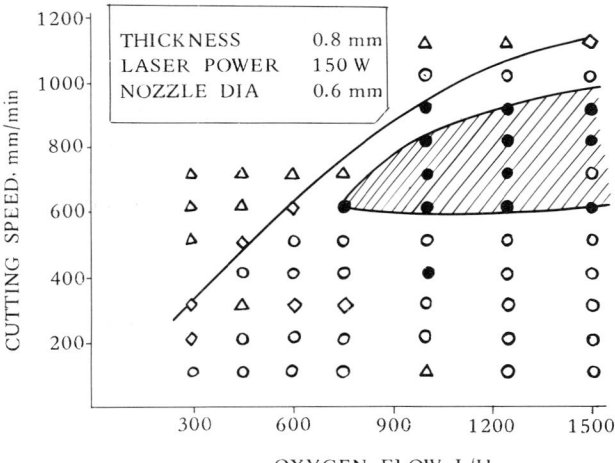

Fig. 20 (d)

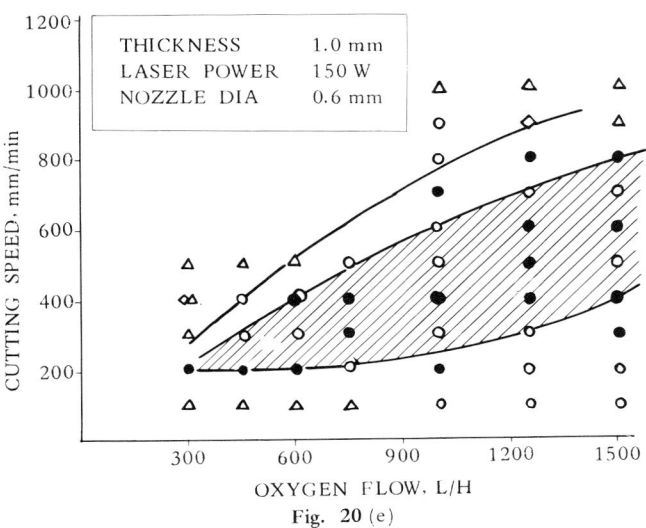

Fig. 20 (e)

Fig. 20   Influence of cutting speed and oxygen flow on the quality of cuts. Laser power; 150W.

Fig. 19 shows the relation between the laser power output and thickness possible to be cut.   Thickness of rough surface increases more rapidly than of fine surface with increasing the laser power output.   If the laser power output is too high for the thickenss, molton slag tends to freeze out on the cutting edge providing a rough cut and wide kerf.

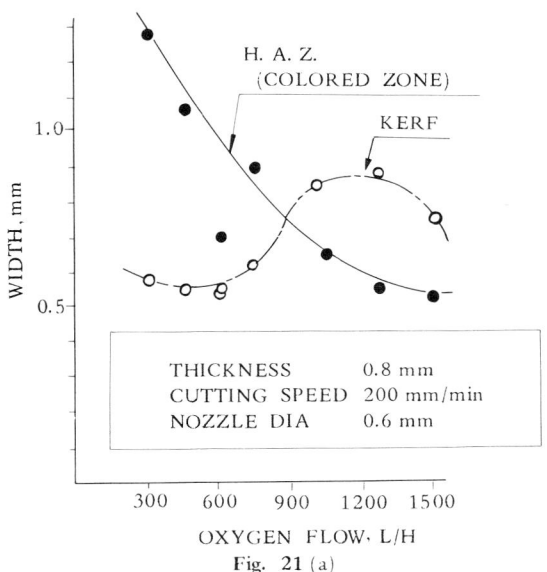

Fig. 21 (a)

122

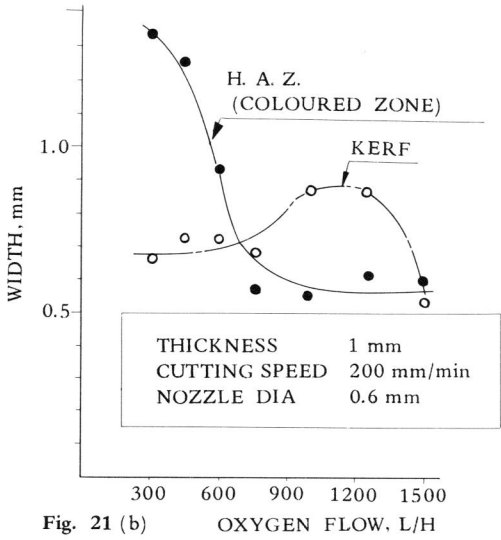

Fig. 21 (b)   OXYGEN FLOW, L/H

Fig. 21   Influence of the oxygen flow on the widths of kerf and heat affected zone.

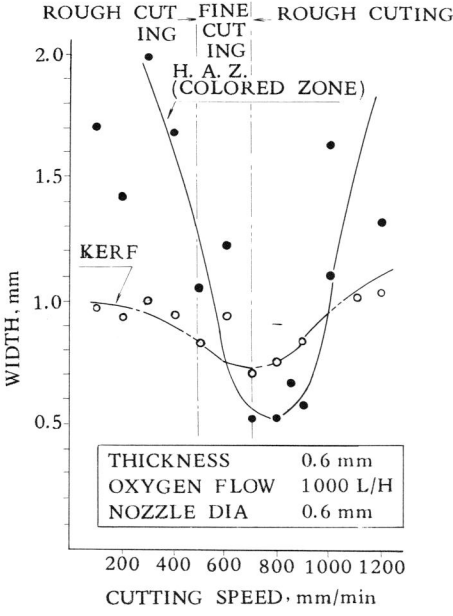

CUTTING SPEED, mm/min

Fig. 22   Influence of the cutting speed on the widths of kerf and heat affected zone.

Fig. 20 shows the influence of cutting speed and oxygen flow on the quality of cuts. Fig. 21 shows an influence of the oxygen flow on a kerf width and colored zone which corresponds to a zone heated above 200°C. Fig. 22 shows an influence of cutting speed on a kerf width and the colored zone. The most narrow kerf width of 0.5mm has been obtained. Fig. 23 shows examples of cuts corresponding to each region in Fig.20(c).

(2) Laser welding

Up to now lasers capable of welding were ruby laser or glass laser that only discrete

Fig. 23   Examples of cuts at each region shown in Fig. 20 (c).
            Thickness;   0.6mm
            (a)   poor oxygen flow
            (b)   suitable oxygen flow (fine cut)
            (c)   excess oxygen flow

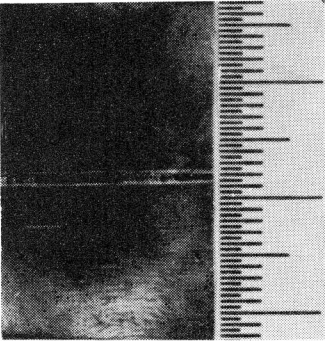

Fig. 24   Weld on 0.5mm thick Zirconium.
            Pretreatment; surface oxidized at 600°C for two hours in air..
            Laser power; 50W
            Welding speed; 120mm/min

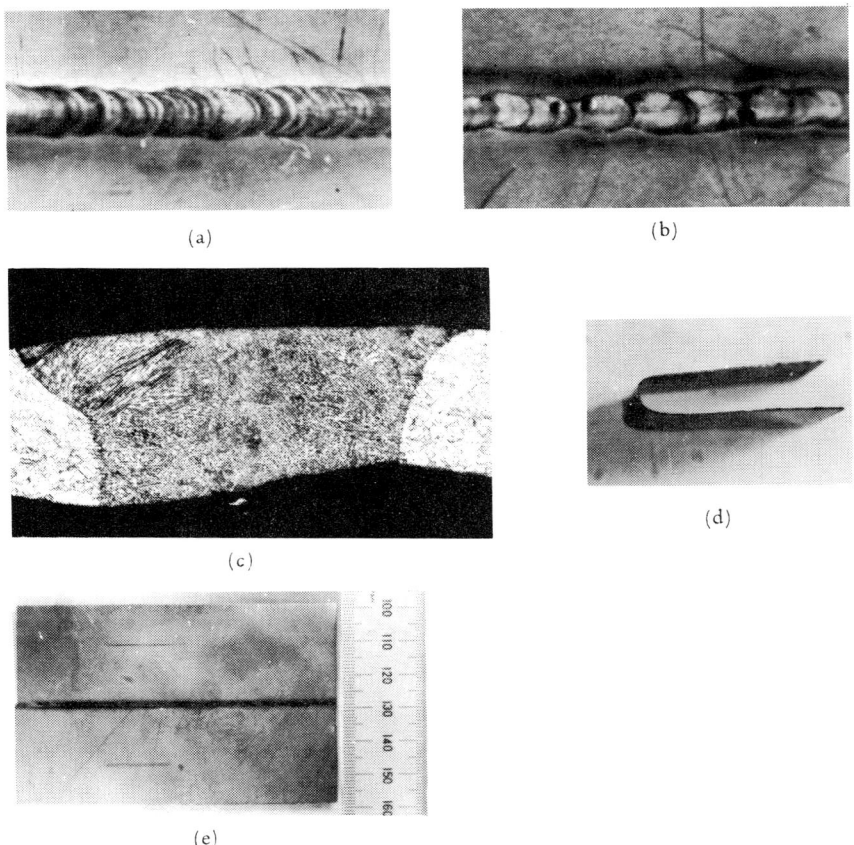

(a)

(b)

(c)

(d)

(e)

Fig. 25   Weld made on 18-8 stainless steel with a thin smear of fine powder.
Thickness; 0.3mm   Laser power; 300W   Welding speed; 800mm/min.
Argon shielding

(a)   Top view (x15)   (b) Bottom view (x30)
(c)   Cross section of base metal/weld metal interface.
(d)   Bending test
(e)   Weld of no shielding. Welding speed; 1000mm/min.

spot welding [6] could be made for the sake of potential ability of intermittent output. But continuous and stable welding are now possible by developement of high power CW CO_2 lasers. In laser welding, when the laser beam impinges upon the metal it delivers its heat to the surface and further penetration beneath the surface relies upon the thermal conduction. One of the most important problems for CO_2 laser welding is that only under 10% of the beam can be absorbed by polished pure metals as previously described. Therefore one must rise up the absorption coefficient of the laser beam by various methods mentioned previously to improve the weldability.

Fig. 24 shows an example that the reducing the reflectivity by means of oxidizing permits the welding with low power output. Metals whose oxides have high absorption coefficient of the laser beam can be welded without any treatment for reducing the reflectivity; i. e. once a molton pool are formed beam absorbable oxide is formed ahead the molton pool.

To improve the weldability furthermore, it is recommended to smear thinly a fine powder on the surface which simultaneously refine the molton metals and reduce the reflectivity. For carbon steel and stainless steel of 0.3mm thickness, comparably high speed welding about 100cm/min can be accomplished at the power output of 300W. Fig. 25 shows an example of the butt weld of the stainless steel of 0.3mm thickness. Shielding is necessary to prevent weld contamination and embritllement by exposure to air at the sacrifice of welding speed slightly. Sound welds can be obtained without surface treatment, but welding speed is very low as shown in Fig. 26 for example.

At present these welding methods which involve the treatments to reduce the reflectivity tend to contaminate the weld bead appearance. Therefore to avoid this disadvantage beam absorbable powder which is readily pealed off should be developed.

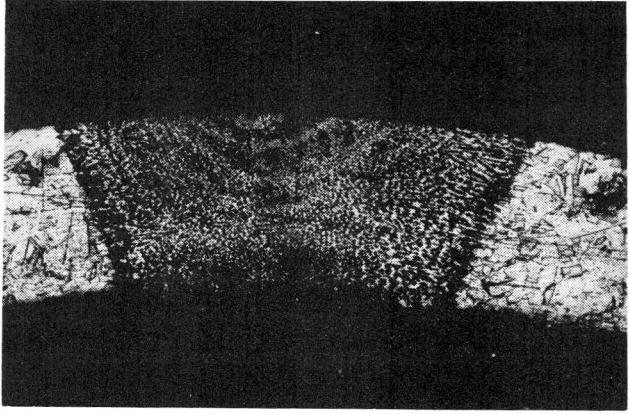

Fig. 26   Weld made on 18-8 stainless steel without any treatment. Welding speed; 15cm/min. Argon shielding.

References

1)   C.K.N. Patel, Appl. Phys. Letters 7, 15 (1965)

2)   G.Moeller and J.D. Rigden, Appl. Phys. Letters 7, 274 (1965)

3)   W.J. Wittman, Phillips Research Reports, 22, 73 (1966)

4)   D. Rosenberger, Phys. Letters, 15, 520 (1966)

5)   Y. Arata and I. Miyamoto, Technol, Repts. Osaka Univ., 17 285, (1967)

6)   R.H. Fairbanks and C.M. Adams, Welding Journal, 43, 97−s (1964)

A.O. Schmidt and T. Hoshi, Welding Journal, 44, 481−s (1965)

A.R. Pfluger and P.M. Maas, Welding Journal, 44, 264−s (1965)

7)   D.R. Whitehous, Laser Tech. July, A6 (1967)

High Power Electric Discharge CO₂ Lasers

A discharge in high flow[1] gas has made possible extremely high power compact CO_2 laser, as compared with the power obtained from the conventional type with very low gas flow. An uniform and stable glow discharge in high flow gas is one of the most important factors in obtaining such high power. When an initially uniform electric power is deposited in a high flow gas where there is a non-uniformity of the velocity, the gas preferentially is heated in the zone where flow speed is lower from flow calorimatry consideration to cause non-uniform temperature distribution, and hence non-uniform electrical conductivity. This non-uniform conductivity leads to a constricted discharge, eventually to an arc. It is necessary not only to make the gas flow distribution as uniform as possible but also to disperse the heat locally concentrated in order to prevent such non-uniformity. The authors test-produced two types of high flow CO_2 laser, and some primary experimental data are shown in this paper. These laser systems consist of circulating pump, heat exchanger discharge tube, and gas mixtures of CO_2, N_2 and He.

In the first type of the laser, the directions of the laser beam, gas flow and discharge perpendicularly intersect each other. The electrode distance is about 3 cm and the maximum gas flow is about 30 m/sec. Under such conditions even without convective action it is comparatively easy to maintain discharge curtain-like over the range of about one meter between an anode and many copper cathodes having their own ballast resistor. The laser power is coupled out from Ge substrate and is nearly proportional to the gas flow rate.

Figure 1 shows the relationship between the discharge current, discharge voltage, and output power at the flow rate of 30 m/sec. The maximum power was about 1.6 KW with 10% efficiency at 11A, 1.5 KV and 30 torr.

Generally as the electrode distance becomes longer, it becomes more necessary to disperse heat

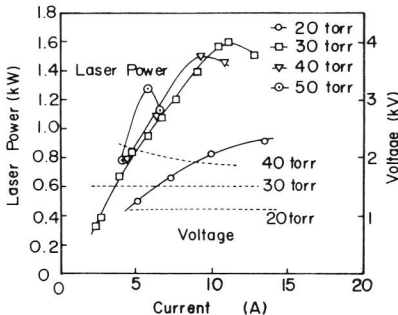

Fig. 1. Relationship between the current, voltage and the output power for the laser shown in Fig. 1 for various gas pressure. ($CO_2 : N_2 : He = 1 : 6 : 3$).

by convective action for the purpose of maintaining an uniform discharge. In the second type of the laser shown in **Fig. 2,** the directions of laser beam, gas flow and discharge are parallel with each other, the discharge cross section is comparatively small, and the electrode distance is long. The maximum gas flow in the discharge tube of 74 mm in diameter was 110 m/sec. The orifice type cathode was set at the upper stream in order to produce convection and consequently make the discharge uniform. A cylindrical tube was used for an anode at downstream. Both electrodes have the same internal diameter as laser tube. The laser power was taken out from the coupling hole of 20 mm in diameter. Various forms of the discharge tube were used and it was found that the

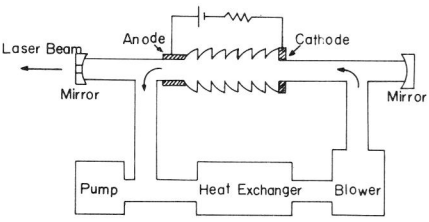

Fig. 2. Schematic diagram of high flow CO_2 laser.

structure of modified tube shown in Fig. 2 was most effective for producing convection. In a straight discharge tube, the gas flow is inclined to be slower near the tube wall than around the center, and as the result the positive column is apt to concentrate on the low flow area, which causes non-uniform discharge. The convection effect at the orifice also weakens this tendency a little, but uniform discharge cannot be maintained any more under high electric input.

Figure 3 shows the limit of uniform discharge for various gas pressures in both straight and modified tubes. The maximum current of the modified tube is larger by 30～50% than that of the straight tube. The convection effect of the modified tube became larger with an increase in gas pressure.

The relationship between the discharge current and the laser power is shown in **Fig. 4,** where solid curves correspond to the modified tube, and a dotted curve the straight tube. Comparing data for both tubes at 15 torr, the laser power of the modified tube is higher than that of the straight tube for each current value. The current of maximum laser power for each curve nearly coincides with the limit current shown in Fig. 3. The maximum power was about 420 W with the specific input power of about 4 KW /kg/sec, which was considerably lower than that shown in Fig. 1. This is because the internal diameter of orifice-type cathode is so large that the initial convection effect at upper stream is insufficient. Therefore it is assumed that input power can be enhanced by bending the discharge path and then by making the internal diameter small.

Photograph 1 shows an example of the cross section of bead welded by 1.5 KW beam obtained from the laser of the first type. The beam was focused through a concave spherical mirror with the radius of curvanture 310 mm with the incident angle to the mirror at about 5°. This apparatus emitted the laser beam with a comparatively large divergent angle, so that the beam was not concentrated well through the mirror used in this experiment, and deep penetration welding was observed, although not so obviously. However, the use of the optical system with short focal length increases the energy density, and thereby deep penetration welding may be achieved in much more obvious form.

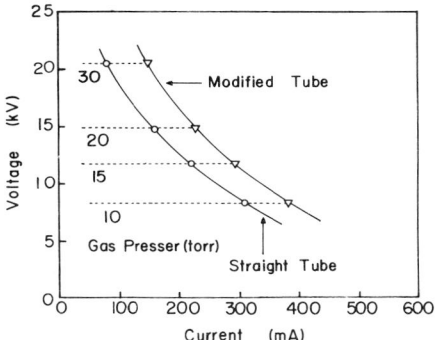

Fig. 3. Limit of uniform discharge for straight and modified tubes with discharge length of 75 cm.

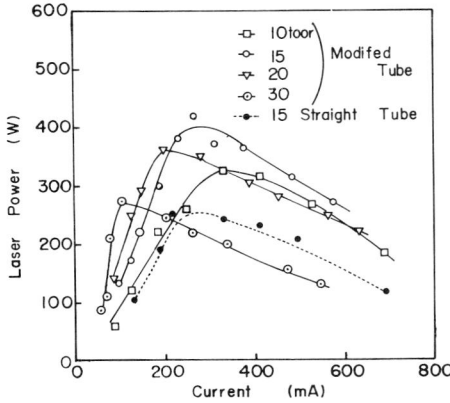

Fig. 4. Relationship between the current and the output power (CO₂ : N₂ : He ＝ 1 : 24 : 25).

Photo. 1. An example of weld bead of AISI 304 stainless steel, 0.8 mm in thickness. Laser power＝1 KW, welding speed＝1 m/min.

Reference

1) W. B. Tilang, R. Targ and J. D. Foster: Kilowatt CO₂ Gas-Transport Laser, Appl. Phys. Letters, Vol. 15 (1969), No. 3
A. E. Hill: Uniform Electrical Excitation of Large Volume High Pressure Near Sonic CO₂-N₂-He Flowstream, Appl. Phys. Letters, Vol. 18 (1971) No. 3.

Plasma Beam

Magnetic Control of Arc Plasma and Its Application for Welding
Y. Arata and H. Maruo
[Tech. Report Osaka Univ. **22** (1972), 135. IIW Doc. IV−53−71.]

Generation of Point Arc and Its Characteristics
Y. Arata and K. Inoue
[Trans. JWRI **3** (1974), 201.]

Fundamental Characteristics of Stationary Plasma Arc in Gas Tunnel
Y. Arata and A. Kobayashi
[Trans. JWRI **13** (1984), 173.]

Development of Gas Tunnel Type High Power Plasma Jet
Y. Arata and A. Kobayashi
[J. High Temp. Soc. **11** (1985).]

Magnetic Control of Arc Plasma and Its Application for Welding

Abstract

In an ordinary plasma arc process, plasma arc column is restricted principally by the water-cooled metallic nozzle. For the purpose of realizing more intensified energetic arc plasma and of achieving high speed welding or cutting, the influence of cusp type magnetic field on tungsten inert gas arc (TIG arc) is investigated.

When the cusp magnetic field is imposed upon the TIG arc, arc column changes its cross-section from circular into elliptical in accordance with the magnitude of field applied, and all of arc voltage, potential gradient in arc column, and current density at the anode are observed to increase respectively. Heat delivered to the anode increases, too, but its ratio to the electrical input is remained unchanged around 75%.

These elliptical arc plasma yield the similar characteristics to a linear distributed surface heat source.

In welding 18−8 stainless steel plate with use of magnetically constricted arc, weld width was observed to decrease and penetration to increase as a function of magnetic field applied.

1. Introduction

In the field of welding technology, the problem of magnetic control of welding arc has become of interest recently. Many researchers[1]−[6] have reported on the effect of external additional magnetic field on a welding arc.

Almost of them were, however, concerned with the effect of transverse (magnetic flux is perpendicular to the direction of electrode travel), parallel (magnetic flux is parallel to the direction of electrode travel) or longitudinal (magnetic flux is parallel to the axis of arc) magnetic field.

When a transverse magnetic field is, for example, inposed upon the arc, arc column could be deflected either forward or backward with respect to the electrode. Deflected welding arc is, however, not so changed in essence from the view point of arc physics.

It seems there are two way of thinking magnetic control of welding arc, one is to control the arcing path between the electrodes, and another is to change some specific properties of arc itself.

If it can be possible to constrict the welding arc magnetically, the average arc current density varies inversely as a cross-sectional area of the arc, and so the conditions govering heat flow to a work piece from the arc can be expected to change. Accordingly, it may be possible to regulate the shape and size of weld bead.

It is, however, impossible to constrict the arc uniformely toward the center of them, because there is no way to produce such an external magnetic field except one dueing to the arc current itself.

For the alternative way to increase the energy density of welding arc, the effects of cusp type magnetic field upon the TIG arc was investigated. When the cusp magnetic field is imposed upon the arc, arc column was restricted laterally from both sides, and its cross section was changed from circular to elliptical. All of the arc voltage, potential gradient of arc column, and current density at the anode were observed to increase respectively. Heat delivered to the work piece had, also, increased.

As a consequence of these results, it may be possible to consider that the elliptical arc plasma yields similar characteristics to those of linear distributed surface heat source. In welding 18—8 stainless steel plate with above elliptical arc, it was made clear that weld bead width became narrower and penetration became deeper as a function of magnitude of cusp magnetic field applied. Cross-sectional area of the bead was still remained unchanged.

2. Experimental Equipments and Procedure

In order to producing the cusp type magnetic field, two types of electromagnet were used in this study. First is a pair of air-cored solenodis (O.D. 480 mm, I.D. 34 mm, Length 40 mm) each of which is enclosed in a austenitic stainless steel casing and is water-cooled. Each solenoid is capable to produce a magnetic field of 8000 gauss at a central point of solenoid with a coil current of 300 A.

Cusp field is produced in such a alignment as the magnetic flux made by each solenoid is equall in magnitude and opposit in direction to each other. These magnets were, however, used in a preliminary experiment to seek the general behavior of TIG arc in a cusp type magnetic field.

On the basis of preliminary results, other sets of electromagnet were developed, considering practical manipulation. Figure 1 illustrates the final set of iron-cored electromagnet, which is designed so small in size that can be installed to the TIG arc torch. A general view of the apparatus is shown in Fig. 2.

The magnetic flux was brought to the arcing zone by two "L"—shaped soft iron yokes and pole pieces. Surface of the pole piece exposed to the arc was covered with a

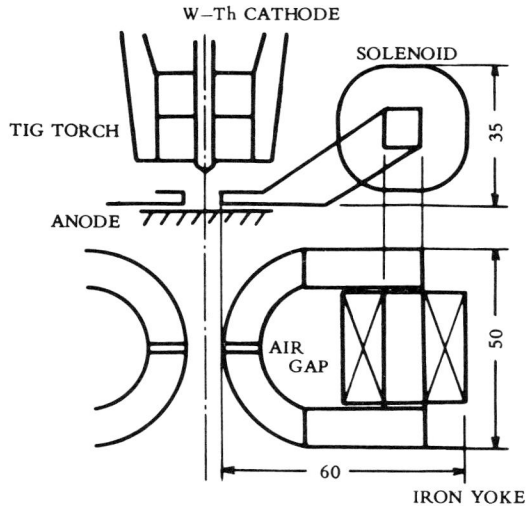

Fig. 1   Iron-cored electromagnet used for magnetic control of welding arc.

Fig. 2   Experimental apparatus.

water cooled copper plate to protect them from thermal damage.

This magnet is capable of producing the magnetic field of 3000 gauss with a field exciting current of 8 A. Strength of magnetic field in the arcing zone is propotional to the exciting current up to 10 A.

Measurement of magnetic field strength was carried out using gauss meter with a transverse Hall probe having a 2% accuracy rating. At the same position, components of

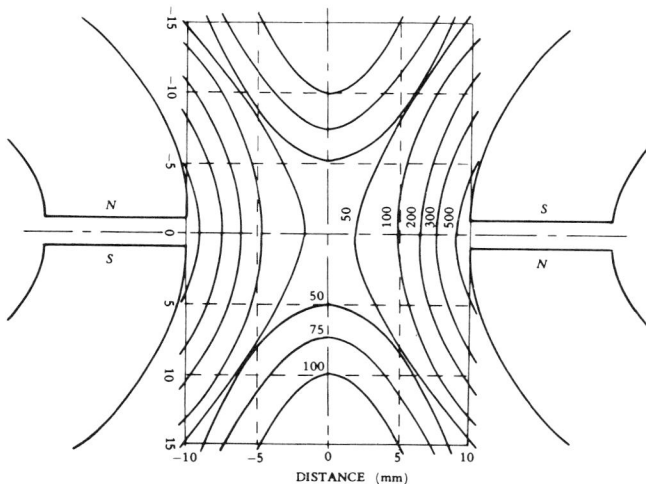

Fig. 3   Equi-strength line of cusp magnetic field made by iron cored electromagnet. Field exciting current, I_f = 1.5A

134

magnetic flux B_x and B_y were measured by rotating the sensing probe. An example of magnetic field strength distribution is given in Fig. 3 which represents the equi-strength line of magnetic field, B, $(=\sqrt{B_x^2+B_y^2})$.

Welding arc was ignited vertically between the thoriated tungsten cathode and water cooled copper anode in such a manner as the axis of arc is being perpendicular to that of cusp field.

A high frequency arc starter is used to facilitate initial break down of the electrode gap. Arc currents is provided by the 40 KVA ordinary DC arc welder having the drooping characteristics.

Measurements of the distribution of arc currents at the anode were accomplished with use of split anodes, each of which was insulated each other by inserting mica plate, 0.05 mm in thickness, as illustrated in Fig. 4. Arc currents flowing into the half anode were measured and recorded continuousely as a function of displacement of the arc. The arc deformation was determined by taking picture of the arc plasma.

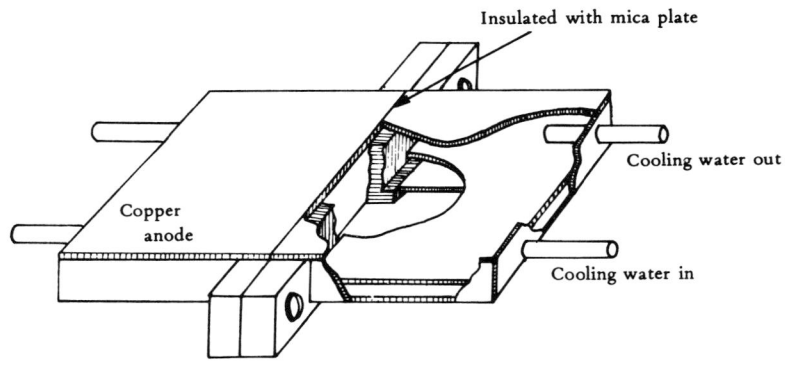

Fig. 4   Detail of split half anodes.

All the welds made were bead-on-plate type with no filler metal using direct current straight polarity. The materials used in this investigation were typical 18—8 austenitic stainles steel plate, 6 mm and 10 mm in thickness. The cross-sectional area of each weld bead was determined from photomacrograph of the cross section prepared at X10 magnification.

3.   Results and Discussions

3.1   General behavior of arc in a cusp magnetic field

1.   Appearance of arc in a cusp magnetic field

When the transverse magnetic field was imposed upon the arc, arc could be deflected either forward or backward with respect to the traveling electrode under the Fleming's left hand rule. When the cusp type magnetic field is used, arc will be deformed in two directions, outward and inwards, depending to the direction of Lorentz force act on each portion of arc column, as shown in Fig. 5.

Figure 6 shows typical appearance of TIG arc in a cusp magnetic field made by air-cored solenoids. The free burning arc (open arc) is, naturally, in rotationally symmetric about its axis, but when the cusp field is imposed upon it, arc plasma is compressed towards its center in one direction and is stretched along perpendicular direction. Deformation of arc column becomes, of course, higher as the strength of magnetic field increases.

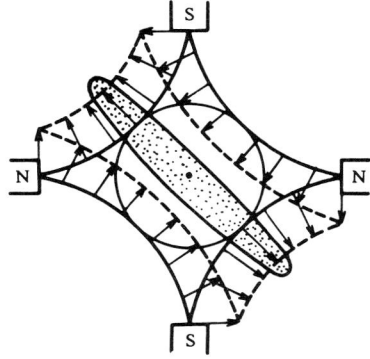

Fig. 5   Deformation of arc plasma in the cusp magnetic field (schematic illustration, two dimensionally). Arrows indicate the direction of Lorentz force act on arc.

It was also observed that some portion of arc was flowing to diagonally upward direction and it had became bright and distinct as the magnetic field was increased. These plasma stream was rather easily observed when the arc distance was prolonged over about 5 mm. Cross-section of the arc column was observed to be elliptical, and there was no longer axial symmetry as observed in an open arc. The trace left on anode plate, shown in Fig. 6, will confirm the validity of this observation.

Iron-cored electromagnet can easily generate the comparable strength of cusp magnetic field to that of air-cored solenoids. Distribution of magnetic fluxes in arcing zone made by such iron-cored electromagnets are, however, somewhat different from those of air-cored solenoids, i.e., transverse components of the magnetic flux in arcing zone are very strong and, accordingly, constricting effects upon the arc from its both side are considered to be much highly.

Typical appearances of TIG arc between the thoriated tungsten cathode and copper anode are shown in Fig. 7 for the both cases magnetic field exists or not. When the magnetic field was applied, arc column was constricted strongly at the midplane between both pole pieces of magnet. If the field exciting currents of each magnet are not the same, arc is always deflected to the weaker side of field.

When the arc was constricted evenly, it was observed that the deformed arc column had a very bright, distinct and thin high temperature plasma region, and its

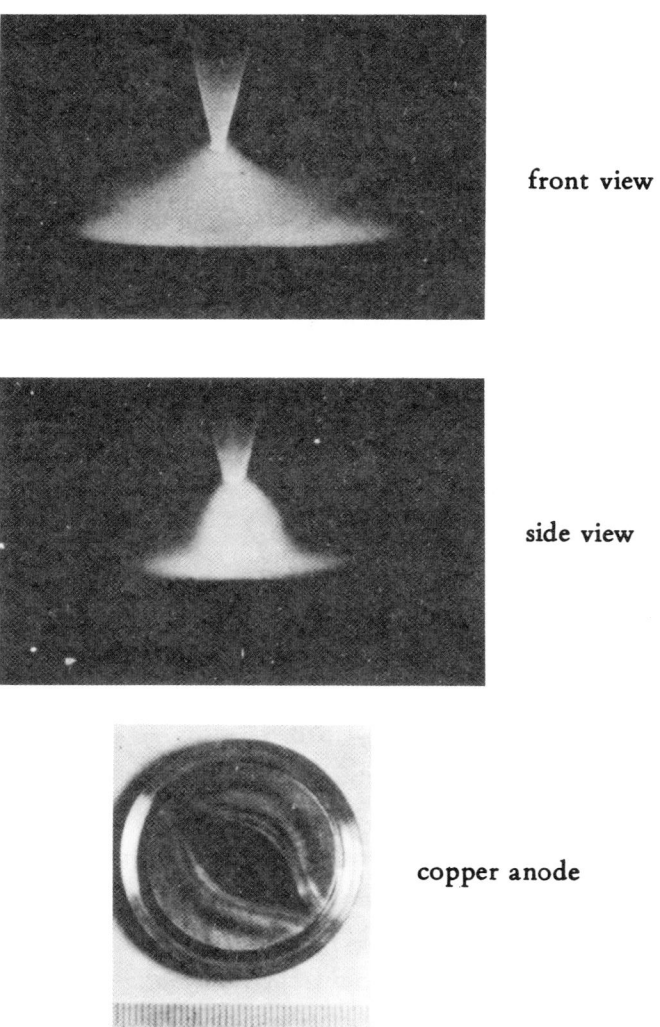

front view

side view

copper anode

Fig. 6   Appearances of arc in the cusp magnetic field. Arc current: 300A, arc length: 6mm, arc voltage: 21.2V, field coil current: 240A.

 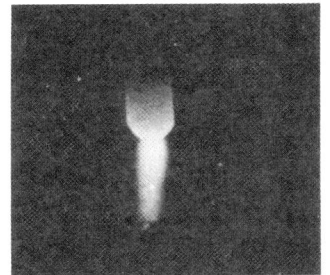 

open arc   arc in cusp field. l_f = 7 amp

Fig. 7   Appearance of TIG arc in cusp magnetic field made by iron-cored electromagnet.
Arc current: 150A, arc voltage: 17.2V, field coil current: 7A.

cross-section was also elliptical.

It has also found that the vertex angle of cathode tip affected on a stability
of arc in the magnetic field. If a round shaped cathode tip was used, bright arc
column was apt to deflect toward an indefinite direction, and arc became almost
unstable. It may caused by the fluctuation of the directional cathode flame. When the
vertex angle was, however, prepared at 80 degree or less, arc was stable in general at
any arc current and magnetic field strength.

2.   Influence of positioning arc in the cusp field

In a cusp type magnetic field, magnitude and direction of the magnetic flux is
different of all spatial positions, and so welding arc in cusp field will be subjected to
different magnetic action in accordance with its location.

Figure 8 represents the collective results of the influence of arc position in the
cusp magnetic field, in which arc voltage, shape of anode spot, rotational angle ,θ, of
major axis of anode spot with respect to the axis of magnetic field, and stable range of
arc are indicated. In this case, arc length was kept at 6 mm through the whole
experiments. So fat as arc was stable, appearance of arc was remained unchanged in
general, but whenever positioning the arc was located too much above or below with
respect to the axis of cusp field, arc had became unstable. This instability has arisen in
consequence of successive increase of twist component in magnetic force act on the arc
plasma.

From these experiments, it was found preferable to locate the arc at the central
portion of cusp magnetic field, where the transverse component of magnetic flux has
dominant effect upon the arc rather than the longitudinal component of cusp magnetic
flux.

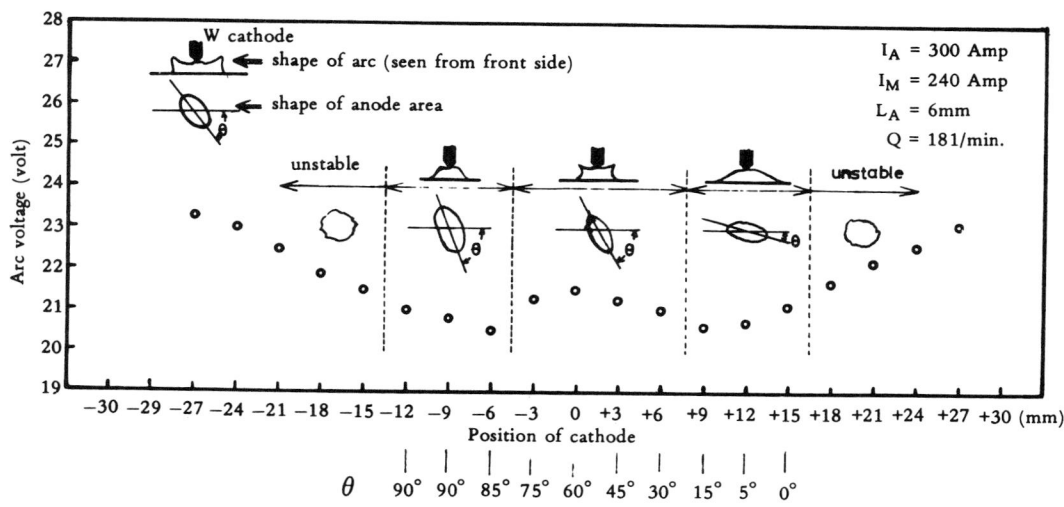

Fig. 8   Effect of positioning the arc in a cusp magnetic field on arc voltage, shape of arc, rotational angle θ, and on stability of arc.

Arc current: 300A, arc length: 6mm, field current: 240A, argon flow rate: 18 l/min.

3.2   Electrical properties of magnetized arc

For the typical influences of cusp type magnetic field upon the welding arc, change in electrical characteristics of the arc was, firstly, measured.

At any arc currents, arc voltage rised in accordance with the increase of magnetic fiedl strength, as in Fig. 9. Similar results were obtained for each arc length of 5.2 mm, 6.0 mm and 6.9 mm. Figure 10 shows the apparent potential gradient in arc column numerated from above results. As can be seen in this figure, the rate of increase of potential gradient was almost constant untill the field exciting current reached at around 7 A, but was somewhat reduced after the field current exceeded 7 A.

Immediate reason for this phenomena may be considered as the fall of field strength at arcing zone where arc plasma is compressed to. When the magnetic field is intensified gradually, arc column is compressed toward the central region of the magnetic field, but its field strength falls inversely as the distance from the pole pieces. Interaction between the arc and magnetic flux becomes, therefore, more weakly as the arc is compressed into a central portion of field. Relative confinement effect is thus lessened at a higher range of field exciting current, because the arc has already compressed sufficiently.

139

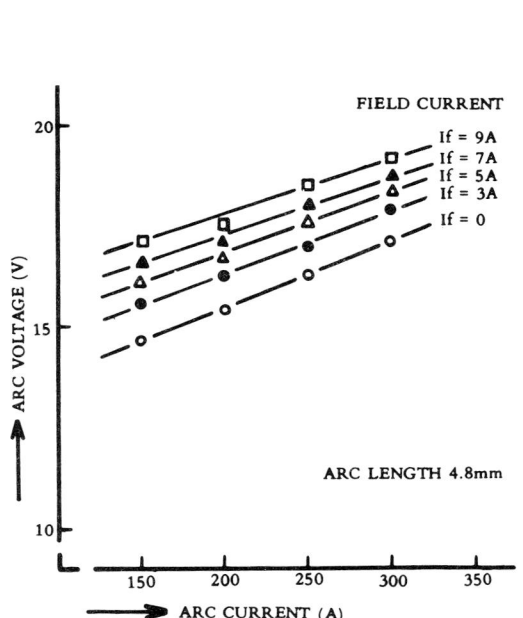

Fig. 9　Arc voltage-arc current characteristics of TIG arc in the various strength of cusp magnetic fields.

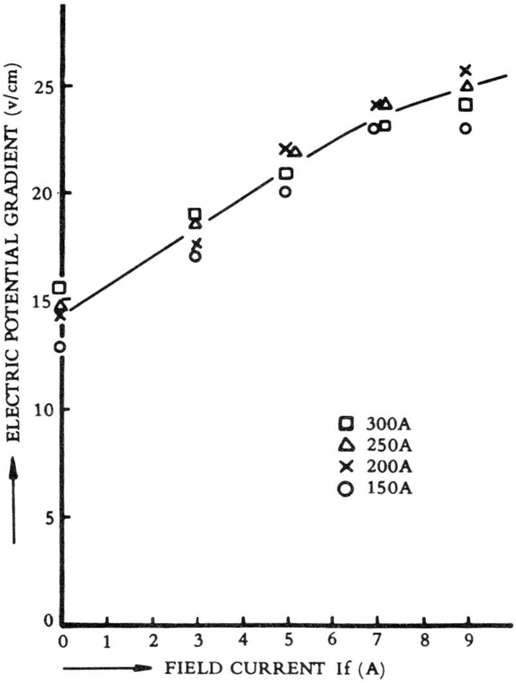

Fig. 10 '　Electric potential gradients in arc column under various strength of cusp magnetic fields.

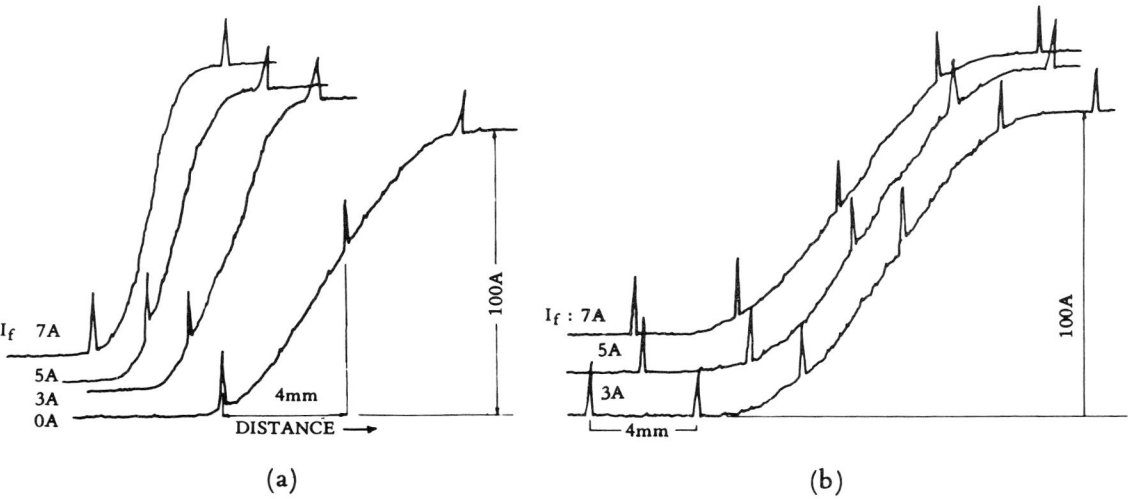

(a)　　　　　　　　　　　　(b)

Fig. 11　Arc current flowing into half anode in direction of (a) minor axis and (b) major axis. Arc current: 100A.

3.3 Distribution of arc currents

The cross-section of the arc column was observed to be elliptical under the influence of cusp magnetic field. In order to confirm this visual observation and to know actual distribution of the arc currents, a series of experiments was carried out with use of split copper anodes. Arc currents investigated were 100 A and 150 A, and arc length was 6 mm.

As an arc traveling across the boundary between the both half anodes from left to right, arc current flowing into the right-hand anode increases gradually untill the whole arc current is sustained. Typical changes of arc current flowing into half anode is illustrated in Fig. 11, in which signals indicating travel distance of 4 mm are superimposed upon the curve of arc current. Traveling velocity was fixed at

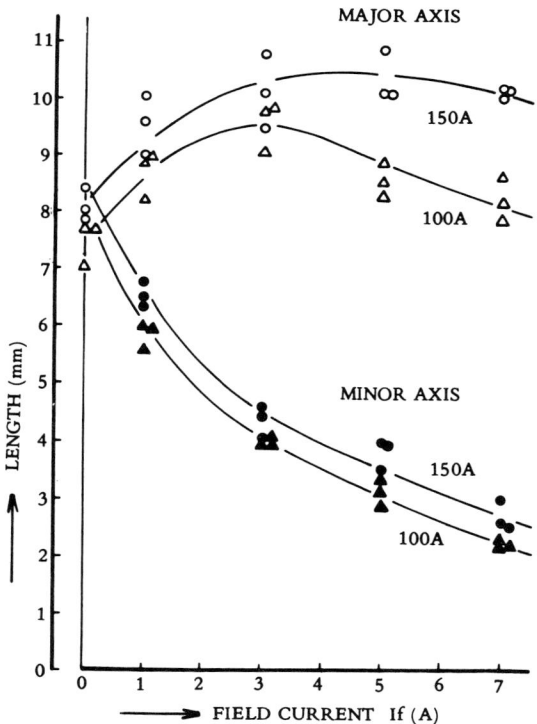

Fig. 12   Effective measure of arc column width in which
90% of arc current is included.

20 mm/min after preliminary inspections at various speeds.

If the major axis of elliptical arc is provided as parallel to the boundary of half

141

anode, the change of currents flowing into right-hand anode will be ended within a relatively short distance. Traveling distance across the anode spot is corresponded to the effective diameter of arc. It can be seen from Fig. 11 that the thickness of arc becomes thiner as the magnetic field is intensified. When the minor axis of the arc is set parallel to the anode boundary, marked change in width of arc cannot be observed even with high field current of 7 A or more. Figure 12 gives an effective measures of arc column in which 90% of the whole arc currents is flowing.

For a rotationally symmetric arc, lateral magnitude of arc current is converted to radial current density using the Abel inversion equation;

$$j(r) = -\frac{1}{\pi} \int_r^R \frac{I(x)dx}{\sqrt{x^2 - r^2}} \tag{1}$$

where $j(r)$ is local current density at radius r, and $I(x)$ is the first derivative of the measured lateral current. R is the effective radius of the arc from its axis to fringe, r is an arbitary radial distance from the arc axis in the range zero to R, while x is the distance to the axis from the center of the chord for which the lateral current is being measured.

Assuming that the current distribution inside the magnetized arc is completely elliptical as well as its appearance, it may be possible to evaluate the current density. The above assumption seemed to be valid from fact that the fusion line of weld crater was practically elliptical.

The Abel inversion was performed by a numerical calculation. In this calculation, elemental cross-sectional area was chosen as the area enclosed by the crossing lines by which major axis or minor axis is divided into ten equally length.

Distribution of arc current densitites at the anode region for the arc currents of 100 A are shown in Fig. 13. Stereographic illustrations of the current densities for the arc of 150 A are given in Fig. 14.

Current densities estimated for an open arc was not so high over a whole region of anode, and its maximum value at the center was still the order of 8 A/mm^2, but in case of magnetized arc, current densities increased extensibly. They reached up, for example, as much high value as 60 A/mm^2 for the arc current 150 A and the field current 7 A. Similar experiments were carried out for the arc above 150 A, but satisfactory result was not obtained, since the thermal damage of anode plate due to high heat influx was unavoidable.

It is of importance to note that the application of cusp magnetic field bring about not only change of arc shape, but the considerable increase of current density. Average current density over a whole area of arc is, of course, not so high as its peak value, but is still fairly high as twice or more compared with that in an open arc.

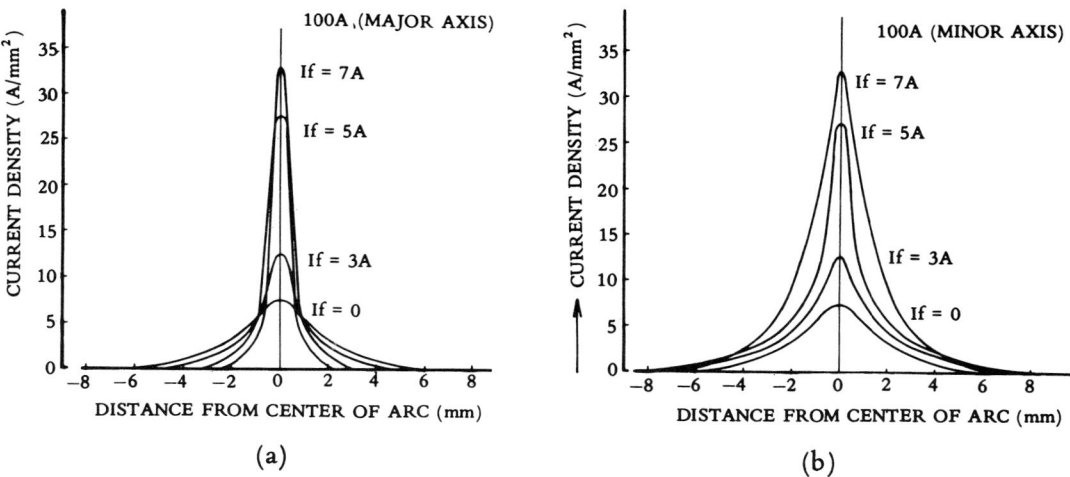

Fig. 13   Distribution of arc current density in each direction of major axis and minor axis under various strength of cusp magnetic fields. Arc current: 100A.

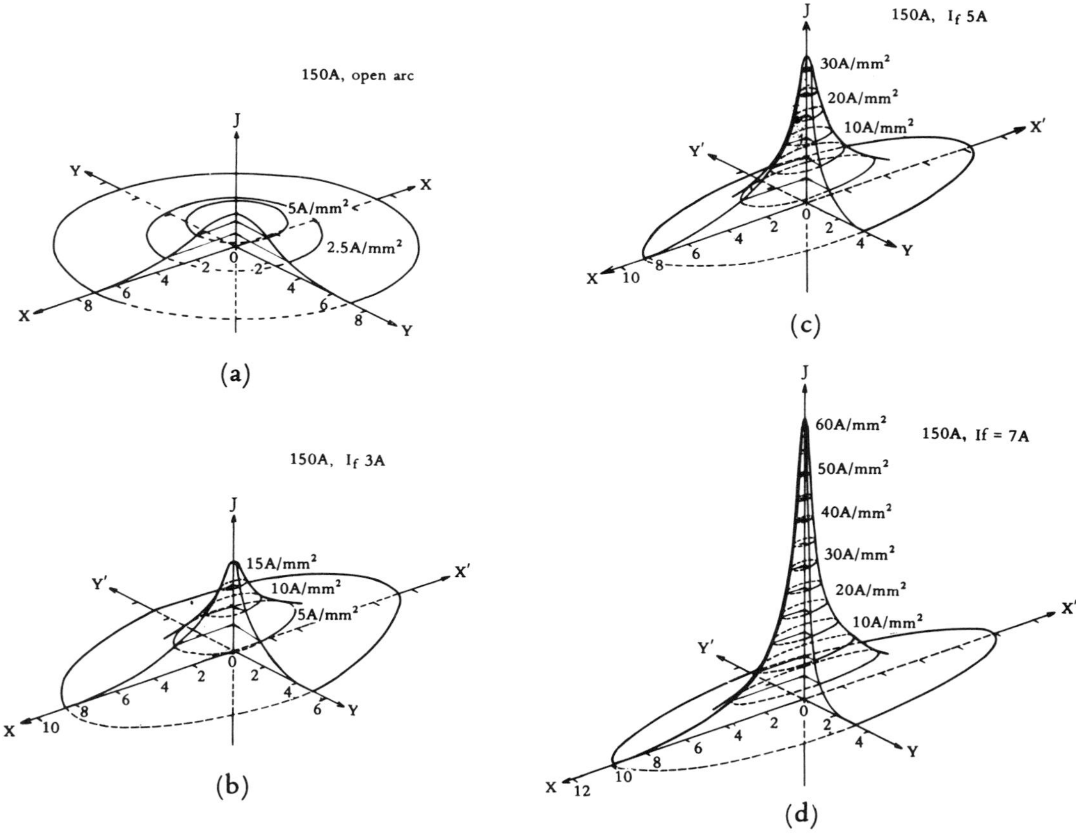

150A, open arc

(a)

150A, I$_f$ 3A

(b)

150A, I$_f$ 5A

(c)

150A, If = 7A

(d)

Fig. 14   Stereographic illustrations of arc current density of the open arc and of magnetized arc. Arc current: 150A.

4 Energy density of magnetized arc.

As previously described, both the potential gradient in arc column and the current ensity had fairly increased.

From empirical values of potential gradient, maximum or average current density nd of area of anode region, difference of energy density between the magnetized arc nd the open arc can be evaluated.

Relative values of energy densities in the magnetized arc is shown in **Fig. 15** as a unction of exciting current of electromagnet. In this expression, energy density of pen arc was taken as standard unit. It may be clear that the average power density, E^*, increased extensively as the magnetic field has become stronger.

As the cross-sectional shape of arc in a cusp magnetic field was elliptical, its eformation parameter can be defined as b/a, where a is the length of major axis, and is of minor axis. The practical values of a and b are able to find in **Fig. 12**. Then, ie relative power density of magnetized arc can be plotted with various deformation arameter, a/b, as shown in **Fig. 16**.

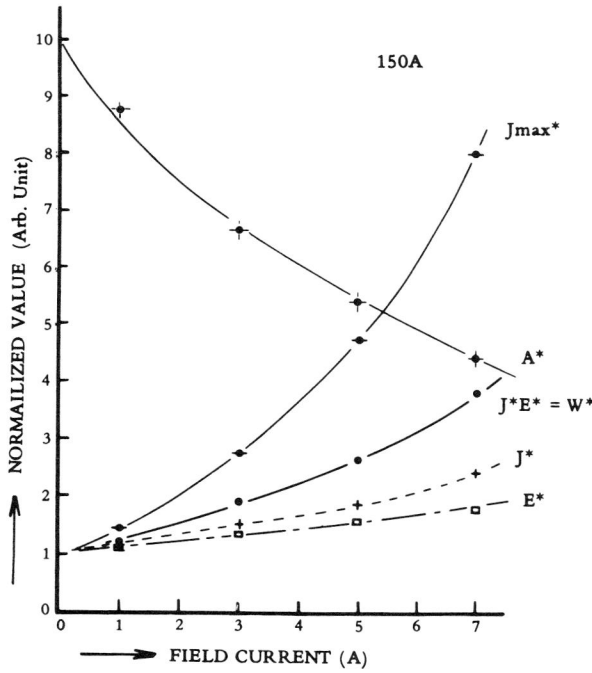

Fig. 15　Effect of magnetic field strength upon the relative values of cross sectional area of arc colum, A^*, average current density J^*, potential gradient, E^*, average power density, $J^* \cdot E^* = W^*$, and of peak current density, J_{max}^*.

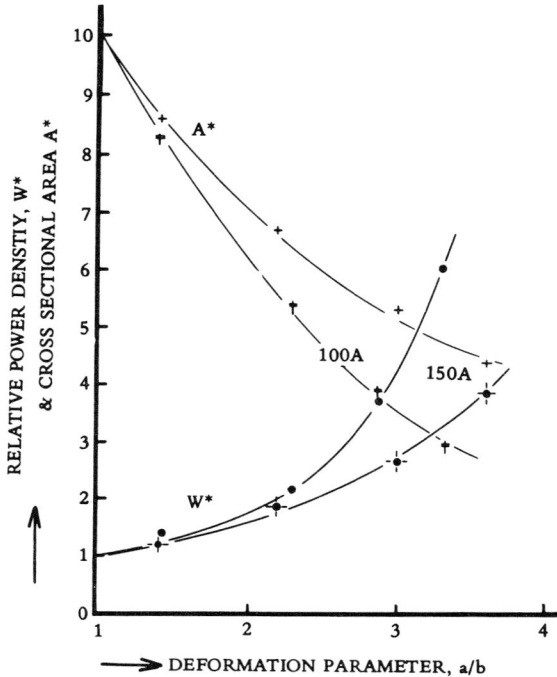

Fig. 16   Relationships among the relative power density, W*, cross
sectional area, A* and deformation parameter, a/b.

The shape of the arc column is determined by such many complex forces act on arc as electric, magnetic and also gas dynamic forces inside and/or outside the arc. It may be extremely difficult to establish the formula governing magnetic deformation of arc.

While, magnetized arc seemed to behave like an elastic body, since it recovers quickly to the initial state (open arc) when the magnetic field vanished. From this point of view, it may be feasible to analogize the energy accumulated in elastic body.[7] The following assumptions have been made in this treatment; (a) The arc is regarded as a straight conductor having an elliptical cross-section, (b) Arc column is assumed as a gaseous conducter having properties of an elastic solid, (c) Force acting on the arc column is concentrated at its focus.

Then, increment of elastic energy, ΔW, is described as

$$\Delta W \propto \frac{1}{2} k_a f^2 = \frac{1}{2} k_a a^2 e^2 \tag{2}$$

where k_a is elastic coefficient of arc, f is distance between origin and focus, a is length of major axis, and e is eccentricity of ellipse. Since the force is proportional to a

145

product of current, I, and magnetic field strength, B, equation (2) may be considered as equivalent to

$$\Delta W \propto \frac{1}{4} I \cdot B \cdot f = \frac{I \cdot B}{4} \cdot a \cdot e$$

The circumferential length of oval arc was observed to be nearly constant. Length of major axis, a, can be, therefore, connected to the initial radius, r_0, of the open arc,

$$a = \left(\frac{4}{3 + 3\sqrt{1 - e^2} - 2\sqrt[4]{1 - e^2}} \right) r_0 \tag{3}$$

Combination of equation (2) and (3) lead to

$$\Delta W \propto \frac{1}{4} \pi r_0^3 \left(\frac{4}{3 + 3\sqrt{1 - e^2} - 2\sqrt[4]{1 - e^2}} \right)^3 (1^2 - e)^{1/2} \cdot i \cdot B = \text{const} \cdot F(e) i \cdot B \tag{4}$$

where i is the average current density and is found to be proportional to the magnitude of magnetic field applied. Value of $F(e)$ can be calculated for various deformation parameter, a/b. The elastic coefficient, k_a, is denoted from equation (2) and (3) as

$$k_a = \frac{I \cdot B}{2f} = \frac{I \cdot B}{2a \cdot e} = \frac{\pi r_0}{2} \left(\frac{4}{3 + 3\sqrt{1 - e^2} - 2\sqrt[4]{1 - e^2}} \right) \left(\frac{1 - e^2}{e^2} \right)^{1/2} i \cdot B$$

$$\propto \text{const} \cdot f(e) \cdot B^2 = \text{const}$$

This equation means that $f(e)$ is inversely proportional to B^2.

Figure 17 shows the change of ΔW as a function of deformation parameter, and it exhibits similar trend as well as Fig. 14.

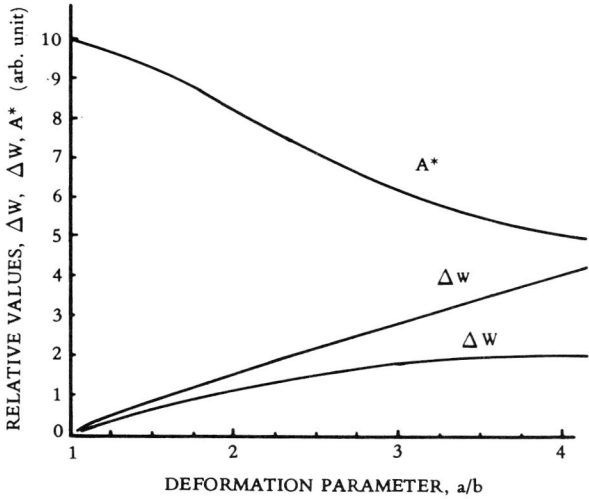

Fig. 17 Change of relative values of power density ΔW and Δw, and cross-sectional area A^* derived from elastic strain analysis.

146

3.5 Welding with cusp magnetized arc

All the weld made in tests were bead on plate type with no filler metal in straight polarity. Non-magnetic austenitic stainless steel plate were used in order to eliminate the change of magnetic field at the arcing zone.

Figure 18 shows the weld bead made with arc current of 150 A and traveling speeds of 60 mm/min and 160 mm/min under various strengthes of magnetic field. Traveling of the welding arc was quite stable at any field strength and traveling speed. As can be seen in figures, bead width and shape of weld crater were strongly influenced, that is, width of the crater was reduced by 40% when the magnetic field current was around 7 A, and longitudinal length of crater was somewhat prolonged by

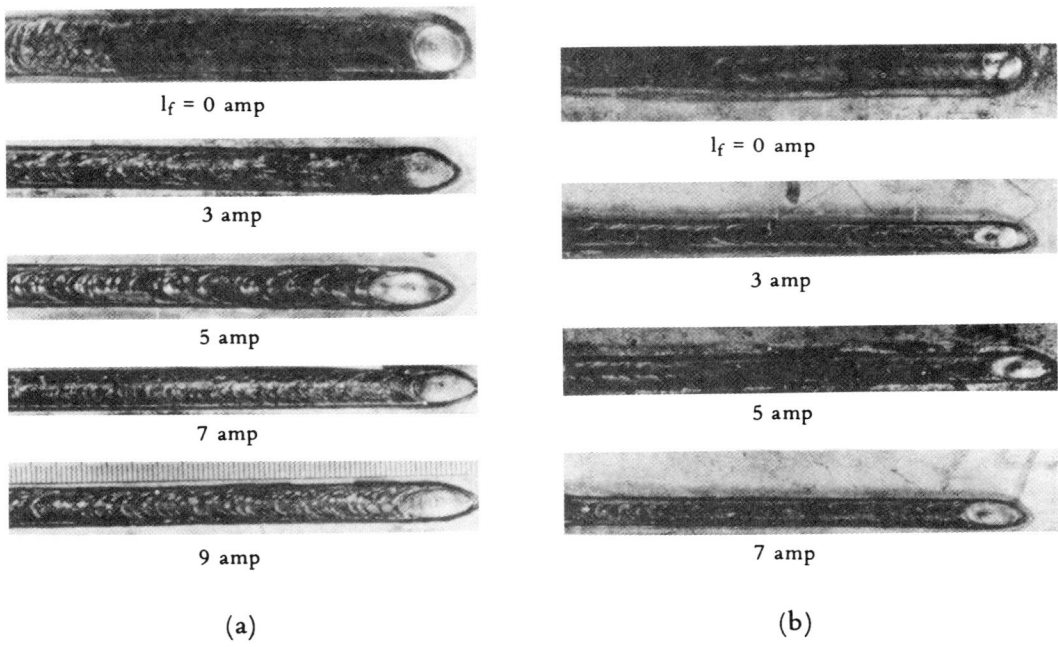

(a)   (b)

Fig. 18   Appearance of weld bead on 18–8 stainless steel plate, 10mm in thickness.
Arc current: 150A, travel speed: (a) 60mm/min, (b) 160mm/min.

the magnetic constriction.

Longitudinal and transverse cross-sections of weld bead are illustrated in Fig. 19. As can be seen in these macrophotographes, shape of transverse cross-section made by magnetized welding arc was different largely from that by an open arc, and became alike a deep coffee-cup shape.

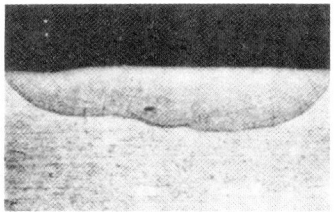

150A, l_f : OA, V = 120mm/min.

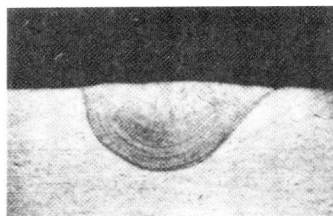

150A, 7A, 120mm/min.

Fgi. 19   Transverse and longitudinal cross-sections of weld bead.

Bead width observed under various welding conditions were summarized as Fig. 20. in which they were plotted as a function of arc energy per unit length of weld length. It may be easily understood that there is large difference between magnetized weld bead and non-magnetized one.

The area of fused metal in weld bead was, however, remained unchanged so far as arc energy supplied to unit length of weldment did not changed as indicated in Fig. 21.

From the view point of welding heat source, cusp magnetized arc may regarded as linear distributed surface heat source. Formation of a semi-circular transverse cross section (Fig. 19) can be explained well if surface linear heat source were existed along the central line of weld bead. The results obtained in current distribution measurement have also confirmed the validity of above explanation.

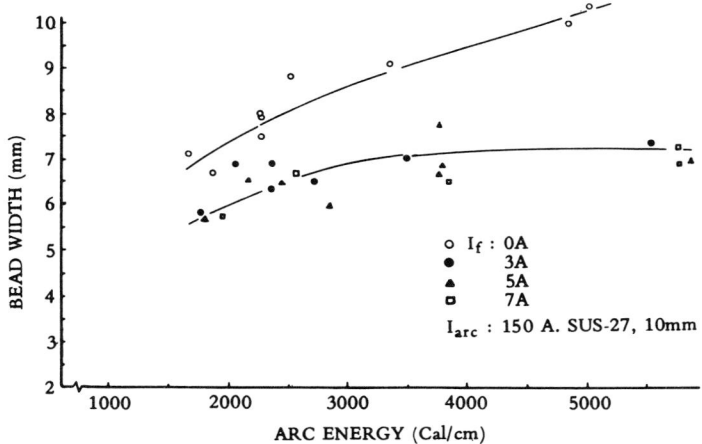

Fig. 20   Effect of cusp magnetic field on the weld bead width at various arc energy per unit length of weld

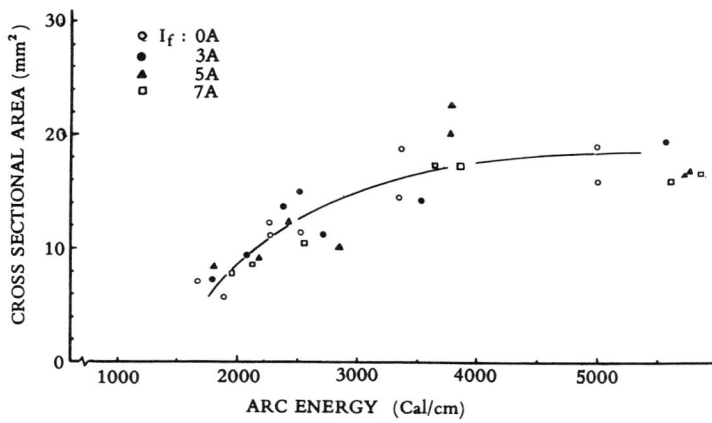

Fig. 21   Cross-sectional area of weld bead at various arc energy per unit length of weld.

4. Conclusion

Experimental results obtained in this series of experiments are summarized as follows;

(1)   Welding TIG arc changes its appearance by the superposition of cusp type magnetic field. Its cross-section varied from circular to elliptical according to the magnitude of cusp field applied.

(2)   Effect of magnetic field on electric properties of arc is investigated. Both arc voltage and potential gradient of arc column increase as the magnetic field becomes stronger.

(3)   Distribution of arc current at the anode was measured. Cusp magnetic field is much effective to raise the current density, and its peak values were as much high as 4—8 times of an open arc. Maximum current density was evaluated as 60 A/mm^2 for the cusp type magnetized arc of 150 A, while it was only 8 A/mm^2 for the open arc of same currents.

(4)   Energy density of arc is also increased by an application of cusp magnetic field. Change of energy density of arc was discussed analogically from view point of elastic strain energy.

(5)   The effect of cusp magnetic field on weld bead formation was investigated. By virtue of cusp magnetic field, bead width becomes narrower as the magnetic field becomes higher, but penetration depth becomes deeper.

(6)   Cross-sectional area of weld bead was, however, remained unchanged. In the formation of weld bead, cusp magnetized arc gives similar effect as the linear heat source distributed on the surface.

Referencess

1)   G.K.Hicken and C.E.Jackson, Welding Journal, Nov. 515s—524s (1966)

2)   I.A.Vachelis, Svar. Proiz. 7, 8—10(1963) Trans.BWRA, Svar. proiz. 1, 17—19(1965) Trans.BWRA

3)   I.M.Kovalef, Svar.Proiz. 10, 4—6(1965) Trans.BWRA

4)   V.S.Levakov and K.V.Lyubavskii, Svar.Proiz, 10, 9—12(1965) Trans.BWRA

5)   A.I.Akulov, L.K.Martinson and I.M.Kovalev, Svar.Proiz, 11, 43—45(1969) Trans.BWRA

6)   R.J.Perry and Z.Paley, Welding Journal, Sept. 389s—394s(1970)

7)   Y.Arata, Preprint of the Annual Meeting of Japan Physical Society, Oct. 8—11, 3, 9a—M—5 (1969)

Color Plates

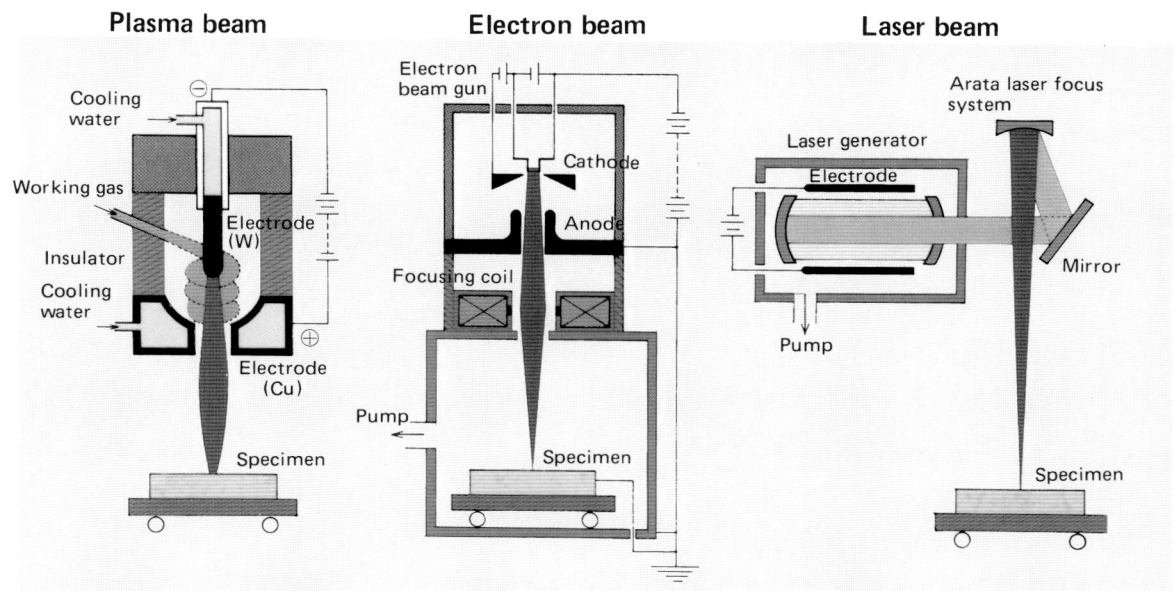

Fig. I Conceptual drawings of various ultra high energy density heat sources

Fig. II Relationship between beam energy density and welding penetration depth

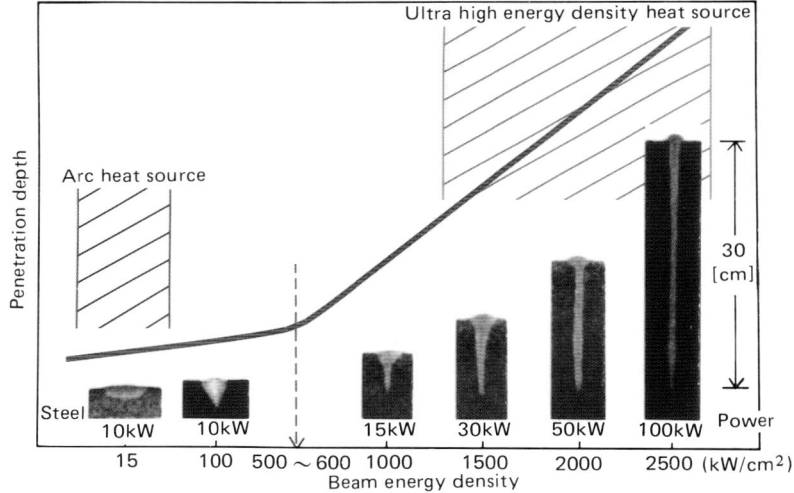

Fig. III Energy density and power (Continuous output heat sources)

	Heat sources	Energy density (kW/cm²)	Max. power (kW)
Light	Sunlight focus beam	1~2	—
	Arc light focus beam (Xenon lamp)	1~5	10
Arc	Arc (1 atm)	~15	50
	Plasma jet	50~100	200
	Point arc	~1000	—
High Energy Density Beam	Plasma beam (Gas tunnel plasma)	500~1000	500 (5,000)
	Electron Beam (Vacuum)	over 1000	300
	Laser Beam (1 atm)	over 1000	100

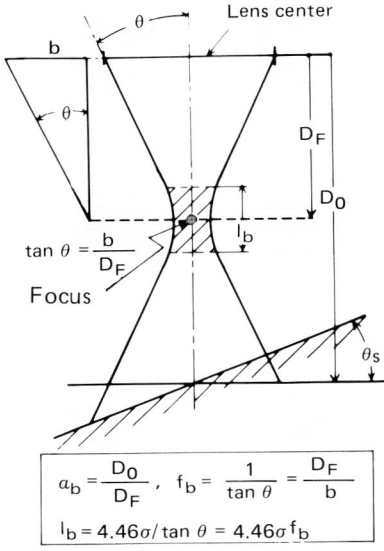

$$a_b = \frac{D_0}{D_F}, \quad f_b = \frac{1}{\tan \theta} = \frac{D_F}{b}$$

$$l_b = 4.46\sigma/\tan \theta = 4.46\sigma f_b$$

$$\tan \theta = \frac{b}{D_F}$$

Fig. IV   High energy density beam and its parameters. (refer to p. 194)

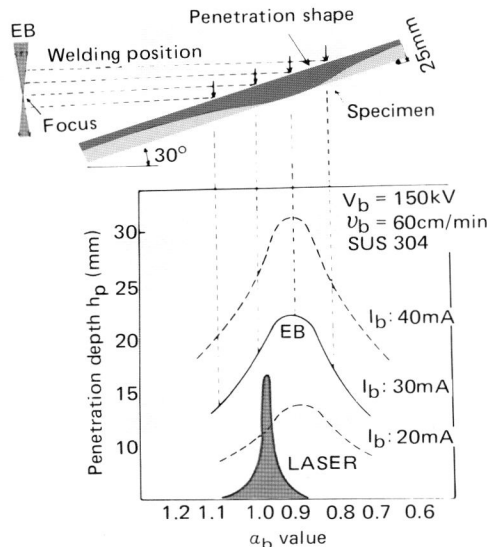

Fig. V   Relation between active parameter a_b and penetration depth h_p for various current I_b. (refer to P. 8)

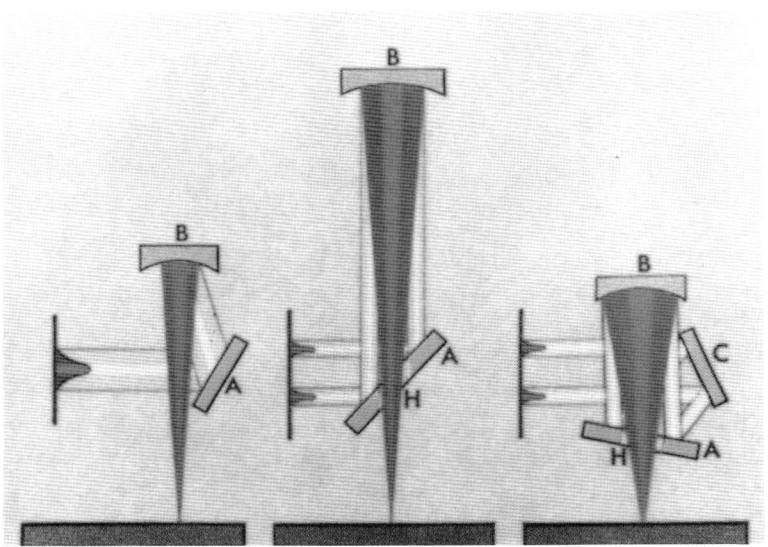

Fig. VI   Arata laser focus system (refer to p. 7 and 411)

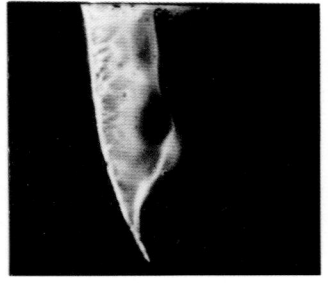 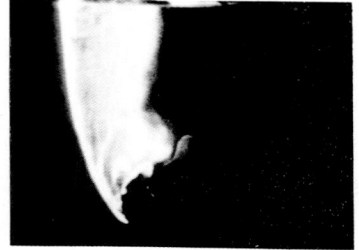 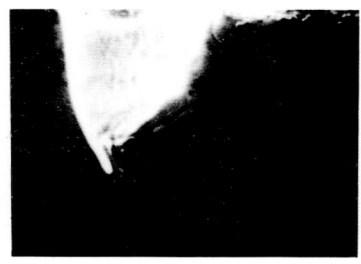

v_b = 100 cm/min v_b = 150 cm/min v_b = 250 cm/min

B) **High speed photograph of beam hole by optical method**
[Laser beam welding of glass, atmospheric pressure, 8000 frame/s]

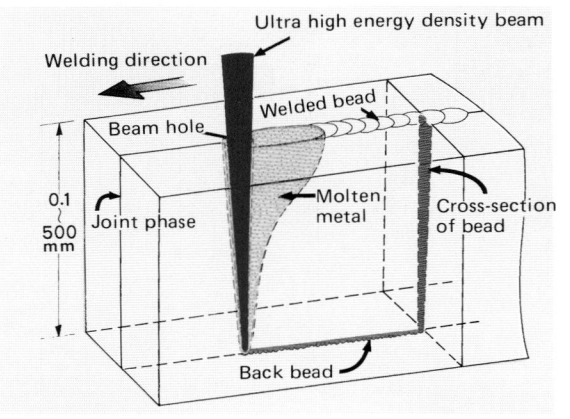

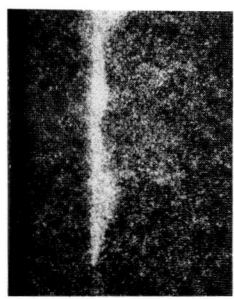   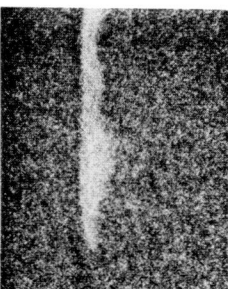

Electron beam welding   Laser beam welding

C) **High speed photographs of beam hole by the X-ray method (300 frame/s)**
[11 kW, v_b = 10 cm/min, HT80, p = 10^{-4} Torr]

A) **Principle of ultra high energy density beam welding**

Fig. VII   Dynamic behavior of beam hole during ultra high energy density beam welding

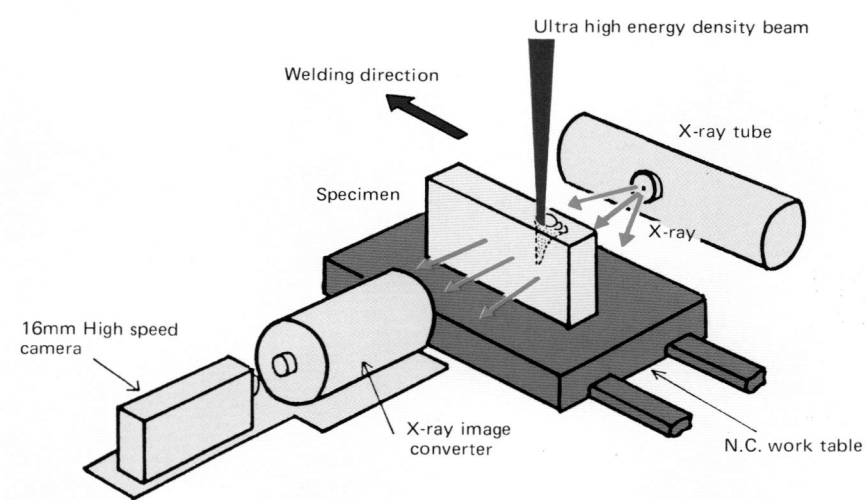

Fig. VIII   Dynamic observation of beam hole by X-ray method (refer to p. 411)

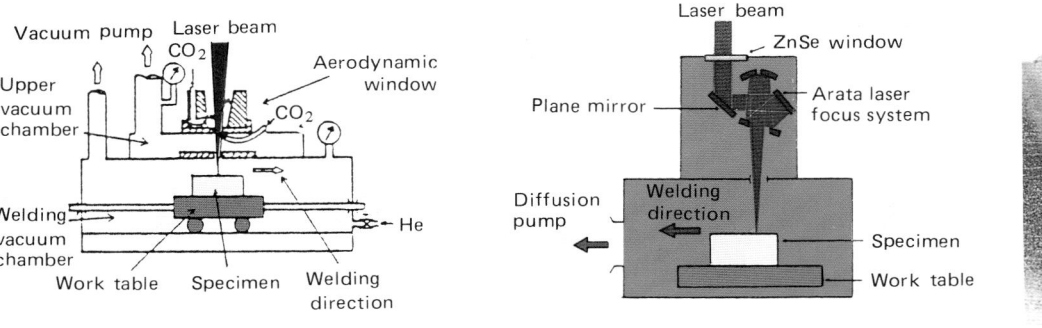

A) Partial vacuum type with aerodynamic window.

B) High vacuum type

C) Cross section of laser welding bead (SUS304, HT80 etc.)

Fig. IX Experimental apparatus of vacuum laser welding (refer to p. 519 and p. 522)

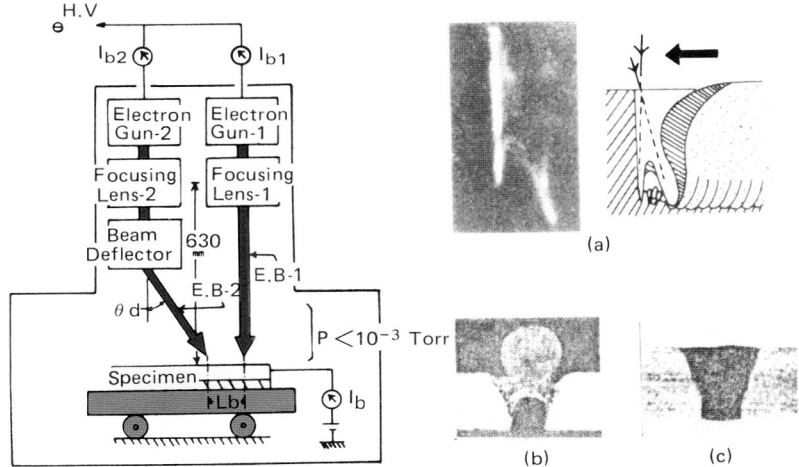

(a) Pinhole photograph and its explanation of beam hole X-ray

(b) Humping bead occurred in conventional high speed electron beam welding

(c) Sound bead obtain by tandem electron beam welding

Fig. X Tandem electron beam welding (refer to p. 393, 482 and 487)

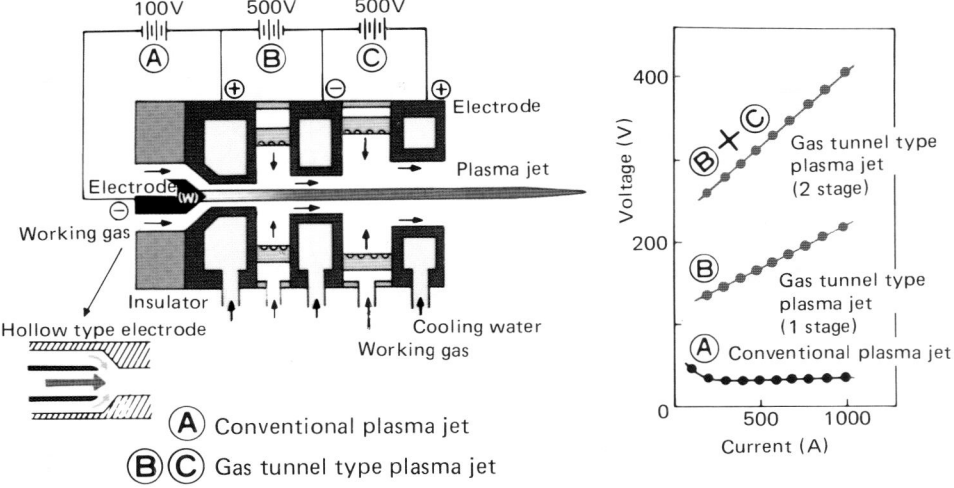

(A) Conventional plasma jet

(B)(C) Gas tunnel type plasma jet

Fig. XI Gas tunnel type high power plasma jet (refer to p. 167 and 168)

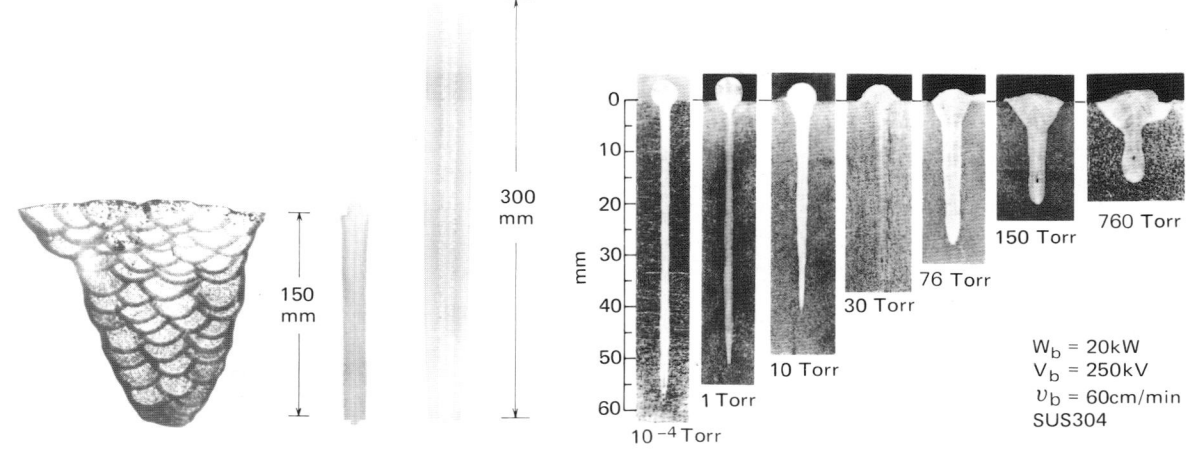

150 mm

300 mm

Arc welding (55 pass)

Electron beam welding
(1 pass)

W_b = 20kW
V_b = 250kV
v_b = 60cm/min
SUS304

Fig. XII Comparison of cross section profile of welded zone

(Steel of pressure vessel, thickness 150,300mm)

Fig. XIII Cross section of electron beam welding bead at various pressures

(refer to p. 102)

Fig. XIV Local vacuum type electron beam welding for HT60 steel pipe

⎡HT60, diameter: 1400mm⎤
⎣thickness: 50mm   ⎦

Fig. XV High pressure body of deep-submarine exploror (Electron beam welding)

⎡10Ni-8Co, diameter: 2100mm ,⎤
⎣ thickness: 80 mm   ⎦

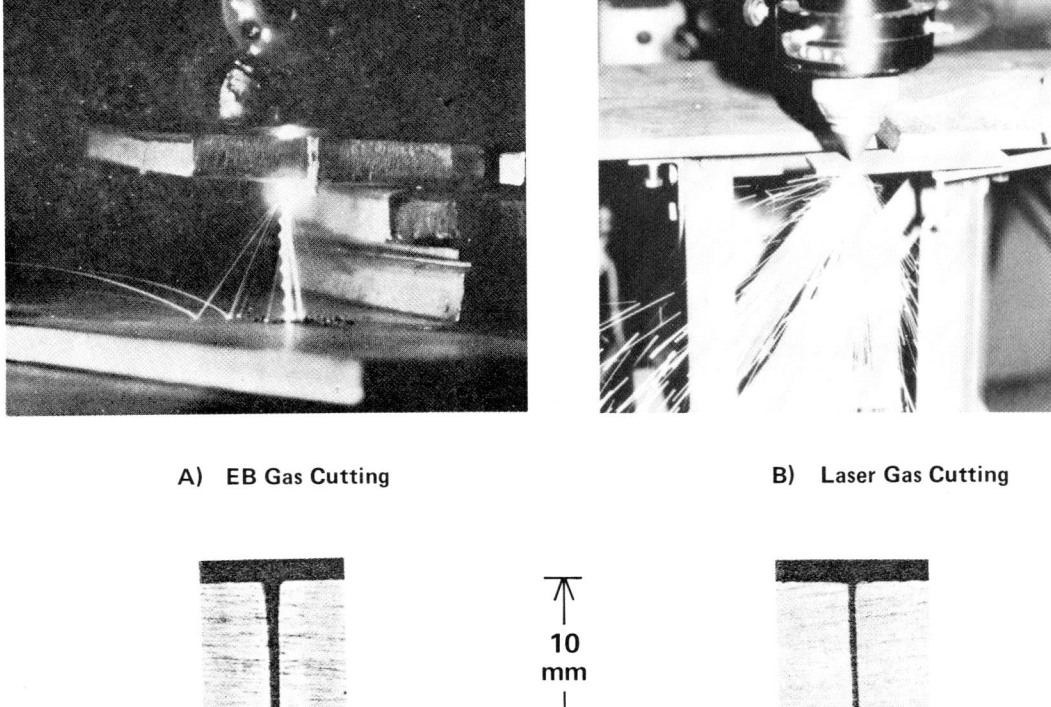

A) EB Gas Cutting

B) Laser Gas Cutting

10 mm

Cross section
(Stainless steel)

Cross section
(Stainless steel)

Fig. XVI High Energy Density Beam Gas Cutting (refer to p. 36)

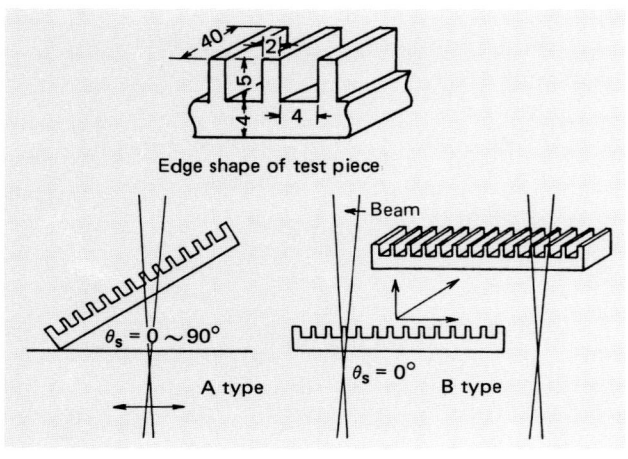

A) **Schematic drawing of "AB test"**

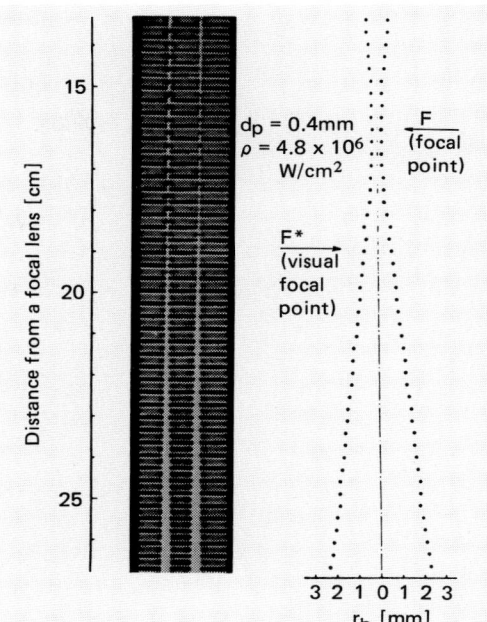

B) **Measured beam shape**
$[V_b = 150kV, I_b = 40mA, v_b = 60cm/min]$

Fig. XVII AB test (Arata Beam Test Method on beam shape decision)
(refer to p. 9, 177 and 197)

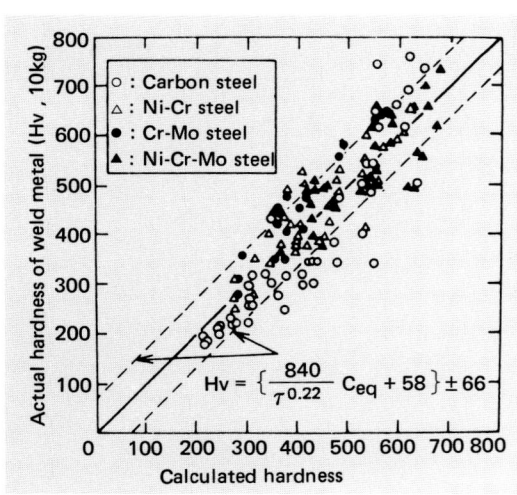

$$H_v = \frac{840}{\tau^{0.22}} C_{eq} + 58$$

$$[C_{eq}] = [C] + \frac{[Mn]}{2.4} + \frac{[Si]}{24} + \frac{[Ni]}{14} + \frac{[Cr]}{16} + \frac{[Mo]}{60}$$

$$\tau_{800\to500} \fallingdotseq 3.8 \times 10^{-2} \left[\frac{0.8 I_b V_b}{v_b h_p}\right]\left[\frac{1}{(500-T_o)^2} - \frac{1}{(800-T_o)^2}\right]$$

A) **Arata electron beam weldability**
(refer to p. 18, 326 and 333)

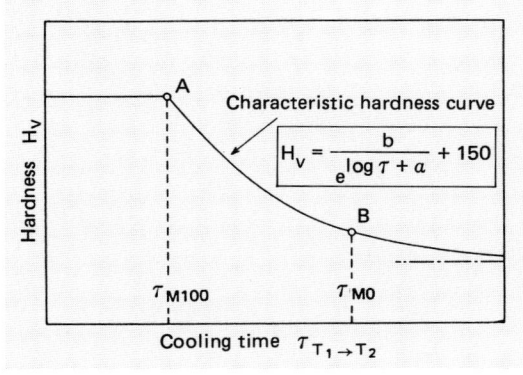

A: $H_v(\tau_{M100}) = 835[C] + 287$

$\log\tau_{M100} = 2.55([C] + \frac{1}{6.3}[Mn] + \frac{1}{3.6}[Si]) - 0.92$

B: $H_v(\tau_{M0}) = 273([C] + \frac{1}{13}[Mn] + \frac{1}{9.7}[Si]) + 133$

$\log\tau_{M0} = -0.37([C] - \frac{1}{1.1}[Mn] - \frac{1}{0.44}[Si]) + 1.02$

B) **Characteristic hardness curve**
(refer to p. 18 and 334)

Fig. XVIII Arata Weldability by hardness prediction

Generation of Point Arc and its Characteristics

Abstract

A new method for constricting arc plasma is proposed. Extermely high current density and high power density of arc plasma can be obtained continuously by this method, in which the dynamic pressure, the thermal pinch effect and the "strip effect" of cool gas are utilized.

The structure of such condensed anode zone is clarified and the·existence of the "arc ball" or "micro arc column" whose steady current density can be easily elevated about 10^5 [A/cm²] or more and·power density becomes 4.5×10^6 [Watt/cm³] is confirmed. This arc ball is steady and stable. The authors named such arc, in which the exposured arc ball or micro arc column is existing steadily and stably, the "point arc", and distinguish it from an ordinary arc.

The point arc can be applied to heat processings such as welding, cutting and so on as a new heat source, and one example is shown.

1. Introduction

The current density in open arc is kept constant even if arc current is increased, but it is increased condiderably by constricting with a water cooled nozzle of metal. It is so called plasma arc.

The current density of open arc column is usually $10 \sim 10^3$ [A/cm²], that of plasma arc is $10^2 \sim 10^4$ [A/cm²], and that of Gerdien Arc, whose current is 1500 [A], is 5×10^4 [A/cm²].[1] The current density of anode spot is $30 \sim 10^3$ [A/cm²], which is slightly higher than that of arc column, in many cases.

However, the various contrivance for generating method of arc makes its current density higher. For example, the steady arc with discontinuous travelling anode spot has current density of $10^3 \sim 10^4$ [A/cm²].[2] Considerably high current density of $10^4 \sim 10^6$ [A/cm²] is obtained as current density of cathode spot with the atmospheric pulse arc of $1 \sim 20$ [μsec] pulse width.[3] Many attempts have been made to increase the current density of plasma arc. Authors obtained the good result by applying steady magnetic cusp field to plasma arc.[4, 5] Furthermore, they succeeded in obtaining much higher current density by generating the "point arc". That is reported as follows.

2. Generation of Point Arc and Its Characteristics

A usual method for generating arc is shown in **Fig. 1.** Investigations on cathod spot, anode spot and the structure of arc including them have ever been made to the arc generated in this manner.

One of the authors proposed the method shown

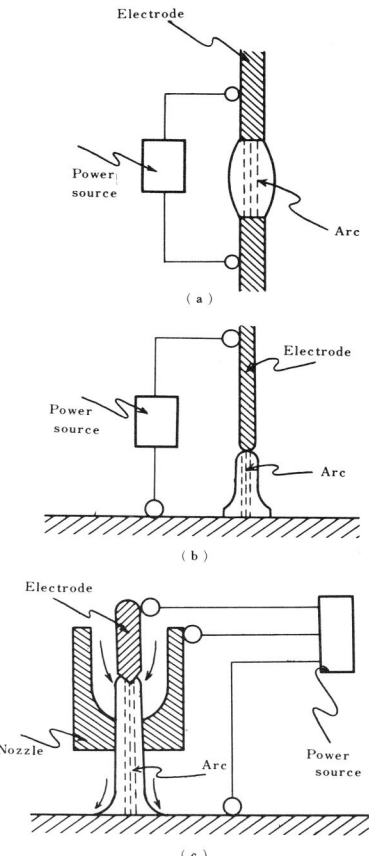

Fig. 1. Usual method to generate open arc ((a), (b)) and plasma arc ((c)).

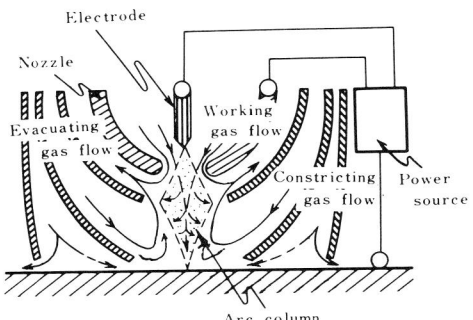

Fig. 2. Principle for generating "point arc".

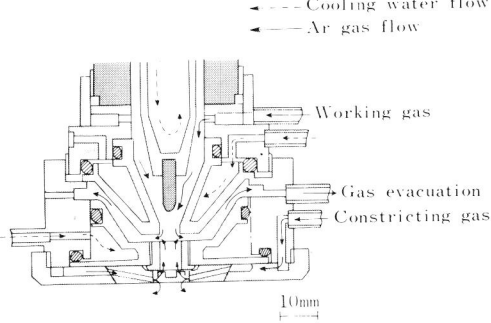

Fig. 3. Cross sectional view of point arc torch.

in **Fig. 2,**[6] to generate the arc of much higher order current density, made the device - the torch - as a trial and experimented for generation of arc to some degree with his co-werkers in 1965.[7] The principle of this method to be described referring to Fig. 2.

Cool inert gas is blown against the arc column and the electrode zone of ordinary plasma arc centripettally and spirally, which generates between the rod and the plate electeode and is constricted by the metal nozzle, from the circumferential inlet and forms the constricting gas flow. Gas evecuation is made radially through the circumferential outlet arranged near by the inlet at the same time. The part of the constricting gas flow, together with the part of the working gas flow supplied along the electrode into the arc column, change into the evacuating gas flow by the evacuation action and forms the flow reverse to the working gas flow. These gas flows are shown with solid lines in Fig. 2. The arc column is cooled and affected by the thermal pinch effect remarkably, and the gass flow strips the arc column of its thermal sheath and plasma flame, so called the "strip effect" occurs, under this condition for the gas flow. The compression effect due to the dynamic pressure of the constricting gas flow comes accompanied with furthermore. It is expected to obtain an extra fine arc column and the higher order current density, owing to the multiplication effect of these. Authors made the torch, whose structural appearance is shown with the principal cross sectional view in **Fig. 3,** and tried to generate the arc in conformity to the above mentioned principle.

The straight polality connection, in which the rod electrode of the torch is the negative pole and the plate electrode (the work) is the positive pole, was adapted, and the investigation on the structure and the behavior of the anode spot and its neibouring zone was made as a first step. The gas flow, presumed in the inner part of the torch at the arc

generation, is shown with solid lines in Fig. 3. (The water flow is also shown with broken lines.) It proves strongly effective on the stabilization of the arc to blow the constricting gas against the arc column to its tangential direction, as shown in **Fig. 4,** from the heading part of the torch. A conventional vacuum cleaner was used for the gas evacuation through the outlet. The gas flow rate was adjusted with the needle valve installed on the route pass to the outlet.

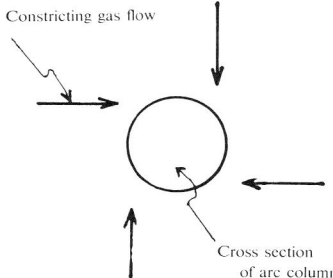

Fig. 4. Constricting gas flow.

Various aspects of the arc which was generated on the water cooled copper plate with the torch of trial made are shown in **Photo. 1,** in which **(a)** corresponds to an ordinary plasma arc without the constricting and the evacuating gas flow and **(b), (c), (d) (e), (f)** and **(g)** correspond to the case of the evacuating gas flow formation, with the constricting and the evacuating gas flow.

The "strip effect" is not sufficient in (b) of these photographs, for lack of the evacuating action, and not sufficient in (c), in which the arc ball growing up is not realized. The existence of the complete, steaby and stable "arc ball" or "micro arc column" can be confirmed in (d), (e), (f) and (g) in Photo. 1. The arc

152

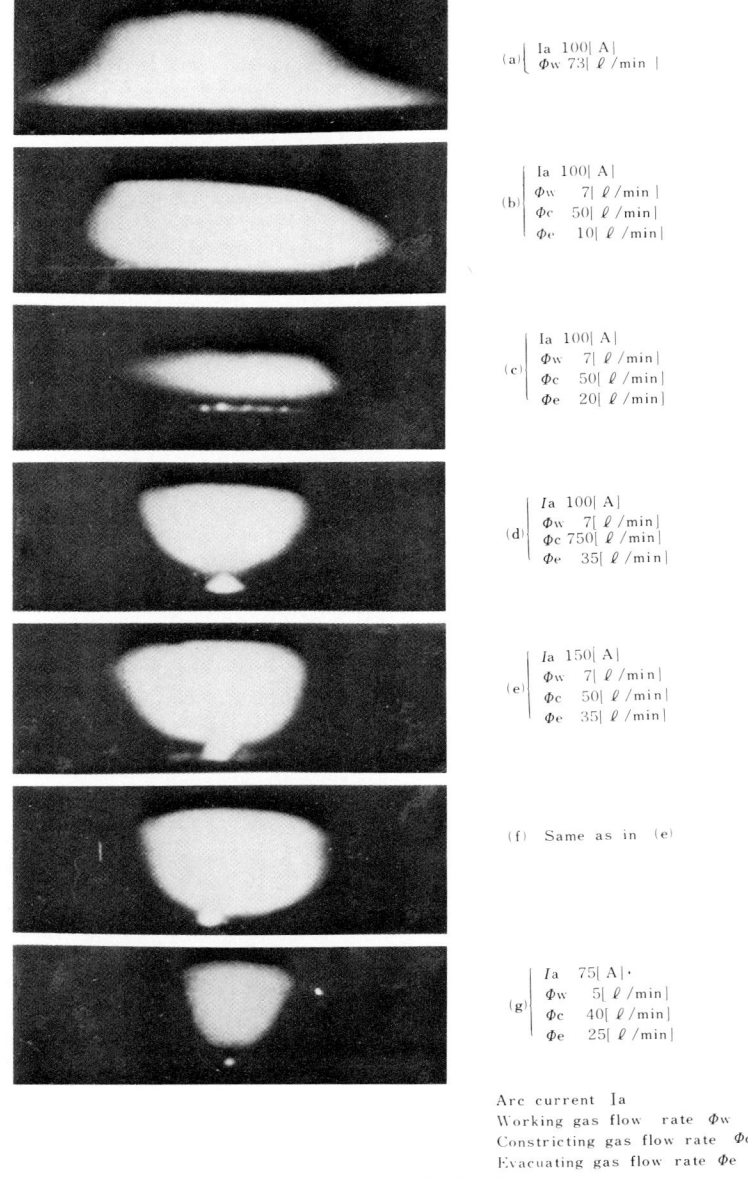

(a) | Ia 100[A]
      Φw 73[ℓ /min]

(b) | Ia 100[A]
      Φw 7[ℓ /min]
      Φc 50[ℓ /min]
      Φe 10[ℓ /min]

(c) | Ia 100[A]
      Φw 7[ℓ /min]
      Φc 50[ℓ /min]
      Φe 20[ℓ /min]

(d) | Ia 100[A]
      Φw   7[ℓ /min]
      Φc 750[ℓ /min]
      Φe 35[ℓ /min]

(e) | Ia 150[A]
      Φw 7[ℓ /min]
      Φc 50[ℓ /min]
      Φe 35[ℓ /min]

(f) Same as in (e)

(g) | Ia 75[A]·
      Φw 5[ℓ /min]
      Φc 40[ℓ /min]
      Φe 25[ℓ /min]

Arc current Ia
Working gas flow rate Φw
Constricting gas flow rate Φc
Evacuating gas flow rate Φe

Photo. 1. Various aspects of arc constricted by gas flow.

current is higher in (e) than in (d), the rotating axis of thé gas is not accord with the center axis of the arc current is somewhat low in (g) in these photographs. We can affirm that the "arc ball" appears at the final growing up stage of the anode spot, judging from look of a series of these photographs. The evacuating

gas flow rate is proper if it is about eighty percent of the total flow rate of the working gas and the constricting gas, more evacuation than this is not only effective on the constriction for the arc, but also causes the arc to be unstable, furthermore, to be in danger to damage the torch, by air inclusion and promotion of oxidation and exhaustion of the electrode due to it. The anode spot grows up to the arc ball of high brightness and sharply outlined and its shape and size, can be measured fairly with accuracy as a result of the "strip effect", as is seen in (d), (e), (f) and (g) of Photo. 1. It is presumed that the current density is elevated to 10^5 [A/cm^2] or more if the most of arc current concentrates on the arc ball. Authors named such arc the "point arc" in order to distinguish from an ordinary arc.

The I-V characteristics of the point arc are shown in **Figs. 5** and **6.** The I-V characteristic of the point arc is compared with that of the plasma arc in Fig. 5. The arc voltage of the former is about 10 [V] higher than that of the latter. It will be proper to suppose that the most of the increment voltage is loaded on the arc ball. The electric power density of the arc ball becomes $4.5 \sim 10^6$ [Watt/cm^3] if we assume the arc ball is spherical in shape, its diameter is 0.75 [mm], the arc current is 100 [A] and the voltage of 10 [V] is loaded on it.

The same I-V characteristics as in Fig. 5 are shown for several kinds of values of the working gas flow rate in Fig. 6. The constricting effect of the gas flow decreases, the arc voltage for the same arc current goes down and the external appearance of the arc changes from that corresponding to (d) of Photo. 1 for that of (a) gradually, as the working gas flow rate increases. The ratio of the constricting gas flow rate to the working gas flow rate is the important factor which controls the constricting effect.

An example for the macro etching of the weld base cross section is shown in **Photo. 2 (a),** it is obtained from the bead on plate of SUS 304, 6 [mm] in thickness. Weld was made under the condition written in addition to Photo. 2.

The dynamic pressure of the working gas flow against the anode spot is considerably high in case of an ordinary plasma arc welding, the plasma gas stream penetrates the work, and blows out through the key hole, and the burn through would occur due to too high heat input per unit weld length if weld was made under the same welding condition as for the point arc, written in Photo. 2. However, the dynamic pressure against the anode spot scarcely exists owing to the reverse direction flow to the working gas in case of the point arc welding, and the burn through does not occur.

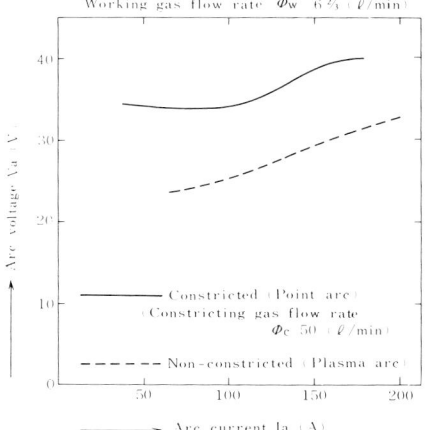

Fig. 5. Comparison of I-V characteristic of point arc and of plasma arc.

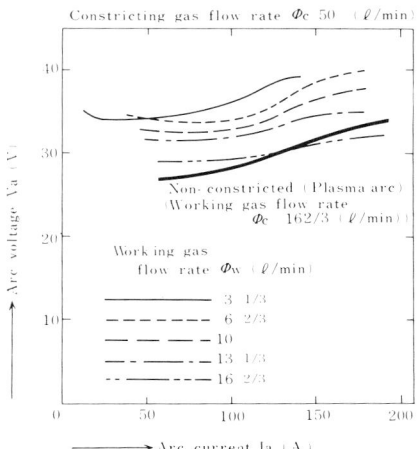

Fig. 6. Effect of working gas flow rate on I-V characteristic of point arc.

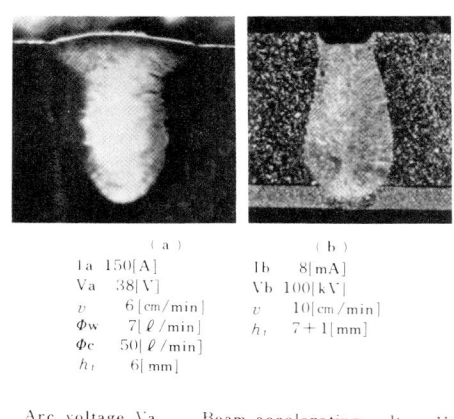

(a)		(b)	
Ia	150[A]	Ib	8[mA]
Va	38[V]	Vb	100[kV]
v	6[cm/min]	v	10[cm/min]
Φ_w	7[ℓ/min]	h_t	7+1[mm]
Φ_c	50[ℓ/min]		
h_t	6[mm]		

Arc voltage Va   Beam accelerating voltage V$_b$
Welding speed v Beam current   I$_b$
Thickness of work h_t
Material of work : Stainless steel SUS 304

Photo. 2. (a) Bead cross section by point arc welding.
　　　　 (b) Bead cross section by small output atmospheric electron beam welding.

The bead cross section by the point arc welding is about 3 [mm] in width and 5 [mm] in depth, it resembles that of the atmospheric electron beam welding of small power in shape and size, as is seen in Photo. 1 (a) and (b).[8]

The shape of the weld bead cross section in the photograph is compared with the theoretical contour line map calculated on the basis of the heat conduction theory. The calculation is made for the generalized travelling line heat source which has a certain heat input distribution along the vertical direction (Z direction) to the plate of definite thickness to obtain peak temperature T_m at any place in the medium (the plate).

The contour line map for T_m can be figured according to the result of the calculation. Authors give the distribution function for the line heat source as,[9]

$$Q^*(Z^*) = (1-Z^*)^\alpha \text{------------ (1)}$$

where

Q^* ; relative heat source distribution function

Z^* ; dimensionless depth

h_t ; plate thickness [mm]

α ; heat source distribution index

under the welding condition in Photo. 2, the dimensionless welding speed is

$$v^* = \frac{v \cdot h_t}{2k_D}$$

where

k_D; thermal diffusivity [cm²/sec]

v ; heat source travelling speed (welding speed) [cm/sec]

The obtained contour line maps for this value of v^* and a few values of α is shown in **Figs. 7, 8** and **9.** Each value of α is written in addition to these figures. The abscissa Y^* in these figures expresses the dimensionless distance for the vertical to the heat travelling direction on the plate from the heat source.

The contour line in Fig. 9 is most similar to the shape in the photograph among these figures, but the heat distribution along the Z direction corresponding to the α value in Fig. 9 is rather different from a typical point source which we can see in Photo. 1 (d), (e), (f) and (g) and it is expressed by the equation $Q^* = \sqrt{1-Z^*}$ as a line heat source. The reason why the heat distribution is like as in Fig. 9 can be considered as follows.

1) The arc ball entered into the work (the plate) to some extent when it is melted.

2) The heat is not only trasmitted from the arc ball directly, but from the plasma flame by convection and

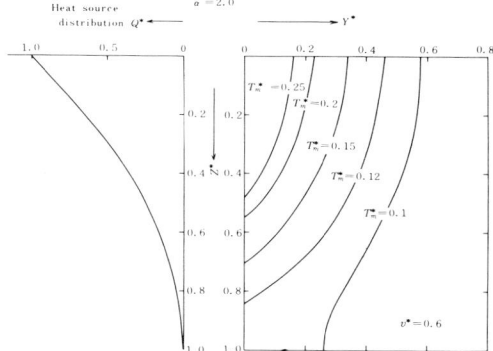

Fig. 7. Contour line map of peak temperature due to generalized line heat source.

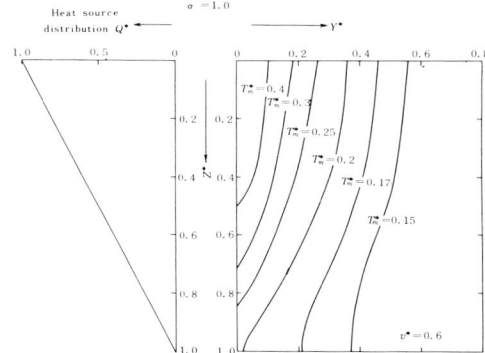

Fig. 8. Contour line map of peak temperature due to generalized line heat source.

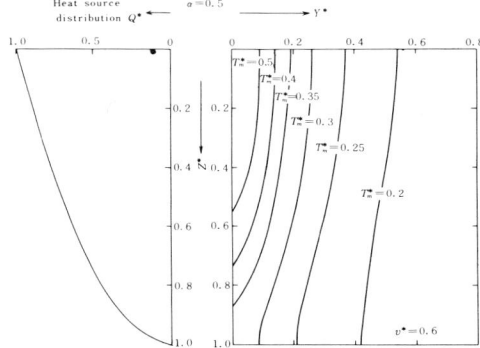

Fig. 9. Contour line map of peak temperature due to generalized line heat source.

radiation.

3) The downward stream of the molten metal caused as the reaction to the metal vaporization transports the input heat.

3. Conclusion

1) The singular arc is obtained by the action of the constricting and evacuating gas flow which are produce by the specially made arc torch.

2) The anode spot of this arc grows up to the "arc ball" or the "micro arc column" which is highly bright and definite in shape and size of 0.5~1.0 [mm] in diameter owing to the "gas dynamic pressure effect" the "thermal pinch effect" and the "ship effect" of the above mentioned gas flow. These effects are much influenced by the supplied gas flow rate. Authors named this arc the "point arc".

3) The voltage of the point arc is about 10 [V] higher than that of an ordinary plasma arc for the same arc current and the same clearance between the electrodes.

4) The power density put into the arc ball is estimated to be 4.5×10^6 [Watt/cm^3] if we assume that all the voltage increment is loaded on the arc ball whose shape is spherical and diameter is 0.75 [mm] and that the arc current is 100 [A].

5) The shape of the bead cross section by the point arc welding resembles that by the atmospheric electron beam.

6) The line heat source distribution function is obtained by using heat conduction theory and by the trial method comparing the calculated result with the weld bead configuration.

7) Consideration is made on the reason why the line heat source distributes in that manner.

8) The authors dare to presume that the dark part we can see between the arc column and the anode in the photograph of the point arc corresponds to the anode dark space at the discharge in vacuum, but the detailed investigation on it is the problem in the next step.

References

1) H. Gerdien and A. Lotz: Z. f. techn. Physik, 4, 157 (1948).

2) W. B. Kouwenhoven and T. B. Jones: J. A. W. W, 27, 470-S (1948).

3) J. M. Somerville and W. R. Blevin: Phys. Rev., Vol. 67, 982 (1949).

4) Y. Arata: Preprint of the Anuual Meeting of Japan Phys. Soc. Oct. 8-11, 3, 9-M-S (1969).

5) Y. Arata and H. Maruo: Tech. Reppts. Osaka University, Vol. 22 No. 1039 (1972). : JWRI, Vol. 1, No. 1 (1972).

6) Y. Arata: Private report, Patent 567-605 (1965) (in Japanese).

7) Y. Arata and K. Inoue: Preprint of the Annual Meeting of Japan Weld. Soc. P. 109–200 (1967) (in Japanese).

8) Y. Arata and M. Tomie: Tech. Repts. Osaka University, Vol. 12, No. 6 (1967). : JWS, Vol. 1, No. 1 (1970).

9) Y. Arata and K. Inoue: JWRI, Vol. 2, No. 1 (1973).

"Fundamental Characteristics of Stationary Plasma Arc in Gas Tunnel"

Abstrast

The stationary plasma arc in gas tunnel has been investigated and the characteristics are clarified experimentally.

At first the mechanism of the formation of gas tunnel by strong vortex flow with big flow rate was studied theoretically. And on the pressure distribution in the vortex chamber, the results showed good agreement with the experimental results in the annular region of the vortex chamber.

The plasma arc produced in this gas tunnel has high temperature and high electron density as compared with conventional plasma arc, because of strong thermal pinch effect. The characteristics of this plasma arc has been mainly carried out by spectroscopic measurement. As the result almost fully ionized plasma has been obtained in gas tunnel at even small current of 200 A, by means of shortening the vortex chamber length.

KEY WORDS: (Plasma Arc) (Vortex Flow) (Pressure Gradient) (Stationary) (Fully Ionized)

1. Introduction

Study on plasma arc at an atmospheric pressure has been done by many workers up to this time[1,2]. And, for the stabilizing methods of plasma arc are used a lot of kind of methods including gas rotation[3], water stabilizing[4], wall stabilizing[5] and so on. Now, among those methods, the stabilization by gas rotation is most popular[6]. But on the conventional method, the flow rate of the working gas is a little (about 50 ℓ/min) and the pressure gradient of radial direction in the cylindrical chamber is small, and the influence of the pressure gradient on plasma has not been investigated in detail.

On the other hand, in this study, by gas jet having big flow rate, more than 300 ℓ/min, from fine nozzles, a strong vortex flow is formed in the cylindrical chamber. And using gas divertor nozzles at both ends of the chamber, the pressure-wall having sharp pressure gradient in radial direction is produced. The pressure at the center region surrounded by the pressure-wall is very low and the region was named vortex gas tunnel or "gas tunnel"[7,8]. In this gas tunnel, plasma arc is produced and investigated in detail. By this method, two effects are given to plasma arc. One of them is that the ionization of the working gas is easy due to the low pressure in the center region, and the other is that the stability of plasma beam is improved remarkably by the sharp pressure gradient in the radial

direction.

Therefore, in this study, the mechanism of the formation of "gas tunnel" produced in the center region of vortex flow having sharp pressure gradient is clarified. And the characteristics of the plasma arc in the gas tunnel are investigated experimentally. In addition, for the purpose of production of stationary fully ionized plasma beam whose temperature is more than a few ten thousands Kelvin degree, the experiment has been carried out by means of the increase of electric input.

2. Mechanism of Formation of Gas Tunnel

In this section the mechanism of the formation of the vortex gas tunnel has been investigated theoretically and experimentally.

In the case that the gas as fluid is moving as vortex motion in the cylindrical chamber, two models which are shown in **Fig. 1** have been considered in this study. One of those is ordinary type of vortex motion, "model I", so called "forced vortex". And the other is vortex motion, "model II" which has a strong sink in the center axis of the vortex chamber[9].

In model I, by the viscosity of the fluid, the gas rotates with a constant angular velocity from the chamber wall to the center, as a rigid body. Then the pressure distribution in the chamber is presented as follows[9].

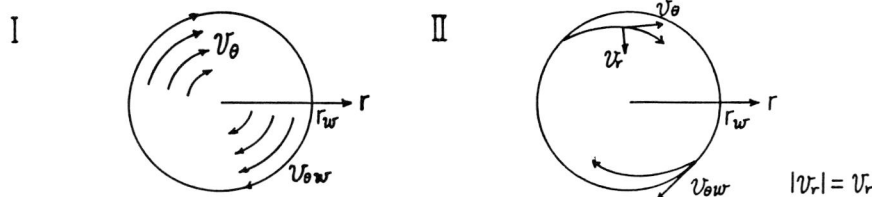

Fig. 1 Two models of vortex flow in the cylindrical chamber, where I is "model I" and II is "model II".

$$\frac{p}{p_w} = \exp\left[\frac{1}{2}\frac{v_{\theta w}^2}{RT}\left\{\left(\frac{r}{r_w}\right)^2 - 1\right\}\right] \quad (1)$$

where p is pressure, p_w is the pressure at the chamber wall, $v_{\theta w}$ is the velocity in the tangential direction at the wall, r is radius, r_w is the radius at the wall, R is gas constant, and T is temperature.

On the other hand, in the case of the vortex flow of model II, the gas is drawn into the sink in the center of the chamber and therefore, the radial velocity is generated. Here, the viscosity of the working gas can be neglected. And, the pressure distribution is presented by the following equation[9].

$$\frac{(p/p_w)^2 - (v_{rw}/v_{\theta w})^2}{1 - (v_{rw}/v_{\theta w})^2}$$
$$\doteqdot \exp\left[-\frac{1}{2}\frac{v_{\theta w}^2}{RT}\left\{\left(\frac{r_w}{r}\right)^2 - 1\right\}\right] \quad (2)$$

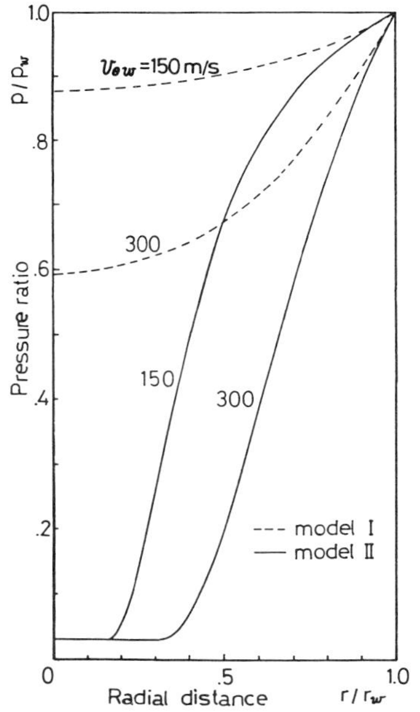

Fig. 2 Theoretical pressure distributions in the radial direction

These theoretical results of the both models are shown in **Fig. 2**. Thus, the characteristics of the two models, model I and model II, are very different each other. In model I, the pressure distributions are like a parabola, and those bottoms are broad. But, in model II the pressure gradient is very sharp, besides the pressure of the center p_o is very small value. Anyway, in both types, the pressure gradient is sharper in the annular region, as the gas velocity at the wall is higher.

In order to realize these two types of vortex flow, the simple experiments are carried out using two types of the

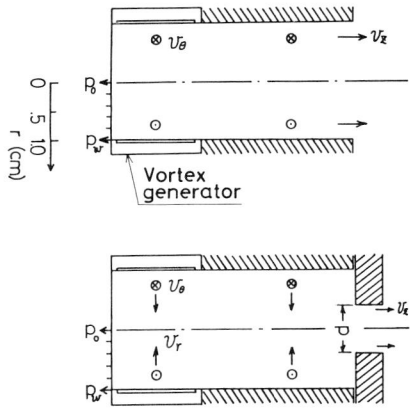

Fig. 3 Experimental apparatus for realizing two types of vortex flow, model I and model II

apparatus shown in **Fig. 3**. The diameter of cylindrical chamber is 20 mm, and the length is 40 mm, respectively. And vortex generator is located at the chamber wall. Air, the working gas, flows into the chamber in tangential direction from the vortex generator, and forms vortex flow in the chamber. The one side of the chamber is closed and the pressure is measured at these side points. In model I, the other side is open, but in model II, there is a nozzle called gas divertor nozzle.

The results in model I are shown in **Fig. 4**. This shows that the velocity at the wall is 150 m/s. In comparison with theoretical results, experimental data of pressure appear a little high in the center region, and the velocity v_θ is different from constant angular velocity. But the characteristics of pressure distribution show good coincidence with theoretical result.

In **Fig. 5** the results of experimental pressure distribution in the case of model II are shown. Those plots show experimental data, where each diameter of gas divertor nozzle is 13 mm, or 6 mm. Those curves shown by dotted lines, are good coincidence respectively with theoretical line in the annular region. In the case of small diameter of gas divertor nozzle, $v_{\theta w}$ is small value, and the coincident annular region becomes larger. In this case, the value of

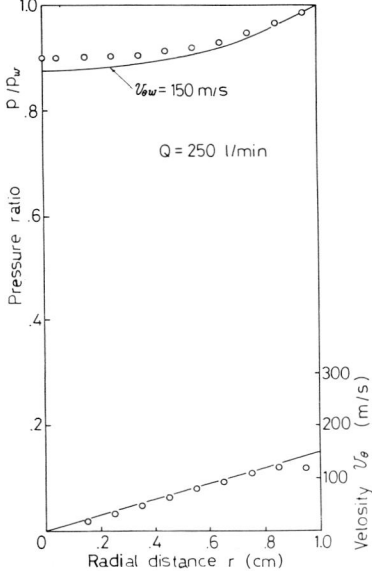

Fig. 4 Experimental pressure distribution in model I

the pressure at the axis is very small, and this low pressure region including the axis, surrounded by the gas wall having large gradient has been named "vortex gas tunnel" or simply "gas tunnel" as already described.

In **Fig. 6**, the each pressure at the center axis and the wall of the chamber is shown corresponding to gas flow rate. Here, the diameter of gas divertor nozzle is 8 mm. As the flow rate is bigger, large pressure difference of Δp ($= p_w - p_o$) can be obtained. But, in the case that the pressure outside of gas divertor nozzle is an atmospheric pressure, "without pump", the gas flows into gas tunnel in the axial direction from the gas divertor nozzle. Therefore, the value of the pressure at the axis is limited, about 300 Torr in the case of 300 ℓ/min.

But, the pressure at the axis can be lower value by

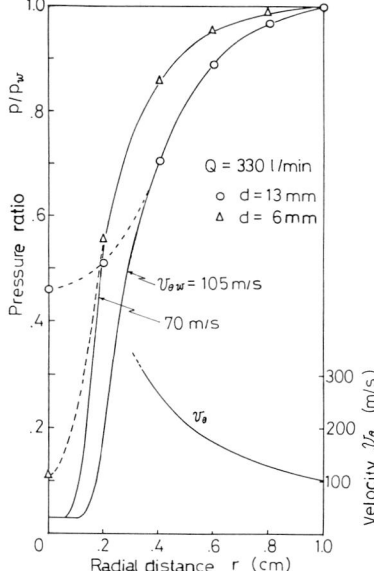

Fig. 5 Experimental pressure distributions in model II

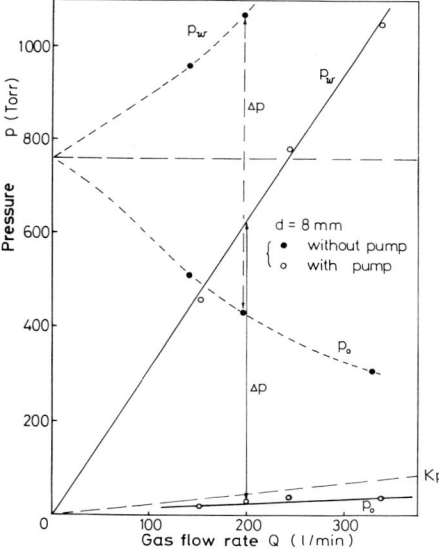

Fig. 6 Dependences of pressure in the gas tunnel upon gas flow rate

means of preventing the counter flow using vacuum pump. This case is shown in Fig. 6 as "with pump", and the pressure at the axis is about half of the capacity (characterstic line, Kp) of the vacuum pump. Besides in this case the pressure difference is almost the same in comparison with "without pump".

The typical pressure distribution is shown in **Fig. 7**. In this case, the gas flow rate is 340 ℓ/min, the diameter of gas divertor nozzle is 8 mm, and this experiment has been carried out by using a vacuum pump. The pressure at the axis is less than 40 Torr against the pressure at the wall, 1050 Torr, and the ratio p_o/p_w is about 0.04.

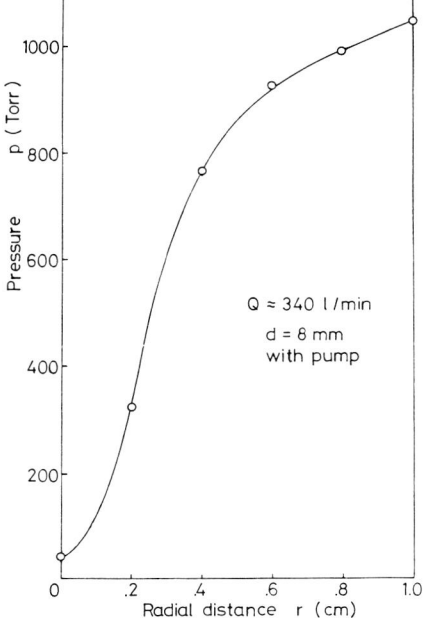

Fig. 7 Typical pressure distribution in the radial direction

These characteristics of the gas tunnel are very useful for the plasma production and stabilization, and thermal pinch of plasma arc.

3. Experimental Apparatus

The schematic diagram of experimental apparatus applied in this study is shown in **Fig. 8**. The discharge tube whose diameter is 20 mm is so called "cascade type" and constituted from seven segments insulated electrically. The "vortex generator" which supply the working gas to the chamber in the tangential direction is located at the center of the discharge tube. And at the both sides, gas divertor nozzles made of boron-nitride, whose diameter is 8 mm, are located.

We call the area which surrounded by the wall of the discharge tube and two gas divertor nozzles, "vortex chamber". The location of gas divertor nozzles can be easily changed. So for the length of vortex chamber i.e. chamber length is selected each value 7.4 cm, 3.4 cm,

1.4 cm. As the electrodes, tungsten is used for the cathode, copper for the anode. The distance between those electrodes is long sufficiently in comparison with plasma diameter. Besides, the applied power source is direct current type and the load voltage is 120 V and electric current is 500 A.

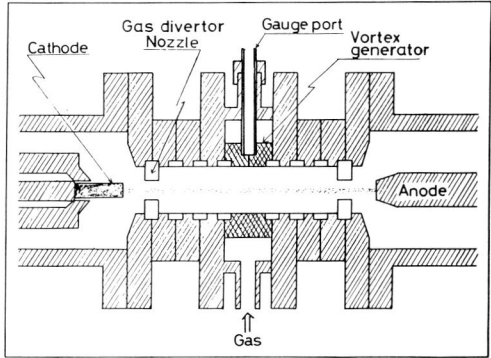

Fig. 8 Schematic diagram of plasma generator

Next we explain the concrete methods of the experiment of this study. At first, argon gas spouting in the vortex chamber through the vortex generator's nozzles forms a vortex flow with high velocity. The gas, as rotating with high velocity, is exhausted to the both outsides of the vortex chamber through the gas divertor nozzles by a vacuum pump having large capacity. Then the vortex gas tunnel at low vacuum pressure as shown in Fig. 7 is produced. At this time, the tungsten cathode is inserted in the vortex chamber, and a high-frequency discharge starts plasma arc. After production of the plasma arc, the cathode is pull out to the certain location, and the gas flow rate and electric current are adjusted proper values.

The measurement for physical parameters of this plasma has been done through the guage port and through windows located at the center of the vortex chamber. As the probes, thermocouple, pitot tube, and so on, are used in order to measure the distribution of temperature, electric potential and pressure at the region surrounding plasma. And the measurement of plasma parameter has been done by spectroscopic method, which is indicated by block diagram in **Fig. 9**. Where electron density of plasma is decided from the stark broadening of hydrogen spectral lines of Balmar series, H_α, H_β. The plasma temperature is obtained by Saha's equation assuming thermal equilibrium of the plasma, and by the relative intensity ratio of spectral lines of argon ion ArII. Besides the plasma diameter is decided by the half width of spatial distribution of ArII line intensity.

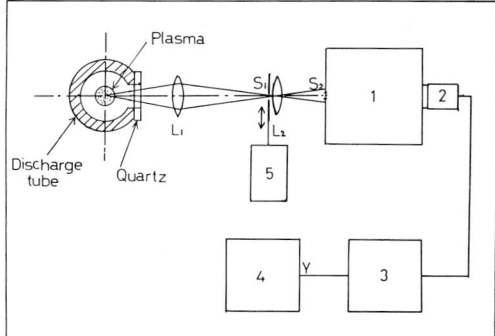

Fig. 9 Block diagram of spectroscopic measurement
1: Spectrometer, 2: Photomultiplier, 3: Amplifier,
4: X Y recorder, 5: Slit scanner

4. Production of Fully Ionized Plasma

4.1 Characteristics of plasma beam in gas tunnel

Fig. 10 shows heat losses from the plasma beam in the both cases, conventional method by gas rotation ("without gas divertor nozzle") and this method by strong vortex flow ("with gas divertor nozzle"). These measurement values are obtained by the increase in temperature of the cooling water for each segment of the plasma apparatus. In the case of "without nozzle", about 28% of the total heat loss is heat loss in radial direction. On the other hand, in the case of "with nozzle", radial heat loss is a little, and the rate against total heat loss is about 4%, very small value. This reason is that the strong vortex flow

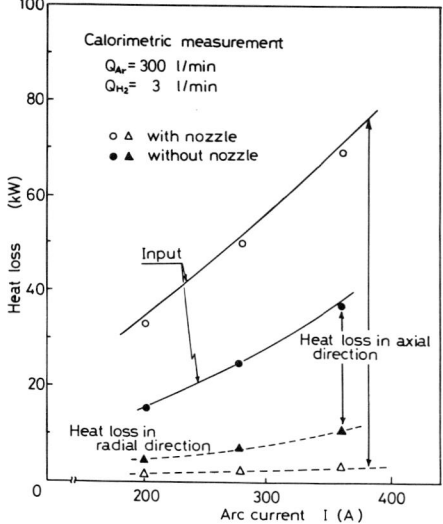

Fig. 10 Characteristics of heat loss from plasma beam

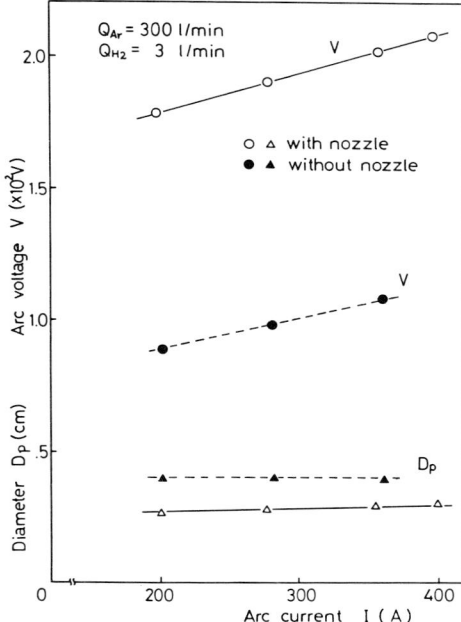

Fig. 11 Dependences of plasma diameter and arc voltage on arc current

has a large radial velocity component. This velocity component remarkably suppresses radial heat loss by thermal conduction and generates strong thermal pinch effect. As the result, high temperature and high electron density plasma whose heat loss is dominated by convection, and whose diameter is very fine, has been produced.

Then, the plasma beam diameter and/or "plasma diameter" defined by the half width of spatial distribution of ArII is shown in **Fig. 11** as a dependence on electric current. It is clear that the plasma diameter in this study using the gas divertor nozzles is finer than that by means of the conventional method at the same current. At 400 A the diameter Dp is 2.5 mm. In Fig. 11, the characteristics of the arc voltage are also shown. In the case of "with nozzle", the arc voltage is about two times of that of "without nozzle".

Then, **Fig. 12** shows the comparison of plasma temperature and electron density, between the both cases. Dotted lines show the values in the case of "without nozzle", and real lines, "with nozzle". As the electric current increases, the electron density of the plasma increases in both cases, but in the latter case, the value is about two times bigger than that in the former case. And in the case of "with nozzle", the plasma temperature increases largely. This result shows that the thermal pinch effect for the plasma is stronger in the case of "with nozzle".

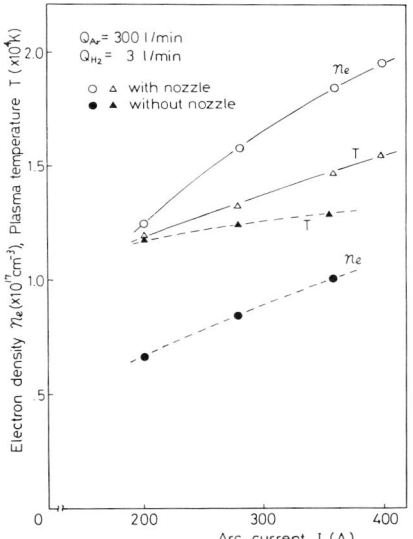

Fig. 12 Dependences of plasma temperature and electron density on arc current.

It is considered that these results are due to the strong vortex flow. So, the pressure in the region surrounding the plasma was measured by a pitot tube. The results is shown in **Fig. 13**. In the case of "without nozzle", the radial

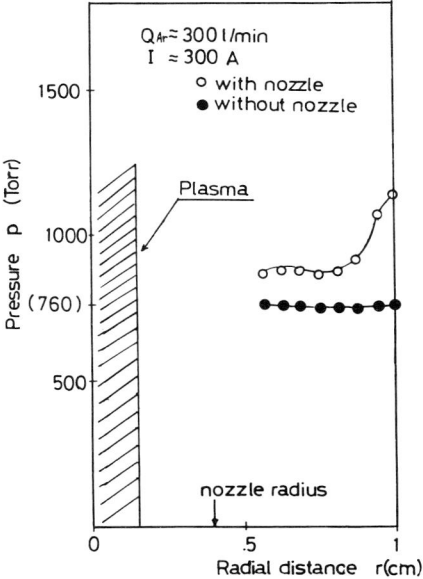

Fig. 13 Results of pressure measurement by pitot tube

pressure gradient is almost zero in the region surrounding plasma. On the other hand, in the case of "with nozzle", as shown in the same figure, the pressure gradient near the wall of vortex chamber is maintained even though plasma exists. Consequently, the plasma stability shows a good result, and the thermal pinch effect is very strong. So, plasma having high electron density and high temperature can be obtained easily.

Now, as described above, according to the increase of argon flow rate, dp/dr increases, and then the role of vortex flow for the stabilization of the plasma increases, and at the same time, the thermal pinch effect is stronger. **Fig. 14** shows the experimental results of the characteristics of the plasma parameters. At this time the electric current is a constant value, 240 A, and the flow rate of argon gas is changed in a range of $100 \sim 300$ ℓ/min. According to the increase of gas flow rate, the thermal pinch effect by the strong vortex flow is stronger, and the plasma diameter gradually decreases. On the other hand, the arc voltage increases, and as the result, the electron density and the temperature of the plasma increase. Corresponding to this result, ArII intensity increases as the gas flow rate increases.

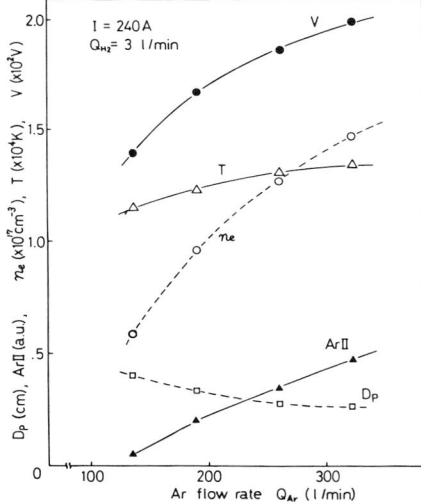

Fig. 14 Dependences of plasma parameters on gas flow rate

4.2 Influence of chamber length

Here we mention about the influence of chamber length i.e. the distance between two gas divertor nozzles, on the plasma parameters. So, the radial pressure distribu-

162

tion has been measured by the pitot tube in the case that plasma doesn't exist. The result is shown in **Fig. 15.** For the chamber length can be selected the values, 1.4, 3.4, 7.4 cm. The experimental results of the pressure measurement show that the shorter the chamber length is, the bigger the pressure gradient is. In this case the pressure at the center axis is not correct by means of the disturbance of the gas flow by the pitot tube. The real pressure at the center axis seems to be less than 40 Torr.

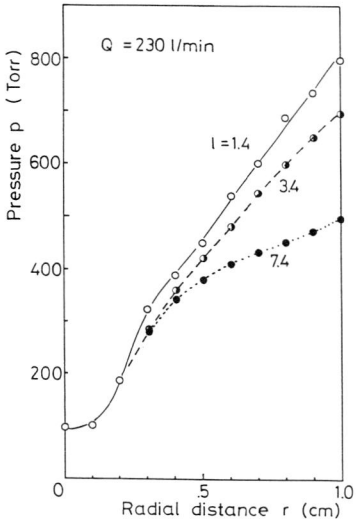

Fig. 15 Radial pressure distributions in various chamber length

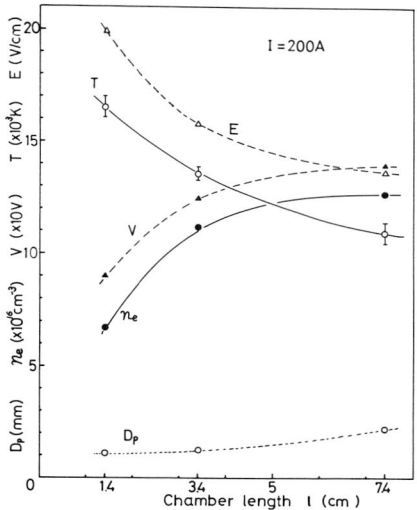

Fig. 16 Characteristics of plasma parameters against chamber length

Fig. 16 shows the characteristics of the plasma parameters for the chamber length. Here, the electric current value is constant, that is 200 A, and the distance between two electrodes is also constant, 10 cm. As the chamber length is shorter, the electric potential difference E increases, but the arc voltage decreases. Because the electric potential difference at the outside of the vortex chamber is rather small value, $4 \sim 8$ V/cm. The thermal pinch effect by the vortex flow is strong, when the chamber length is short. As the result, the plasma diameter is very fine, 1.2 mm, and the plasma temperature increases, and reaches about 17000 K in the case of l = 1.4 cm. On the other hand, the electron density of the plasma decreases gradually, the value is 6×10^{16} cm^{-3} in the case of l = 1.4 cm. Because, the shorter the chamber length becomes, the lower the pressure at the gas tunnel is, by means of the effect of the strong vortex flow.

As mentioned above, in order to obtain more stable and higher temperature plasma, the apparatus of the plasma generator whose chamber length is shorter is appropriate. Then, for instance, when the chamber length is 1.4 cm, fully ionized plasma has been produced in the gas tunnel at 200 A.

5. Conclusion

The stationary plasma arc in the gas tunnel has been investigated and the characteristics are clarified experimentally. The results obtained in this study are as follows:

(1) The mechanism of the formation of gas tunnel by the strong vortex flow with big flow rate, has been studied theoretically. And on the radial pressure distribution in the vortex chamber, the results showed good agreement with the experimental results in the annular region of the vortex chamber.

(2) The plasma in the gas tunnel has high temperature and high electron density as compared with the conventional plasma arc, because of the strong thermal pinch effect. The effect appears remarkably in the case of rather larger flow rate, and the velocity in the radial direction suppresses thermal conduction loss of the plasma.

(3) The chamber length has important role for the characteristics of the plasma parameters. As the result, almost fully ionized plasma has been obtained in the vortex gas tunnel at even small current of 200 A, when the chamber length is 1.4 cm.

163

Development of Gas Tunnel Type High Power Plasma Jet

Abstract

A new type of high power plasma jet were developed and its fundamental characteristics were clarified.

High energy density plasma beam can be generated in a "gas tunnel" produced by a strong vortex flow with a high flow rate. And it can be ejected as a high power plasma jet from the torch. It is easier to obtain extremely high power plasma jet with this "gas tunnel type" than with the conventional type.

The characteristics of the gas tunnel type of plasma jet are follows

1) high voltage
2) high thermal efficiency
3) high energy density

1. Introduction

The plasma jet, currently being put to various industrial applications, was not developed until the late 1950s[1, 2, 3]. In the initial stage of studies, water was employed as the working fluid, causing excessive consumption of electrodes, making it hard to obtain stable plasma. For this reason, it was not tried as a new high-temperature heat source, being confined within the range of academic research.

However, the inert gas sealed arc using tungsten electrodes (TIG arc), which makes it easy to obtain stable plasma, had already found applications in welding. And, a high-temperature, high-velocity plasma jet obtained by fitting a water-cooled copper nozzle to the tip of this TIG arc torch was tried to apply to cutting metal materials[4, 5]. The plasma ejected from the torch by generating an arc between the electrode and the nozzle electrode, constricted by the nozzle wall and the working gas, is called "plasma jet". This plasma jet is characterized by being a high-temperature heat source with high energy density as well as an extremely high velocity.

The remarkable progress made with the plasma jet since then has allowed versatile engineering applications including melting & processing (cutting, welding, metallizing, etc.) of metal materials, and high-temperature chemical reaction[6, 7, 8, 9, 10, 11, 12].

The plasma jet, due to its easy operation in addition to reduced heat generating costs, high power, high temperature and high thermal efficiency, is certain to be applied in an ever widening range of fields. However, with the conventional plasma jet it was difficult to obtain any higher output due to such problems as the damage of electrodes. The plasma jet most generally used has an output of only 100kW or less.

In this study, we attempted to obtain a high-temperature, high energy density, high power plasma jet by generating a plasma beam in a "gas tunnel" designed to provide extremely powerful thermal pinch effects, and investigated its basic characteristics.

2. Experimental apparatus

Fig. 1 shows the block diagram of the experimental apparatus for the high power plasma jet. Power was provided by six DC power sources with a rated current of 1500A and load voltage of 50V, allowing both serial and parallel arrangement by simply switching over the wiring, allowing a maximum 450kW output.

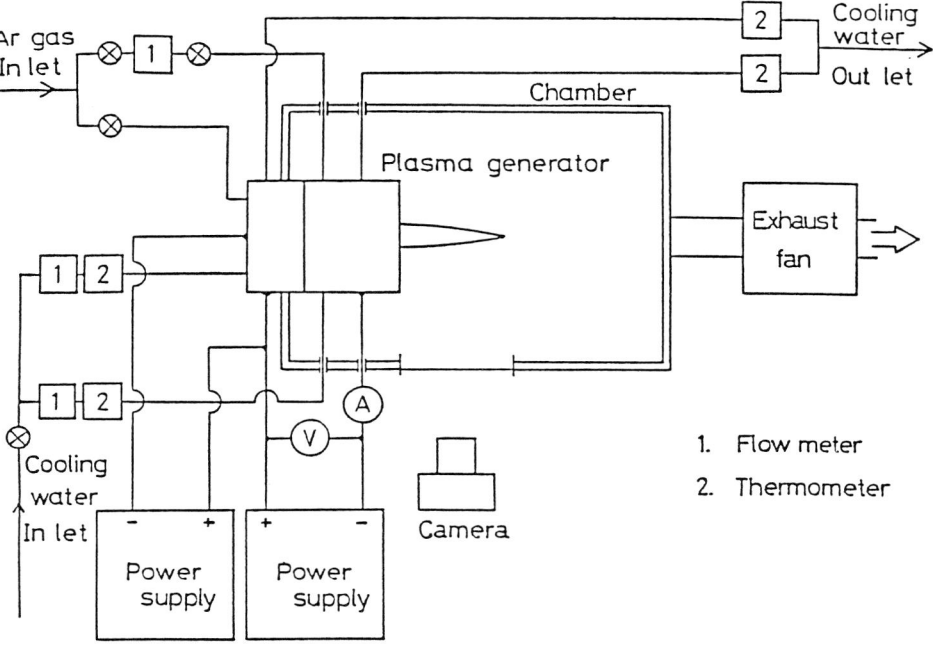

Fig. 1 Block diagram of experimental apparatus for high power plasma jet

Fig. 2 is a sectional view of the high power plasma jet developed for this study. In this figure, (A) is the plasma jet gun for igniting the high power plasma jet, which features a similar construction and function as that of the conventional plasma jet. (B) is the gas tunnel generator, consisting of the vortex generator and gas divertor nozzle (electrode). This vortex generator, provided with numerous small holes in its circumference, forms a vortex with excellent axial symmetry. The large volume of gas forced from these small holes (vortex generator nozzle) is provided with radial velocity by the gas divertor nozzle, forming pressure distribution with an sharp gradient inside the vortex chamber. This forms a short of gaseous wall, i.e. "gas wall", which generates a low-vacuum (vortex) gas tunnel in the vicinity of its inner central axis[13, 14]. Discharge inside this gas tunnel is called "gas tunnel discharge".

The plasma jet is generated as follows. Argon gas is fed into the water-cooled plasma jet generator in order to form the gas tunnel at the central axis of the vortex generating unit. Then, the plasma jet is initiated and led into the gas tunnel by the conventional method; on applying high voltage between the two electrodes, a high current flows, starting the gas tunnel discharge, generating the high-energy, high-power plasma jet.

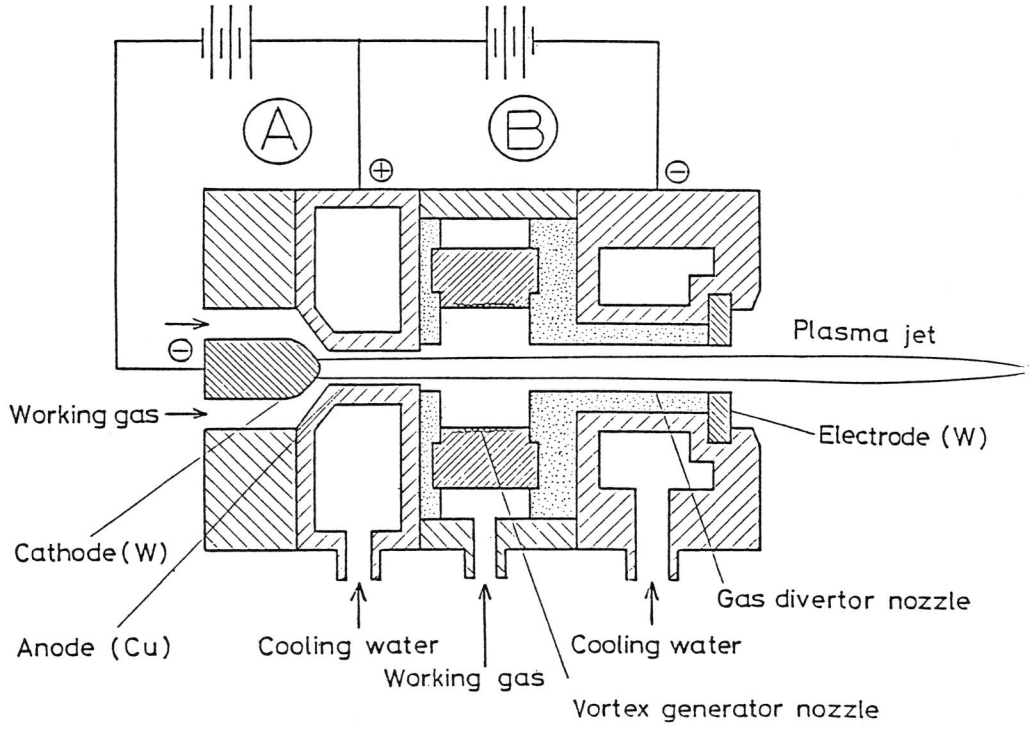

Fig. 2 Schematic diagram of plasma jet generator

In this study, the gas divertor nozzle (electrode) was used as cathode; but by changing the power source wiring, it may also be used as anode. The voltage between electrodes varies depending on the inner diameter of the gas divertor nozzle, the length of the vortex chamber, gas flow rate and pressure level of the vortex working gas. For this reason it is extremely easy to increase plasma jet output. Incidentally, in this study, the distance between electrodes was kept a constant 4.6 cm.

Compared to the conventional plasma jet, the plasma jet generated by this method guarantees a high-energy plasma column with optimum stability and an extremely high electrical potential difference, due to the formation of the gas tunnel by strong vortex.

3. Basic characteristics of the high power plasma jet

3-1 Current-voltage characteristics

The current-voltage characteristics of the conventional plasma jet, though varying depending on the gas flow rate, the torch shape, the type of the gas used etc., generally show decreasing voltage characteristics. Under high current, it shows constant-voltage characteristics, the voltage being about 50V. The gas tunnel (discharge) type plasma jet developed through this study, however, shows current-voltage characteristics as shown in **Fig. 3**, in which the voltage tends to increase from a fairly low-current range; voltage increases linearly with the increase in current. The figure shows an example of argon as working gas, gas flow rate of 400ℓ/min and gas dirertor nozzle

167

diameter of 8mm; at the current level of 900A, voltage exceeds 200V. The electrical potential difference of the plasma jet is estimated to be about 45V/cm. The current-voltage characteristics of the conventional plasma jet when using argon as working gas is also shown in Fig. 3, where the voltage is as low as 40-50V and electrical potential difference is 20V/cm or less, presenting typical dropping characteristics.

Fig. 4 shows the voltage characteristics when gas divertor nozzle diameter varies. The gas flow rate at this time is a constnat 400ℓ/min. The figure shows voltage at the current level of 500A and 1000A — at both values, as nozzle diameter decreases, the voltage shows a sudden rise, since the thermal pinch effect of the strong vortex is enhanced as the gas divertor nozzle diameter decreases. As a consequence, the plasma jet diameter is reduced, while the electrical potential difference is increased. If gas divertor nozzle diameter remains constant, linear current-voltage characteristics as shown in Fig. 3 are obtained.

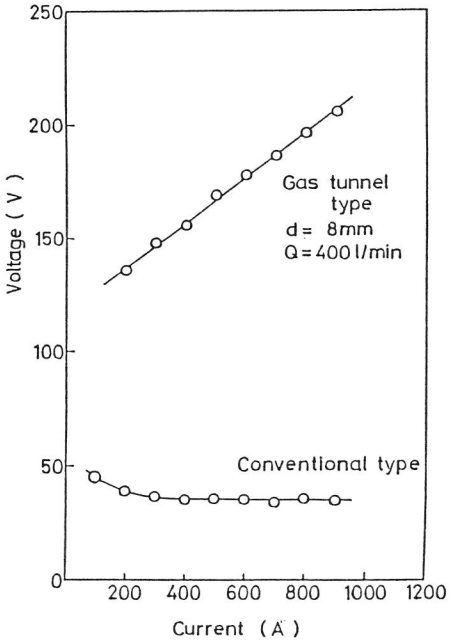 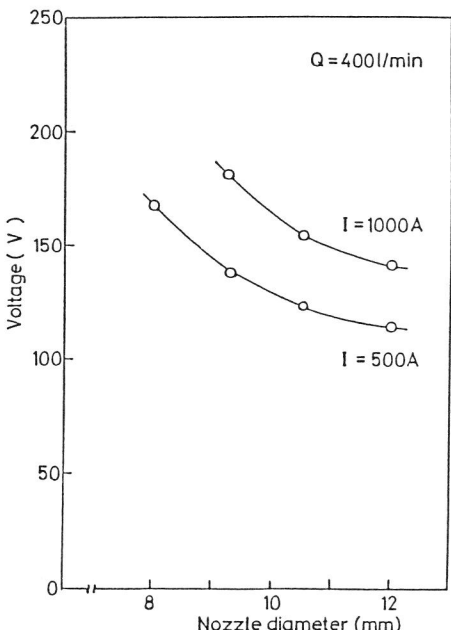

Fig. 3 Current-voltage characteristics of plasma jet **Fig. 4** Dependences of voltage on nozzle diameter

The correlation between gas flow rate and arc voltage is shown in **Fig. 5** for gas divertor nozzle diameter of 8mm and current level of 200A and 400A. As gas flow rate increases, the plasma jet voltage increases linearly at either current level. The increase rate of voltage to gas flow rate, dV/dQ, remains constant, independent from current level at 12.5 V/100ℓ/min). The fact that the voltage rises as gas flow rate increases is due to the enhanced thermal pinch effects of the strong vortex, just as when the gas divertor nozzle diameter is reduced.

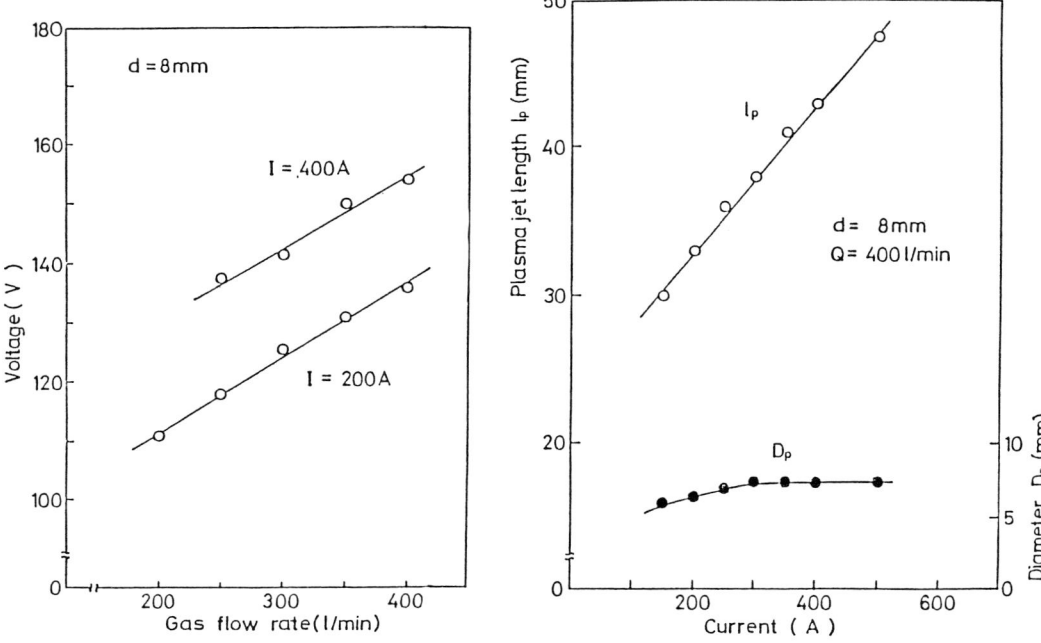

Fig. 5 Dependences of voltage on working gas flow rate

Fig. 6 Dependences of plasma jet length and its diameter on current

3-2 Shape of the plasma jet

The high power plasma jet increases in output as the current increases, raising the plasma light luminance significantly higher. Therefore, to photograph the plasma jet, we used very strong filters. The shape of the plasma jet was found with this photo, and its length and diameter were determined by the size of the highly luminous white core flame.

Fig. 6 shows the correlation found between the length and flame of the plasma jet and the current, when gas divertor nozzle diameter is 8mm and gas flow rate is a constant 400ℓ/min. As current increases, the length of the plasma jet length increases linearly at the rate of about 6mm per 100A; at higher current levels, a plasma jet as long as 50mm or more can be obtained. The conventional plasma jet is about 30mm when input is 50kW, but this plasma jet was 43mm long, almost 1.5 times longer, at the same input; with high-temperature plasma ejection, thermal efficiency is enhanced as explained below.

The plasma jet diameter is also shown in Fig. 6, which shows that below 300A, as the current is raised, the diameter gradually increases, while at a higher current level, the plasma jet attains the diameter of 7mm, slightly smaller than the gas divertor nozzle diameter of 8mm, remaining almost constant in relation to the increase in current. This is due to the remarkable thermal pinch effect of the strong vortex.

3-3 Thermal efficiency of the plasma jet

The electrical input for the plasma jet is given by the following equation:

$$P = IV = (V_a + V_k + V_P)$$

where, I; current, V; voltage, V_a; anode drop voltage, V_k; cathode drop voltage and V_p; the so-called anode column voltage.

The electrical input for this plasma jet is ejected to the outside as the thermal energy of the plasma jet, except energy loss at the electrode and thermal loss into the torch wall from plasma column. The torch thermal loss can be experimentally determined by the temperature rise and the flow rate in the cooling water. The thermal efficiency of the plasma jet is determined by ((electrical input − torch thermal loss)/electrical input) x 100%.

Fig. 7 shows the thermal efficiency of the high power plasma jet at varying current levels. These results are with the gas divertor nozzle diameter of 10.5mm and gas flow rate of 400ℓ/min; in that case, the thermal efficiency reaches a maximum at the current level of 300A, gradually decreasing as the current rises.

Under the experimental condition range, the thermal efficiency of the high power plasma jet of this study proved to be 80% or more, an extremely high value. This is due to the fact that the thermal loss was largely suppressed by the vortex flow with radial velocity, and that the plasma jet diameter was reduced by the thermal pinch effect of the strong vortex.

Fig. 8 shows thermal loss at varying current levels of the torch during the measurement of the thermal efficiency shown in Fig. 7. The thermal loss of the torch comprises that of anode and

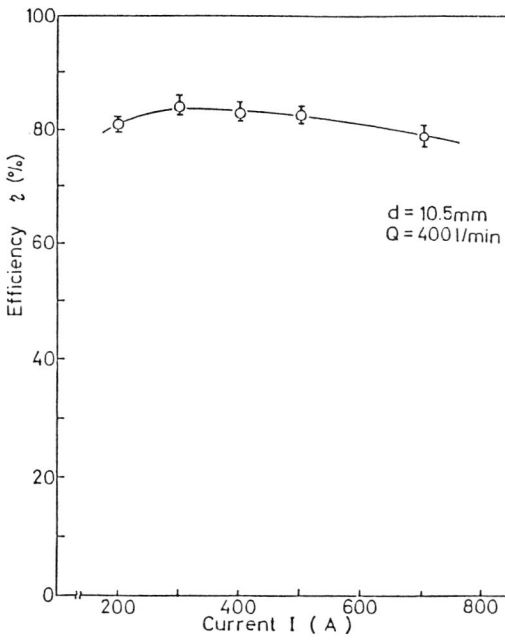

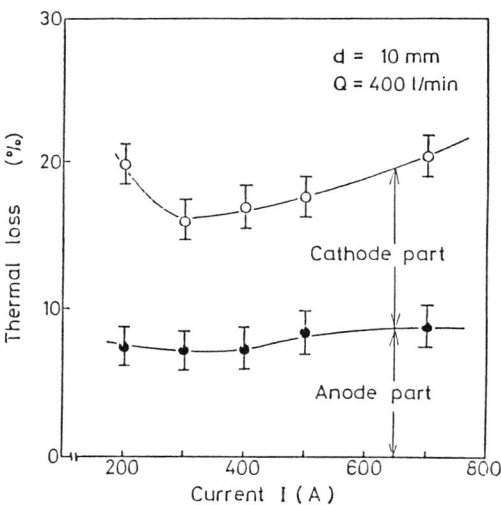

Fig. 7 Dependence of thermal efficiency of plasma jet on current

Fig. 8 Characteristics of thermal loss from plasma jet against current

cathode, the minimum being about 16% at 300A. The high power plasma jet uses inverse polarity — the gas divertor nozzle electrode as cathode — as shown in Fig. 2, resulting in thermal loss more or less balanced between anode and cathode, for an extra design advantage.

Compared to the thermal efficiency (40 − 50%) of the conventional plasma jet, the high power plasma jet has nearly twice the thermal efficiency.

4. Discussion

In order to clarify the characteristics of the high power plasma jet developed in this study, comparison was made with the conventional plasma jet; the results are shown in **Table 1**. The experiments carried out in this study have clearly shown the high power plasma jet to have high electrical potential difference, high thermal efficiency and high energy density. These phenomena have been attained by suppressed the plasma jet through thermal pinch effects and the protection of the electrode by means of the special strong vortex.

The plasma jet obtained by gas tunnel discharge, unlike the conventional plasma jet, is a high-voltage type with an extremely large electrical potential difference, which allows extremely high power to be obtained. Moreover, by introducing the multi-stage layout, a super-high-power plasma jet generator capable of producing as high as several **MW** may also be feasible.

The thermal pinch effects of the special strong vortex constricts the plasma jet, resulting in high temperature and longer plasma. Moreover, since the loss in radial thermal conductivity is suppress-ed, thermal efficiency extremely higher that that of the conventional plasma jet is guaranteed, which should prove highly advantageous in actual application.

The high power plasma jet developed in this study features extremely optimum stability, making versatile engineering applications possible by taking advantage of the characteristics of such a high power plasma jet.

The use of this high power plasma jet featuring extremely high-temperature and high energy density should prove to be highly efficient (higher velocity and greater efficiency than the conventional plasma jet) for melting metals with high melting points, melting ceramics and other insulating materials with high melting points, refining metals, and the like. Moreover, by using a large volume of working gas, any sort of high-temperature gas or reduction gas may be directly and easily obtained in a large quantity without using the current method of passing it through heat exchangers.

Table 1 Performances of high power plasma jet as compared with conventional plasma jet

	High power plasma jet	Conventional plasma jet
Electrical potential difference	40 ∼ 50 V/cm	10 ∼ 20 V/cm
Output	200 kW	<100 kW
Plasma temperature	20000 K	10000 K
Thermal efficiency	80 %	50 %

Far higher quality and work efficiency can be attained in processing, surface treatment or metallizing of materials, if applied to metal materials with high melting points or various metal materials, ceramics, etc. The high-temperature chemical reaction based on the high temperature of the plasma jet may also open roads to its application to the production of various synthetic substances, formation of crystals, production of spherical powders of high-purity alumina, degrading of poisonous substances, etc.

5. Conclusions

The high power plasma jet obtained by the gas tunnel discharge developed through this study is characterized by an extremely high electrical potential difference due to the high thermal pinch effect of the strong vortex, also featuring positive current-voltage characteristics. This in turn makes it extremely easy to give the plasma jet higher power. Besides, this gas tunnel discharge can be made in multiple stages, bringing about the possibility of a device to generate a super-high-power plasma jet of several MW.

This gas tunnel type plasma jet also has optimum stability; the plasma jet can be constricted through the thermal pinch effects of the special strong vortex, realizing higher temperature and longer plasma. This vortex suppresses the loss in the radial thermal conductivity, having extremely high termal efficiency when compared to the conventional plasma jet.

By making the best of the characteristics of the high power plasma jet, experimentally clarified through this study, the high power plasma jet may be used in a far wider range of applications as a high-temperature heat source. And this plasma jet featuring high energy density will make possible its use in various new fields impenetrable for the conventional plasma jet.

Acknowledgements

The author should like to take this opportunity to express his sincere gratitude to Mr. Yasuhiro Habara, Nippon Metal Industry, for his generous cooperation in performing the necessary experiments in this study.

REFERENCES

 1) R. Wiess: Z. Phys., 138 (1954), 170
 2) G. M. Giannini: Scientific American Aug. (1957)
 3) J. A. Browning: Welding Journal, Sep. (1959)
 4) R. M. Gage: U.S. Patent No. 2, 806, 124 (1957)
 5) J. A. Browning: Welding Journal, May (1962)
 6) H. W. Leutner and C. S. Stokes: Ind. Eng. Chem., 53, May (1961)
 7) R. A. Cresswell: Welding Metal Fabrication, 30 (1962), 435
 8) J. E. Anderson and L. A. Case: I & EC Proces s Design and Development, 1 No. 3 (1963)
 9) M. L. Levin: British Welding Journal, May (1964)
10) M. Okada and H. Maruo: Technol. Repts. Osaka Univ. 14 (1964)
11) M. Okada, Y. Arata: "Plasma Engneering", Nikkan Kogyo Shinbunsha (1965), 337

12) A. M. Howatson: "An Introduction to Gas Discharges", Pergamon Press (1976), 216
13) Y. Arata: J. Phys. Soc. Japan, 43 (1977), 1107
14) Y. Arata and A. Kobayashi: Trans. JWRI 13 (1985)

CHAPTER 3

FUNDAMENTAL CHARACTERISTICS OF ULTRA HIGH ENERGY DENSITY HEAT SOURCE

Focusing

Heating of Material

High Energy Density Plasma Beam

Focusing

Evaluation of Beam Characteristics by the AB Test Method
Y. Arata
[IIW Doc. IV–340–83.]

Shape Decision of High Energy Density Beam
Y. Arata, M. Tomie, K. Terai, H. Nagai and T. Hattori
[Trans. JWRI **2** (1973), 130. IIW Doc. IV–114–73.]

Focusing Characteristics of High Energy Density Beam
Y. Arata, T. Ishimura and I. Miyamoto
[Trans. JWRI **2** (1973), 1. IIW Doc. IV–110–73.]

Some Fundamental Properties of High Power Laser Beam as a Heat Source (Report 1: Beam Focusing Characteristics of CO_2 Laser)
Y. Arata and I. Miyamoto
[Trans. JWS **3** (1972), 1. IIW Doc. IV–88–72. Tech. Report Osaka Univ. **22** (1972), 1033.]

Wall-Focusing Effect of Laser Beam
Y. Arata and I. Miyamoto
[Proc. 2nd Int. Symp. of JWS (1975).]

Evaluation of Beam Characteristics by the AB Test Method

In welding with high energy density beams such as electron beam and laser beam, the clarification of characteristics of the beam itself, such as beam diameter and focal position, is not only important in viewing the relations with welding phenomena but also indispensable for indicating common welding conditions. Although various measurements of energy distribution and beam diameter have been conducted so far by electrical and/or optical methods, they have required large-scale equipment and extreme care as well as a long period of time for measuring. In addition, in the case of such high output beams as those generally employed in practical welding, the measurement itself tends to be very difficult.

For this reason, the JIW. IV Commission has studied how to simply, efficiently and inexpensively measure the characteristics of the beams employed in practical welding.

With regard to electron beam welding, at the 1973 annual meeting of the IIW, Dr. Arata and his co-researchers proposed the AB test method utilizing a cutting phenomenon at the metal edge[1]. Since then this method has been generally adopted (at least in Japan) and employed broadly[2]. The appropriateness of the AB test method for electron beam has been verified as mentioned above thanks to the electrical method for measuring electron beam characteristics directly (measuring energy distribution by using slits or the like). Recently, however, measurement by the pinhole method with higher precision than the slit method has been tried. That is why we rechecked the validity of the AB test method.

As a result, we have confirmed that if we can obtain a sharp cut section which does not cause a secondary melting phenomenon at the edge or the surface of a standard shaped sample in the AB test for electron beam, then the measurement results are almost identical to those of the pinhole method, thus proving the appropriateness of the AB test method.

To date, no considerable proposal has been made about laser beam. Thus we investigated the possibility of also using for laser beam the AB test method for electron beam. In the method in which an inclined sample of standard shape is shifted under the beam, the measurement of strength distribution at the beam focal position was inaccurate because the plasma produced at the beam application point absorbed and scattered the laser, preventing direct incidence of beams to the sample and making it unusable.

Considering the above difficulties, we propose here a method using acrylic resin as the AB test for laser beam because plasma is not so easily produced by this method and the energy of the applied beam is consumed mostly in the evaporation of the material. There the beam strength distribution is directly displayed in the shape treated by evaporation.

Hereinafter in Report 1 and Report 2 we describe in detail the method, results, and so on of the tests to confirm the appropriateness of the AB test method for electron beam and laser beam.

Report 1: AB Test Method for Electron Beam

1. Introduction

To indicate welding conditions in electron beam welding, the values of accelerating voltage, beam current, welding speed, beam focusing condition (focus lens current) and working distance have usually been used. Among these, however, lens current does not indicate welding condition correctly. This is not only because lens current can not take a common value depending on the individual welding machine, but also because even on the same welding machine and at the same focus current there is the possibility that focal position and beam diameter vary due to changes of the electron gun with time (wear and tear of electron gun components and deposit of metallic vapor). Consequently, in indicating common welding conditions, it is essential to describe directly the characteristics of the electron beam itself such as beam focal position, beam diameter, etc. instead of lens current.

In order to measure electron beam characteristics directly, many researchers have tried to measure energy distribution and beam diameter by various electrical techniques[3]~[8]. Correct information may be obtained by these methods if measurement is conducted with adequate accuracy and extra care. Generally, however, it is difficult to measure high output electron beam, and it requires a long measuring time as well as high equipment cost. Then adoption of electrical methods to welding machines on the production site is extremely difficult.

For this reason, even with accuracy spoiled to some extent, if a simple and inexpensive method to find beam diameter and/or focal position were put into practice, it would be very significant from the viewpoint of guaranteeing the quality of electron beam welds.

For such a simple and inexpensive measuring means, Dr. Arata and his co-researchers have once proposed a method for testing the beam diameter and focal position by cutting of metal edge with electron beams (AB test method = Arata Beam Test Method)[1],[2].

In this paper we present a review of the proposal of the AB test, discuss the appropriateness of the test and requirements for its specimens, and also describe the results of its application to various welding machines in Japan.

2. Review of the Proposal of the AB Test

To obtain a simple means for measuring electron beam diameters and focal positions, Dr. Arata and his co-researchers have proposed a method —— the AB test based on numerous experiments[1].

According to the principle of this measurement method, electron beam is applied directly to a metal to cut it, and the effective diameter at which the electron beam melts each material is determined by the cutting width. In this study, they measured how thin plates were cut by ultra-high output beam, and also measured the cutting width at the edge of projection, with various projections as shown in Fig. 1. Then they proposed the AB test as follows:

(1) As for the configuration of specimen, a comb-shaped test-piece provided with teeth having a square cross section as shown in Fig. 2 is desirable. Currently in Japan a specimen of g = 4 mm, δ = 2 mm is normally used.

(2) When welding is done on a specimen inclined at the proper angle θ_s, the distribution of electron beam energy in the axial direction can be measured at one time. Incidentally, if $\theta_s = 30°$ is selected, the distance in the axial direction can be easily calculated.

(3) As for the welding direction, down-slope welding i.e. from each tooth root to surface, or from right to left in Fig. 2, is preferable because then the secondary melting due to the deposit of molten metal is less likely to occur and sharper cutting can be achieved.

(4) Stainless steel is desirable for the test piece material. The measurement results of beam dia-

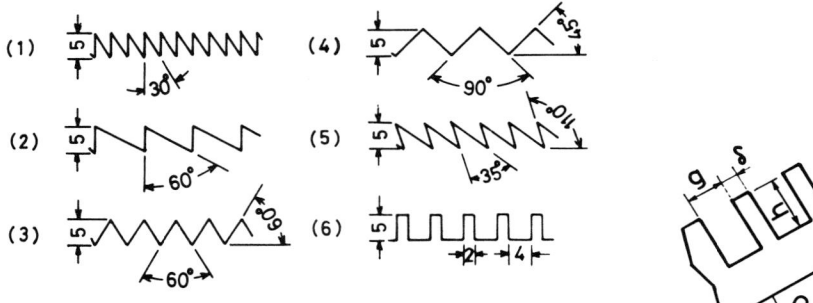

Fig. 1 Various edge shapes used for measuring E.B. diameter

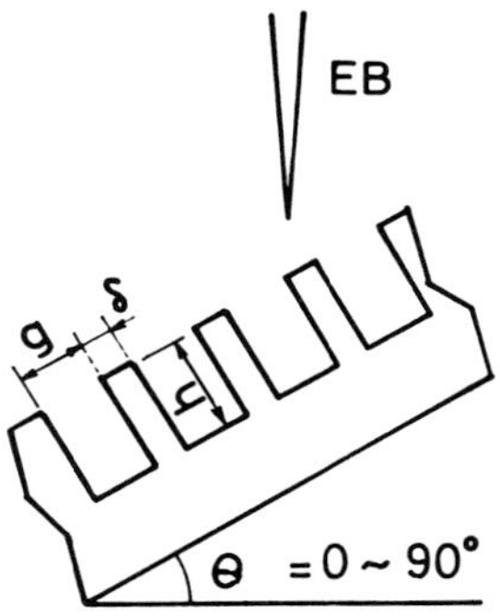

Fig. 2 Configuration of the comb-shape AB test piece

meters using stainless steel test pieces were almost constant in a wide range of welding speed, and practically no secondary melting was observed. On the other hand, secondary melting was conspicuous on other materials (copper and its alloys, aluminum alloy, titanium) and good results were not obtained.

3. Effectiveness of the AB Test and its Application Criteria

The comb-shaped AB test proposed in the preceding section is based on the results of weldings using high voltage electron beam welding machines. So we applied the test method to a low voltage welding machine and tried to compare the results with electrically measured beam diameters, thus investigating the effectiveness of the AB test and its application criteria. The conclusion we have reached is that under the working conditions where sharp cutting independent of welding speed is obtained, the AB test is capable of measuring beam diameters almost equal to real ones and determining focal positions as well.

The following are details of the experiments.

3.1 Beam Diameter Measuring Method

To measure the electron beam energy distribution electrically, we adopted the method shown in Fig. 3 (called the pinhole method). In this method a tungsten die having in its center a pinhole 0.1 mm in diameter was buried in a water-cooled copper plate with great heat capacity. The beam was oscillated horizontally by deflection coils, so that the electron beam center would pass over the pinhole. The value of the electron beam current which passed through the pinhole at this time was measured by the Faraday cup located below and the energy density was calculated.

179

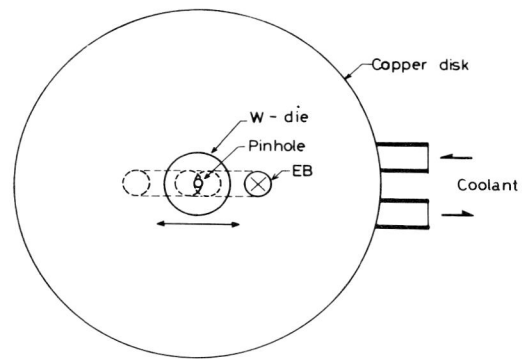

Fig. 3 Top view of electron beam current distribution measuring equipment (pinhole method)

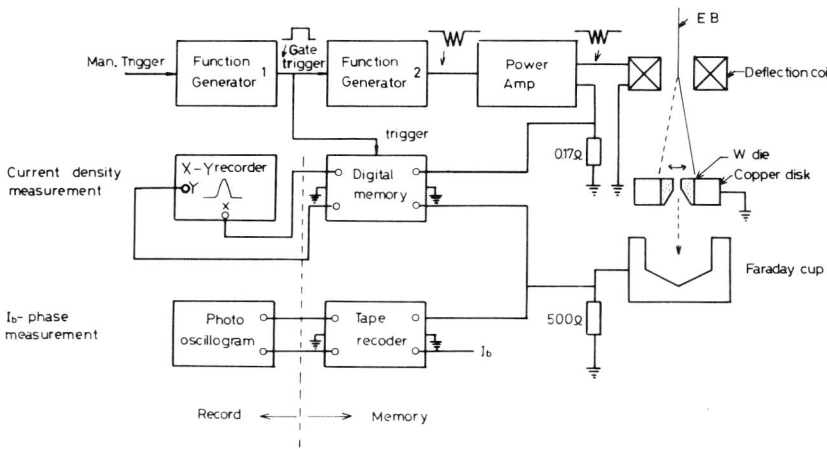

Fig. 4 Electron beam energy density distribution measuring circuit

The measurement circuit is shown in Fig. 4. A digital memory with high sampling speed was used for the measurement. The deflection width of the electron beam and the current value of the Faraday cup were recorded synchronously and reproduced on an X-Y recorder.

Fig. 5 shows a typical reproduced wave-form. To decide the beam diameter from such a wave-form, various proposals have been made so far. For example, a method of measuring the half-value width, the width value at $1/e$ of the peak value (beam diameter per energy), a method of deciding the diameter from the standard deviation (3σ method), and a method of measuring the diameter at $1/10$ of the peak value (ignoring the effect of energy density less than $1/10$ on welding) have been available. The first three methods are effective as long as the energy distribution is gaussian. In this test, however, various wave-forms were present at the same time, and so we determined the beam diameter by the following method. As shown in Fig. 5, a considerable amount of noise was contained in the measuring system. So we brought the mean value of noise to the zero level, and measured the beam diameter for the value there. That beam diameter was larger by 10% at most in comparison with the value at the $1/10$ level.

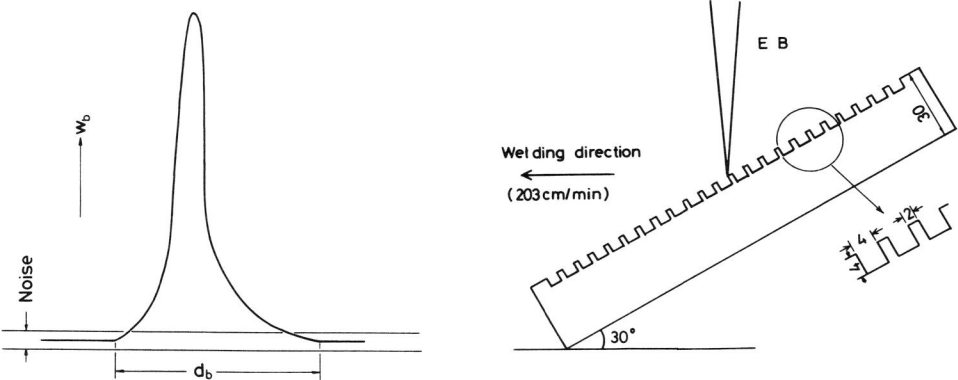

Fig. 5   A typical result of electron beam energy density distribution measurement

Fig. 6   An AB test piece made from low carbon steel

With the variation with time of electron beam taken into consideration, measurement was made on 10 samplings at each point.

In the AB tests conducted at the same time, low carbon steel plates (JIS SM 50 grade steel) 30 mm thick as shown in Fig. 6 were used. The electron gun of the welding machine was made by the Sciaky Co. in the U.S. (60 kV − 500 mA).

3.2 Test Results

The test results are shown in Figs. 7 and 8. In the figures, the left side shows the beam diameter distribution by the pinhole method and by the AB test method respectively, and the right side shows the energy density distribution wave-forms observed by the pinhole method.

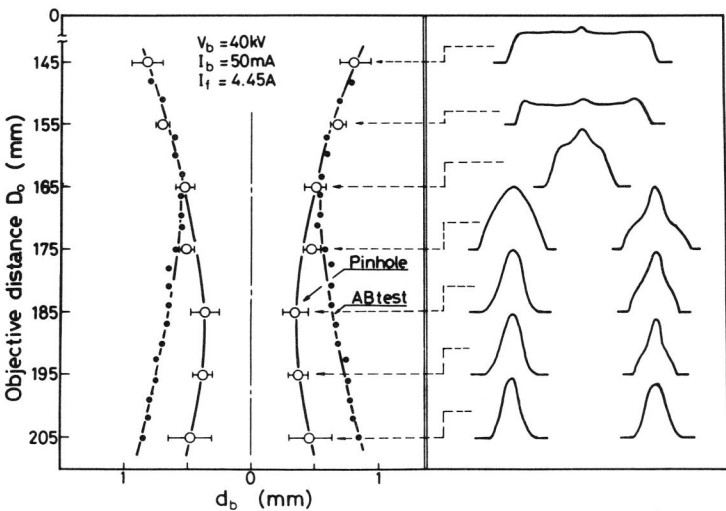

Fig. 7   Measurement results of electron beam diameter and energy density distribution

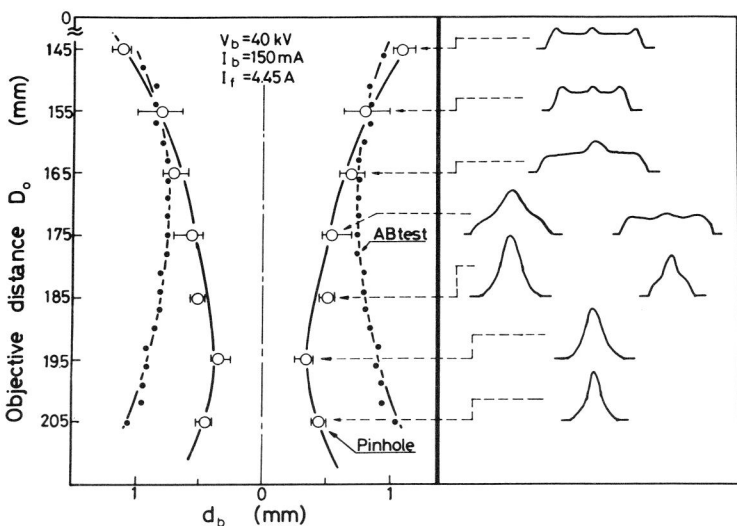

Fig. 8 **Measurement results of electron beam diameter and energy density distribution**

Roughly, the energy density distribution wave-forms with the welding machine used in this test were trapezoid in the vicinity of the focus lens, but as the distance from the lens increased, they turned out to be in the gaussian mode. (At the points extremely far away from the lens the wave-forms deviated from the gaussian mode.)

The focal point in the AB test was near the point where the trapezoid distribution changed into the gaussian mode. While the diameters in the trapezoid mode measured by the AB test method and by the pinhole method respectively matched with each other very well, in the gaussian mode the values obtained in the AB test were fairly greater than the results of the pinhole method. The focal point by the pinhole method was about 30 mm away from the lens compared to that in the AB test.

To investigate why such differences were caused, we examined the melting or cutting mechanism in AB test specimens. The results are shown in Fig. 9. As clearly seen from the figure, in the trapezoid mode, a good cutting essential to the AB test was achieved in the welded zone, whereas in the gaussian mode, molten metal remained in the welded zone and the molten width increased conspicuously near the tooth surface due to secondary melting. Therefore, in Fig. 10 we plotted the molten widths at three locations in each tooth. As clearly seen from the figure, compared to the focal points measured on the surface, the focal points in other locations were about 30 mm away from the lens, well corresponding to the results of the pinhole method. It indicates that even in the gaussian mode, if a sharp cutting without secondary melting could be achieved, the AB test method would be a very useful measure to determine beam diameter and focal point. Also the above results lead to the possibility of measuring beam diameter and focal point by cutting the welds of steel plates and measuring its molten width. The test results regarding that will be described in detail in the appendix A.

182

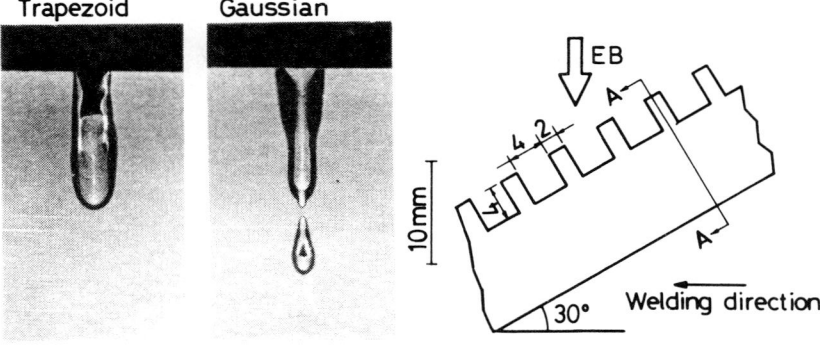

AB test : SM50-steel, V_b=50kV, I_b=200mA, v_b=200 cm/sec

Trapezoid Gaussian

A-A section

Fig. 9 Molten states in low carbon steel AB test piece

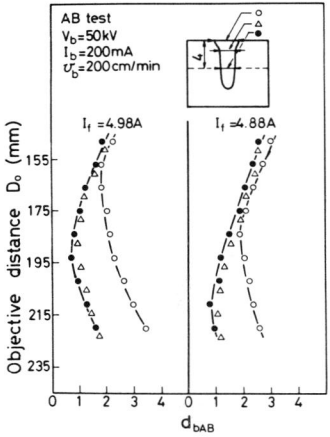

Fig. 10 Measurement results of molten width of low carbon steel AB test piece

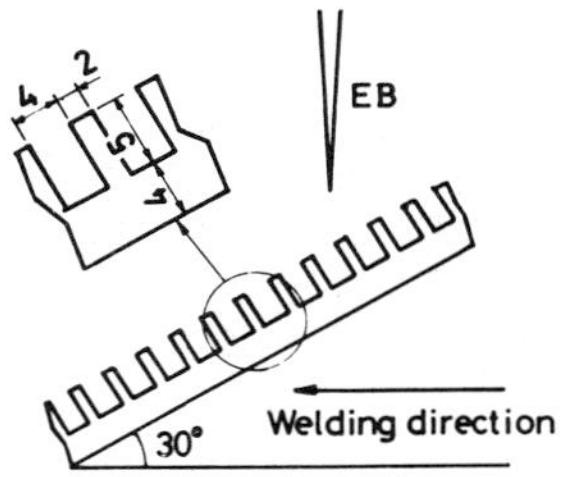

Fig. 11 An AB test piece made from stainless steel

3.3 Application Criteria of the AB Test Piece to the Welding Machine

As shown in Fig. 6, the AB test piece used in the preceding section was low carbon steel and underwent a partial penetration welding. Following the proposal in Section 2, we prepared a specimen of stainless steel and, as shown in Fig. 11, reduced its plate thickness and conducted a full penetration welding so that the molten metal could be removed more easily. The measurement results are shown in Fig. 12 in comparison with the results obtained with carbon steel. As seen from the figure, the results of both AB test pieces matched with each other without any remarkable difference. Fig. 13 shows the molten cross section of stainless steel. The parallel zone of the molten width of the stainless steel specimen was hard to judge compared to that of the carbon steel

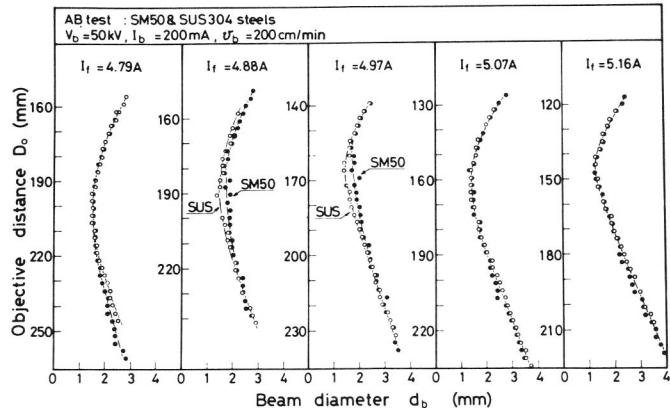

Fig. 12 Comparison of beam diameters of low carbon steel and stainless steel AB test pieces

AB test : SUS304 steel
$V_b=50kV$, $I_b=200mA$, $v_b=200cm/min$

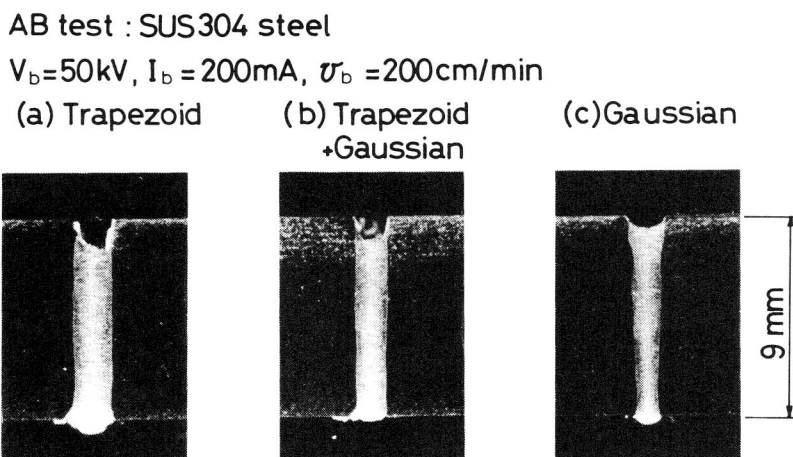

(a) Trapezoid (b) Trapezoid +Gaussian (c)Gaussian

Fig. 13 Molton states in stainless steel AB test piece

specimen, but there was no trace of secondary melting after melting under the trapezoid distribution (a), showing a good result. On the other hand, after welding under the gaussian distribution (c), an increase in width of about 55% on the surface due to secondary melting was observed, and a good result was not achieved.

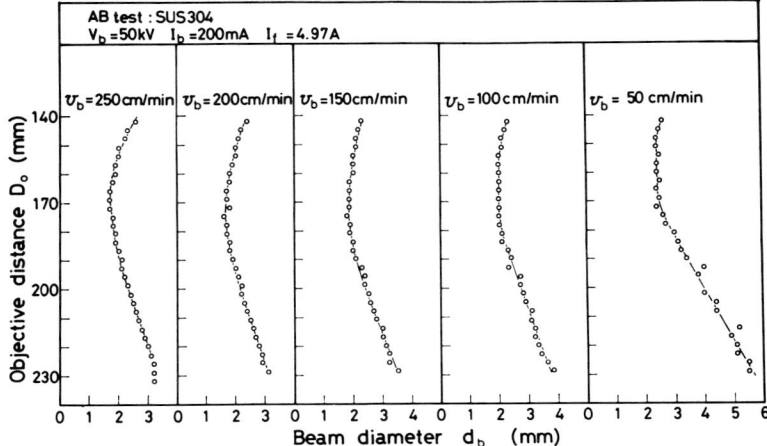

Fig. 14 Speed dependency of beam diameter in AB test

Next we examined the welding speed dependency. The results are shown in Fig. 14. As clearly seen from the figure, in the trapezoid mode (e.g. D_o = 150 mm) where a good cutting could be made, the beam diameter was constant at $v_b \geqq 100$ cm/min, whereas in the gaussian mode (e.g. $D_o \geqq 170$ mm) the beam diameter increased according to the decrease in welding speed.

From the above results we define the AB test application criteria as follows:

(1) Any steel can be applied.

(2) The cutting width should not vary in a wide range of welding speed.

(3) Although it is desired that no trace of secondary melting be found near the surface, even some traces to an ignorable extent can be allowed. Taking these into account, **h, g,** δ in Fig. 2 and plate thickness may be selected for the AB test piece size.

(4) Furthermore, if and when secondary melting is present at any D_o, the focal point can be determined as far as the secondary melting is to the same extent. It will be assumed that even then there is no speed dependency of the focal point.

4. Examples of AB Test Results in Japan

Table 1 shows some examples of AB test results per electron beam welding equipment in Japan.

185

Table 1 Examples of the AB test result in various electron beam welding equipments

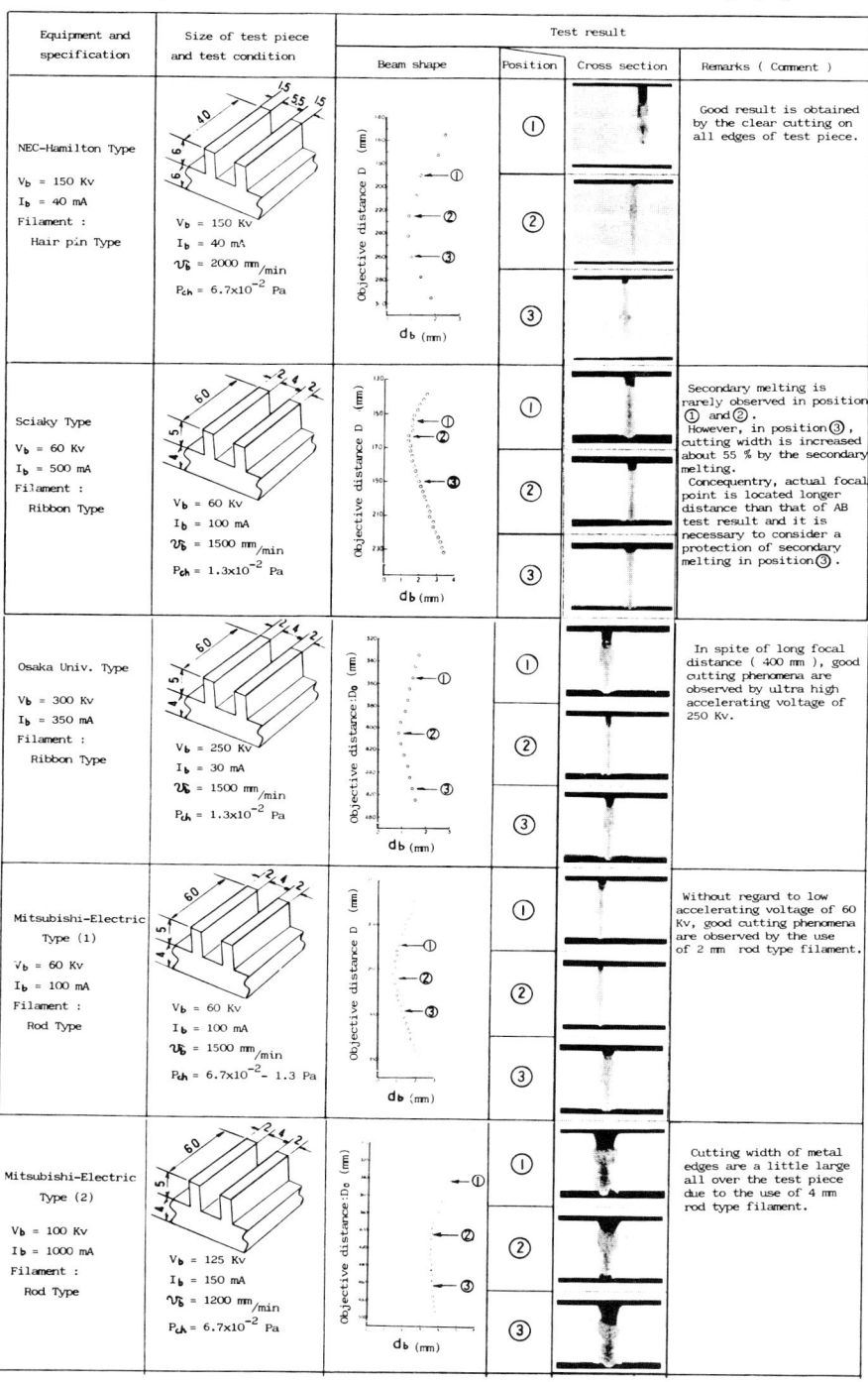

Appendix A.   Decision of Beam Characteristics by Measurement of the Weld Width

As mentioned in Section 3.2, beam characteristics can be known to some extent by measuring the molten width at specific locations in the weld when carbon steel is welded. From this viewpoint we describe our test results in this appendix.

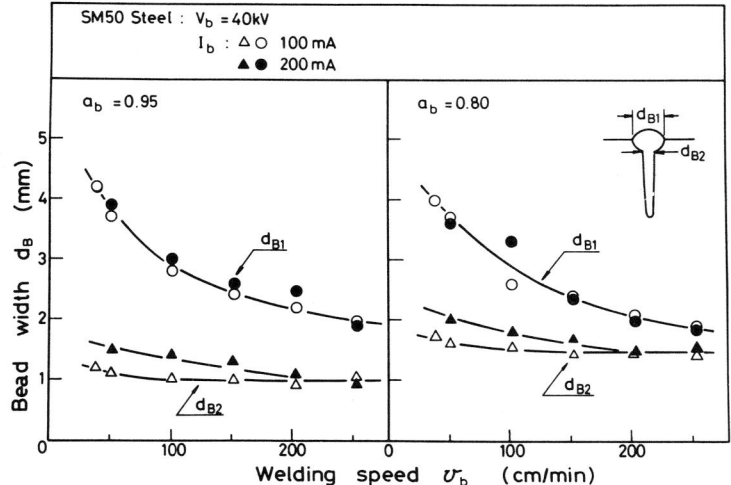

Fig. 1A   Welding speed dependency of penetration form

Fig. 1A shows welding results under various conditions with use of the welding machine employed in Section 3.

The width d_{B1} on the surface was highly dependent on welding speed and varied rapidly according to the change of welding speed. On the other hand the width d_{B2} in the neck portion of the weld was relatively less dependent on welding speed and hardly varied at high speeds. From this we consider that the increase of d_{B2} caused by secondary melting was not so great but that it corresponded with the electron beam diameter.

**Table 1A   Correlation between molten zone width and
beam diameter**

v_b (cm/min)	Pattern	Δd (mm)			
		$d_{B1} - d_b$	$d_{B1} - d_{bAB}$	$d_{B2} - d_b$	$d_{B2} - d_{bAB}$
203	Trapezoid	+ 1.22	+ 1.04	− 0.14	− 0.32
	Gaussian	+ 2.21	+ 1.11	− 0.13	− 1.15
102	Trapezoid	+ 2.35	+ 2.19	+ 0.11	− 0.10
	Gaussian	+ 3.53	+ 2.41	+ 0.27	− 0.85

+ : $d_B > d_b$
$V_b = 40kV$, $I_b = 50 \sim 200mA$

So we conducted welding under various conditions and examined the difference $\Delta \tilde{d}$ between beam diameter by the pinhole method (d_b), beam diameter by the AB test (d_{bAB}) and widths d_{B1}, d_{B2}. The results are shown in Table 1A where classification has been made by the mode or pattern of electron beam energy density distribution. With this test equipment (60 kV − 500 mA), as mentioned in Section 3.2, a difference was found between d_b and d_{bAB} when the mode of density distribution was gaussian, but when one compares d_b with the molten width, one notices how small the difference between d_b and d_{B2} is. From the result it is possible to catch the beam characteristics by measuring the molten width in the neck portion at relatively high welding speeds.

Next, we made a similar test, using a high voltage welding machine (300 kV − 350 mA) that enabled the focal point rather easily detectable by the AB test method no matter what the mode of energy density distribution is. The results are shown in Fig. 2A. As clearly seen from the figure, the focal point obtained from the molten zone width d_{B2} in the neck portion agreed very well with the focal point obtained in the AB test. Although there was a slight difference between beam diameters in both cases, judging from the results in Table 1A, both were expected to agree with each other by increasing the welding speed. Furthermore the molten zone width d_{B1} on the surface where secondary melting was present increases naturally in comparison with d_{B2} and d_b, and the focal point calculated from d_{B1} approached considerably to the lens as seen in the results described in Section 3.2.

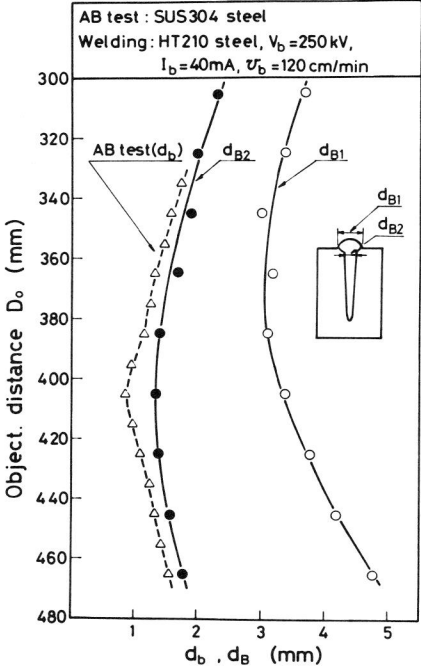

Fig. 2A Beam diameter and penetration form in ultra-high voltage welding machine

Appendix B. On the a_b Parameter

Using the electron beam characteristics measured as described above, to indicate the positional relationship between beam focal point and welded material in electron beam welding, the a_b parameter defined by the following formula is used in Japan as already announced[1, 2].

$$a_b = \frac{D_o}{D_F}$$

where D_o is the distance from objective lens center to welded material surface (objective distance) and D_F is the distance from objective lens center to focal point (focal distance).

References

1) Y. Arata, M. Tomie, K. Terai, H. Nagai and T. Hattori: IIW Doc. IV-114-73 (1973), Trans. JWRI, 2 (1973) 2, 1

2) K. Watanabe, T. Shida, M. Suzuki and H. Okamura: J. JWS, *43* (1974) 678 in Japanese

   Y. Arata, I. Futami, K. Terai, H. Nagai, S. Shimizu and T. Aota: J. JWS, *43* (1974) 981 in Japanese

   Y. Arata, F. Matsuda, Y. Shibata, S. Hozumi, Y. Ono and S. Fujihira: J. JWS, *44* (1975) 1011 in Japanese

   H. Nagai, S. Shimizu, T. Aota and K. Satoh: The 2nd Inter. Symp. of JWS, August, (1975)

   T. Nagao, S. Kosuge and S. Kobayashi: The 2nd Inter. Symp. of JWS, August (1975)

   M. Katsumura, M. Hamasaki and Y. Arata: J. High Temp. Society, 2 (1976) 2, 87 in Japanese

   Y. Arata, K. Terai, H. Nagai, S. Shimizu and T. Aota: Trans. JWRI, *5* (1976) 2, 119: J. High Temp. Society, 4 (1978) 2, 72 in Japanese

   Delegation of Germany, Fed. Rep.: IIW, Doc. IV-199-76 (1976)

   Y. Arata, M. Tomie and A. Kohyama: Trans. JWRI, *5* (1976) 1, 11

   H. Irie, T. Hashimoto and M. Inagaki: JWS, *46* (1977) 642 in Japanese

   Y. Suezawa, H. Kuroda, H. Kobayashi, M. Takada and T. Hase: J. JWS, *47* (1978) 418 in Japanese

   G. Sayegh: IIW, Doc. IV-276-79 (1979)

   J. Tanaka and S. Kosuge: J. JWS, *49* (1980) 628 in Japanese

   M. Nakanishi, J. Furusawa, S. Yasunaga, M. Tomie and Y. Arata: Sumitomo Metals, 33 (1981) in Japanese

   S. Tsukamoto, H. Irie, T. Hashimoto and M. Inagaki: J. JWS, *51* (1982) 286 in Japanese

   Thesis for Dr. of Eng, applying AB Test Method
   S. Matsuda (1974), M. Osumi (1976), H. Nagai (1976), Y. Shibata (1977), S. Shimizu (1977), H. Irie (1978), M. Tomie (1978), S. Satoh (1982)

3) P. Dumonte et G. Sayegh: IIW Doc. IV-131-73 (1973)

4) M.J. Adams: Brit. Weld. J., *15* (1968) 451

5) A. Sanderson: Brit. Weld. J., *15* (1968) 509

6) J. Sandstrom, J.F. Buchen and G.S. Hanks: Weld. J., *49* (1970) 293S

7) P. Drews, D. Schumacher and B. Spies: IIW Doc. IV-144-74 (1974)

8) H. Irie, T. Hashimoto and M. Inagaki: Trans NRIM, *22* (1980) 95

1. Introduction

The position of the focal point of electron beam in welding is determined by the AB test method in which a standard-shaped inclined test-piece is moved and heated so that the change in the width of molten bead width (or slot width) is examined as mentioned above.

However, in the case of laser beam heating of metals, laser plasma is produced at the beam irradiation point and the laser beam is absorbed and dispersed thereby so that direct incidence of beams on the test-piece is prevented and beam intensity distribution measurement is thus made inaccurate. That is why a substitute test method needs to be developed in laser beam welding. The method proposed here is a new one to employ non-metallic materials on which plasma is not produced.

The most recommended material for AB testing in laser welding is acryl resin. As listed below, this material has properties suitable for this testing method.

(1) Low thermal conductivily
(2) Sublimation (unfusable)
(3) Highly evaporable
(4) Low reflectivity
(5) No laser plasma

Since metallic materials are used as test-pieces in the electron beam a_b value determining method, measured beam intensity distribution are affected by heat conductivity, molten metal flow, etc. On the other hand, if an acryl resin, having the above-mentioned properties, is employed, intensity distribution of focussed laser beams can be obtained from an evaporated shape since most energy is consumed for evaporating the material when the laser beam is irradiated.

Hereinafter we propose a method using acryl resins to determine beam focal point and beam intensity distribution.

2. Measuring Method

Beam intensity distribution can be obtained from the shape of the evaporated portions in a static specimen placed for a short period or by scanning of the test-piece at a constant speed under focussed laser beams. In these two methods the latter is recommended for its higher reproducibility.

The broken line in Fig. 1 shows an example of cross-sectional shape of the evaporated groove when an acryl specimen was scanned perpendicularly to the focussed CO_2 laser beam axis. The solid line in this figure shows a profile of the focussed beam at the same power level and focal position as the dotted line measured by using a knife-edge slit (40μm wide). Both curves are in good agreement, indicating that the method using acrylic resin makes it possible for beam intensity distribution to be obtained simply and accurately.

To determine beam intensity distribution of focussed laser beam and to determine the focal point, using acryl resins, the test-piece is inclined at angle θ and scanned at a constant speed in the same manner as in the metal AB test shown in Fig. 2. Then the test-piece is cut at the fixed intervals, for transverse cross sections, and the change in the cross-sectional shape of the evaporation groove is examined.

Then the following conditions must be satisfied for the measurement.

(1) Angle of inclination (Fig. 2)

Since a laser beam is focussed normally by short-focus lens, its depth of focus is extremely shallow. For this reason, to improve the accuracy of measurement, θ should be smaller than a

190

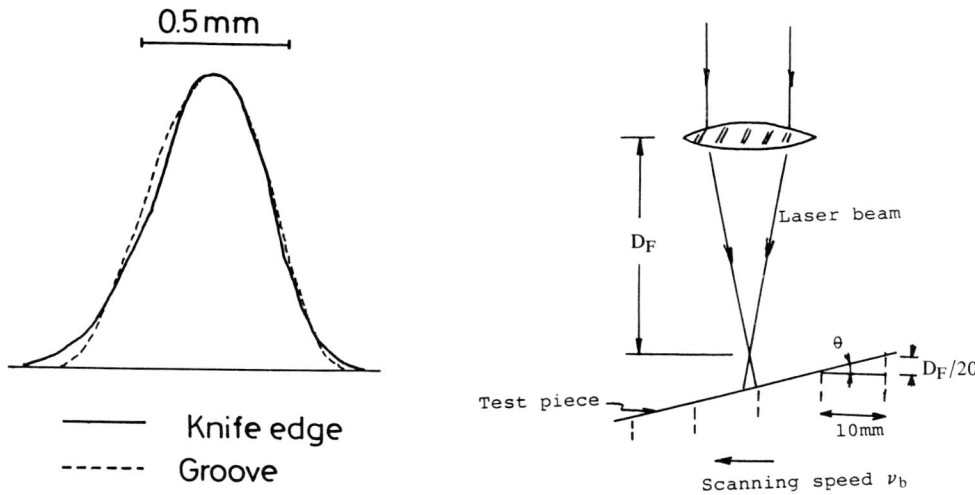

0.5mm

——— Knife edge
------ Groove

Fig. 1 Comparison of evaporated groove with beam intensity profile measured with knife edge slit

D_F

Laser beam

Test piece

θ

$D_F/200$

10mm

Scanning speed ν_b

Fig. 2 Inclination angle θ of test piece for AB testing

certain value according to the focal distance of lens.

Now focal distance of the lens is supposed to be D_F. The following values are recommended as standard for measurement.

1) Measurement precision of a_b value: 1/200

2) Cutting interval of test-piece: 10mm

A plane acrylic plate can be used as the test-piece.

Then,

$$\tan \theta = \frac{\frac{D_F}{200}}{10} = \frac{D_F}{2000} \quad \cdots\cdots\cdots\cdots\cdots (1)$$

will serve as the desired inclination.

(2) Scanning speed

At too low a scanning speed, the beam incident angle on the wall surface of the groove will increase due to deep groove formed and so will the reflection factor as well (Fig. 3). Then due to the wall-focusing effect[1] of reflection beams concentrating to the groove bottom, the groove shape does not represent beam intensity distribution any more. The wall-focusing effect can be ignored when the groove width is approximately equal to the groove depth. This gives the maximum groove depth allowed. On the other hand, you have to magnify groove cross-section for observation. Then the deeper the groove, the more accurate the measurement. Therefore, the sweep speed is recommendable at which the depth and width of the slot are equal.

The groove depth is in proportion to the laser power, and inversely proportional to the beam diameter and moving speed. For a given beam, a groove whose ratio of depth to width is approximately one can be obtained easily by varying the ν_b and cutting and trying. Figure 4 shows the calculated curves giving the relationship between beam diameter and sweep speed for this groove shape.

191

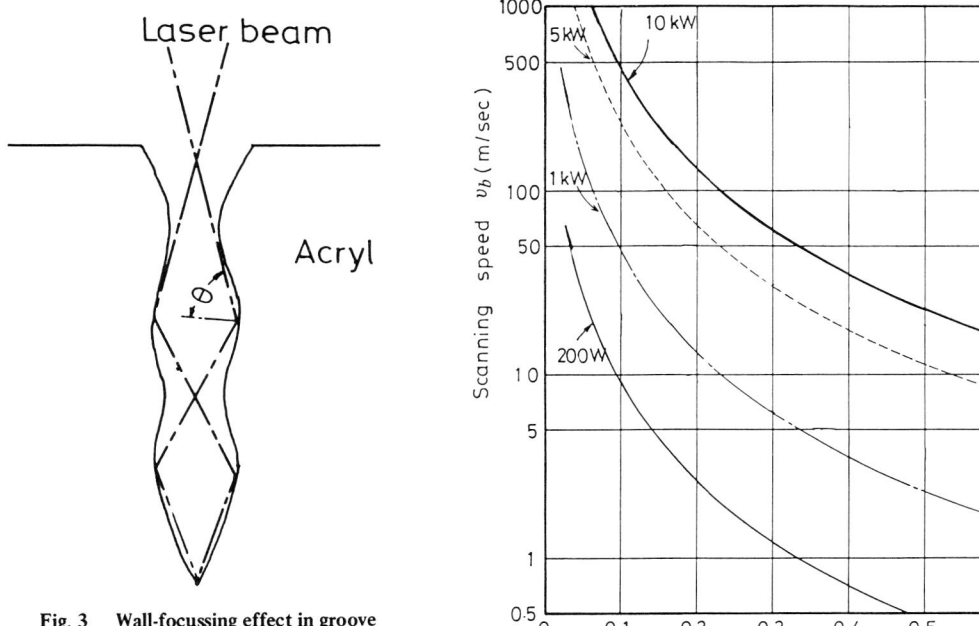

Fig. 3   Wall-focussing effect in groove

Fig. 4   Recommended scanning speed for various laser power and beam diameter

(3) Atmospheric gas

Generally measurements are almost independent of the atmospheric gas used. To protect the lens, however, it is desirable to remove material vapor from the beam irradiating part by using assist gas.

3. Necessity of Real Time Measurement

When a lens is irradiated by strong laser beams, there is the problem of the focal point variation with irradiation time. As exemplified in Fig. 5, the focal distance decreases in accordance with beam irradiation time and tends to saturate after a certain period of irradiation. That is because the edge of the lens are cooled whereas their central part is expanded by heating. This problem matters when the beam absorption coefficient of the lens material is large, or when the lens surface is diteriorated by sputtering. On such occasions, real time measurement of focal position is required.

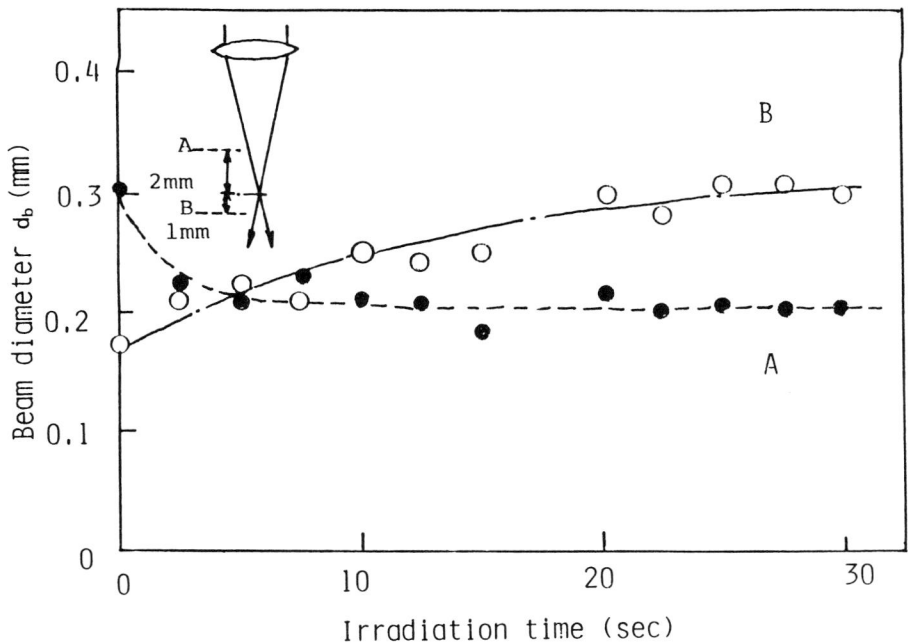

Fig. 5   Time change in beam diameter

4. Example of Real Time Measurement Method

As shown in Fig. 6, using a smoke emitting substance like an incense rod, we delivered smoke around the focal point to visualize the beam. An example of the image thus obtained is shown in Fig. 7. Clearer images are expected to be attained by selecting the smoke emitting substance and by improving the smoke emitting method.

Real time detection of beam profile is considered achievable by picking up such an image with a TV camera and by using microcomputer or the like in the image processing system.

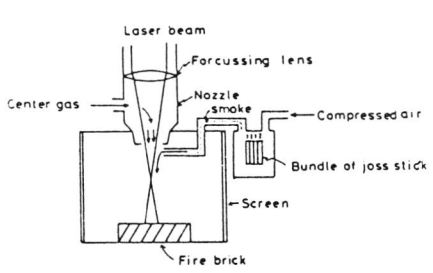

Fig. 6 Real Time Measurement Method of Beam Locus

Fig. 7 An example of beam image

Reference

1) Y. Arata and I. Miyamoto: Trans. JWRI, *2* (1973) 2, 19.

Shape Decision of High Energy Density Beam

Abstract

A new principle and device of measuring accurate beam shape, focal point and energy density of a beam was proposed by using the same beam (beam voltage, beam current, welding speed, gas pressure in vacuum chamber and a_b value) as used in the practical welding.

This method can be applied to any high power and high energy density beam and to all the weldable materials. The authors recommended stainless steel, killed steel or ceramics as a testing material which gives superior quality obtaining beam diameter and is convensional material. With application of this method, the interaction between electron beam and its beam plasma becomes strong around the vacuum of $(0.3—1)\times10^{-1}$ Torr, and phenomena of expanding beam diameter ("beam expansion") and marked decrease of energy density of beam were obviously observed.

1. Introduction

In the high energy density beam welding such as electron beam and laser beam welding, there appears a parameter which represents the location of the beam focal point for a workpiece besides beam voltage, beam current and welding speed as the fundamental welding parameter. This parameter, a_b value, acts as an important factor in giving considerable effects on penetration depth, width of bead, shape of bead and appearance of weld defects, and the definition and characteristics of which have reported in detail by Arata and co-workers[1~5]. As shown in **Fig. 1**, the beam zone in the vicinity of focal point beam active zone, having high energy density gives a shape of bead that shows the most remarkable characteristics of this welding method featuring narrow width of bead and deep and sharp penetration. On the contrary, however, it easily gives rise to such weld defects as porosity, spike and cold shut. While, the beam zone off around the focal point gives a shape of bead having wide width of bead and relatively shallow penetration depth that serves to prevent those weld defects.

In order to clarify these fundamental phenomena and mechanism of bead penetration, it should be required to know the accurate beam shape, that is, beam diameters varied continuously in relation to the distance from the focusing coil and its energy density particularly those at the focal point.

Despite various methods have been tried, no satisfactory method is yet being established due to its extreme difficulty in measuring the shape of beam having high power and high energy density as used in practical welding process, which differs from the measurement of beam diameter having a low power that no melting action is given to the test piece by the irradiation of electron beam as used for the electron microscope. Furthermore, no experiment has so far been tried on the method to determine the shape of beam and focal point during the process of practical welding.

In this technical paper, therefore, the method of measuring the shape of high power and high energy density beam during the practical welding process is proposed and its usability proved by the many experiments, and more the feature of the beam is studied.

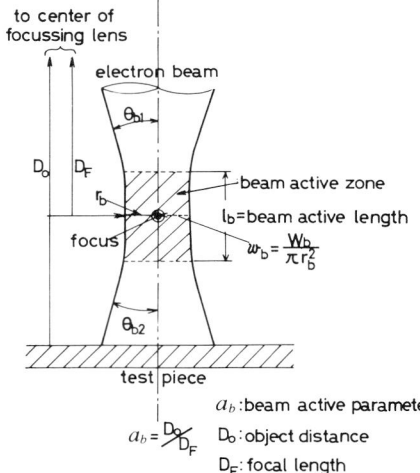

a_b : beam active parameter

$a_b = \dfrac{D_o}{D_F}$ D_o : object distance

D_F : focal length

Fig. 1. Schematic diagram illustrated both shape of electron beam and beam parameters.

2. Preliminary Research

The measuring methods of the shape of beam are largely classified one with melted beam irradiation point and no melted point. For the former, use of sheet material, utilization of burn through phenomenon and use of easily vaporized material are considered as essential means. However, there has been lack of reappearance and reliability in the conventional testing methods and functions of testing materials. For the latter, measurements have been made by the linear or circular movement of the cooling plate or bar[6]. Since application of this method is limited to the range within the low beam power in which the cooling plate is not melted, it is not a suitable method to measure the shape of high power and high energy density beam as used in welding.

In this paper, the authors describe the measuring methods of effective beam diameter for the practical electron beam and its usability is studied through various experiments.

Firstly, effects of various plate thickness on the burn through phenomenon at the weld zone were examined by using the step type workpiece having its thickness from 2 mm to 20 mm ($40^{(W)} \times 240^{(L)}$, 40 mm length for each thickness) as shown in **Fig. 2**. In this

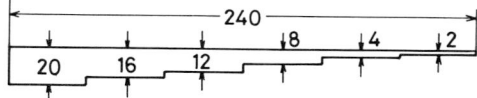

Fig. 2. Step type test piece.

test, effects of welding speed were also examined with the speed varied in four steps as the only parameter. Austenitic stainless steel AISI 304 was used as a test piece, and the bead was given by the electron beam welding machine 150 KV-40 mA type (6 KW). As the result, kerf or "beam groove", moving zone of beam hole where molten metal is not filled up, on both side and its width are shown in **Photo. 1** and **Table 1**.

The conclusion from the facts described above is as follows.
1) Marked variation of kerf or beam groove by plate thickness change despite a value remained unchange.
2) Key hole action as seen in plasma arc welding appears as the plate thickness becomes smaller, molten metal residued without burn through, which resulted in appearance of "secondary melting phenomenon" giving varied kerf or beam groove.
3) This method is considered not practical because a

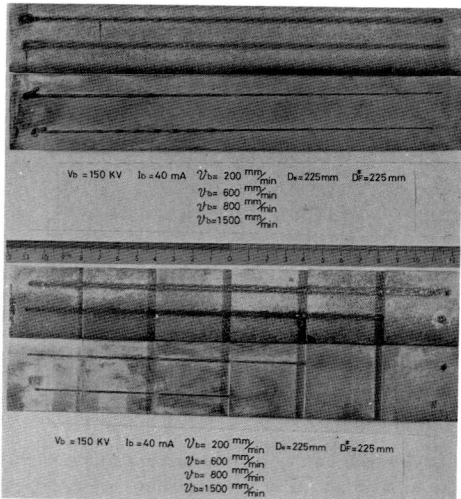

Photo. 1. Surface and reverse bead profile for various welding speed.

Table 1. Bead width for various test piece thickness.
(V_b=150 KV, I_b=40 mA, v_b=200 mm/min).

thickness (mm)	20	16	12	8	4	2
bead width (mm)	0.81	1.05	0.88	0.95	1.50	1.70

large number of test pieces is required to observe continuous change of the shape of beam along its axis.

Secondly, "slope welding" method was then applied as shown in **Fig. 3** by using the test piece with much smaller plate thickness for the purpose of clarifing corelation between kerf width and beam diameter when beam cutting was performed, and continuous change of the shape of beam. In this test, stainless steel (AISI 304) with 1 mm plate thickness, nickel with 0.4 mm and monel metal with 0.2 mm were used.

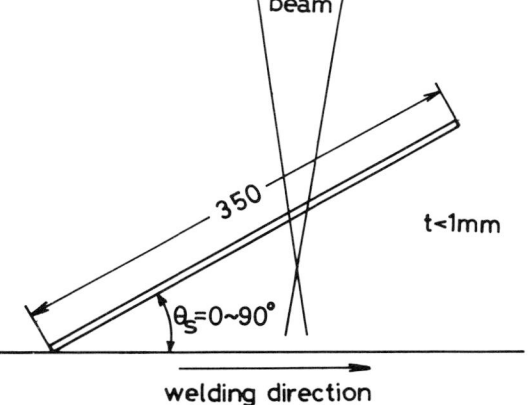

Fig. 3. Schematic drawing of slope welding with thin plate.

195

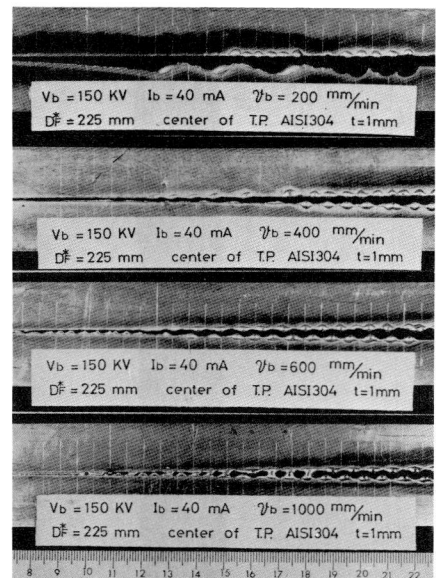

Photo. 2. Secondary melting effect of condensation of molten metal by surface tension for various welding speeds (stainless thin plate).

The results are, as shown in **Photo. 2,** that secondary melting effect of molten metal decreases as welding speed increases, and the kerf width which makes it possible to estimate continuous change of beam diameter can be obtained. On the contrary, adhesion phenomenon of molten metal to the kerf wall by surface tension is further promoted. This phenomenon appears, as shown in **Photo. 3,** even when different materials with much smaller plate thickness are used. This is a substantial phenomenon and cannot be avoided. It has been proved, however,

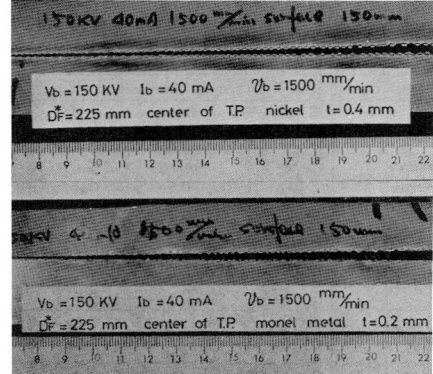

Photo. 3. Condensation of molten metal by surface tension in case of changing plate thickness.

that the slopewelding method used here was extremely effective in observation of the shape of beam at a glance by the use of one test piece.

From the above facts, it has been concluded that the slope welding is the most suitable as a testing method of the shape of beam (simply called "beam test method" or "beam test") which can be observed the shape of beam in a wide range, and for the test piece, it should be necessary to have enough thickness to function rapid and effective removal of molten metal.

Accordingly, the authors have found that such function of the test piece can be given by the application of "edge effect" of a thick plate. Properties of a sharp edge of the thick test piece with profitable angle may be easily estimated as follows.

1) The tip will be vaporized or melted by the slight beam energy.

2) Excess beam energy given around the tip is absorbed rapidly in the base zone, cool and large heat absorber expanding under the tip in the form of heat energy. This action is so powerful that condensation of molten metal by the surface tension will not be allowed to occur, as shown in Photos. 2 and 3.

3) In the case of powerful beam having such excess power that even melts the heat absorber as in welding, most part of the energy is absorbed in the form of melting heat absorber. And at the same time, such melted heat absorber will be removed from the edge zone rapid enough not to occur secondary melting. The authors call such property "edge effect". The test piece with such edge effect therefore is used for the beam test performed by the authors. And, in order to obtain a test piece to put to practical use as well as to meet the purpose, examination was made on edge effects with various shapes and sizes as shown in **Fig. 4.** As a result, test piece with edge shape as shown in **Fig. 5** is selected from more than 10 types of test piece to be used for the beam test methods A type and B type.

A type, as formerly described, is so called "slope welding" using a test piece having slope angle $\theta_s \approx 0°$ ~90° which makes one-dimensional movement in horizontal axial direction, and B type is a method

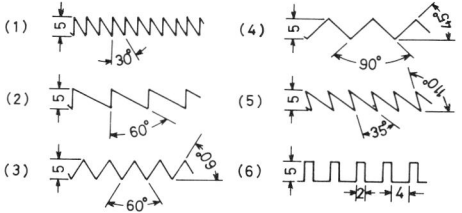

Fig. 4. The kinds of edge shapes.

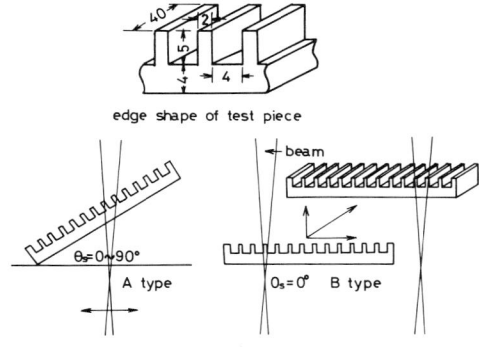

edge shape of test piece

experimental method

Fig. 5. Schematic drawing of beam test methods.

simply called "glide-welding" by the authors using a test piece having $\theta_s = 0°$ which makes oblique movement (two-dimensional movement) along the gradient with glide angle θ_s.

Fig. 6 is the fundamental drawing of the test device. The slope angle is determined by the length of beam zone to be measured (length of test piece required for the purpose) and the condition of fluid of liquid metal from the molten pool. When beam power becoms large or speed is slow, it is necessary to have small $\theta_s = \pm(15°\sim45°)$ is most frequently used. In this test $\theta_s = \pm 30°$ was used.

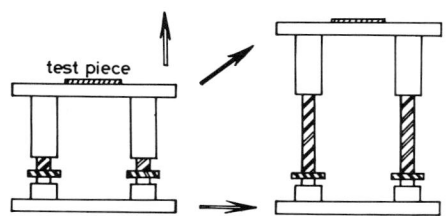

Fig. 6. Schematic drawing of glide beam test method device.

The symbol, $+$, shows upslope-welding which gives a welding bead going toward upside along a slope with θ_s, and symbol, $-$, shows downslope- welding which gives a welding bead going toward downside along the same sloping surface. As shown in **Fig. 7,** when $\theta_s = +30°$ the molten metal is rather widened in the vicinity of the most narrowly focussed beam active zone (focal point) compared to that in case of $\theta_s = -30°$. This is because, in case of upslope-welding, liquid metal in the molten pool tends to flow only downward and to gather in the beam groove near the surface. Such tendency becomes more and more conspicuous with larger θ_s and slower welding speed at the narrower beam groove, i. e., near the focal point to occur secondary melting.

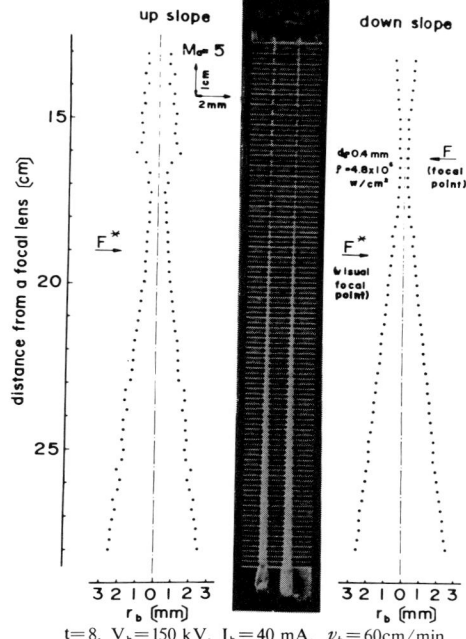

$t = 8$, $V_b = 150$ kV, $I_b = 40$ mA, $v_b = 60$cm/min

Fig. 7. Difference of measured beam shape between up slope welding and down slope welding.

Accordingly, $\theta_s = -30°$ was used for any A type test with edge shape of a test piece as shown in Fig. 5. It is natural that, for B type, no such action needs to be considered because $\theta_s = 0°$.

3. Shape Decision of High Energy Density Beam

3-1 Used materials and test device

Table 2 shows materials and their chemical composition used for test. The test piece has the edge shape shown in Fig. 5 from the results of Preliminary

Table 2. Chemical composition of test piece.

material	mark	chemical composition								
stainless steel	AISI 316L	C <0.08	Si <1.00	Mn <2.00	P <0.04	S <0.03	Ni 12.00 16.00	Cr 16.00 18.00	Mo 2.00 3.00	Fe Bal
phosphor bronze	PBC-1	Sn 10.0 13.0	P 0.05 0.15		Pb+Zn+Fe <1.5			Cu Bal		
copper	DCuP2 -1/2		P 0.004 0.040					Cu Bal		
titanium	KS-40	C+Fe+N ~0.60			O <0.300	H <0.010			Ti Bal	
aluminum alloy	5083	Cu <0.1	Si <0.4	Fe <0.4	Mn 0.3 1.0	Mg 3.8 4.8	Zn <0.1	Ti <0.2	Cr <0.5	Al Bal
aluminum alloy	7075	Cu 1.2 2.0	Si <0.4	Fe <0.5	Mn <0.3	Mg 2.1 2.9	Zn 5.1 6.1	Ti <0.2	Cr 0.18 0.35	Al Bal

Research with the size $40^{(W)} \times 350^{(L)}$. The welding device is 150 KV-40 mA Hamilton type 6 KW EB welder which has variable gas pressure in vacuum chamber from 10^{-1} Torr to more than 5×10^{-4} Torr.

3-2 Test methods

The test adopts the shape of test piece and beam test methods based on the results of Preliminary Research. For test procedures, firstly austenitic stainless steel AISI 316L was selected as a representative test piece, which was fixed with a jig at slope angle 30° based on Fig. 5 A type test method, and downslope welding was performed to examine effects of various factors (beam voltage: V_b, beam current: I_b, welding speed: v_b, a_b value and gas pressure in vacuum chamber: P_{ch} on the shape of beam. The same method was applied on test piece of titanium, copper and its alloy, aluminium alloy to examine effects of materials on the shape of beam.

In order to examine effects of vaporization on the measurement of the shape of beam, A type beam test method was applied by using 7075 Al alloy, in which easily vaporized element (zinc, etc.) is contained, and quartz to measure the shape of beam. Quartz test piece used here was a thin plate having 2 mm$^{(t)}$ $\times$ $40^{(W)} \times 300^{(L)}$.

A similar test to A type was further performed on AISI 316L steel and titanium based on B type beam test method as shown in Fig. 5 using a device as shown in Fig. 6 to compare with the results of A type test method. Prior to welding, grease was removed from all the test pieces by methyl-ethyl ketone. For the determination of measured value of beam diameter, average of three points, i. e., the measured point, before and behind it, was calculated to be adopted as measured value.

3-3 Test results

3-3-1 Effects of welding factors on the shape of beam

AISI 316L stainless steel was used to examine effects of various welding factors on the shape of beam. **Figs. 8~12** show the results. When burn through region of each edge for the test piece, i. e., its kerf width or beam groove represents the beam diameter, these figures show the changes of such beam diameter by the distance from the focusing lens.

Fig. 8 shows how the beam diameter becomes larger as beam voltage decrease, Fig. 9 shows how the beam diameter becomes larger as beam current increases. These tendencies can easily be predicted when considering velocity of electrons and action of space charge based on the beam current value.

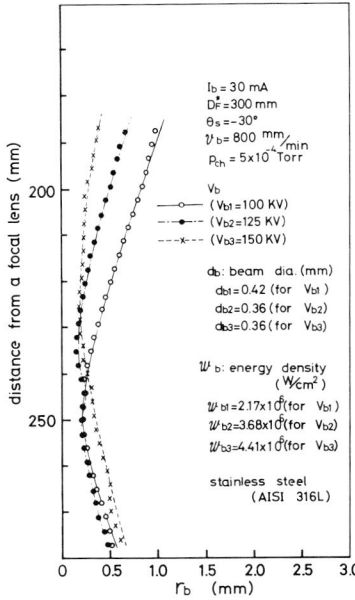

Fig. 8. Relation between beam voltage and beam shape (A type test).

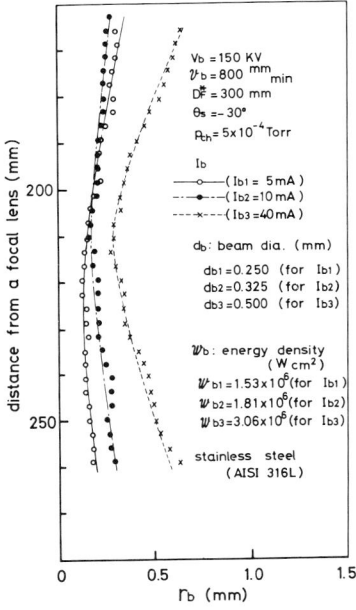

Fig. 9. Relation between beam current and beam shape (A type test).

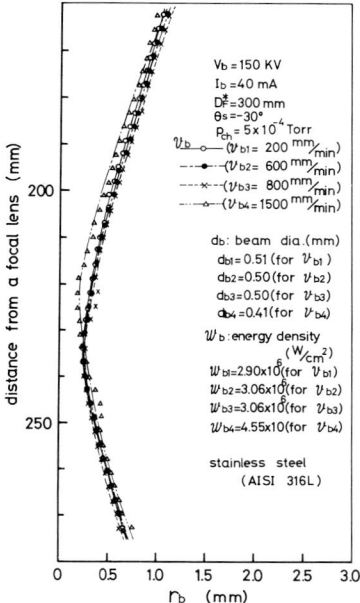

Fig. 10. Relation between welding speed and beam shape (A type test).

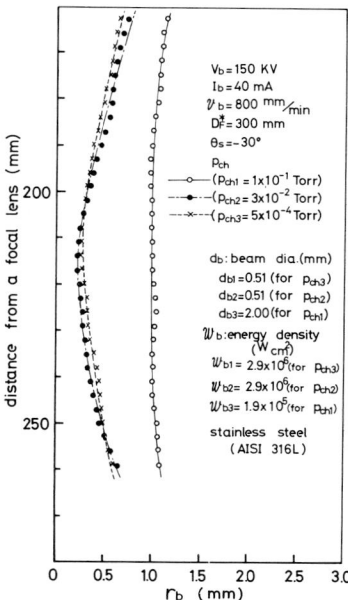

Fig. 12. Relation between gas pressure in vacuum work chamber and beam shape (A type test).

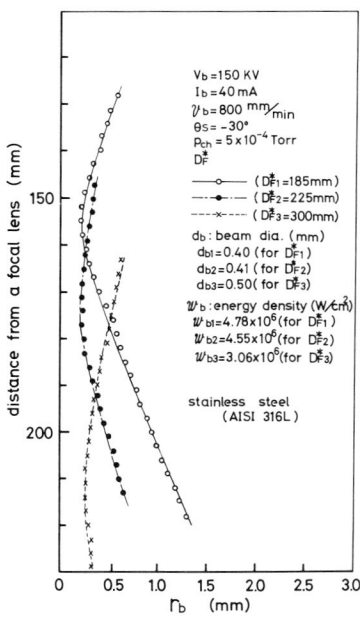

Fig. 11. Relation between visual focal length and beam shape (A type test).

Fig. 10 shows effects of welding speed, which indicates that such change of speed does not give any effect on the shape of beam for AISI 316L stainless steel.

Fig. 11 shows effects of visual focal point on the beam diameter. The visual focal point defines as the target position where beam spot is the smallest, its brightness the highest and is observed visually when the beam of small current ($1 \sim 2$ mA) is irradiated on the target of movable tungsten. The results of Fig. 11 prove that the true focal point of the practical beam welding is located considerably upper from each visual focal point (for example, true focal points for Hamilton type 6 KW electron beam welding machine are located approximately 3 cm, 5 cm, 7.5 cm upper from each visual focal point 18.5 cm, 22.5 cm, 30.0 cm.). Since these facts play extremely important role, true focal point of the welding beam by the beam power must be measured prior to practical welding.

The minimum beam diameter tends to be widened slightly as visual focal point becomes longer.

Fig. 13 shows relation between visual focal point D^*_F, penetration depth h_p and object distance D_o (distance from focusing coil to test piece). Such relation is more clearly expressed by visual beam active

199

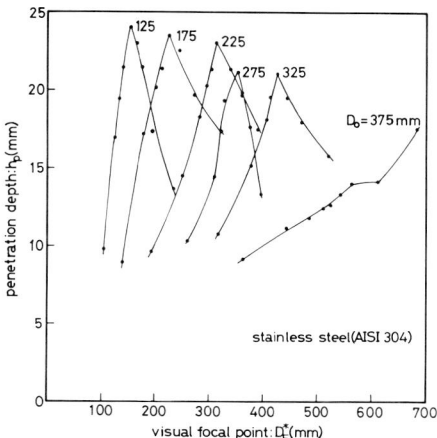

Fig. 13. Relation between penetration depth and visual focal point for various object distance, (V_b = 150 KV, I_b = 30 mA, v_b = 500 mm/min.).

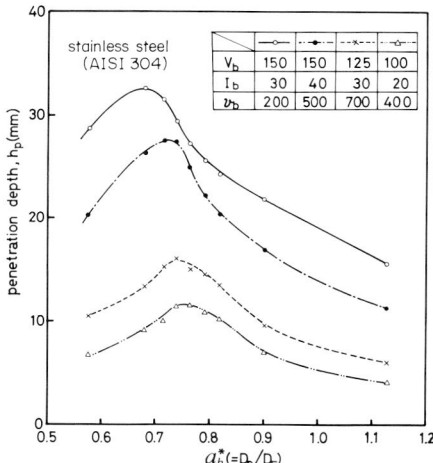

Fig. 14. Relation between penetration depth and visual active parameter: a_b^*.

parameter a_b^* ($\equiv D_o/D_F^*$) as shown in **Fig. 14.** The object distance at the visual focal distance corresponding to each maximum penetration depth as shown in Fig. 13, is located near the real focal distance obtained from Fig. 11 under same visual focal distance above mentioned, which can be easily explained in consideration of its energy density[7]. From these facts described above, penetration depth reaches the maximum when focal point comes near the immediate below the test piece surface as shown in **Fig. 15.**

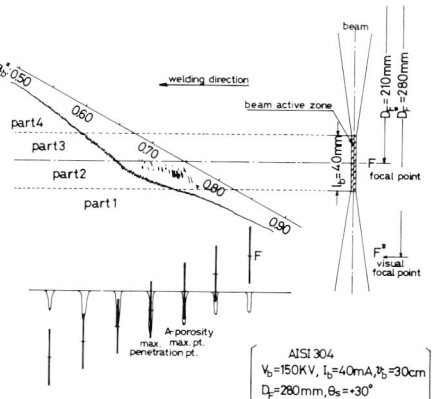

Fig. 15. Schematic explanation of mutual relation on situation between occuring range of porosity. and beam.

Fig. 12 shows effects of gas pressure in vacuum work chamber, P_{ch} on the beam diameter. It seems that the beam diameter measured vacuum chamber having gas pressure from high vacuum condition to 3×10^{-2} Torr receives so little effect of P_{ch} that is can be disregarded. The beam diameter rather looks slightly smaller around 10^{-2} Torr than high vacuum is considered may be due to the action of plasma lens force. When the degree of vacuum lowers to 1×10^{-1} Torr, it is observed that the beam diameter suddenly widened largely. This suggests that interaction between electron beam and its beam plasma becomes abruptly strong in the vacuum range of $(0.3-1) \times 10^{-1}$ Torr, and change in mechanism of collision. Such vacuum range is called "beam expansion vacuum". This new fact has obviously discovered for the first time by A type beam test device, which indicates abrupt decrease of the beam energy density in this vacuum range. It is considered that the above fact will give extremely important contribution to understanding and application of welding phenomenon. For example, well known relation between gas pressure in vacuum chamber and penetration depth as shown in **Fig. 16** is remarkably well explained[7].

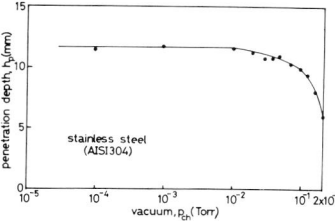

Fig. 16. Relation between penetration depth and gas pressure in vacuum work chamber.

3-3-2 Effects of variation in material on the shape of beam

The same method as that for AISI 316L stainless steel was applied to titanium, copper, phosphor bronze and 5083 Al alloy to examine effects of materials on the shape of beam. The results are shown in **Figs. 17~21.** These results show similar tendency to AISI 316L stainless steel for the effect of each

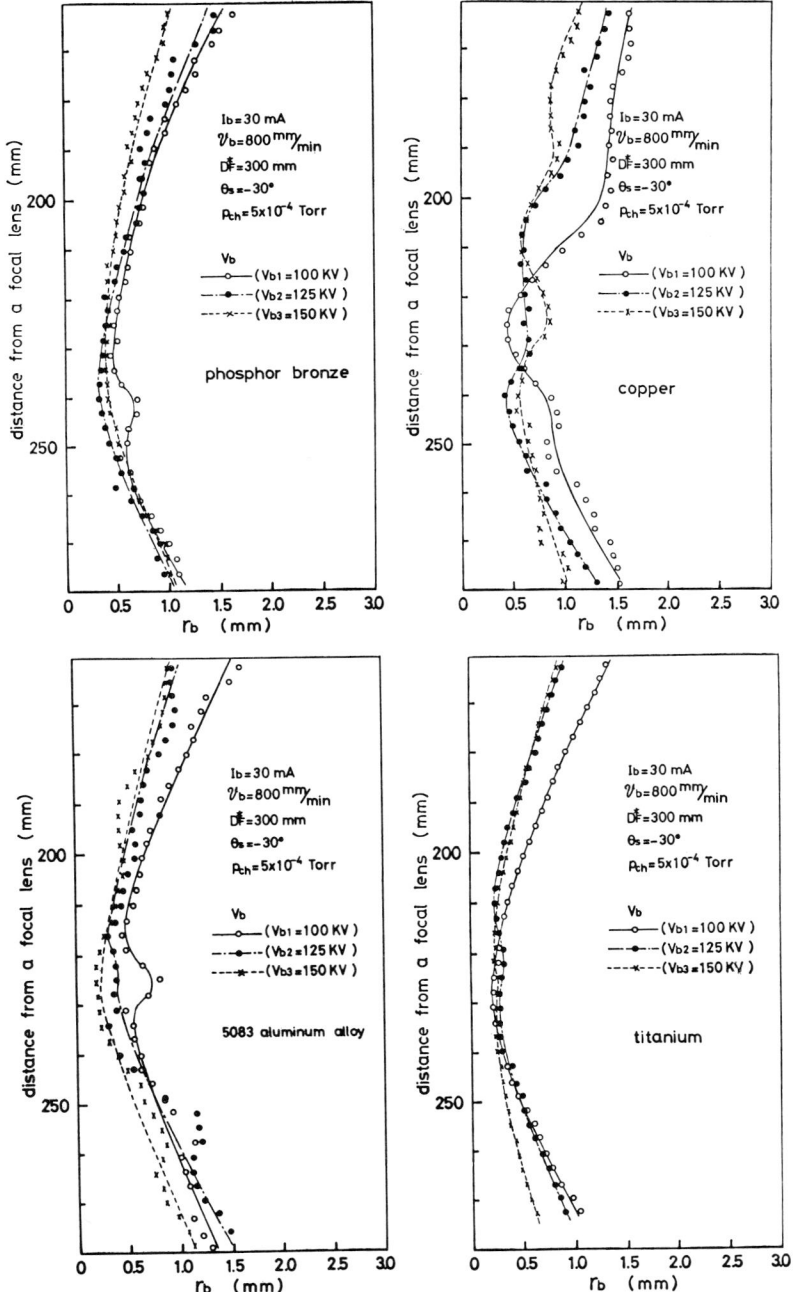

Fig. 17. Relation between beam voltage and beam shape for various materials (A type test).

201

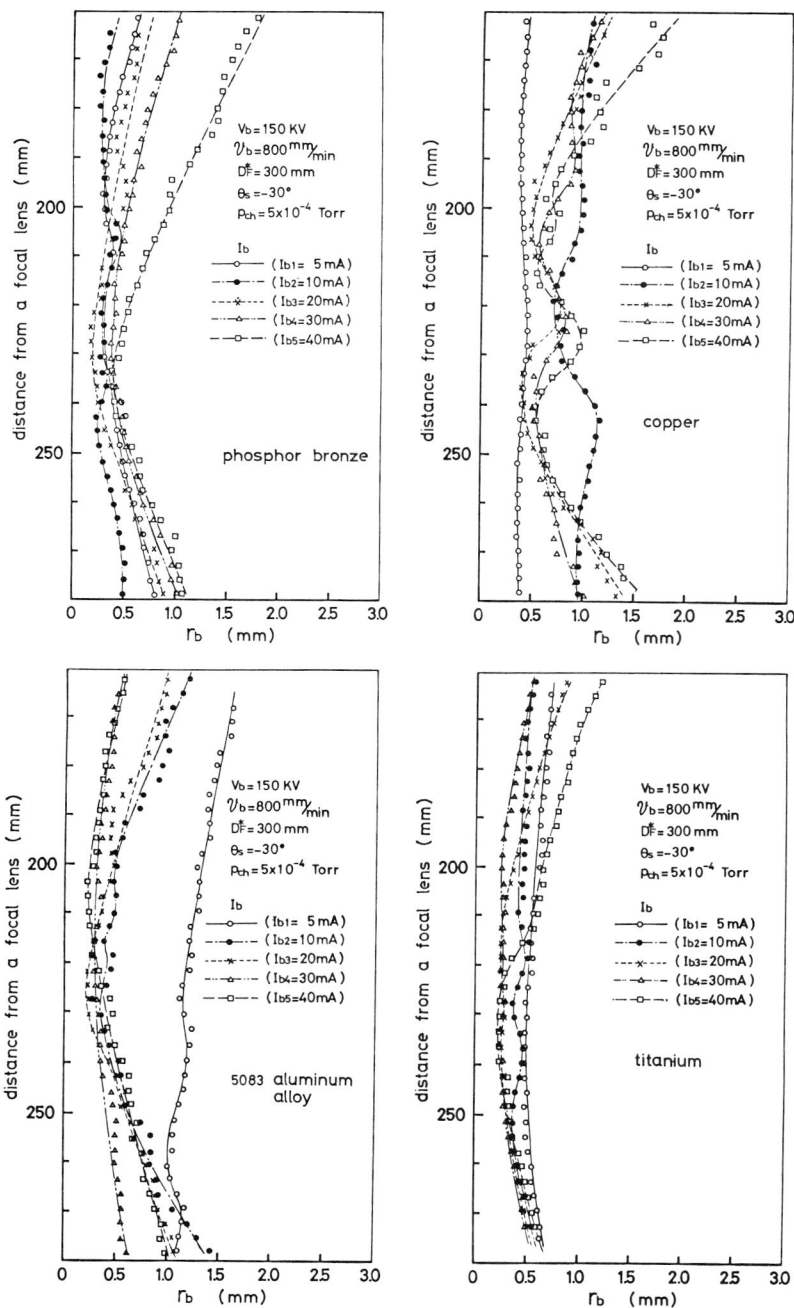

Fig. 18. Relation between beam current and beam shape for various materials (A type test).

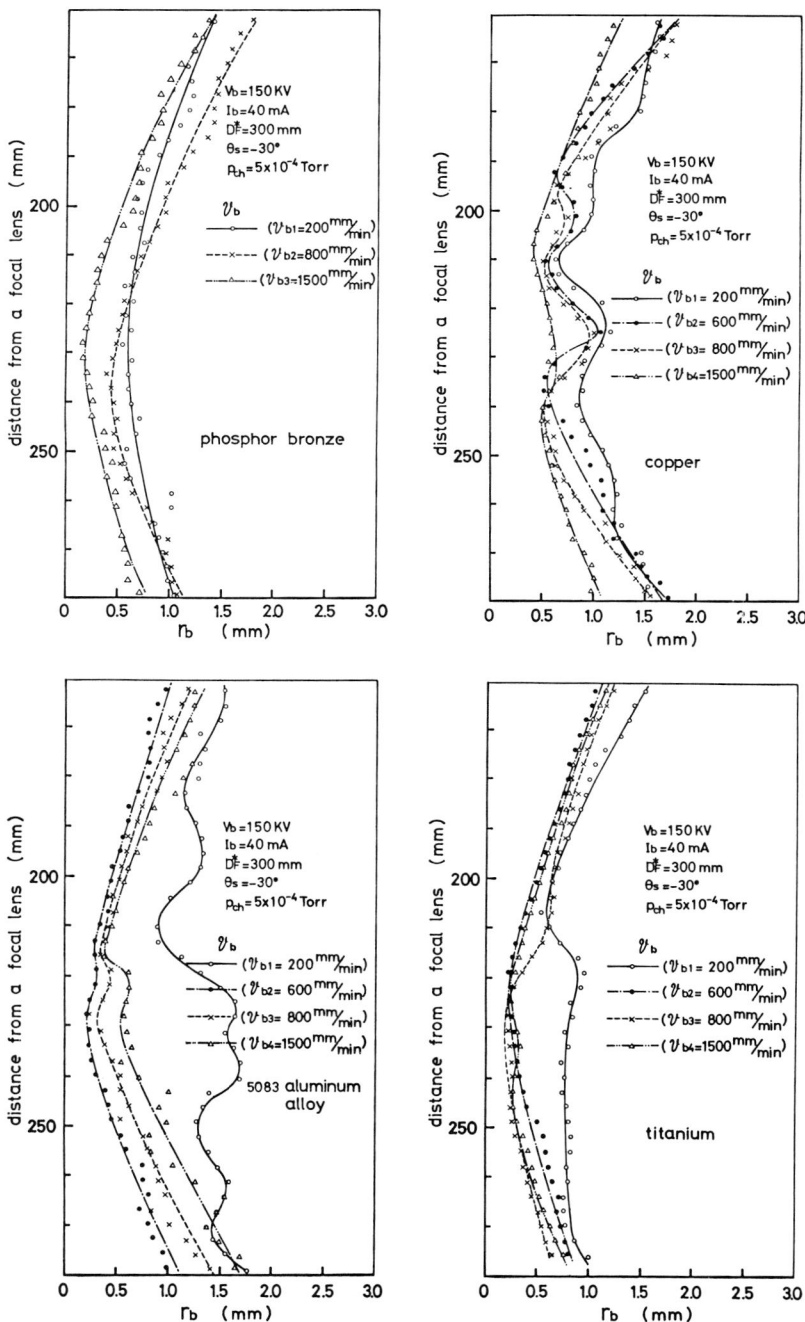

Fig. 19. Relation between welding speed and beam shape for various materials (A type test).

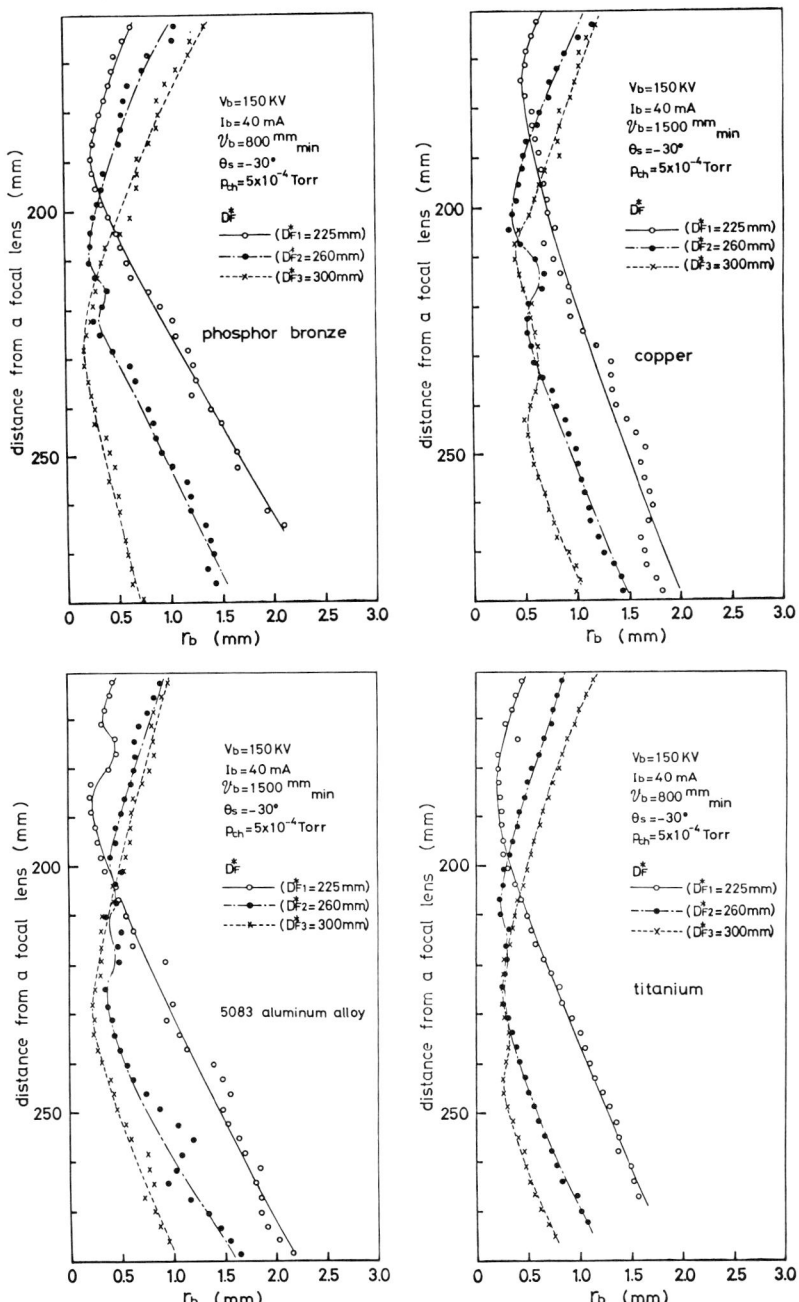

Fig. 20. Relation between visual focal length and beam shape for various materials (A type test).

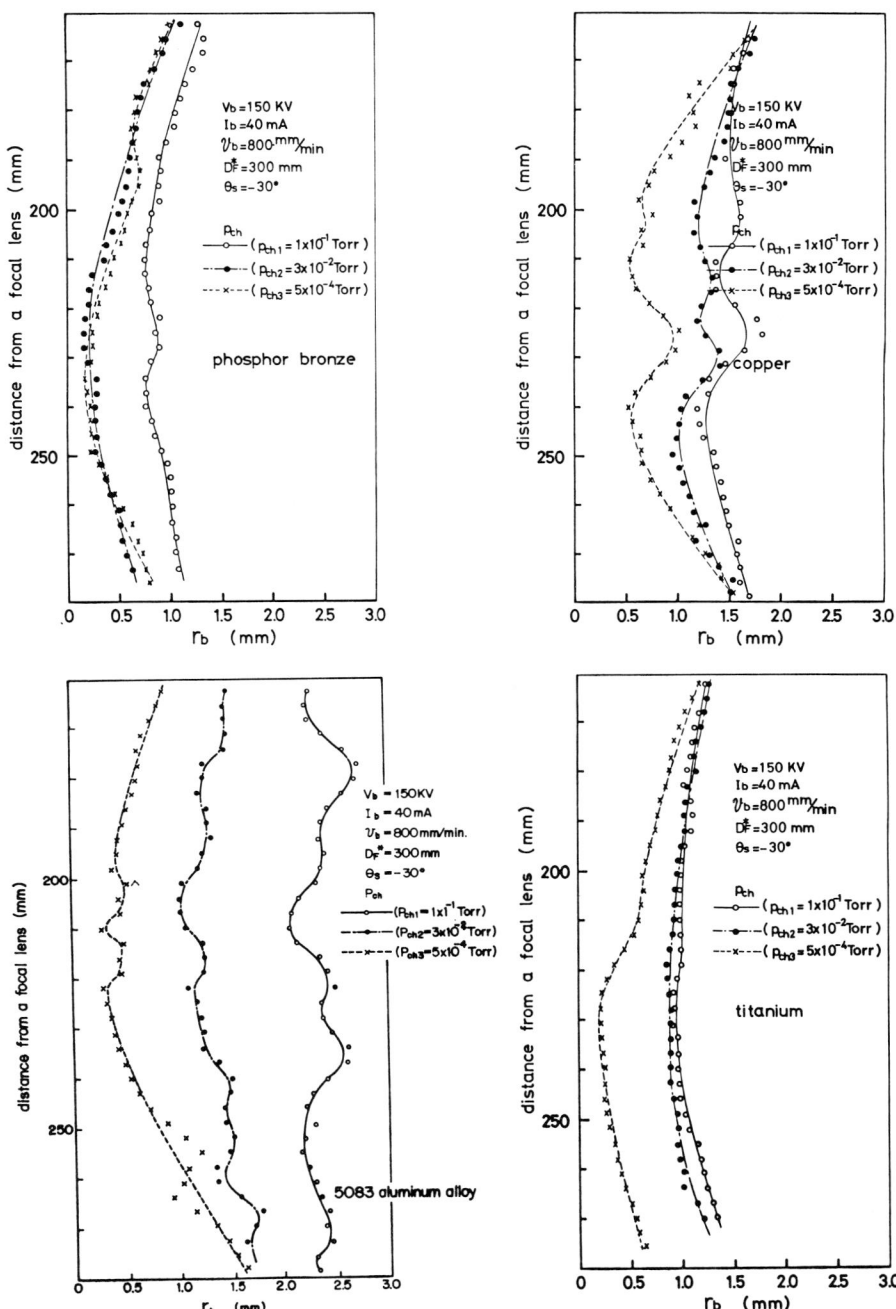

Fig. 21. Relation between gas pressure in vacuum work chamber and beam shape for various materials (A type test).

welding parameter on the shape of beam. As shown in Fig. 19, however, welding speed has a marked effect on the shape of beam for copper and 5083 Al alloy. Particulary, when welding speed becomes slow, secondary melting effect appears as shown in **Photo. 4** so that actual beam diameter is not indicated around the focal point. Also, the result in Fig. 18 shows widened beam diameter in small current range, which may give rise to the beam acted as surface heat source producing slightly melted surface, concerning both effects of large heat conductivity of the material and lower beam power, and molten metal was remained at the edge without burn through as shown in **Photo. 5.**

However, if the limited condition is given to these materials (copper and 5083 Al alloy, etc.), for example, exceedingly fast welding speed with large beam power is used, similar results to those for other materials can be obtained. It has proved, therefore, that measurement of the shape of beam by this method can be applied to any material.

3-3-3 Effects of vaporization on the shape of beam

In order to examine effects of vaporization on the shape of beam, slope welding was performed using 7075 Al alloy in which asily vaporized element (Mg and Zn, etc.) was contained. Effect of gas pressure in vacuum work chamber was examined under the fixed welding conditions ($V_b = 150$ KV, $I_b = 40$ mA, $v_b = 1500$ mm/min, $D^*_F = 300$ mm). The results show, as shown in **Fig. 22,** the similar shape of beam to other materials described above is obtained, and fairly good results are obtained under low vacuum of 10^{-2} Torr. On the contrary, when gas pressure in vacuum work chamber is decreased and vaporization becomes vigorous, they make it impossible to obtain the correct shape of beam. It is natural that such disorder should appear, and determination of gas pressure in vacuum work chamber in consideration of vapor pressure for easily vaporized element contained in the material and beam expansion vacuum is required for a material with remarkable vaporization occurs in the vicinity of focal point. As shown in **Fig. Fig. 23,** when quartz (2 mm$^{(t)}$ × 40 mm$^{(w)}$ × 300 mm$^{(L)}$) which brings extremely vigorous vaporization was used as a test piece, similar beam shape to that of an ordinary metal material was obtained though slightly narrow. It is considered that this result is caused by the effect of space charge which is charged up on the test piece.

To prevent such space charge effect, authors used parallel-arranged ceramic strip fixed by jig of steel, as show in **Fig. 24,** to make space charge

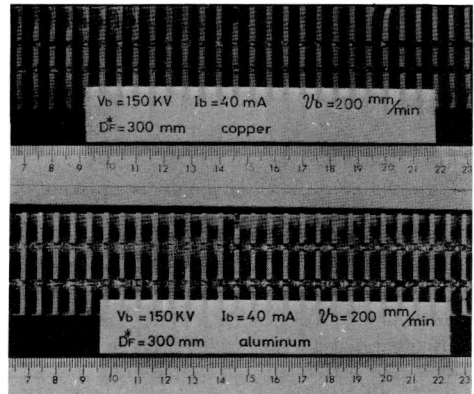

Photo. 4. Swelling of measured beam diameter in the vicinity of focal point.

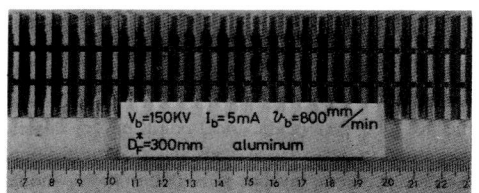

Photo. 5. Bead profile in case of using small beam current.

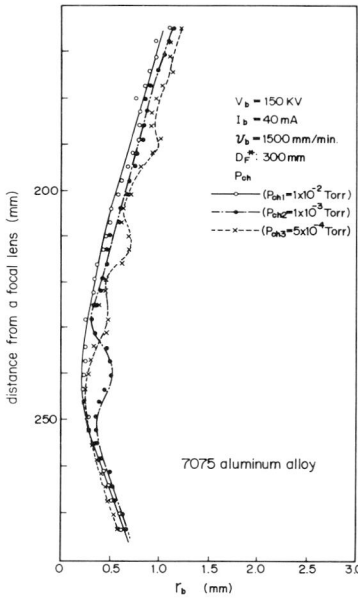

Fig. 22. Relation between gas pressure in vacuum work chamber and beam shape for 7075 aluminum alloy (A type test).

206

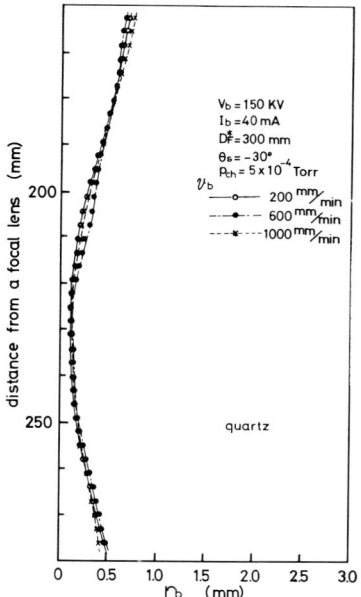

Fig. 23. Relation between welding speed and beam shape for (A type test).

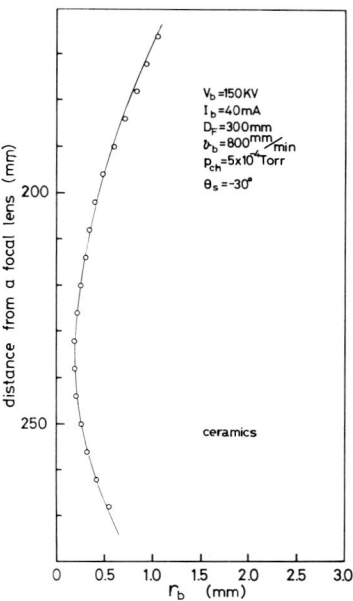

Fig. 25. Beam shape in test piece with parallel-arranged ceramic strip fixed by jig of steel.

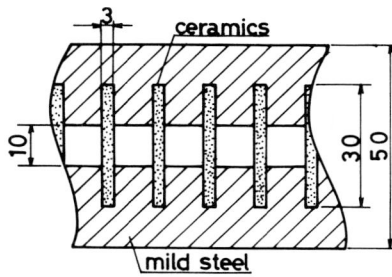

Fig. 24. Schematic drawing of test piece with parallel-arranged cermic strip fired by jig of steel.

escape to the fixed jig of steel through plasma generated around the ceramic specimen. As this result, beam shape is approximately corresponded to the one gained in the specimen of AISI 316L stainless steel as shown in **Fig. 25.**

3-3-4 Tests on reliability of A type testing method

The problems of A type testing method in which downslope welding was used, i. e., effects of flow of molten metal with the test piece placed obliquely on the measured value of beam diameter were examined by using B type testing method ($\theta_s = 0°$) in which glide welding was used as shown in Fig. 5.

As a test piece, AISI 316L stainless steel and

titanium, by which relatively smooth beam shape was obtained in A type testing method, were adopted, and the test was conducted under the same conditions as A type. The results are shown in **Figs. 26 ~29,** which agree well with those of A type testing method on the beam shape and minimum beam diameter.

3-3-5 Summary

From the above results, proper beam shape for each type of material is summarized in **Fig. 30.** From Fig. 30, since tendency and minimum beam diameter for each material almost agree except a few materials (Cu and quartz), this measuring method can be applied to any type of material. In the case of Cu, it is considered that the same results can be obtained if the welding is performed with a large beam power at much faster welding speed. It is therefore allowed to use cheaper materials instead of high priced materials used for welding in affirmation of shape of beam and energy density prior to practical welding. The most suitable material of the test piece that can be recommended from such point of view is stainless steel followed by killed steel. Practically speaking, such beam properties (shape of beam, focal point, energy density) vary oftened in accordance with the use of filament and its replacement, so that it is better these should be measured before welding.

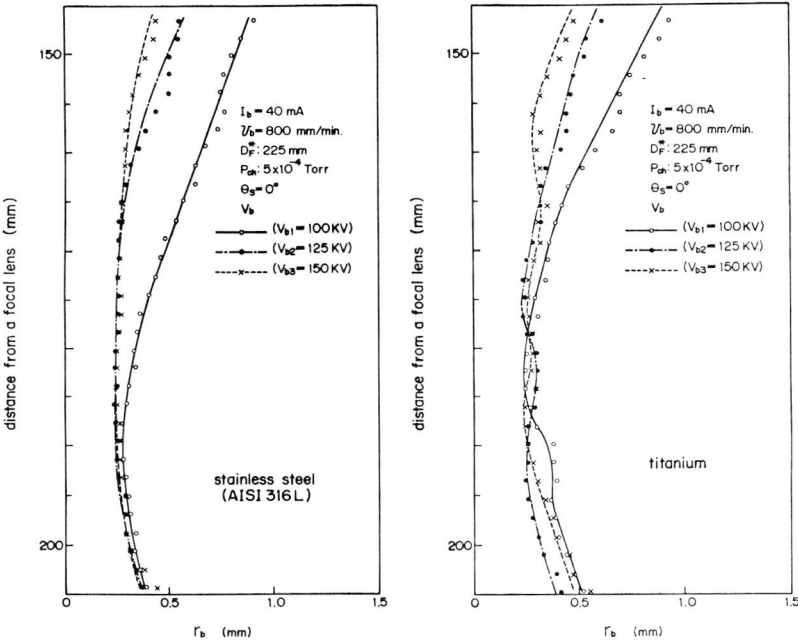

Fig. 26. Relation between beam voltage and beam shape (B type test).

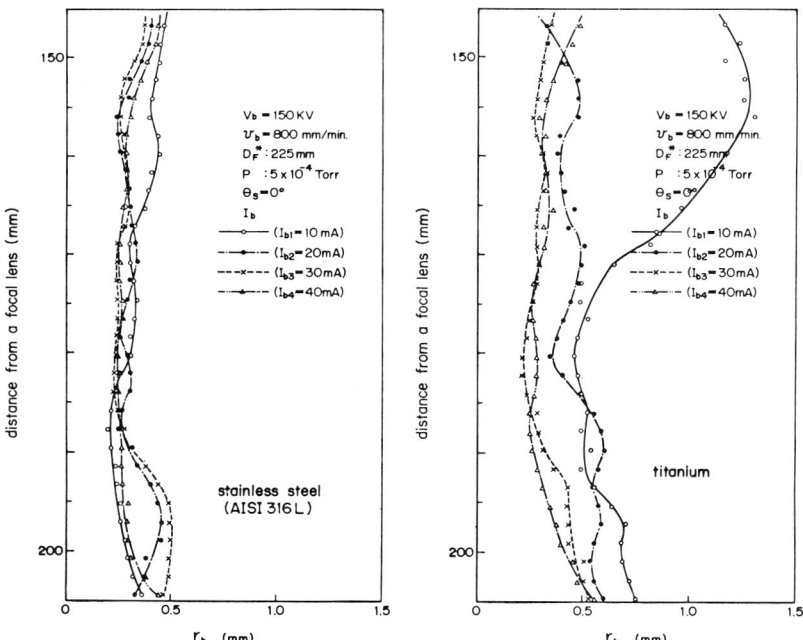

Fig. 27. Relation between beam current and beam shape (B type test).

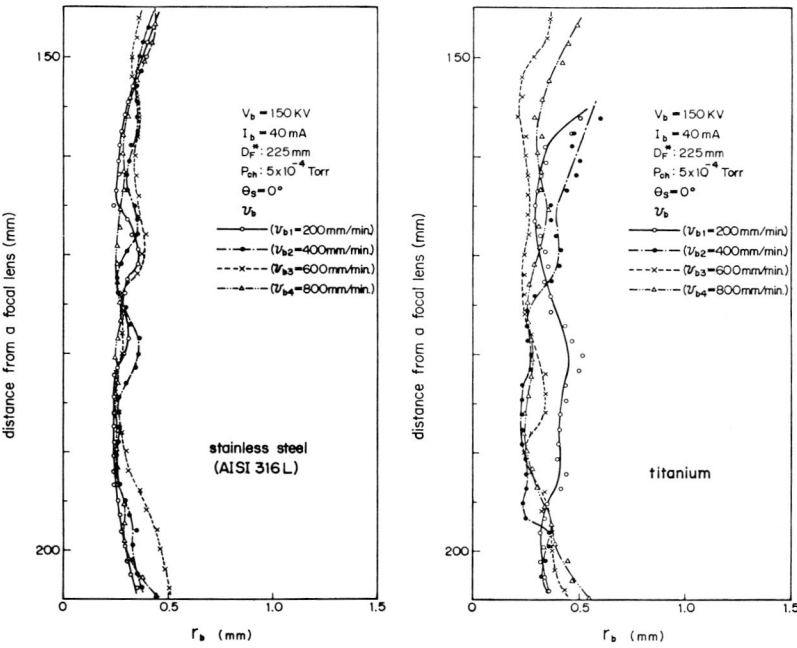

Fig. 29. Relation between gas pressure in vacuum work chamber and beam shape (B type test).

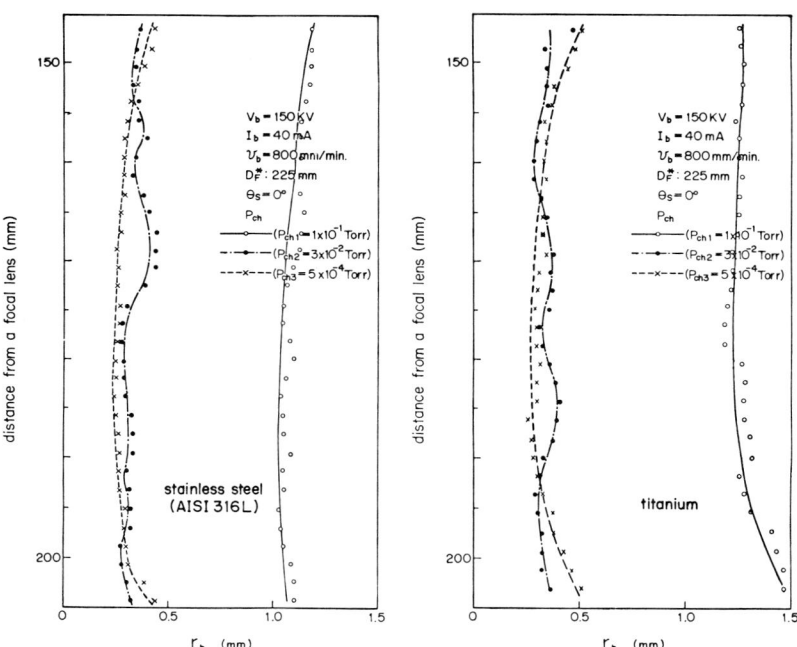

Fig. 28. Relation between welding speed and beam shape (B type test).

209

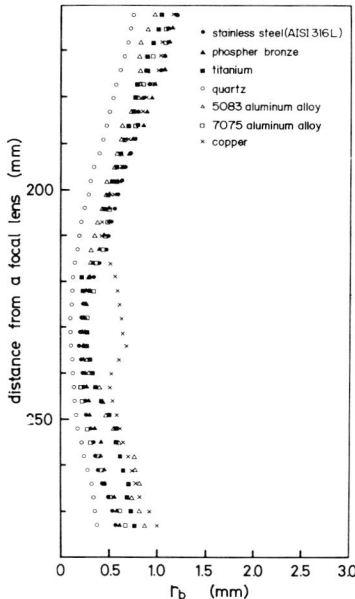

Fig. 30. Summarization of proper beam shape for each type of materials.

4. Conclusion

From the test results described above the following conclusion can be drawn.

1) The new beam test method has been achieved by using 2 types of welding methods ("slope welding", particularly "downslope welding" and "glide welding") and a test piece having function of edge effect.

2) Since this method can be performed under the same conditions as the practical welding, namely, the same beam conditions as during the welding (beam voltage, beam current, shape of beam, welding speed, gas pressure in vacuum chamber and a_b value) and the test piece of the same material as welds can be used, effective beam diameter and energy density during the practical welding can be measured.

Only the difference between this testing method and the practical welding is that the test piece has function of edge effect.

3) This method allows to observe effective profile at a glance quite easily and correctly: focal point of beam and its effective energy density under any welding condition can be measured correctly.

4) It was proved that, in vacuum range $(0.3-1) \times 10^{-1}$ Torr, interaction between electron beam and its beam plasma became vigorous, and energy density abruptly decreased due to abrupt expansion of beam diameter ("beam expansion"). From the above fact, it was also proved that so-called "beam expansion vacuum" exists for powerful electron beam and relation between gas pressure in vacuum work chamber and penetration depth, which was already measured but remained unexplained, could be properly explained.

5) This testing method can be applied to all the weldable materials. Practically, it can be recommended that stainless steel is the best as a test piece followed by killed steel.

6) Welding conditions as a test piece to obtain effective beam diameter for such materials as stainless steel and steel having inferior heat conductivity and good fluid of liquid metal are rather easy in restrictions whereas those for such material having good heat conductivity as Cu and Al or containing a large amount of element of vaporosity with high vapor pressure are severe in restrictions.

7) Remarkable secondary melting phenomenon may appear on some materials under certain welding conditions and burn through effect disappears. In such a case, true beam diameter does not show. On the contrary, such phenomenon will help to study property of the material in high temperature.

8) In an insulating material, beam shape is approximately corresponded to the one with AISI 316L stainless steel by use of parallel-arranged ceramic specimen fixed by jig of steel or other metals.

References

1) Y. Arata: "Characteristics of Electron Beam Heat Souce and View of Development on Electron Beam Welding Technology", J. Japan Welding Society, Vol. 41, No. 11 (1972).

2) Y. Arata, M. Tomie and Y. Katoh: "Some Properties of 30-KW class Electron Beam for Welding", Vol. 2, No. 1 (1973).

3) Y. Arata, K. Terai and S. Matsuda: "Study on Characteristics of Weld Defect and Its Prevention in Electron Beam Welding (Report 1)", Vol. 2, No. 1 (1973).

4) Y. Arata, M. Tomie, K. Terai, H. Nagai and T. Hattori: "Shape Decision of High Energy Density Beam (Report 1)", Document of Committee of Electron Beam Welding, JWS, No. EBW-69-72 (1972).

5) Y. Arata, M. Tomie, K. Terai, H. Nagai and T. Hattori: "Shape Decision of High Energy Density Beam (Report 2)", Document of Committee of Electron Beam Welding, JWS, No. EBW-76-73 (1973).

6) H. Suzuki, T. Hashimoto and F. Matsuda: "Characteristics of Electron-Beam for welding", .J. Japan Welding Society, Vol. 32, No. 5 (1963).

7) K. Terai, T. Toyooka and H. Nagai "Effects of Process Parameter on the Penetration Depth in High Voltage Electron Beam Welding", Transaction of Japan Welding Society, Vol. 3, No. 1 (1972).

Focusing Characteristics of High Energy Density Beam

Abstract

Mathematical formulations which give power density of a focused beam are proposed. The calculated values are compared with experimental data obtained from CO₂ laser and electron beams. It is shown that both are considerably coincidental.

1. Introduction

In the heat processing such as welding, cutting and drilling, phenomena occurring in the process differ according to energy density and its distribution of the heat source used, thereby affecting the quality of the processed goods. Laser and electron beams, which can be well concentrated, producing considerably higher energy density than conventional heat source, have not only improved the quality but even enable the processing which has been impossible. However, knowledge of the energy densities and distributions of the sources such as laser and electron beams, which are concentrated by a lens, is not sufficient, in spite of their importance, because of the difficulty of experimental and theoretical treatments.

In general, as the beam has a certain divergency which varies with many factors including beam species, quality, energy density and so on, it is almost impossible to propose the precise energy density of the focused beam. In this paper the energy density distribution is proposed theoretically by assuming the beam divergence to be uniform in all portions for simplicity, and is compared with the experimental data obtained from laser and electron beams.

2. Energy Density Distribution of Focused Beam

In order to calculate the energy distribution of the focused beam the following two assumptions are introduced:

1) The beam radiated from a point diverges in a cone like shape of which axis is in the radiated direction.
2) Its beam energy density in the plan perpendicular to the axis is given by a Gaussian curve.

Then the beam power $f(x, y, z)$ in the xy plane at (x, y, z) due to the limited beam $q \cdot ds$ radiated from the origin in the direction to the z axis is given by

$$f(x, y, z) = \frac{q}{2\pi\sigma_0^2 z^2} \, exp\left(-\frac{x^2+y^2}{2\sigma_0^2 z^2}\right), \quad \text{------------- (1)}$$

where σ_0 is the standard deviation of the Gaussian curve at $z=1$.

Here the power distribution of the beam around the origin is proposed under condition the beam $q = q(X, Y)$ in the $z=Z$ plane is directed toward the origin as shown in **Fig. 1**. The beam $q \cdot dx \cdot dy$ from a small area $dx \cdot dy$ at point $S(X, Y)$ provides the beam power density in the plane perpendicular to the z axis at point $A(x, y, z)$ given by the following equation:

$$f(x, y, z; X, Y) = \frac{q \, dx \, dy}{2\pi\sigma_0^2 l^2} \, exp\left(-\frac{m^2}{2\sigma_0^2 l^2}\right) \cos\beta. \text{ --(2)}$$

Where

$$l = \overline{SB} = \frac{X^2 + Y^2 + Z^2 - (xX + yY + zZ)}{\sqrt{X^2 + Y^2 + Z^2}},$$

$$m = \overline{AB} = \left[\frac{(x^2 + y^2 + z^2)(X^2 + Y^2 + Z^2) - (xY + yX + zZ)^2}{X^2 + Y^2 + Z^2}\right]^{\frac{1}{2}} \quad (3)$$

and

$$\cos\beta = \frac{Z - z}{\sqrt{(X-x)^2 + (Y-y)^2 + (Z-z)^2}}.$$

(β represents the angle between $\overline{AS}$ and the z axis). Putting $x = r\cos\theta$, $y = r\sin\theta$, $X = R\cos\Theta$ and $Y = R\sin\Theta$, l, m and $\cos\beta$ may be written as follows;

$$l = \frac{R^2 + Z^2 - (rR\cos\varphi + zZ)}{\sqrt{R^2 + Z^2}},$$

$$m = \left[\frac{(r^2+z^2)(R^2+Z^2)-(rR\cos\varphi+zZ)^2}{R^2+Z^2}\right]^{\frac{1}{2}} \quad ---- (4)$$

and

$$\cos\beta = \frac{Z-r}{\sqrt{R^2+Z^2-2rR\cos\varphi-2zZ+r^2+z^2}},$$

where $\varphi = \theta - \Theta$.

When the beam power is constant in a circlar region of radius R_0;

$$q(R,\ \Theta) = \left\{\begin{array}{ll} q_0, & R \leq R_0 \\ 0, & R > R_0 \end{array}\right\}, \quad ---------------------- (5)$$

and is focused through a lens of the focal length $Z = D_F$, the beam power in a plane vertical to the z axis at location $A(x,\ y,\ z)$ is given by

$$w_b(x,\ y,\ z) = \int_0^{R_0}\int_0^{2\pi} f(x,\ y,\ z;\ R,\ \Theta)\ R\ dR\ d\Theta. \quad (6)$$

Now the following assumptions are introduced;
1) Radius of the beam source in $z = D_F$ plane is sufficiently smaller than the focal length; $R_0 \ll D_F$.
2) The values r and z treated here are limited to a sufficiently small region in comparison with source radius R_0 and focal length D_F, respectively, $r \ll R_0$ and $z \ll D_F$.
Then it becomes

$$l^2 \approx z^2,$$
$$m^2 \approx \frac{r^2 D_F^2 + z^2 R^2 - 2rzRD_F\cos\varphi}{D_F^2} \quad ----------------- (7)$$

and

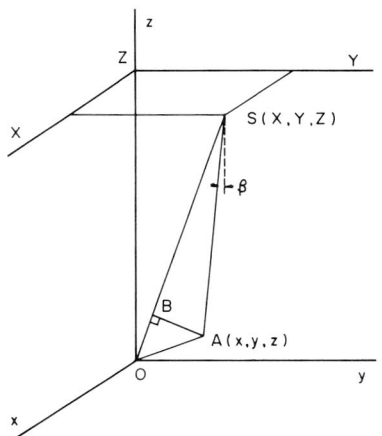

Fig. 1. Beam radiated from point $S(X,\ Y,\ Z)$.

$$w_b(r,\ \theta,\ z) = \frac{q_0}{2\pi\sigma_0^2 D_F^2}$$
$$\times \int_0^{R_0}\int_0^{2\pi} exp(-\frac{r^2 D_F^2 + z^2 R^2 - 2rzRD_F\cos\varphi}{2\sigma_0^2 D_F^4})$$
$$\times R\,dR\,d\varphi = \frac{q}{\sigma^2}\int_0^{R_0} exp(-\frac{r^2 D_F^2 + z^2 R^2}{2\sigma_0^2 D_F^4})$$
$$\times I_0(\frac{rzR}{\sigma^2 D_F^3})\ R\,dR \quad ------------------ (8)$$

where $I_0(a)$ represents the modified Bessel function of the first kind of order zero. Putting

$$Q = \pi R_0^2 q_0,$$
$$\sigma_F = \sigma_0 D_F, \quad -------------------------------------- (9)$$
$$r^* = r/\sigma$$

and

$$z^{*\prime} = \frac{Z}{\sigma_F}\cdot\frac{R}{D_F},$$

it may be written

$$w_b(r^*,\ z^*) = \frac{Q}{2\pi\sigma_F^2}\cdot\frac{2}{z^{*2}}\int_0^{z^*} exp(-\frac{z^{*\prime2}+r^{*2}}{2})$$
$$\times I_0(z^{*\prime}r^*)\ z^{*\prime}dz^{*\prime} \quad ------------------ (10)$$

where $z^* = (z/\sigma_F)(R_0/D_F) = z/\sigma_F \tan\theta_b$ and $2\theta_b$ represents convergence angle. At focal point, $r^* = z^* = 0$, it becomes

$$w_b(0,\ 0) = \frac{Q}{2\pi\sigma_F} \equiv w_{bm} \quad -------------------- (11)$$

Normalizing the value of $w_b(r^*,\ z^*)$ by that of focal point w_{bm},

$$w_b^*(r^*,\ z^*) = \frac{2}{z^{*2}}\int_0^{z^*} exp(-\frac{z^{*\prime2}+r^{*2}}{2})\cdot I_0(z^{*\prime}r^*)$$
$$\times z^{*\prime}dz^{*\prime} \quad -------------------------------- (12)$$

Writing $I_0(a)$ in the form

$$I_0(a) = \sum_{n=0}^{\infty}\frac{a^{2n}}{2^{2n}(n!)^2}, \quad ------------------------------- (13)$$

Eq (12) becomes

$$w_b^*(r^*,\ z^*) = \frac{2}{z^{*2}}\int_0^{z^*} exp(-\frac{z^{*\prime2}+r^{*2}}{2})$$
$$\times \sum_{n=0}^{\infty}\frac{z^{*\prime2n+1}r^{*2n}}{2^{2n}(n!)^2}dz^{*\prime} \quad ----------------- (14)$$

As the results of the partial integration of Eq (14),

$$w_b{}^*(r^*,\ z^*)=\frac{2}{z^{*2}}\left[1-exp\left(-\frac{z^{*2}+r^{*2}}{2}\right)\right.$$
$$\left.\times\sum_{n=0}^{\infty}\ \sum_{m=0}^{n}\ \frac{1}{n!m!}\left(\frac{z^*}{2}\right)^{2m}\left(\frac{r^*}{2}\right)^{2n}\right]\ \text{------}(15)$$

In special cases;

1) in the focal plane, $z^*=0$,

$$w_b{}^*(r^*,\ 0)=exp\left(-\frac{r^{*2}}{2}\right)\ \text{------------------}(16)$$

2) on the beam axis, $r^*=0$,

$$w_b{}^*(0,\ z^*)=\frac{2}{z^{*2}}\left[1-exp\left(-\frac{z^{*2}}{2}\right)\right]\ \text{----------}(17)$$

In Figs. **2** and **3** nomalized beam power density in the planes perpendicular to the z axis and parallel to the z axis is plotted. On the beam axis the beam power density falls to e^{-1} of $w_b\ (0,\ 0)$ at $z^*=2.23$. This corresponds to $z=2.23\sigma/\tan\theta_b$. Recently one of the authors named the length along the beam axis between $w_b\ (0,\ 0)/e$ points above and below focal point as "beam active length", l_b given by

$$l_b=4.46\frac{\sigma}{\tan\theta_b}\ \text{------------------------------}(18)$$

where $2\theta_b$ represents the convergence angle of the beam toward the focal point.

3. Comparison with Experimental Results

3.1 Experimental Setup and Method

Two kinds of beams—laser and electron beams were used in order to check the theoretically analysed results described in the last section. One beam was obtained from the CO_2 laser which consisted of a water-cooled discharge tube 74 mm i. d., 12 m long, placed between gold evaporated plane and 29.2 m radius curvature mirrors (Ref. 1). The multimode laser power with a wavelength 10.6 μ was coupled out from the cavity through 15 mm diam hole at the center of the plane mirror. The coupled out beam was focused through a concave spherical mirror 2.0 m away from the laser head. Another beam was produced from electron beam apparatus, 150 KV-40 mA type (Hamilton Standard Division).

The test piece put on the slant, which moved horizontally at a constant speed v_b as shown in **Fig. 4,** was irradiated by the beam and the width of the fused bead or groove was measured. The contour of the width represents the equi-power line if there is no heat

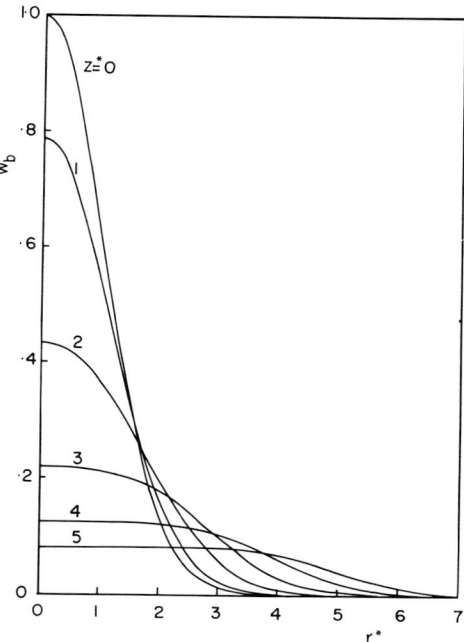

Fig. 2. Relation between $w_b{}^*$ and r^* for various z^*.

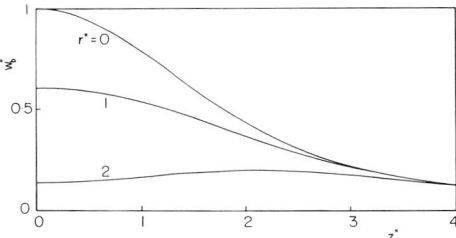

Fig. 3. Relation between $w_b{}^*$ and z^* for various r^*.

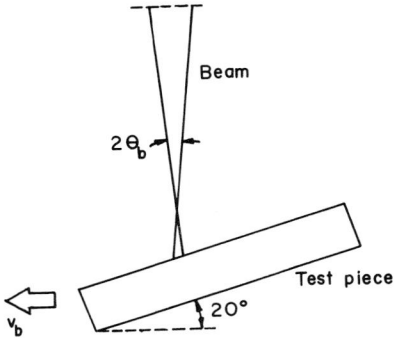

Fig. 4. Schematic diagram of experimental setup.

Table 1. Composition of SUS316L steel (%).

C	Si	Mn	P	S	Ni	Cr	Mo
0.03	1.0	2.0	0.04	0.03	12—16	16—18	2—3

conduction. Here the materials with low thermal conductivity—acryl for laser beam and austenitic SUS316L steel were used to minimize the effect of the thermal conduction. The composition of SUS316L steel was shown in **Table 1.**

3.2 CO_2 Laser Beam

The multimode CO_2 laser beam coupled out from the hole of diameter 15 mm was focused by means of a concave spherical mirror of focal length $D_F = 155$ mm at distance about 2 m from the coupling hole in the same way as mentioned in Ref. 1; in this way the incident beam to the mirror was off-axial, but the effect of the astigmatism could be disregarded by making the incident angle small. Energy density distribution in the plane vertical to the beam axis at the focal point was measured and it was found that the profile coincided quite well with the Gaussian curve of standard deviation $\sigma_F = 0.177$ mm $(\sigma_0 = 1.14 \times 10^{-3})$ (Ref. 1).

Assuming that for the beam reflected from the spherical mirror Eq. (1) is valid and the beam power density is uniform in the circular region of the effective radius R_0, for simplicity, the beam source assumed in section **2** may be regarded to be at the mirror surface. In order to obtain the effective source radius R_0, a film was placed at mirror surface and irradiated by the laser beam for a short time. The radius of the scoached area of the film was about 9.8 mm. Then r^* and z^* are given as follows:

$$r^* = \frac{r}{\sigma_F} = 5.65\ r$$
$$z^* = \frac{z}{\sigma_F}\frac{R_0}{D_F} = 0.36\ z\ .$$

As the moving slant test piece shown in **Fig. 4** continuously irradiated by the laser beam, wedge-shaped deep groove was formed, and width of the groove at its surface was measured. In **Fig. 5** the experimental data of the half width r are plotted in the cases of $v = 2$ cm/sec, 5 cm/sec and 6 cm/sec. As values r in mirror side and the opposite side were almost symmetric with respect to the focal point, the mean value of both was plotted. In this figure solid lines show theoretical value and agree well with the experimental data. Since the laser power level is 85W, energy density w_{bm} at the focal point, $r^* = z^* = 0$, is about 430 W/mm^2. For example, 4.3 W/mm^2 is ob-

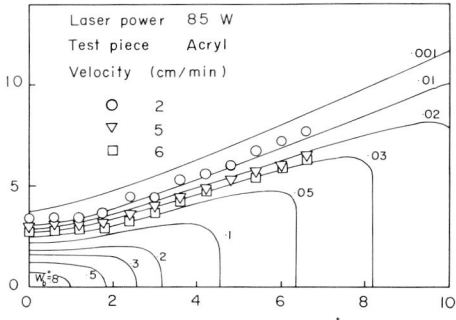

Fig. 5. A comparison of theoretical values with experimental data obtained from laser beam.

tained at $w_b{}^* = 0.01$. The thermal conductivity of acryl is exceedingly small, of the order 10^{-4} cal/sec °C cm so that energy density at the groove edge divided by the velocity of the test piece v_b may be regarded to be constant. It corresponds to 217 (W/cm^2) (cm/sec) in both cases of $v_b = 2$ cm/sec and 6 cm/sec as shown in **Fig. 5.**

3.3 Electron Beam

It has been experimentally made clear by A. Sanderson that energy density distribution can be approximated by a Gaussian curve (Ref. 2). According to his data standard deviation of the Gaussian curve σ_F is $0.06 \sim 0.07$ mm when accelerating voltage $V_b = 30$ KV and beam current $I_b = 10$ mA, and here 0.06 mm was adopted expediently.

A series of experimens was performed under various conditions and the bead width was measured (Ref. 3). In **Fig. 6** the bead half width is plotted against the distance from the focal lens for SUS316L steel, for example. The contour of the bead was not symmetric with respect to the focal point; the width at lens side was somewhat smaller than that at the opposite side.

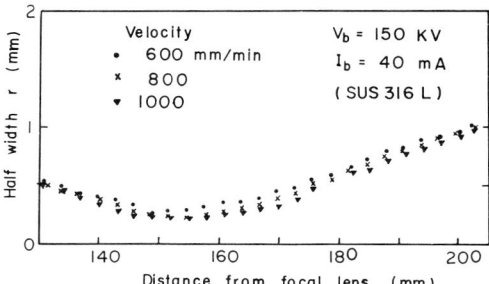

Fig. 6. Half beam width plotted against the distance from a focal lens.

214

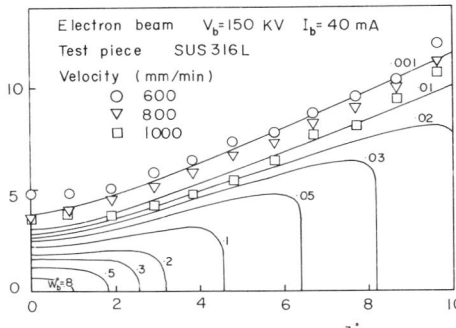

Fig. 7. A comparison of theoretical values with experimental data obtained from electron beam.

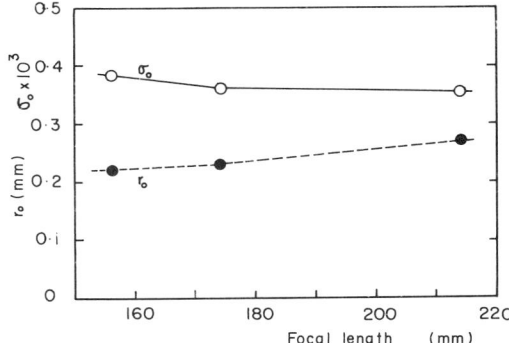

Fig. 8. The values of r_0 and σ_0 plotted against focal length.

In **Fig. 7** the data are compared with the equi-power lines in r^*-z^* co-ordinates obtained from theoretical analysis. When data were converted into non-dimensional values, r^* and z^*, $\sigma_F=0.06$ mm and $\tan\theta_b$ $=1/52$ were used. The convergency angle θ_b was proposed from the inclination of the linear part of the curve imagined from the data, where the beam source was assumed to locate at the center of the lens in the same manner as the laser beam. For the half bead width the average value of the lens side and opposite one was adopted. According to $\tan\theta_b=1/52$, the effective radius of the beam source R_0 became about 3 mm. The experimental data are a little bigger than the theoretical value around the focal point, but it can be said that both are considerably coincidental. This may be because electron velocity is not constant so that the beam does not converge toward a point. As the energy density at the focal point, $r^*=z^*=0$, w_{bm} $=I_b \cdot V_b/2\pi\sigma_F^2=2.56\times10^5$ W/mm^2, energy density at $w_b^*=0.001$, for example, corresponds to 2.65×10^2 W/mm^2.

The standard deviation, which represents the measure of the divergency of the electron beam, is affected by various factors. Here the effect of the focal length and the pressure in the chamber on the standard deviation is discussed, compared with the case of focal length 156 mm, in which σ_F equals 0.06 mm.

By controlling the current of lens coil the focal length D_F was varied. In **Fig. 8** half bead width at focal point r_0 is plotted for each focal length 0.22 mm, 0.23 mm and 0.27 mm. Neglecting the effect of the heat conduction, each point has the same energy density. The standard deviation of the Gaussian curve, which represents the energy density distribution,

is equal to 0.06 mm as already mentioned. Thus the energy density at $r_0=0.22$ mm is given by

$$w_b\left(\frac{0.22}{\sigma_F},\ 0\right)=w_{bm}\ exp\left(-\frac{r_0^2}{2\sigma_F^2}\right).$$

Therefore the standard deviations σ_1 and σ_2 in case of $D_F=174$ mm and 214 mm are respectively given by

$$w_{b_1}\left(\frac{0.22}{\sigma_{F1}},\ 0\right)=w_{b_2}\left(\frac{0.27}{\sigma_{F2}},\ 0\right)=w_{b_i}\left(\frac{0.22}{\sigma_F},\ 0\right). \text{-----}(19)$$

Eq (19) provides $\sigma_{F1}=0.063$ and $\sigma_{F2}=0.076$ mm. Dividing each σ_F by the focal length, the standard deviation per unit length σ_0 is obtained, and is plotted in **Fig. 8**.

As the figure shows, there is such a tendency that σ_0 decreases a little with increase of the focal length. But the decrease is very small, so σ_0 can be regarded as constant. This result does not contradict the assumption described in section **2**.

The pressure in the chamber is usually maintained 10^{-4} torr order or less, and it is considered that as the gas pressure increases σ_0 varies because of scattering by the gas particles and interaction between electrons and plasma produced by the collision of the electrons with the gas particles. Assuming that there is a beam spread with the appearance of Gaussian distribution and there is no energy loss even if there are the scattering and the interaction, the standard deviation for each case can be calculated based on the experimental data having different chamber pressures.

Table 2 shows the data of half bead width at focal point and the values of standard deviation σ_0. When the pressure p increases slightly from 5×10^{-4} torr to 3×10^{-2} torr, shows a decrease of about 20 %, which is

215

Table 2. The values of σ_0 for different pressure ($D_F = 214$ mm).

Pressure (torr)	5×10^{-4}	3×10^{-2}	1×10^{-1}
r_0 (mm)	0.23	0.22	0.27
σ_0	3.55×10^{-4}	2.95×10^{-4}	2.0×10^{-1}

supposed to be caused by the lens effect of plasma (Ref. 4) produced by the beam collision.

From the figure it is seen that the energy density at the focal point has increased by about 1.5 times that of 5×10^{-4} torr. In case of 1×10^{-1} torr, the beam is remarkably spread by the scatter, and energy density shows a decrease of about 3 % that of 5×10^{-4} torr.

References

1) Y. Arata and I. Miyamoto: "Some Fundamental Properties of High Power Laser Beam as a Heat Source—Beam Focusing Characteristics of CO_2 Laser—", Trans. Japan Welding Society, Vol. 3, No. 1 (1972).

2) A. Sanderson: "Electron Beam Delineation and Penetration", British W. J., Oct. 1968.

3) Y. Arata, H. Nagai and T. Hattori: "The Shape Decision of High Energy Density Beam", To be published.

4) Y. Arata and M. Tomie: "Non-vacuum Electron Beam (1)", Tech. Rept. Osaka Univ. Vol. 17, No. 773 (1967).

5) Y. Arata: "Characteristics of the Electron Beam Heat Soorce and View of the Development on Its Welding Technology", J. Japan Welding Society, Vol. 41 (1972) No. 11.

Some Fundamental Properties of High Power Laser Beam as a Heat Source (Report 1)

— Beam Focusing Characteristics of CO_2 Laser —

Abstract

Beam focusing characteristics of multimode laser which is constructed for a heat source are discussed. The size and shape of the focused laser beam spots are measured at various locations on the chief ray under various focusing conditions using a simple method. The method of measuring beam energy distribution at focal point and focal line is developed. Its profile obtained by a concave spherical mirror with a radius of curvature 310mm may be approximated by a Gaussian curve having 0.5 mm diameter at e^{-1} power point. It is shown that deep hole can be obtained due to "self-focusing effect" of the laser beam in the drilled hole.

1. Introduction

Since laser produces a beam of highly collimated coherent light that is nearly monochromatic, it can be focused to very small spot with an extremely high density which can be used as a heat source for precise processing such as welding, cutting, drilling, plasma heating and so on. In order to achieve an exact control of the heat processing, it is very important to know shape, size and energy distribution in the focused beam spot on the work-piece. In case of completely plane parallel and monochromatic coherent light beam, it may be theoretically focused to a diffraction limited spot by a good lens. Beam spot, however, obtained from actual laser beam is somewhat larger than theoretical one due to imperfection of optical system in laser apparatus and imperfect monochromaticity.

In the present investigation, as the first problem, optical systems to converge the high power CO_2 laser beam, which is the best of the other lasers as a continuous heat source at present, are discussed and energy distribution in focused beam spot is measured in comparison with theoretical one.

For this purpose, a method for measuring the profile of the beam energy density at focal point was developed, and using this method focusing characteristics of the laser were studied experimentally with the aid of expedient method with which variation of size and shape of the beam spot along the chief ray was able to be obtained easily.

In the case focused laser beam is used as a processing heat source of various materials, it is important to clarify the feature of the processed portion in relation with focusing condition. Since the focused laser beam is not parallel but has appreciably large convergent angle determined by the optical system, the quality of the laser-processed material may be affected by the size and shape of the laser beam penetrated to the material, especially in case of thick material. Thus, in the second problem, interaction between focused laser beam and various material is discussed from point of view of beam focusing characteristics of laser.

2. Optical system

The CO_2 laser apparatus used in this experiment, which is similar to the one described in Ref. 1, consists of 74 mm diameter water cooled discharge tube, 12 m long, placed between two gold coated mirrors of 29.2 m radius of curvature spaced 13 m apart. There are six sections to the tube, each section of 2 m long having its own electrode with a series ballast resister. The multimode laser power was coupled out from the cavity through a 15 mm diameter hole perforated at the center of the one of the mirror.

Figure 1 shows a schematic diagram of optical system to move the converged beam spot to an arbitrary location. Two-dimensional motion in a given plane

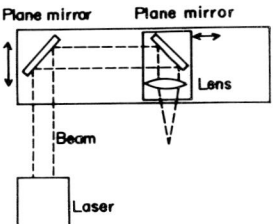

Fig. 1. Schematic diagram of beam focusing system.

is shown for simplicity in this figure, but it can be extended easily to three-dimensional motion based on the same principle. Since the laser beam has an excellent directivity and is not absorbed in the air, it can be focused even at distant place from the laser apparatus without energy loss.

The laser beam can be converged by means of a converging lens or a concave mirror. Figure 2 shows the schematic diagram of optical systems to converge

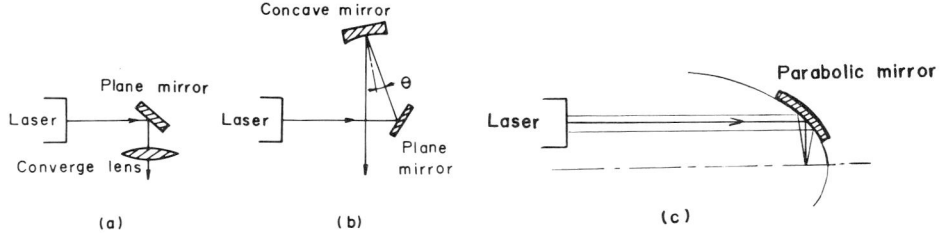

(a) **(b)** **(c)**

Fig. 2. Beam focusing methods. In order to converge the beam downward, optical systems nsing converging lens, spherical mirror which are conbined by a plane mirror or parabolic mirror can be used.

Table. 1. Comparison of window and lens materials for CO_2 lasers.[2]

Material	Index of Refra cti on	Fresnel Reflectivity	Fabry-Perot Reflectivity	Absorptivity	Hygroscopic
NaCL crystal	1.49	0.039	0.145	$\ll 0.01 cm^{-1}$	Yes
KCL crystal	1.454	0.034	0.127	$\ll 0.01$	Yes
BaF_2 crystal	1.42	0.030	0.113	0.077	Slightly
KRS–5	2.38	0.167	0.490	0.02	Slightly
Irtran–2	2.19	0.139	0.429	≈ 0.55	No
Irtran–4	2.4	0.170	0.497	≈ 0.1	No
Germanium	4.0	0.360	0.779	≈ 0.07	No

the laser beam downward on a horizontal plane. In case of converging lens system (a), it should be chosen from 10.6 μ transparent materials, which are shown in Table 1[2] for example, but materials are limited due to hygroscopic, unsuitable mechanical properties or large beam absorption coefficient. In addition, the inevitable optical absorption, no matter how small, eventually leads to destruction for high power laser beam, since the cooling is limited only to edge water-cooling for non-hygroscopic material or surface air cooling but their thermal conductivity is poor, the cooling is not sufficient.

In case of concave mirror made of glass or metallic substrate, the mirror surface is coated with vacuum deposited gold which is the best due to the high reflectance, near 100% at 10.6 μ, and chemical inertness in the air. Furthermore, when such metallic mirror is used, it is sufficiently cooled by water in comparison with infrared transparent materials. Thus authors proposed an optical system shown in Fig. 2 (b) and (c) for high power laser beam[1]. In this ex-

periment (b) system was adopted.

When laser beam is converged using the concave spherical mirror, the angle between incident laser beam and mirror axis, θ, cannot become zero. The result is that two mutually perpendicular images are formed instead of a point image for $\theta = 0$ in geometrical optics. This effect is known as astigmatism and is illustrated by perspective diagram in Fig. 3. The reflected rays in tangential plane RALT are seen to cross at T and the rays in sagittal plane JAKS across at S. If a screen is placed at E and moved toward the mirror, the image is generally elliptical in shape, but especially becomes a vertical line at S, a circular disk at M and horizontal line at T. The images at T, M and S correspond to the first focal line, circle of least confusion and second focal line, respectively. The location and length of the focal line can be calculated from the ray tracing based on the geometrical optics.

Figure 4 shows the ray tracing of the parallel incident light with an incident angle θ in tangential plane RALT. Let the axes ξ and η be in direction of the mirror axis and direction perpendicular to it,

Fig. 3. Astigmatic images of off-axis object point at infinity, as formed by a concave spherical mirror. The focal lines T and S are perpendicular to each other.

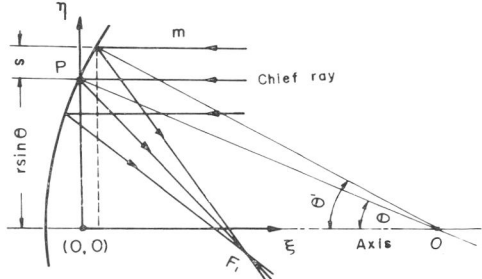

Fig. 4. Geometry for ray tracing through a single spherical surface with off-axis parallel incident light.

218

respectively.

The reflected chief ray is given by

$$\eta = R \sin \theta - \xi \tan 2\theta \qquad (1)$$

where R is the radius of curvature of the concave spherical mirror. The reflected ray m is given by

$$\eta = R \sin \theta' - (\xi - s \tan \theta)\tan 2\theta', \qquad (2)$$

where s is a radius of circular aperture stop and θ' is given by

$$\theta' = \sin^{-1}\left\{\frac{R \sin \theta + s/\cos^2\theta}{R + s \sin \theta/\cos^2\theta}\right\}. \qquad (3)$$

From Eq(1) and Eq(2), the solution of ξ is given by

$$\xi_1 = \frac{R(\sin \theta' - \sin \theta) + s \tan 2\theta' \cdot \tan \theta}{\tan 2\theta' - \tan 2\theta}.$$

Therefore the distance from the mirror to F_1 along the chief ray $\overline{PF_1}$ is given by

$$\overline{PF_1} = \frac{\xi_1}{\cos 2\theta}, \qquad (4)$$

where F_1 is the intersection between m and the chief ray and is the location of the first focal line. Let the F_2 is the location of the second focal line, $\overline{FP_2}$ is given by

$$\overline{PF_2} = \frac{R}{2 \cos \theta}. \qquad (5)$$

Let f_1 and f_2 are length of the first and second focal line respectively, they may be written by

$$f_1 = \frac{2s}{\overline{PF_1}}F(\theta) \qquad (6)$$

$$f_2 = \frac{2s}{\overline{PF_2}}F(\theta) \qquad (7)$$

where $F(\theta)$ is the magnitude of the astigmatism and is given by

$$F(\theta) = \frac{R}{2 \cos \theta}\left\{1 - \frac{R(\sin \theta' - \sin \theta) + s \tan 2\theta' \cdot \tan \theta}{\tan 2\theta' + \tan 2\theta}\right\} \qquad (8)$$

The calculation based on geometrical optics gives rise to line or point spot, but it is of course contradict with the actual. An exact dimension of the beam spot should be obtained from physical optics. Diffraction pattern of plane parallel waves focused by a circular lens with a circular aperture consists of bright central disk, known as Airy's disk, surrounded by a number of fainter rings. The intensity at radial distance r from beam axis in focal plane is given by

$$q(r) = q_0\left\{\frac{2J_1(\pi r/A)}{\pi r/A}\right\}^2 \qquad (9)$$

where $A = f\lambda/D_0$,
f = focal length of the converging lens,
λ = wavelength of the beam,
D_0 = aperture of the lens,
$q_0 = \pi P/A^2$,

P = total beam power and
J_1 = Bessel function first kind one order.

3. Beam focusing characteristics

(1) Divergent angle of the laser beam

The divergent angle of the laser beam was measured by a simple method; a film was placed in a plane vertical to the laser beam at a given distance from the laser head, and was heated by the laser beam for a given time. The diameter of the scoached area in the film was measured. Figure 5 shows the relation between the diameter of the scoached spot and the distance from the laser head to the film L. The divergent angle

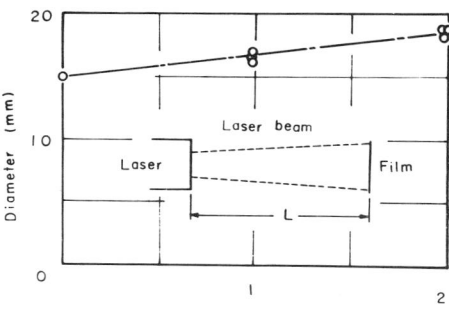

Fig. 5. Oscillated beam diameter vs. distance from laser head.

calculates from Fig. 5 was

$$\Delta\theta = 1.8 \times 10^{-3}$$

(2) Spot size and shape

Spot size and shape of the CO_2 laser formed by the optical system shown in Fig. 2 (b) were measured for various incident angles as a function of the distance from the concave shperical mirror along the chief ray L. Experimental setup used in this measurement is shown in Fig. 6. A thin film was placed in horizontal

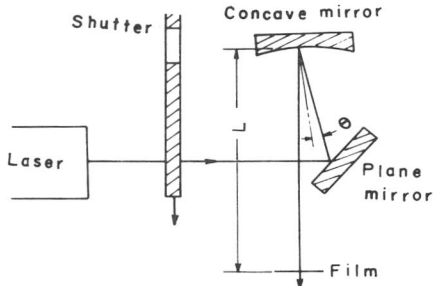

Fig. 6. Experimiental setup of measuring the spot size shape using film.

plane at the vertical distance L from the mirror, and was heated by the laser beam which was coupled out from 15 mm diam hole with the power level of 40 W. A shutter consisted of a rectangular plate with a

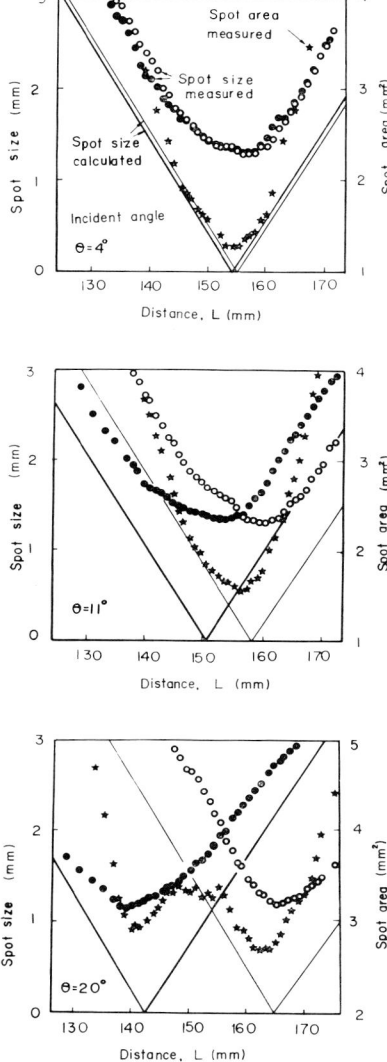

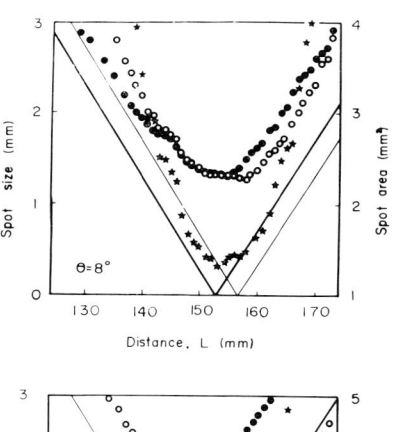

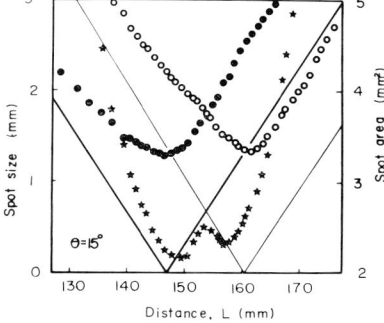

Fig. 7. Spot size and spot area vs. distance from concave spherical mirror to the film along chief ray (focal length: 155 mm, beam diameter: 15 mm).

lines for $\theta < 15°$.

Figure 8 shows the minimum minor diameter of elliptical beam spot and major diameter-to-minor diameter ratio at focal line plotted as a function of θ. The minimum minor diameter was constant value 1.3 mm which was the focusing limit of the laser beam

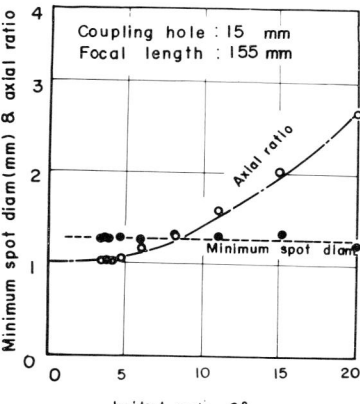

Fig. 8. Minimum minor spot diameter and major diameter-to-minor diameter ratio vs. incident angle.

window which falls freely, and determined the heating time, 1/50 sec. The focal length of the concave shperical mirror was 155 mm. The film was scoached elliptically in shape by the laser-heating, and both diameters in major and minor axes which were contained in the sagittal or tangential plane were measured. Figure 7 shows the two diameters crossing with the sagittal and tangential plane and the spot area as a function of the distance L for various incident angles. In each case open symbols indicate data points of the spot diameter measured in the sagittal plane and filled symbols indicate data points of the spot diameter measured in tangential plane. The spot area became minimum at location of circle of least confusion for incident angle $\theta < 15°$, while it became minimum at location of focal

converged with this optical system. This method, however, measuring the spot diamter using film is expediental and spot size obtained from this measurement is of rough estimation. An exact spot size should be determined from quantitative measurement of energy distribution in the beam spot as described later. The ratio at focal line was near unit and the astigmatism was negligible for $\theta < 6°$. It, however, increased gradually with increasing θ for $\theta > 6°$.

Figure 9 shows the minimum spot area plotted as a function of θ. It was almost constant for $\theta < 6°$, but it increased with increasing θ for $\theta > 6°$. In order to obtain small beam spot by means of this optical system, θ must be less than 6° in which the astigmatism is negligible small.

(3) Energy distribution at focal point and focal line

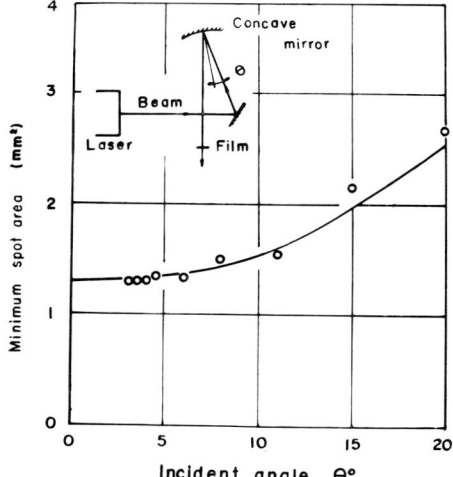

Fig. 9. Minimum spot area vs. incident angle (focal length: 155 mm, beam diameter: 15 mm)

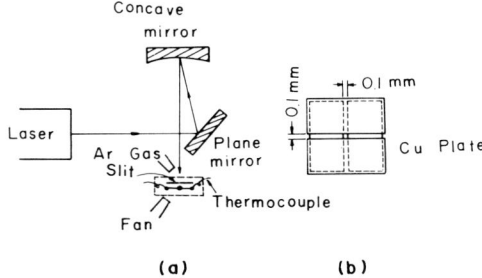

Fig. 10. (a) Experimental setup of measuring the energy distribution at focal point and focal lines.
(b) Detail of the slit.

The method measuring the beam spot size using film is practically convenient, but the data obtained by this method are not precise because the diameter is affected by the laser power and its heating time. The precise spot size should be determined based on the energy distribution profile in the beam spot as described above.

Figure 10 is the schematic diagram for measuring the energy distribution at focal line or focal point. A slit with a small square hole (0.1 mm × 0.1 mm) which consisted of four copper foils (0.2 mm in thickness) was moved in the horizontal plane. The beam energy passed through the slit was measured by means of a thermocouple and recorded in relation to the location in the beam spot. In order to protect the thermal oxidation, the top surface of the copper foil was polished smoothly and shielded with argon gas during laser heating. The thermocouple and the back surface of the slit were cooled by a fan.

Figure 11 (a) is an example of energy distribution profile recorded at focal point in the tangential and sagittal planes which correspond to X-X and Y-Y, respectively in case of $\theta = 5°$. Both energy distribution

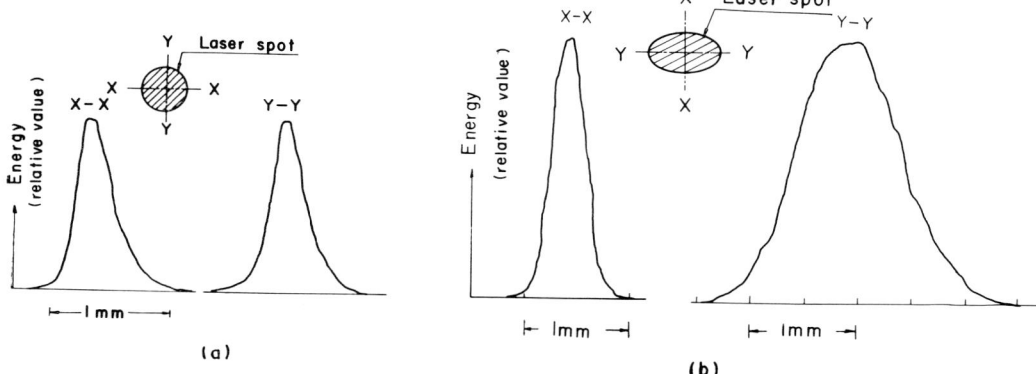

Fig. 11. Energy distribution profile at focal point and focal line(focal length: 155 mm, beam diameter: 15 mm).
(a) At focal point, $\theta = 5°$.
(b) At focal line, $\theta = 20°$.

221

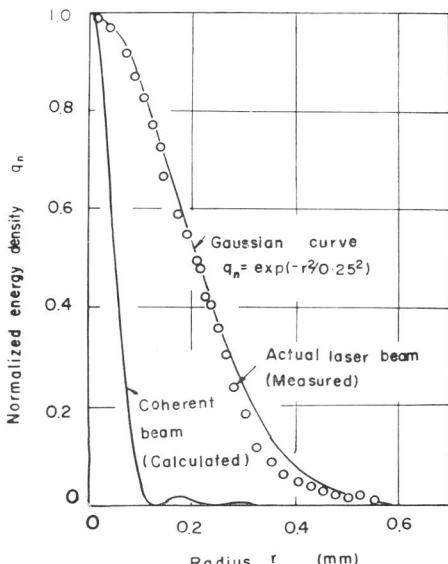

Fig. 12. Comparison between actual laser beam measured and coherent beam (focal length: 155 mm, beam diameter: 15 mm).

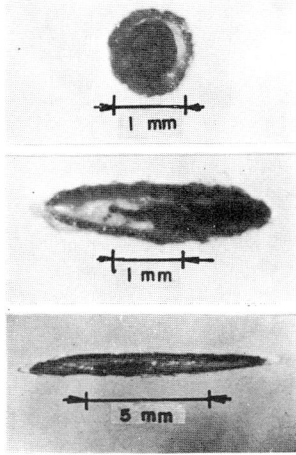

Photo. 1. Examples of the spot shape of focal point and focal lines (focal length: 155mm, beam diameter: 15 mm).
(a) $\theta=5°$, at focal point
(b) $\theta=20°$, at focal line
(c) $\theta=45°$, at focal line.

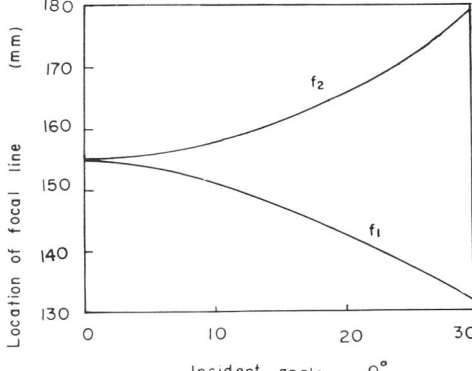

Fig. 13. Locations of the first focal line and the second focal line vs. incident angle.

profiles in X-X and Y-Y were almost same. It is found that intensity in these profiles falls to almost zero at the spot diameter 1.3 mm which was measured by means of the film. Figure 11 (b) is the energy distribution at second focal line in case of $\theta=20°$. Both energy distributions at focal line in minor axis X-X for $\theta=20°$ and at focal point for $\theta=5°$ had the almost same profile.

Figure 12 shows the comparison of the measured energy distribution profiles at focal point and the theoretical one calculated from the plane parallel beam with a wavelength 10.6 μ. The actual distribution profile may be well approximated by a Gaussian curve

$$q_n=\exp(-r^2/a^2)$$

where r is the radial distance from the center in the focal plane, a is radius of $1/e$ power point at the center and q_n is the normalized intensity at radial distance. As shown in this figure, a is equal to about 0.025 cm, which is about four time as large as theoretical radius in the Airy's disk. When the power level of the laser beam is 100 W for example, the power density at center $r=0$ becomes about 5.6×10^4 w/cm².

(4) Control of the beam spot shape

In the conventional heat source such as arc, plasma jet and electron beam, it has been important problem not only to converge them to a small spot, but to deform the spot by the magnetic field etc., according to their purpose. In case of laser beam, it has an advantage that the beam spot can be continuously and widely deformed by a simple optical system. For example, when laser beam is converged by a concave shperical mirror, a long and narrow spot can be obtained at the focal line by use of its astigmatism.

Photograph 1 shows examples of deformed beam spot obtained at focal point and focal line. Figure 13 shows the location of the first and second focal line formed by concave spherical mirror with a radius of curvature 310 mm plotted as a function of the incident angle.

4. Self-focusing effect of laser beam

It has been described earlier that the laser beam is useful for precise heat processing such as welding, cutting and drilling because it can be focused to a very small spot with very high energy density. However, since the laser beam focused by optical system is not

222

parallel but has an appreciably large convergent angle, it is natural to inquire whether a laser processing of thick material is possible or not. In this section several examples of laser processed materials are illustrated, and feature and possibility of laser processing of thick material are discussed on the basis of these results.

Photograph 2 shows an example of the laser-drilled acryl for various beaming times. In this experiment the laser beam with the power level of about 250 W was focused just upon the acryl by means of the optical

(a) (b) (c) (d) (e) (f) (g) (h) (i) (j) (k) (l) (m) (n) (o)

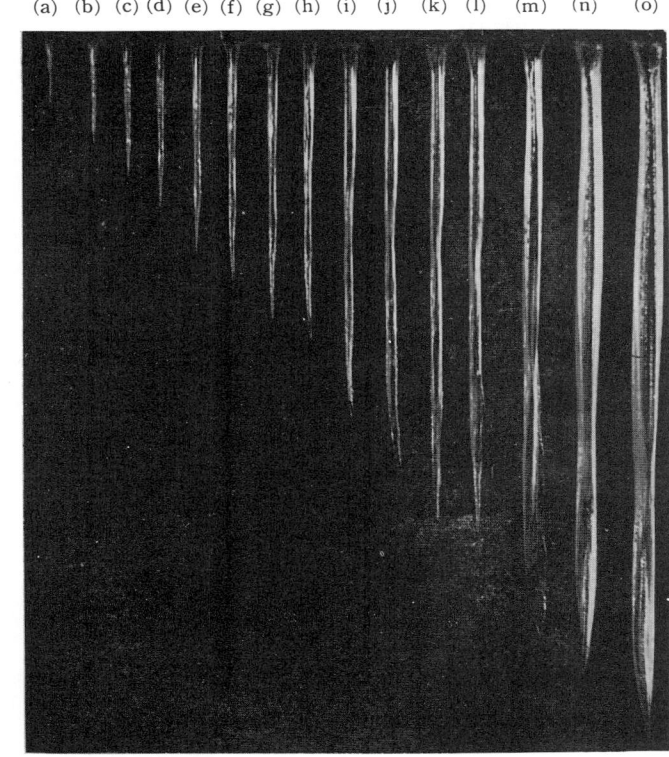

10mm

Photo. 2. Example of laser drilled acryl (Laser power=250 W). Beaming time (sec):
(a) 0.02, (b) 0.04, (c) 0.06, (d) 0.08, (e) 0.13, (f) 0.18,
(g) 0.23, (h) 0.28, (i) 0.28×2, (j) 0.28×3, (k) 0.28×4,
(l) 0.28×5, (m) 0.28×10, (n) 3.0, (o) 5.0.

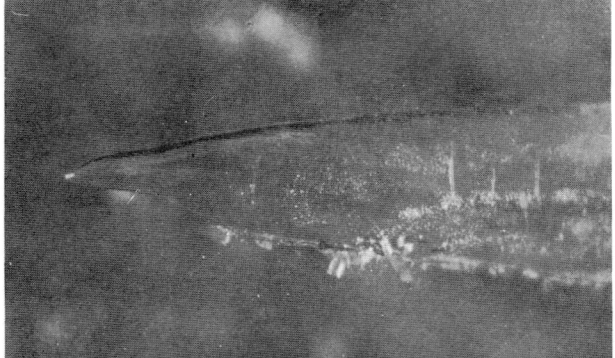

Photo. 3. Magnified photograph of Photo. 2(m)

system shown in Fig. 2 (b) (incident angle $\theta = 5°$). The drilling was carried out with considerably higher speed, initially about 25 cm/sec. The depth of the drilled hole was remarkably larger incomparison with length whose spot diam was regarded as constant at focal point in the optical system, and the bottom edge of the hole always had wedge shape having a very acute angle even at far off-focal point as shown in Photo. 3.

In most case the diameter of the hole which was somewhat larger at the top surface varied with the depth having a peak value at two locations. From these facts, drilling mechanism by the focused laser beam can be shown as follows; At first, incident laser beam drills a hole with small diameter. As the following beam incomes the hole with very large incident angle with the wall of the hole, about 90°, it reflects with almost 100% reflectance.

Then the reflected beam will be re-focused at the bottom edge as shown schematically in Fig. 14 giving acute edge angle at bottom of the hole. The authours named this phonomenon as "self-focusing effect" of the laser beam. From Fig. 14, the incident beam seems to be reflected about two times to reach the bottom edge by the wall of the hole resulting the two peak portions of the hole diameter at B and C, because the wall of the drilled hole which is not completely smooth, absorbs the beam energy much larger at main reflected points (B and C) than other part. Figure 15 shows the

relation between beaming time and depth of the drilled hole. Drilling speed computed from this figure was about 25 cm/sec till beaming time 30 msec as described above but it decreased gradually with beaming time. This is due to the increase of the consumed beam energy at the wall and so the hole becomes gradually fat with beaming time as shown in Photo. 2.

To confirm the self-focusing effect of the laser beam, two acryl plates of 5 mm in thickness were heated by the focused laser beam with different distances between two plate. In this test, the beam was focused on the top surface of the upper plate. The shape of the drilled holes is shown in Photo. 4. When the distance d was zero, the beam diameter at the bottom surface of the upper plate, ϕ_{1b}, was same as that at the top surface of the lower plate, ϕ_{2t}. The remarkable features in case of $d \neq 0$ were that a discrepancy between ϕ_{1b} and ϕ_{2t} increased with increasing the distance d and the hole of the lower plate became quasi-conical in shape. From d-value and the difference between ϕ_{1b} and ϕ_{2t}, divergent angle about 5° of the beam passed through the hole of the upper plate was obtained. It is, however, noteworthy that the beam diameter at bottom surface of the lower plate, ϕ_{2b}, was considerably smaller than that of the upper plate (ϕ_{1b}). It is clear from these facts that the beam is re-focsued inside the laser-drilled hole, namely the self-focusing effect exists in laser drilling, and the effect is not available without

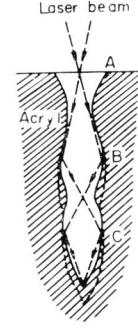

Fig. 14. Schematic diagram showing "self-focusing action". Laser beam is re-focused to the bottom edge of the hole.

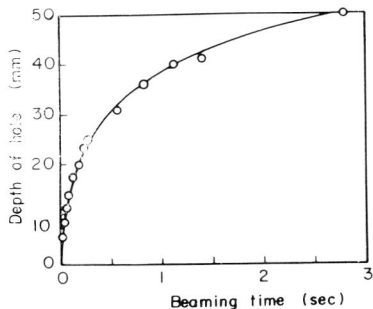

Fig. 15. Relation between beaming time and depth of hole (Acryl, laser power = 250 W).

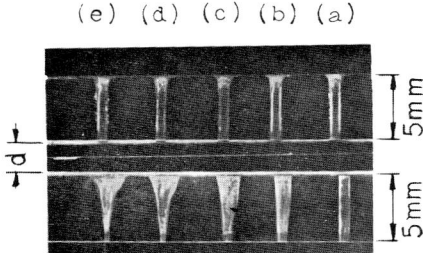

Photo. 4. Shape of drilled holes for acryl with various distance d. Distance d (mm): (a) 0, (b) 3, (c) 6, (d) 9, (e) 12.

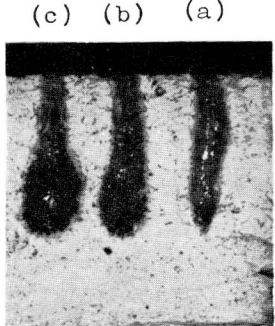

Photo. 5. Laser drilled firebrick (Laser power = 250 W). Beaming time (sec): (a) 5.0, (b) 10.0, (c) 20.0.

224

surrounding wall. In general, such the self-focusing phenomenon of the high power laser beam should occur in heating of any material involving solid and liquid if the heating was accompanied by the drilling action. Since the drilling action becomes remarkable as the laser beam power and its density increase, the self-focusing action depends upon the beam power and its density. Therefore the self-focusing action becomes effective in the material having lower thermal conductivity, easily sublimated or vaporized property and small latant heat of the vaporization or the sublimation such as acryl. Photograph 5 indicates the cross section of the laser drilled firebrick. In this case until the beaming time was about 5 sec, the bottom edge of the drilled hole possesed an acute angle due to the self-focusing action. And then the depth of the hole reached the limit and did not increase any more. Such limiting depth depends upon the beam focusing condition, beam power and material. After reaching the limiting depth at which the vaporization was supressed or stopped, the bottom wedge shape of the hole became gradually round shape so that the self-focusing action disappeared.

As described above, the self-focusing action occured also for liquid material. Photograph 6 illustrates the self-focusing action for water. In this case the bottom

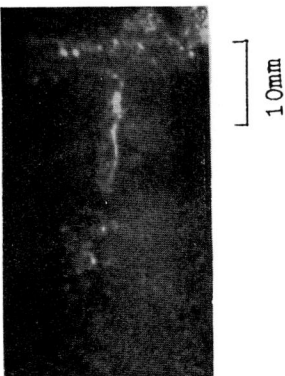

Photo. 6. Laser drilled water (Laser power=250 W).

edge was somewhat round in comparison with above two examples. This is due to both surface tension and vaporized pressure of the water.

In case of metal, the self-focusing phenomenon should also take place in higher laser beam power which induced the drilling action. From these results it can be predicted that so called "deep penetration"

which has been typical in electron beam welding only is also obtained in high power laser beam welding. Consequently it was clarified that laser beam processing involving deep penetration welding, cutting and drilling of thick material, became also possible as well as electron beam heat source by introducing new developed idea, self-focusing effect of the laser beam.

5. Conclusions

Optical system to converge the high power CO_2 laser beam was discussed and its focusing characteristics obtained by a concave spherical mirror was analized. The spot size and shape of the laser beam were measured expediently using film and the method of measuring the energy density profile in the beam spot was developed.

Conclusions obtained are summerized as follows;

(1) The divergent angle of the CO_2 laser beam coupled out from 15 mm diam hole was about 1.8 milliradians.

(2) In an optical system to converge the high power CO_2 laser beam, a concave reflecting mirror is recommendable rather than a lens.

(3) The astigmatism in the optical system using a concave spherical mirror with a radius of curvature 310 mm was negligible when the incident angle $\theta < 6°$.

(4) The energy distribution profile at focal point may be approximated by a Gaussian curve and the diameter of the focused beam at $1/e$ power point was about 0.5 mm, which was approximately four times the diffraction limit.

(5) The shape of the focused beam spot can be widely controled and a long and narror spot can be obtained.

(6) The laser beam is re-focused to the bottom edge of the laser-drilled hole due to "self-focusing effect" of the laser beam when power level of the laser beam is enough high to arise the drilling action.

(7) The self-focusing action occurs in high power laser-heating of any material involving solid and liquid, and the high power laser processing such as deep penetration welding, cutting and drilling of thick material, is possible due to the self-focusing effect even though the convergent angle of the focused laser beam is considerably large.

(8) When the self-focusing action disappears, the limiting depth arises in the processing.

References

1) Y. Arata, I. Miyamoto and M. Kubota: I.I.W. Doc. IV-4–69, 1969
2) D. R. Whitehouse: Laser Tech., July, A6, 1967.

Wall-Focusing Effect of Laser Beam

I.   Introduction

     Since the absorption by non-metallic materials of infrared radiation is generally very good, CW CO_2 lasers may be most widely used for processing the materials. In laser drilling or cutting, deep hole or groove is obtained even using short focal length lens. Some mechanisms, such as the generation of a type of shock wave (1) and a liquid ejection (2), have been proposed for such a material removal in hole drilling. These may be, however, predominant in pulsed laser having extremely high peak power, but are considered to be not a major factor in CW CO_2 laser.

     In this paper it is shown that a wall-focusing effect, which is caused from reflection of the incident beam in the hole or groove, is predominant in CW CO_2 laser processing. In order to appreciate the effect, a transparent acrylic resin was used as a test material since it made the theoretical treatment and observation of the hole shape easy. The laser used in this work was mainly of conventional design.

II.   Some aspects of CO_2 laser drilling

     When materials are irradiated by the laser beam, whether focused or not, with enough high power density to vaporize them, deep and narrow holes are formed as shown in Fig. 1. These results can be explained well by internal multiple reflections of the beam having large incident angles to the wall in the hole as schematically shown in Fig. 2.

     In general the reflectance from non-metallic material of infrared radiation corresponding to the large incident angle is very large, though that in normal incidence is very low. Relation between the incident angle θ of CO_2 laser and reflectance R from acrylic resin measured at sufficiently low power density is shown in Fig. 3, where R is no less than 0.5 for $\theta=85°$, for instance.

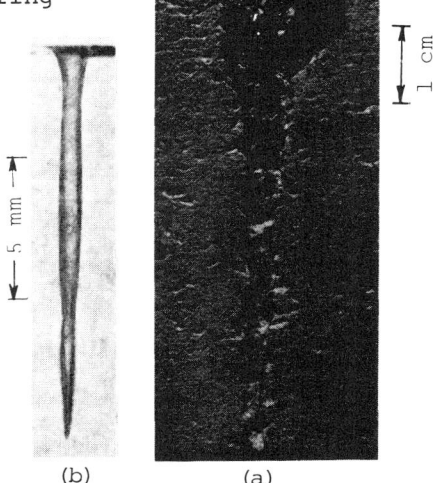

(b)   (a)

Fig. 1 Examples of CO_2 laser drilled hole. (a) A drilled fire brick by unfocused beam (3KW). (b) A drilled acrylic resin by focused beam (f=100mm).

226

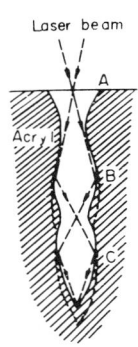

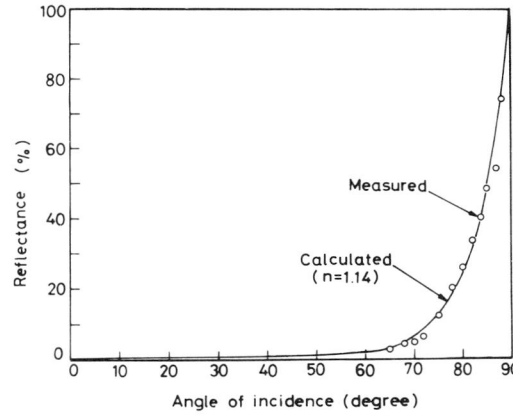

Fig. 2 Schematic diagram
        illustrating multi-
ple internal reflections
of the incident beam.

Fig. 3 Relation between incident
        angle of CO_2 laser beam
to acrylic resin and reflectance.

Thus one can expect that the power density in the hole is larger than one formed by the optics used as the result of the internal reflection at high incident angle θ, and the authors have named this effect, which increases the power density and transports the beam energy successfully to the top of the hole, a "wall-focusing" effect . It may be said that a wall-focusing becomes the more effective because the refractive index of the material is large or the convergent angle of the optics used is small.

Volume of the hole drilled by CW CO_2 laser in the acrylic resin was measured precisely for various values of the input energy. The volume was fairly proportional to the input energy as shown in Fig. 4. No material removal by liquid or solid ejection was observed from the high speed movies.

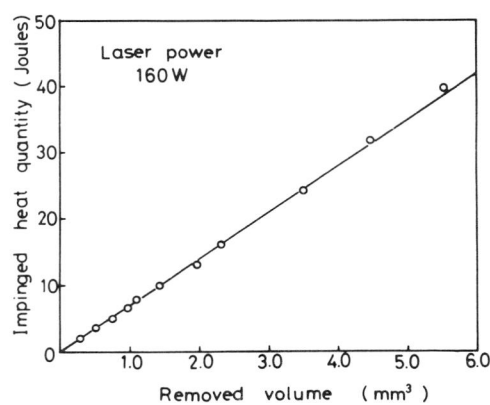

Fig. 4 Relation between
        removed volume of
acrylic resin and input
energy.

From these results it can be concluded that the bulk removal of acrylic resin is only by evaporation, and that the absorption by the vapor is negligible. In order to appreciate the degree of the wall-focusing effect, the power density in the laser drilled hole is calculated

227

based on the theory developed in the following section and is
compared with one formed by optics alone.

III.   Theory

Figure 5 illustrates schematically
laser drilled holes corresponding to
irradiation time t and t+dt. Assuming
that r(z,t) is radius at z and l is
depth of the hole for irradiation time
t, the volume contained from z=Z to
z=l is given by

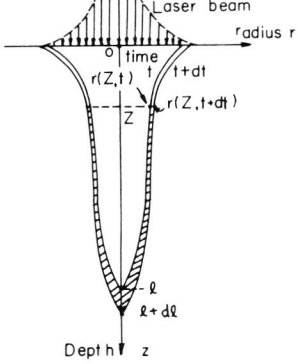

$$V(Z,t)=\pi\int_{Z}^{1} r^{2}(z,t)dz. \quad ....(1)$$

Energy amount passed through the z=Z
plane from time t to t+dt is proportional to the hatched volume dV(Z,t) in
Fig. 5, and is given by

$$dJ=KdV(Z,t) \quad ....(2)$$

where a proportional constant K is the
energy required to vaporize unit volume
of the material, and numerical value of
K for acrylic resin is given in Fig. 4.
Thus the beam power W(Z,t) passing through z=Z at time t is
written

Fig. 5   Schematics of
laser drilled
holes.

$$W(Z,t)=2\pi K\int_{Z}^{1} r(z,t)\frac{\partial}{\partial t}r(z,t)dz, \quad ....(3)$$

and the average power density $\overline{w}_{v}$ in the plane perpendicular to
the z-axis is

$$\overline{w}_{v}(Z,t)=\frac{2K}{r^{2}(Z,t)}\int_{Z}^{1} r(z,t)\frac{\partial}{\partial t}r(z,t)dz. \quad ....(4)$$

On the other hand, the power density $\overline{w}_{s}$ absorbed in unit
area of the side wall increasing the hole radius is given

$$\overline{w}_{s}(Z,t)=-\frac{1}{2\pi r(Z,t)}\frac{\partial}{\partial t}W(Z,t)$$

$$= K\frac{\partial}{\partial t}r(Z,t). \quad ....(5)$$

The proportional constant K and r(z,t) are necessary to obtain
$\overline{w}_{v}$ and $\overline{w}_{s}$.

Power density distribution of the beam focused by the
optics at the surface of the work piece may also be calculated
using these results.   The radial distribution of the power density at the focal point is generally given by a Gaussian curve

$$w(r)= \frac{W}{\pi\alpha^{2}}\exp(-r^{2}/\alpha^{2}) \quad ....(6)$$

228

where W is power of the beam and α radius at 1/e power point. Then the mean value of the density in the circle of radius r is written by

$$w(r) = \frac{W}{\pi r^2}[1 - \exp(-r^2/\alpha^2)]. \quad \dots\dots\dots\dots\dots\dots\dots(7)$$

In the vicinity of the surface where the effect of the reflection from the side wall may be neglected, the value α is obtained from r(Z,t) and W(Z,t) for arbitrary time t as follows:

$$\alpha = \frac{r(Z,t)}{[-\ln\{1-W(Z,t)/W\}]^{0.5}} \quad \dots\dots\dots\dots\dots\dots\dots(8)$$

The power density at the top of the hole may also be obtained from the drilling velocity v as

$$w = Kv. \quad \dots\dots\dots\dots\dots\dots\dots(9)$$

IV. Power density in the hole

Figure 6 shows the holes drilled by CW CO_2 laser in acrylic resin with power level 160 w of multimode. It was concentrated through a GaAs lens of focal length 60 mm, and focal point was just on the work piece surface.

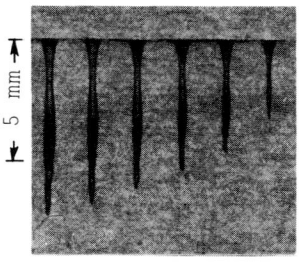

a   b   c   d   e

Fig. 6 Laser drilled holes in acrylic resin. Irradiation time (msec): (a) 60 (b) 50 (c) 40 (d) 30 (e) 20 (f) 10.

Because the value α obtained from Eq(8) was rather scattered using r(Z,t) and W(Z,t) at the surface, it was extraporated from the values obtained near the surface. The extraporated value α was about 0.19 mm and the resultant maximum density formed by the lens was about 1.5×10^5 w/cm^2. When defocusing amount was larger than 3-4 mm, the hole corresponding to short irradiation time was so shallow in shape that the beam incident angle was small enough to neglect the wall-focusing effect. In such a case the power density of the optics becomes the value given by Eq(9). The power density of optics along the beam axis thus obtained is plotted in Fig. 7.

On the other hand transient power density for defocusing amount L=0, which is also shown in Fig. 7, has been two or three times as large as the value produced by the optics alone. Difference between two values, however, became smaller as the hole formation progressed. Increase in power density due to the wall-focusing effect was also obtained except for L=0, but the power density for L=0 was the highest. The time required for the initiation of the appreciable wall-focusing decreased with decreasing the defocusing amount, and it was 1 msec or less for L=0 according to the high speed movies.

In Fig. 8 the mean power density $\overline{w}_v$ in the hole is plotted against the depth z. Each curve has its own peak at the depth 1-1.5 mm, which corresponds to the waist of the hole. The maximum value of $\overline{w}_v$ obtained in this test was about 2.6×10^5 w/cm^2, which agreed well with the maximum value shown in Fig. 7.

The value $\bar{w}_s$, which was absorbed by the wall to increase the hole radius, had minimum at the waist of the hole, and increased rapidly until showing a decrease near the top of the hole. The peak value of $\bar{w}_s$ became smaller and the depth z corresponding to the peak of $\bar{w}_s$ became larger, as the irradiation time was longer.

These results illustrate that as the hole formation progresses, the incident beam is transported successfully to the top of the hole by the internal reflections providing deep hole even using short focal length lens.

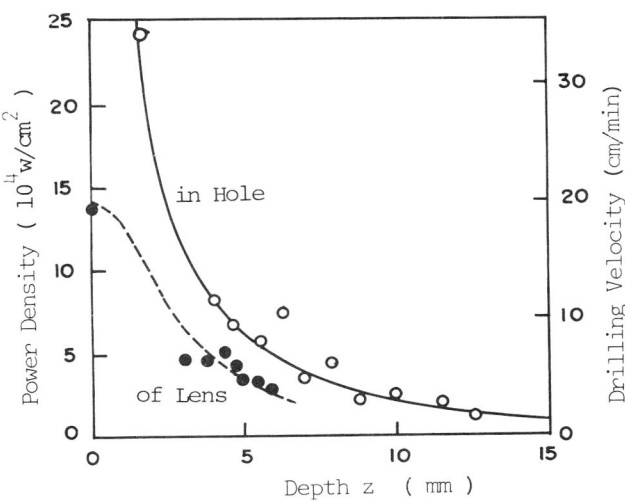

Fig. 7 Comparison of power density in the hole for L=0 with one obtained by the lens used.

V.   Conclusion

Results obtained are summerized as follows:
(1)   The bulk removal of acrylic resin by CW CO_2 laser is only by evaporation.
(2)   A method obtaining power densities at surface and in the hole of the work piece by measuring the dimension of the hole and drilling velocity has been proposed.

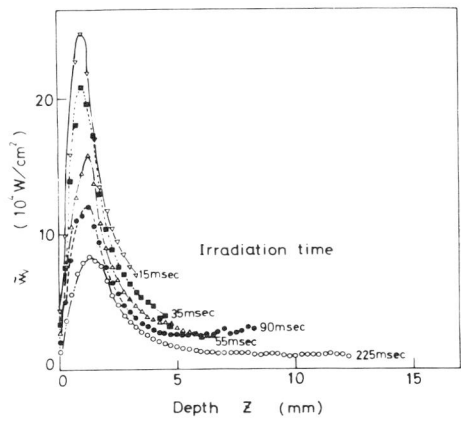

Fig. 8 Relation between z and $\bar{w}_v$.

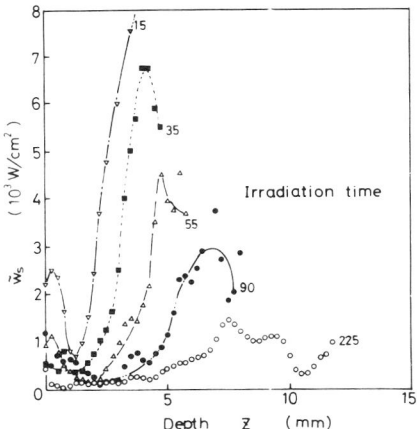

Fig. 9 Relation between z and $\bar{w}_s$.

(3) Power density in the hole has been higher, 2 or 3 times in this test, than that produced by the lens because of a wall-focusing effect, which is caused from internal reflections so as to transport successfully the beam power tward top of the hole.

(4) The effet becomes more evident with decreasing the defocusing amount and increasing the refractive index of the material.

Acknowledgments

The authors are grateful to K.Nakamura for his assistance in the laboratory.

REFERECE

(1)J.F. Ready:"Effects Due to Absorption of Laser Radiation", J. Appl. Phys., Vol. 36, 1522 (1965).

(2)M.K. Chun and K. Rose:"Interraction of High Intensity Laser Beams with Metals", J. Appl. Phys., Vol. 41, 614 (1970).

Heating of Material

Some Fundamental Properties of High Power Laser Beam as a Heat Source (Report 2: CO_2 Laser Absorption Characteristics of Metal)
Y. Arata and I. Miyamoto
[Trans. JWS **3** (1972), 152. IIW Doc. IV–50–71. Tech. Review FINO-MECHANIKA.]

Some Fundamental Properties of High Power Laser Beam as a Heat Source (Report 3: Metal Heating by Laser Beam)
Y. Arata and I. Miyamoto
[Trans. JWS **3** (1972), 163. IIW Doc. IV–51–71.]

Theoretical Analysis of Weld Penetration Due to High Energy Density Beam
Y. Arata and I. Miyamoto
[Trans. JWRI **1** (1972), 11.]

Investigation of Moving Line Heat Source—Numerical Calculation of Temperature Distribution
Y. Arata and K. Inoue
[Trans. JWRI **2** (1973), 41.]

Some Fundamental Properties of High Power Laser Beam as a Heat Source (Report 2)

− CO₂ Laser Absorption Characteristics of Metal −

Abstract

The absorption characteristics of metallic specimen in various surface conditions were evaluated in the wavelength 10.6 μ: transient absorptance as well as relation between the absorptance and some parameters involving the power level and traveling velocity of the specimen were obtained when focused and unfocused CO_2 laser beam impinged upon the specimen. It was found that the surface treatment was the promising method to enhance the absorptance, which was so low in polished surface condition that a CO_2 laser had been less desirable as a heat source for metal processing such as welding. No bad effect of the surface treatment on the metallurgical and mechanical properties was observed when CO_2 laser was applied for welding of metal. A welding method was presented in order to make possible or easy the welding of thick metal.

1. Introduction

In the case CO_2 laser is applied as a heat source for processing metal, in particular for thin metal, the effect of its absorptance in wavelength of CO_2 laser, 10.6 μ, on the processing is very large. In general the infrared absorptance of metal with smooth surface is very small due to its good electrical conductivity. For example, the absorptance of highly polished gold does not reach only 1% and so it can be used as a reflecting surface film evaporated on the mirror in CO_2 laser systems. It seems to have contradicting content that it is necessary to have the absorptance as high as possible in metal processing by means of CO_2 laser which contains metallic mirrors of which absorptance should be extremely small, and so this has made CO_2 laser less desirable as a heat source for the metal processing up to date.

Fortunately, however, the absorptance is significantly affected by its surface condition. Thus the authors wish to propose to improve the absorptance of metal by means of the surface treatment. The present investigation was undertaken to study the effect of surface treatment on the absorptance at wavelength of 10.6 μ.

Many workers[1)~4)] have measured the absorptance in the extreme condition, namely the absorptance of metal in highly polished, clear and stress-free surface condition or of evaporated metal film on the super smooth surface in high vacuum. In practice, however, the absorptance in the surface condition supplied by commercial vender or of various surface finished is necessary rather than extreme condition described above, but these measurements have not yet been undertaken.

When a specimen coated with thin layer produced by the surface treatment is appropriately heated by high power focused laser beam, condition of this layer in heated region varies with time due to melting, evaporating or spattering so that the absorptance also varies with time. In general the absorptance is not steady at all, but it should be a function of a heating time as well as power level of the laser beam, optical and thermal properties of the specimen and its surface treatment.

In the case the power level becomes high furthermore, so called "drilling action" due to reaction of metallic vaporization becomes remarkable as may be the case with electron beam, and then a main amount of incident high laser power may be trapped within the drilled hole due to "self-focusing effect"[5)] of the beam which occurs as a result of its multiple reflections by the wall of the hole so as to re-focus the beam to the bottom of the hole. In this situation the absorptance increases in excess of 50~60%, sometimes up to near 100%, and it does not almost depend upon the optical properties of the metal or its surface condition any more.

When thick metal is welded by CO_2 laser beam, such a high absorptance condition may be realized not only by increasing the laser power, but by devicing the shape of the weld joint. In this paper it will be exhibited the absorptance is enhanced by the latter even at lower laser power without the sufficient drilling action. In the case of thin metal in comparison with the diameter of focused laser beam,[5)] it must be emphasized that the surface treatment is especially important to enhance the absorptance.

This paper consists of five sections: Section **2** shows our CO_2 laser apparatus and experimental setups for measuring the absorptance of the specimen, and section **3** introduces a theory for optical properties of metal. Section **4** gives our experimental rsults of the absorptance of mettallic specimen in various surface conditions. The final section gives several conclusions obtained in this study.

2. Experimental Apparatuses and Procedures

The CO_2 laser apparatus used in this experiment, which is similar to the one described in Ref. 6, consists of a 74 mm diameter water cooled discharge tube, 12 m long, placed between two gold coated mirrors of 29.2 m radius of curvature spaced 13 m apart. There are six sections to the tube, each section of 2 m long having its own electrode with a series ballast resister. The laser power was coupled out from the cavity through a 15 mm diameter hole perforated at the center of the one of the mirror. The present laser apparatus was operated from an ac electric power supply of 60 Hz, giving laser pulsed output of 120 Hz.

The laser power output measurements were carried out with power meter shown in Fig. 1.

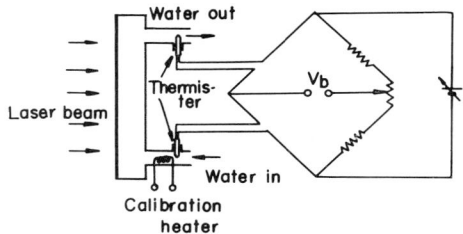

Fig. 1. Schematic diagram of power meter.

Water is flowed through the power meter at a constant rate which is monitored by a flow meter, and the temperature difference of input and output water is measured by a pair of thermisters which makes the temperature measurements very precise. Then a detecting voltage of the bridge, V_b, is recorded on charts. To make the measurements accurate, a heater is incorporated for calibrating the V_b-value with an electrical input of known power level. The absorption plate of the power meter consists of thin copper plate (0.3 mm in thick), of which surface is chemically treated to have an appropriate magnitude of uniform absorptance. During measurement of the absorptanace, power output of CO_2 laser is monitored by the power meter to keep it constant. The absorptance of metal at room temperature was obtained by measuring the absorbed power of an another power meter which was similar to one described above except that the absorption plate of the power meter was replaced by the specimen of thin metal with unknown absorptance. In this case, unfocused radiation beam, 15 mm in diameter, with power level of approximately 200 W was supplied to the specimen. A thin metal, 0.3 mm in thick, was used to reduce the temperature gradient within the specimen in direction of the thickness, and then its surface temperature was kept nearly at room temperature during measurement since its opposite surface was cooled with water.

Figure 2 shows a schematic diagram of the experimental setup for measuring the absorptance of molten metal.

The molten pool was produced by the laser heating

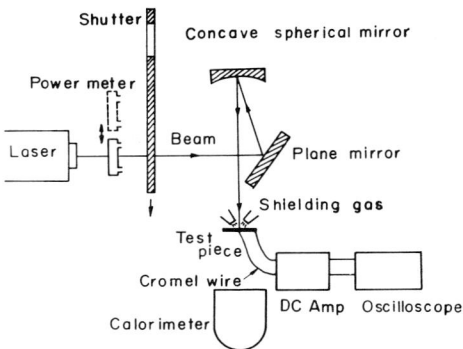

Fig. 2. Experimental setup for measuring the beam absorptance of the molton metal.

of the specimen using the beam focused through the optical system containing a concave mirror with a radius of curvature of 310 mm as indicated in Fig. 2. The optical system was adjusted so as to have its focal point on the surface of the specimen. Energy distribution profile at the focal point may be approximated by a Gaussian curve, of which diameter at $1/e$ power point of the center is about 0.5 mm.[7] A shutter consists of a rectangular plate with a window which falls freely, and determins the beaming time. Each specimen heated by the laser was immediatly thrown into the calorimeter shown in Fig. 3 and the calorimetric gain was measured.

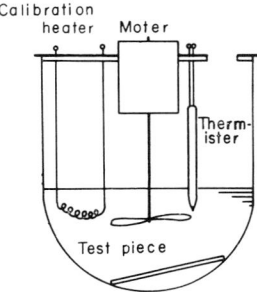

Fig. 3. Schematic diagram of calorimeter.

A heater and a thermister were incorporated to the system, and the similar circuit as shown in Fig. 1 was used; in this case one of the thermisters was replaced by a solid resister. The relation between electrical input to the heater, and change of V_b-value was checked beforehand, and thus in order to obtain the energy absorbed by the specimen, the V_b-value was recorded before and after the specimen was thrown in calorimeter. The absorbed energy, in calories, is plotted as a function of beaming time, and the absorptance may be computed from the incident laser energy, in calories, to the specimen and slope of a curve obtained from these experimental values.

In this case the absorptance varied sharply with

time, the temperature rise at location on the bealy axis of the back surface was also measured to make sure the absorptance. In order to measure the rapid temperature change, a thin Chromel wire, 0.1 mm in diameter, was welded to the back surface of the specimen, 0.3 mm in thick, using the condenser-discharge type spot welder. Its welded point was then aligned so as to put just on the beam axis, and the thermoelectromotive-force between a Chromel and the specimen was recorded by a oscillograph.

3. Optical Properties of Metal

The two most characteristic optical properties of metal are the opacity and the high reflecting power of its polished surface. One valence electron in metal which interacts with the light is equivalent to the system containing a free electron with the effective mass, m^*, and some of harmonic oscillators with a given specific frequency, but in the infrared the absorptance of metal is dominated by conduction absorptance due to free electron.

It is well known that for the electromagnetic wave with the frequency, ω, the electric permitivity of metal may be obtained by replacing that of insulater, ε (ε_0 in vacuum space), having real quantity with the complex quantity, $\bar{\varepsilon}$;

$$
\begin{aligned}
\bar{\varepsilon} &= \varepsilon + i\frac{\sigma}{\omega} \\
\text{or} \quad \bar{\varepsilon}^* &= \varepsilon^* + i\frac{\sigma}{\omega\varepsilon_0} = \varepsilon^* + i\sigma^*,
\end{aligned}
\tag{1}
$$

where σ is the electrical conductivity of metal, seting $\varepsilon = \varepsilon_0 \cdot \varepsilon^*$, ε^* is dielectric constant and dimensionless parameter. And then the optical constants, n and k, are illustrated in terms of the electric parameters, ε^* and σ^* ($\equiv 1/\eta^*$);

$$
\begin{aligned}
k &= -\varepsilon^* \eta^* + [\varepsilon^{*2}\eta^{*2} +]^{1/2} \\
n^2 k^2 &= \frac{1}{2}\{-\varepsilon^* + [\varepsilon^{*2} + \sigma^{*2}]^{1/2}\}
\end{aligned}
\tag{2}
$$

The reflectance at surface with complete smoothness of metal in the case of the normal incident is given by Fresnel's equation,

$$
\begin{aligned}
R &= \left|\frac{\bar{n}-1}{\bar{n}+1}\right|^2 \\
&= \frac{(n-1)^2 + n^2 k^2}{(n+1)^2 + n^2 k^2}
\end{aligned}
\tag{3}
$$

where $\bar{n} = n(1+ik)$.
This may be obtained by substituting Eq. (2) into Eq. (3).
In general, since σ and σ^* (or η^*) are the function of ω, it should be obtained from the free electron theory of Drude. However, when the wavelength of light is sufficiently long, giving slowly varying electric field, the current in metal can be treated as being steady which is proportional to the eledtric field. If σ_0 or $\sigma_0^*(=\sigma_0/\omega\varepsilon_0)$, which is determined from experiments on

stationary of slowly varying current, is used for σ or σ^* value corresponding to the infrared. For instance, for copper $\sigma_0 = 5.8 \times 10^7$ ohm^{-1} m^{-1} and taking $\omega = 1.78 \times 10^{14}$ sec^{-1} corresponding to the CO_2 laser beam, 10.6 μ, $\sigma_0^* = 37000$ may be obtained with $\varepsilon_0 = 8.85 \times 10^{-12}$ farad·m^{-1}. A ε^* is of the order of unity and therefore is small compared with the σ_0^* value computed above. Therefore, Eq. (2) gives a good approximation;

$$
\begin{aligned}
k &= 1 & \text{(a)} \\
n^2 &= \frac{1}{2}\sigma_0^* & \text{(b)}
\end{aligned}
\tag{4}
$$

Substituting Eq. (4) into Eq. (3), the infrared absorptance of metal may be written

$$
\begin{aligned}
A &= 1 - R \\
&= \frac{2}{n} = \sqrt{8\eta_0^*}
\end{aligned}
\tag{5}
$$

For CO_2 laser beam, Eq. (5) yields

$$
A = 112.2\sqrt{\eta_0}
\tag{6}
$$

where η_0 is dc resistivity of metal, σ_0^{-1}. This is well-known Hagen-Rubens's equation.

4. Results and Discussion

(1) Absorptance of polished metal
As the absorptance of metal is markedly affected by its surface condition, a sufficient care is necessary for measuring the absorptance. Some measurements[2,4] have been carried out for the infrared absorptance of metal which is evaporated on clean and super smooth material, such as fused quartz or cleavage surface of NaCl, in high vacuum. It was shown that the infrared absorptance measured was reasonably in agreement with the theoretical value, Drude equation, calculated from the relaxation time of conduction electrons and dc conductivity, which was somewhat lower than the value calculated from Eq. (6).
However, in practical metal processing using CO_2 laser, the absorptance of metal in the surface condition supplied by a commercial vender is necessary rather than that in the extreme condition mentioned above. The sheet metal, 0.3 mm in thick, was recieved in the form of highly polished strip, and was sheared into 3 cm × 3 cm. Then the specimen was rinsed with a neutral detergent, and then with benzen. Though smoothness, purity, removal of residual stress and cleanness of the specimen are not so well comparison with other worker,[2,4] the authors call these specimens "polished metal".
The absorptance of various metals for high power CO_2 laser beam was measured at both room and fusion temperatures using calorimeter. In Fig. 4 the absorptance of polished metals at both temperatures is plotted vs square root of dc resistivity, in $(\mu\Omega\cdot cm)^{1/2}$. The experimental absorptance and theoretical one calculated from Eq. (6), which is shown by a solid line on this figure, are reasonably in good agreement

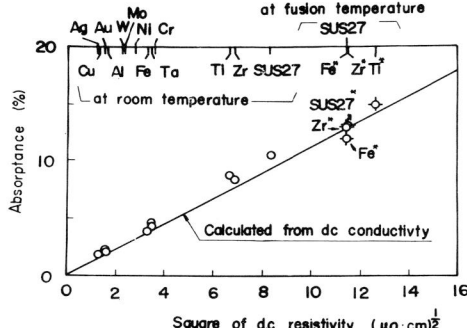

Fig. 4. Absorptance of metals at room temperature and fusion temperature marked with symbol*. SUS27 standarized by JIS corresponds to AISI 304 stainless steel.

Table 1[8]. Relation between the abrasive grain size and number of sand paper

number	240	280	320	400	500
abrasive grain size (μ)	80	67	57	40	34

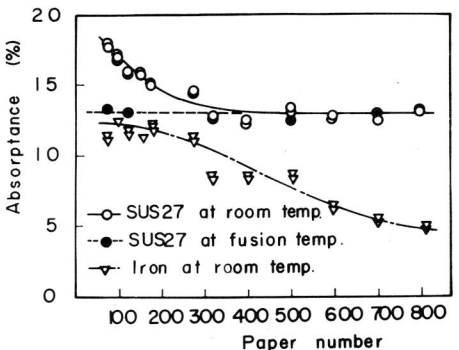

Fig. 5. Absorptance of roughened metallic surface.

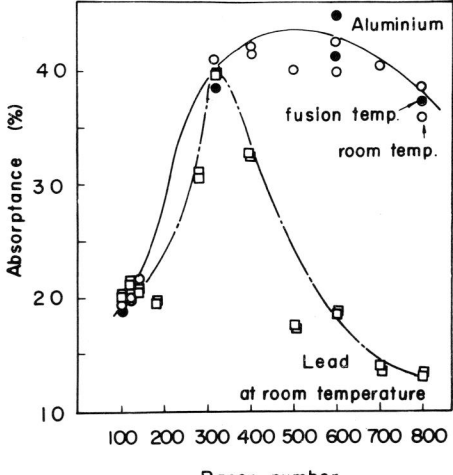

Fig. 6. Absorptance of roughened metallic surface.

at both temperatures. At room temperature, SUS27*) stainless steel has the highest absorptance and copper has the lowest one of any measured metal. At the fusion temperature the absorptance is large in comparison with one at the room temperature due to increase in the resistivity of metal, but it does not exceed only 15%, so that most incident laser energy is lost due to reflection.

(2) Surface treatments

The absorptance of metal is so low even after melting as shown in Fig. 4 that more than 85% of impinged laser beam is reflected out. This makes the CO_2 laser processing of metal inefficient, sometimes even impossible. The absorptance, however, fortunatelly is markedly affected by the surface condition, and therefore in order to improve the absorptance surface treatment was employed.

Many surface treatments were examined and a series of tests was carried out to evaluate their effect on the absorptance. Surface conditions are divided into four groups according to their absorptance characteristics:

(a) roughened, (b) coated with non-metallic thin film, (c) coated with fine metallic and non-metallic powder and (d) created by the irradiation of focused laser beam.

(3) Absorptance of roughened specimen

In order to analize the relation between the surface roughness of the specimen and the absorptance, the specimen was roughened with sand papers with various grain size. This treatment is practical and wide range of roughness is easily obtained with reproducible properties. The absorptance was measured after the roughened specimen was rinsed with benzen.

In Figs. 5 and 6 the absorptance is plotted vs the roughness in term of the number of the sand paper. The relation between the abrasive grain size of the sand paper and its number[8] is listed in Table 1.

As shown in Fig. 5, the absorptance at room temperature (open symbols) has a tendency to increase with

increasing the surface roughness for iron and SUS27. However, once the surface melts, the absorptance (filled symbols) decreases down to a constant value (for example in case of SUS27, about 13%) which agrees with the polished one.

On the other hand, as shown in Fig. 6, the absorptance curves of both aluminium and lead have appearant maxima at #300~#500 of roughness region at room temperature. The maximum values

---

*) The SUS27 standarized by JIS corresponds to AISI 304 stainless steel.

are considerably larger than polished; 45% for Al, which is larger by nearly a factor of 20 than the polished, and 40% for Pb, which is larger by a factor of 4—5 than polished. Both specimens treated with sand paper were blackened, especially at maxima of the absorptance curves. These phenomena did not occur for iron and SUS27. The increasing in the absorptance of both Al and Pb seems to be caused by the fine particles of the sand paper attached on their soft roughened surface rather than by roughening. Namely, the high absorptance measured at room temperature is maintained after melting because the fine particles remain on the surface, whereas the effect of the roughening alone is lost on melting as shown in SUS27.

It was found that the surface treatment with sand paper gives the high absorptance for both Al and Pb, after all, though the roughening treatment leaves the fine sand particles on their base surface. This treatment, however, seems to belong to the category of the treatment of coating with non-metallic films rather than that of the surface roughening.

(4) The absorptance of specimen coated with non-metallic thin film
(a) The absorptance at room temperature

Since the reason for which the polished metal has very low absorptance is that its electric conductivity is very high as descrived previously, the absorptance may be increased by coating its surface with non-metallic thin film. Many kinds of coatings produced by thermal oxidation, anodization, chemical treatment etc. were examined. To study the effect of non-metallic film coating on the absorptance, Fe_2S_3 and Al_2O_3 films were formed on the SUS27 and Al by the sulphurizing and anodizing respectively. These films have sufficient uniformity and reproducible properties.

In Fig. 7 the absorptance of sulphurized SUS27 specimens at room temperature is plotted vs time of the sulphurizing treatment. The absorptance increases with increasing treatment time, but it reaches a constant value, about 80%, at 7 minutes and does not increase any more. The weight gain was measured as a function of sulphurizing time, and it also saturated at about 7 minutes. It seems that this is due to the protective properties of Fe_2S_3 film.

On the other hand, the anodized film produced in dilute sulphuric acid can grow up to sufficient thickness providing 100% absorptance at room temperature because of its porous structure. Figure 8 shows the relation between absorptance and anodizing time. Anodizing was carried out in a 15% solution of H_2SO_4 at a temperature of 20°C and anode current was 1.2 A/dm².

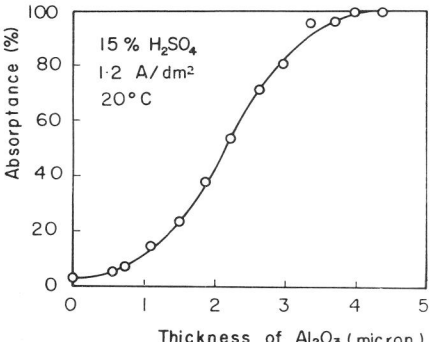

Fig. 8. Thickness of Al_2O_3 vs. absorptance at room temperature.

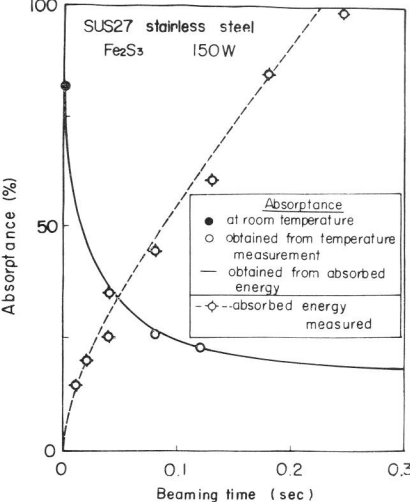

Fig. 9. Relation between absorptance and beaming time for SUS27 stainless steel (0.3 mm thick).

(b) Change of the absorptance with time

In order to study the absorption properties of the specimen coated with non-metallic thin film, change of absorptance with time during irradiation of focused high density laser beam on the specimen was measured. In Fig. 9 absorbed energy is plotted as a function of beaming time for SUS27 coated with Fe_2S_3 film with

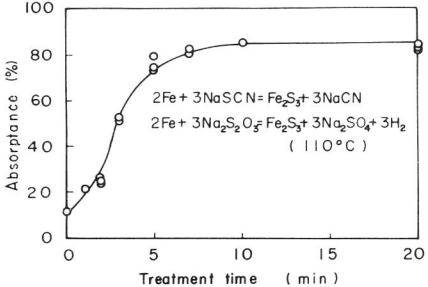

Fig. 7. Treatment time vs. absorptance at room temperature.

150 W focused laser power.

A solid curve shows the absorptance computed by differentiating the absorbed energy curve with respect to time. The absorptance obtained from the temperature measurement, which is shown by open circle symbols in this figure, agrees with solid curve precisely.

This absorptance curve decreases rapidly with time at first, and then its decreasing rate becomes gradual. The absorptance seems to fall down to that of polished, about 13%. Photograph 1 shows the surface irradiated by focused laser beam for duration of 40 msec and 150 msec with power level of 150 W.

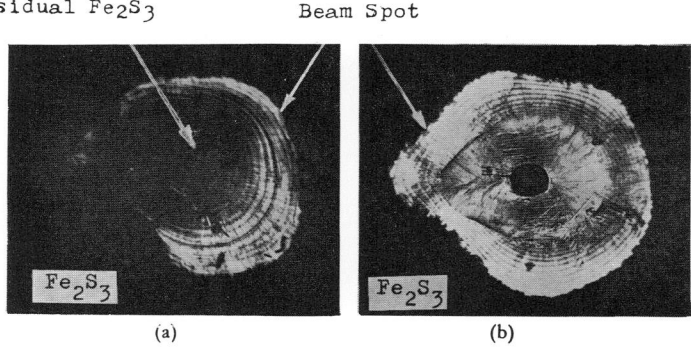

Photo. 1. Appearance of the surface irradiated by focused laser beam (SUS27 stainless steel with Fe₂S₃ film). Beaming time: (a) 40 msec (b) 150 msec.

It can be seen that Fe_2S_3 film still remains on the surface for 40 msec, but dissapears from the irradiated region for 150 msec. These results show that the decrease in absorptance is due to removal of Fe_2S_3 film via evaporation.

Figure 10 shows the change of the absorptance with time when focused laser beam with power level of 150 W impinged upon 0.3 mm Al specimens coated

with Al_2O_3 films taking the anodizing times; 11, 6.5 6 and 5.5 minutes.

They have 100, 64, 61 and 48% absorptance respectively at room temperature. The absorptance decreased sharply down to constant value, about 55%, though it was much higher than 3.7% for polished, during initial 50 msec of beaming time for the three specimens corresponding to 100, 64 and 61% absorptance at room temperature. It does not, however, change for the specimen corresponding to 48% absorptance at room temperature.

These results show that Al_2O_3 film keeping maximum 55% absorptance at room temperature can still remain on the 0.3 mm Al specimen when 150 W-focused laser beam impinges upon the specimen. According to Fig. 8, it is estimated that Al_2O_3 film having 2.2 μ in thickness can remain.

In order to confirm these results, the specimens were weighed both before and after the irradiation of

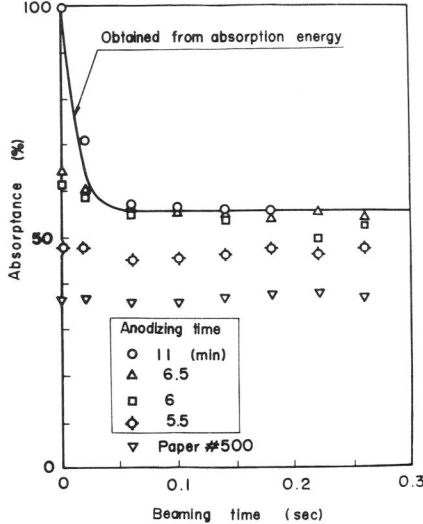

Fig. 10. Relation between absorptance and beaming time for anodized aluminium (0.3 mm thick).

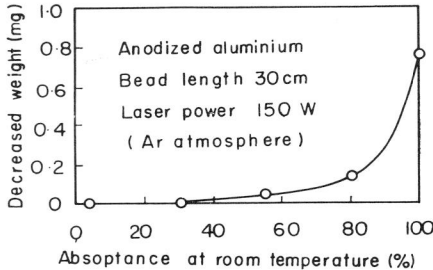

Fig. 11. Decreased weight vs. absorptance at room temperature (bead length=30 cm).

239

150 W-focused laser beam under the same condition. Since the decrease in weight is very small, long bead, 30 cm long, was formed. The decreased weight, in mg is plotted as a function of the absorptance at room temperature in Fig. 11. There were almost no decrease in weight for specimens having lower absorptance than 55% at room temperature, though appreciable decrease occured for the specimen having the higher absorptance than 55%.

The computation of the residual thickness was made for the specimen having the 100% absorptance at room temperature under the following assumptions: (1) Al_2O_3 film has a density of 2.5 g/cm³ [9]. (2) The film is removed in the rectangular shape of 0.5 mm wide by 30 cm long with a certain uniform thickness. (3) The film has an uniform thickness of 4 μ before laser heating. (4) The residual film is of uniform thickness and its density does not change after melting.

Since the computed thickness of removed Al_2O_3 film was about 2 μ, the thickness of the residual film was equal 2to μ, which was in good agreement with the thickness (2.2 μ) corresponding to the constant absorptance (55%) in Fig. 8. It was verified that the decrease in the absorptance of anodized Al was due to removal of the Al_2O_3 film via evaporation, and that

a certain thickness of the film remains so as to keep considerably higher absorptance.

This constant value to which the absorptance of such an anodized Al falls decreases with increasing the focused laser power as shown in Fig. 12, because the amount of Al_2O_3 film removed due to evaporation increases with increasing the laser power.

Namely, during laser heating it seems that the top surface of the Al_2O_3 film will reach its boiling point or above, and that the bottom surface, which contacts with the top surface of the aluminium sheet, should have the same temperature as the aluminium top surface having the considerably lower temperature in comparison with that of top surface of the film. Therefore the steep gradient in temperature should be produced within the film in the direction of the thickness, and thus a part of film heated over its boiling point is removed.

It was found that the absorption properties of anodized Al was quite distinct from that of sulphurized SUS27, i.e. the film in the former remained and contributed to enhance the absorptance, but in the latter did not remain at all when focused laser beam impinged upon the specimen for relatively long time. Since the bottom surface has the same temperature as the top surface of the base metal as described already, whether the film remains or not at the surface depends on the surface temperature of the base metal and the boiling point of the film. In the case of Al, its surface temperature is not so high because of its high heat conductivity, keeping the temperature of the film's bottom surface (contacted to the base metal) lower than its boiling point, for the given focuced laser power. On the other hand, in the case of SUS27 the top surface temperature of the base metal may exceed the boiling point of the surface-treated film due to the low heat conductivity.

(c) Effect of the traveling velocity and power level of laser beam on the absorptance

The absorption characteristics described already have been limited to the case there are no relative motion between the specimen and the focused laser beam. It is, however, practically important to analize the absorption characteristics of moving specimen or moving laser heat source for the laser processing such as welding. A series of bead-on-plate tests were carried out on both anodized Al and sulphurized SUS27, and the absorptance was measured as a function of the traveling velocity of the laser heat source and the laser power level, using the calorimeter.

Figure 13 shows examples of relation between the traveling velocity and the absorptance. In general, the absorptance of metal specimen coated with non-metallic films is inclined to increase with increasing traveling velocity.

Figure 14 shows examples of the relation between the impinged laser power and the absorptance at given constant traveling velocities. The absorptance has a tendency to decrease with increasing laser power because of the remarkable removal of the films owing to the evaporation.

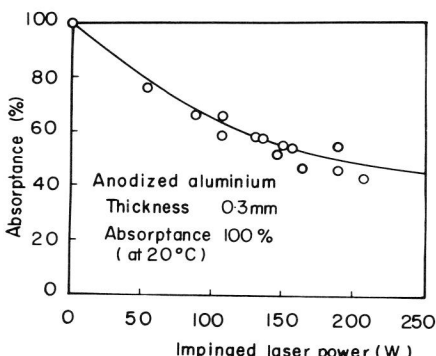

Fig. 12. Impinged laser power vs. absorptance of anodized aluminium in steady state ($v=0$).

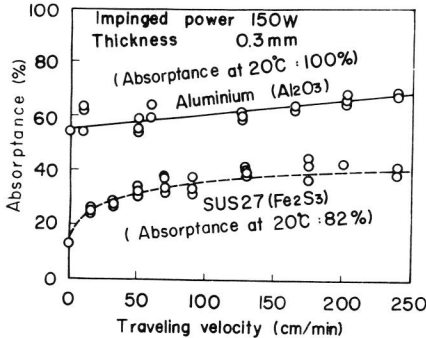

Fig. 13. Effect of traveling velocity on the absorptnace.

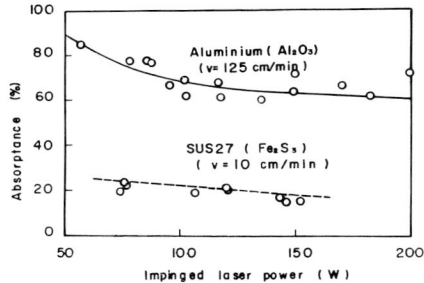

Fig. 14. Effect of impinged laser power on the absorptance.

Photo. 2. Appearance of the bead coated with paint.

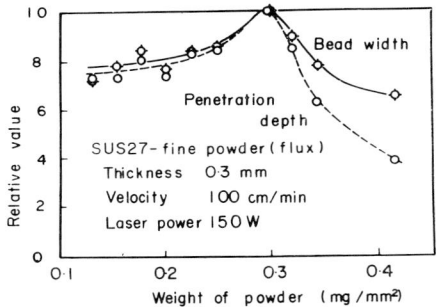

Fig. 15. Effect of the weight of powder.

The effect of surface treatment on the absorptance of Al is higher than that of SUS27 as shown both in Figs. 13 and 14, though Al has much lower absorptance than SUS27 under polished surface condition.

When the film coated on the metal surface has very low boiling point in comparison with base metal, the effect of the film on the absorptance cannot be expected. Although the specimen coated with a paint has nearly 100% of absorptance at room temperature, at the moment the focused laser beam impinges upon the specimen, the paint film evaporates completely as shown in Photo. 2 and has little effect on the absorptance.

(5) The absorptance of the specimen coated with fine powder

Metal specimen coated with metallic or non-metallic fine powder had neary 100% absorptance at room temperature. Fine powder (>200 mesh) was dissolved with proper concentration in alcohol, and it was spread on the surface of metal so as to form a uniform layer and then dried in the air. Whether impinged laser energy is effectively delivered to the base metal or not, depends upon the thickness of the powder, and to obtain the high absorptance a considerable thick layer is neccesary. In this case, since the absorptance was difficult to measure by means of the calorimetric method, a bead-on-plate test was carried out, and then the penetration depth and bead width were measured as shown in Fig. 15. These were plotted as a function of weight of the powder and there were maxima at 0.3 mg/mm^2 in these curves. They decrease sharply at weight above 0.3 mg/mm^2, because laser beam to be transmitted to the base metal is cut off by the thick layer of powder and is wasted for the melting or evaporating the powder.

(6) Increase in the absorptance caused by irradiation of the laser beam (for example, effect of the thermal oxidation on the absorptance)

The surface-treated specimens mentioned already have lower absorptance for focused laser beam than for unfocused, because the focused laser beam evaporates the thin surface layer so as to decrease the absorptance. However, when the metal is heated by focused laser beam in the air, depending on the metal, it is possible to make the absorptance higher than that at room temperature by forming the thin oxide film on the base surface. In the metal such as steel whoes oxide film is dense and adheres closely to the base surface, the oxidation may be limitted to the thin layer.

Figure 16 shows the relation between traveling velocity of laser heat source and the absorptance for SUS27, laser-heated in the air. The absorptance is inclined to increase with decreasing the traveling velocity for polished specimen because of proceeding of oxidation. There is a maximum in the absorptance curve at about 50 cm/min in the case sulphurized SUS27 is heated in the air because of multiple effects of Fe_2S_3 film (shown by a chain curve) and the

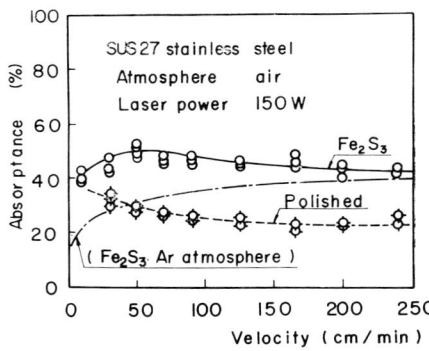

Fig. 16. Effect of the velocity on the absorptance (0.3 mm thick).

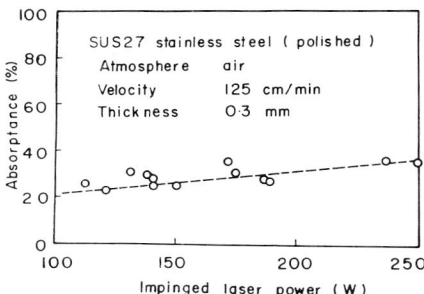

Fig. 17. Effect of laser power on the absorptance.

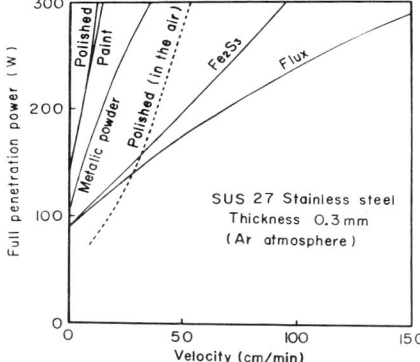

Fig. 18. Velocity vs. full penetration power. (SUS27 stainless steel).

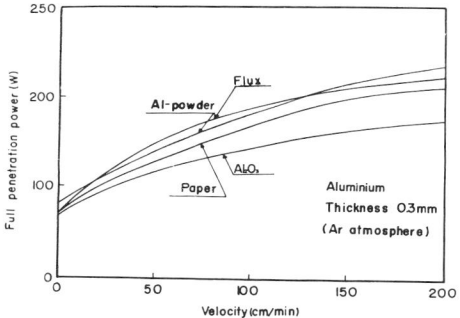

Fig. 19. Velocity vs. full penetration power (aluminium).

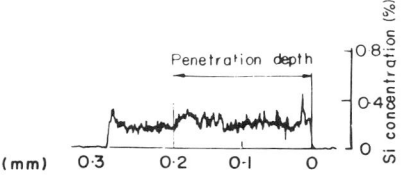

Fig. 20. Distribution of Si concentration (SUS27 stainless steel, non-metallic fine powder coating).

oxidation.

Figure 17 shows the relation between the impinged laser power and the absorptance for polished SUS27, laser-heated in the air. The absorptance increases linearly with increasing the laser power.

(7) Comparison between surface treatments

In previous sections, it was shown that the absorptance was a function of the traveling velocity of the laser heat source and its power, when the surface-treated metal specimens were heated by focused laser beam. Here, in order to compare the effect of each surface treatment on the absorptance under same condition, a full penetration power for various metals with a given thickness is measured under various surface conditions.

The data for 0.3 mm SUS27 are shown in Fig. 18. There is little differnce between polished specimen and one coated with paint: a film made of material of low boiling point such as paint has little effect on the absorptance. There was considerably larger difference of the effect of the absorptance in the surface treatments. Coating with non-metallic fine powder was the most efficient surface treatment among the other. On the other hand, coating with metallic fine powder had not so large effect as to non-metallic fine powder, becauce once metallic powder melted it acted as a polished metal. But since non-metallic fine powder melted to form the non-metallic thin film, it had high absorptance even after melting.

The data for aluminium, 0.3 mm in thick, are shown in Fig. 19. Under polished surface condition, even the surface of the specimen did not melt with power level of laser beam used in this experiment because of its extremely high refrectance. The absorptance was remarkably improved by the surface treatments, and there were only a little differnce of the effect on the absorptance in the surface treatments in the case of Al.

It should be noted that coating treatment with aluminium fine powder had nearly the same effect as non-metallic fine powder in contrast to the case of SUS27. It is believed that full penetration of Al specimen may be obtained at the moment when the layer of Al fine powder, which has high absorptance only at solid state, melts down, due to its high thermal conductivity and small thickness, 0.3 mm. Specimen coated with paint film, of which data are not indicated in this figure, does not melt even at the surface, because the paint evaporates easily since its boiling point is much lower than the base metal.

(8) Effect of the surface treatment on the welds

In order to study the effect of the surface treatment on the welds, concentration measurement of the element involved in the surface-treated layer and mechanical test were made. Distribution of the concentration in the weld bead was measured for a laser-heated SUS27 stainless steel specimen coated with non-metallic fine powder. Figure 20 shows the silicon content which is the principal contsituent of the non-metallic fine powder. Silicon content was negligible in molten metal. Non-metallic powder seems to vaporize away and have no bad effect on the bead.

242

Photo. 3. (a) V-joint weld of SUS27 stainless steel (0.3 mm thick).

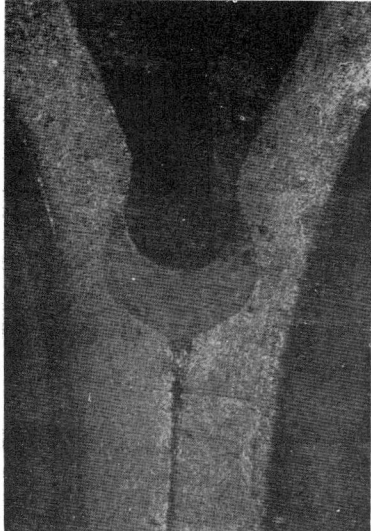

(b) The cross section.

Similar results were obtained from other kinds of surface treatment. Tensile test transverse to the direction of welding was made for SUS27 stainless steel specimen. The surface conditions used were (1) polished, (2) coated with metallic fine powder, (3) coated with non-metallic fine powder, and (4) sulphurized. Test specimens failed usually in the base metal or heat affected zone in the cases of surface conditions (1) and (2), and sometimes in the weld bead in (3) and (4). There were, however, almost no difference in tensile strength between both cases, and the weld joint efficiency ranged from 90~100%.
(9) Welding method of thick metal

It has been shown in Report 1[5] that the beam is reflected by the wall of the hole so as to re-focus it to the bottom. As the result, the hole has an acute angle of wedge shape at the bottom and the incident beam may be almost trapped within the hole showing the high beam absorptance.

Photograph 3 shows an example of V-joint weld as a result of improving the beam absorptance based on this idea. In this case, the absorptance about 60% was measured. Such a condition having high beam absorptance may be also realized by the drilling action of the high power laser beam as described above. But the depth of the drilled hole and the thickness of weld are limited by the laser power. In case of V-joint welding, there are no limiting thickness in principle.

5. Conclusions

Absorptance of various metals in polished surface condition at wavelength 10.6 μ was evaluated at both room and fusion temperatures, and as the results it was emphasized that the absorptance was too low. The authors proposed to improve the absorptance by surface treatments. The surface treatments were divided into four groups; (a) roughened, (b) coated with non-metallic thin layer, (c) coated with fine powder and (d) created by the focused laser beam heating.

Conclusions obtained may be summarized as follows:

(1) The absorptance of metal sheet in the surface condition supplied by a commercial vender, which was highly polished, was proportional to a square root of the dc resistivity and agreed well with the theoretical values calucurated from Hagen-Rubens's equation at both room and fusion temperatures.

(2) Surface treatments such as superficial roughening or coating having a material of lower boiling point than base metal had little effect on the absorptance.

(3) When the focused laser beam impinged upon the aluminium specimen in the surface condition (b) or (c) for a long time, a part of the surface-treated layer remained so as to keep the considerably high absorptance. However, in the case of SUS27 stainless steel specimen the absorptance decreased with beaming time, at last down to the polished one due to complete removal of the surface-treated layer.

(4) In the case of SUS27 there was large difference between the absorptance in different surface conditions, and coating with non-metallic fine powder had the largest effect on the absorptance. On the other hand, in the case of Al there was little difference between the absorptance in different surface condition except for coating with a material of very low boiling point.

(5) Though, in polished surface condition, Al was very difficult for CO_2 laser processing in comparison with SUS27, surface treatment made the processing easier for Al than SUS27.

(6) In general, the absorptance of metal had a tendency to increase with increasing the traveling velocity of the laser heat source and decreasing its power level in the surface conditions (b) and (c), but the tendency was completly contrary in the surface condition (d).

(7) Surface treatments had no bad effect on the weld bead.

(8) In case of welding a thick metal, V-joint welding was favourable.

References

1) E. Hagen and H. Rubens: Ann. Physik, Vol. 11 (1903). 873.
2) L. G. Schulz: "An Experimental Confirmation of the Drude Free Electron Theory of the Optical Properties of Metals for Silver, Gold and Copper in the Near Infrared", JOSA Vol. 44 (1954), No. 7, 504–545.
3) H. E. Bennett, M. Silver and E. J. Ashley: J. Opt. Soc. Am. Vol. 53 (1963), 1089.
4) J. M. Bennett and E. J. Ashley: "Infrared Refrectance and Emittance of Gold Evaporated in Ultrahigh Vacuum", Appl. Optics Vol. 4 (1965), No. 2, 221–224.
5) Y. Arata and I. Miyamoto: "Some Fundamental Properties of High Power Laser Beam as a Heat Source (Report 1)", Transactions of the Japan Welding Society Vol. 3 (1972), No. 1.
6) Y. Arata, I. Miyamoto and M. Kubota: "Some Fundamental Properties of High Power CW Laser Beam as a Heat Source", I. I. W. Doc. IV-4-69, (1969)
7) Y. Arata and I. Miyamoto: "Studies of High Power CO_2 Gas Laser as a Heat Source (Report 1)", Journal of the Japan W. S. Vol. 39, (1970) in Japanese, No. 12, 1307–1314.
8) JIS R6252-1966 Abrasive paper.
9) Aluminium Tashenbuch, 12, Anflage S. 576.

Some Fundamental Poroperties of High Power Laser Beam as a Heat Source (Report 3)

— Metal Heating by Laser Beam —

Abstract

Mechanism of CO_2 laser procssing was discussed on the basis of heat conduction theory and experiments. A generallized heat conduction theory for moving surface heat source with non-uniform energy distribution and any waveform of power supply was developed, and temperature distributions were computed in relation to time and heating conditions. A laser-weldability was discussed on the basis of the heat conduction and optical properties of the work piece and metals were clasified according to it. Interaction between focused laser beam and surface-treated metal was observed using high-speed camera and also discussed from the view point of the heat conduction theory.

1. Introduction

Authors have made close investigation on a CO_2 laser absorption properties[1] of surface-treated metal and suggested that they depend not only on surface treatment but on surface temperature of metal. Therefore it is important to know a precise temperature distribution of metal specimen in relation to time and heating condition in order to make suitable laser processing as well as to reveal the absorption properties.

In this paper, interaction between laser beam and surface-treated metal specimen and its temperature distribution are discussed on the basis of heat conduction theory and observations.

Some workers[2]–[4] have studied heat conduction problems involved laser heating, but an actual heat source has been approximated by an idealized simple model because of mathematical difficulty and experimental difficulty to estimate the actual heating condition. These treatments, however, were not in general enough to know precise temperature distribution, especially in the vicinity of the heat source.

This paper begins by describing our CO_2 laser apparatus and experimental setups for measuring energy distribution profile of focused laser beam and its power output waveform. Section **3** shows the features of laser processing and section **4** develops a heat conduction theory for generallized surface heat source. Section **5** gives a comparison between computed and experimental results to confirm the theory, and section **6** discusses a laser-weldability of various metal based on the heat conduction theory and their optical properties. Section **7** deals with an interaction between focused laser beam and surface-treated metallic specimen, and section **8** discusses a convection driving force on the basis of the observed results. The final section gives main conclusions obtained from this study.

2. Properties of CO_2 Laser Beam

The authors will explain briefly our CO_2 laser apparatus and fundamental properties of the output beam. This CO_2 laser apparatus, whoes schematic diagram is shown in Fig. 1[5], consists of a water-cooled

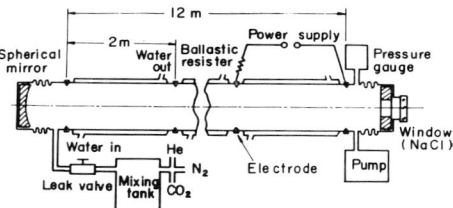

Fig. 1. Schematic diagram of 12 m CO_2 laser apparatus.

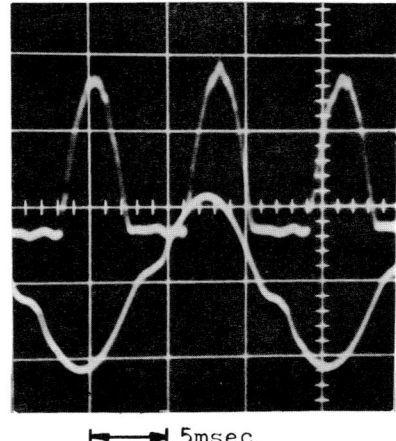

$\vdash\!\!\longrightarrow\!\!\dashv$ 5msec

Photo. 1. CO_2 laser output waveform detected with Ge:Au (upper) and discharge current waveform (lower).

discharge tube 74 mm i.d., 12 m long, placed between gold evaporated plane and 29.2 m radius curvature mirrors. The laser power with a wavelength 10.6 μ was coupled out from the cavity through a 15 mm diam hole at the center of the plane mirror. There are six sections to the tube, each section with 2 m long having its own cathode with a ballast resistor. A cold cathode discharge, either ac or dc, was run on each section in flowing gas mixture of CO_2, N_2 and He.

Power output measurements were carried out with a power meter and a Au-doped germanium detector. In this experiment, the laser was operated from 60 Hz electric power supply which gives a pulsed output of 120 Hz as shown in Photo. 1.

The output laser beam was focused through the optical system which consisted of a plane mirror and a 310 mm radius curvature mirror. Figure 2[1)6)] shows the beam energy profile at focal point which may be approximated by a Gaussian curve with a diameter of 0.5 mm at the $1/e$ power point.

3. Some Features of Laser-Heated Workpiece

The authors have shown that CO_2 laser beam absorptance of metal, which is very low in the polished surface condition, may be remarkably enhanced by appropriate surface treatment of the metal. Photograph 2 shows a typical bead cross section of surface-treated sheet metal, SUS27*) stainless steel. Surface-treated thin layer on the sheet metal consists of Fe_2S_3 produced by a sulphurizing. The shape of the bead cross section is heat conduction type as is evident from this photograph. Computing the temperature distribution from ideal point heat source, for example, a width-to-depth ratio of the fusion isotherm should be 0.5 for semi-infinite thickness plate and less than 0.5 for thin plate with a thickness of the order shown in Photo. 2. It contradicts, however, with the fact that actual width-to-depth ratio is considerably larger than 0.5 as shown in Photo. 2. This discrepancy indicates that the focused laser beam cannot be approximated by an ideal point source at least, but its energy distribution should be taken into account in the heat conduction problems.

Photograph 3 shows an another example of the bead cross section. This fusion isotherm has a slightly depressed contour at its center in comparison with Photo. 2. Such a fusion isotherm, which is a special case, has been obtained only in the case the base metal is SUS27 stainless steel and surface-treated layer consists of non-metallic fine powder as well, within the limits of our experiments. It seems to be, however, quite all right to consider that shape of bead cross section shown in Photo. 3 is also heat conduction type on the whole.

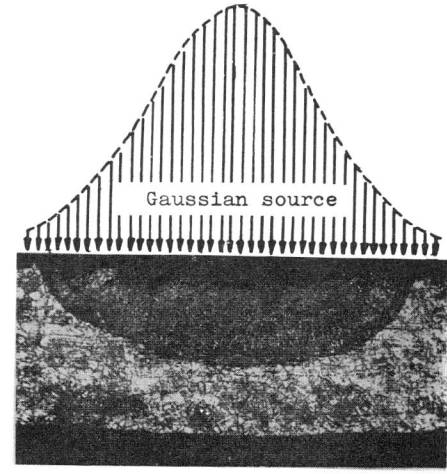

Photo. 2. Typical bead shape heated by CO_2 gas laser beam (SUS27 stainless steel with Fe_2S_3film, 0.3 mm thick).

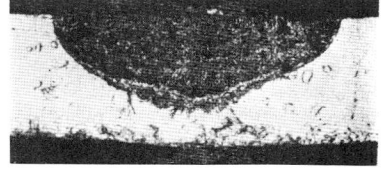

Photo. 3. Cross sections of the bead of SUS27 stainless steel (coated with non-metallic fine powder).
Thickness: 0.3 mm
Velocity: 20 cm/min
Laser power: 170 W

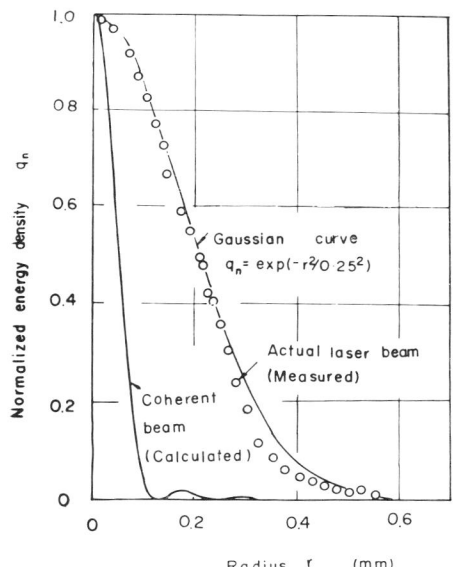

Fig. 2. Comparison between actual laser beam measured and coherent beam (f=155 mm, beam diam=15 mm)[6)]

*) SUS27 standerized by JIS corresponds to AISI 304 stainless steel.

Photograph 4 shows examples of the typical bead appearances at top surface, a periodic scale-shape bead, which differs essentially from one obtained by arc welding. A pitch of the periodic scale-shape bead increases with increasing the traveling velocity. In Fig. 3 the pitch is plotted as a function of traveling velocity. In this case experimental results agree closely with a solid line computed with 120 Hz which corresponds to the frequency of the laser output waveform shown in Photo. 1. These show that at least near the surface the temperature distribution is affected by variation of power with time caused by an excitation from ac electric power supply. Photograph 5 shows the typical bead cross section affected by the output waveform of laser beam: it is obvious that both melting and solidification occur in each period of the 120 Hz laser output.

In order to analize the effect of periodic change of the laser power on the temperature distribution of the specimen, the temperature rise at location on the axis of the impinged focused laser beam on the bottom surface was measured. To measure the rapid change of its temperature, a fine Chromel wire, 0.1 mm in diameter, was welded to its bottom surface by means of condenser-discharge type of spot welder, and potential difference generated between the Chromel wire and the specimen was recorded by oscillograph. Figure 4 shows examples of temperature records for Al and SUS27, 0.3 mm in thick. For SUS27, a fluctuation of temperature rise curve owing to the periodic change of laser power is very small and may be neglected at the bottom surface. On the other hand, the fluctuation of the Al specimen is appreciably large because of its large thermal diffusivity.

From experiments descrived above, it is found that for 60 Hz ac-excited focused CO_2 laser beam with

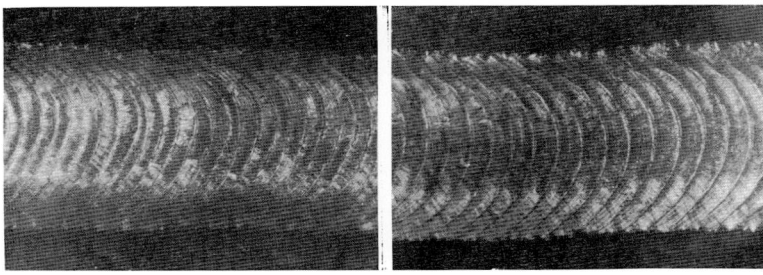

Photo. 4.   Appearance of bead (SUS27 stainelss steel).
(a) Velocity: 40 cm/min
(b) Velocity: 60 cm/min

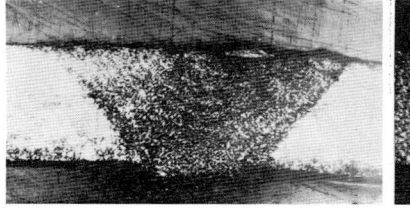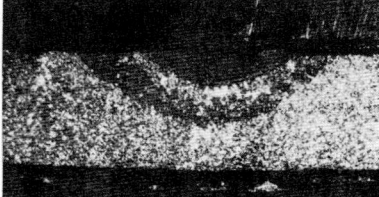

Photo. 5.   Cross sections of the bead (0.3 mm)
(a) SUS27 stainless steel
(b) Aluminium.

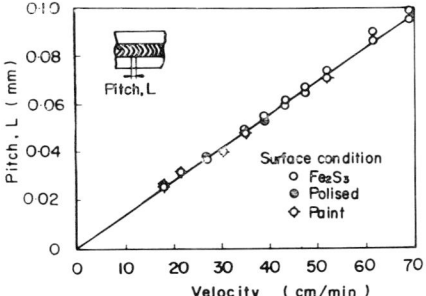

Fig. 3.   Velocity vs. pitch L in the case of ac excited laser heating.

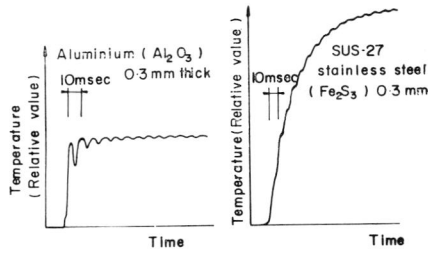

Fig. 4.   Examples of thermo e.m.f. records corresponding back surface temperature, θ_b.

power level of the order of several hundreds watts, cross sectional bead shape formed is a heat conduction type, and that treating the heat conduction problem, both energy distribution profile at the surface and waveform of the laser beam as indicated in Photo. 1 and Fig. 2 must be taken into account.

In the follwing section, a generallized heat conduction theory will be developed and applied to analize the experimental results described above.

4. Heat Conduction Theory for Surface Heat Source

Heat conduction problem in laser heating may be reduced to the problem to solve the temperature distribution for moving surface heat source with non-uniform intensity distribution of which power supply per unit time varies with time. In order to treat such a complicated problem, it is convenient to start from rather simplified model, the source with constant power and next to develop into the case power supply per unit time varies with time using Duhamel's integral. In this treatment variation of the physical coefficients of material, radiation heat loss and convection in the molten pool are neglected.

(1) Temperature distribution due to a stationary surface heat source with constant power

When heat concentrated to a point is liberated at origin of a cylindrical coordinate system located on the surface of semi-infinite solid at constant rate q per unit time, the temperature rise at (r,z) at time t due to the heat delivered from 0 to t is given by

$$\theta(r,z;t) = \frac{q}{2\pi k \sqrt{z^2 + r^2}} \text{erfc} \frac{\sqrt{z^2 + r^2}}{\sqrt{4k_D t}} \quad \cdots(1)$$

where r is radial distance measured parallel to the surface from the origin, z vertical distance measured perpendicular to the surface from the origin, t time measured from the initiation of heat delivery, k thermal conductivity of the material, k_D thermal diffusivity of the material and

$$\text{erfc } X = \frac{2}{\sqrt{\pi}} \int_X^\infty \exp(-v^2) dv$$

In the case a heat liverated has a rotational symmetric distribution with respect to z axis: its power density at r' is given by $q(r')$, the temperature rise at (r, z) at time t can be written

$$\theta(r,z;t) = \frac{1}{k\pi} \int_0^\pi \int_0^\infty \frac{q(r')}{R} \text{erfc} \frac{R}{\sqrt{4k_D t}} \cdot r' \cdot dr' \cdot d\varphi \quad \cdots(2)$$

where $R = (r'^2 - r^2 - 2r' \cdot r \cdot \cos\varphi + z^2)^{1/2}$.

For instance, in case of steady superficial heat source with an uniform flux density of q_0 over a circular region on the surface of a semi-infinite solid, temperature rise at (r,z) at t is given by App. 1)

$$\theta(r,z;t) = \frac{2q_0}{\pi k} \sqrt{k_D t} \int_0^{\varphi_0} \left[\text{ierfc} \frac{\sqrt{z^2 + l_2^2}}{\sqrt{4k_D t}} - \text{ierfc} \frac{\sqrt{z^2 + l_2^2}}{\sqrt{4k_D t}} \right] d\varphi \quad \cdots(3)$$

where $|l_1| = r \cdot \cos\varphi + (a^2 - r^2 \cdot \sin^2\varphi)^{1/2}$,

$$\text{ierfc} X = \int_X^\infty \text{erfc } u \cdot du$$
$$= \exp(-X^2)/\sqrt{\pi} - X \cdot \text{erfc } X,$$

and φ_0 and $|l_2|$ is given by

$$\varphi_0 = \sin^{-1} a/r$$
$$|l_2| = r \cdot \cos\varphi - (a^2 - r^2 \sin^2\varphi)^{1/2}, \text{ when } r > a$$
$$\varphi_0 = \pi$$
$$l_2 = 0, \text{ when } r < a$$

At location on the axis, $r = 0$, Eq. (3) yields

$$\theta(O,z;t) = \frac{2q_0}{k} \sqrt{k_D t} \left[\text{ierfc} \frac{z}{\sqrt{4k_D t}} - \text{ierfc} \frac{\sqrt{z^2 + a^2}}{\sqrt{4k_D t}} \right] \quad \cdots(4)$$

At the origin, the center of the circular source, the temperature rise becomes simple form:

$$\theta(O,O;t) = \frac{2q_0}{k} \sqrt{k_D t} \left(1.1284 - \text{ierfc} \frac{a}{\sqrt{4k_D t}} \right) \cdots(5)$$

and at steady state, $t = \infty$, Eq. (5) yields:

$$\theta(O,O;\infty) = \frac{q_0}{k} \quad (\equiv \theta_m) \quad \cdots(6)$$

This is a limiting temperature at center of the uniform circular heat source.

Since Eq. (2) posseses singularity in $r'^2 + r^2 - 2r \cdot r' \cdot \cos\varphi + z^2 = 0$, we represents the temperature rise using Eq. (3) in order to remove the singularity. Assuming $\Theta_\rho(r,z;t)$ represents $\theta(r,z;t)$ in Eq. (3), under condition $a = \rho$ and $\pi\rho^2 q_0 = 1$, Eq. (3) may be written by

$$\theta(r,z;t) = \int_0^\infty \pi\rho^2 \left[-\frac{d}{d\rho} q(\rho) \right] \Theta_\rho(r,z;t) \cdot d\rho \quad \cdots(7)$$

For the heat source having Gaussian distribution with power density at center $(\rho = 0)$ q_0 and radius of $1/e$ power point at center a given by

$$q(\rho) = q_0 \cdot \exp(-\rho^2/a^2) = \frac{Q_0}{\pi a^2} \exp(-\rho^2/a^2), \quad \cdots(8)$$

substitution of Eq. (8) into Eq. (7) yields

$$\theta(r,z;t) = \frac{Q_0}{\pi a^4} \int_0^\infty \rho^3 \cdot \exp(-\rho^2/a^2) \Theta_\rho(r,z;t) \cdot d\rho \quad \cdots(9)$$

where Q_0 is the total heat delivered per unit time and is equal to $\pi a^2 q_0$. At location on the axis $(r = 0)$, Eq. (9) is expressed in the form:

$$\theta(O, z;t) = \frac{\sqrt{k} Q_0}{\sqrt{\pi^3 k}} \int_0^t \frac{\exp(-z^2/4k_D\tau)}{\sqrt{\tau} (a^2 + 4k_D\tau)} \cdot d\tau \quad \cdots(10)$$

By writting

$l^* = l/a$ (l is a quantity with a dimension of length)

$t^* = (2/a) \sqrt{k_D t}$ \hfill (11)

$\theta^* = \theta/\theta_m$

Eq. (10) is expressed in non-dimensional units as follows:

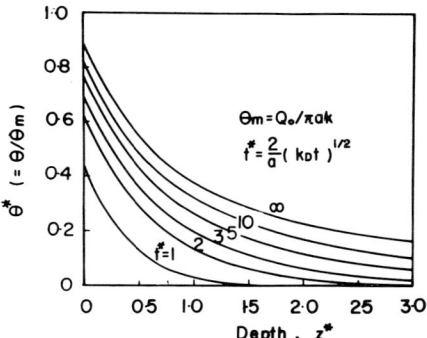

Fig. 5. Relation between z^* and θ^* for semi-infinite thickness plate (Gaussian source, $r^*=0$).

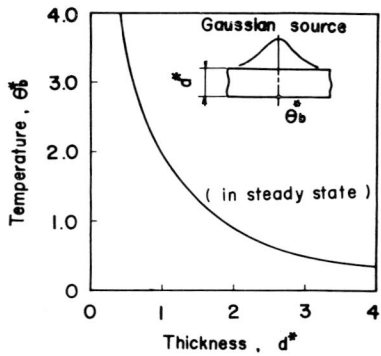

Fig. 6. Thickness vs. temperature θ_b^* in steady state.

$$\theta^*(O,z^*;t^*)=\frac{1}{\sqrt{\pi}}\int_0^{t^*}\frac{\exp(-z^{*2}/\tau^{*2})}{1+\tau^{*2}}\cdot d\tau^* \cdots\cdots(12)$$

where $z^*=z/a$ and $r^*=r/a$. Simplified equation may be obtained in special case: at center of the heat source $(r^*=z^*=0)$,

$$\theta^*(O,O;t^*)=\frac{1}{\pi}\cdot\tan^{-1}t^* \cdots\cdots(13)$$

and in the steady state $(t^*\to\infty)$ at (O,z^*),

$$\theta^*(O,z^*;t^*)=\frac{\sqrt{\pi}}{2}\cdot\exp(z^{*2})\cdot\mathrm{erfc}\ z^* \cdots\cdots(14)$$

Figure 5 shows the relation between z^* and θ^* $(O,z^*;t^*)$ for various values of t^* in case semi-infinite plate.

In the case the plate has a limited thickness d giving rise to a series of reflection terms, Eq. (12) is written by

$$\theta^*(O,z^*;t^*)\frac{1}{\pi}\int_0^{t^*}\frac{1}{1+\tau^{*2}}\cdot\sum_{n=-\infty}^{+\infty}\exp\left[-\frac{(2nd^*+z^*)^2}{\tau^{*2}}\right]\cdot d\tau^*$$
$$\cdots\cdots(15)$$

where $d^*=d/a$. Figure 6 shows the relation between d^* and θ^* at location (O,d^*) in steady state in the case Gaussian source acts on the surface of the plate having limited thickness.

(2) Temperature distribution due to moving surface heat source with constant power

Temperature distribution due to the ideal point heat source having constant power and moving at constant speed has been presented by Rosenthal[7] in quasi-steady state. Expanding his results for a non-uniformly distributed surface heat source, temperature rise at point (r,φ,z) in the system of cylindrical coodinates attached to the heat source, r and φ in the surface and z extending perpendicular downward, in quasi-steady state is given by

$$\theta(r,\varphi,z)=\frac{1}{2\pi k}\int_0^{2\pi}\int_0^\infty\frac{q(r')}{R'}\cdot\exp\left[-\frac{v}{2k_D}(R'+r\cdot\cos\varphi\right.$$
$$\left.-r'\cdot\cos\varphi')\right]r'\cdot dr'\cdot d\varphi' \cdots\cdots(16)$$

where $R'=[r'^2+r^2-2r\cdot r'\cos(\varphi-\varphi')+z^2]^{1/2}$ and $v=$ traveling velocity of the heat source. At location on the axis, $r=0$, Eq. (16) yields

$$\theta(O,\varphi,z)=\frac{1}{2\pi k}\int_0^{2\pi}\int_0^\infty\frac{q(r')}{\sqrt{r'^2+z^2}}\cdot\exp\left[-\frac{v}{2k_D}(r'^2+z^2\right.$$
$$\left.-r'\cdot\cos\varphi'\right]\cdot r'dr'd\varphi$$
$$=\frac{1}{k}\int_0^\infty\frac{r'q(r')}{\sqrt{r'^2+z^2}}\cdot\exp\left(-\frac{v}{2k_D}\sqrt{r'^2+z^2}\right)\cdot I_0\left(\frac{vr'}{2k_D}\right)\cdot dr'$$
$$\cdots\cdots(17)$$

where $I_0=$ Bessel function first kind zero order.

If heat delivers uniformly over a circular region with a radius a, Eq. (17) becomes

$$\theta(O,\varphi,z)=\frac{q_0}{k}\int_0^a\frac{r'}{\sqrt{r'^2+z^2}}\exp\left(-\frac{v}{2k_D}\sqrt{r'^2+z^2}\right)$$
$$\cdot I_0\left(\frac{vr'}{2k_D}\right)\cdot dr'.$$

Putting $y^2=r'^2+z^2$, this may be written

$$\theta(O,\varphi,z)=\frac{q_0}{k}\int_z^{\sqrt{a^2+z^2}}\exp\left(-\frac{vy}{2k_D}\right)\cdot I_0\left(\frac{v}{2k_D}\sqrt{y^2-z^2}\right)dy.$$

Now putting $y^*=y/a$, this is expressed in non-dimensional units as follows:

$$\theta^*(O,\varphi,z^*)=\int_{z^*}^{\sqrt{1+z^{*2}}}\exp(-v^*y^*)\cdot I_0(v^*\sqrt{y^{*2}-z^{*2}})\cdot dy^*$$
$$\cdots\cdots(18)$$

where $v^*=av/2k_D$. For Gaussian heat source shown in Eq. (8), Eq. (17) becomes

$$\theta^*(O,\varphi,z^*)=\int_{z^*}^\infty\exp(-y^{*2}+z^{*2}-v^*y^*)\cdot$$
$$I_0(v^*\sqrt{y^{*2}-z^{*2}})\cdot dy^* \cdots\cdots(19)$$

Figures 7–9 show examples of isothermal lines in both transverse and longitudinal cross sections computed[App2] in cases of ideal point source, circular source with uniform intensity and Gaussian source in quasi-steady state $(d^*=1.2, v^*=0.25)$.

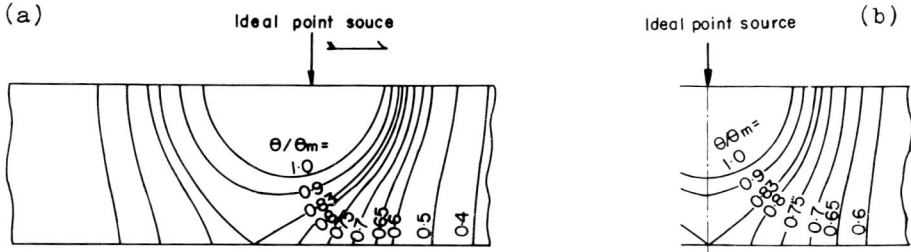

Fig. 7. Isothermal lines for ideal point source in quasi-steady state ($d^*=1.2$, $v^*=0.25$).
(a) Longitudinal cross section
(b) Transverse cross section
(maximum value).

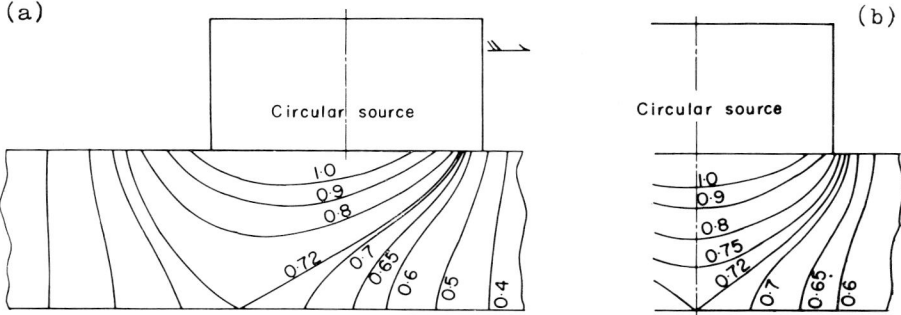

Fig. 8. Isothermal lines for circular source in quasi-steady state ($d^*=1.2$, $v^*=0.25$).
(a) Longitudinal cross section
(b) Transverse cross section
(maximum value).

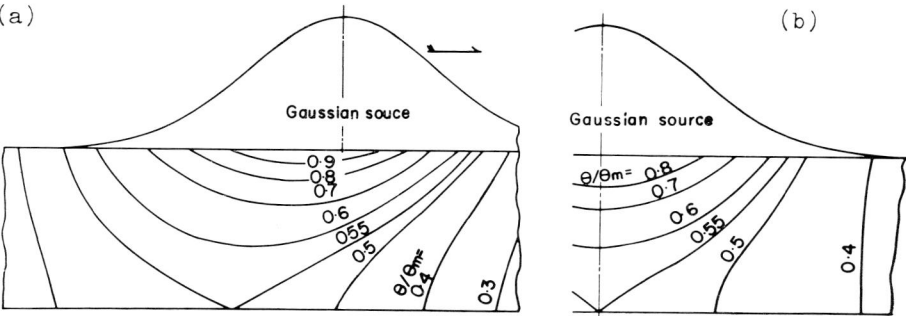

Fig. 9. Isothermal lines for Gaussian source in quasi-steady state ($d^*=1.2$. $v^*=0.25$).
(a) Longitudinal cross section
(b) Transverse cross section
(maximum value).

Figure 10 shows relation between v^* and maximum temperature at top and bottom surfaces of sheet ($d^*=1.2$) in non-dimensional form, θ_s^* and θ_b^* respectively in both cases of ideal point source and Gaussian source (θ_s^* for ideal point source is not indicated on this figure because it becomes infinity). A ratio, θ_s^*/θ_b^*, for Gaussian source is also indicated

by a chain line on this figure: it increases with increasing v^*.

In order to know the effect of heat distribution on the temperature rise, θ_b^* for each source was compared. Figure 11 shows the relation between v^* and ratio of θ_b^* for ideal point source and circular source with uniform intensity to θ_b^* for Gaussian source.

250

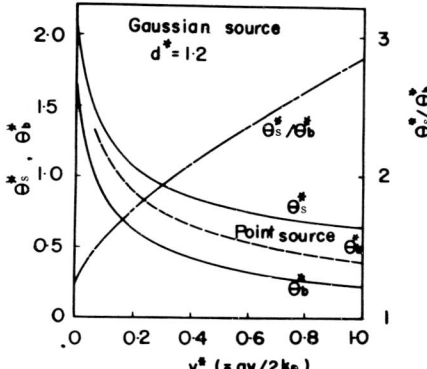

Fig. 10. Relation between θ_s^*, θ_b^* and v^* in quasi-steady state (0.3 mm).

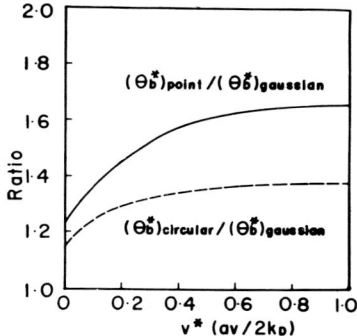

Fig. 11. Comparison of normalized temperature between point, circular and Gaussian sources.

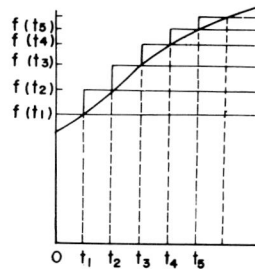

Fig. 12. Approximation of $Q_0(t)$ by step function.

It may be seen that difference of the temperature rise between Gaussian source and both ideal point and circular sources has a tendency to increase with increasing v^*.

(3) Temperature distribution due to surface heat source with the varing power with time

Here heat conduction problems are treated under condition a heat source whoes total power varies with time but its flux distribution profile does not vary, moves with a constant speed. A situation where a rate

of power supply of source varies with time or heat absorbability of work piece varies with its temperature, surface condition or surrounding is occasionally produced in laser as well as other conventional source, and it will be important to know the temperature distribution in such a situation. Let us calculate its distribution by dividing the calculation into two steps.

In the first step transient temperature distributions in workpiece are calculated under conditions heat flux distributes non-uniformly, the heat source is stationary at origin and its power $Q(t)$ is constant:

$$Q(t) = 1, \text{ when } (t \geq 0) \left. \right\}$$
$$Q(t) = 0, \text{ when } (t < 0) \quad\quad \cdots\cdots(20)$$

And at the second step the transient temperature distribution in the case the source power varies with time and the source moves with a given speed are calculated using results obtained in the first step by Duhamel's integral.

Temperature rise at (r, z) at time t due to heating by its source, which is stationary at origin, having non-uniform flux distribution with a total power per unit time shown in Eq. (20) may be obtained from Eq. (2), putting

$$2\pi \int_0^\infty q(r') \cdot r' \cdot dr' = 1$$

We shall write the temperature rise at (r, z) at time t as $\Theta(r, z; t)$.

Let us approximate the heat supply per unit time by step function as shown in Fig. 12. Writing

$$n \cdot \Delta t = t_n \quad (n = 1, 2, 3, \cdots\cdots) \quad\quad \cdots\cdots(21)$$

for small time element Δt, and replacing $Q(t)$ shown in Fig. 12 by $Q(t_n)$ for time duration $t_{n-1} < t < t_n$, temperature rise at (r, z) at time t due to heating from 0 to t_n is given by

$$\begin{aligned}
\theta(r, z; t) &= (t_1) \cdot \Theta(r, z; t) \\
&\quad + [Q(t_2) - Q(t_1)]\Theta(r, z; t - t_1) \\
&\quad + \cdots\cdots \\
&\quad + [Q(t_{n-1}) - Q(t_n)] \cdot \Theta(r, z; t - t_{n-1}) \\
&= Q(t_1) \cdot [\Theta(r, z; t) - \Theta(r, z; t - t_1)] \\
&\quad + Q(t_2) \cdot [\Theta(r, z; t - t_1) - \Theta(r, z; t - t_2)] \\
&\quad + \cdots\cdots \\
&\quad + Q(t_n) \cdot [\Theta(r, z; t - t_{n-1}) - \Theta(r, z; t - t_n)] \\
&= \sum_{i=1}^n Q(t_i)[\Theta(r, z; t - t_{i-1}) - \Theta(r, z; t - t_i)] \\
&= \sum_{i=1}^n \frac{\Delta\Theta(r, z; t)}{\Delta t} Q(t_i) \cdot \Delta t \quad\quad \cdots\cdots(22)
\end{aligned}$$

In the limit, $\Delta t \to 0$ Eq. (22) yields

$$\theta(r, z; t) = \int_0^{t_n} \frac{\partial}{\partial t} \Theta(r, z; t - t') \cdot Q(t') \cdot dt' \quad\quad \cdots\cdots(23)$$

Now we consider the case such a source moves with a constant speed v. Temperature rise at (x, y, z) at t in the system of Cartesian coordinate attached to the center of the source, x extending tword moving direction of the source, y in a surface and z downward, is given by

$$\theta(x,y,z;t)=\sum_{i=1}^{n}\frac{\Delta\Theta[r(t_i),z;t]}{\Delta t}\cdot Q(t_i)\cdot\Delta t \qquad \cdots\cdots(24)$$

where

$$r(t_i)=\sqrt{[x-v(t-t_i)]^2+y^2}$$

At limit, $\Delta t\to 0$, Eq. (24) becomes

$$\theta(x,y,z;t)=\int_0^{t_n}\frac{\partial}{\partial t}\Theta[r(t'),z;t-t']\cdot Q(t')\cdot dt' \quad\cdots(25)$$

where

$$r(t')=\sqrt{[x-v(t-t')]^2+y^2}.$$

5. Comparison Between Theoretical and Experimental Temperature Distribution

In this section the computed temperature distribution is compared with experimental through a micro-examination.

General shape of typical bead cross section in laser heating shown in Photo. 2 agrees satisfactorily with isotherms computed on the basis of heat conduction theory developed in the last section. However, only in case of SUS27 stainless steel coated with non-metallic fine powder, fusion isotherm obtained from

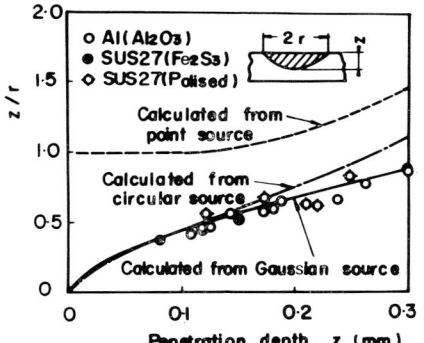

Fig. 13. Relation between z and z/r.

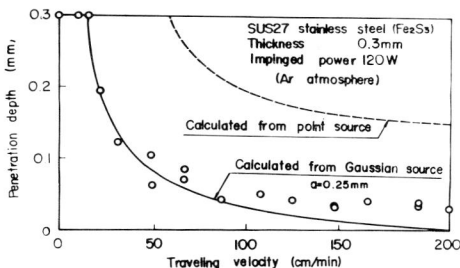

Fig. 14. Relation between velocity and penetration depth.

micro-examination has a slightly depressed contour at the center comparing with computed isotherm as shown in Photo. 3. This is assumed to be due to convection in the molton metal during laser heating. The effect of the convection will be discussed later and here the authors will consider about the case except for the SUS27 specimen coated with non-metallic fine powder.

A bead width $2r$ is measured for various bead depth z in order to make the more detailed comparison. In Fig. 13 a r-to-z ratio is plotted as a function of z. Calculations[**] of z/r as a function of z were carried out for ideal point, circular and Gaussian sources assuming that each power is constant. The experimental values agree with calculated from Gaussian source indicated by a solid line as shown in Fig. 13. Agreement between experimental relults and calculated from circular source is sufficient in smaller z, but it becomes gradually insufficient as z increases. In case of ideal point source, the large difference is recognized.

In Fig. 14 penetration depth is plotted as a function of traveling speed for sulphurized 0.3 mm SUS27. A solid curve was obtained by calculation for the Gaussian source with constant laser power. In this calculation, effective heat input entered actually into the specimen, is obtained on the basis of the absorptance measurements as a function of traveling velocity. Agreement between experimental and theoretical results is quite well in the lower traveling velocity, but the experimental values become larger than the calculated in the higher traveling velocity where the actual penetration depth is less than 0.05 mm. This disagreement is due to the fact that the waveform of the laser power output used is not constant but pulsive.

Thus it seems to be quite well that the laser power output waveform is regarded as constant except for the upper part of the sheet of SUS27. Inversely, the absorptance can be obtained by dividing the calculated full penetration power by experimental one. This is only an effective method to obtain the absorptance because the absorptance of the specimen coated with fine powder cannot be measured using calorimeter.[1] As described previously, the shape of the bead cross section differs slightly from heat conduction type in the case of SUS27 specimen coated with non-metallic fine powder, but deviation from the theory is negligible small. In Fig. 15 the calculated absorptance is plotted as a function of the traveling velocity.

(2) Effect of the pulsive waveform on the penetration depth

As previously described, the effect of the pulsive waveform of laser output is large and it cannot be neglected even at the bottom surface of the sheet specimen with a thickness of the order of 0.3 mm, in the case of Al due to its high thermal diffusivity.

Figure 16 shows an example of temperature distribution computed on the bead center line at the bottom surface of 0.3 mm Al, in the case of Gaussian laser source with the pulsive waveform as shown in Photo. 1

[**] In the calculation, it was assumed that diameter of circular source and diameter at $1/e$ power point of Gaussian source were 0.5 mm.

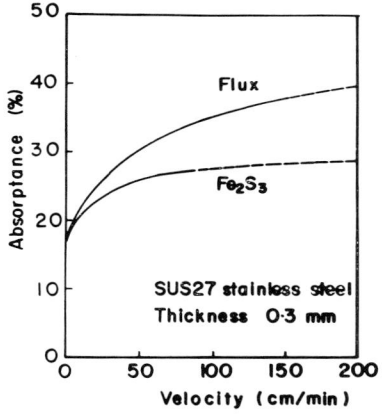

Fig. 15. Absorptance vs. velocity in the case of full penetration.

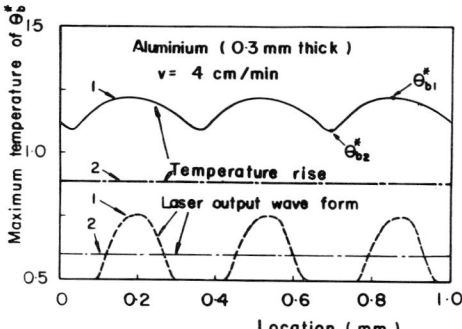

Fig. 16. Maximum temperature rise by pulsive and equivalent constant heating (0.3 mm thick).

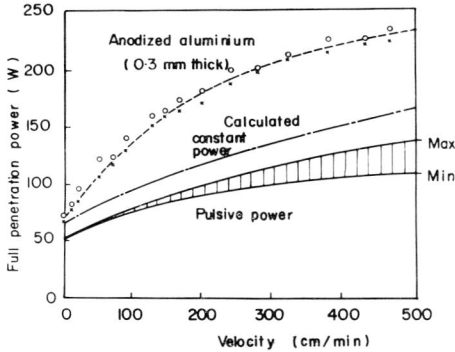

Fig. 17. Relation between absorptance and velocity.

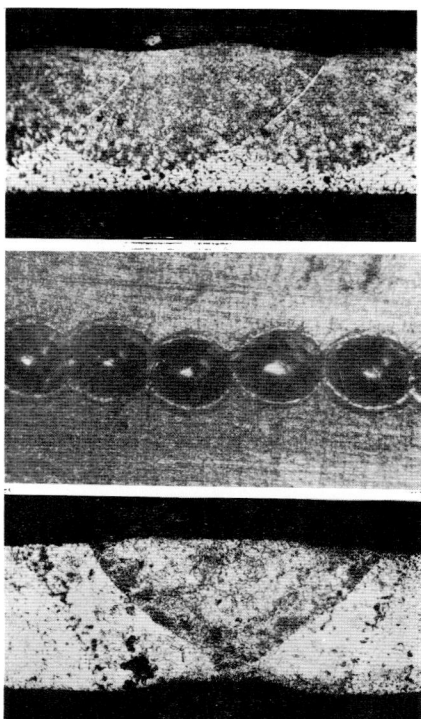

Photo. 6. Anodized aluminium heated by focused laser beam.
(a) Longitudinal cross section
(b) Bottom view
(c) Transverse section.
Thickness: 0.3 mm, Velocity: 2.4 m/min

which started from infinite distance point. Let θ_{bm} be maximum temperature rise at the point under consideration while the source is passing through the point, θ_{bm} depends upon the phase of the laser waveform at the point. In Photo. 6, θ_{bm} on the bead center line at bottom surface, expressed in non-dimensional unit, is plotted in the case of 4 m/min. Let θ_{b1}^* and θ_{b2}^* be maximum and minimum temperature of θ_{bm} on the bottom bead center line respectively, even θ_{b2}^* is higher than the maximum temperature corresponding to equvalent constant heating. Therefore, this indicates that power level required to obtain full penetration is lower for the pulsive waveform than for the equivalent constant heating.

Photograph 6 (a) shows the longitudinal bead cross section of 0.3 mm Al specimen at the traveling velocity 2.4 m/min. It is found that solidification has been completed during off-part of cycle and that there are deepest and shallowest penetration points corresponding to θ_{b1}^* and θ_{b2}^* respectively. When the laser power becomes slightly higher, positions corresponding to θ_{b2}^* come to melt as shown in Photo. 6 (b). Photograph 6 (c) shows the transverse bead section.

Figure 17 shows the effective full penetration power

in 0.3 mm Al computed from the actual laser wave-form and equivalent constant power supply. In the case of the actual waveform, difference of the full penetration power between θ_{b1}^* and θ_{b2}^* points increases with increasing the traveling velocity. The full penetration power, however, even for θ_{b2}^* points was smaller than for equivalent constant laser power supply. Measured full penetration power in θ_{b2}^* points for 0.3 mm anodized Al is also plotted in Fig. 17.

Since the amount of the computed effective full

penetration power divided by the measured gives rise to the absorptance of the specimen, in order to check the calculated values, the computed and measured absorptances are compared. Figure 18 shows the computed and measured absorptance for 0.3 mm anodized Al: a chain curve represents the measured absorptance and a dotted and solid curves are com-puted from equivalent constant and actual waveform power, respectively. Experimental values agreed well with computed from the pulsive waveform.

6. Laser-Weldability

When material is heated by focused CO_2 laser beam, its temperature rise depends upon the energy distri-bution profile and power level impinged as well as the beam absorptance and heat conduction properties of material. Therefore, it is convenient to estimate the CO_2 laser-weldability of material in term of the minimum power level required to melt the material from its optical and heat conduction properties.

When focused CO_2 laser beam, having Gaussian flux distribution with radius of $1/e$ intensity at center a, impinges upon the surface of semi-infinite material, of which beam absorptance is A, temperature rise at location on the beam axis in steady state becomes, by Eq. (14), in simplified form:

$$\theta(O,z;\infty) = \frac{AQ_0}{2\sqrt{\pi}k}\exp(z^2/a^2)\cdot\mathrm{erfc}(z/a) \quad \cdots\cdots(26)$$

where Q_0 is the total laser power per unit time. At center of the source, $(0,0)$

$$\theta(O,O;\infty) = \frac{AQ_0}{2\sqrt{\pi}ka} = \frac{\sqrt{\pi}}{2}A\theta_m. \quad \cdots\cdots(27)$$

Minimum power required to melt surface of semi-infinite solid is given by

$$Q_f = \frac{2\sqrt{\pi}ka}{A}\theta_f \quad \cdots\cdots(28)$$

where θ_f is fusion temperature of material.

In the case of polished metal, the absorptance at fusion temperature is given by[1]

$$A_f = 4\sqrt{\frac{\pi c\varepsilon_0\eta_f}{\lambda}} \quad \cdots\cdots(29)$$

where c is velocity of light, ε_0 the electric permativity of vacuum, η_f dc resistivity of metal at fusion temper-ature and λ wavelength of the beam. Substitution of Eq. (29) into Eq. (28) yields

$$Q_f = \frac{1}{2}ka\sqrt{\frac{\lambda}{c\varepsilon_0\eta_f}}\theta_f \quad \cdots\cdots(30)$$

For CO_2 gas laser beam, $\lambda = 10.6~\mu$. With $c = 3 \times 10^{10}$ cm/sec and $\varepsilon_0 = 8.85 \times 10^{-14}$ farad/cm, and taking $a = 0.025$ cm, Eq. (30) gives

$$Q_f = 7.9 \times 10^{-4}\frac{k\theta_f}{\sqrt{\eta_f}} \quad \cdots\cdots(31)$$

If the absorptance of material is enhanced up to 100%,

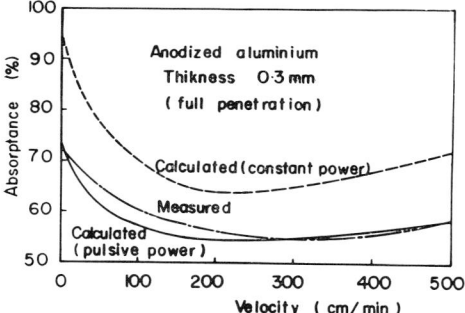

Fig. 18. Relation betwen velocity and full penetration power.

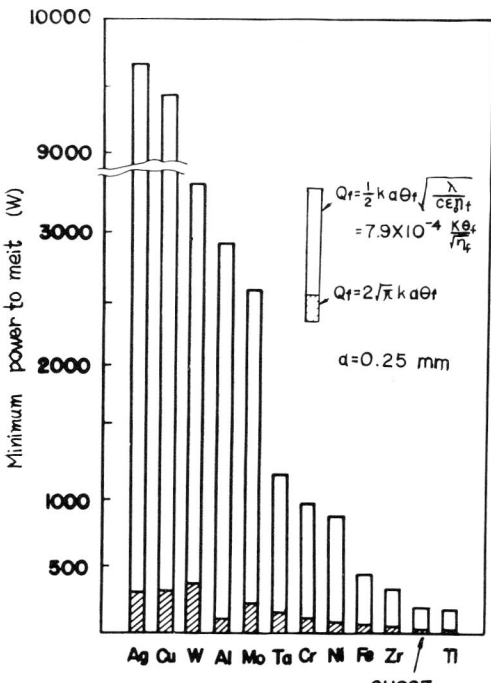

Fig. 19. Minimum laser power to melt various metals with infinite thickness. In this figure η_f represents dc resistivity at fusion temperature (constant power, Gaussian distribution).

Eq. (28) yields

$$Q_f = 2\sqrt{\pi}\, ka\theta_f \qquad\qquad \cdots\cdots(32)$$

Figure 19 shows Q_f for various metals of semi-infinite thickness in both cases, polished and of 100% absorptance. As it is clear from this figure, iron, zirconium, titanium, SUS27 etc. may be relatively easily processed by CO_2 laser in polished surface condition: Q_f is about 200 W for titanium and SUS27 and 300–400 W for iron and zirconium. On the other hand both silver and copper in polished surface condition are most difficult; Q_f of both metals is extremely high value,

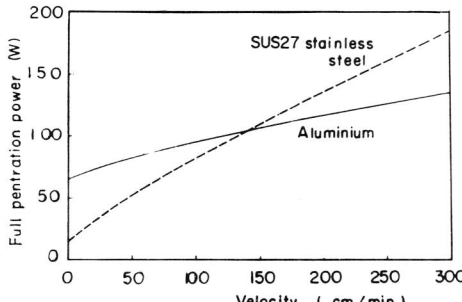

Fig. 20. Velocity vs. effective power for full penetration of polished metals with 0.3 mm thickness.

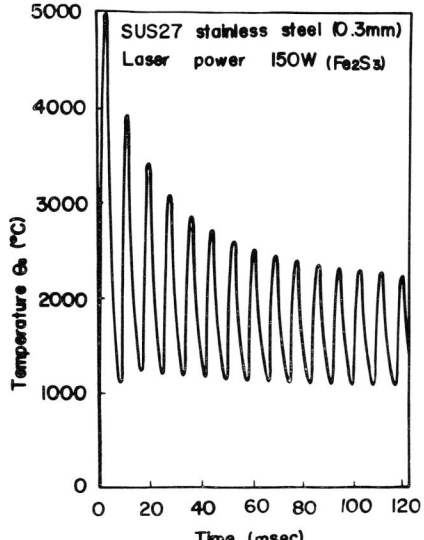

Fig. 21. Calculated surface temperature rise at location on the axis ($r=0$) by focused laser beam (150 W) Material: SUS27 stainless steel (Fe_2S_3) Thickness: 0.3 mm

9–10 KW. Thus in polished surface condition, difference of Q_f in metal is large and comparatively higher laser power is required.

On the other hand, when the absorptance is enhanced up to 100%, difference of Q_f in metal becomes small and the magnitude of Q_f falls to less than 200 W[***].

Figure 20 shows relatio bnetween the traveling velocity, in cm/min, and effective full penetration power computed in quasi-steady state for Al and SUS27, 0.3 mm in thick. It was found that in the velocity range lower than about 150 cm/min, SUS27 can be welded in lower power level than Al, but in the range higher than 150 cm/min, the relation becomes inverse due to the relation, $v^* \propto \dfrac{1}{k_D}$.

7. Interaction Between Laser Beam and Surface-treated Specimen

The authors have shown that an appropriate surface treatment is necessary to increase CO_2 laser beam absorptance of metal in order to make the metal processing by CO_2 laser efficient, and that Al and SUS27, which differ in thermal diffusivity and melting point, differ in the absorption properties: when focused

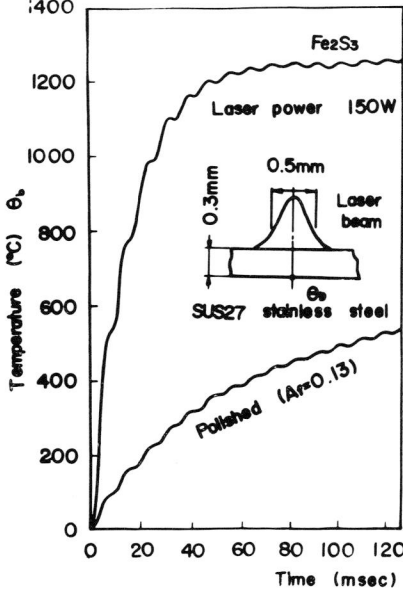

Fig. 22. Calculated bottom temperature rise at location on the axis ($r=0$) by focused laser beam (150 W) Material: SUS27 stainles ssteel Thickness: 0.3 mm

[***] According to the law of Wiedermann-Franz, electrical conductivity of pure metal is proportional to its thermal conductivity. Then for polished metal, Eq. (31) yields
$$Q_f \propto k^{3/2}$$
and for 100% absorptance metal, Eq. (32) becomes
$$Q_f \propto k$$

laser beam impinges upon metallic specimen coated with thin layer produced by an appropriate surface treatment, the laser beam absorptance varies as the evaporation of the surface layer progresses. It seems to be quite well to consider that the bottom surface of such thin layer has the same temperature as the top surface of the base metal which contacts with this layer, though there is a steep temperature gradient within the layer in the direction of the thickness due to its poor heat conductivity. Therefore it is assumed that condition of the layer is affected by the top surface temperature of the base metal. Thus the authors have computed the transient temperature distribution of the specimen on the basis of heat conduction theory described in section 4 in order to discuss the relation between the temperature distribution of the base metal and the absorptance.

Transient temperature distribution of the base metal can be computed from Eq. (22) or Eq. (23) when focused laser beam impinges upon the specimen coated with thin layer at rest. Energy distribution of focused laser beam, its output waveform and transient beam absorptance, which are necessary in this computation, have been already measured by the authors[1]: since $Q(t)$ in these equation represents the absorbed power, this is given by the product of the absorptance and impinged laser power.

(1) SUS27 stainless steel specimen

Figures 23[1] shows the relation between time duration of laser heating and the absorptance of the sulphurized 0.3 mm SUS27 specimen heated by 150 W-focused laser beam. Both top and bottom surface temperature rises at location on the beam axis were computed based on this result shown in Fig. 21 and 22. The top surface temperature varies considerably with time according to the laser power output waveform with a period of

1/120 sec, having maximum of temperature for each pulse. This is the reason for which the bead obtained by laser heating exhibits a periodic scale-shape appearance described in section 2. Though a maximum temperature corresponding to the first pulse is the highest, after that the maximum temperatures decrease shraply with time in spite of continuous heating because of decrease in the absorptance. In the first and second pulses, it is clear that the surface temperature becomes enough high for Fe_2S_3 film to evaporate, and thereafter

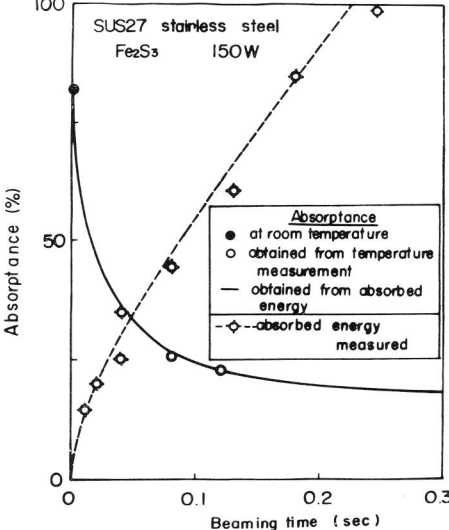

Fig. 23. Relation between absorptance and beaming time for SUS27 stainless steel (0.3 mm thick)[1]

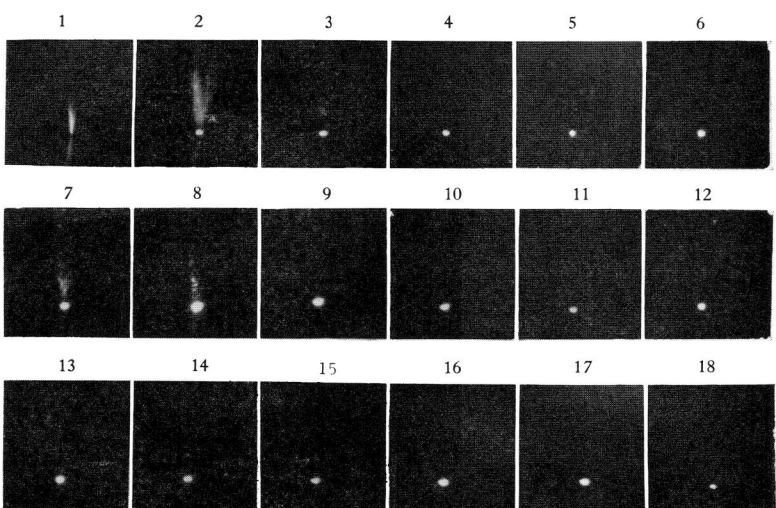

Photo. 7. Irradiation of focused laser beam to SUS27 stainless steel with Fe_2S_3 film. Laser power: 150 W, Velocity: 0 (Argon atmosphere) (700 frames/sec)
1—6: first pulse. 7—12: second pulse. 13—18: third pulse.

256

the surface temperature is also enough high. Thus it was confirmed from the theoretical analysis of the temperature rise that Fe_2S_3 film cannot remain on the base metal so that the absoptance decreases down to the value in the polished surface condition, infered in Ref. 1.

Photograph 7 shows the high speed photographs of the surface comdition during laserea hting. In the first pulse an evaporation is observed clearly, in the second pulse it becomes relatively weeker and there-after it may be no longer observed. But in the third pulse and after, a week evaporation seems to continue. These observation agrees with the tendency infered from the analysis of temperature rise described above.

On the other hand, there was little fluctuation in computed temperature curve of 0.3 mm sulphurized SUS27 at the bottom surface at location on the beam

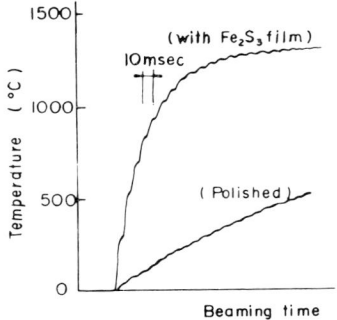

Fig. 24. Temperature records at location on the axis of the surface (SUS27 stainless steel, 0.3 mm thick).

axis. Its temperature is elevated sharply within initial 5–10 pulses and subsequently it does not almost increase. The temperature rise corresponding to polished one ($A_f = 13\%$) is also shown in this figure. The temperature of polished specimen is elevated very slowly in comparison with the sulphurized one.

Figure 24 shows measured temperature rises of sulphurized and polished 0.3 mm SUS27 under the same condition. Experimental values in both polished and sulphurized agree well with the theoretical results.

In the case CO_2 laser is applied to the welding of SUS27, in order to obtain full penetration the response time of temperature rise at bottom surface is fairly long due to its poor thermal diffusivity. Therefore a situation under high beam absorptance must be maintained for a long time. This means that surface layer must remain on the base metal for a long time. But in order to melt the bottom surface, it is inevitable that temperature at the top surface becomes so high that the surface layer evaporates away as the temperature gradient within the bese metal in the direction of the thickness becomes large due to its poor thermal diffusivity.

From the consideration based on the heat conduction theory, one can make the follwing statement: in order to enhance the absorptance, it is necessary for the thin surface layer to remain at the top surface of base metal under high temperature condition. Photograph 8 shows high speed motion pictures of non-metallic fine powder. In the first pulse distinct evaporation as well as spattering of the surface layer is observed and evaporation is still continued for longer time than the case of Fe_2S_3 film coating. During the evaporation continues, it seems that the surface layer attaches

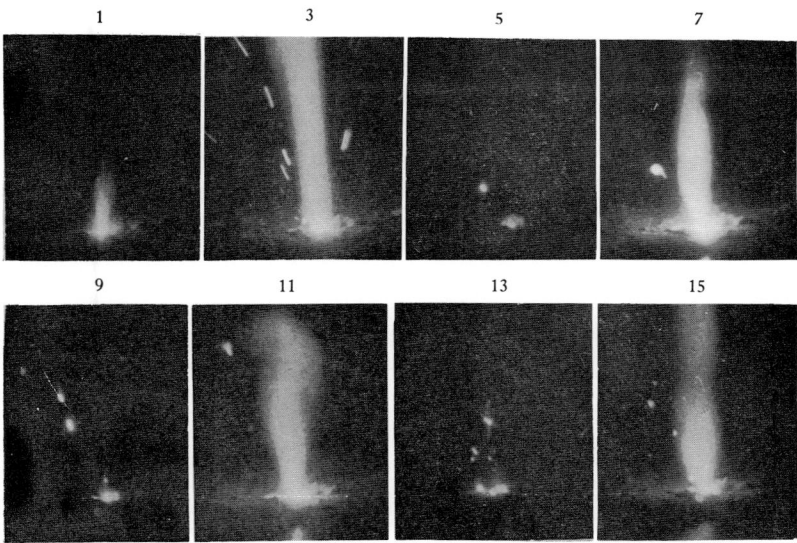

Photo. 8. Irradiation of focused laser beam to SUS27 stainless steel (500 frames/sec)
Laser power: 150 W, Surface treatment: flux (non-metallic fine powder)
1—3: first pulse. 5—7: second pulse, 9—11: third pulse 13—15: fourth pulse.

closely to the base metal in the form of thin film as a result of its melting, and enhances the absorptance. In case of SUS27 the fact that the specimen coated with a non-metallic fine powder which is indicated in Ref. 1 has the highest absorptance can be explained from these observations.

(3) Aluminium specimen

Figure 25 shows the computed temperature rises of anodized 0.3 mm Al sheet on the beam axis. It was assumed that the anodized Al had 100% absorptance

at room temperature and that its transient absorptance curve was given by Fig. 26.[1] In the same manner as the SUS27 specimen, at the top surface the temper-

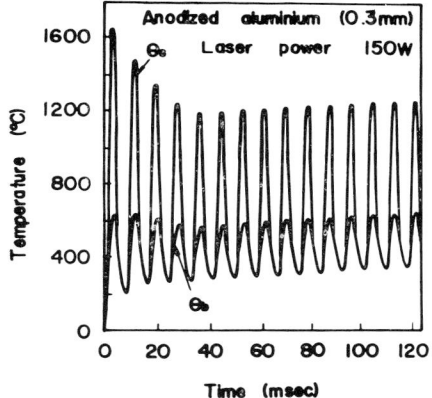

Fig. 25. Calculated temperature rise on the axis ($r=0$) of aluminium plate, 0.3 mm in thick. Transient absorptance is shown in Fig. 26.

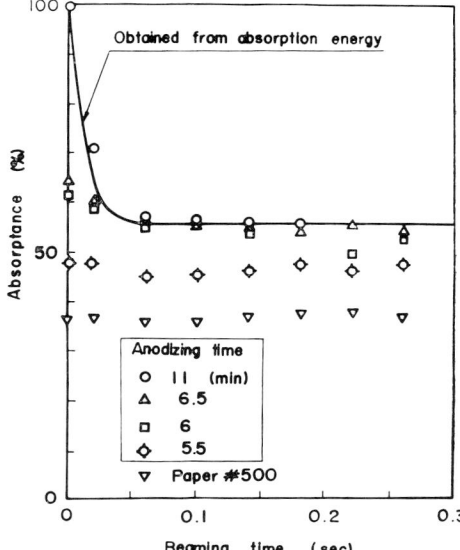

Fig. 26. Relation between absorptance and beaming time for anodized aluminium (0.3 mm thick)[1].

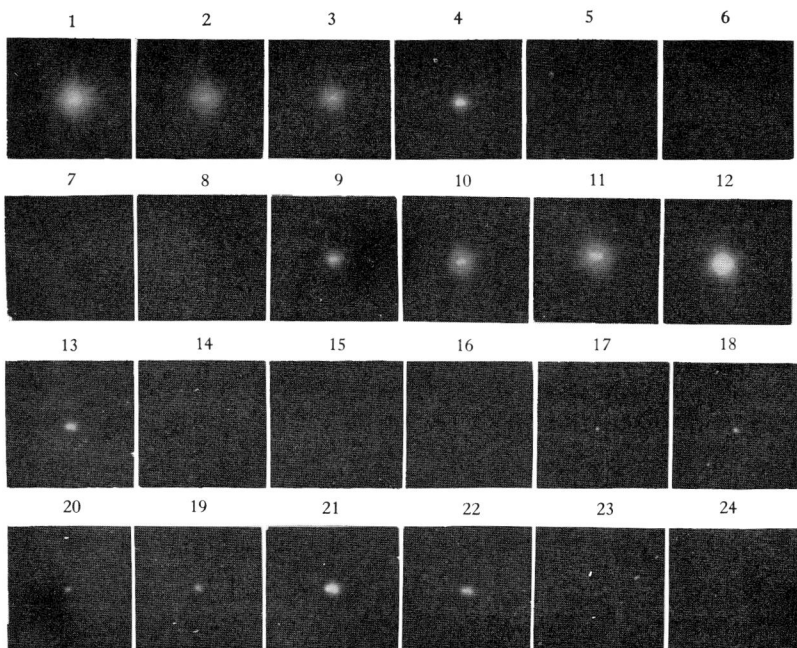

Photo. 9. Irradiation of focused laser beam to anodized aluminium, 0.3 mm in thick (1000 frames/sec) Laser power: 150 W (stationary) 1~8: first pulse. 9~16: second pulse. 17~24: third pulse.

ature rise in first pulse is the highest. However, it should be noticed that the surface temperature of the base metal is low, 1600°C at the highest, which is considerably lower than the boiling point of Al_2O_3 coated on the base metal. Therefore the Al_2O_3 film can remain even after the anodized specimen is heated and melted by focused laser beam for a long time, and enhances the beam absorptance.

The bottom temperature also becomes the highest during the first pulse. It can be assumed from computation that both melting and solidification of specimen at both top and bottom surface occur in each pulse duration because of its high thermal diffusivity. High speed motion pictures were taken to confirm the computed results. As shown in Photo. 9, which is experimented under the same condition as the computation, indicates that relults on the basis of observation agree qualitatively with computed.

While anodized 0.3 mm Al specimen was being heated by focused laser beam, its bottom temperature at location on the beam axis was recorded as shown in Fig. 27. The absorptance of the specimen used was 100% at room temperature and the laser beam impinged at same location four times. In this case each beaming time (150 W) was 200 msec. The absorptance corresponding to the first heating with time duration of 200 msec is assumed to vary accorcing to a solid curve shown in Fig. 26: at the end of the heating period, the absorptance is already constant value, 55%, and thereafter further heating, which involves second, third and fourth heating, does not change the absorptance. Actual temperature measurements con-

firm that the constant absorptance value is kept from the second heating. Figure 28 shows behavior of the calculated temperature corresponding to the experimental results indicated in Fig. 27 (b), (c) and (d). The computed results corresponding to Fig. 27 (a) is shown in Fig. 25. Both experimental and computed results are qualitatively in good agreement. Comparison between Fig. 22 and Fig. 25 or Fig. 22 and Fig. 28 shows how fast the bottom surface temperature of Al responses in comparison with SUS27.

8. Convection in the Molten Metal

As previously described, only in case of SUS27 coated with non-metallic fine powder, fusion isotherm obtained from micro-examination has slightly depressed contour at the center in comparison with one predicted from heat conduction theory. When metallic specimen coated with non-metallic thin layer is heated by focused laser beam, some evaporation occurs as previously described. The vapor pressure and repulsion of the evaporation will exert mechanical influence on the liquid metal around the laser irradiated zone so as to thrust downward vertically and thus the fusion isotherm is depressed at the bead center. Evaporation is especially violent in the case of the specimen coated with non-metallic fine powder. However, there are still problems because the depressed fusion isotherm obserbed in the case of SUS27 specimen was not obtained for Al specimen coated with non-metallic powder.

And so, to study the convection behaviour, distribution of elements contained in the surface layer was measured by means of X-ray micro-analizer. In this test, SUS27 and Al specimens were coated with non-metallic fine powder mixed with copper and zinc

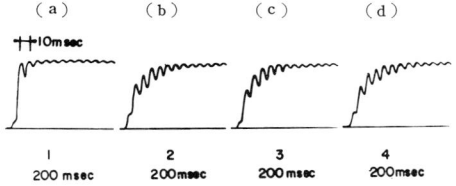

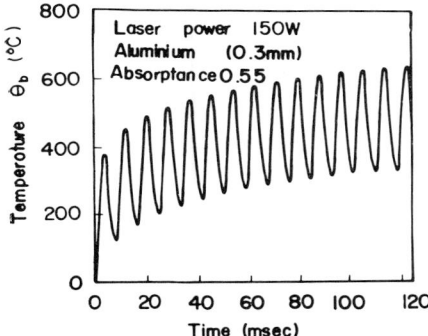

Fig. 27. Temperature records at location on the axis of (back surface) anodized aluminium, 0.3 mm in thick ($v=0$).

Fig. 28. Calculated bottom temperature rise on the axis ($r=0$) of aluminium plate, 0.3 mm in thick, with constant absorptance (55%).

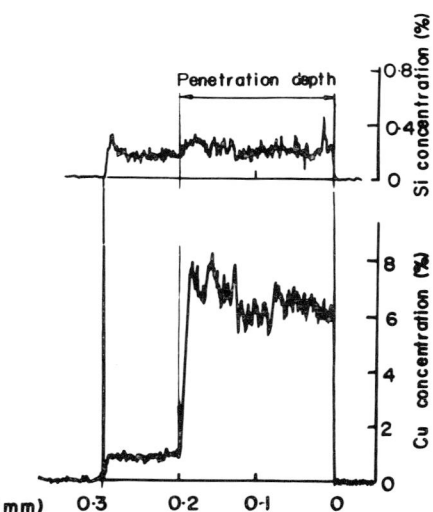

Fig. 29. Concentration of Cu and Si (0.3 mm thick). SUS27 stainless steel (Cu powder+flux), velocity=70 cm/min, 250 W.

fine powder, respectively. The concentration of copper on the beam axis across the SUS27 sheet is shown in Fig. 29. The copper content was higher in the bottom part of the bead than of the upper part. This indicates that a convection occurs in the molten metal. Figure 29 also shows the silicon content which is a principal constituent of the non-metallic powder. The silicon content was negligible in molton metal as well as base metal. Non-metallic powder seems to vaporize away.

The concentration of zinc on the beam axis across the Al sheet is shown in Fig. 30. The zinc distributed only in the upper part of the molten metal. This indicates that the convection occurs only in the upper part of

the molten metal. It was found that for given penetration depth the convection is more violent in SUS27 than in Al.

It is believed that this difference in convection behavior is caused by the differnce in the thermal diffusivity of both metals: in case of Al specimen, molten metal is thrusted during only "on" part of the cycle and completely solidifies during "off" part of the cycle as shown in Fig. 25. Thus the convection becomes discrete so that it occurs only in upper part of the molten metal due to its higher rate of the heat conduction.

In case of SUS27, the convection will be almost continuous through one cycle, so that copper coated on the surface can reach to the bottom of the bead. However, the effect of convection on the shape of fusion isotherm is very small and a heat conduction is still predominant, since convection driving force is discrete.

In higher power level, the evaporation will be violent and the effect of the thrust due to evaporation will be remarkable. To study the behaviour of a liquid which is more violently evaporating, focused laser beam was irradiated on the surface of water, and the photographs near the focal point were taken with high-speed motion camera. In order to make the observation easy, water was put in a glass container and was slightly coloured. Photograph 10 shows the photographs of the water irradiated locally by the laser beam. It is clearly observed that a hole with a depth of about 1.5 cm is formed in the water and bubbles move downward. This shows that in case of sufficient laser power, the repulsion due to the evaporation and local vapor pressure make a deep hole and induce an action to convect the liquid material. The depressed surface becomes flat as shown in frame number 17 in Photo. 10, which corresponds to "off" part of the cycle, due to pulsive laser power supply, and so that the maximum depth is limited to the depth perfolated during "on" part of one cycle.

When the energy supply is continuous as an electron beam, deeper hole will be sustained in the liquid

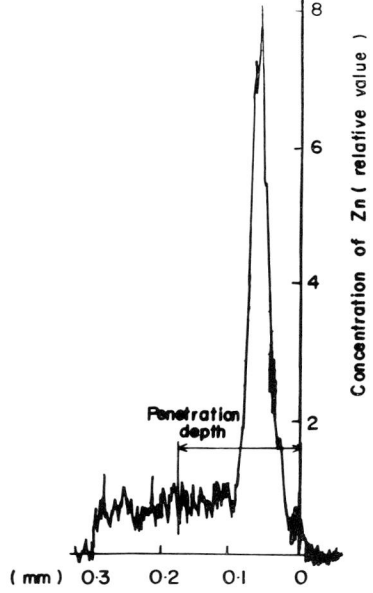

Fig. 30. Concentration of Zn (0.3 mm thick). Aluminium (Zn powder + flux), $v=250$ cm/min, 200 W.

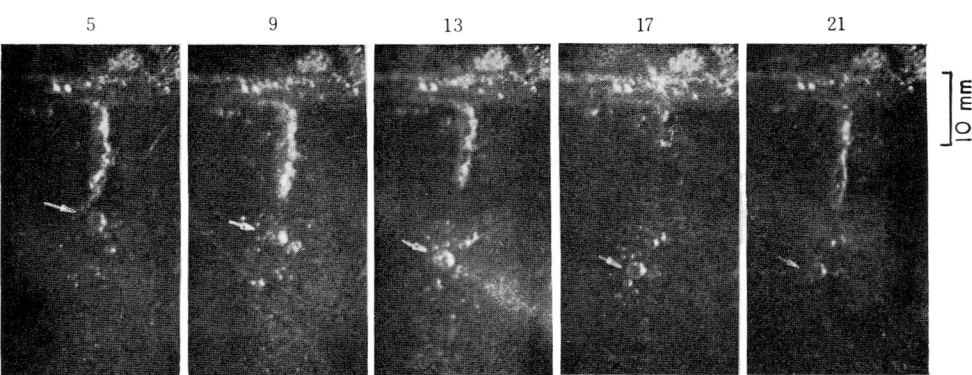

Photo. 10. Irradiation of focused laser beam to water.
Laser power: 250 W (500 frames/sec)

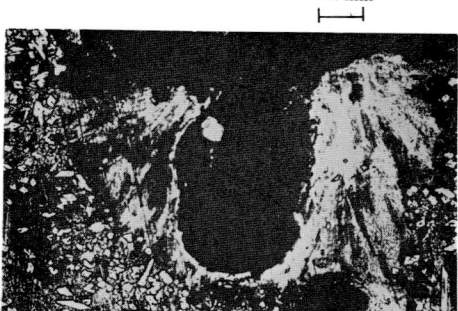

0.1 mm
⊢━━━┤

Photo. 11.　Cross section of Zinc (Flux)
Thickness: 0.5 mm, Velocity: 0, Beaming
time: 10 sec, Laser power: 200 W.

material, and "deep penetration" will be obtained in the case of metal.　In such situation, incident laser beam will be reflected by the side wall of drilled hole toward its bottom and cannot escap from the hole, providing almost 100% beam absorptance.

A deep hole is already obtained in the case of low boiling point material under several hundreds watts as shown in Photo. 11.

9.　Conclusions

Generellized heat conduction theory in respect to a surface heat source was developed and interaction between focused CO_2 laser beam and surface-treated metal were discussed on the basis of heat conduction theory and some experimental results.

Conclusions obtained may be summarized as follows:

(1)　Generalized heat conduction equation for a surface heat source, where the source moves with a constant speed, its intensity distribution is non-uniform and its power waveform is not constant but pulsive (120 Hz), was derived as shown in Eq. (25).

(2)　The computed transient temperature distributions were in good agreement with the results obtained from micro-examination and temperature measurements in the CO_2 laser heating.

(3)　When the actual focused laser beam, having a Gaussian distribution, is approximated by a circular or ideal point sources, the deviation from the actual temperature distribution increases with increasing the traveling velocity of the source, as shown in Fig. 11.

(4)　In the case of SUS27 stainless steel, an effect of the pulsive waveform (see Photo. 1) on the temperature rise may be negligible except for at location in the vicinity of the surface, and then the laser beam may be regarded as a equivalent constant power source. However, for metal with high thermal diffusivity such as aluminium, the effect is large even at the bottom surface of the sheet having thickness of the order of 0.3 mm.

(5)　A CO_2 laser-weldability is the best in case of titanium, zirconium, SUS27 stainless steel etc., and

it was worst in silver, copper etc. When its absorptance is enhanced sufficiently, aluminium sheet is the most suitable metal for high speed welding.

(6)　When melting point of the thin surface layer coated on the base metal is higher than the metal's one, the layer with appreciable thickness can remain on the metal surface to enhance the absorptance.

(7)　The absorptance of sheet metal having higher thermal diffusivity and lower melting point can be effectively enhanced by the surface treatment and depends hardly on the surface treatment.

(8)　Non-metallic fine powder coating induces an appreciable convection in the liquid metal, and slightly affects the shape of fusion isotherm in the case of low thermal diffusivity metal such as SUS27.

(9)　Under sufficient laser power, a deep pentration should be obtained as well as the case of electron beam. But when the laser is excited by using ac (of the order of 60 Hz) electric power supply, the deep penetration may be considerably suppresed.

Appendix
App 1)

In the case of circular source with uniform intensity, it is convenient to introduce l and φ defined by following equations as shown in Fig. 31:

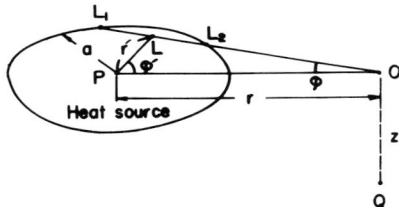

Fig. 31.　Circular heat source with an uniform intensity
($\mathrm{OL_1}=l_1$, $\mathrm{OL_2}=l_2$, $\mathrm{OL}=l$).

$$r' \cdot \sin \varphi' = l \cdot \sin \varphi$$

$$r' \cos \varphi' + l \cdot \cos \varphi = r$$

In the region $r > a$, Eq. (2) may be written as

$$\theta(r, z; t) = \frac{q_0}{\pi k} \int_0^{\varphi_0} \int_{l_2}^{l_1} \frac{1}{\sqrt{z^2 + l^2}} \operatorname{erfc} \frac{\sqrt{z^2 + l^2}}{\sqrt{4k_D t}} \cdot l \cdot dl \cdot d\varphi$$

$$\cdots\cdots (\mathrm{A1})$$

where $l_1 = \overline{OL_1}$ and $l_2 = \overline{OL_2}$ given by following equations, assuming L is a point at any arbitrary distance $r < a$ from P and L_1 and L_2 the points of intersection of the line $\overline{OL}$,

$$\left. \begin{array}{l} |l_1| = r \cdot \cos \varphi + \sqrt{a^2 - r^2 \cdot \sin^2 \varphi} \\ |l_2| = r \cdot \cos \varphi - \sqrt{a^2 - r^2 \cdot \sin^2 \varphi} \end{array} \right\} \qquad \cdots\cdots (\mathrm{A2})$$

φ_0 is $\overline{OL}$ in the case line $\overline{OL}$ is the tangent to the circle, and given by

$$\varphi_0 = \sin^{-1} \frac{a}{r} \qquad\qquad \cdots\cdots (\mathrm{A3})$$

Putting

$$R = \frac{\sqrt{z^2 + l^2}}{\sqrt{4k_D t}}$$

we have

$$\frac{l \cdot dl}{\sqrt{z^2 + l^2}} = dR \sqrt{4k_D t}$$

Thus Eq. (A1) yields

$$\theta(r,z;t) = \frac{q_0}{\pi k} \sqrt{4k_D t} \int_0^{\sin^{-1} a/r} \int_{R_2}^{R_1} \mathrm{erfc}\, R \cdot dR \cdot d\varphi$$

$$= \frac{q_0}{\pi k} \sqrt{4k_D t} \int_0^{\sin^{-1} a/r} \left[\int_{R_2}^{\infty} \mathrm{erfc}\, R \cdot dr \right.$$

$$\left. - \int_{R_1}^{\infty} \mathrm{erfc}\, R \cdot dR \right] d\varphi$$

where

$$R_1 = \frac{\sqrt{z^2 + l_1^2}}{\sqrt{4k_D t}} \quad \text{and} \quad R_2 = \frac{\sqrt{z_1^2 + l_2^2}}{\sqrt{4k_D t}}$$

Writing

$$\int_X^{\infty} \mathrm{erfc}\, u \cdot du = \mathrm{ierfc}\, X,$$

we have

$$\theta(r,z;t) = \frac{q_0}{k\pi} \sqrt{4k_D t} \int_0^{\sin^{-1} a/r} (\mathrm{ierfc}\, R_2 - \mathrm{ierfc}\, R_1) \cdot d\varphi$$

$$\cdots\cdots(A5)$$

On the other hand, in the region $r < a$, since

$$|l_1| = r \cdot \cos\varphi + \sqrt{a^2 - r^2 \cdot \sin^2\varphi},$$

$$|l_2| = 0 \quad \text{and}$$

$$\varphi_0 = \pi$$

Eq. (2) yields

$$\theta(r,z;t) = \int \frac{q_0}{\pi k} \cdot \sqrt{4k_D t} \int_0^{\pi} \left(\mathrm{ierfc}\, \frac{z}{\sqrt{4k_D t}} \right.$$

$$\left. - \mathrm{ierfc}\, \frac{\sqrt{z^2 + l^2}}{\sqrt{4k_D t}} \right) \cdot d\varphi \qquad \cdots\cdots(A6)$$

App 2)

Temperatures at (r,z,φ) for point, circular and Gaussian sources in quasi-steady state are given by following equations:

(1)　Point source:

$$\theta(r,z) = \frac{Q}{2\pi k} \cdot \frac{1}{\sqrt{r^2 + z^2}} \exp\left[-\frac{v}{2k_D} (r \cdot \cos\varphi \right.$$

$$\left. + \sqrt{r^2 + z^2}) \right] \qquad \cdots\cdots(A7)$$

(2)　Circular source

$$\theta(r,z) = \frac{q_0}{\pi k} \int_0^{\pi} \int_0^a \frac{l}{R} \cdot \exp\left[-\frac{v}{2k_D} \cdot (R + r \cdot \cos\varphi \right.$$

$$\left. - r' \cdot \cos\varphi') \right] \cdot r' \cdot dr' \cdot d\varphi' \qquad \cdots\cdots(A8)$$

(3)　Gaussian source

$$\theta(r,z) = \frac{q_0}{k\pi} \int_0^{\pi} \int_0^{\infty} \frac{1}{R} \cdot \exp\left[-\frac{v}{2k_D} \cdot (R + r \cdot \cos\varphi \right.$$

$$\left. - r' \cdot \cos\varphi') - \frac{r^2}{a^2} \right] \cdot r' \cdot dr' \cdot d\varphi' \quad \cdots\cdots(A9)$$

where

$$R = [r'^2 + r^2 - 2 \cdot r' \cdot r \cdot \cos(\varphi - \varphi') + z^2]^{1/2}$$

In the case the plate thickness is finite value d, Eqs. (A7)–(A9) become

$$\theta_{\varphi,d} = \sum_{n=-\infty}^{+\infty} \theta_{\varphi}(r, 2nd-z) \qquad \cdots\cdots(A10)$$

References

1) Y. Arata and I. Miyamoto: "CO$_2$ Laser Absorptance Characteristics of Metal", I. I. W. Doc. IV-50–71, (1970) Y. Arata and I. Miyamoto: "Some Fundamental Properties of High Power Laser as a Heat Source (Report 2)", Transactions of the Japan Welding Society, Vol. 3 (1972), No. 1.

2) R. H. Fairbanks and C. M. Adams: "Laser Beam Fusion Welding", W. J., Vol. 43 (1964), No. 3, 97s–102s.

3) A. O. Sdhmidt and T. Hoshi: "An Evaluation of Laser Perfomance in Microwelding", W. J., Vol. 44 (1965), No. 11, 481s–488s.

4) R. Guenot and J. Racinet: "Heat Conduction Problem in Laser Welding", British W. J., Vol. 14 (1967), No. 8, 427–435.

5) Y. Arata and I. Miyamoto: "Some Fundamental Properties of High Power CW Laser Beam as a Heat Source", I. I. W. Doc. IV-4-69, (1969).

6) Y. Arata and I. Miyamoto: "Some Fundamental Properties of High Power Laser Beam as a Heat Source (Report 1)", Transactions of the Japan Welding Society, Vol. 3 (1972). No. 1.

7) D. Rosenthal: "Mathematical Theory of Heat Distribution during Welding and Cutting", W. J. Vol. 20 (1941), No. 5, 220s–234s.

262

Theoretical Analysis of Weld Penetration Due to High Energy Density Beam

Abstract

Heat flow due to band and rectangular sources moving in direction perpendicular to the source plane is mathematically analysed. New mathematical formulations which give bead depth in high energy density beam welding such as electron and laser beam weldings as a function of beam parameters, physical constants of the material and welding speed are proposed.

The theoretical weld depth (bead penetration depth) agrees well with the published data obtained not only from electron beam welding but from laser welding.

1. Introduction

Many works have been reported in order to correlate the weld depth with the parameters in the electron beam welding. The theoretical treatment[1], however, to explane the data on a physical basis has not been sufficient, though it is important not only to predict the weld depth but also to clarify the welding mechanism.

In the typical electron beam welding, it is well known that the beam penetrates into the workpiece so as to sustain a deep hole which moves along the intended joint producing the straight sided narrow weld zone. It has been shown that the similar weld also may be obtained by means of recently developed CW CO_2 laser with high power level. In these weldings, the incident beam energy to the workpiece is consumed mainly for melting, evaporation and surface-reflecting which is important only in laser welding. In order to estimate the energy for melting the workpiece, which is the most predominant of all[2], a moving line source model has been used in the heat conduction treatment. However, this model which provides an infinite temperature at location of the source leads to noticeable error in temperature estimation at distant point from the beam axis compared with the beam diameter. In this model the size of the beam spot cannot be taken into account in calculating the temperature rise in spite of the fact that the weld depth is considerably affected by the beam diameter.

In the present paper, the authors propose to introduce a new model, band or rectangular plane heat source of uniform intensity, which moves in the direction perpendicular to the source plane, and of which

width corresponds to the hole diameter. The temperature and bead penetration depth are described by a set of non-dimensional variables in order to establish an universal relationship independent of parameters such as physical constants of material, welding speed, source diameter, beam current and acceleration voltage. The calculated values of the penetration depth are compared with the published data which have been obtained in electron beam welding[3,4] as well as laser welding[5]. In these treatments variation of the phyisical constants of material, radiation heat loss and convection in the molten pool are neglected. Results obtained from the band source model may be also applied to the analysis of the gas cutting.

In sections 2–4 the theory of rectangular and band plane sources with uniform intensity moving with constant velocity in the direction perpendicular to the source plane is dealt with. Section 5 gives the theoretical weld depth due to these sources. In section 6 the theoretical weld depth is compared with the published data obtained from electron and laser beam weldings. The final section gives the summery of this study.

2. Fundamental solution

The evaporation of the material seems to play an important role to sustain a deep hole drilled in the electron beam welding from the view point of the dynamics. However the energy needed for evaporating the material is considerably smaller[2]. Therefore it is assumed that the heat deliveres to the workpiece is consumed only for melting.

Due to the drilling action of the high energy

density beam, heat delivered to the workpiece shows three-dimentional distribution profile, which varies with the physical properties of metal and the welding variables. But here in order to simplify the problem it is assumed that heat distributes two-dimensionally in a plane S of a certain shape perpendicular to the moving direction and steady temperature distribution is formed.

The temperature distributions due to such a source are obtained by integration of the solution of the moving point source in steady condition[6]. Assuming the moving direction is paralell to the x-axis and heat is liberated at the rate of q per unit time per unit area at (x', y', z'), the steady temperature at the point (x, y, z) at time considered in an infinite solid, initially zero temperature is

$$\theta = \frac{1}{4\pi k} \exp\left\{ -\frac{v}{4k_D}(x-x') \right\} \iint_S \frac{q}{R}$$
$$\times \exp\left(-\frac{vR}{4k_D} \right) dy' \, dz' \tag{1}$$

where $R = \left| (x-x')^2 + (y-y')^2 + (z-z')^2 \right|^{\frac{1}{2}}$, $k =$ the heat conductivity of the metal, $k_D =$ the heat diffusivity of the metal and $v =$ moving velocity of the heat source. In the following sections the temperature rises due to a "rectangular heat source" and a "band heat source" with an infinite length are considered.

3. The steady temperature due to the moving rectangular source

Here the temperature distribution of infinite body due to the moving rectangular source as shown in **Fig. 1** is treated; the heat is liberated uniformly at the rate q per unit time per unit area over the rectangle of sides $2a$ parallel to the y-axis and $2b$ to the z-axis, which moves with velocity v along the x-axis, and the

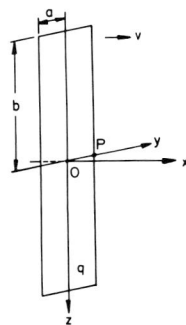

Fig. 1. Rectangular source with uniform intensity in the direction perpendicular to the source plane.

center of the rectangle is at origin. In order to derive the universal equation independent of physical properries of the material, moving velocity and the beam parameters, dimensionless parameters are introduced as follows:

$$\left.\begin{array}{l} l^* = l/a, \\ v^* = av/2k_D, \\ \theta^* = \theta/\theta_m = 2k\theta/aq = 4k\theta b/Q \text{ and} \\ \theta_m = aq/2k \end{array}\right\} \tag{2}$$

where l is a quantity with the dimension of length, and Q is total beam power per unit time. At the point (x, y) in $z=0$ plane, Eq (1) gives rise to

$$\left.\begin{array}{l} \theta^* = \theta_1^* + \theta_2^*, \quad \text{for} \quad |y^*| \leq 1, \\ \theta^* = \theta_1^* - \theta_2^*, \quad \text{for} \quad |y^*| > 1, \end{array}\right\} \tag{3}$$

where

$$\theta_{1,2}^* = \frac{\exp(-v^* x^*)}{\pi v^*} \left\{ \exp(-v^* |x^*|)(\alpha_{1,2} + \beta_{1,2}) \right.$$
$$\left. - \int_0^{\alpha_{1,2}} \exp(-v^* R_{\alpha_{1,2}}^*) d\varphi - \int_0^{\beta_{1,2}} \exp(-v^* R_\beta^*) d\varphi \right\}$$

$$\alpha_1 = \tan^{-1} \frac{b^*}{1+y^*}, \qquad \alpha_2 = \tan^{-1} \frac{b^*}{|1-y^*|}$$

$$\beta_1 = \tan^{-1} \frac{1+y^*}{b^*}, \qquad \beta_2 = \tan^{-1} \frac{|1-y^*|}{b^*}$$

$$R_{\alpha_1}^* = \left\{ x^{*2} + \frac{(1+y^*)^2}{\cos^2\varphi} \right\}^{\frac{1}{2}}, \quad R_{\alpha_2}^* = \left\{ x^{*2} + \frac{(1-y^*)^2}{\cos^2\varphi} \right\}^{\frac{1}{2}}$$

and $\quad R_\beta^* = \left\{ x^{*2} + \frac{b^{*2}}{\cos^2\varphi} \right\}^{\frac{1}{2}}$ \hfill (4)

For points in $x^*=0$ plane, Eq. (4) gives

$$\theta_1^* = \frac{1}{\pi v^*} \left[(\alpha_1 + \beta_1) - \int_0^{\alpha_1} \exp\left\{ -\frac{v^*(1+y^*)}{\cos\varphi} \right\} d\varphi \right.$$
$$\left. - \int_0^{\beta_1} \exp\left(-\frac{v^* b^*}{\cos\varphi} \right) d\varphi \right]$$

$$\theta_2^* = \frac{1}{\pi v^*} \left[(\alpha_2 + \beta_2) - \int_0^{\alpha_2} \exp\left\{ -\frac{v^* |1-y^*|}{\cos\varphi} \right\} d\varphi \right.$$
$$\left. - \int_0^{\beta_2} \exp\left(-\frac{v^* b^*}{\cos\varphi} \right) d\varphi \right] \tag{5}$$

At the edge of the rectangle P, $x^*=0$ and $y^*=1$, it becomes

$$\theta^* = \frac{1}{\pi v^*} \left\{ \tan^{-1} \frac{b^*}{2} + \tan^{-1} \frac{2}{b^*} - \int_0^{\tan^{-1}\frac{b^*}{2}} \right.$$
$$\left. \exp\left(-\frac{2v^*}{\cos\varphi} \right) d\varphi - \int_0^{\tan^{-1}\frac{2}{b^*}} \exp\left(-\frac{v^* b^*}{\cos\varphi} \right) d\varphi \right\} \tag{6}$$

4. The steady temperature due to moving band source

In this section steady temperature distribution of infinite body due to moving band source in which heat is liberated uniformly at the rate q per unit time per unit area over an infinite strip parallel to the z-axis and of width $2a$ along the y-axis is presented.

Supposing that the center of the band which moves with the velocity v in the direction to the x-axis is at origin at the instance considered, the temperature rise at (x, y) is obtained by putting $b^* \to \infty$ in Eq. (3) and (4). For the present only the temperature in the plane of the source, $z=0$, will be discussed. Putting $x^*=0$ and $b^* \to \infty$, Eq. (3) becomes

$$\theta^* = \frac{1}{v^*} \left[1 - \frac{1}{\pi} \int_0^{\frac{\pi}{2}} \left\{ \exp\left(-\frac{v^*(1+y^*)}{\cos\varphi} \right) \right. \right.$$
$$\left. \left. + \exp\left(-\frac{v^*(1-y^*)}{\cos\varphi} \right) \right\} d\varphi \right], \quad |y^*| \leq 1,$$

$$\theta^* = \frac{1}{v^*} \int_0^{\frac{\pi}{2}} \left\{ \exp\left(-\frac{v^*(y^*-1)}{\cos\varphi} \right) \right.$$
$$\left. - \exp\left(-\frac{v^*(y^*+1)}{\cos\varphi} \right) \right\} d\varphi, \quad |y^*| > 1,$$

(7)

At the edge of the source, putting $y^*=1$ in Eq. (7) it becomes

$$\theta^* = \frac{1}{\pi v^*} \left\{ \frac{\pi}{2} - \int_0^{\frac{\pi}{2}} \exp\left(-\frac{2v^*}{\cos\varphi} \right) d\varphi \right\}$$

(8)

Numerical values θ^*/θ_0^* in the source plane, that is, the ratio of θ^* to that at the center of the source, are shown in **Fig. 2** for various values of $v^*=av/2k_D$. For large v^*, the integral in Eq. (8) may be approximated by $(\pi/2)\exp(-2v^*)$, and then it becomes

$$\theta^* = \frac{1}{2v^*} \left\{ 1 - \exp(-2v^*) \right\}$$

(9)

Eq. (8) also represents the temperature due to a semi-circular plane source at the center which moves in the direction perpendicular to the plane and of which center is at the location considered. For $v^* \gg 1$,

$$\theta^* = \frac{1}{2v^*}$$

(10)

The temperature due to the band source also may be obtained by integration of the solution of the moving line source[7]. Then Eq. (8) may be written in the another form:

$$\theta^* = \frac{1}{2\pi v^*} \int_0^{2v^*} K_0(u)\,du$$

(11)

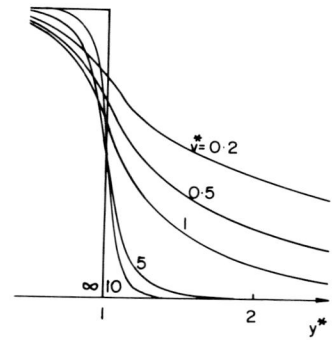

Fig. 2. Relation between θ^*/θ_0^* and y^* in the $x^*=0$ plane for band source

where K_0 is the modified Bessel function of the second kind of order zero. In the case $v^* \ll 1$, it becomes

$$\theta^* = \frac{2}{\pi} (1.1159 - \log_e 2v^*)$$

(12)

5. Penetration depth

Since a liquid-enveloped deep hole moves along the intended joint in high power electron or laser beam welding, the temperature at the edge of the rectangular or band source may be somewhat higher than the fusion temperature of the material θ_f. But here this temperature is approximated by θ_f to simplify the problem.

Putting $\theta = \theta_f$ in Eq. (2),

$$\theta_f^* = \frac{4k\theta_f b}{Q} = \frac{4ka\theta_f}{Q} b^*$$

(13)

The solution of θ_f^* in the case of the rectangular source is obtained by equating Eq. (6) to Eq. (13), which gives the welding conditions. Relation between θ_f^* and b^* for the rectangular source is shown by the solid curves in **Fig. 3**. The values in the limit $b^* \to \infty$ in this figure correspond to that of the band source given by Eq. (8).

On the other hand θ_f^* in Eq. (13) is a linear function of b^* with gradient $4ka\theta_f/Q$, which is plotted as a function of Q for various materials for $a=0.5$ mm in **Fig. 4**. It may be easily found that the solution θ_f^* in case of the rectangular source, which is given by the coordinate of the intersection of Eq. (13) and Eq. (6), approaches to that of the band source as Q increases, or k, a or θ_f decreases. The dotted lines in **Fig. 3** represent the relation given by Eq. (13) for various materials in the case of $Q=5$ KW and $a=$

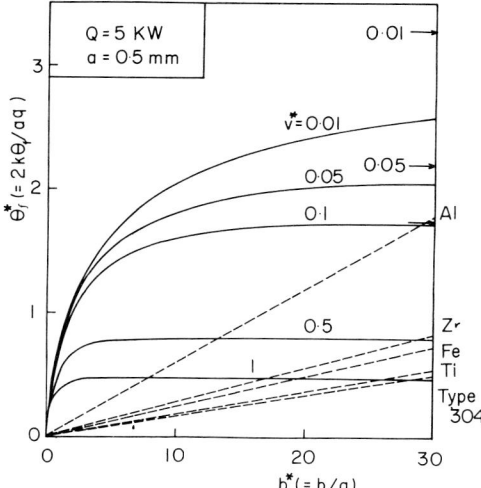

Fig. 3. Relation between θ_f^* and b^* for rectangular source.

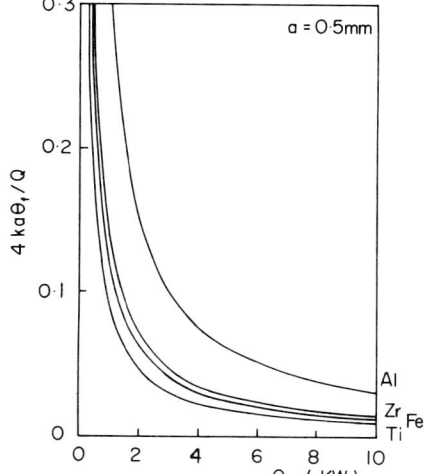

Fig. 4. Relation between Q and $4ka\,\theta_f/Q$.

0.5 mm, for an example. For low heat conduction material such as titanium, zirconium and type 304 stainless steel, θ_f^* for the rectangular source agrees well with that of the band source. Even for aluminium with large k, it seems to be quite well to consider that θ_f^* for both sources agrees well for $v^* > 0.05$. Thus Eq (8) is available for estimating the penetration depth in typical electron beam welding, in which the beam power is usually 5 KW or more, because the contribution of a spurious part of the band source from

$z^* = b^*$ to ∞ may be neglegible. In order to expand the problem to the case of generalized high power beam involving laser and electron beam, and to make the estimation more accurate, taking the effects of the reflectance loss of the beam and latant heat of the fusion into consideration, the final result of θ^* is given by

$$\theta_f^* = \frac{Q}{AQ - 2abHv} \cdot \frac{1}{\pi v^*} \left\{ \frac{2}{\pi} \right. \tag{14}$$
$$\left. - \int_0^{\frac{\pi}{2}} exp \left(-\frac{2v^*}{\cos\varphi} \right) d\varphi \right\}$$

where A is the beam absorptance of the material and H the latant heat of fusion in the form cal/cm³. In the case of the electron beam A is equal to unit, but in the laser beam it is usually less than unit because the beam reflectance for normal incidence is higher. However, the absorptance approaches to unit as the drilled hole becomes deep, because the beam is re-focused to the bottom of the hole through the self-focusing effect of its wall[9] trapping the beam within the hole effectively.

6. Comparison with experimental data

In the calculation of the bead penetration depth in electron or laser beam welding one of the most difficulty to be overcome is the estimation of the diameter of the beam-drilled hole, which corresponds to the width of the band or rectangular source. In general the diameter varies not only with the beam parameters which involve the beam current, the acceleration voltage, the current of the focusing coil and work distance but also with the physical properties of the workpiece and welding velocity. Furthermore, it is impossible or very difficult to measure the hole diameter during welding. Here the diameter of the hole during welding will be approximated by that of the fusion isotherm obtained from micro-examination for convenience. In general the electron beam welding produces a penetration profile with a certain deviation from the ideal rectangle as shown in **Fig. 5**. In well-controlled electron beam

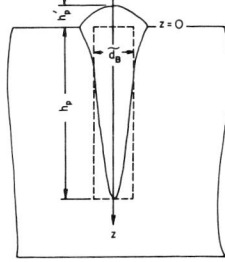

Fig. 5. Equivalent rectangle having same area and depth with actual bead.

266

welding the deviation is not appreciable, and then the penetration profile may be approximated by an equivalent rectangle having same depth and area with it. The width of the equivalent rectangle is given by

$$\widetilde{d_B} = \frac{1}{h_p} \int_{-h_p'}^{h_p} d(z)\,dz \qquad (15)$$

where $d(z)$ is the diameter at depth z, h_p penetration depth. As the result putting $2a = \widetilde{d_B}$ and $b = h_p$, v^* and θ_f^* are written in the forms

$$v^* = \frac{\widetilde{d_B}\,v}{4k_D} \qquad (16)$$

$$\theta_f^* = \frac{2k\theta_f\,\widetilde{d_B}}{Q}\,\theta^* \qquad (17)$$

The experimental data to be compared with the theory are obtained from references 3, 4 and 10 in which the photographs of the penetration profile are illustrated. Consequently, the data give the variety to the welding parameters involving beam current, accelerating voltage, focal current, work distance, welding speed and physical properties of material as shown in **Table 1.** Relation between $\dot{\theta}_f^*$ and v^* reduced from the experimental data is shown in **Fig. 6.** The solid curve in this figure represents the theoretical value calculated from the band source model, which has been found to be reasonable for analysing the concerned problems from the study in section 5 because the beam power is higher or equal to 5 KW. Close agreement between the data and theoretical over a wide range of v^* from

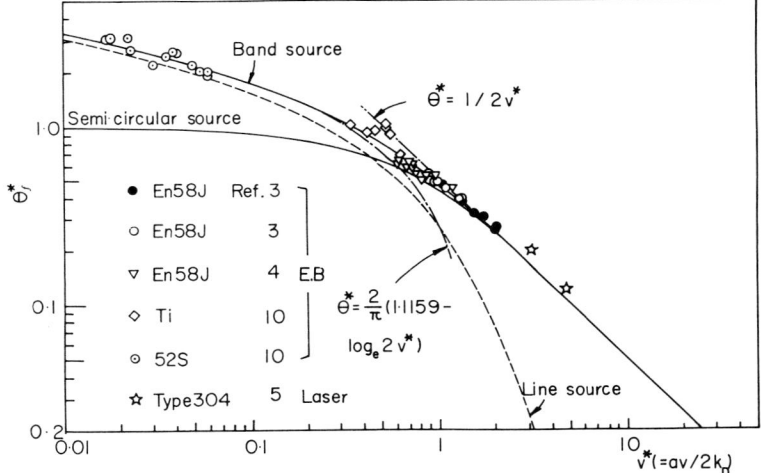

Fig. 6. A comparison of the theoretical equations with experimental data.

Table 1.

Reference	3		4	10		5
Material	En58J**		En58J	52S*	Ti	Type 304
k (cal/cm. sec. °C)	0.07		0.07	0.4	0.065	0.07
k_D(cm²/sec)	0.05		0.05	0.76	0.08	0.05
	Electron Beam					Laser
Beam	130 KV	27~33 KV	30 KV	150 KV		20 KW
	40 mA	150 mA	100 mA	30 mA		
v (cm/min)	120~300	66	66	50		126~430
		const.	const.	const.		
Parameter	Velocity	Beam focusing current	Beam focusing current and focus position	Work distance		Velocity

  * 52S (Mg-2.4 %, Cu-0.04 %, Fe- 0.18 %, Si-0.1 %, Cr-0.17 %, Al-remainder)
  ** En58J (C-0.06 %, Ni-10.89 %, Cr-17.65 %, Mo-2.85 %, Fe-remainder)

267

about 0.01 to 2.0 is obtained.

Additional experimental data are obtained from the CO_2 laser welding[5]. The laser beam absorptance A is regarded as 100 % in the estimation of θ_f^*, because the extremely high laser power, 20 KW, provides deep penetration, about 15 mm, in which the incident laser beam may be trapped without any loss. The data agree well with the theory, though the data are not sufficient in number because only a few data of the CO_2 laser welding have yet been published.

7. Summary

Heat flow in electron and laser beam welding has been mathematically analysed by introducing rectangular and band heat source model, and new mathematical formulations avairable for estimating the weld depth as a function of beam parameters, physical constants of the material and welding speed have been derived. In the power level of several kilowatts or above, it has been found from the mathematical analysis that a value obtained from the rectangular source may be approximated by that from the band source. The theoretical values have been compared with published data obtained from laser and electron beam welds. The actual bead has been replaced by an equivalent rectangle with the width $\widehat{d}_B$ having same depth and area with the actual, and then agreement between theoretical bead depth and experimental one has been obtained.

Acknowledgment

We wish to thank Kawasaki Heavy Industry Co. Ltd. for providing the experimental data of electron beam welding.

References

1) T. Hashimoto and F. Matsuda: "Effect of Welding Variables and Materials upon Bead Shape in Electron-Beam Welding", Trans. of NIRM Vol. 7 (1965) No. 3.
   B. T. Lubin: "Dimensionless Parameters for the Correlation of Electron Beam Welding Variables", Weld. J., Vol. 47 (1968) No. 3.

2) T. Hashimoto and F. Matsuda: "Penetration Mechanism of Weld Bead in Electron-Beam Welding", Trans. of NIRM Vol. 7 (1965) No. 5.

3) M. J. Adams: "Low Voltage Electron Beam Welding: Effect of Process Parameters", British Weld. J., Vol. 15 (1968) No. 3.
   M. J. Adams: "High Voltage Electron Beam Welding" British Weld. J., Vol. 15 (1968) No. 11.

4) A. Sanderson: "Electron Beam Delineation and Penetration", British Weld. J., Vol. 15 (1968) No. 10.

5) E. L. Locke, E. D. Hoag and R. A. Hella: "Deep Penetration Welding with High-Power CO_2 Lasers", IEEE Journal of Quantum Electronics, Vol. QE-8 (1972), No. 2.

6) D. Rosenthal: "Mathematical Theory of Heat Distribution during Welding and Cutting", Weld. J., Vol. 20 (1941) No. 5.

7) D. Rosenthal: "The Theory of Moving Source of Heat and Its Application to Metal Treatments", Trans. ASME, Vol. 68 (1946).

8) Y. Arata and I. Miyamoto: "Some Fundamental Properties of High Power Laser Beam as a Heat Source-CO_2 Laser Absorption Characteristics of Metal -", Trans. Japan Welding Society, Vol. 3 (1972) No. 1.

9) Y. Arata and I. Miyamoto: "Some Fundamental Properties of High Power Laser Beam as a Heat Source-Beam Focusing Characteristics of CO_2 Laser -", Trans. Japan Welding Society, Vol. 3 (1972) No. 1.

10) Kawasaki Heavy Industrial Co. Ltd., private communication.

11) C. Y. Ho and R. E. Taylor: "Thermal Conductivity", Plenum Press. New York (1969).

12) Y. S. Touloukian, R. W. Powell, C. Y. Ho and P. G. Klemens: "Thermal Conductivity", IFI/Plenum, New York-Washington (1970).

Investigation on Moving Line Heat Source
—Numerical Calculation of Temperature Distribution—

Abstract

Calculation based on heat conduction theory was made for moving line heat source of non-uniform input energy distribution and temperature distribution in the vicinity of the heat source was obtained.
These results are expressed in isothermal contour line maps.
These can be compared with the actual welding results such as arc welding, plasma welding, electron beam welding and so on.

1. Introduction

The heat input source during actual welding is neither a point, nor a line along which input energy is uniformly distributed, though these two cases have been used on theoretical treatment for the heat conduction problem of welding. The actual heat input source seems to lie midway between both cases.

When we measure the temperature at a position considerably distant from the heat source, it has enough accuracy for the purpose to use a point heat source approximation or a "so called" line heat source approximation.

When we discuss, however, the temperature surrounging the heat source, we frequently get unsatisfactory results from these extreme approximations. In this report, a medium case between a point heat source approximation and a line heat source approximation is formulized by introducing a dimensionless quantity α, and the temperature distribution is calculated numerically.

2. Equation for Temperature Distribution

2.1 Notation

We take the position of the heat source of the plate surface as the origin, O, the moving direction of the heat source as the X co-ordinate axis, the perpendicular directions to the X co-ordinate in the horizontal and the vertical plane as the Y and the Z co-ordinate respectively and assume the co-ordinate system which moves with the line heat source as shown in **Fig. 1.**

The other notation is as follows,

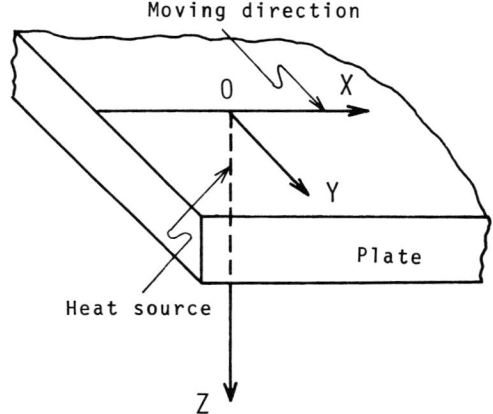

Fig. 1. The co-ordinate system.

Q : input energy per unit length along Z axis.

k : heat conductivity.

k_D: thermal diffusivity (or temperature conductivity) $= k/c \cdot \rho$

c : specific heat

ρ : density

v : moving speed of heat source

D: plate thickness

and the asterisk symbol $*$ expresses a dimensionless quantity in all cases.

2.2 Method of Calculation

The temperature T_p at the point (X, Y, Z) in the infinite medium due to the moving point heat source $Q(Z_1)dz_1$ can be calculated with eq. 1,

$$T_p = \frac{Q\ dZ_1}{4\pi k_D c \rho R}\ e^{-\frac{v}{2k_D}(X+R)} \quad \text{------------------------(1)}$$

where Z_1 is the position of the point heat source on the Z co-ordinate (only in this case, we can take the origin at the arbitrary position on the Z axis)

$$R^2 = X^2 + Y^2 + (Z - Z_1)^2$$
$$= R_1^2 + (Z - Z_1)^2$$
$$R_1^2 = X^2 + Y^2$$

We introduce the following dimensionless quantities,

$$X^* = \frac{X}{D}, \quad Y^* = \frac{Y}{D}, \quad Z^* = \frac{Z}{D}$$

$$Z_1^* = \frac{Z_1}{D}, \quad R^* = \frac{R}{D}, \quad R_1^* = \frac{R_1}{D}$$

$$T_\Theta = \frac{Q(0)}{4\pi k} = \frac{Q(0)}{4\pi k_D c \rho}, \quad Q^*(Z_1^*) = \frac{Q(Z_1^*)}{Q(0)}$$ ----- (2)

$$v^* = \frac{v \cdot D}{2 k_D}$$

and obtain the dimensionless temperature T_p^* $(= \frac{T_p}{T_\Theta})$

$$T_p^*(v^*, X^*, R_1^*, Z^* - Z_1^*, Z_1^*) = \frac{Q^*(Z_1^*) dZ_1^*}{R^*}$$
$$\times e^{-v_*(X^* + \sqrt{R_1^{*2} + (Z^* - Z_1^*)^2})}$$ ---------------------- (3)

In case the medium is of finite thickness, it is necessary to calculate the dimentionless temperature T_{pl}^* by the following equation.

$$T_{pl}^*(v^*, X^*, R_1^*, Z^* - Z_1^*, Z_1^*)$$
$$= T_p^*(v^*, X^*, R_1^*, Z^* - Z_1^*, Z_1^*) + T_p^*(v^*, X^*,$$
$$R_1^*, Z^* + Z_1^*, Z_1^*) + \sum_{n=1}^{\infty} \{T_p^*(v^*, X^*, R_1^*,$$
$$2n - z^* - Z_1^*, Z_1^*) + T_p^*(v^*, X^*, R_1^*, 2n + Z^* -$$
$$- Z_1^*, Z_1^*) + T_p^*(v^*, X^*, R_1^*, 2n - Z^* + Z_1^*, Z_1^*)$$
$$+ T_p^*(v^*, X^*, R_1^*, 2n + Z^* + Z_1^*, Z_1^*) \}$$
$$= e^{-v^* X^*} T_{p2}^*(v^*, R_1^*, Z^*, Z_1^*) dZ_1^*$$ --------- (4)

When it is required to treat the line heat source whose input energy distribution function $Q^*(Z_1^*)$ changes along the Z co-ordinate axis, we need to sum up all the contribution of $Q^*(Z_1^*) dZ_1^*$, then, to obtain the integral as

$$T_{line}^*(\alpha, v^*, X^*, Y^*, Z^*) = e^{-v^* X^*} \int Q^*(Z_1^*)$$
$$\times T_{p2}^*(v^*, R_1^*, Z^*, Z_1^*) dZ_1^* = e^{-v^* X^*}$$
$$\times T^*(\alpha, v^*, R_1^*, Z^*)$$ ---------------------- (5)

where $Q^*(Z_1^*)$ may be chosen arbitrarily.

Considering the actual heat input situation in welding process, we put

$$Q^*(Z^*) = (1 - Z^*)^\alpha$$ -------------------------------- (6)

where α is the index to the input energy distribution. Then, at $\alpha = 0$,

$$Q^*(Z^*) = 1$$ -- (7)₁

and it corresponds to the line heat source of uniform input energy distribution, so called "line heat source" which has been employed so far.

at $\alpha \rightarrow \infty$
$$\left. \begin{array}{l} Q^*(Z^*) = 1 \quad (Z^* = 0) \\ \quad\quad\quad = 0 \quad (Z^* \neq 0) \end{array} \right\}$$ ------------------------------ (7)₂

and corresponds to the point heat source as is well known. Dependency of $Q^*(Z^*)$ on Z^* is shown in **Fig. 2,** when α is in a certain value between both limits.

Substituting eq. (6) into eq. (5), $T^*(\alpha, v^*, R_1^*, Z^*)$ was calculated first, because on calculation of T_{line}, it is profitable to obtain T^*. Calculations were made on various combinations of α and v^* values. Calculations for integral in eq. (5) were performed numerically, using Simpson's formula.

Then, the temperature on the path of the heat source $T_t^*(\alpha, v^*, X^*, O, Z^*)$ was calculated for $X^* < 0$.

From the obtained $T^*(\alpha, v^*, R_1^*, Z^*)$, $T_{line}^*(\alpha, v^*, X^*, Y^*, Z^*)$ was calculated and $T_m^*(\alpha, v^*, Y^*, Z^*)$ was obtained from eq. (8).

$$T_m^*(\alpha, v^*, Y^*, Z^*) = \text{Max.} \{T_{line}^*(\alpha, v^*, X^*, Y^*, Z^*)\}$$
$$= T_{line}^*(\alpha, v^*, X_m^*, Y^*, Z^*)$$ -- (8)

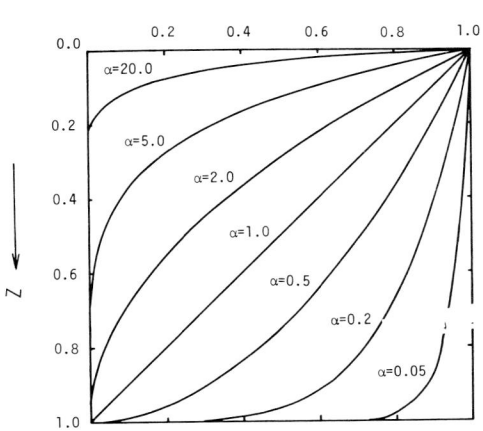

Fig. 2. The input energy distribution function $Q^*(Z)$.

270

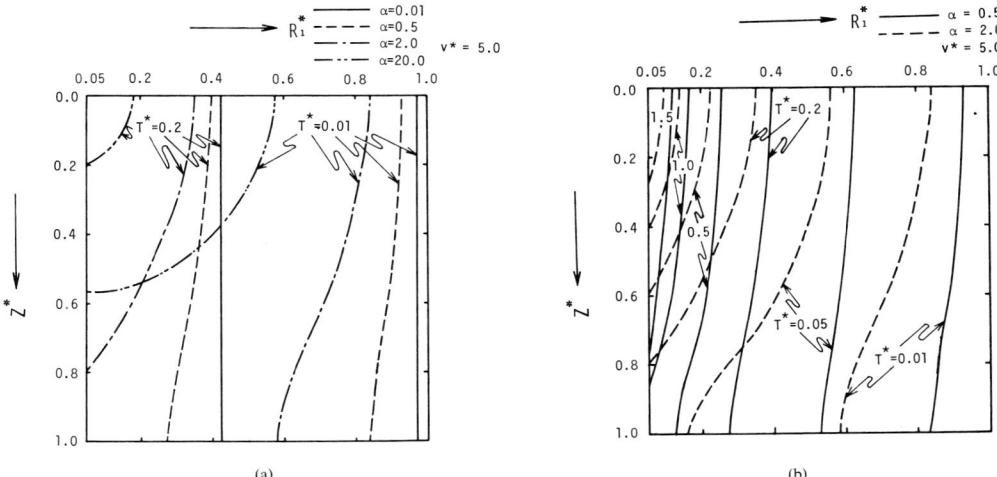

Fig. 3. The contour map of T^* on the $R_1{}^*$—$Z_1{}^*$ plane.

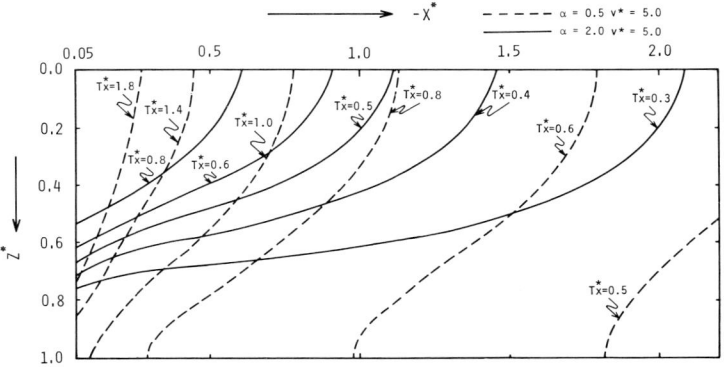

Fig. 4. The contour map of $T_x{}^*$ on the X^*—$Z_1{}^*$ plane

where the value X^* which makes $T_{line}{}^*$ maximum is put to $X_m{}^*$.

These calculated results of T^*, $T_x{}^*$ and $T_m{}^*$ were plotted on the $R_1{}^*$—Z^*, the X^*—Z^* and the Y^*—Z^* plane respectively and the isothermal contour line maps were figured.

3. Results of Calculations

Figure 3 (a), (b) shows the contour line maps of T^* for the constant value of $v^*(=5.0)$ and a few values of α.

It is seen from this figure that the pattern of the temperature distribution varies from that of a line heat

source to that of a point heat source as α increases. An example of the $T_x{}^*$ contour line map is shown in **Fig. 4** in which α and v^* are the same values as in **Fig. 3 (b)**.

The distribution of the maximum temperature $T_m{}^*$ on the Y^*—Z^* plane and its dependency on α are shown in **Fig. 5 (a), (b), (c)** and **(d)**. These are arranged in the same way as in **Fig. 3** for the constant value of v^* and a few values of α. The dependency of the T^* and the $T_m{}^*$ distribution on the value of α are shown in **Fig. 6** and **Fig. 7**.

The values of T^* are plotted on the R^*—Z^* plane for the constant value of $\alpha(=1.0)$ and a few values of v^* in **Fig. 6 (a), (b)** and **(c)**. The change of T^*

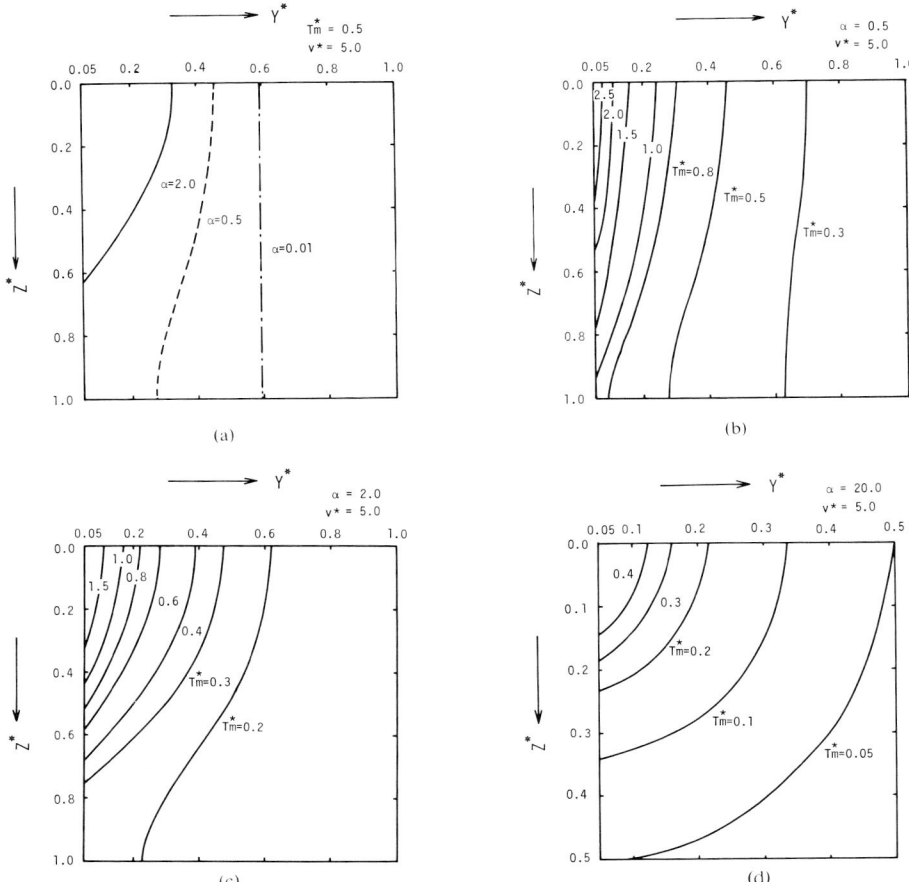

Fig. 5. The contour map of T_m^* on the Y^*—Z^* plane.

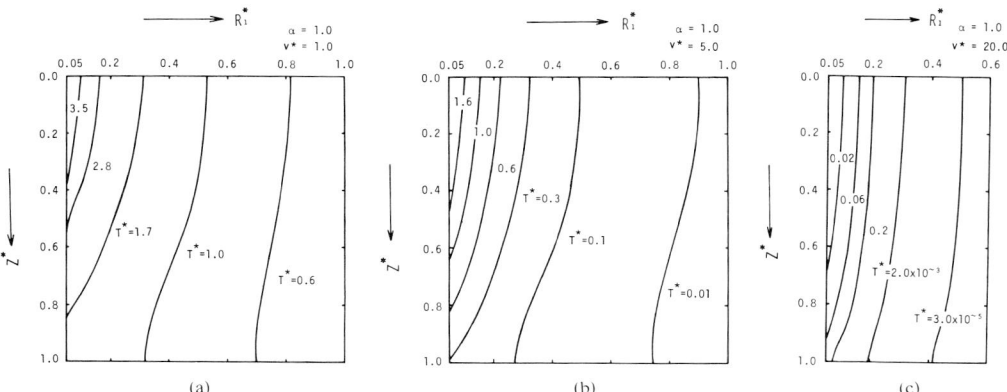

Fig. 6. The contour map of T^* on the R_i^*—Z^* plane.

272

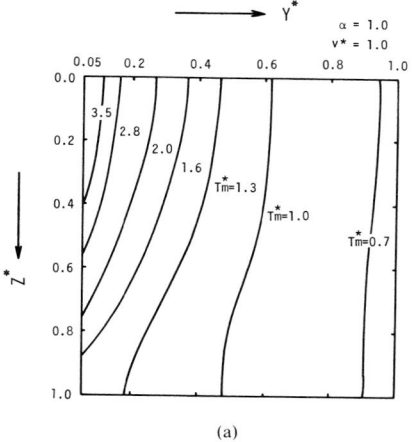

(a)

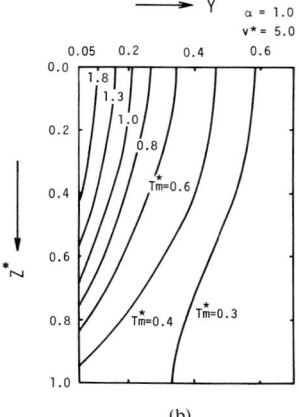

(b)

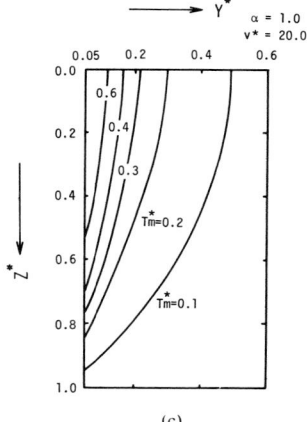

(c)

Fig. 7. The contour map of T_m^* on the Y^*—Z^* plane.

becomes remarkable as the value v^* increases in this figure. The value of T_m^* are plotted on the Y^*—Z^* plane in the same way as in **Fig. 6** as shown in **Fig. 7 (a), (b)** and **(c).**

In this figure, the remarkable tendency as is in **Fig. 6** is not seen.

4. Summary

Calculations, made in this report, will prove their worth when comparing with various welding results.

This is the problem in which α is to be determined under the given value of v^* Some experimental approaches should be undertaken for this purpose.

273

High Energy Density Plasma Beam

Achievement of High Energy Density for Plasma Beams
Y. Arata
[J. High Temp. Soc. **11** (1985).]

Magnetic Control of Plasma Arc Welding—On the Pressure Distribution of Plasma and its Effect on Key-Hole Action
Y. Arata, H. Maruo, Y. Hirata and Y. Horio
[Trans. JWRI **3** (1974), 127.]

Concept of Vortex Gas Tunnel and Application to High Temperature Plasma Production
Y. Arata
[J. Phys. Soc. Japan **43** (1977), 1107.]

High Power Microwave Discharge in Atmospheric Hydrogen Gas Flow
Y. Arata, S. Miyake and A. Kobayashi
[J. Phys. Soc. Japan **44** (1978), 998.]

Pulsed High Current Heating of a Microwave Plasma at Atmospheric Pressure
Y. Arata, S. Miyake, M. Ushio and Y. Yoshioka
[J. Phys. Soc. Japan **44** (1978), 1483.]

Achievement of High Energy Density for Plasma Beams

Abstract

The basic thermal properties that govern the generation of "high energy density plasma beams" are investigated. The arc state, which is important in the utilization of plasma beams as a high energy density heat source, is studied by separately considering the two regions of the arc: the arc column and an electrode (anode or cathode) region. The mechanism of the "thermal pinch" phenomenon is clarified in the arc column and the role of various parameters which enhance the functioning of the "thermal pinch" phenomenon is examined; including the effect of the column configuration and the roles played by both thermal conductivity and the temperature gradient. The conditions necessary to obtain a plasma column with a high temperature and high energy density are described and actual examples are given. It is also shown that a "point arc" of high energy density can be generated in the region near the electrodes by the supply of a special vortex gas flow. The "arc ball" that appears in the anode region of the "point arc" is found to have an energy density as high as 10^6 W/cm² , and can be stably sustained in a fixed position.

It is usually more difficult to generate a high energy density in a plasma beam than in laser beams or electrically charged particle beams such as electron and ion beams. Laser beams can be focused up to the diameter of the wavelength by means of an optimally designed lens system. A high energy density can be achieved relatively easily in charged particle beams by such technical means as accelerating individual particles and strongly focusing the beams.

Plasma beams, however, can be neither focused like laser beams nor accelerated like charged particle beams. For the energy per specific volume of a plasma at pressure p, temperature T and number density n, the following equation holds, known as the equation of state,

$$p = nkT(=\rho R_0 T), \quad \dotfill \quad (1)$$

where k is the Boltzmann's constant, R_0 is the gas constant and ρ is the mass density ($\rho = mn$ and m is the mass of a particle). When the pressure is constant

$$nT = p/k = \text{constant}. \quad \dotfill \quad (2)$$

For example, in a high power hydrogen plasma jet to obtain the highest possible temperature at atmospheric pressure, we can derive the following values from Eq. (2).

$$T = 3 \times 10^4 \text{ (K)} \sim 2.6 \text{ (eV)}, \quad \dotfill \quad (3a)$$
$$n = 2.4 \times 10^{17} \text{ (cm}^{-3}). \quad \dotfill \quad (3b)$$

In case of a nuclear fusion plasma, as yet unachieved, we obtain at a pressure of a few atmospheres

$$T = 10^8 \text{ (K)} \sim 10^4 \text{ (eV)}, \quad \dotfill \quad (4a)$$
$$n = 2.4 \times 10^{14} \text{(cm}^{-3}). \quad \dotfill \quad (4b)$$

In the former case the energy of a plasma particle acting on a test piece surface at atmospheric pressure is less than a few electron volts and the particle density is only about 1×10^{23}/cm² per second. The energy density obtained by this heat flux is 6×10^4 W/cm² , smaller than that of either a laser or a charged particle beam. In order to obtain an energy density of more than 1×10^6 W/cm² which is the same level as the laser or charged particle beam heat sources, we must resort to the high temperature plasmas created by nuclear fusion at one hundred million degrees Kelvin.

276

Thus, it is very difficult to obtain the same high energy density as laser and charged particle beams in a plasma beam which is governed by Eq. (1). There is a method of utilizing plasma beams as a high energy density heat source by making it at an arc state, whose characteristics differ from those of other heat sources in the fact that it behaves in the same way as a gas fluid. For instance, 1) it is easy to obtain dynamic pressures, 2) shock waves are generated at the speed of sound. They can be either an advantage or a disadvantage when used as a heat source.

As described above, one of the simple methods of obtaining high energy density plasma beams is to create an "arc state" of the plasma i.e., to generate joule heat by the flow of electrical current through the plasma. When it is used as a heat source, however, we must consider the following problems;

(A) the achievement of a high energy density in the region where joule heat is generated,

(B) the achievement of a high energy density in the region near the electrode.

(A) The Achievement of a High Energy Density in the Region Where Joule Heat is Generated.

The region is generally called an "arc column". In order to maintain a temperature T in this arc column, a certain amount of energy (the input energy W_a) per second and per unit length of the arc column must be input. At the same time, there is an energy output (energy loss W_L), resulting in the following energy balance.

$$W_a = W_L \qquad \dots\dots\dots\dots\dots\dots\dots\dots\dots\dots\dots\dots\dots\dots (5a)$$
$$= W_{LT} + W_{LR} + W_{LO}, \qquad \dots\dots\dots\dots\dots\dots\dots\dots\dots\dots (5b)$$

where W_{LT} is the energy loss related to the increase in temperature and is the most important factor in the generation of a high energy density. W_{LR} is the radiant energy, and W_{LO} is all energy loss other than W_{LT} and W_{LR}, as described below.

The energy loss W_{LT} is carried out through the surface of the arc column. This loss is thus proportional to the surface area S_L, to the temperature gradient from the center to the surface, and to the thermal conductivity κ.

When the plasma column has a temperature gradient in the radial direction and heat is generated only by the flow of electrons, the thermal flux density, $i_T = \kappa E_T = -\kappa \,\mathrm{grad}\, T$, has the same form as in a simple Ohm's law, $i = \sigma E = -\sigma \mathrm{grad} V$. Therefore, the energy W_{LT} that is transferred outside the arc column through the surface area S_L is

$$W_{LT} = S_L\, i_T\ (\equiv I_T), \qquad \dots\dots\dots\dots\dots\dots\dots\dots\dots\dots\dots\dots\dots (6)$$
where
$$i_T = -\kappa \,\mathrm{grad}\, T \qquad \dots\dots\dots\dots\dots\dots\dots\dots\dots\dots\dots\dots\dots (7a)$$
$$= \kappa E_T, \qquad \dots\dots\dots\dots\dots\dots\dots\dots\dots\dots\dots\dots\dots\dots\dots (7b)$$
$$E_T = -\mathrm{grad}\, T. \qquad \dots\dots\dots\dots\dots\dots\dots\dots\dots\dots\dots\dots\dots\dots (8)$$

Thermal conductivity κ is primarily dependent on temperature T, as shown in the following equations. For a fully ionized hydrogen plasma.[1]

$$\kappa = 4.67 \times 10^{-10}\, T^{5/2}(C^*\ell n V)(\mathrm{cal/deg.\ m.\ sec}) \qquad \dots\dots\dots\dots\dots\dots\dots (9a)$$
$$\cong 2 \times 10^{-11}\, T^{5/2}\ (\mathrm{cal/deg.\ m.\ sec}) \qquad \dots\dots\dots\dots\dots\dots\dots\dots (9b)$$
$$(\text{in hydrogen plasma})$$

where $C^* \equiv k^*{}_L/Z$, Z is the charge number and $k^*{}_L$ is the correction term of collision, such as ion recoil due to collisions of electrons or ions. C^* is 0.225 for a high power hydrogen plasma jet with $Z = 1$, and the plasma parameter $\ln \Lambda$ is 5-6 (this is 15-20 in nuclear fusion plasma), thereby resulting in Eq. (9b).

S_L and i_T have a very close relationship which forms the basis for the generation of a high energy density in an arc column. If S_L is constant, i_T is a function of the temperature only and becomes heat loss due solely to electrons. Therefore the energy loss W_{LT} or total heat flux I_T increases dramatically as the temperature increases. In other words, the greater the increase in I_T, the higher the temperature of the arc column and the higher the energy density that can be obtained. This is the important point of the so-called "thermal pinch" phenomenon[2], and studies have been made on how to cool the arc column from the "Gerdien arc"[3] up to the present day plasma jet. This problem must be studied on in more detail by examining various factors to increase i_T.

(1) The Effect of S_L.

S_L is an important parameter which plays a fundamental role in increasing i_T, dominating the functioning of the "thermal pinch'. phenomenon. There are two ways of reinforcing the function of this phenomenon by increasing i_T.

a) The Increase in i_T Due to a Reduction in S_L

When the cross-section of an arc column is circular and the plasma behaves in a self-sustaining manner according to the "minimum energy principle", reduction in S_L will result in an increase in i_T. Conventional arc columns which are governed by the thermal pinch effect belong to this category. Their examples are a plasma jet and "gerdiem arc", where S_L is written by

$$S_L = 2\pi r \quad \dots \dots \dots \dots \dots \dots \dots \dots \dots \dots \dots \dots \dots \dots \dots \dots \quad (10)$$

per unit length of the column and r is the column radius. The gas tunnel type plasma beam,[4] or plasma jet,[5] developed by the authors, also belongs to this category. It is important to obtain practical methods to decrease S_L as low as possible or to make it at a constant value even at an increase in the arc current.

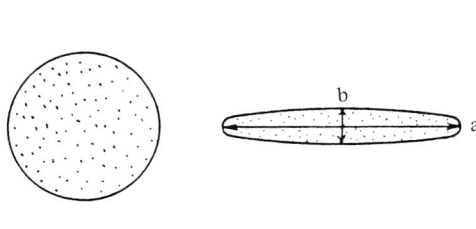

Fig. 1 Column cross-sections of open arc and magnetized sheet arc

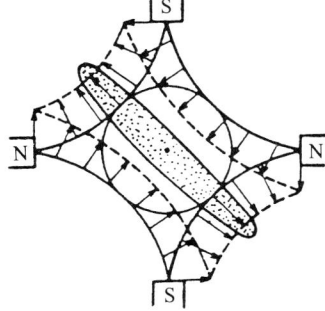

Fig. 2 Block diagram of cusp field

b) The Increase in i_T Due to an Increase in S_L.

As described already, an arc column autonomously forms a profile which minimizes energy loss. Since the development of arc by Davy in 1801 and the application of the carbon arc heat source to welding by Benardos and Olszewski in 1885, it has been considered that the energy density of the arc column is very difficult to control. Only the plasma jet, which follows a Gerdien arc or a wall stabilized arc, etc., were examples overcoming this difficulty, although they still follows the autonomous variations of the arc column.

The authors have tried to overcome the autonomous characteristic of the arc column and examined to increase S_L. As shown in Fig. 1, we thought of flattening the circular cross-section of the arc column,[6] and succeeded in obtaining such a configuration by applying the cusp magnetic field shown in Fig. 2.[7] The i_T increased dramatically and W_{LT}, or I_T were enlarged thus creating a high energy density. This has been named a "magnetized sheet arc" by the author. Fig. 3 shows a comparison of the current density of this arc and a conventional arc.

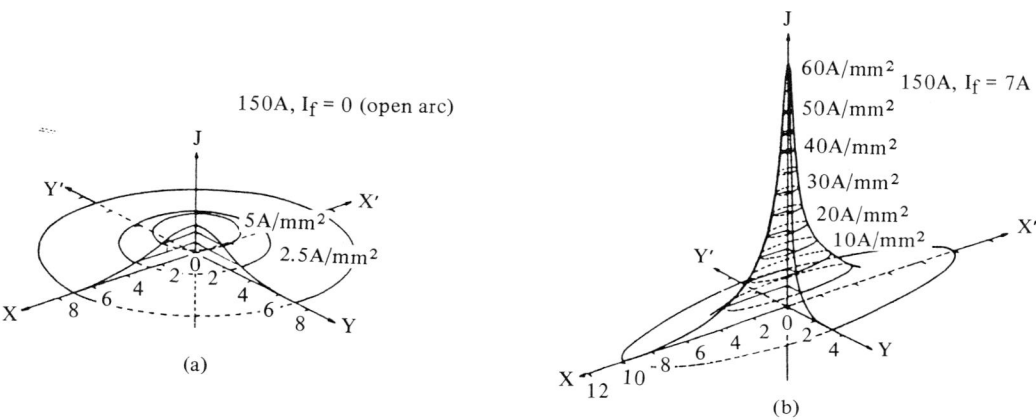

Fig. 3 Current density distributions of open arc and arc in cusp field

(2) The Roles of Thermal Conductivity and Temperature Gradient.

The thermal flux density i_T is determined by both κ and $-\text{grad } T (\equiv E_T)$ as given in Eq. (7). It has been interpreted that the "thermal pinch" phenomenon strengthens whenever there is a large energy loss by cooling of the arc column. The reason why this happens is more clearly understood if we briefly look at the role of the parameters which determine energy loss and dominate the generation of a high energy density. For example, in the following equation.

$$i_T = \kappa_T E_T \dots\dots\dots\dots\dots\dots\dots\dots\dots\dots\dots\dots\dots\dots\dots\dots (11)$$

an increase in E_T and not in κ_T will generate a high energy density by the increase in i_T. When $\kappa_T = \kappa$, as shown in Eq. (9), κ_T can be determined only by the temperature T. While in case of $\kappa_T = \kappa_{eff}$ as descided later, the situation becomes more complex and generally κ_{eff} is larger than κ and sometimes $\kappa_{eff} \gg \kappa$.

On the other hand, if we suppose the existence of an arc column which has the property of $\kappa_T \approx 0$ and which is thermally insulated from the outside, we can imagine that the i_T loss is remarkably small and E_T is very large. It is considered that this thermally-insulated state is very useful in generating both a high temperature and a high energy density. In other words, E_T is the most important parameter affecting whether or not a high energy density is generated. Thus increasing i_T by increasing E_T is a very effective way to enhance the functioning of the "thermal pinch".

Moreover, if a special phenomenon of $\kappa_T \approx 0$ as described above occurs, for example if κ approaches zero on the surface of the arc column, the energy loss become extremely small and a state of thermal insulation will occur. Even though there is a cold wall outside the arc column, due to the state of thermal insulation, both a high temperature and a high energy density can be generated inside the column. If such an arc column can be created, it will have great advantages over a conventional plasma in a high-vacuum and will be a preferable method of producing high temperature plasma for bringing about nuclear fusion. The author is very interested in what level the "magnetized arc column" can approach to this special condition. The problem is how to reduce the magnetized thermal conductivity, κ_{mag} compared to κ.

As an example of making κ_T be very small with a large E_T in case without external magnetic field, there is so-called "Kapitza plasma"[8]. It was reported that a high temperature hydrogen plasma of 10^6 K was obtained steadily in an atmospheric pressure by using a high power microwave energy. The author was so much interested in this result and made a study of confirmation. But we obtained a plasma beam with a temperature of only about 1×10^4 K.[9] Therefore, we tried to use a pulse discharge to heat this plasma beam adiabatically. We super-imposed a pulsed high current of a few kA on this plasma beam, resulting in obtaining a fully ionized plasma beam with a temperature of 5×10^4 K at atmospheric pressure during a few microseconds.[10]

This experiment was carried out without an external magnetic field. If a strong external magnetic field is applied upon this plasma beam with the supply of a rapidly rising pulsed high current to obtain an effective self magnetic pinch, a stable heating would be carried out more effectively, inducing the production of ultra-high temperature plasma. It will eliminate the "wall-problem" in a high-vacuum type nuclear fusion plasma and are sure that it will lead to studies into new types of plasma production for nuclear fusion.

(3) Forced cooling of the arc column

The arc column can be cooled by an electrically insulating fluid (gas or liquid). If cooled by a liquid, the energy loss from the arc is mainly the energy of vaporization, and in case of a gas cooling the loss appears in the form of dissociation and/or ionization energy. This energy is equal to W_{LO} in Eq. (5b). It is the energy loss not directly related to the temperature rise.

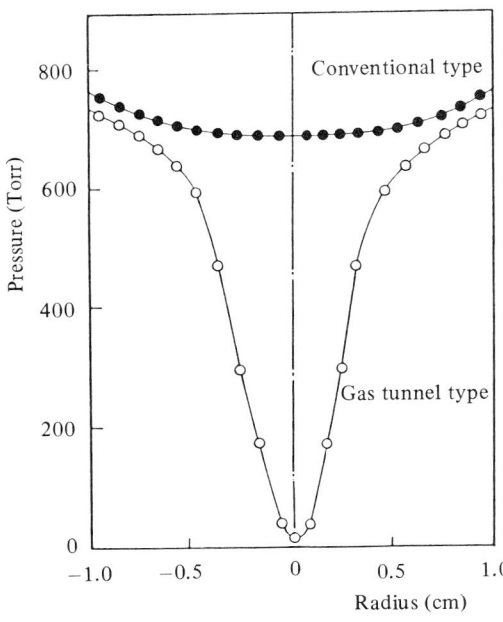

Fig. 4 Pressure distribution in a vortex chamber

Working gas: air
Gas flow rate: 250 ℓ/min
Velocity at wall: 150 m/s

a) Forced Cooling by a Laminar Type Vortex Flow.

Forced cooling of the arc column by a laminar vortex flow maintains an outside temperature to a fixed valve and does not make it increase. Therefore, an increase in the joule heating of the arc column will correspondingly raise the arc temperature. In this case, the heat flux within the arc column i_T follows the Eqs. (6) – (9).

One of the most effective example of this is the "gas tunnel"[11] type plasma beam developed by the authours. Fig. 4 compares the pressure distribution in the special vortex flow of a gas tunnel type plasma beam and that in the conventional type of vortex flow used for typical plasma jets. In the former a vacuum of only a few Torr can be maintained under appropriate conditions along the central axis against a gas pressure of 1 atmosphere at the circumference. When an arc column is generated in this gas tunnel, V-I characteristic of the discharge is different from that of a conventional plasma jet and typically displays "a positive characteristic"[5]. As a result, it is easy to obtain high temperatures of a few tens of thousands of degrees Kelvin.

b) Forced Cooling by Fluid Flowing through the Arc Column.

A strong cooling effect can generally be achieved by means of turbulence caused by convection between the inside and the outside of the arc column. This cooling phenomenon is easily generated in an actual arc column. The following two methods in particular are used, allowing gas flowing from outside through the arc column to pass along and/or across the column. The former method is used for conventional plasma jets, while the latter is realized in "high-speed linear running arcs" by the magnetic drive, and "rotating arcs," or in a "arc in high-speed fluid". For these methods, it is necessary to consider not only Eqs. (6) – (9), but also the new factors of energy loss. As described above in Eqs. (6) – (9), the current of thermal electrons plays the main role in the heat loss of i_T or I_T, which are necessary for increasing the temperature. This increase in heat loss, however, is extreme in cases where the arc column is cooled by fluid flowing through the column because the fluid carries energy away from it corresponding to the fluid velocity. In other words, the time that a plasma particle in the arc column stays in the column, i.e., the "average lifetime" τ_{ap}, becomes very short, and new cold particles enter into the column. Thus the heat loss from the arc column becomes greater, and the thermal conductivity changes to the so-called "effective thermal conductivity", κ_{eff} which is larger than κ in Eq. (9). κ is affected solely by the electron flow, while κ_{eff} includes the infuluence of various types of particles – electrons, ions and neutrals. Consequently,

$$i_T = -\kappa_{eff} \text{ grad } T \ (\equiv \kappa_{eff} E_T), \quad \dots \dots \dots \dots \dots \dots \dots \dots \dots \dots \dots (12a)$$
$$I_T = S_L \ i_T \ (=W_{LT}). \quad \dots \dots \dots \dots \dots \dots \dots \dots \dots \dots \dots \dots \dots (12b)$$

The energy loss by Eq. (12) is larger than Eq. (6), but it does not always lead to a higher temperatures or a higher energy density, as described in section (2). Therefore, it is necessary to decide on an actual case by case basis which way is more advantageous in practical use of this energy loss.

(B) High Energy Density at the Point of an Electrode.

In many cases, the region in the vicinity of the electrode (around the anode or cathode drop) of the arc column is utilized actually as the practical heat source. Typical examples are arc welding, plasma arc welding, arc spraying, and discharge processing in liquid. Usually the electrode drop of arc heat sources is usually ten volts or less and the current density is about $10^2 - 10^4$ (A/cm^2). The energy density is higher than that of an arc column. But the conventional method used for generating high energy density plasma beams has disregarded to make use the electrode spots.

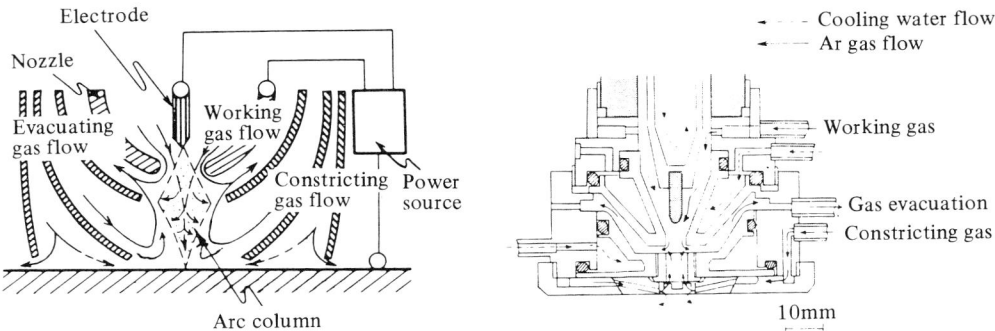

Fig. 5 Principle for generating "point arc".

Fig. 6 Cross sectional view of point arc torch.

The energy density of an anode spot is generally smaller than that of a cathode spot. Using the vortex flow of a gas fluid, the authors were able to generate a high energy density at the anode spot[12]. The method and a cross-section of the apparatus used are shown in Figs. 5 and 6. A high energy density anode spot was obtained, which the author has named an "arc ball,"[13] as shown in Fig. 7. The current density of this arc ball is an extremely high value of 10^5 (A/cm^2) or greater, and the energy density is as high as even over 10^6 (W/cm^2). This state can be constantly and stably maintained at a fixed position. The authours call this arc a "point arc." This arc has a superior character as a heat source for processing. We have obtained using this arc a weld bead cross-section with a similar penetration as the one obtained by non-vacuum EB welding as shown in Fig. 7(C).

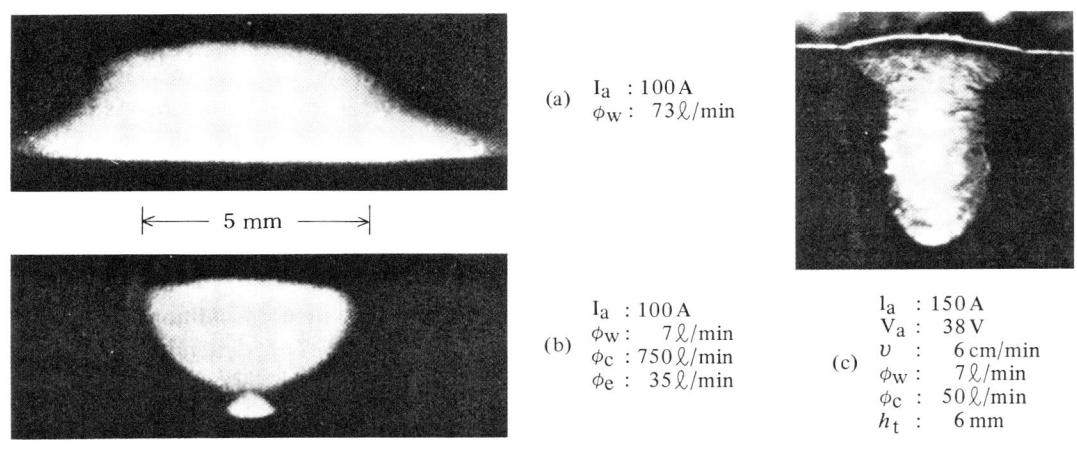

(a) I_a : 100 A
ϕ_W : 73 ℓ/min

(b) I_a : 100 A
ϕ_W : 7 ℓ/min
ϕ_c : 750 ℓ/min
ϕ_e : 35 ℓ/min

(c) l_a : 150 A
V_a : 38 V
v : 6 cm/min
ϕ_W : 7 ℓ/min
ϕ_c : 50 ℓ/min
h_t : 6 mm

Fig. 7 Point arc and its application to welding

(a) (b); Various aspects of arc constricted by gas flow.
I_a : Arc current
ϕ_W : Working gas flow rate
ϕ_c : Constricting gas flow rate
ϕ_e : Evacuating gas flow rate

(c); Bead cross section by point arc welding.
V_a : Arc voltage
v : Welding speed
h_t : Thickness of work
Material of work: Stainless steel SUS 304

Summary

This paper describes in general the thermal characteristics of a plasma beam as well as some properties, which enable the achievement of a high energy density. The author hopes that the results shown in this work will help to open a new area of study in the field of plasma heat sources.

References

1. L. Spitzer: "Physics of Fully Ionized Gases", Interscience Publishers, Inc., NY (1956)
2. M. Okada, Y. Arata: "Plasma engineering", Nikkan Kogyo Shinbunsha (1957) 347
3. H. Gerdien and A. Lotz: Z. f. Techn. Physik **4** (1923) 157
4. Y. Arata and A. Kobayashi: J. High Temperature Society, **11** No. 1 (1985) 18
5. Y. Arata and A. Kobayashi: J. High Temperature Society, **11** No. 3 (1985) 124
6. Y. Arata: Lecture of Phys. Soc. Japan, 9a-M-5 (1969) Oct.
7. Y. Arata and H. Maruo: Kakuyugokenkyu, **22** No. 4 (1973): Tech. Rept. Osaka Univ. **22** No. 1039 (1972) 135: IIW Doc. IV-53-71 (1971): IV-85-72 (1972)
8. Y. Arata, H. Maruo, Y. Hirata and Y. Horio: Trans. JWRI. **3** No. 2 (1974) 127
9. P.L. Kapitza: Sov. Phys. TETP **30** No. 6 (1970) 973
10. Y. Arata, S. Miyake and A. Kobayashi: J. Phys. Soc. Japan **44** (1978) 998
11. Y. Arata, S. Miyake, Y. Ushio and Y. Yoshioka: J. Phys. Soc. Japan **44** (1978) 1483
12. Y. Arata: J. Phys. Soc. Japan **43** (1977) 1107
13. Y. Arata: Private report (1965), (Patent 567605, Japan)
14. Y. Arata and K. Inoue: Trans. JWRI **3** No. 2 (1974) 201

Magnetic Control of Plasma Arc Welding
—On the Pressure Distribution of Plasma and its Effect
on Key-Hole Action—

Abstract

When the cusp type magnetic field is imposed on the plasma arc, its cross section becomes oval. By this magnetic control, feasible range for key-hole welding is expanded markedly. Key-hole is, however, observed to be larger than conventional one. In order to make clear this phenomena, plasma arc pressure and pressure due to surface tension of molten metal are taken into account to determine the equilibrium key-hole diameter.
Calculated value agrees with experimental results.

1. Introduction

In an ordinary plasma arc welding of medium thick plate, key-hole technique is widely used. Key-hole formation is, however, not always stable, but is fairly sensitive to any change of welding conditions such as arc current, working gas flow rate, traveling velocity and so on. At the same time, bead formation in a key-hole process depends on the intrinsic physical properties of material to be welded such as surface tension, viscosity and density of molten metal at elevated temperature.

In general, feasible range of welding operation for thicker plate exceeds 9 mm is comparative narrow to that for medium thick plate of around 6 mm. Burnthrough or lack of penetration will often come arise from small fluctuation of welding parameter.

As described in our previous report,[1, 2] on the other hand, when cusp magnetic field imposed to the plasma arc, feasible range of welding operation could be broadened markedly, and narrow, deep penetrated weld bead was obtained as a result of magnetic deformation of arc plasma. It was also pointed out then to be formed somewhat larger key-hole than that in ordinary plasma welding process.

Presented in this paper are series of analysis on the relation between the pressure of plasma arc at the key-hole region and the diameter of key-hole.

2. Experimental Apparatus and Materials

Plasma arc torch system used in present investigation is composed of ordinary plasma torch, electromagnets to generate a cusp typed magnetic field and auxiliary shield cover. The torch and electromagnets used are almost the same with those in previous report. Their structures are as shown in **Fig. 1** and **Fig. 2**. The pole pieces of magnet were made of soft iron, 1 mm in thickness, and magnetic flux density obtained with field coil current of 3 amp. was about 300 gauss at the site located 5 mm from the nose of pole piece. Field strength varies proportionally as field coil current not exceeds 3 ampare.

Nozzle diameter of plasma torch was 3 mm and arc current was changed over a range from 100A to 200A. Argon was chiefly employed both for plasma working gas and shield gas. Materials used in welding experiment were 304 type stainless steel plate, 6 mm and 9 mm in thickness.

For the pressure measurement of plasma arc on the flat anode plate, a water-cooled copper anode having a pressure tap tube (0.5 mm in diameter) was used. Furthermore, special copper anode having a "key-hole" simulated to the practical one was employed in order to know the actual pressure acting upon the key-hole and neighborhood of them.

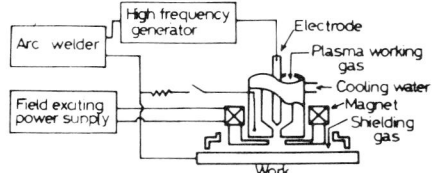

Fig. 1. Schematic diagram of magnetic control of plasma arc.

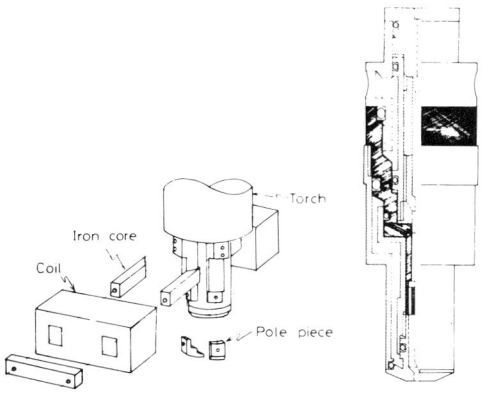

Fig. 2. Electromagnets and plasma arc torch.

3. Experimental Results and Discussion

3.1 Magnetic control of plasma arc with cusp magnetic field

Typical deformation of plasma arc in a cusp type magnetic field is as shown in **Photo. 1.** Each picture of them was taken at the same moment from both direction, longitudinal and transverse. It will be found that the plasma arc column was restricted from both sides and expanded somewhat to the perpendicular direction, and its cross section has deformed to an elliptical as a result. When the working gas flow rate is low, deformation of plasma column can be clearly

recognized, but measurement of current flowing into the anode indicated that its cross section takes an oval shape even at the case with higher working gas flows.[2]

3.2 Feasible range of plasma arc welding

Similarly as described in previous report, all of welding with cusp magnetic field were carried out in key-hole method as well as ordinary one. An example of weld bead made with magnetic field is given in **Photo. 2.** As can be seen in photograph, width of weld bead made by such magnetically controlled plasma arc was fairly narrow comparing with those in conventional method. On the other hand, bottom bead width was observed to be increased rather. It should be noted that undercut free weld bead was obtained even in the case it would be sure to form if no magnetic field was applied.

Consequently, feasible range to give a satisfactory weld has been expanded by an application of magnetic field, as shown in **Fig. 3.**

3.3 Appearances of crater and key-hole

When the cusp magnetic field was applied, weld crater take a distinctive feature and fairly different from those in conventional process. Its front solid wall descends by easier grade and falls into the key-hole. All of molten metal at the front portion were swept away towards rear side of crater in a similar manner as observed in conventional plasma welding,

  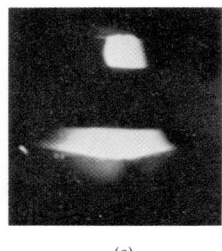

(a)   (b)   (c)

Photo. 1. Appearances of plasma arc, 150 A, 41 V, 13 mm.
(a) I_f; 0 A   (b), (c) I_f; 3 A.

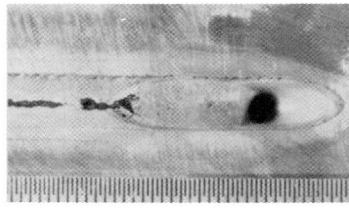 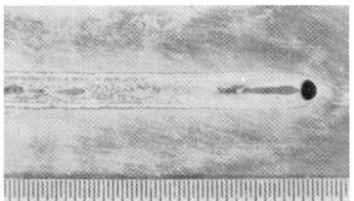

Photo. 2. Weld bead made by magnetic control of plasma, 304 Stainless Steel, 6 mm, 150 A, 35 V, 20 cm/min, I_f 3 A, 4 l/min Argon.

285

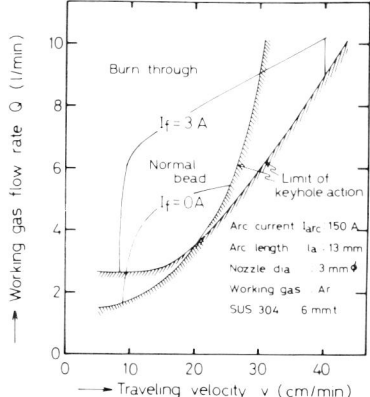

Fig. 3. Feasible range of key-hole welding for both plasma.

but they formed a long tailed oval molten puddle.

Diameter of key-hole was observed to increase certainly whenever magnetic field was applied. Its diameter was determined by taking a photograph of effluent plasma through out the key-hole. **Figure 4** and **5** give the change of key-hole diameter when the

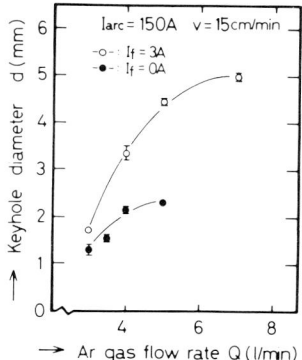

Fig. 4. Effect of working gas flow rate on key-hole diameter.

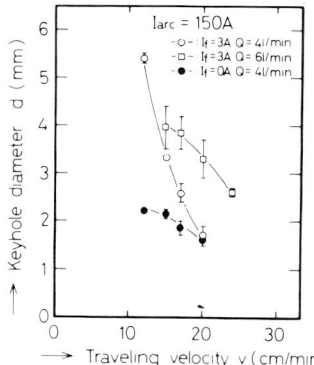

Fig. 5. Effect of welding speed on key-hole diameter.

working gas flow rate or traveling velocity was altered.

Schematic illustration of longitudinal cross section of weld bead is given in **Fig. 6.** Solid-liquid interface could be determined by removing the molten metal. This was carried out by impinging heavy weight hummer upon the work piece so as to rotate it very quickly.

The experiment showed that the heat input per unit length of weld does not so much change the length of crater, λ_0. It, however, prolonged certainly by an application of magnetic field, as shown in **Fig. 7.**

Most closest diameter of "solid" wall around the the key-hole, d_0, was observed to be remained unchanged, while key-hole diameter changed in response to the welding parameter as mentioned above.

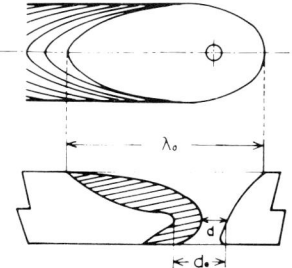

Fig. 6. Longitudinal cross section of molten pool.

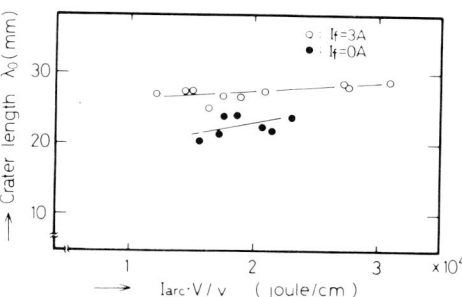

Fig. 7. Relationship between heat input and length of crater.

3.4 Dynamic pressure due to plasma arc

When the fully penetrated key-hole welding is progressing, certain balance should be established at the bottom of weld bead among the forces acting upon the molten metal. They are upward force due to surface tension of liquid metal, downward gravity force, and dynamic force due to plasma stream.

Balance of these forces is, in principle, very delicate, and each force is also influenced by many factors.

286

Figure 8 shows an experimental result obtained in the measurement of pressure due to plasma arc, which was carried out with devices as shown in **Fig. 9.** The effect of magnetic field is seen more clearly when the working gas flow rate is relative low. When the gas flow rate became higher, difference could be hardly seen. Their peak value increased, of course, with working gas flows.

These results are, however, obtained with flat anode, and is insufficient to elucidate the reasons why large diameter of key-hole is formed by an application of magnetic field.

In order to know actual pressure distribution while the key-hole is formed, another experiment was carried out using water cooled copper anode having a simulated key-hole. Cross-section of this anode plate is shown in **Fig. 10.**

In practical welding, axis of torch and center of key-hole does not meet each other, and torch is usually shifting forward some distance. So, measurement was made by changing the distance of torch axis measured relative to that of key-hole, ξ, in steps. Diameter of key-hole was also changed as 0, 1, 2, 3 and 5 mmϕ.

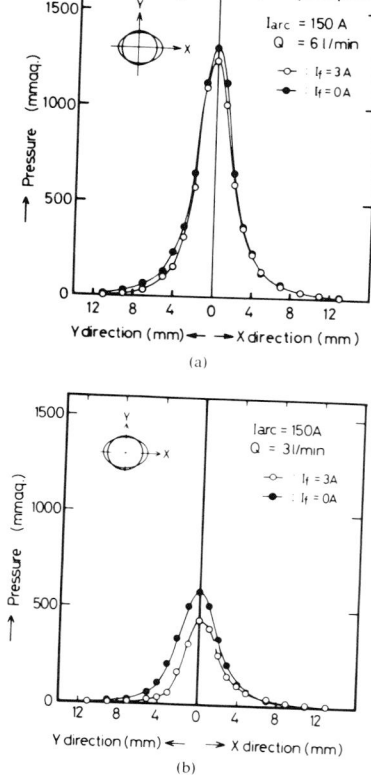

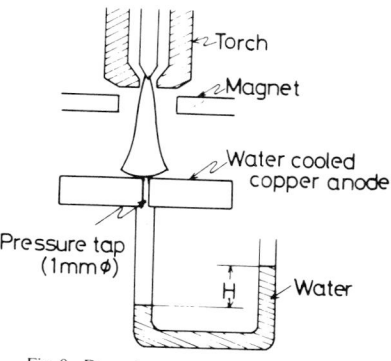

Fig. 8. Pressure distribution on the flat anode.

Fig. 9. Dynamic pressure measurement devices.

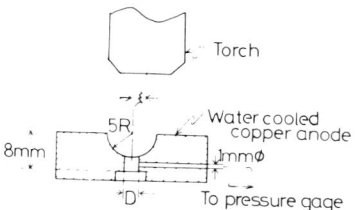

Fig. 10. Water cooled copper anode for the measurement of static pressure at the key-hole.

In **Fig. 11** is given the results obtained for the case that key-hole is not formed. Stagnation pressure at the center of crater is seemed to decrease by an application of magnetic field. As can be seen in lower range of working gas flows in Fig. 3, key-hole is still formed if no magnetic field is applied. For this reason, it will be sufficient to consider only stagnation pressure.

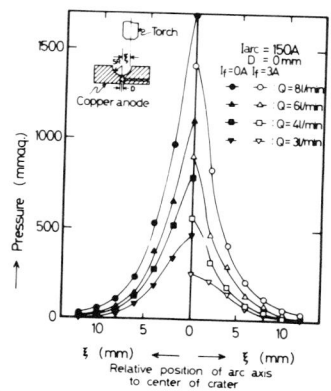

Fig. 11. Relationship between positioning of arc and total pressure at the bottom of crater.

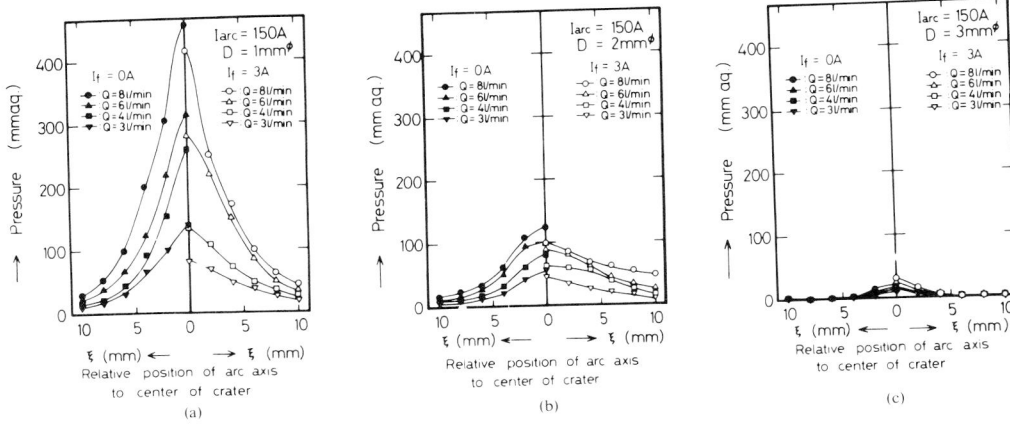

Fig. 12. Effect of arc positioning and key-hole diameter on static pressure at the key-hole.

Static pressure at the midpoint of key-hole was measured for various diameter of key-hole, and effect of lag distance (ξ) on static pressure was investigated. Experimental results are summarized as **Fig. 12**. Static pressure increased with working gas flow rate, but decreased rapidly as a diameter became larger. In present experiment, static pressure was no longer appreciable when diameter of key-hole exceed 3 mm.

As obvious from these results, lag distance, ξ, has much effects on the static pressure at the key-hole region. When the lag distance is not so large, static pressure inside of key-hole produced by ordinary plasma arc is higher than those of magnetized plasma arc, but this correlation is the reverse for the larger value of ξ as shown in **Fig. 13**. This characteristics plays an important role on occasion to determine the key-hole diameter, as described later.

In order to evaluate the actual pressure act on molten pool, similar measurements were carried out.

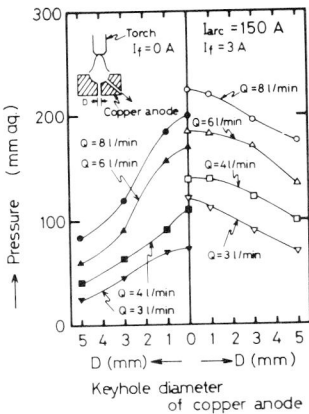

Fig. 14. Dynamic pressure act on molten pool.

Its result is represented in **Fig. 14**. In this experiment, plasma torch was set just above the key-hole. It was found that pressure act on molten pool was enhanced by an application of magnetic field, and this trend was confirmed at any key-hole diameter. These strong pressure would be produced by the change of plasma stream as a result of magnetic constriction.

Relationship between key-hole diameter and static pressure inside key-hole is as shown in **Fig. 15**.

3.5 Pressure due to surface tension of molten metal

As described in former section, it is presumed that molten metal is sustained by the balance of forces, which are dynamic force of plasma stream, gravity head and pressure difference at the surface of molten metal due to surface tension.

If key-hole is assumed to be hollow cylinder

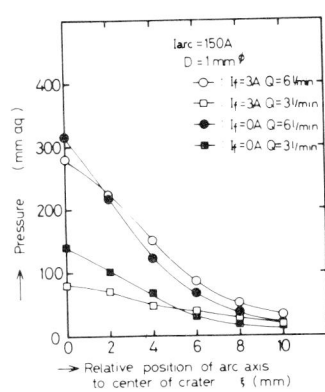

Fig. 13. Relation between static pressure and arc position.

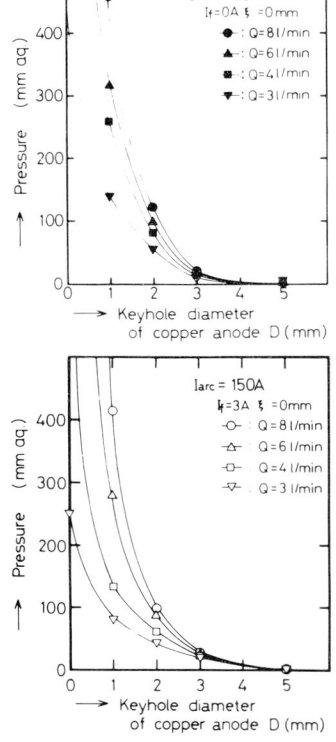

Fig. 15. Effect of key-hole diameter of copper anode on the static pressure of key-hole plasma.

having a diameter of d, difference of internal pressure ΔP is given after Laplace equation,

$$\Delta P = \gamma \left(\frac{1}{R_1} + \frac{1}{R_2} \right) \tag{1}$$

ΔP : Difference of internal pressure

γ : Surface tension

R_1, R_2 : Radii of curvature at the point referred, which are perpendicular each other

Substituting $R_1 = d/2$ and $R_2 = \infty$ to above equation (1), following relation is obtained.

$$\Delta P = \frac{2\gamma}{d} \tag{1'}$$

From equation (1'), it is easily deduced that if the diameter of key-hole, d, becomes larger, the static pressure of plasma stream becomes inversely small, which is necessary to balance with ΔP. Actual key-hole is not, of course, hollow cylinder, but likely to be such shape as in Fig. 6. Then, it may possible to consider the model as shown in **Fig. 16**, and we

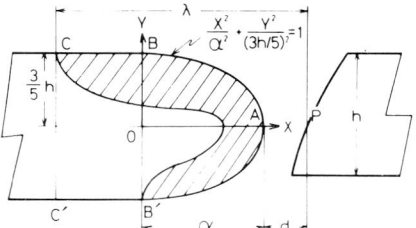

Fig. 16. Molten pool model for calculation.

assume that;

1) there is no molten metal at the front area of crater

2) key-hole is always being circle, and its diameter, d

3) envelope of longitudinal cross section of molten metal at the rear part of key-hole is elliptical curve

In Fig. 16, X axis was set at 3h/5, considering real shape of cross-section and bottom bead, too. Surface of molten metal between the lines B–B' and C–C' is assumed to be flat.

If length of molten pool, λ, is given, α can be calculated so as to not change the area of CBAB'C', since volume of molten metal should be ever constant.

If λ is invariant, we can obtain a following equation,

$$\frac{6}{5} h \cdot (\lambda - d - \alpha) + \frac{\pi}{2} \cdot \frac{3}{5} h \cdot \alpha = \frac{\pi}{2} \cdot \frac{3}{5} h \cdot \lambda$$

$$\alpha = \lambda - \frac{4}{4 - \pi} d \geq 0 \tag{2}$$

From above equation, we can be obtain a following inequality.

$$0 \leq d \leq \frac{4 - \pi}{4} \lambda$$

This equation suggest that key-hole diameter, d, has a maximum value with respect to the value of λ, which is corresponding to $\Delta P = 2\gamma/d$ in Eq. (1'). If d is beyond this maximum value, it is no longer easy to maintain the molten metal as it was, · and burn through will be occured. This result does not contradict with previous statement, that is, d will take a larger value as λ becomes larger.

Now at a point A in Fig. 16, if its curvatures $1/R_1$, $1/R_2$ are known, internal pressure difference required to maintain the key-hole at a diameter of d can be determined using equation (1).

In general, curvature $1/r$ of the curve, which is expressed $x = f(t)$ and $y = g(t)$ in x-y coordinate with parameter t, can be shown as following equation.

$$\frac{1}{r} = \frac{\dfrac{df}{dt} \cdot \dfrac{d^2g}{dt^2} - \dfrac{dg}{dt} \cdot \dfrac{d^2f}{dt^2}}{\left\{\left(\dfrac{df}{dt}\right)^2 + \left(\dfrac{dg}{dt}\right)^2\right\}^{\frac{3}{2}}} \quad (3)$$

Then, for the key-hole in Fig. 16, curvatures $1/R_1$, $1/R_2$ can be expressed as follows;

$$1/R_1 = d/2, \quad 1/R_2 = 25\alpha/9h^2$$

Using equation (1) and (2), following equation is led.

$$\Delta P = \frac{\gamma}{9(4-\pi)h^2} \cdot \frac{100d^2 - 25(4-\pi)\lambda d + 18(4-\pi)h^2}{d} \quad (4)$$

In **Fig. 17** represented the results calculated for various parameter λ, plate thickness h=6mm, surface tension of molten metal γ=1500 dyne/cm.[3] From this figure, it can be seen that the static pressure necessary to maintain the key-hole of d may be enough low, if length of molten pool λ is so long.

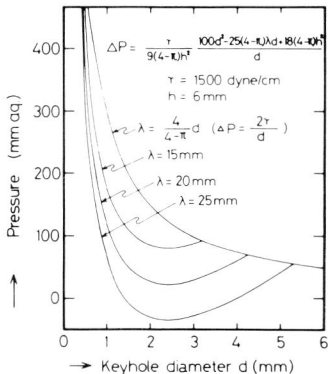

Fig. 17. Pressure due to molten metal calculated.

3.6 Equilibrium diameter of key-hole

Static pressure of plasma stream at the key-hole is already obtained experimentally. Pressure required to maintain the key-hole could be obtained. Then, it becomes possible to consider the equilibrium diameter of key-hole.

For example, if the key-hole is presumed to be a hollow cylinder with a diameter of d, we can find a intersecting point on both curves, as shown in **Fig. 18**. Then it may be possible to consider that the key-hole will be formed at this pressure and with this diameter. If key-hole becomes larger than d*, it will be soon restricted by sorrounding molten metal wall, since

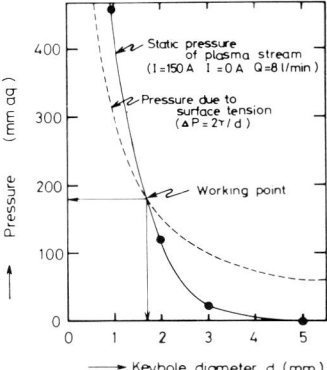

Fig. 18. Pressure due to molten metal calculated.

pressure due to the surface tension exceeds the static pressure of plasma stream. The reverse is also true.

In practical welding, plasma arc torch was in advance of key-hole, and their actual length was measured for both plasma arc processes with and without magnetic control. As can be seen in **Fig. 19,** distance of arc measured relative to the center of key-hole increased almost lineary to the traveling velocity, and usually the arc axis of controled plasma arc located closer to key-hole axis.

The effect of welding parameter on this equilibrium diameter of key-hole is also possible to evaluate.

When the working gas flow rate is altered, key-hole will change its diameter as indicates in **Fig. 20.** From this result, it may be suggested that the key-hole in controlled plasma arc welding would be larger than that in conventional one.

An example of the effect of traveling speed on diameter of key-hole is shown in **Fig. 21.** As the traveling speed becomes fast, key-hole diameter

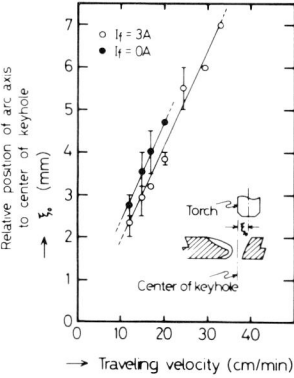

Fig. 19. Position of plasma torch relative to key-hole at various traveling speed.

290

decreases gradually, but it does not take a continuous value to the point of d=0.

In practical welding, it was observed frequently that the key-hole disappear abruptly after diminishing its diameter to certain value.

Comparison of experimental results and calculated value of key-hole diameter is as shown in **Fig. 22.** Both diameter agree well within a experimental error.

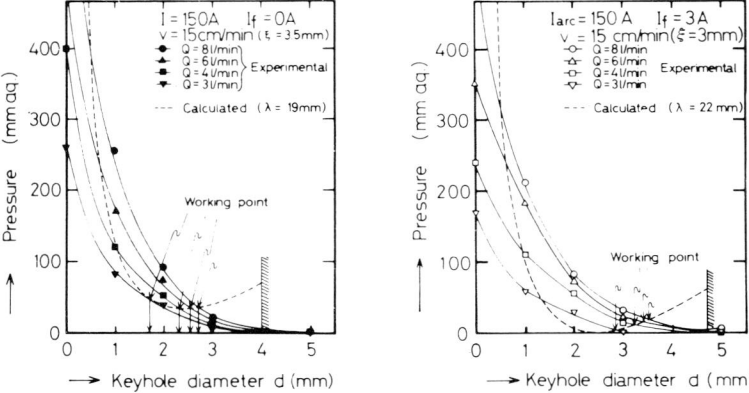

Fig. 20. Determination of key-hole diameter at various working gas flow rate.

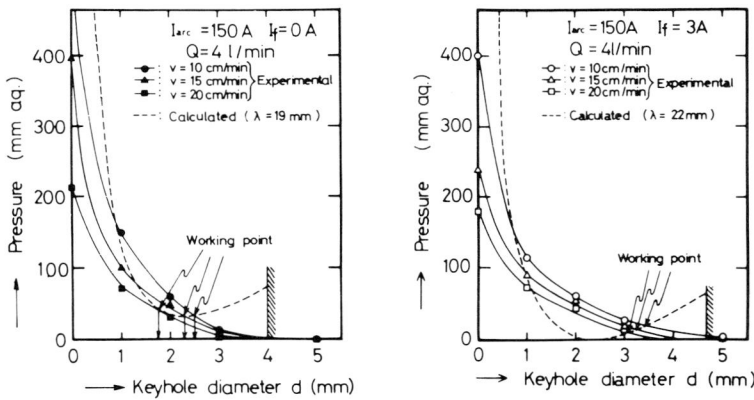

Fig. 21. Determination of key-hole diameter at various traveling speed.

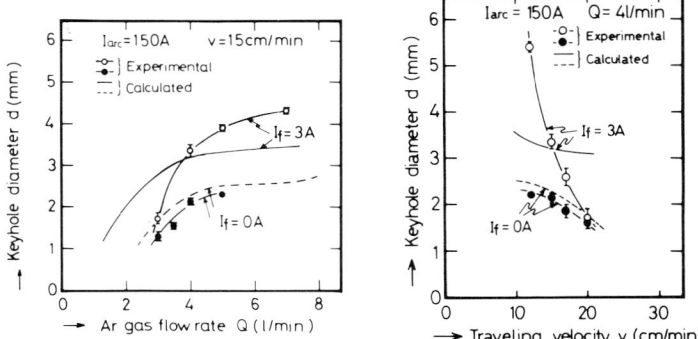

Fig. 22. Comparison of key-hole diameter calculated with experimental value.

4. Conclusion

The results obtained in this series of experiments are summarized as follows;

(1) When the cusp typed magnetic field is imposed on the plasma arc, plasma arc column changes its cross section from circular to oval.

It is possible to weld a stainless steel and others in a key-hole technique with this magnetically controlled plasma arc.

(2) By the magnetic control, feasible range to give a fully penetrated satisfactory weld bead was expanded towards both higher working gas flow rate region and higher traveling velocity side.

(3) No remarkable difference was observed in the total pressure distribution on the flat plane anode, between the both plasma arc processes with or without magnetic field.

(4) Pressure distribution for the case that key-hole was formed was investigated. Static pressure at the key-hole is much influenced by the magnetic field, key-hole diameter and torch positioning relative to the key-hole.

(5) At the key-hole, inward pressure due to surface tension of liquid metal was considered as one of the force to maintain a key-hole.

(6) Equilibrium key-hole diameter was calculated by assuming that pressure due to plasma stream balanced with that due to the surface tension of molten metal.

(7) Effects of working gas flow rate and traveling speed on the diameter of key-hole formed was evaluated. These results agreed well with experimentals.

References

1) Y. Arata and H. Maruo; Magnetic control of arc plasma and its application to welding, Welding in the World, vol. 10 no. 7/8, 1972, " (2nd report) IIW Doc. IV-85-72, 1972.

2) Y. Arata, H. Maruo and K. Yasuda; Some properties of magnetically controlled plasma arc, Transaction JWRI, vol. 2 No. 1, 1973.

3) Franz Cech; A memorandum on surface tension, Preliminary report 71—11.

Concept of Vortex Gas Tunnel and Application to High Temperature Plasma Production

To obtain a high temperature plasma insulated by gas of atmospheric pressure, a newly developed gas flow system is proposed which induces an extremely larger pressure difference in the radial direction (over several hundreds Torr) compared with the conventional one (several Torrs). This concept of "gas tunnel" in strong vortex gas flow has been verified by the experiment. Several points of excellence on the production of a high temperature plasma in the "gas tunnel" are emphasized.

Recently plasma-wall interaction has been closed up as a very troublesome item for the nuclear fusion research. To obviate such difficulty as much as possible, concepts of gas insulation[1] and gas blanket[2] have been proposed. But their insulation effects have not been fully proved experimentally.[3-6] In this letter are described a newly developed gas flow system and its application to the production of a high temperature plasma.

The key point of the new method is to keep the vicinity of the central axis of vortex gas flow in as much a vacuum condition as possible, resulting in a state of tunnel. For this purpose let us examine the performance of a strong vortex gas generator G_V whose ends are open to the atmosphere as shown in Fig. 1 by using air flow. The air is jetted tangentially from supersonic nozzles along the inner wall of the cylinder over 1 atmospheric pressure. The air accordingly forms a vortex flow resulting in a minimum pressure on the C–C' axis, as is well known, but an interesting point is that a "cylindrical inversion layer" of r_R

radius is formed in the flow vector along the axial direction as is shown in the figure. Within this layer, air from the outer atmosphere flows into violently along its axis while on the exterior of this layer a vortex flow is seen to go outside smoothly. When a disc of r_R radius is placed at both ends ($\overline{ACB}$ and $\overline{A'C'B'}$) of G_V, the ingress of open air is prevented and a pressure on the C–C' axis is remarkably reduced. Furthermore, the pumping of air within this inversion layer, instead of its closing with the disc, would make it possible to form a more effective vacuum condition. In order to materialize this understanding, the author has introduced a concept of "gas divertor".

Figure 2 shows a schematic diagram of the gas divertor D_G which has two kinds of exhaust system. One is D_B directly connected to the suction side of a blower and the other D_P is connected to a vacuum pump. The vortex flow gas, which acts as a gas wall, is exhausted outside of the system through D_B, while the gas within the inversion layer is pumped out at D_P.

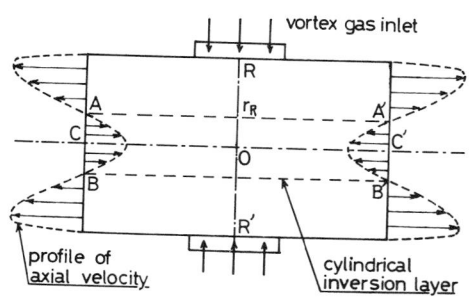

Fig. 1. Schematic diagram of vortex gas generator with radial profiles of axial velocity at both ends.

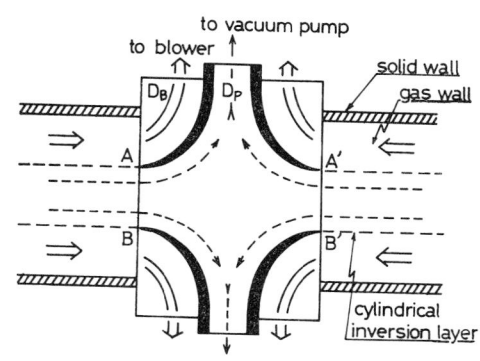

Fig. 2. Schematic diagram of gas divertor, D_G.

293

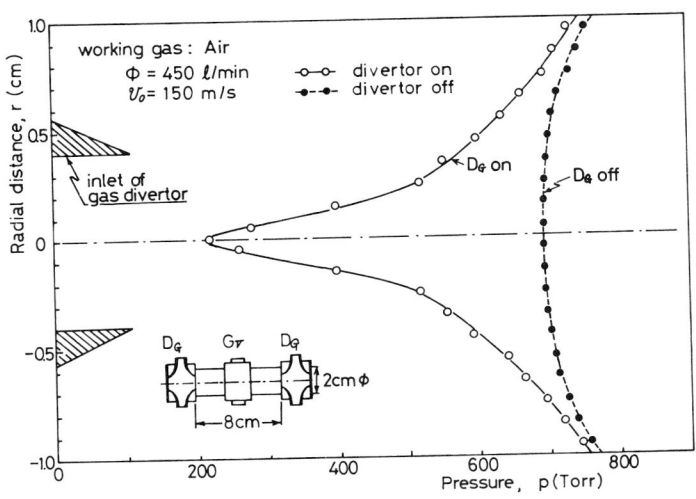

Fig. 3.   Radial static pressure distribution obtained in a vortex gas generator.

Using such a method, a pressure within the cylindrical inversion layer is easily reduced from 1 atmosphere to a vacuum condition near 100 Torr or below. While a pressure on the inner surface of the solid wall is still found at a level over an atmospheric pressure. This condition is considered to be a kind of tunnel state, and the author, thinks that it can be called "gas tunnel" or "vortex gas tunnel".

Figure 3 shows one of the results obtained from preliminary experiment carried out under the above understanding. The air of 450 l/min is jetted at a constant speed ($v_0 = 150$ m/sec; smaller than the sound speed!) along the inner wall of G_V (3 cm in length, 2 cm in dia), whose ends are connected with quartz tube and D_G as shown in the figure. In case of no operation of D_G a pressure distribution shown by full circles and the inversion layer of 4–5 mm in radius were obtained. By operating D_G with its entrance radius of 4 mm efficient gas wall ($p_{00} \simeq$ 760 Torr) and gas tunnel ($p_0 \simeq 200$ Torr) were formed with a steep pressure gradient in the radial direction as shown by white circles. Moreover $p_0 = 160$ Torr was obtained by increasing p_{00} to 1000 Torr. It is easy to get $p_0 < 100$ Torr with $p_{00} > 1000$ Torr by improv-

ing experimental conditions.

For the high temperature plasma generated in this gas tunnel, the plasma-wall interaction is attributed to the plasma-gas wall interaction and will be obviated with high reliability by gas insulation effect. Moreover it can intensify remarkably the high temperature plasma stability, since the pressure gradient is by far the larger, as compared with the case of a conventional vortex flow. It was preliminary proved by experiment that a plasma beam in the gas tunnel has higher temperature, and is more stable compared with the one in a conventional vortex flow and/or still gas discharge conditions, in which all researchers have performed experimental studies until now.

References

1)   H. Alfvén and E. Smars: Nature **188** (1960) 801.
2)   B. Lehnert: Nuclear Fusion **8** (1968) 173.
3)   P. L. Kapitza: Soviet Physics-JETP **31** (1970) 199.
4)   Y. Arata, S. Miyake, A. Kobayashi and S. Takeuchi: J. Phys. Soc. Japan **40** (1976) 1456.
5)   O. Klüber: Naturforsch. **22a** (1967) 1599.
6)   F. C. Schüller, for the RINGBOOG team: Report on the International Symposium on Plasma-Wall Interaction, Jülich, F.R.G. 18–22, Oct. '76, E$_2$E–8.

High Power Microwave Discharge
in Atmospheric Hydrogen Gas Flow

A microwave plasma in helical flow of atmospheric hydrogen gas is experimentally investigated. Because of a very small diameter of the plasma beam (about 1 mm) the ambipolar diffusion effect induces the deviation in the temperatures of electrons and heavy particles ($T_e = 12000$ K, $T_g = 9000$ K at $P_i = 25$ kW, $\phi = 800$ l/min), and underpopulations of excited states with the electron deficiency are clarified.

§1. Introduction

In a previous paper[1] properties of atmospheric nitrogen plasma in a high power microwave discharge were experimentally investigated. Concerning the discussions on the so called Kapitza plasma,[2] however, one should study the hydrogen and/or helium plasma, because the extremely high temperature state of 10^6 K is insisted in a filamentary discharge of such a light element.

In Kapitza's paper it was demonstrated that the plasma core of 10^6 K was guaranteed by the anomalous skin effect and the formation of electric double layer. But Dymshitz[3] threw doubts on the existence of such a high temperature by reanalyzing his spectroscopic data taking into account of recombination radiation. Batenin et al.[4] studied spectroscopically a microwave plasma in hydrogen gas with a coaxial type plasmatron[5] and obtained a weakly ionized low temperature plasma below 10^4 K. To compare with Kapitza's result, however, their experiment has a disadvantage that the plasma is very short in length and fixed to the water-cooled electrode. In such a configuration strong cooling of the plasma and destructuion of electric double layer might occur at the electrode region.

In this paper an electrodeless hydrogen plasma produced at atmospheric pressure by a 30 kW-class CW microwave is studied and some comparisons with Kapitza plasma are described.

§2. Experimental Precedures

Figure 1 shows the schematic diagram of experimental apparatus. Configuration of plasma generator (1) is similar to the one in the previous paper[1] and has a rectangular waveguide section. A closed circuit of the gas flow is produced by compressor (2), heat exchanger (3), flow meter (4) and rotary pump (5). Several gas species can be used (H_2, He, Ar, N_2 etc.) with the flow rate below 1600 l/min. The monochrometer (Shimazu GE-100, 1st order dispersion 8.3 Å/mm) used for optical measurement has the calibrated instrumental half width of 0.2 Å with the exit slit of 25μ. The standard lamp (EPLEY, Type-EP) was used for the sensitivity calibration of the optical system.

As for the discharge tube dimension the quartz pipe of 20 mm in diameter was used to obtain a higher plasma temperature than the case of 40 mm[1] due to thermal pinching effect.

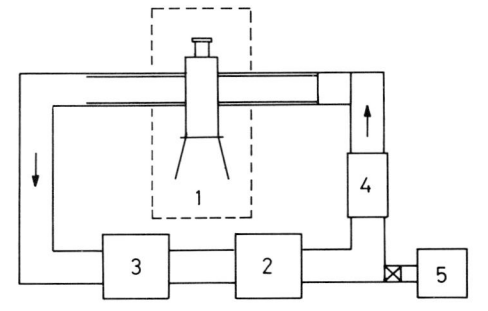

Fig. 1. Schematic diagram of experimental apparatus.

§3. Experimental Results

In the study of nitrogen plasma[1] it was confirmed that the absorbed power P_a to the plasma in this type of plasma generator was quite high and reached to about 80% of the incident one P_i. The reflected power was below several percents, so that there are unknown losses. This character does not change even when the plasma production is made in hydrogen or helium gas. While the plasma diameter D is lowered with the increase in hydrogen content to a value of about 1 mm as is shown in Fig. 2. The plasma length L shows a little decrease (below 10%) by the hydrogen content and is about 20 cm at $P_i = 25$ kW. While electron density N_e and plasma temperature T increase from 6×10^{14} cm^{-3}, 7.6×10^3 K to 5×10^{15} cm^{-3} and 8.8×10^3 K respectively in the change of hydrogen mixing rate γ from 20 to 100%. In pure hydrogen gas dependence of N_e and T on P_i is given in Fig. 3. At $P_i = 25$ kW, that is $P_a = 20$ kW, T is about 9×10^3 K and the plasma is still in a weakly ionized state. It should be noticed here that D keeps a constant value of about 1 mm. As for the influence of the change in the gas flow rate between $500 \sim 1600$ l/min a higher flow rate gives a little increase in T in place of the decrease in D. The plasma parameters D, N_e, T in Figs. 2–3 were estimated as follows; D from a half width of the ob-

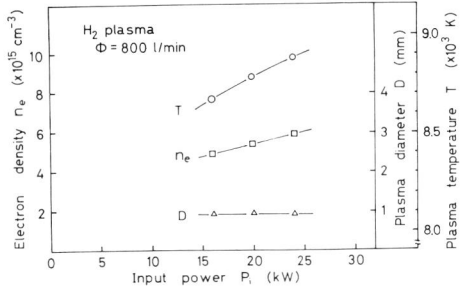

Fig. 3. Dependence of plasma parameters on input power at $\phi = 800$ l/min.

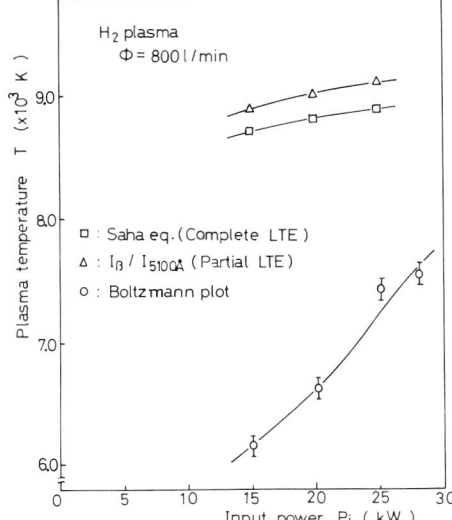

Fig. 4. Dependence of plasma temperature on input power decided from various methods at $\phi = 800$ l/min.

served intensity distribution of H_β line across the plasma column, N_e from its Stark broadening and T by using Saha equation with the assumption of complete LTE.

Figure 4 shows the plasma temperature T obtained with various methods. The line with squares is T from Saha equation and the one with triangles from the intensity ratio of H_β line to the continuum at $\lambda = 5100$ Å (wavelength interval of 80 Å) on the partial LTE assumption. Both gives an approximately equal temperature for various P_i. But the temperature from the Boltzmann plot of the intensity of Balmer lines (H_α, H_β, H_γ, H_δ) results in a very low value. In Fig. 5 the absolute intensities of the recombination and bremsstrahlung continua on the axis $(r = 0)$ behind the tail of Balmer series are shown

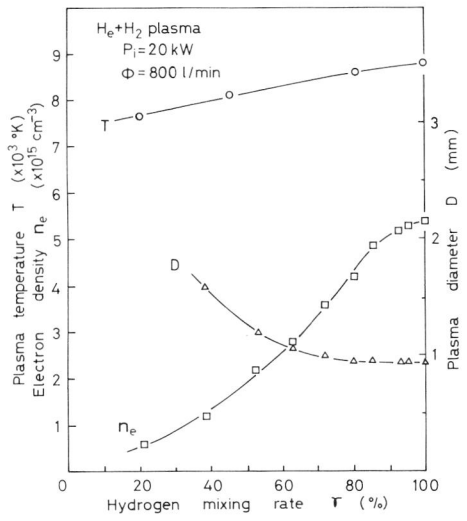

Fig. 2. Change of plasma parameters on hydrogen mixing rate at $P_i = 20$ kW, $\phi = 800$ l/min.

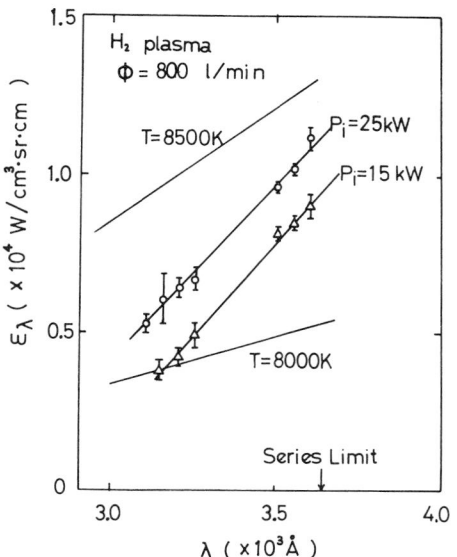

Fig. 5. Absolute intensities of continuous spectra behind the tail of Balmer series for two input powers at $\phi = 800$ l/min.

indicating the plasma temperature of about $8.5 \sim 8.0 \times 10^3$ K. Lines with $T = 8500$ and 8000 K are the calculated ones by Roberts and Voigt[6] on the assumption of complete LTE. True radial temperature distribution obtained by Abel inversion of the emission intensities can be seen in Fig. 6.

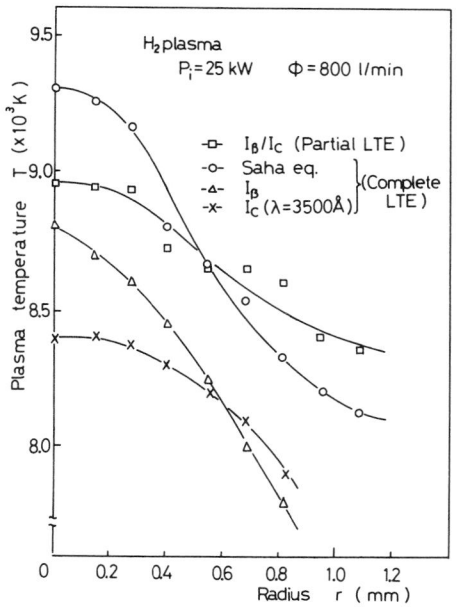

Fig. 6. Radial temperature distributions obtained by various methods at $P_i = 25$ kW, $\phi = 800$ l/min.

§4. Discussion

As for the accuracy of the electron density estimation Stark broadening of H_β and H_γ lines were compared and both gave 6.0×10^{15} cm^{-3} at $P_i = 25$ kW, $\phi = 800$ l/min. This density satisfies the criterion[7] of partial LTE of an optically thin and homogeneous steady state hydrogen plasma for the principal quantum states of $n \geq 2$. So that the population densities emitting Balmer lines may be expected to follow Boltzmann distribution and an equilibrium would be established between free electrons and excited atoms over $n \geq 2$. But in Figs. 2–3 a very low temperature is obtained from the data of Boltzmann plot of Balmer series. Some nonequilibrium effect such as strong nonuniformity of the plasma should be taken into account to answer this discrepancy, since the above criterion is for the homogeneous case. Drawin and Emard[8] have shown that the usual partial LTE condition is pushed to so high quantum levels by the diffusion dominated nonuniformity effect. The characteristic dimension[9] of the plasma in which the diffusion effect is dominant is $L_d = \sqrt{D_a/\alpha N_e}$, where D_a is the ambipolar diffusion coefficient and α is the recombination coefficient. Taking the data of D_a from Devoto[10] and α from Drawin,[11] we obtain $L_d \simeq 0.5$ mm at $T = 9000$ K, $N_e = 6 \times 10^{15}$ cm^{-3} with $D_a = 160$ cm^2/sec, $\alpha \simeq 1.5 \times 10^{-11}$ cm^3/sec. Indeed the radial plasma dimension is 0.5 mm and comparable with L_d. Then the strong diffusion effect might occur in our hydrogen plasma, which led to the uncorrect estimation of the exciting temperature from the Boltzmann plot as a result of electron deficiency.[12]

Moreover that effect will bring about the temperature difference of electron and heavy particles. Biberman et al.[13] have shown a method of the determination of electron density N_e and its temperature T_e by measuring the populations of three excited states in a nonequilibrium plasma. Applying this method to H_α, H_β and H_δ lines we obtained $T_e = 12000$ K at $P_i = 25$ kW, $\phi = 800$ l/min and N_e was calculated to be 6.1×10^{15} cm^{-3} which well agreed with the experimental result in Fig. 3. Figure 7 is the population densities experimentally obtained from the intensities of Balmer series (H_α, H_β, H_γ, H_δ). In the figure

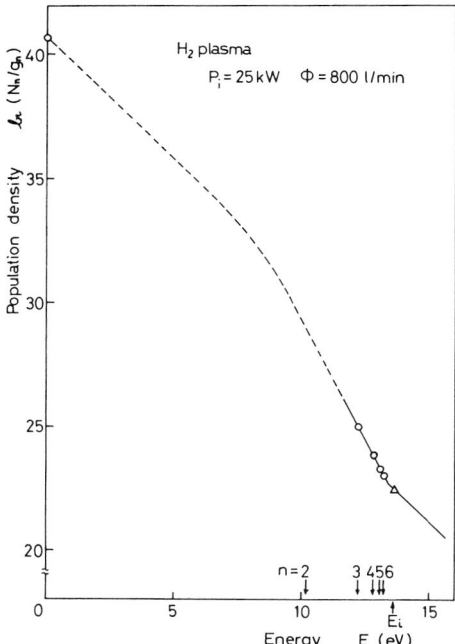

Fig. 7. Population densities n_i/g_i of atomic hydrogen over excited levels at $P_i=25$ kW, $\phi=800$ l/min.

free electrons are characterized[13] by the population $(N_e^2/2U_i)(h^2/2\pi m_e kT_e)^{3/2}$ and is given at $E=E_i$ with a triangle; where $U_i=$ partition function and equal to 1, $h=$ Plank constant, $m_e=$ electron mass, $k=$ Boltzmann constant and $E_i=$ ionization energy of atomic hydrogen. The underpopulations of excited atoms emitting Balmer lines and electron deficiency are clear and this result satisfactorily explains the lower temperatures obtained from the measurement of the absolute intensity of H_β line or the continuum at $\lambda=3500$ Å on the assumption of complete LTE (Figs. 5 and 6).

The observed difference in T_e and T is also connected with the heating power or the microwave electric field strength. When the wave field E_{rf} with the frequency ω is larger than the plasma field $E_p(=4.2\times10^{-10}$ $\sqrt{\delta_{eff}T(\omega^2+\nu_{eff}^2)}$ V/cm),[14] T_e becomes larger than T, where $\delta_{eff}=$ the fraction of electron energy transferred to heavy particle by collisions and ν_{eff} is the electron collision frequency. By using the data for δ_{eff} and ν_{eff} from Ginzburg[14] we obtain $E_p=24$ V/cm at $T=9000$ K. While E_{rf} is estimated from the relation $P_a'=\sigma\bar{E}_{rf}^2$, where P_a' is the wave power

absorbed to a unit volume of the plasma and σ is the electrical conductivity and $\bar{E}_{rf}$ is the wave field averaged over plasma volume. By using the values of $P_a=0.8P_i$ and the plasma dimension ($=\pi D^2L/4$), we obtain $E_{rf}\simeq90$ V/cm at $P_i=25$ kW and $E_{rf}/E_p\simeq0.38$, which results in $T_e\simeq1.14$ $T=10300$ K for $T=9000$ K. The wave field can have a remarkable contribution to the electron overheating, but it is weaker than the diffusion effect. Other factor which induces temperature nonequilibrium is the energy exchange time[9] τ_{ex} of electrons to heavy particles. At $T=9000$ K, $\tau_{ex}=3/4$ $\delta_{eff}\nu_{eff}$ and equal to 1.2×10^{-10} sec. While the time of the field repetition is $2\pi/\omega=1.1\times10^{-9}$ sec, so that electrons have enough time to exchange energy with heavy particles in one cycle of the wave.

From these results we can conclude that a very small value of the plasma diameter, that is, the strong nonuniformity of the plasma induces the difference in T_e and T accompanying underpopulations of excited states and electron deficiency. The electron temperature of 12000 K is the averaged value over plasma diameter so that if we want to know the radial distributions of T_e and $T=T_g$ we must solve the rate equations for hydrogen atoms together with the energy equation for electrons. Such a problem is studied for example in an arc plasma[15] and our next step is the numerical study of our hydrogen plasma based on the two temperature model.

§5. Comments on Kapitza Plasma and Comparison with Our Result

Let's first consider the differences of experimental conditions between Kapitza plasma (case-I) and ours (case-II).

The wave frequency ω in case-I is 10^{10} sec^{-1} and in case-II 5.7×10^9 sec^{-1}. It will give little difference in both plasma parameters, since the skin depth is proportional to $\omega^{-1/2}$. While the radius of the cold wall R is 10 cm in case-I and 1 cm in case-II. This difference comes mainly from that of the operated wave mode. Case-I is operated at TM$_{01}$ mode in a cylindrical cavity resonator with a high Q-value. While case-II has TE$_{01}$ mode in a rectangular cavity with a small Q, by which we have taken in this experiment away the coupling window and the tuning

298

plunger. The latter makes a much higher efficiency of the power absorption (about 80%) than the former (about 30%[16]). To increase the absorbed power over 20 kW it was necessary[17] in case-I to alter the cavity dimension (also the wave frequency) due to the change in the cavity mode, while in our case we have only to increase the output of the magnetron over 25 kW.

Figure 8 shows the radial distributions of the static pressure difference ΔP_s from 1 atm, axial gas velocity V_z and azimuthal velocity V_θ in helical hydrogen flow of 800 l/min without plasma. They were obtained from the data of Pitot tube of 0.6 mm in diameter with a lateral hole having a diameter of 0.3 mm. The gas velocities V_z and V_θ have a large value on the outer part of the plasma column, forming a static pressure gradient of about 1.5×10^2 dyn/cm³. These conditions of the gas flow and the small tube radius ($R=1$ cm) made a slender and long filamentary plasma of $D \leq 1$ mm, from which the nonequilibrium state by the ambipolar diffusion took place as was stated in §4. In case-I the flow condi-

tion is not shown clearly at all. But the plasma diameter D increases with the input power and changes from 2 to 6 mm for $5 \leq P_a \leq 15$ kW at $p=1$ atm.[1] In the large installation $4 \leq D' \leq 26$ mm for $15 \leq P_a \leq 40$ kW, where D' is defined as the distance where N_e decreases to 0.1 of the axial one.[17] So that in Kapitza plasma the nonuniformity effect leading to the non-LTE state will be weaker compared with our situation.

Furthermore it is described in case-I that the plasma density decrease with the input power accompanying the enlargement of the plasma volume. This character can be explained on the assumption of LTE by solving the energy equation (Elenbaas=Heller equation or eq. (1.3) in Meierovich's paper[18]). It is principally a nonlinear differential equation, so that we have numerically calculated it for a hydrogen plasma in a microwave field of $\omega=10^{10}$ sec⁻¹ or 5.7×10^9 sec⁻¹. Figure 9 shows the calculated dependence of the plasma temperature T and the plasma radius r_p on the microwave power per unit length Q_0 for various cold wall radii R. Meierovich has treated the problem analytically and obtained a similar curve in case of Kapitza plasma, where Q_0 ranged around 1 kW/cm with $R=10$ cm. In the figure the plasma radius r_p is defined as the radial distance at which N_e decreases to a half value of the central one. It is clear from this figure that for the cases of $R \geq 1.5$ cm T decreases with Q_0 around 1 kW/cm followed by the increase in r_p. The decrease in T below 10^4 K surely results in that of N_e, which coincides with Kapitza's result. This region corresponds,

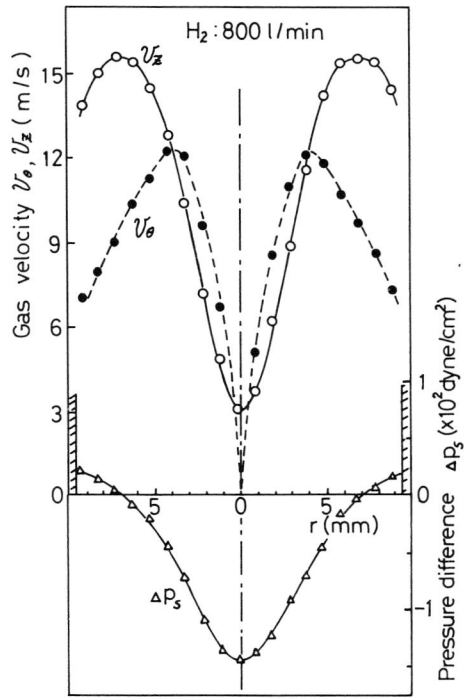

Fig. 8. Radial distributions of the gas velocities V_z and V_z and the static pressure difference ΔP_s from 1atm in case of no plasma at $\phi=800$ l/min.

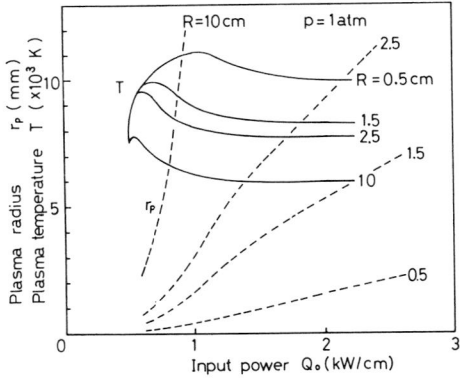

Fig. 9. Theoretical dependence of plasma temperature and its radius on input power per unit length for various wall radii on the assumption of LTE.

as is described by Meierovich, to the one where the microwave energy is spent only to increase the plasma volume due to the strong skin effect on its surface. The experimental N_e in case-I approximately corresponds to the above temperature with similar values of r_p.

§6. Conclusion

By using 30 kW-class CW microwave energy a hydrogen plasma was studied in the helical gas flow of $500 \sim 1600$ l/min. Typical plasma parameter was $T = 9000$ K, $N_e = 6 \times 10^{15}$ cm^{-3} for $P_i = 25$ kW and because of the small tube diameter and a high gas flow rate a non-equilibrium state was induced with the formation of a very slinder plasma beam of about 1 mm in diameter and 20 cm in length. The ambipolar diffusion effect caused the under-populations of excited atoms and the electron deficiency. From the temperature determination by measuring the populations of three excited states $T_e = 12000$ K was obtained at $P_i = 25$ kW.

Some comparisons of our results with the so called Kapitza plasma were performed.

The authors would like to express their appriciation to Dr. Fujiie on the arrangement of spectroscopic measurement.

This work was supported in part by Japan Securities Scholarship Foundation.

References

1) Y. Arata, S. Miyake, A. Kobayashi and S. Takeuchi: J. Phys. Soc. Japan **40** (1976) 1456.
2) P. L. Kapitza: Soviet Physics-JETP **30** (1970) 973.
3) B. M. Dymshitz and Ya. P. Koretskii: Optics and Spectrosc. **32** (1972) 17.
4) V. M. Batenin, V. S. Zrodnikov, V. K. Roddatis and V. F. Chinnov: Teplofiz. Vys. Temp. **13** (1975) 270 [in Russian].
5) S. Miyake, S. Takeuchi and Y. Arata: Japan. J. appl. Phys. **13** (1974) 296.
6) J. R. Roberts and P. A. Voigt: J. of Research of NBS **75A** (1971) 291.
7) H. W. Drawin: Z. Phys. **228** (1969) 99.
8) H. W. Drawin and F. Emard: Z. Naturforsch. **28a** (1973) 1289.
9) V. M. Goldfarb: *Physics and Techniques of Low Temperature Plasma*, ed. S. V. Dresvin (Atomizdat, 1972) [in Russian].
10) R. S. Devoto: J. Plasma Phys. **2** (1968) 617.
11) H. W. Drawin: EUR-CEA-FC-510 (1969).
12) V. S. Vorobév and M. B. Zheleznyak: Optics and Spectrosc. **35** (1973) 361.
13) L. M. Biberman, V. S. Vorobév and I. T. Yakubov: Soviet Physics-USPEKHI **15** (1973) 375.
14) V. L. Ginzburg and A. V. Gurevich: Usp. Fiz. Nauk. **70** (1960) 201.
15) J. F. Uhlenbusch and E. Fischer: Proc. of IEEE **59** (1971) 578.
16) C. D. Bogomolov: private communication.
17) E. A. Tishchenko and V. G. Zatzepin: Soviet Physics-JETP **41** (1975) 268.
18) B. E. Meierovich: Soviet Physics-JETP **34** (1972) 1006.

Pulsed High Current Heating of a Microwave
Plasma at Atmospheric Pressure

A plasma heating is investigated in an atmospheric hydrogen and/or helium gas mixture, by the superposition of a pulsed high current on a long plasma beam produced by a high power CW microwave energy. An efficient ionization and Joule heating are achieved and the plasma of $T=5.5 \times 10^4$ K and $N_e = 1.8 \times 10^{17}$ cm^{-3} is obtained in He+40%H$_2$ gas mixture. About 20% of the input energy to the discharge is dissipated to increase the thermal energy of the plasma. Stabilizing effect of MHD instability by an admixture of a heavy gas is clarified and the correlation between the instability growth and the heating efficiency is demonstrated.

§1. Introduction

Positive column of a discharge in a high pressure gas environment is well known as arc plasma and has been investigated for a long time. Alfven et al.[1] proposed in 1960 with the name "gas insulation" that arc plasma could be the high temperature core of controlled thermonuclear reactor (CTR) when it was immersed in a strong axial magnetic field. On the experimental study of gas insulation around atmospheric pressure, the plasma in a high power microwave reported by Kapitza[2] gave much interest, since the high temperature core of 10^6 K was demonstrated without a magnetic field at a power per unit length of around 1 kW/cm. Indeed it was a very good example of efficient thermal insulation by the cold gas. Dymshitz,[3] however, stated that the paper by Kapitza included some questions in the analysis of the spectroscopic data and threw doubts on the achievement of 10^6 K. We have also studied[4,5] experimentally a high pressure microwave discharge but it was impossible to obtain such a high temperature with the input power of about 1 kW/cm and only a plasma of around 10^4 K could be obtained.

While in case of low pressure gas blanket concept developed by Lehnert,[6] an experimental study has been performed by Ringboog team.[7] However they have not succeeded in obtaining the impermeability of a plasma against neutral gas.

It can be said at present that the concept of gas insulation or gas blanket are not yet clarified experimentally. To obviate many unfavorable interactions of a hot plasma with the solid wall, it is by far the better to study this problem for a plasma in a high pressure environment. At the same time it is first of all necessary to obtain a quasi-steady plasma core with a temperature over 10^5 K for the efficient thermal insulation and impermeability against neutral gas.[8] In the gas pressure around 1 atm the electron density which satisfies the criterion of impermeability is obtained quite easily but the plasma temperature over 10^5 K has not been reported until now.

In studying a gas insulated plasma it must be long enough to be able to neglect the end loss compared with the radial one. Because the interaction of the plasma and the neutral gas is important in the region where the magnetic field has little effect, we should clarify in advance a fully ionized non-magnetized plasma character in such a light gas as hydrogen or helium. The hydrogen plasma[5] we have studied is not fully ionized, but it is slender and long enough ($D \simeq 1$ mm, $L \simeq 20$ cm). It has also a high stability in the steady state with an electrical conductivity of about 1 ℧/cm with a temperature of about 10^4 K. When a high current is supplied to this plasma efficient Joule heating can be achieved with good reproducibility and a fully ionized plasma is obtained quite easily.

In this paper plasma characters of pulsed high current discharge are reported in hydrogen and helium gas mixture at atmospheric pressure. Problems of stability and heating efficiency are discussed, and a concluding remark is given as to the further experimental plan of a gas insulated plasma production in a high pressure gas environment.

§2. Experimental Apparatus and Methods

The experimental apparatus is quite similar to the one in ref. 5. Two copper electrodes (10 mm in dia.) having cone-shaped tungsten tips are inserted into a quartz tube of 40 mm in diameter in the axial direction to contact with the microwave plasma beam. The electrode distance was 22 cm. A pulsed high current is supplied from these electrodes with a charging voltage below 10 kV. The operating conditions are shown in Table I. The plasma produced by the microwave acts as an initial filamentary conductor enclosed by cold high pressure gas and a condenser bank (10 kV, 5 kJ) is used as a pulse current source. The experimentally obtained current half cycle was 40 μs.

Schematic diagram of diagnostic system is shown in Fig. 1. The time behavior of the voltage V_p supplied to the plasma is measured by the divider of 1:1000 and the plasma current I_p is detected by Rogowsky coil. For spectroscopic measurement two monochrometers are used in addition to the optical multichannel analyzer OMA (PAR, model 1205). Typical specifications of monochrometers are $f=1$ m with 8.0 Å/mm in 1st order dispersion for SG-12D-10B (Mizojiri) and $f=25$ cm with 33 Å/mm in dispersion

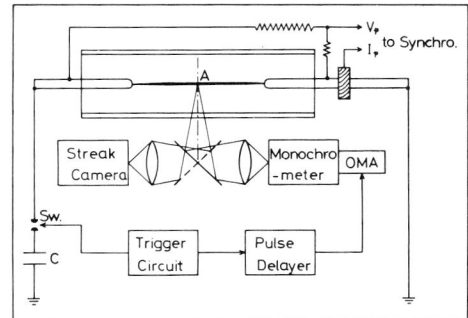

Fig. 1. Schematic diagram of diagnostic system.

for JE-25E(NJA). By the gated operation of OMA we can obtain emission intensities over wide wavelength range at a certain time after the start of Joule heating. The width of the gating time was selected to be 1 μs.

§3. Experimental Results

Figures 2–4 show some streakphotographs of the plasma heated in various gas species. In Fig. 2 the helium plasma has a very weak fluctuation and expands almost symmetrically

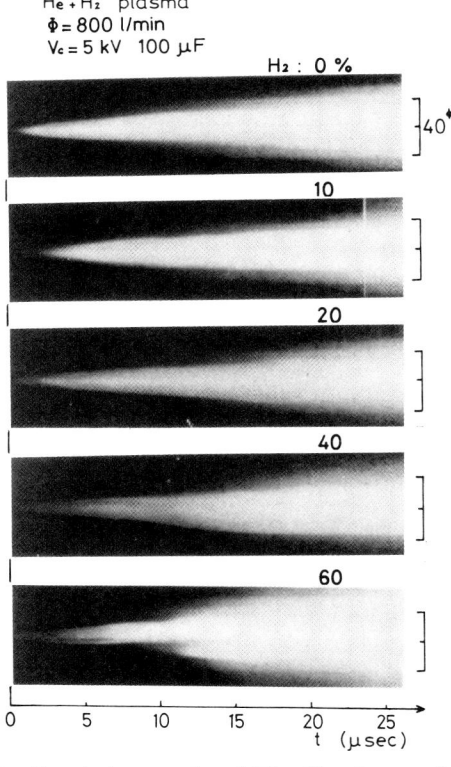

Fig. 2. Streakphotographs of He+H$_2$ plasma with various hydrogen mixing rate.

Table I. Operating conditions.

Microwave	$f=915$ kHz
	$p_l \leqslant 30$ kW, CW
Gas	He, H$_2$, Ar, N$_2$ and their mixtures
Flow rate	$\phi=500\sim1600$ l/min
Initial plasma	$L=20\sim30$ cm
	$D=1\sim3$ mm
	$N_e=10^{15}\sim10^{16}$ cm^{-3}
	1 ℧/cm, $T\lesssim10^4$ K
Pulse current source	10 kV, 100 μF, 5 kJ

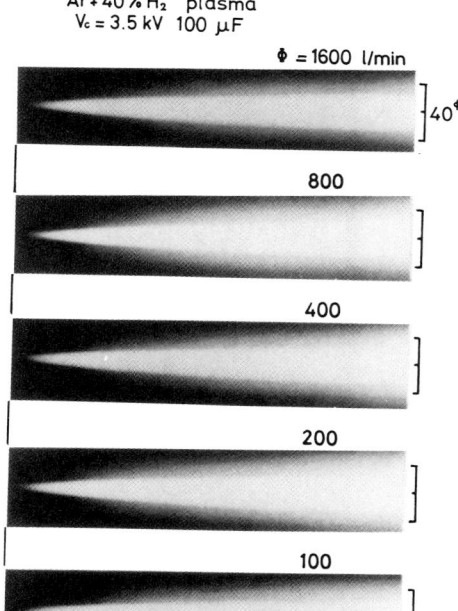

Ar + 40% H₂ plasma
V_c = 3.5 kV 100 μF

Φ = 1600 l/min

800

400

200

100

Fig. 3. Streakphotographs of Ar+40%H₂ plasma.

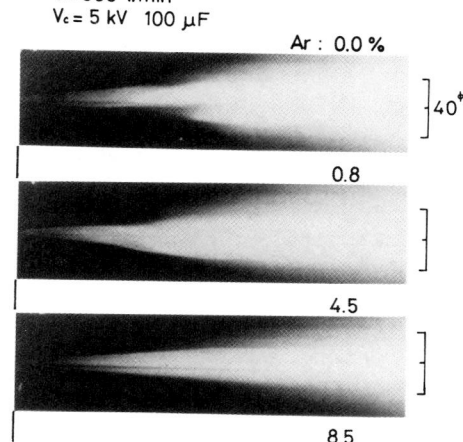

He + 60% H₂ + Ar plasma
Φ = 800 l/min
V_c = 5 kV 100 μF

Ar : 0.0 %

0.8

4.5

8.5

Fig. 4. Streakphotographs of He+60%H₂ plasma with addition of a small bit of Ar.

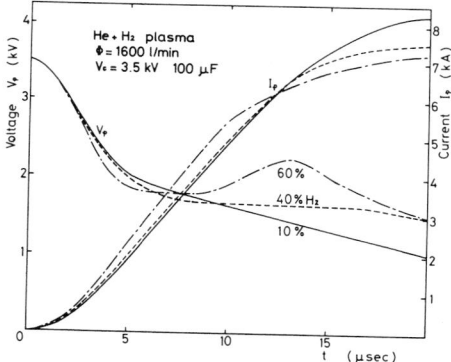

Fig. 5. Typical V–I characteristics for various hydrogen mixing rate to He gas.

in the radial direction. With the increase of hydrogen content, however, it becomes unstable and at 60%H₂ content it has no symmetry after $t \geq 5$ μs with the strong deflection towards the tube wall. While Ar+40%H₂ plasma (Fig. 3) is quite stable and not influenced by the gas flow rate. Contribution of a heavy gas to the plasma behavior is seen in Fig. 4. Addition of several percentages of Ar to the unstable He+60%H₂ plasma brings forth stable symmetrical expansion of the plasma up to the time of the maximum current ($t = 20$ μs). This character was also preserved in case of N₂ admixture. It was experimentally clarified that the instability did not occur when the equivalent mass M_e of the discharge gas was over 5 in relative atomic mass unit.

Figure 5 shows typical voltage-current (V–I) characteristics in He+H₂ plasma. With the increase in H₂ content anomaly in I_p and V_p appears. This corresponds to the unstable plasma behaviors in Fig. 2. By using these V–I characteristics electrical conductivity $\bar{\sigma}_p$ averaged over plasma diameter was calculated and is given in Fig. 6. The plasma diameter

was decided from streakphotographs. When the plasma is unstable as in case of 60%H₂, $\bar{\sigma}_p$ decreases strongly after $t \geq 5$ μs on the contrary to that of He+10%H₂ plasma. It should be noticed moreover that by the addition of 7% Ar to the unstable plasma the decay of $\bar{\sigma}_p$ after $t \geq 5$ μs becomes quite weak.

Figure 7 shows an example of spectroscopic data detected by JE-25 and OMA for He+40%H₂ plasma at $V_c = 3.0$ kV. During the first quarter cycle of the heating current ($0 \leq t \leq 20$ μs) the intensity of ion line (HeII

303

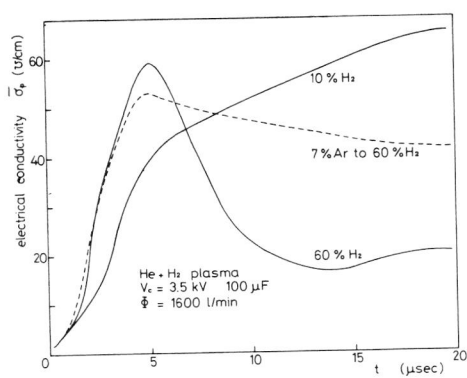

Fig. 6. Temporal changes of averaged plasma conductivity σ_p for stable and unstable plasma parameters.

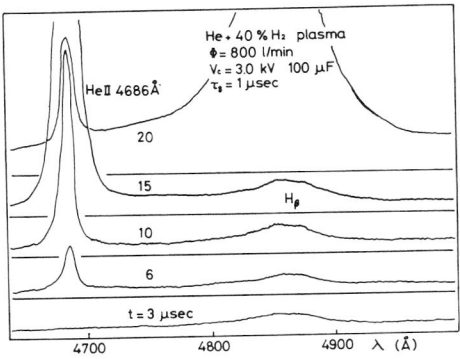

Fig. 7. Typical visible light emissions from He+40%H$_2$ plasma at various times after the beginning of Joule heating.

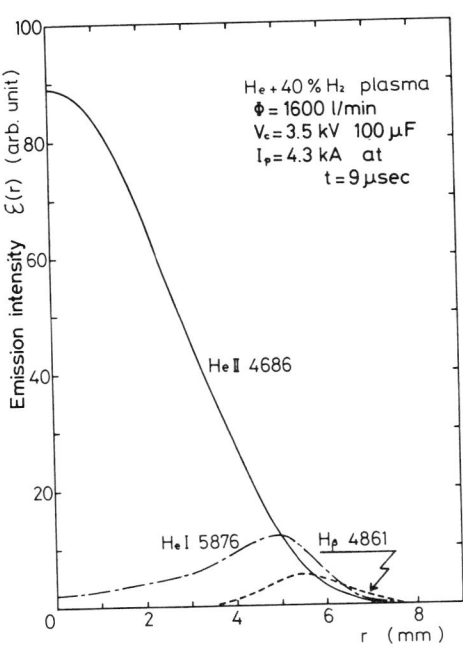

Fig. 8. Radial intensity distributions of emitted visible lines at the time of $I_p=4.3$ kA for He+40%H$_2$ plasma.

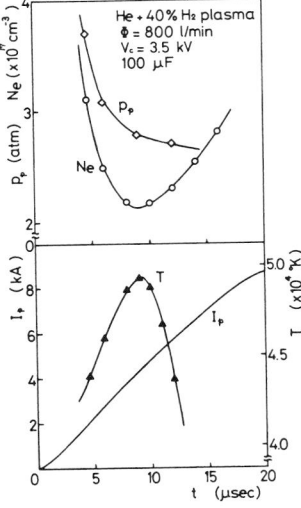

Fig. 9. Temporal changes of plasma pressure, plasma temperature and its density in case of He+40%H$_2$ plasma.

4686 Å) increases with time until $t=15\ \mu$s whereas that of neutral line (H_β 4861 Å) keeps almost a constant value. This suggests an efficient heating of the plasma core accompanying the emission of H_β line on the plasma surface. Radial intensity distributions measured with two monochrometers and OMA are shown in Fig. 8 after Abel inversion. The ion line (HeII 4686 Å) is emitted from the central core of the plasma and the neutral lines (HeI, H_β) are strong on the boundary region. From these data we can decide plasma temperature and electron density. The electron density N_e was estimated from Stark broadening of HeII and H_β lines. The plasma temperature T was decided from the intensity ratio of HeII and HeI lines on the assumption of local thermodynamic equilibrium (LTE) state of the plasma.

The temporal behavior of the plasma parameters averaged over its diameter is seen in Fig. 9 for He+40%H$_2$ gas at $V_c=$ 3.5 kV. With the increase in I_p, T grows up to 4.8×10^4 K until $t=9\ \mu$s. The steep decay in $p_p (=2N_e kT)$ before this time corresponds to the plasma expansion involving its efficient

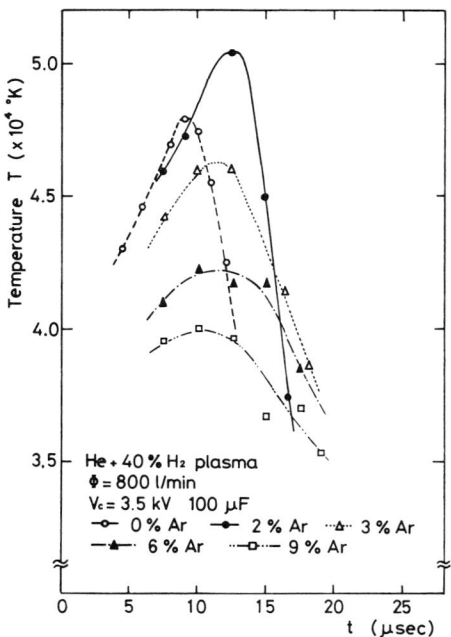

Fig. 10. Temporal behaviors of plasma temperature for the change in the percentage of Ar additive to He+40%H₂ plasma.

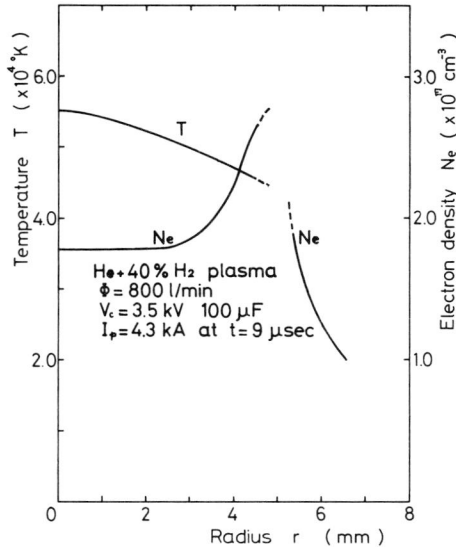

Fig. 11. Radial temperature and density distributions of He+40%H₂ plasma at the time of $I_p=4.3$ kA.

heating. The effective charge Z_{eff} is calculated to be below 1.2 and neglected in the estimation of p_p. After $t>9$ μs T is rapidly lowered with a slight decrease in P_p. The temporal change of T in this plasma with the addition of several percentages of Ar was measured as shown in Fig. 10. The maximum plasma temperature decreases with Ar additive but its decay after $t \gtrsim 10$ μs becomes slower.

The radial distribution of T and N_e at $t=9$ μs obtained from the data of Fig. 8 is shown in Fig. 11. The maximum plasma temperature is 5.5×10^4 K with the density of 1.8×10^{17} cm^{-3}. Similar profiles of T and N_e was obtained also at $t=5$ μs, $I_p=2.15$ kA with the maximum plasma temperature of 4.7×10^4 K.

§4. Discussion

4.1 Contribution of the microwave plasma to heating experiment

In this experiment the initial stable microwave plasma plays the role of a preionized conductor over 20 cm in length. Its diameter changes for different gas species but becomes very slender and keeps almost a constant value of about 1 mm when the hydrogen content is larger than 40%.[5] Therefore it expands symmetrically by the supply of a pulsed high current and is heated from the central region with a good reproducibility. It is very important in our opinion to have such a stable and long initial plasma beam for the detailed study of a pulse heating experiment at a high pressure and little is stated[9] on this importance until now. Usually a thin conducting wire is used to initiate the heating in a long plasma column, but it is apt to be contaminated by heavy impurity elements. Another way[10] of preionization by a pulsed high intensity light source such as laser is hard to obtain an axially uniform plasma column.

4.2 Heating process and its efficiency

When a pulsed high current is superposed to the initial microwave plasma beam, the rapid increase in its internal energy led to the radial expansion with a velocity V_r of about 1 km/sec. And it was followed by the efficient heating to about 4×10^4 K (Fig. 9) at $3 \sim 4$ μs after the current start. The plasma pressure decided from the relation[11] $p_p=0.9\rho_0 \bar{V}_r^2+p_0$, where ρ_0 and p_0 are mass density and static pressure of the initial cold gas respectively, gives $p_p=2.9$ atm and approximately coincides with the one from $p_p=2N_e kT$. We note here that the time of the current penetration is estimated[12] from $\tau = 4\mu_0\sigma r_p^2/\pi^2$, where σ is the Spitzer conductivity and r_p is the plasma radius that is decided from the radial distribution of T and N_e, not from the streak-photograph. For the data of Fig. 9 $\tau \simeq 5 \times 10^{-7}$ sec with $Z_{\text{eff}} \simeq 1.2$, so that the heating current can flow in the central region during the heating time.

When the plasma is stable and quasi-adiabatic assumption is satisfied until the time of the current maximum, the heating efficiency is limited principally by the plasma expansion. So that at a certain time the increase of T stops, after which it begins to decay gradually. When an instability is developed the anomaly of expansion appears accompanying the deflection of the plasma column and steeper decay of T will result. This process is well understood from Fig. 10. With the addition of 2% Ar the time of instability growth is delayed from 9 μs to 12 μs after which the plasma temperature drops abruptly. By the further increase in Ar additive the temperature decay becomes weaker keeping the stable symmetrical expansion.

Let's discuss here the thermal properties of He+40%H$_2$ plasma. The plasma can be considered in the state of LTE when the following criteria are satisfied,[13,14] making use of experimental value of $T = 4.5 \times 10^4$ K, $N_e = 2 \times 10^{17}$ cm^{-3}: i) the density requirement for the establishment of complete LTE gives $N_e \gtrsim 3 \times 10^{17}$ cm^{-3} and it is nearly satisfied. ii) the characteristic time for complete LTE in a transient plasma becomes 1.3×10^{-9} cm^{-3} and it is much smaller than the heating time. iii) as for the time for kinetic equilibrium between electrons and atoms (τ_k^a) or ions (τ_k^i), τ_k^i is important since the heated plasma is almost fully ionized and $\tau_k^i \simeq 5 \times 10^{-9}$ sec. This condition is also quite easily satisfied experimentally. iv) the electric field E_p over which is induced a difference in temperatures of electrons and heavy particles is about 5×10^3 V/cm and the applied electric field is below 10^2 V/cm (Fig. 5). This field is negligible compared with E_p. v) the effect of the plasma nonuniformity is characterized by the effective diffusion lengths for atoms λ_a and ions λ_i. It is estimated to be $\lambda_a \lesssim 10^{-3}$ cm and $\lambda_i \simeq 0.1\lambda_a$. The radial plasma

dimension $2r_p$ is over 0.5 cm already at $t = 5$ μs so that this effect is of no importance.

From above estimations we can satisfactorily consider that the heated plasma is in LTE state and the determination of the plasma temperature by the relative intensity ratio of HeII and HeI lines is guaranteed well.

Next we estimate the transfer rate of input energy to the heating of the plasma. The liberation processes of the energy W_p fed into the plasma are works of dissociation W_d, ionization W_i, expansion W_e and the work to increase thermal energy W_t of the plasma. In case of Fig. 9 the energy W_p during the first 5 μs can be estimated to be 7.8J from V–I characteristics eliminating inductive energy. While we obtain $W_d = 0.15$J, $W_i = 3.8$J, $W_t \simeq W_e = 1.54$J. The summation of these energies gives $W_a = 7.03$J. Thus W_a/W_p becomes 0.9 with $W_t/W_a = 0.22$. During the time of 9 μs, $W_p = 26.3$J is obtained and $W_a = 21$J. Similarly $W_a/W_p = 0.8$, $W_t/W_a = 0.24$. Thus only $0.1 \sim 0.2 W_a$ is attributed to the energy loss including radiation, and about 20% of W_p is used to increase the thermal energy of the plasma until the growth of the instability. As was stated in the introduction we consider that the low radiation loss with $Z_{\text{eff}} \simeq 1$ and the moderately high heating efficiency are maintained because of the long plasma dimension ($L/2r_p \simeq 10$) and of the formation of a well-defined initial current path, by a high power microwave energy in the high pressure gas of light element.

4.3 Stability of expanding plasma

Generally speaking a current-carrying plasma column inherently suffers from MHD instabilities due to the interaction of the current and its self magnetic field. They are well known as kink or sausage instability[15] in a Z pinch plasma and helical instability[16,17] in a Joule heated arc type discharge. Their growth rates are usually connected with the current density and the mass density. The instability observed in this experiment is helical one which is easier to be excited by the existence of heating electrodes. Among many stabilizing factors[16-19] of these instabilities at a high pressure, the following two factors are strongly coupled to the mass density; i) pressure gradient by vortex gas flow,[19] ii) inertial force of neutral

The static pressure distribution of cold neutral gas in the radial direction was measured with Pitot tube and $|\partial p_0/\partial r|$ at $r=r_p$ showed a value of 6×10^3 dyn/cm^3 for He$+40\%$H$_2$ gas with $\phi = 1600$ l/min, where p_o is the neutral gas pressure. It is smaller by about one order than $H_\theta^2/8\pi r_p$ ($=8 \sim 16 \times 10^4$ dyn/cm^3 for $r_p = 2.5 \sim 5$ mm), where H_θ is the self magnetic field. The experiment also shows even in a low gas flow of 100 l/min in Ar$+40\%$H$_2$ gas the plasma expands quite stably (Fig. 3) with $|\partial p_0/\partial r| \leq 10^3$ dyn/cm^3. Thus the vortex flow in this experiment does not play a decisive role to the plasma stability, though it becomes in general a very active stabilizing method in a more refined fashion of the flow condition.

Now the growth rate γ_g of the helical instability that is affected by condition ii) is $\gamma_g = j\sqrt{\mu_0 F(y)/\rho_0}$,[17)] where j is the current density decided from $I_p/\pi r_p^2$ and μ_0 is the magnetic permeability. The variable $F(y)$ is the numerical factor with $y = 2\pi r_p/\lambda$ and λ is the wavelength of the perturbation. In case for He$+40\%$H$_2$ plasma ($M_e = 3.2$), j changes from 1.2×10^4 ($Z_{\text{eff}} \simeq 1.0$) to 8×10^3 A/cm^2 ($Z_{\text{eff}} \simeq 1.2$) during $4 \leq t \leq 10$ μs. The substitution of ρ_0 and μ_0 gives $1/\gamma_g = 13 \sim 20$ μs with $F(y) = 5.2 \times 10^{-2}$ and $y = 0.6$ at which the maximum growth rate is given. This result shows that the helical instability of $\lambda = 2.6 \sim 5.2$ cm exponentially develops in a time of $13 \sim 20$ μs for He$+40\%$H$_2$ plasma. By the addition of 9% Ar to this gas M_e and Z_{eff} change to 6 and 1.5, respectively. Thus $1/\gamma_g = 19 \sim 29$ μs with the reduced current density due to the increase in Z_{eff}. It assures the stable expansion until the time of the current maximum. These results well correspond to the experimental data of Figs. 4 and 10.

Thus the development of the plasma instability in this experiment is remarkably governed by the inertial effect of the outer cold gas and the change of Z_{eff}. At the temperature level of 4×10^4 K Ar (or N$_2$) gas is multiply ionized[20)] with a nearly equal density of Ar^{2+} and Ar^{3+}. So that a considerable energy is spent for ionization, even a small quantity of Ar (or N$_2$) is mixing. In Fig. 10 the electron density at the time of the maximum temperature showed a similar value of 2×10^{17} cm^{-3} for all cases. It then shows the decrease

of W_t and W_e by Ar admixture. So that the newly introduced energy dissipation results in the reduction of the plasma temperature. Actually for the case of 9% Ar additive in Fig. 9, this additional energy ΔW_i is about 1.9J. While decreases of W_t and W_e are summed to be $\Delta W = 1.6$J. They are in good agreement each other ($\Delta W_i \simeq \Delta W$).

From above result it can be said that the heating experiment in the light gas is better than in the heavy one to obtain a higher temperature but the onset and the growth of an instability are easier. So that the stabilization by the strong pressure gradient of neutral gas is recommended for the further study of a gas insulated plasma production, although it was very difficult in such a conventional vortex flow method as this experiment. We will in turn apply a new type gas flow system[21)] with the name "vortex gas tunnel", in which a much stronger pressure gradient than the conventional method is obtained. The new experimental result will be reported in the next paper.

§5. Conclusion

To study a fully ionized hydrogen and/or helium plasma at atmospheric pressure, Joule heating experiment was performed initiating the discharge by a high power CW microwave energy.

For example in He$+40\%$H$_2$ plasma a plasma core of $T = 5.5 \times 10^4$ K and $N_e = 1.8 \times 10^{17}$ cm^{-3} on the axis was obtained with $I_p = 4.3$ kA at $t = 9$ μs after the start of the current supply. The plasma length was fixed to be 22 cm and the core diameter was about 1 cm at that time.

Estimations of thermal properties of the heated plasma certified its LTE state and about 20% of the input energy was liberated to increase the thermal energy of the plasma without a large radiation loss.

Discharges in various gas species made it clear that a heavy gas stabilized the heated plasma accompanying the decrease in the heating efficiency. The observed instability was attributed to the helical one and the stability effect of a heavy gas was well explained by the cold gas inertial force and an increase of Z_{eff}. At the same time reduction of the heating efficiency was due to the increase of the energy dissipation to multiple ionization.

A further experiment in "vortex gas tunnel" is planned, which has a potential to stabilize and insulate a high temperature core.

A part of this work was supported by Japan Securities Scholarship Foundation.

References

1) H. Alfvén and E. Smars: Nature **188** (1960) 801.
2) P. L. Kapitza: Soviet Physics-JETP **31** (1970) 199.
3) B. M. Dymshitz and Ya. P. Koretskii: Optics and Spectrosc. **32** (1972) 17.
4) Y. Arata, S. Miyake, A. Kobayashi and S. Takeuchi: J. Phys. Soc. Japan **40** (1976) 1456.
5) Y. Arata, S. Miyake and A. Kobayashi: J. Phys. Soc. Japan **44** (1978) 998.
6) B. Lehnert: Nuclear Fusion **8** (1968) 173.
7) F. C. Schüller, for the RINGBOOG team: Reports on the International Symposium on Plasma-Wall Interaction, Jülich, F.R.G. 18–22, Oct. '76 E{E-8.
8) B. Lehnert: *Proc. IIIrd Int. Symp. on Toroidal Plasma Confinement, Garching, March 26–30* (1973) C-7.
9) A. V. Pyshnov: Teplofiz. Vys. Temp. **13** (1975) 279 [in Russian].
10) J. R. Vaill, D. A. Tidman, T. D. Wilkierson and D. E. Koopman: Appl. Phys. Letters **17** (1970) 20.
11) S. I. Braginskii: Soviet Physics-JETP **34** (1958) 1068.
12) H. H. Woodson and J. R. Melcher: *Electro-mechanical Dynamics, Part III* (John Wiley and Sons, 1968).
13) H. R. Griem: *Plasma Spectroscopy* (McGrow-Hill Book Comp., 1964).
14) H. W. Drawin: Z. Phys. **228** (1969) 99.
15) M. Kruskal and M. Schwartzschild: Proc. Roy. Soc. **223A** (1954) 348.
16) J. Mentel: Z. Naturforsch. **26a** (1971) 526.
17) K. Ragaller: Z. Naturforsch. **29a** (1974) 556.
18) H. E. Wilhelm: *Proc. Vth. Int. Conf. on Ioni, Phoenom, in Gases, Munich, Aug. 28–Aut. 1* (1961) p. 2233.
19) A. G. Elfimov: Soviet Physics-Tech. Phys. **17** (1972) 592.
20) H. N. Olsen: J. Quant. Spectrosc. Radiat. Transfer **3** (1963) 305.
21) Y. Arata: J. Phys. Soc. Japan **43** (1977) 1107.

CHAPTER 4

PHYSICAL CHARACTERISTICS OF ULTRA HIGH ENERGY DENSITY BEAM WELD ZONE

Hardness and Defect of Weld Zone

Fracture Path Transition Temperature

Dynamic Behaviour of Welding and Cutting

Hardness and Defect of Weld Zone

100KW Class Electron Beam Welding Technology—Characteristics of Deep Penetration Bead and Its Analysis
Y. Arata, M. Tomie and A. Kohyama
[Trans. JWRI 5 (1976), 11. IIW Doc. IV–193–76.]

Quench Hardening and Cracking in Electron Beam Weld Metal of Carbon and Low Alloy Hardenable Steels
Y. Arata, F. Matsuda and K. Nakata
[Trans. JWRI 1 (1972), 39. IIW Doc. IV–109–72.]

Weldability Concept on Hardness Prediction
Y. Arata, K. Nishiguchi, T. Ohji and N.Kohsai
[Trans. JWRI 8 (1979), 43. IIW Doc. IV–263–79.]

Study on Characteristics of Weld Defect and Its Prevention in Electron Beam Welding—Characteristics of Weld Porosities
Y. Arata, K. Terai and S. Matsuda
[Trans. JWRI 2 (1973), 103. IIW Doc. IV–114–73.]

100KW Class Electron Beam Welding Technology
—Characteristics of Deep Penetration Bead and its Analysis—

Abstract

The welding of thick plates was carried out by the 100 KW class electron beam welder as shown in Ref.(1). As a result, the characteristics of deep penetration welding could be studied, in relation to the application of high power electron beam to several kinds of thick steel plates.

Furthermore, investigation was made into welding defects peculiar to the welding electron beam and the structure of weld metal.

1. Introduction

For the one pass welding of steel plates over 100 mm thickness with a short time, only the electron beam energy with very high levels of the power and its density is available.[1],[3],[4] Using such electron beam welding, in the case of small bead width, this welding will have quick heating and cooling rate. In this instance, however, welding defects, which appear concurrently with solidification as peculiar to the electron beam welding, are often observed dangerously.

Concerning the weldability of the above, several reports are available about a case of a few centimeters penetration at a low power, but no detailed report is made public about a case of a high power of 50 KW and upwards. Under the circumstances, study was made about the weldability of mild and high tensile steel plates as using a high power electron beam welder,[1] which plates were extensively used for large structures.

2. Materials and welding conditions

The test specimens used were taken from steel plate (SS41) for general structure, steel plate (SM41) for welding structure and high tensile steel plates (HT 60 and HT80) for welding structure, all available in the market. Their chemical composition is shown in Table 1. The electron beam power of the welder is 50KW and 100KW (V_b = 100 KV constant).

Table 1. Chemical composition of material used.

	C	Si	Mn	P	S	Cu	Cr	Ni	Mo	V	B	Wt(%) ↔ ppm	
												O	N
HT80	0.11	0.32	0.92	0.0008	0.006	0.26	0.46	1.03	0.47	0.003	0.0012	10	34
HT60	0.14	0.30	0.88	0.0008	0.008	—	0.26	0.45	0.23	0.05	—	20	33
SM41	0.18	0.47	0.71	0.015	0.010	—	—	—	—	—	—	36	68
SUS304	0.05	0.74	1.74	0.030	0.010	0.12	19.5	10.9	0.16	—	—	75	360
SS41 (KILLED)	0.21	0.29	1.25	0.016	0.011	—	—	—	0.008	—	—	20	36
SS41 (SEMI KILLED)	0.18	0.03	0.87	0.009	0.015	0.23	0.26	—	—	—	—	230	42

By use of the AB-test method,[2] the electron beam was preliminarily checked in respect of its form, and the focus (D_F) of the beam as applied for welding was taken at 500 mm. Also, the active beam parameter (a_b) was fixed at 0.9 of a value which ensures a maximum penetration depth. The beam was deflected at a set-back angle of 10 degrees, in order to restrain the arcing of the electron beam gun due to metal vapour as generated during welding.

3. Test results and observation

Fig. 1 shows an effect of welding speed (v_b) upon penetration depth (h_p) at 50 KW and 100 KW beam outputs, and it is seen from this Fig. 1 that relations of $h_p \propto 1/\sqrt{v_b}$ are approximately established for both beam outputs at welding speed over a certain level. In the case

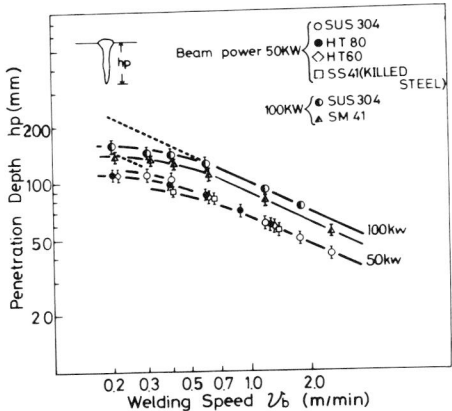

Fig. 1   Relations between v_b and hp

of 50 KW beam output, $h_p \propto 1/\sqrt{v_b}$ can be observed, when the welding speed is higher than 50 cm/min. The penetration depth (h_p), however, begins to be saturated, when the speed is lower than 50 cm/min. This saturation is caused, because molten metal once flowed back in a beam hole formed by deep penetration, is re-heated by the beam and energy necessary for the deep penetration is lost. Photo. 1 shows the longitudinal cross sections of the flat position welding bead on the steel plates (HT80

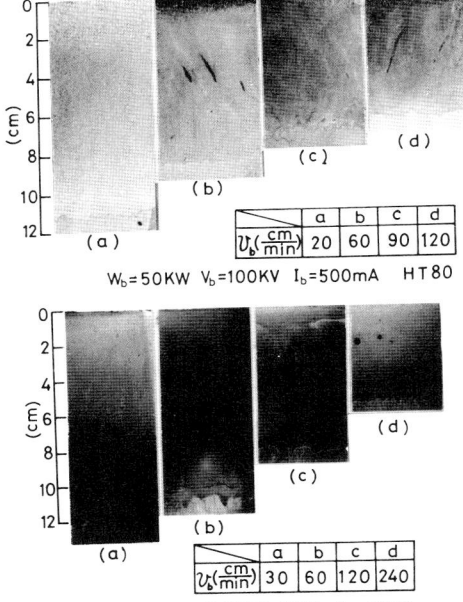

W_b=50KW V_b=100KV I_b=500mA   HT80

W_b=100KW, V_b=100KV, I_b=1000mA, SM41.

Photo. 1 Longitudinal cross sectional view of the bead for HT80 and SM41

and SM41). The penetration is well form and it is observed that the bead appearance deteriorates with a drop in welding speed. As welding defects, A-porosity (active zone porosity) and R-porosity[5] are generally observed. The former frequently occurs around the center of the penetration depth and more along the ripple line on the longitudinal cross section at the bead center line, and the latter is much observed to be surrounded by the dendrite at the vicinity of the bottom or spiking of the bead.

As welding defects liable to be caused by use of high power electron beam, there are (1) blowhole, (2) A-porosity, (3) R-porosity, (4) cold shut and cold shut crack and (5) horizontal crack.

(1) Blowhole

The occurrence of blowholes can be avoided, if a parent material contains a minimum amount of gases, and they do not appear in high tensile and killed steel plates.

Photo. 2 shows a case of blowholes in semi-killed steel. It is seen that beads are abruptly disordered due to

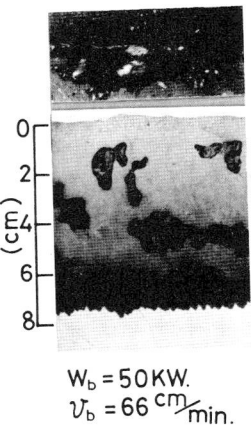

W_b = 50KW.
v_b = 66 cm/min.

Photo. 2 Bead appearance and longitudinal cross section for Semi-killed steel.

the blowing of gases from the parent material. Photo. 3 shows scanning electron micrographs of defective surfaces. The many solidification tip parts of the dendrites can be observed. It is to be noted that gaseous contents of about 50 ppm does not matter in the formation of blowholes, but when the contents are about 100 ppm, the occurrence of blowholes depends upon a welding condition, which behavior is very complicated.

(2) A-Porosity

This porosity is frequently observed especially in the penetration of wine bottle form, but appears with a slight

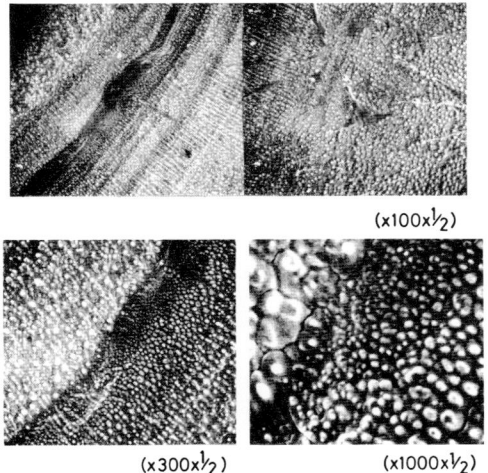

(x100x½)

(x300x½) (x1000x½)

Photo. 3 Scanning electron micrographs of porosity surface
(W_b = 50 KW, v_b = 60 cm/min, Semi-killed steel)

change in bead width, even in the case of the "well-form" penetration. In most cases, A-Porosity occurs in the weld metal containing alloy elements of high vapor pressure

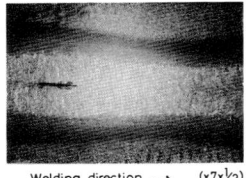

Welding direction ⟶ (x7x½)

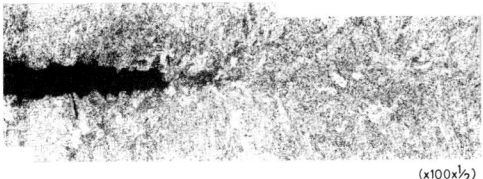

(x100x½)

Photo. 4 Horizontal cross section of HT80 weld metal
(W_b = 50 KW, v_b = 60 cm/min)

and over than certain level of amount of gases. Photo. 4 shows a horizontal cross section of weld metal and the occurrence of A-porosity. In this case, it has a nugget structure and is accompanied with free dendrite formed arround its edge in an anti-welding direction, though the occurrence of another type is recognized.[5] For the mechanism of A-porosity occurrence, It depends on the kinds of parent material and welding conditions. Photo. 5

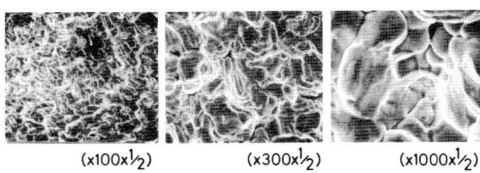

(x100x½) (x300x½) (x1000x½)

Photo. 5 Scanning electron micrographs of porosity surface
(W_b = 50 KW, v_b = 40 cm/min, HT80)

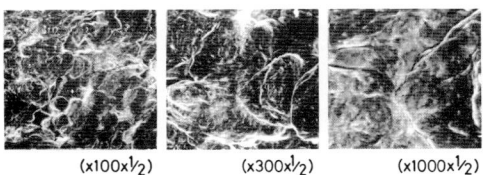

(x100x½) (x300x½) (x1000x½)

Photo. 6 Scanning electron micrographs of porosity surface
(W_b = 50KW, v_b = 20 cm/min, HT60)

and 6 show scanning electron micrographs of the A-porosity surface of high tensile steel. From both of these Photograph, it can be inferred that the defect is formed by a condition similar to shrinkage cavity in casting. Namely, Photo. 5 relates to the central part of the A-porosity and a solidified structure develops at right angles with the steel surface. It is observed that the structure has underwent a slight deformation just before solidification. On the other hand, the A-porosity in Photo. 6 is found in the vicinity of a weld end and the solidified structure, undergoes a substantial deformation. In this respect, the formation of the porosity is attributable to a cause different from the case of a blowhole as formed due to a heavy

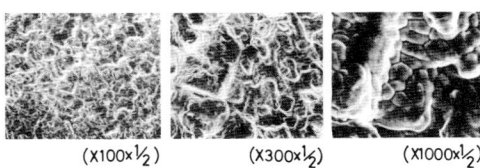

(x100x½) (x300x½) (x1000x½)

Photo. 7 Scanning electron micrographs of porosity surface
(W_b = 50 KW, v_b = 40 cm/min, Killed steel)

existence of gases. Photo. 7 shows a case of mild steel (killed steel) and a phenomenon similar to the case of high tensile steel can be observed. Furthermore, a similar phenomenon can be observed in the case of 100 KW output.

(3) R-porosity

As in the case of A-porosity, R-porosity is frequently surrounded with dendrite extended toward a welding direction, and it is formed when the molten fluid un-fills space which is produced by boiling in the bottom weld metal. Photo. 8 shows scanning electron micrographs of the R-porosity surface, and it is seen that this is a quickly

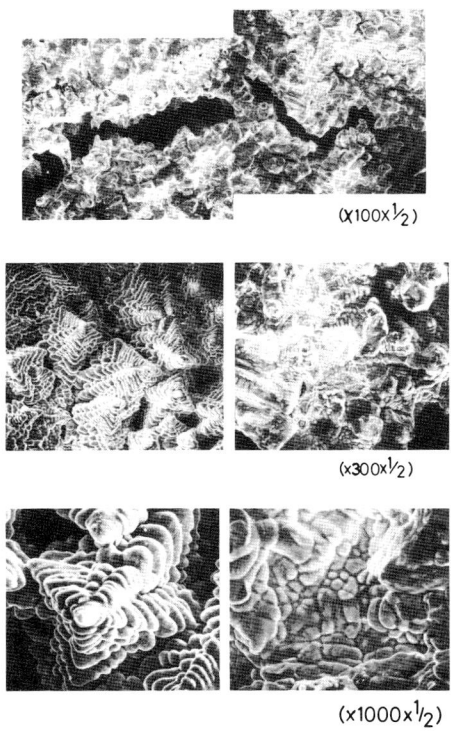

(×100×½)

(×300×½)

(×1000×½)

Photo. 8 Scanning electron micrographs of porosity surface
(W_b= 50 KW, v_b= 40 cm/min, Killed steel)

cooled solidified structure. In Photo. 8, a crack is shown at right angles with the photograph which is enlarged up to 100 times and this is a horizontal crack.

(4) Cold shut and cold shut crack

The formation of a large spike frequently leaves room where the cold shut and its crack which are liable to be caused along the ripple line in the bottom side, and it is not so easy to eliminate completely the cold shut and its crack. Especially when deep penetration weld is applied and the reheating phenomenon of the molten pool is occurring, spike become large and both of cold shut and cold shut crack are liable to take place.

(5) Horizontal crack

As shown in Photo. 1 and 2, in most cases, a horizontal crack occurs concurrently with the A-porosity and the R-porosity. Its occurrence is frequently observed in mild

steel and others of high sulfer content. This crack is considered to be hot cracking due to a deformation at the time of solidification. Also, some small horizontal cracks are considered to be attributable to the insufficient fluidity and amount of molten metal, similar to the case of the A-porosity.

The most basic cause for the occurrence of such defects should generally be found in the stability of a beam hole formed by electron beam. In relation to this fact, weld position should be taken into consideration. In the case of flat position welding by the application of vertical beam, full penetration of good structure can be expected, and more full penetration of better structure is obtained by horizontal beam compared with vertical beam. Especially[6], the beam hole can be maintained at the most stable condition, in the case of upward vertical position welding by the horizontal beam.   It seems that the most effective improvements for the welding defects can be expected in that instance.

Another problem in the deep penetration by the electron beam welding is related to the strength of the welded part. That is, it is concerned with the uniformity of structure in a direction of plate thickness. We have no exact data about this problem as yet.

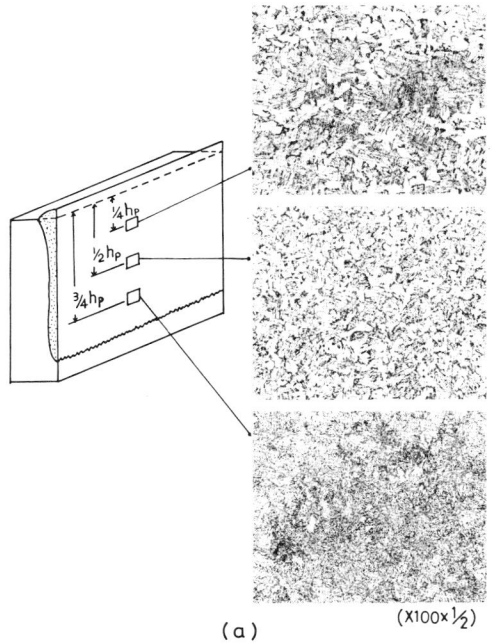

(×100×½)

(a)

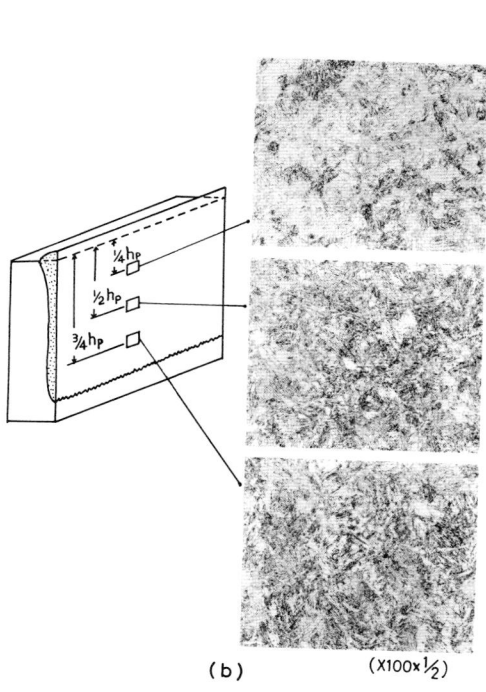

(b)   (×100×$\frac{1}{2}$)

Photo. 9 Microstructure of HT80 weld metal.
(W$_b$= 50KW, (a); υ_b= 20 cm/mun, (b); υ_b= 120 cm/min

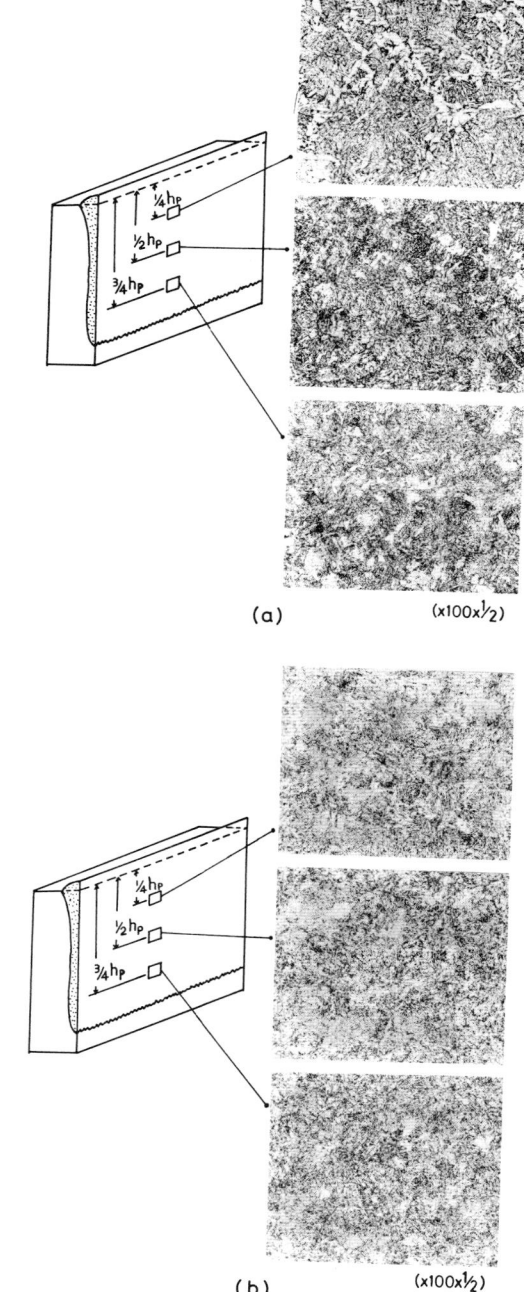

(a)   (×100×$\frac{1}{2}$)

(b)   (×100×$\frac{1}{2}$)

Photo. 10. Microstructure of HT60 weld metal
(W$_b$ = 50KW, (a); υ_b = 20 cm/min, (b); υ_b = 120 cm/min

Photo. 9(a) and (b) show the microstructure of HT80 weld metal, and a difference of structure at ¼ depth, ½ depth and ¾ depth from the surface is shown for two welding speeds of 20 cm/min and 120 cm/min. In the case of (a), ferrite structure with large ferrite as initially noticed is found at ¼ depth and the initial ferrite becomes small at ½ depth, gradually changing into micro ferrite structure. On the other hand, bainite is formed at ¾ depth and it is inferred that a substantial difference of toughness as well as strength is existing at that portion.

Photo. 10(a) and (b) are related to HT60 weld metal and it is seen that the structure gives a broad difference, depending upon the position of beads (direction of penetration), similar to the case of HT80 weld metal.

315

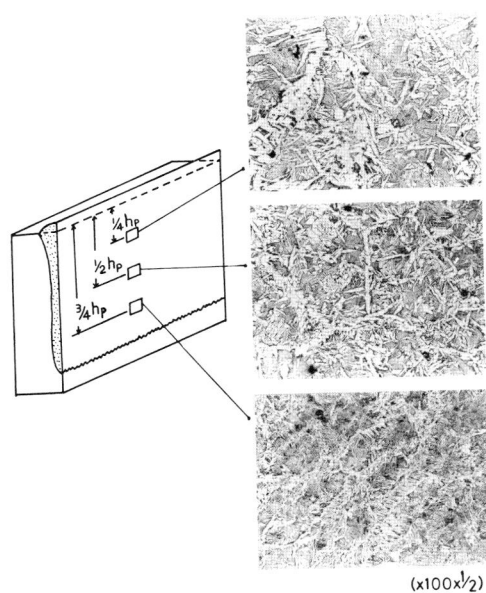

(×100×½)

Photo. 11. Microstructure of SM41 weld metal
(W$_b$ = 100KW, v_b = 20 cm/min)

Photo. 11 shows a case of SM41 weld metal obtained by 100KW electron beam welding. The hardness distribution of the welded part of HT80 is shown in Fig. 2 which indicated that maximum hardness changes conspicuously,

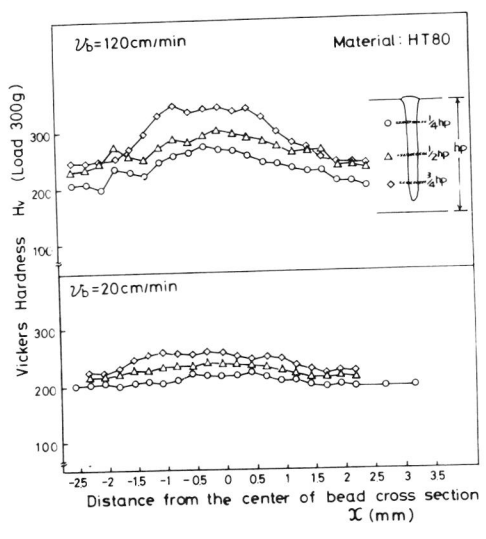

Fig. 2. Hardness curves for welded part of HT80 (W$_b$ = 50KW)

depending upon a position on the cross sectional bead. Also, it is not observed that a very high maximum hardness is usually recognized at the bonded part around 20~30 mm depth from a steel surface. Fig. 3 shows a case of

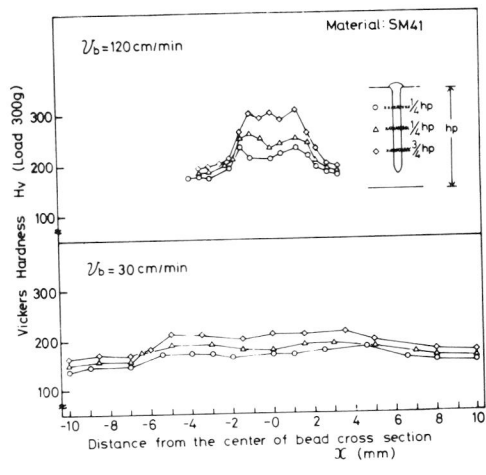

Fig. 3 Hardness curves for welded part of SM41 (W$_b$ = 100KW)

SM41 weld metal. Now, the chemical compositions of HT60 and HT80 weld metals are shown in Table 2. It is seen that only manganese drops in amount and sulfer and

Table. 2 Chemical composition of weld metal.

Mark	x*	C	Si	Mn	P	S	Cu	Cr	Ni	Mo	V	B
HT 80	¼ hp	0.11	0.30	0.70	0.008	0.006	0.24	0.46	1.03	0.47	0.03	0.0012
	½ hp	0.10	0.29	0.76	''	''	''	''	''	''	''	''
	¾ hp	0.11	''	0.81	''	''	0.25	''	''	''	''	''
HT 60	¼ hp	0.14	0.30	0.71	''	0.008	—	0.24	0.45	0.23	0.05	—
	½ hp	0.13	''	0.74	''	''	—	''	''	''	''	—
	¾ hp	''	''	0.80	''	''	—	''	''	''	''	—

x : Distance from specimen surface
WELDING CONDITION Beam Power 50KW(100KV, 500mA)
Welding speed 20cm/min

phosphorus having a high vapor pressure, undergo little change, due to their small absolute amount. In order to make uniform the chemical composition of weld metal, it would be recommended to add an alloy element to the specimen surface and cause a density grade in a direction of plate thickness. For this purpose, the electron beam is impinged upon the usual welding electrode of 2¼Cr−1Mo steel which is placed on the specimen surface, as shown in Fig. 4. In the case of 50 KW beam output and 120 cm/min welding speed, chrome density changes so that it is 0.59% at ¼ depth from the surface and 0.50% at ¾ depth from the surface respectively, as shown in Table 3. Also, the structure of weld metal is identical at ½ and ¾ depth

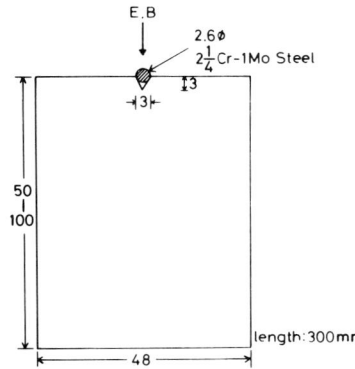

Fig. 4. Specimen preparation for C_r-M_o addition from specimen surface, using usual welding electrode of 2¼ C_r-1M_o steel.

Table. 3 Chemical composition of weld metal.
(Cr-Mo Addition from specimen surface)

x^*	C	Si	Mn	P	S	Cu	Cr	Ni	Mo	V	B
¼ hp	0.08	0.33	0.82	0.007	0.007	0.24	0.59	0.98	0.51	0.02	0.0012
½ hp	"	"	0.83	"	"	"	0.55	"	0.50	"	"
¾ hp	0.10	"	0.87	"	"	"	0.50	1.00	"	"	"
M.M**	0.11	0.32	0.92	0.008	0.006	0.26	0.46	1.03	0.47	0.03	0.0012

(*) Distance from specimen surface
(**)Mother Material
Welding Condition : Beam power 50KW(100KV, 500mA)
Welding speed 120 cm/min
insert metal 2¼ Cr-1Mo steel
Welding rod (2.6Φ)

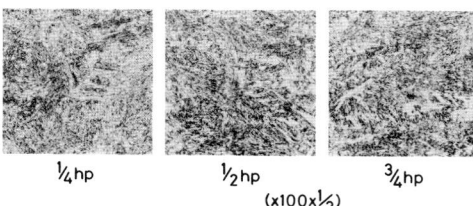

¼ hp ½ hp ¾ hp
(x100x½)

Photo 12 Microstructure of HT80 weld metal.
(Cr—Mo addition from specimen surface
W_b: 50KW, v_b: 120 cm/min)

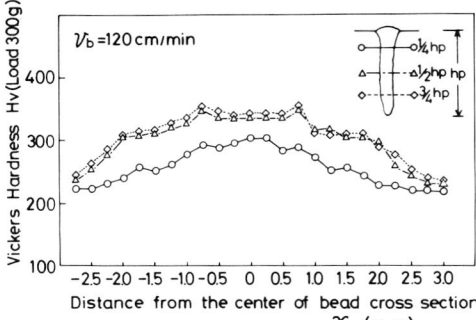

Fig. 5 Hardness curves for welded part of HT80, (W_b=50KW)

from its surface, as shown in Photo. 12. Fig. 5 shows a hardness distribution at the welded part. It is observed that there is no difference of hardness at ½ and ¾ depth from the surface, and a difference between the maximum hardness and hardness at ¼ depth is only less than 50 by H_v-values. Thus, it is concluded that the addition of chrome to the specimen surface is effective to unify about each chemical composition and the structure of weld metal.

4. Conclusion

Study was made about the problems of the deep penetration welding of thick steel plates using the high power electron beam welding method, and the following conclusions were obtained.

(1) Stainless, mild and high tensile steel plates were welded and in the case of the stainless steel (SUS 304), the maximum penetration depth from the surface was 120 mm at a 50 KW beam power and 160 mm at a 100 KW beam power respectively. In the case of high tensile steel, it was 110 mm and 140 mm respectively.

(2) With beam powers of 50 KW and 100 KW, $h_p \propto 1/\sqrt{v_b}$ can be observed, if the welding speed is over 50 cm/min. The penetration depth h_p is approximately constant, if the welding speed is lower than 20 cm/min. In a region of welding speed under 50 cm/min, a phenomenon of reheating molten metal by electron beam is taking place due to the violent "wall-fluctuation" of the beam hole.

(3) Welding defects as observed were a blowhole, A-porosity, R-porosity, a cold shut, cold shut crack and a horizontal crack. For these defects, its surfaces were carefully observed by the help of scanning electron micrographs, to find out a cause for the occurrence of the defects. As a result, it was concluded that the upward vertical position welding by horizontal beam should be adopted and its full penetration weld should be recomended, in order to avoid the occurrence of these welding defects for the thick plates.

(4) In the large power electron beam welding, the bonded part is free from abnormal hardening, but the structure of the welded part gives a substantial change in a direction of plate thickness. It is also clarified that the mechanical properties of the welded part are seriously affected by such structure. This problem needs special attention in a region where a phenomenon of re-heating molten metal by electron beam appears.

(5) In order to minimize the structual change of the welded part in the penetrated direction, investigation was made into the addition of an alloy element

317

(chrome in this report) to the specimen surface. As a result, it was found possible to unify the structure of the welded part in the penetrated direction.

Aknowledgement

The authors would like to express their gratitude to Dr. K. Ito, Prof. Dr. N. Igata and Dr. N. Oda for their encouragements.

References

1) Y. Arata, M. Tomie: "100 KW Class Electrom Beam Welding Technoloty (Report I)", Trans. of JWRI, vol. 2, No. 1 (1973); "100 KW Klasse-Elektronenstrahlen-Schweiβtechnologie (Bericht II)", Trans. of JWRI, vol. 4, No. 1 (1975)

2) Y. Arata, M. Tomie, K. Terai, H. Nagai and T. Hattori: "Shape Decision of High Energy Density Beam" Trans. of JWRI, vol. 2, No. 2 (1973)

3) A. Sanderson; A 75KW Electron Beam Installation for Thick Section Welding, Proc. Conf. on Advances in Welding Processes, Harrogate (1974) The Welding Institute.

4) G. Sayegh, P. Dumonte and T. Nakamura Design and Manufacture of a 100 KW Electron Gun, Advanced Welding Technology. The Second International Symposium of J.W.S., August (1975)

5) Y. Arata, K. Terai and S. Matsuda: Study on Characteristics of Weld Defect and Its Prevention in Electron Beam Welding (Report I)–Characteristics of Weld Porosities–, Trans. of JWRI, vol. 2, No.1 (1973)

6) Y. Arata, M. Osumi, K. Higuchi and K. Noda: Study on Electron Beam Welding of High Strength Aluminum Alloy, Preprints of the National Meeting of J.W.S. No.16 (Spring 1975) (In Japanese).; Advanced Welding Tech., The Second Int. Symp. August (1975)

Quench Hardening and Cracking in Electron Beam Weld Metal of Carbon and Low Alloy Hardenable Steels

Abstract

Using seventeen kinds of hardenable steel of plain carbon, Ni-Cr, Cr-Mo, Ni-Cr-Mo low alloy and partly 13 Cr steels the hardenability was investigated for the electron beam weld metals which were welded with a change in welding parameters and preheating temperature, Then a generalized prediction equation for the hardness of electron beam weld metal was established using the concepts of the carbon equivalent and the cooling time from 800°C to 500°C. Hereafter the hardness of the weld metal can be estimated in advance when the chemical compositions of steel, the welding parameters, the preheating temperature and the penetration depth are given.

Moreover the occurrence of four kinds of carck in weld metal, Horizontal Crack, Vertical Crack (I), Vertical Crack (II) and Cold Shut, was macro-and microscopically investigated. As a result the Vertical Crack (II) is only related to weld metal hardening, while the other cracks occur during solidification of weld metal.

1. Introduction

One of the most advantageous features in electron beam welds results in a deep, narrow and parallel-sided fusion zone in spite of low weld heat input.

The low weld heat input to the parent metal in electron beam welding confines the metallurgical changes to a narrow band on either side of the weld centerline in comparison with any other conventional arc welding. However, as electron beam welding is a fusion welding process, the same metallurgical effects as in any other fusion process occur in both the fusion and the heat-affected zones, although the associated heating and cooling rates and temperature gradients are much higher in the case of electron beam welding. Therefore electron beam welds are also subject to the similar metallurgical basic difficulties as any other fusion welding process including cracking, porosity, lack of fusion, hardening and softening and so forth. With steels which are hardenable by a phase change on cooling, structures which are harder than would be obtained with other fusion processes are produced. This usually due to the formation of a martensitic structure. The hardness reached depends mainly on the carbon and other some alloying elements.

The handness of the conventional arc weld deposit can be controlled since filler additions are made in the welding process. However, with electron beam welding, it is usual to melt only the parent metal although some investigations have been undertaken to assess the value of filler materials in reducing peak hardening.

In certain materials, hardened welds will be prone to quench cracking under conditions of high restraint, necessitating the use of preweld and/or postweld heat treatment.

Moreover the high handness of the weld area are generally undesirable because they constitute a sudden change in material properties in the weld zone.

Therefore with electron beam welding of hardenable steels it is very important to investigate the hardness of weld metal with or without preheating under various welding conditions. Furthermore the relation between hardening and quench cracking in the weld metal is also very important for practical purposes.

Unfortunately few investigation, however, has been done [1-2] only for particular steels within authors' knowledge.

In this report, therefore, authors have treated hardening and cracking in electron beam weld metals of various carbon and low alloy hardenable steels.

Firstly, using seventeen kinds of hardenable steel, the maximum hardness in each weld metal with three electron beam welding conditions with or without preheating has been investigated. Then an experimental equation which can estimate the maximum hardness in these electron beam weld metals in advance of welding has been established. Finally the occurrences of weld crackings which are often found in weld metal have been discussed in relation to welding conditions.

2. Experimental Procedure

2. 1 Materials used

Commercial five carbon steels, the carbon content of which varies from 0.18 to 0.53 %, four nickel-chromium low alloy steels, two chromium-molybdenum low alloy steels, four nickel-chromium-molybdenum low alloy steels and additionally two 13 % chromium alloy steels are used in this experiment. The chemical compositions of these steels used are shown in **Table 1.**

These steels were prepared as a rod shape of 50 to 55 mm in diameter and 500 mm in length for electron beam welding purpose.

2. 2 Welding conditions

For electron beam welding 30 kV—500 mA (maximum 15 kW) type machine was used. A bead-on-plate welding was done longitudinally along the side of each rod under each welding condition.

Three welding conditions [A], [B] and [C] have been principally chosed after the preliminary tests. The welding parameters in [A], [B] and [C] are shown in **Fig. 1.**

The penetration depth in these three welding conditions was given about 5, 10 and 20 mm, respectively.

Moreover preheating conditions of 150 and 300°C other than room temperature (about 20°C) have been

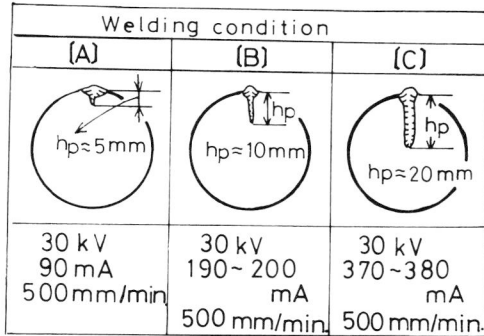

Fig. 1. Welding Conditions used in This Experiment.

adopted for each welding condition.

2. 3 Hardness measurement and microscopic observation

The hardness was measured with Vickers hardness tester with 10 kg load as shown in **Fig. 2 (a)** for weld metal and in **Fig. 2 (b)** for welded joint, and then the hardness of weld metal was defined on an average of the values of three points.

For microscopic observation the solutions of alcohol saturated with picric acid and alcohol saturated with picric acid + (1 g/100 g) wetting agent + 0.5 % $CuCl_2$ + 5 % nital were utilized to reveal the primary

Table 1. Chemical Compositions (wt %) of Specimen used

Material	JIS* SPEC. of material	Chemical composition (wt %)								
		C	Mn	Si	Ni	Cr	Mo	S	P	Cu
Carbon steel	S15C	0.18	0.45	0.23	0.01	0.01	—	0.021	0.009	0.02
	S25C	0.24	0.45	0.24	0.02	0.01	—	0.024	0.007	0.03
	S38C	0.39	0.80	0.29	0.02	0.14	—	0.018	0.022	0.05
	S48C	0.50	0.77	0.26	0.02	0.13	—	0.021	0.017	0.03
	S53C	0.53	0.81	0.28	0.01	0.01	—	0.018	0.014	0.02
Ni-Cr low alloy steel	SNC2	0.31	0.48	0.26	2.68	0.80	—	0.014	0.019	—
	SNC3	0.36	0.52	0.26	3.09	0.69	—	0.026	0.020	0.13
	SNC21	0.13	0.48	0.30	2.19	0.37	—	0.007	0.011	—
	SNC22	0.12	0.48	0.26	3.10	0.88	—	0.023	0.015	0.15
Cr-Mo low alloy steel	SCM4	0.39	0.73	0.29	0.03	1.00	0.18	0.014	0.020	0.02
	SCM21	0.18	0.67	0.30	0.10	1.03	0.17	0.014	0.015	0.17
Ni-Cr-Mo low alloy steel	SNCM2	0.23	0.51	0.24	3.15	1.13	0.18	0.017	0.017	0.10
	SNCM5	0.28	0.57	0.31	2.58	2.66	0.51	0.010	0.020	—
	SNCM8	0.41	0.74	0.26	1.70	0.71	0.17	0.010	0.017	0.13
	SNCM9	0.46	0.73	0.26	1.71	0.80	0.21	0.011	0.017	—
13Cr steel	SUS50	0.09	0.78	0.29	—	11.90	—	0.009	0.024	—
	SUS53	0.32	0.55	0.48	—	13.30	—	0.006	0.025	—

* JIS : Japan Industrial Standard

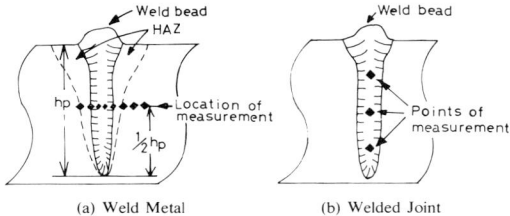

(a) Weld Metal   (b) Welded Joint

Fig. 2. Illustrations for Hardness Measurements.

dendritic structures of solidification and the prior austenitic grain boundaries, respectively.

3. Consideration on Weld Hardenability

3. 1 Hardness in weld metal

As described in the previous report[1], the maximum hardness in electron beam welded joint without filler metal is usually seen in the weld metal where the cooling rate is the maximum. Typical examples of the hardness distribution for welded joints of some steels used in this experiment are shown in **Fig. 3 (a)** and **(b)**.

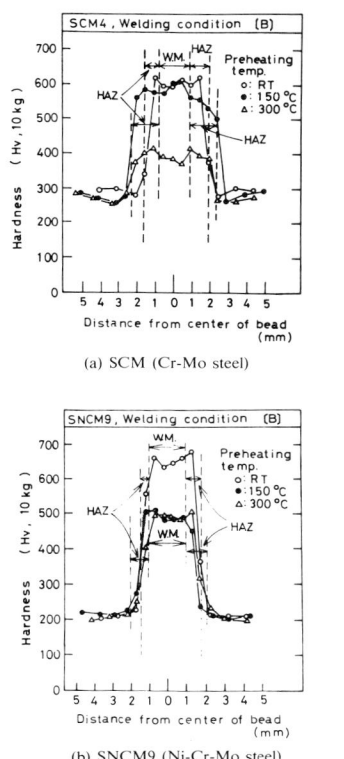

(a) SCM (Cr-Mo steel)

(b) SNCM9 (Ni-Cr-Mo steel)

Fig. 3. Hardness Distribution in Cross-section of Welded Joint.

The heat-affected zone as well as the weld metal shows the maximum hardness in electron beam welds of hardenable steels. The higher the preheating temperature, the softer the hardness in the weld metal and the wider the heat-affected zone in general.

Under the welding conditions of [A], [B] and [C] the relations between the hardness of weld metal and the preheating temperature prior to welding have been investigated for all steels used These results are shown in **Fig. 4 (a)** through **(e)**. In Fig. 4 (a) for carbon steels, the weld metals of low carbon steels did not harden so much even in no-preheating condition, but with an increase of carbon the weld metals became harder. Then in the weld metals of S48C and S53C the hardness reached over than 700 when no-preheating welding was applied. However the validity of preheating, which prolongs the cooling time in the weld metal from 800°C to 500°C, is notable for high carbon

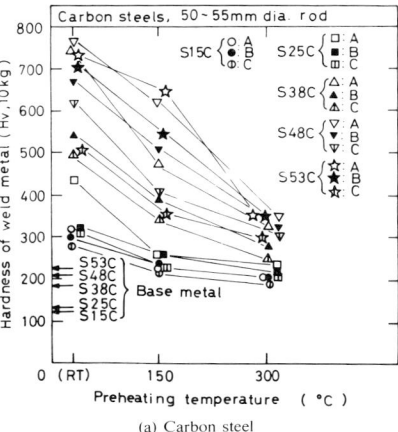

(a) Carbon steel

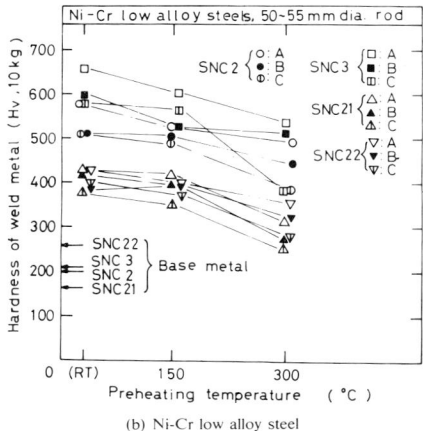

(b) Ni-Cr low alloy steel

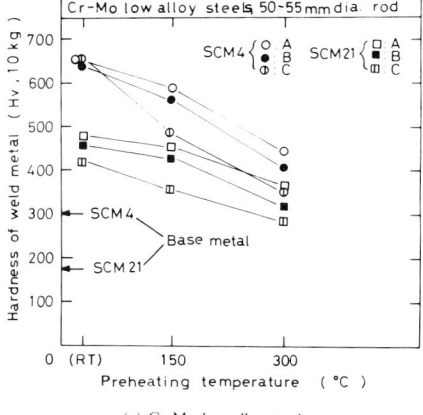

(c) Cr-Mo low alloy steel

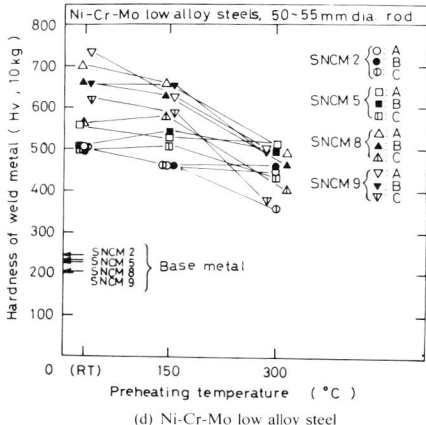

(d) Ni-Cr-Mo low alloy steel

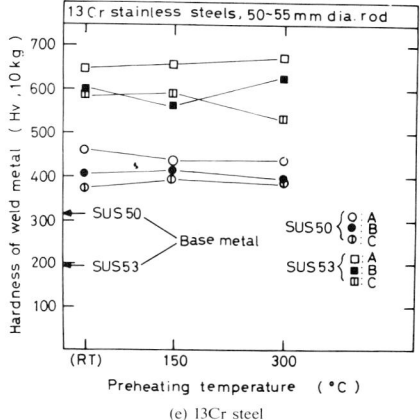

(e) 13Cr steel

Fig. 4. Relations between Hardness of Weld Metal and Preheating Temperature in [A], [B] and [C] Welding Conditions.

steels because the handness decreases promptly with an increase of preheating temperature.

In Fig. 4 (b) for Ni-Cr low alloy steels, the hardness of the weld metals are higher in higher carbon Ni-Cr steels such as SNC2 and SNC3 than for lower carbon steels. Moreover the preheating up to 300°C for these Ni-Cr steels is less effective than that for carbon steels.

The results as to Cr-Mo steels are shown in Fig. 4 (c). The difference in the hardness between SCM21 is depended on the difference of carbon content for which the former is about twice more than the latter. The validity of the preheating is recognized a little and the hardness is reduced to about 400 in the weld metals of SCM4 steel which were preheated at 300°C.

The results as to Ni-Cr-Mo steels are shown in Fig. 4 (d). In spite of lower contents of Ni and Cr in SNCM8 and SNCM9 steels, the hardnesses of these no-preheat weld metals are higher than those of SNCM2 and SMCM5 steels because of higher carbon content in the former. However in the 300°C preheat weld metals of the former steels, the hardnesses are reduced noticeably.

In Fig. 4 (e) the results in 13 Cr stainless steels are shown. The weld metals of SUS53 steel which contains higher carbon shows higher hardness than those of SUS50 steel regardless of preheating temperature. Moreover the preheating up to 300°C is invalid for reducing the hardness.

3. 2 Arrangement for the hardnesses of the weld metals with the C_{eq} which is widely used for arc welding

The definition of Carbon Equivalent (C_{eq}) was widely discussed in the field of arc welding of constructional steels and then various equations for C_{eq} have already been introduced experimentally with respect to the estimation of the maximum hardness in heat-affected zone of arc welded joint[3].

Of the equations the following is widely used for high tensile low alloy steels:

$$C_{eq} = C + Mn/6 + Si/24 + Ni/15 + Cr/5 + Mo/4 \quad (\%) \quad (1)$$

Now authors have tried to obtain the relation between the C_{eq} in eq. (1) and the hardnesses of electron beam weld metals which were given in Fig. 4. These results are shown in **Fig. 5 (a)** for [A] welding condition, **(b)** for [B] and **(c)** for [C]. In each figure, three kinds of mark show the difference in the preheat temperature. From the results in Fig. 5 there is no close relation between them although some relation can be predicted in [C] condition. It must be considered that there is some limitation for the application

322

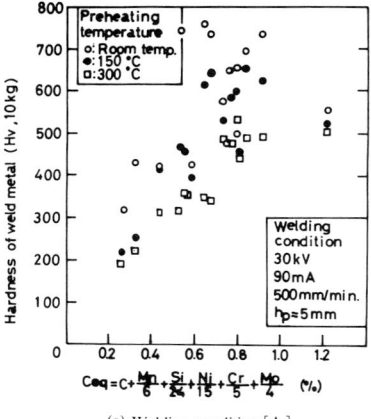

(a) Welding condition [A]

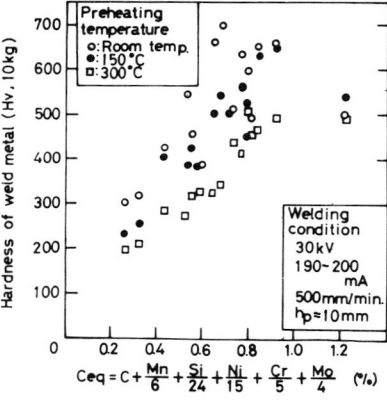

(b) Welding condition [B]

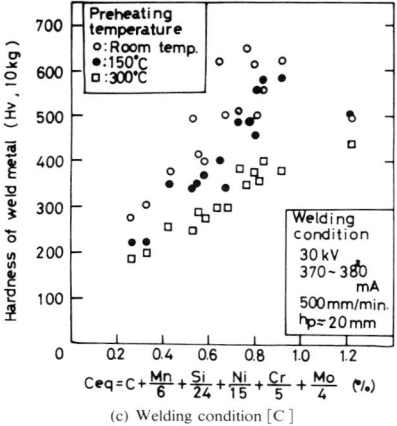

(c) Welding condition [C]

Fig. 5. Relations between Hardness of Weld Metal and Carbon Equivalent for Arc Welding of High Tensile Steels.

of eq. (1) because eq. (1) was obtained from a fixed cooling condition which is given by the defined welding parameters. The cooling condition in the weld metals of the highest heat input of [C] is similar to that in the case of arc welding. Then it seems that hardnesses in [C] welding condition can be fairly put in order with the C_{eq} for arc welding.

3. 3 Prediction for hardness of electron beam weld metal

3. 3. 1 Prediction equation for weld metal hardness to individual welding condition

Using the hardness data obtained, authors have tried to determine new prediction equations for weld metal hardness to individual welding condition by means of the method of least squares. The equations of normal distribution in the method of least squares were computed by means of the iterative process, and FACOM 230-60 is used for this calculation.

In calculation the following two relations were assumed,

$$H_v = a \ C_{eq} + b \tag{2}$$

$$C_{eq} = [C] + \frac{1}{\alpha_1}[Mn] + \frac{1}{\alpha_2}[Si] + \frac{1}{\alpha_3}[Ni] + \frac{1}{\alpha_4}[Cr] + \frac{1}{\alpha_5}[Mo]$$

$$\tag{3}$$

where, H_v: Vickers hardness of weld metal, C_{eq}: Carbon equivalent (%), [C], [Mn], [Si], [Cr] and [Mo]: quantities of carbon, manganese, silicon, nickel, chromium and molybdenum (wt. %), respectively, a, b, $1/\alpha_1$, $1/\alpha_2$, $1/\alpha_3$, $1/\alpha_4$ and $1/\alpha_5$: constants.

Thus values of "a", "b", "$1/\alpha_1$", "$1/\alpha_2$", "$1/\alpha_3$", "$1/\alpha_4$," "$1/\alpha_5$" have been determined for each welding and preheating condition. The results are shown in **Table 2** for each condition.

Moreover relations between hardness and the estimated C_{eq} are shown in **Fig. 6 (a), (b)** and **(c)**.

From these results there are fairly good relations between them.

Therefore the above each equation of the hardness which was calculated has a validity in the fixed welding and preheating conditions.

Moreover as it is understood from Fig. 6 (a) through (e), all prediction equations for the hardness of the weld metal as well as the C_{eq} are approximately similar to each other except the inclinations of these straight lines, although there are some differences in detail. It seems, therefore, that all the hardness of electron beam weld metals will be represented in the same C_{eq} and "b" in eq. (2) with variation of "a", i. e. the inclination of the straight line. The difference in the inclination of the straight line is caused by the

323

Table 2. Values of the Coefficients in the Prediction Equations for Weld Metal Hardness.

Welding condition		Preheat temp. (°C)	α					a	b	σ
			Mn	Si	Ni	Cr	Mo			
(A)	30 kV 90 mA 500 mm/min	RT	2.3	24	34	230	8600	781	88	32
		150	2.4	31	17	23	285	667	38	46
		300	3	18	8.2	5.7	58	387	74	41
(B)	30 kV 190~200 mA 500 mm/min	RT	2.1	24	25	70	400	668	70	34
		150	2.2	26	13	14	6.8	541	53	45
		300	4.3	32	8	8.1	7.9	413	64	32
(C)	30 kV 370~380 mA 500 mm/min	RT	2.6	22	18	18	610	580	91	50
		150	2.7	31	7.6	12	5.2	444	59	47
		300	2.9	14	9.1	5	9.1	272	85	21

$C_{eq} = C + \dfrac{Mn}{\alpha_1} + \dfrac{Si}{\alpha_2} + \cdots\cdots$, σ : Standard deviation of hardness (H_V, 10 kg)

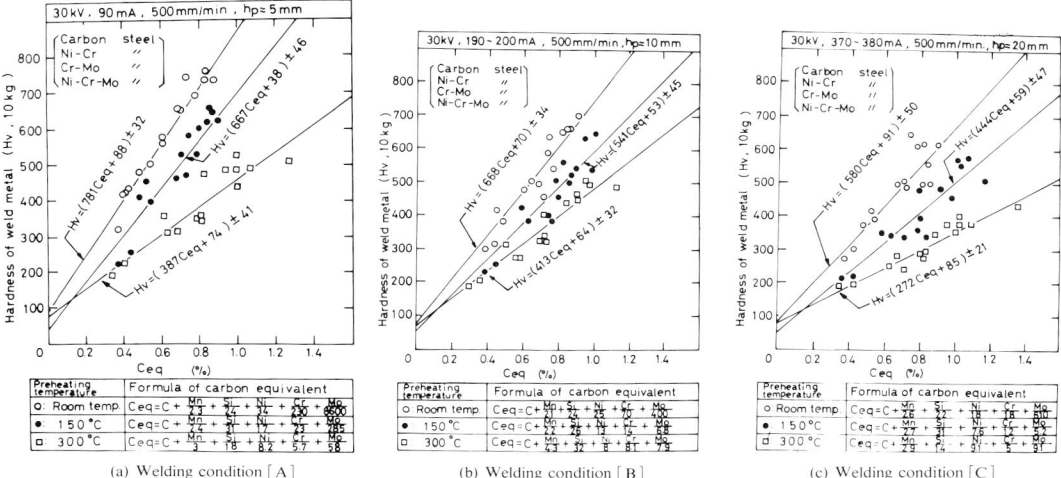

Fig. 6. Relations between Hardness of Weld Metal and Calculated Carbon Equivalent for Individual Welding Condition.

(a) Welding condition [A]   (b) Welding condition [B]   (c) Welding condition [C]

difference in the cooling time passing through the quench temperature range of the steel.

3. 3. 2 A Generalized prediction equation for weld metal hardness in electron beam welding

Authors indicated in the above that the weld metal hardness will be estimated in eq. (2) by taking into consideration for the variation in the value of "a" which is related to the cooling time during quenching. Therefore authors have tried to make a new prediction equation for the hardness of all the weld metals that are welded with electron beam welding.

Authors have reported previously on cooling time

of electron beam weld metal form 800°C to 500°C, $\tau_{800\to500}$, which decides mainly the hardness of weld metal of hardenable steels[1]. That is, in case of electron beam welding the cooling time of weld metal from 800°C to 500°C can be easily estimated from the theory of two-dimensional heat flow from a moving line heat source, the depth of which have a length of the bead penetration. Then the cooling time $\tau_{800\to500}$ was experimentally given by

$$\tau_{800\to500} \doteqdot 3.8 \times 10^{-2} \cdot \left[\frac{V \cdot 0.8\,I}{v \cdot h} \right]^2 \cdot \left[\frac{1}{(500 - T_o)^2} - \frac{1}{(800 - T_o)^2} \right]$$

(sec) (4)

where, V : accelerating voltage of electrons (kV),

I : emitted current of electron from cathode (mA) (the constant 0.8 means the efficiency of the electron beam from the cathode to the workpiece), v : welding speed (cm/sec), h_p: penetration of the weld bead (cm), T_0: initial temperature of the specimen

Moreover authors have established a nomograph eq. (4) as shown in **Fig. 7**.

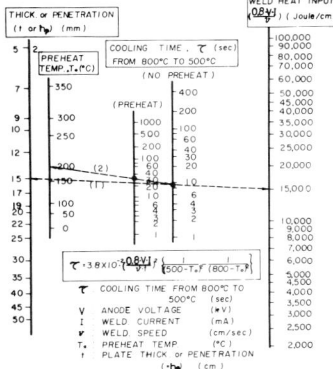

Fig. 7. Nomograph for Estimation of Cooling Time from 800°C to 500°C in Weld Metal

Nextly authors have investigated the relations between the weld metal hardness obtained and the cooling time from 800°C to 500°C which is estimated from eq. (4). These results are shown in **Fig. 8 (a)** and **(b)** in logarithmic co-ordinates.

From Fig. 8 authors have learned that each relation can approximately be represented with a straight line within the range of this experments, while the inclinations of these straight lines repectively show different value within the range of 0.08 to 0.42.

From the grounds of the above argument authors have assumed the following equations as a new generalized prediction equation for the weld metal hardness, that is

$$H_V = A\left\{\frac{1}{\tau_{800 \to 500}^K} \cdot C_{eq}\right\} + B \qquad (5)$$

$$C_{eq} = [C] + \frac{1}{\beta_1}[Mn] + \frac{1}{\beta_2}[Si] + \frac{1}{\beta_3}[Ni] + \frac{1}{\beta_4}[Cr]$$
$$+ \frac{1}{\beta_5}[Mo] \qquad (6)$$

where, A, B, K, β_1, β_2, β_3, β_4, and β_5 : constants, $\tau_{800 \to 500}$: cooling time of weld metal from 800 to 500°C which is estimated by eq. (4) or Fig. 7, C_{eq}: new carbon equivalent, the equation of which is an identical form in spite of welding condition.

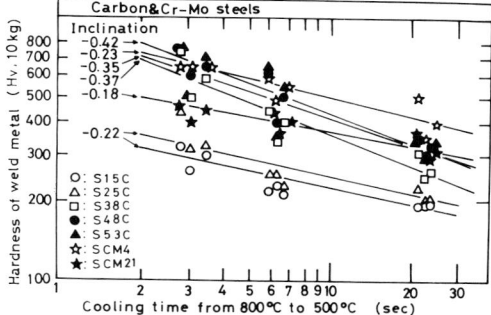

(a) Carbon and Cr-Mo low alloy steels

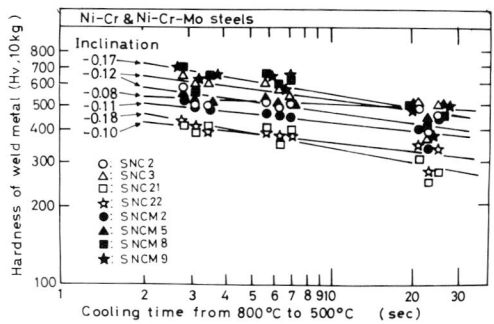

(b) Ni-Cr and Ni-Cr-Mo low alloy steels

Fig. 8. Relations between Hardness of Weld Metal and Cooling Time from 800°C to 500°C in Weld Metal of Various Steels.

On calculation with the computer, K was firstly assumed for nine different values in the range from 0.16 to 0.24, and subsequently the constants were determined for each K using all the hardness obtained.

The calculations showed that the standard deviation of eq. (5) was the minimum at K = 0.22 as shown in **Fig. 9**.

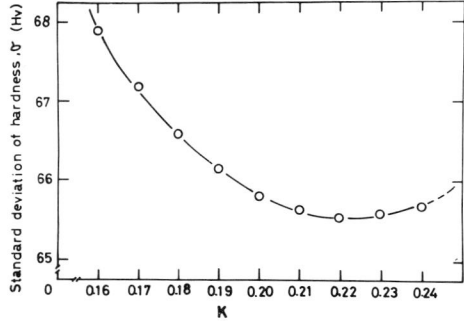

Fig. 9. Relation between K Value and Standard Deviation of Hardness.

325

Therefore authors have adopted this value of K and then the generalized prediction equation for the weld metal hardness was given by

$$H_v = \left[\frac{840}{\tau^{0.22}} \cdot C_{eq} + 58 \right] \pm 66 \qquad (7)$$

$$C_{eq} = [C] + \frac{[Mn]}{2.4} + \frac{[Si]}{24} + \frac{[Ni]}{14} + \frac{[Cr]}{16} + \frac{[Mo]}{60} \qquad (8)$$

where, H_v: Vickers harness of weld metal with load of 10 kg, τ: cooling time from 800 to 500°C which is estimated by eq. (4) or Fig. 7, C_{eq}: carbon equivalent, [C], [Mn], [Si], [Ni], [Cr] and [Mo]: wt. % of carbon, manganese, silicon, nickel, chromium and molybdenum.

Note that eq. (7) is reliable on the carbon and the low alloy steels, the chemical compositions of which are within the limits of 0.1 to 0.55 % C, 0.4 to 0.9 % Mn, 0.2 to 0.3 % Si, less 3.5 % Ni, less 3.0 % Cr, less 0.5 % Mo and less 0.2 % Cu.

The experimental relation between the actual hardness of the weld metals and the predicted hardness from eq. (7) is shown in **Fig. 10.** Most data (about 70 %) are contained within the limits of the standard deviation (± 66), though some of them deviate from both broken lines.

Moreover, in order to confirm the above relation authors have tried to arrange the actual data of hardness, which were obtained in the electron beam weld metals by several researchers,[1, 2, 4, 5] with the predicted

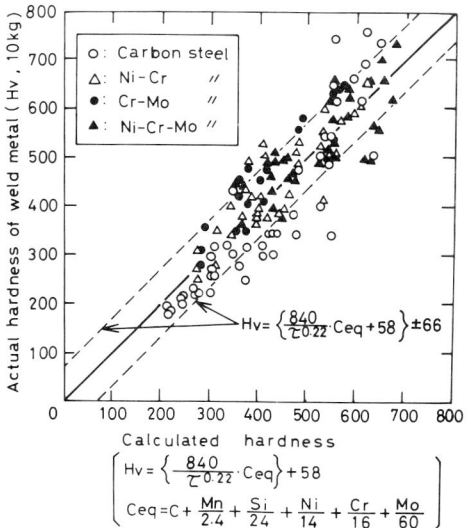

Fig. 10. Comparison of the Actual Hardness with the Calculated Hardness for Electron Beam Weld Metal of Various Steels in This Experiment.

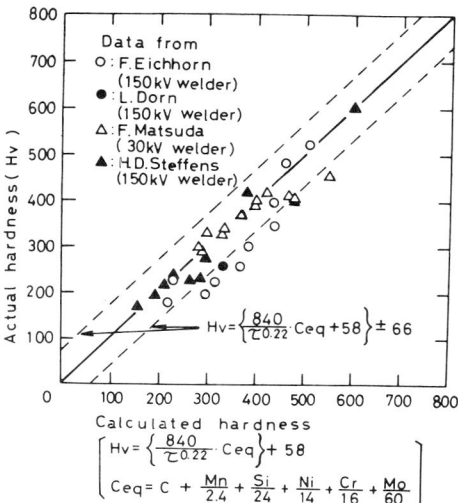

Fig. 11. Comparison of the Actual Hardness of Weld Metals, Which were Measured by Four·Researchers, with the Calculated Hardness

hardness by eq. (7) using the cooling time from the welding parameters and the chemical compositions of the steel used. The result is shown in **Fig. 11.**

These data had been obtained from the different types of welder using some plain carbon, low alloy and high tensile steels.

From a result of Fig. 11 it is considered that equation (7) is fairly valid to predict the hardness of weld metal in electron beam welding of hardenable carbon and low alloy steels.

The hardness of the weld metal, therefore, can be calculated in advance using the welding parameters, preheating temperature, penetration depth predicted and chemical compositions.

The relations between the cooling time of weld metal from 800 to 500°C which is estimated from Fig. 7 and the carbon equivalent in eq. (8) are collectively shown in **Fig. 12.**

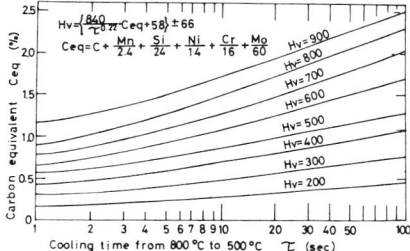

Fig. 12. Estimation Graph for Weld Metal Hardness Which is Determined from Carbon Equivalent and Cooling Time from 800°C to 500°C.

326

4. Consideration on Weld Metal Cracking

4. 1 Occurrence limits of Weld Cracks

There are various kinds of cracks in electron beam weld metal. These cracks are collectively illustrated in **Fig. 13.** Of these weld cracks Vertical Crack (II) is

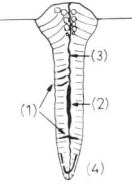

Fig. 13. Schematic Illustration of Various Cracks Which Occur in Electron Beam Weld Metal. (1) Horizontal Crack, (2) Vertical Crack (I), (3) Vertical Crack (II) and (4) Cold Shut.

said to be a cold crack, while the others are said to be hot cracks during solidification.

Some typical examples of the weld cracks which were found in this experiment are shown in **Fig. 14 (a)** through **(e).**

For all 153 weld beads in this experiment authors have investigated the effects of welding condition including preheating on the occurrence of three kinds of crack. These results are collectively shown in **Fig. 15.**

Horizontal Crack occurred in 32 beads (about 20 %), Vertical Crack (I) in 8 (about 5 %) and Vertical Crack (II) in 5 (about 3 %) out of 153 beads. Moreover the Horizontal Crack and the Vertical Crack (I) are abruptly increased in the weld metals with the welding condition $[C]$, the penetration of which shows the deepest in this experiment. On the contrary

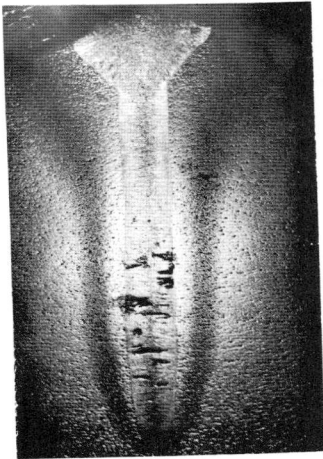
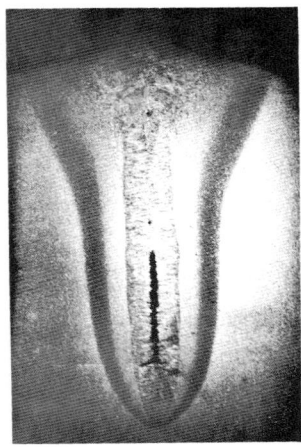

(a) Horizontal Crack (SNC2 steel) (×2.5)   (b) Vertical Crack (I) (SNC22 steel) (×0.9)   (c) Vertical Crack (II) (SCM4 steel) (×1.0)

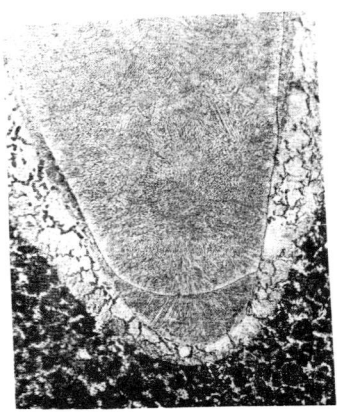
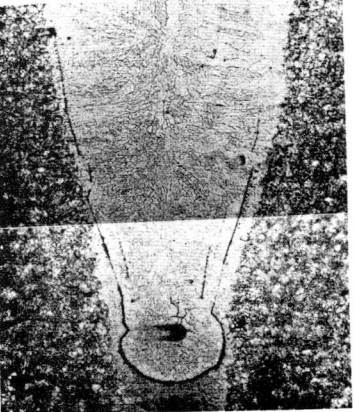

(d) Cold Shut (S53C steel) (×100)   (e) Cold Shut with Crack (SNC2 steel) (×42)

Fig. 14 Typical Examples of Various Cracks Which Occured in Weld Metals in This Experiment.

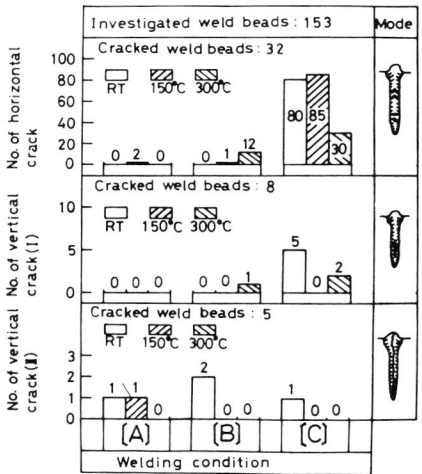

Fig. 15. Relation between Occurrence of Cracks and Welding and Preheating Conditions.

the Vertical Cracks (II) are apt to occur in the weld beads without preheating, the cooling rates of which are usually higher.

It is said in the field of arc welding that the harder in the heat-affected zone, the much susceptible for cold cracking. This will also exist in the field of electron beam welding. Therefore authors have investigated for all weld beads the relation between the hardness of the weld metal and the occurrence of weld cracks. The result is shown in **Fig. 16.** The axes of

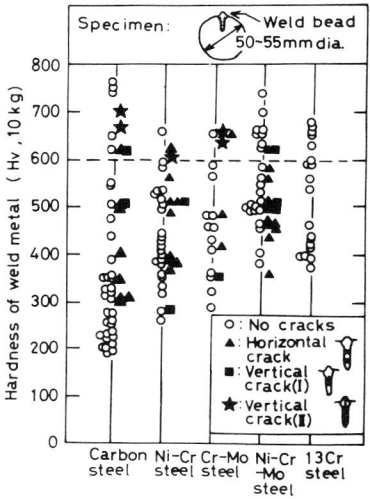

Fig. 16. Relation between Hardness of Weld Metal and Occurrence of Various Cracks.

co-ordinates are the hardness of weld metal and the kinds of steel in which three kinds of crack are shown in three different marks. As a conclusion, the occurrence of the Vertical Crack (II) is only depended on the hardness of weld metal. The limit of the hardness above which the Vertical Crack (II) occurred in the weld metal was placed about 600 in Vickers hardness number. That is to say, the Vertical Crack (II) is much susceptible for electron beam weld metal, the hardness of which exceeds the critical value of about 600 (H_v). Of course it is considered that this critical hardness will be lowered when the restraint of the weld joint is increased.

From the above result it seems to authors that the Vertical Crack (II) is a cold crack, while the other two cracks are incorporated into hot cracks.

4. 2 Metallographic observations of weld cracks

4. 2. 1 Horizontal crack

As shown in Fig. 14 (a) the horizontal cracks occur intercolumnar. This is confirmed with **Fig. 17** which shows the origin of the crack. This kind of crack, occurs during the final stages of solidification usually because of the low melting point compounds formed by impurities such as sulphur. Examples of the sulphur prints for two welds are shown in **Fig. 18 (a)** and **(b)** for SNC2 and SNCM9 steels, respectively.

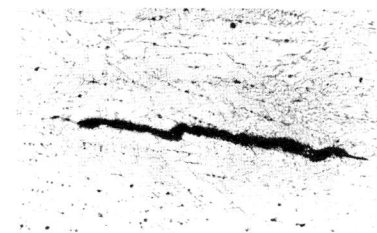

Fig. 17. Horizontal Micro-crack Occurring in Weld Metal of Ni-Cr Steel (SNC3) (×200).

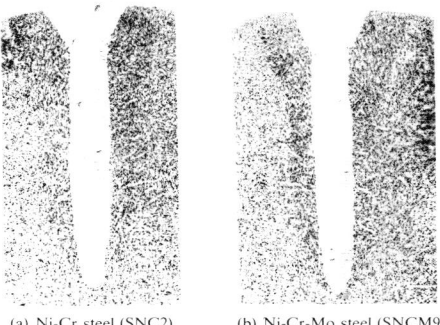

(a) Ni-Cr steel (SNC2)   (b) Ni-Cr-Mo steel (SNCM9)
Fig. 18. Sulphur Print in Cross-section of Welded Joint.

328

It is clear from both sulp' ur prints that sulphur is fairly reduced in electron beam weld metal as compared with base metal because of evaporation. However intercolumnar sulphur bands are sporadically seen, and there the Horizontal crack usually occurs.

4. 2. 2 Vertical Crack (I)

Judging from the crack appearance shown in Fig. 14 (b), this kind of crack also occurs during the final stages of solidification where the grain boundaries at the weld centerline have to take large strains. Therefore this is a solidification crack.

4. 2. 3 Vertical Crack (II)

As previously described, this type of crack is closely related to the hardenability of weld metal. In the weld beads over than 600 in Vickers hardness number this was found as shown in Fig. 16.

Microstructural observations of the typical crack are shown in **Fig. 19 (a)** and **(b)** in panoramic fashion. The solidification structure is revealed in Fig. 19 (a) and the austenitic grain boundaries in (b). The enlarged pictures of the crack are shown in **Fig. 20** for Fig. 19 (a) and in **Fig. 21** for Fig. 19 (b). As a result of the microstructural observations it was made clear that the propagation of this crack does not occur along the dendritic boundaries of solidification structure but occur along the grain boundaries of austenitic secondary structure of the weld metal.

Therefore it seems that this kind of crack is a cold crack which occurs in the temperature below M_s after solidification was completed.

4. 2. 4 Cold Shut

This kind of crack is one of the lack of fusion at or near the tip of penetration as shown Fig. 14 (d) and (e). Examples of the microstructure near the Cold Shut are shown in **Fig. 22 (a)** and **(b).** The growing directions of the substructures on either side of the Cold Shut are apparently different, therefore these grains will be different in their crystallographic orientations.

The Cold Shut will be caused by the intermittent flow of molten metal pushing backward the electron beam which is illustrated in **Fig. 23.**

5. Conclusion

The hardenability and the crack susceptibilty in electron beam weld metals of carbon, Ni-Cr, Cr-Mo and Ni-Cr-Mo low alloy and partly 13Cr alloy steels have been investigated. The main conclusions of this investigation are summarized below.
(1) The electron beam weld metals without preheating

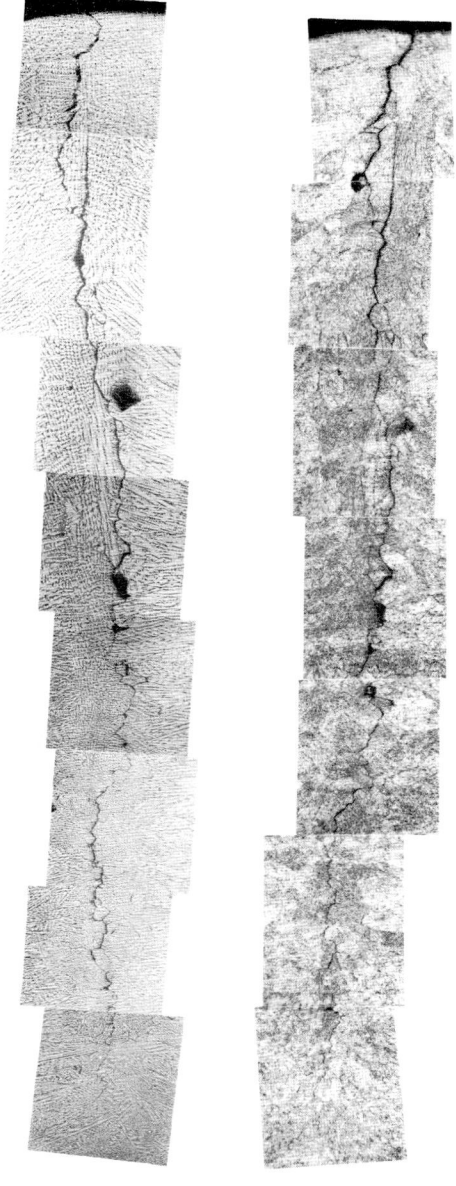

(a) Solidification structure   (b) Prior austenitic grain

Fig. 19. Panoramic Pictures of Vertical Crack (II) in Weld Metal of High Carbon Steel (S53C) (×100).

of high carbon, low alloy and 13Cr steels are so high in hardness that the occurrence of weld cracks or the insufficient ductility in the welded joint are often observed. Therefore preheating and/or post heat treat-

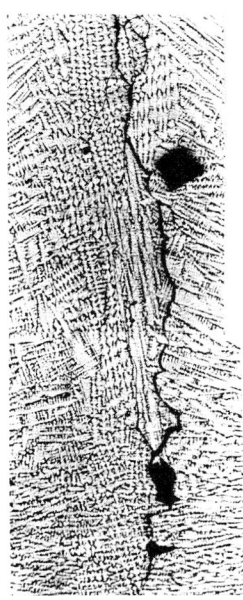

Fig. 20. Microstructure of Cracking Zone in Fig. 19 (a) (Solidification Structure) (×100).

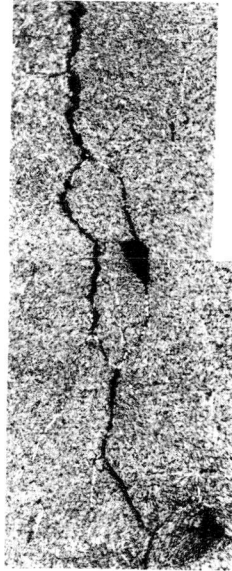

Fig. 21. Microstructure of Cracking Zone in Fig. 19 (b) (Prior Austenitic Grain) (×200).

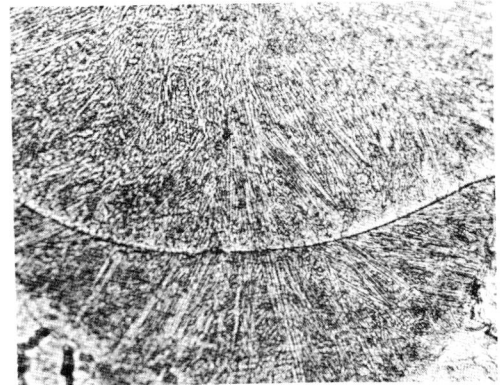

(a) Tip of Cold Shut

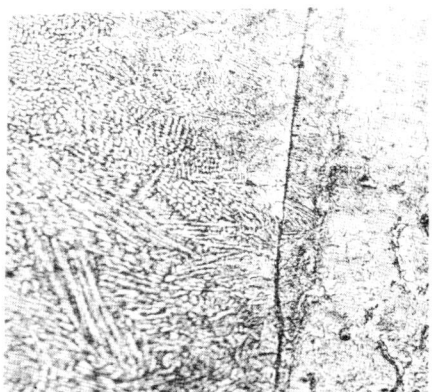

(b) Side of Cold Shut

Fig. 22. Solidification Microstructures on Both Sides of Cold Shut (×200).

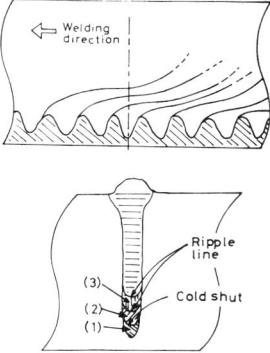

Fig. 23. Schematic Illustration of Cold Shut from Longitudinal and Cross-sections of Welds.

330

ment are required to obtain the sound welds for electron beam welding of these steels.

(2) As to carbon steel in higher level of carbon, the hardness of the weld metal can be reduced to less than 350 in Vickers hardness number with preheating of 300°C, However the hardness of the weld metals of Ni-Cr, Cr-Mo, Ni-Cr-Mo and 13Cr steels can not be reduced so much with preheating up to 300°C.

(3) It is difficult to predict, in general, the hardness of the electron beam weld metals of various hardenable steels, which are welded with various welding parameters, by using the equation of carbon equivalent which was determined and widely used in the field of arc welding.

(4) Therefore in this investigation, a generalized prediction equation for the weld metal hardness was introduced by introducing a concept of the cooling time of weld metal from 800 to 500°C. This is given by

$$H_v = \left[\frac{840}{\tau^{0.22}} \cdot C_{eq} + 58 \right] \pm 66$$

where, H_v: Vickers hardness of the weld metal with 10 kg load, τ: cooling time of weld metal from 800°C to 500°C, C_{eq}: carbon equivalent $[= C + (1/2.4) Mn + (1/24) Si + (1/14) Ni + (1/16) Cr + (1/60) Mo]$. However this equation is reliable on the steels whose chemical compositions are within the limits of 0.1 to 0.55 % C, 0.4 to 0.9 % Mn, 0.2 to 0.3 % Si, less 3.5% Ni, less 3.0 % Cr, less 0.5 % Mo and less 0.2 % Cu.

(5) The above prediction equation is also valid for the prediction on the hardness of weld metals which are welded with the different types of electron beam welder.

(6) Vertical Crack (II) is related to the hardenability of the weld metal, and this crack occurred in the weld metal over than 600 in Vickers hardness number in this experiment.

Moreover this crack propagates through the grain boundaries of the prior austenitic structure, therefore therefore this seems to be a cold crack.

(7) Horizontal Crack, Vertical Crack (I) and Cold Shut seem to occur during solidification of weld metal because the separation occured in the interdendritic boundaries of the solidification structure.

Acknowledgments

The authors wish to aknowledge the co-operative support of Japan Electron Optics Industries Co. Ltd.

References

1) F. Matsuda, T. Hashimoto, Y. Arata: "Some Metallurgical Investigations on Electron-beam Welds", *Trans. Japan Welding Society,* **1**, 1, 72-85 (1970).

2) F. Eichhorn, D. Schumacher, D. Neef: "Investigations on Electron Beam Welding of Preheated Specimens Made of Unalloyed Steels with a Higher Carbon Content", *IIW Doc. IV-56-71-E.*

3) QT Subcomittee JIW IX: "Weldable QT High Strength Steels and Their Applications in Japan", *IIW Doc. IX-673-70.*

4) L. Dorn: "Erfahrungen mit dem Electronestrahlschweissen Allgemeiner Baustähle", *Schw. und Schn.,* **21**, H2 (1969).

5) H. D. Steffens, G. H. Sepold: "Recent Development in Electron Beam Welding of Unalloyed Steels", 4th International Conference on Electron and Ion Beam Science and Technology, 1970, 281-291.

Weldability Concept on Hardness Prediction

Abstract

The maximum hardness is adopted as an index symbolizing the weldability of steels. By newly introducing two parameters (cooling function f(τ) and alloy-element function f(E)) the fundamental equation for the maximum hardness is given. The empirical formula is obtained which is very useful in practice and shows a good agreement with many experimental data.

KEY WORDS: (Weldability) (Maximum Hardness) (Cooling Time) (Alloy-element) (Carbon Equivalent)

1. Introduction

The maximum hardness in the weld part of a metal is one of the most important parameter to estimate the occurrence of "low-temperature crack". The value of this parameter reflects a standardized base for the appropriate selection of the base metal and the welding condition to prevent the low-temperature crack. It is indeed an index symbolizing the "weldability" of a metal and will be formulated in the following expression.

$$H_v = F\{f(\tau), f(E)\} + F_o, \qquad \dots\dots [I]$$

where $F\{f(\tau), f(E)\}$ shows a function of $f(\tau)$ and $f(E)$ which mean cooling function, alloy-element function, respectively and F_o is numerical constant relating to basic hardness of a metal (a certain standard one).

Arata et al.[1] have attempted to solve this formula [I]. By using the carbon-equivalent C_{eq} for $f(E)$ and the cooling function $f(\tau_{T_1 \to T_2})$ for $f(\tau)$ during welding, the maximum hardness H_v is given by

$$H_v = F\{f(\tau_{T_1 \to T_2}), C_{eq}\} + B, \qquad \dots\dots (1)$$

where $\tau_{T_1 \to T_2}$ is the cooling time in seconds from a temperature T_1 to T_2 °C, and B is a numerical constant. T_1 and T_2 are selected to be the ones that characterize the important natures of steels; for example T_1=800°C

and T_2=500°C. Assuming the function F in equation (1) to be the following relation

$$F = \frac{A}{\tau_{T_1 \to T_2}^K} C_{eq}, \qquad \dots\dots (2)$$

where A and K are numerical constants, equation (1) is written as follows by treating many experimental data.

$$H_v = \left\{ \frac{840}{\tau_{800 \to 500}^{0.22}} C_{eq} + 58 \right\} \pm 66, \qquad \dots\dots (3)$$

where $[C_{eq}] = [C] + \frac{[Mn]}{2.4} + \frac{[Si]}{24} + \frac{[Ni]}{14} + \frac{[Cr]}{16} + \frac{[Mo]}{60}$,

$$\tau_{800 \to 300} \doteqdot 3.8 \times 10^{-2} \left[\frac{0.8 I_b V_b}{v_b h_p} \right]$$

$$\left[\left(\frac{1}{(500-T_0)^2} \right) - \frac{1}{(800-T_0)^2} \right]$$

In this equation I_b, V_b, v_b, h_p and T_o are beam current [mA], beam voltage [kV], welding speed [cm/min], penetration depth [cm] and initial temperature of the specimen [°C], respectively. $[C]$, $[Mn]$, $[Si]$, $[Ni]$, $[Cr]$ and $[Mo]$ indicate weight percentages of each element in steels. The standard deviation σ were obtained experimentally to be 66 in the EB-weld part as given in the equation for the many kinds of steels such as carbon and low alloy steels. The result can be illustrated in **Fig. 1** and was confirmed for many kinds of electron beam welds of steels under various welding conditions.[2-3] Such concept is known as Arata Electron beam weldability.[2]

In this paper the generalization of this weldability is intended to be applicable not only to EB welding but also to other welding methods. The maximum hardness is estimated quantitatively with a higher accuracy in the following way: several characteristic values obtained from welding CCT diagrams for many kinds of steels are treated statistically, and by using these values, F function in formula [I] is given by another form instead of equation (2). In this case attention is paid mainly to martensite structure region.

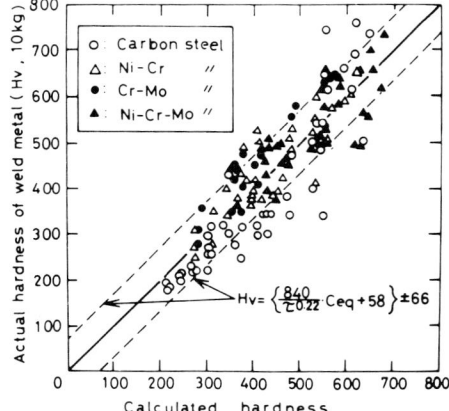

Fig. 1 Comparison of the actual hardness with the calculated hardness for EB-Welds of various steels using eq (3)

2. Formulation of the Problem

As is well known the maximum hardness of HAZ can be estimated[3] from a welding CCT diagram (rapid heating with the maximum heating temperature of 1350°C). **Fig. 2** (a) is a CCT diagram of a typical high strength steel (HT-steel) and (b) shows the relation between the structure area percentage of constituent and the hardness obtained from (a),[4] and it is called CCTSH (Continuous Cooling Transformation Structure Hardness) curve hereafter. In the figure τ_{M100} and τ_{M0} indicate the critical cooling time at which quantity of martensite becomes 100 and 0%, respectively. The solid line in (b) shows the actual hardness curve (CCTSH curve) which enables us to suggest the hardness of HAZ in various welding conditions using any kind of heat source. It is clear from Fig. 2 (b) the hardness is closely connected with the quantity of martensite in the structure. For example in the region of

short cooling time (small τ) with 100% martensite it keeps almost a constant value. While in a long time duration (large τ) it decreases sharply in relation to the martensite.

The solid line in **Fig. 3** shows the approximated H_v-τ curve corresponding to the proposed hardness equation in this research. A and B in the figure are the critical points showing 100% and 0% martensite, respectively. H_v (τ_{M100}) means the hardness at τ_{M100}. C corresponds to the critical ferrite separation point.

The following assumptions are made in obtaining the appropriate hardness curve.
(i) the hardness saturates at $\tau < \tau_{M100}$ with a value $H_v(\tau_{M100}$).
(ii) In the region of $\tau > \tau_{M100}$ the hardness decreases exponentially passing through points B (or B and C) and shows an asymptote with $\tau \to \infty$. Knowing the characteristic values at the critical points A and B (or A, B and C) the necessary hardness curve could thus be obtained. Such H_v-τ curve is called the "characteristic hardness curve" which is approximate to the CCTSH curve (actual hardness curve). The hardness equation (characteristic hardness curve) estimated from such A and B-characteristic values is called as Empirical Formula α and the one from A, B and C-characteristic values as Empirical Formula β. (As for the practical form of the equations, refer to eqs. (16)~(18).)

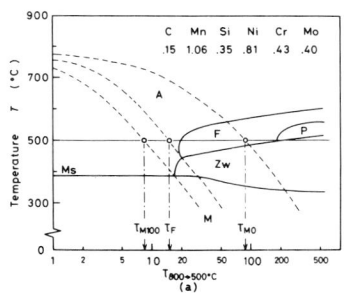

(a): welding CCT (continuous cooling transformation) diagram

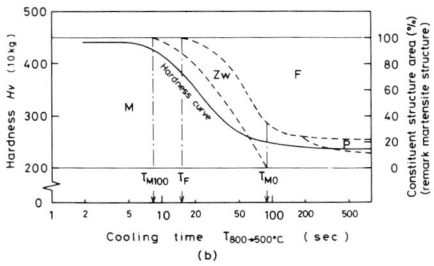

(b): welding CCTSH (continuous cooling transformation structure hardness) diagram

Fig. 2 Relation between welding CCT diagram and hardness curve

333

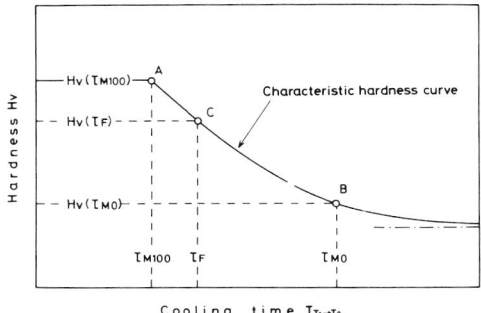

Fig. 3 Relation between A, B, C-characteristic each point, cooling time and hardness (This relation is called characteristic hardness curve which is compared with CCTSH actual curve)

3. Numerical Treatment of the Characteristic Values and the Empirical Formula

3.1 Numerical treatment of the characteristic values

Each characteristic value can be analyzed regressively and in the regression it is assumed that the value depends only on the alloy-elements and each element is a linear independent operator for every characteristic values. A lot of welding CCT diagrams which were obtained by Inagaki et al.[3-11] for about 70 kinds of steels in **Table 1**

Table 1 Chemical composition range of CCT diagram used.

Chemical Composition	Range (%)
C	0.07 ~ 0.53
Mn	0 ~ 1.50
Si	0 ~ 0.60
Ni	0 ~ 3.50
Cr	0 ~ 1.50
Mo	0 ~ 0.60
V	0 ~ 0.15
B	0 ~ 0.01

were adopted for such treatment. In the practical analysis the materials are classified into the conventional welding steels (Si-Mn type steels) and the HT-steels. For the former materials the alloy-elements except C, Si, Mn are included only below 0.1%.

a) Each characteristic value for the conventional welding steels.

A-characteristic value: (it is defined as the one at the critical point A)

$$H_v(\tau_{M100}) = 835[C] + 287 , \qquad \dots\dots (4)$$

$$log\ \tau_{M100} = 2.55([C] + \frac{1}{6.3}[Mn]$$
$$+ \frac{1}{3.6}[Si]) - 0.92 . \qquad \dots\dots (5)$$

B-characteristic value: (defined at B)

$$H_v(\tau_{M0}) = 273([C] + \frac{1}{13}[Mn]$$
$$+ \frac{1}{9.7}[Si]) + 133 , \qquad \dots\dots (6)$$

$$log\ \tau_{M0} = -0.37([C] - \frac{1}{1.1}[Mn]$$
$$- \frac{1}{0.44}[Si]) + 1.02 . \qquad \dots\dots (7)$$

C-characteristic value: (defined at C)

$$H_v(\tau_F) = -277([C] - \frac{1}{12}[Mn]$$
$$- \frac{1}{2.4}[Si]) + 339 , \qquad \dots\dots (8)$$

$$log\ \tau_F = 5.77([C] + \frac{1}{17}[Mn] +$$
$$+ \frac{1}{14}[Si]) - 0.88 . \qquad \dots\dots (9)$$

b) The characteristic value for HT-steels.
A-characteristic value:

$$H_v(\tau_{M100}) = 835[C] + 287 , \qquad \dots\dots (10)$$

$$log\ \tau_{M100} = 5.9([C] + \frac{1}{19}[Mn] + \frac{1}{14}[Si] + \frac{1}{37}[Ni]$$
$$+ \frac{1}{19}[Cr] + \frac{1}{9.1}[Mo] - \frac{1}{49}[V]$$
$$+ \frac{1}{0.31}[B]) - 1.13 . \qquad \dots\dots (11)$$

B-characteristic value:

$$H_v(\tau_{M0}) = 500([C] - \frac{1}{38}[Mn] - \frac{1}{68}[Si]$$

334

$$-\frac{1}{45}[Ni] + \frac{1}{9.0}[Cr] + \frac{1}{9.9}[Mo]$$

$$+\frac{1}{2.1}[V] + \frac{1}{0.48}[B]) + 153 \quad \dots \quad (12)$$

$$log\,\tau_{M0} = -0.20\,([C] - \frac{1}{4.3}[Mn] - \frac{1}{0.40}[Si]$$

$$-\frac{1}{0.58}[Ni] + \frac{1}{0.45}[Cr] - \frac{1}{0.49}[Mo]$$

$$+\frac{1}{240}[V] - \frac{1}{0.0024}[B]) + 1.60 .. (13)$$

C-characteristic value:

$$H_v(\tau_F) = 288([C] + \frac{1}{4.6}[Mn] - \frac{1}{57}[Si]$$

$$+\frac{1}{33}[Ni] + \frac{1}{190}[Cr] - \frac{1}{68}[Mo]$$

$$-\frac{1}{2.0}[V] - \frac{1}{0.43}[B]) + 294, \quad \dots (14)$$

$$log\,\tau_F = 6.18([C] + \frac{1}{17}[Mn] + \frac{1}{35}[Si]$$

$$+\frac{1}{46}[Ni] + \frac{1}{15}[Cr] + \frac{1}{7.0}[Mo]$$

$$+\frac{1}{95}[V] + \frac{1}{0.14}[B]) - 0.93 . \quad \dots (15)$$

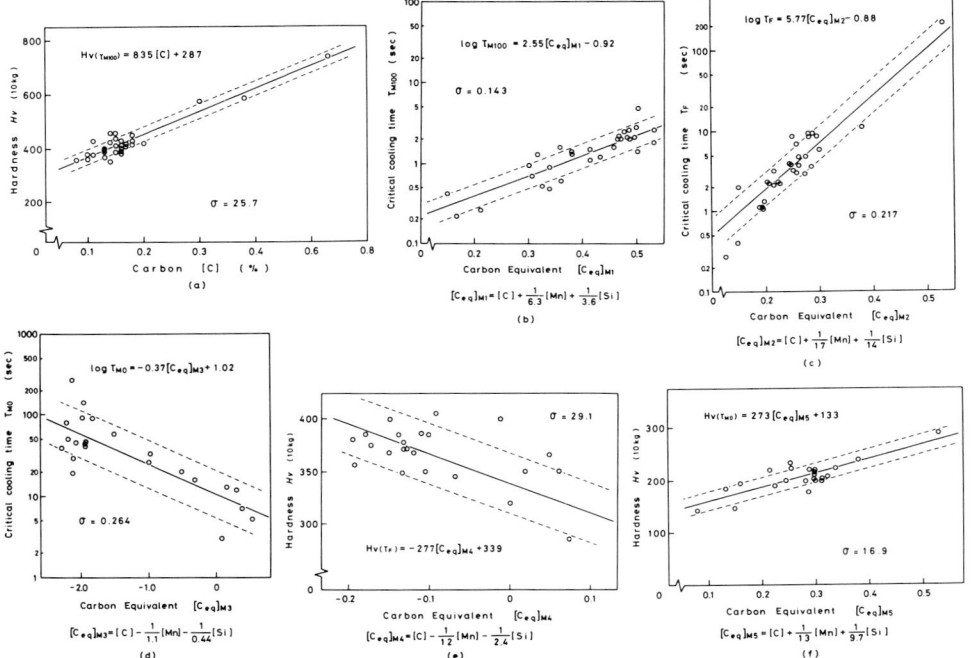

Fig. 4 Characteristic values for conventiond welding steels (Si-Mn type steel)

In the above equations $H_v(\tau_{M100})$ can be treated from the regression equation including only the term of carbon content, because 100% martensite structure hardness depends only on carbon content.[11]

Figure 4(a)~(f) and **Fig.** 5(a)~(f) show the results of eqs. (4)~(15). σ-values in the figures are the standard deviations obtained by using each regression equation. The solid line shows the regression equation and the dotted

one corresponds to σ-value. For example in Fig. 4(a), σ-value of $H_v(\tau_{M100})$ in eq. (4) is estimated to be about ± 26 using 10Kg-Vickers hardness. $[C_{eq}]_{M1}$~$[C_{eq}]_{M5}$ are for some characteristic values of Si-Mn type steels. Indeed not all characteristic values are treated quite well as shown in the figures, but we can satisfactorily draw the hardness curve for any kind of steels using above equations.

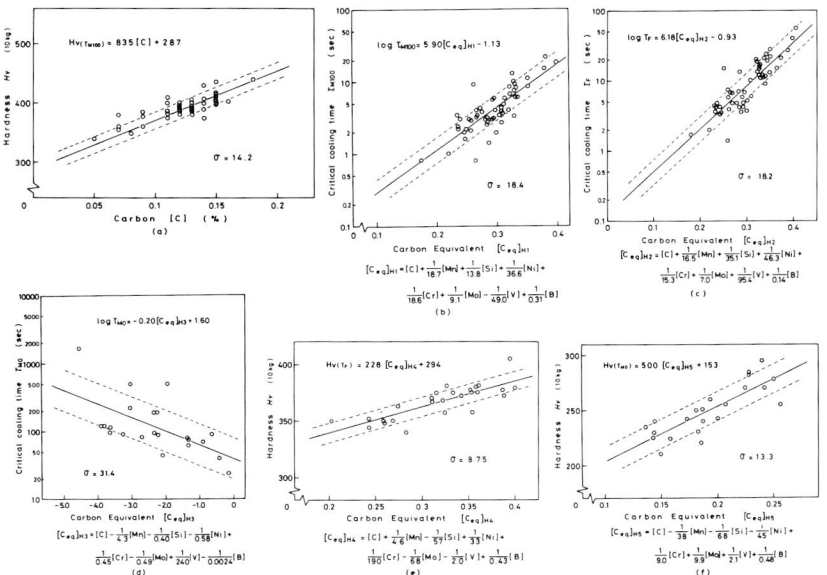

Fig. 5 Characteristic values for HT-steels

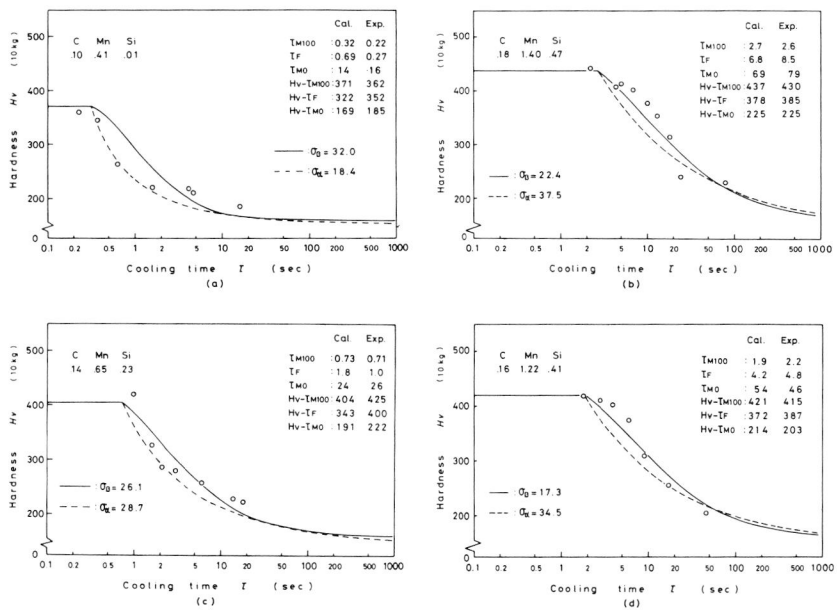

Fig. 6 Comparison between formula α (dashed line), formula β (solid line) and actual hardness (CCTSH; circles), using conventional welding steels

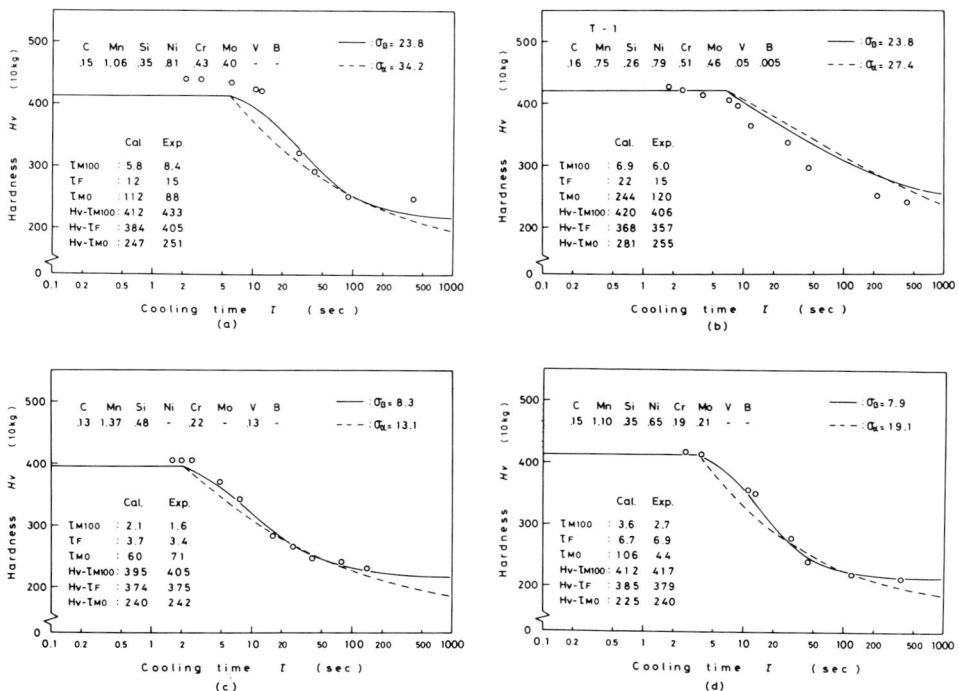

Fig. 7 Same examples as Fig. 6 using HT-steels

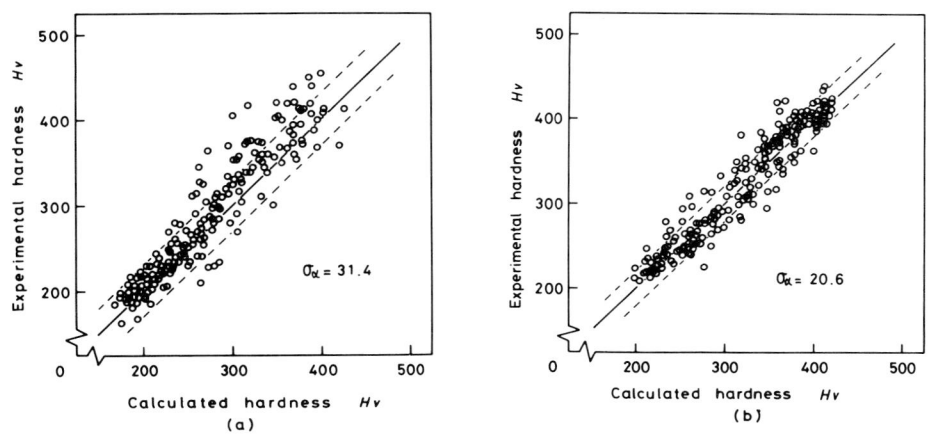

Fig. 8 Summarized result on usability of formula a
(a): Conventional steel (Si-Mn type)
(b): High strength steel

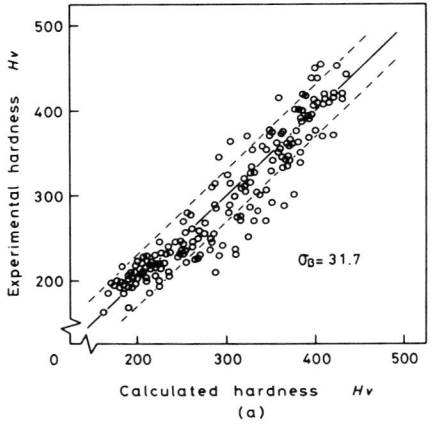

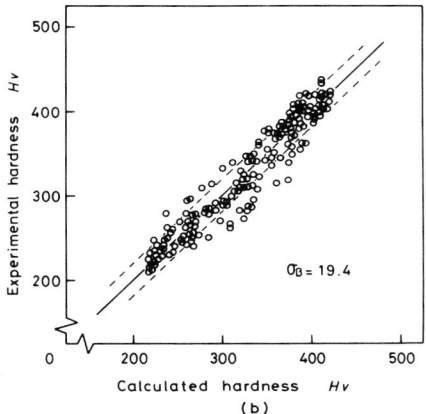

Fig. 9 Summarized result on usability of formula β
(a): Conventional steel (Si-Mn type)
(b): High strength steel

3.2 Empirical formula of the maximum hardness

In this section the characteristic hardness curves as shown in Fig. 3 are obtained by using eqs. (4)~(15) and compared with the actual ones.

The adopted formulas are as follows:

Empirical formula a

$$\tau > \tau_{M100}: \quad H_v = \frac{b}{e^{\log \tau + a}} + 150 \ (160), \quad \cdots (16)$$

$$\tau \leq \tau_{M100}: \quad H_v = 835[C] + 287 . \quad \ldots \ldots (17)$$

Numerical constant 150 (160 in HT steels) in eq. (16)

shows an asymptotic one for $\tau \to \infty$. When a steel is given, critical points A and B are calculated by using its chemical compositions and the characteristic values are substituted into eq. (16). Thus a, b are decided and the empirical equation is obtained.

Empirical formula β

$$\tau > \tau_{M100}: \quad H_v = \frac{1}{e^{c'\log \tau + a'}} + 160 \quad (217 \text{ in HT})$$

$$\ldots \ldots (18)$$

The procedure for obtaining a', b', and c' is quite similar

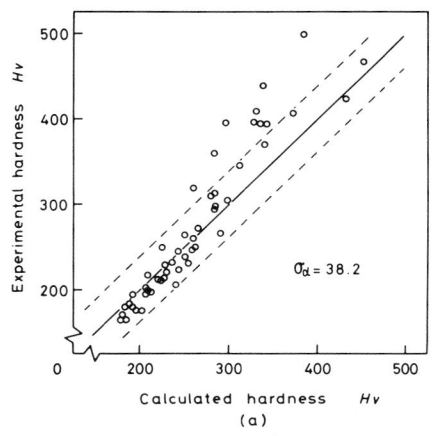

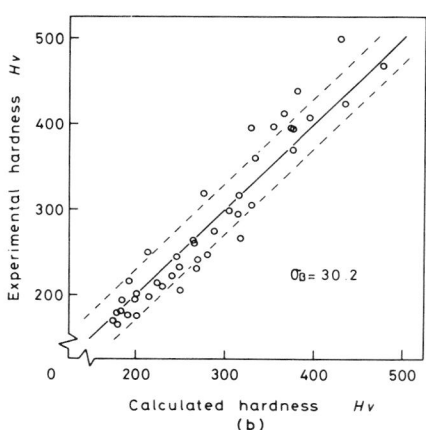

Fig. 10 Check of accuracy of formula a and β for conventional steels

338

to the case of eq. (16). For $\tau \leq \tau_{M100}$, eq. (17) is used.

Figure 6 and 7 show typical examples of comparison of the characteristic hardness obtained from the above method and the measured one by using welding CCT diagram (CCTSH). The standard deviations σ_a, σ_β are obtained by rearranging the measured hardness (circles in the figure) with the formulas a and β, respectively. Clearly the empirical formulas proposed in this paper satisfactorily reflect the actual hardness changes. In **Figs. 8** and **9** are summarized calculated results by using both formulas. For any kind of steels the experimental and the calculated hardnesses agree quite well. We may say there is little difference between the formula a and β judging from values of the standard deviation.

4. Accuracy of Empirical Formulas

First of all the hardness of HAZ is discussed. All of the adopted specimens are heated at 1350°C in the maximum temperature. **Figure 10** shows the comparison of the calculated hardness with the measured ones by Yamamoto et al.[12] for 7 kinds of conventional welding steels. In the figure (a) corresponds to the formula a and (b) to β. There is a larger error in (a). **Figure 11** is the hardness curve for HT-steels obtained from the formula β. The experimental data shown for comparison were reproduced for 50 kinds of steels from welding CCT diagrams made at each steel maker (A~D Co.). Each analyzed result is

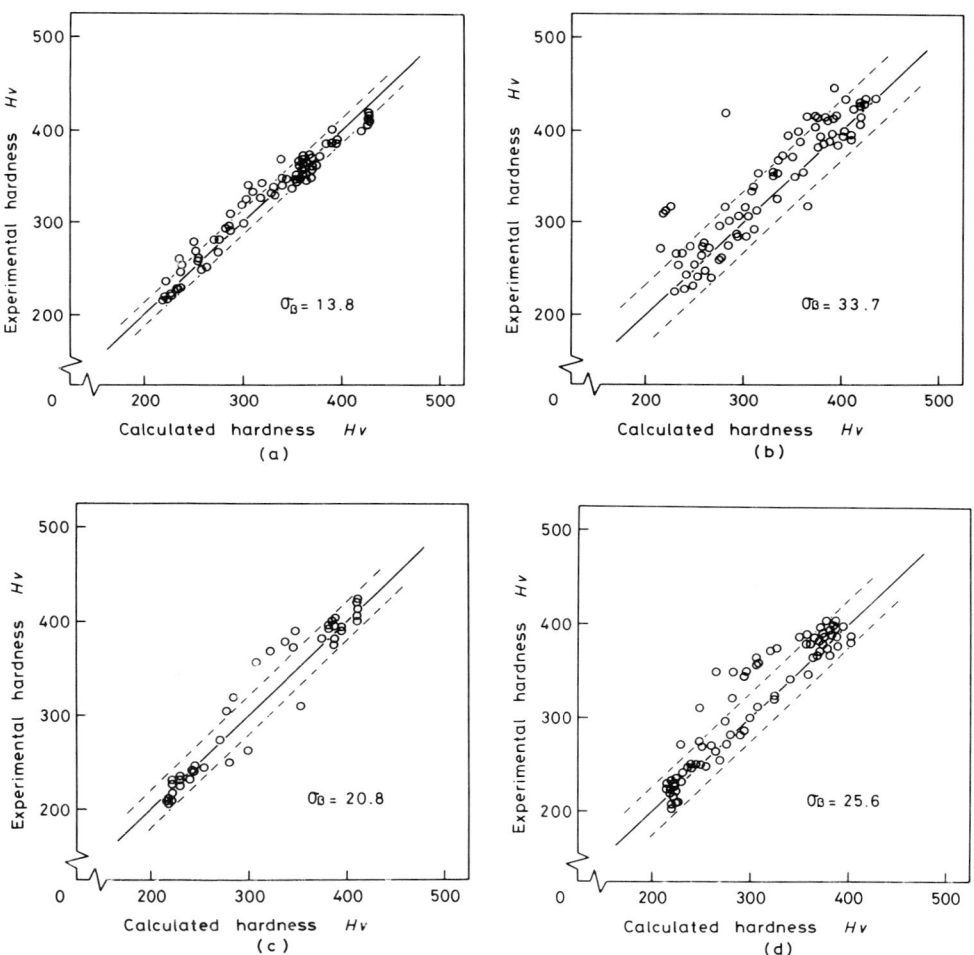

Fig. 11 Check of accuracy of formula β for 50 kinds of HT-steels

drawn with (a)~(d). In **Table** 2 the accuracy of the formula a and β for HT-steel are tabulated by the standard deviations σ_a and σ_β. (See also Fig. 8(b) and Fig. 9(b), σ-values in K of Table 2 were obtained from these figures.)

Table 2 Standard deviation using formula a and β for 50 kinds of HT-steels.

	K	A	B	C	D
σ_α	20.6	31.5	48.3	51.3	43.5
σ_β	19.4	13.8	33.7	20.8	25.6

From these results it can be concluded the hardness curve of CCT diagrams (CCTSH) can be estimated with a high accuracy by using the formula β.

Next, the actual hardness at HAZ are tested by Bessho's data.[13] In his data of the maximum hardness, the welding conditions for 100 kinds of steels (Y-groove inner welding using $1mm\phi$ low hydrogen electrode) are as follows: welding current = 170A; arc voltage = 25; welding speed = 50mm/min. The measured cooling time τ is 4~6 sec. These conditions are typically used in the traditional method for estimating the hardness equation. The traditional method with only one parameter C_{eq} is compared with our result in which chemical composition and cooling time are included. **Figures** 12 (a) and (b) show the examples of the traditional hardness curve by Dearden et al.[14] and Kihara et al.,[15] respectively. They don't give a good result for estimation. While in Fig. 12(c) and (d) our results are shown, where the empirical formulas a and β give by far the more accurate results for wide welding conditions. Comparing (c) and (d) the formula a shows a better agreement than β in contrast to the case of Fig. 10 or Table 2, whose reasons are not

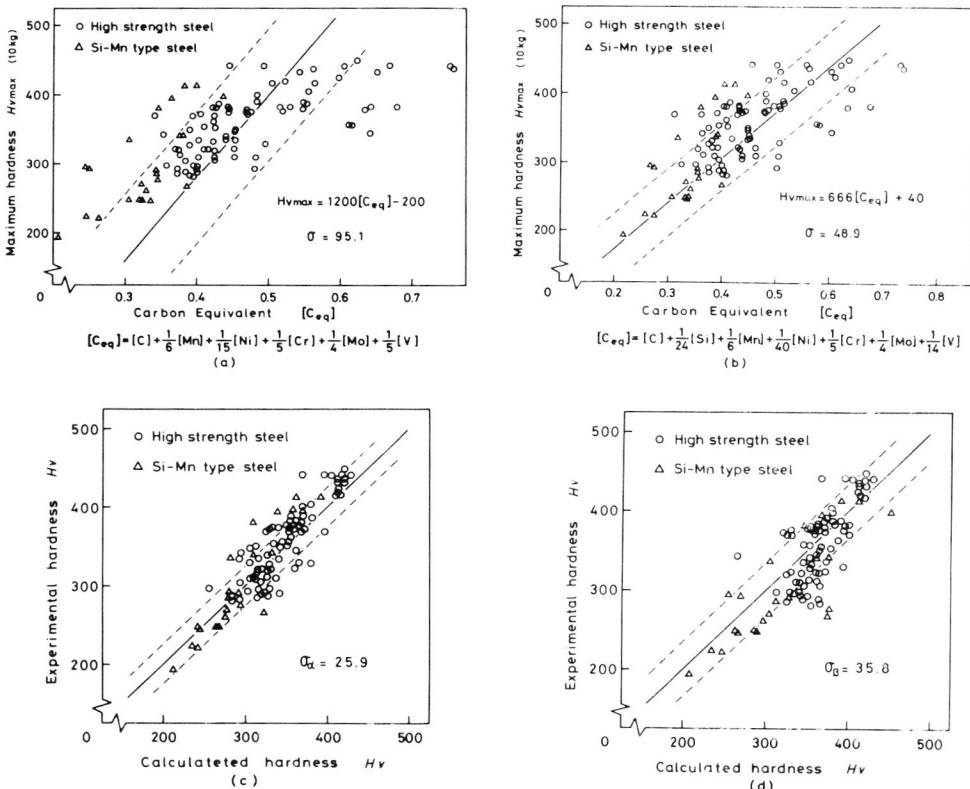

Fig. 12 Comparison between traditional equations and formulas a, β under conventional arc welding condition

clarified.

The result in **Fig. 13** is shown only for reference. The empirical formula (the solid line) gives a considerably good estimation to the experimental hardness[1] of EB-welding part.

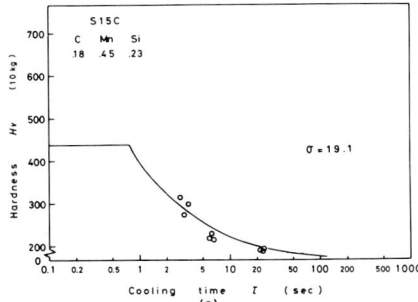

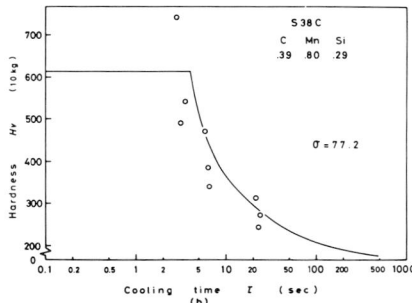

Fig. 13 Comparison between formula α (solid line) and experimental value (circle)

5. Conclusion

Data in CCT diagrams of steels were treated statistically and the estimation of the hardness in the weld part of many kinds of steels were performed. A new index in the evaluation of weldability of steels were obtained by this method.

References

1) Y. Arata, F. Matsuda and K. Nakata; "Quench Hardening and Cracking in Electron Beam Weld Metal of Carbon and Low Alloy Hardenable Steels", JWRI, Vol.1, No.1 (1972); Vol.2, No.1 (1973).

2) M. J. Bibby, J. A. Goldak and G. Burbidge; "Cracking in Restrained EB Welds in Carbon and Low Alloy Steels", Welding Journal, Vol.54, No.8 (1975) 253s-258s.

3) H. Sekiguchi and Inagaki; "Welding CCT Diagram and Its Application (1)", Journal of J.W.S., Vol.29, No.8 (1960) (in Japanese).

4) M. Inagaki and M. Uda; "Welding SH-CCT Diagram for 2H-Super and 2H-Ultra Steels", N.R.I.M. Report, Vol.6, No.1 (1963) (in Japanese).

5) H. Sekiguchi and M. Inagaki; "Welding CCT Diagram and Its Applications (2)", Journal of JWS, Vol.29, No.10 (1960) (in Japanese).

6) M. Inagaki, M. Uda and T. Wada; "A New Apparatus for Determining SH-CCT Diagram for Welding and Its Application to High Strength Steel", Trans. of N.R.I.M., Vol.6, (1964), No.6, 39 54.

7) M. Inagaki, T. Umada and M. Uda; "Welding CCT Diagram for HT-60 Steels", N.R.I.M. Reports, Vol.4, No.2 (1961) (in Japanese).

8) M. Inagaki and M. Uda; "Welding CCT Diagram for HT-70 Steels", N.R.I.M. Reports, Vol.4, No.4 (1961) (in Japanese).

9) M. Inagaki, M. Uda and S. Kanazawa; "Welding CCT Diagram for Various HT Steels", N.R.I.M. Reports, Vol.5, No.3 (1962) (in Japanese).

10) M. Inagaki and T. Kasugai; "Influence of Carbon on Welding SH-CCT Diagram", N.R.I.M. Reports, Vol.14, No.3 (1971) (in Japanese).

11) J. L. Burns, T. L. Moore and R. S. Archer; "Quantitative Hardenability", Trans. of The Americal Society for Metals, Vol.26 (1938), 1~36.

12) S. Yamamoto, T. Sugitaya et. al.; "Study on Embtittlement of HAZ". ND. WM-225-68 (1968) (in Japanese).

13) Bessho; "Study on Index Number of Weld Crack Susceptibility in High Strength Steel", a doctore's thesis in Osaka University (1972) (in Japanese).

14) J. Dearden and H. O'Neill; "A Guide to the Selection and Welding of Low Alloy Structural Steels", Trans. Inst. Weld. (U.K.), Vol.3 (1940), October, 203~214.

15) H. Kihara, H. Suzuki and H. Tamura: "Weldability of High Strength Steels Evaluated by Synthetic Heat Affected Zone Ductility Test", I.I.W. Doc. No.1 IX-288-61.

Study on Characteristics of Weld Defect and Its Prevention in Electron Beam Welding
—Characteristics of Weld Porosities—

Abstract

In this report it performed to investigate upon the types of representative macro weld defects (except crack) and also upon the characteristics of weld porosities which are produced in electron beam weld metal, using eight kinds of materials.

Moreover in these materials, six types of defects such as R-porosity, A-porosity, AR-porosity, unmelted lump, cold shut and spiking, is recognized. The occurring characteristics of these porosities is indicated positively by utilizing I_b-a_b coordinates.

It is proved to be closely related with R-porosity and spiking, with A-porosity and beam active parameter, a_b, and also with AR-porosity and arcing in electron gun.

Additional investigations are made on the effect of gases contained in the materials on the porosities.

1. Introduction

It is well known that the electron beam will be fairly utillized as an excellent heat source for welding.

The most important property of it is to produce an extreme weld deep penetration. It is, however, recognized that the proper defects of the electron beam weld tends to occur in the weld deep zone. Then it is necessary to study the occurring mechanism of these defects and to search its prevention due to extend the industrial application. So it was performed systematically.

So in this report, various investigations of the macroscopic defects such as porosity, unmelted lump, spiking, cold shut and so on except cracks, have been done systematically using eight kinds of metals such as ferrite steel, austenite steel, Aluminum, Titanium and so forth.

2. Experimental Procedure

2.1 Material Used

As shown in **Table 1,** eight kinds of metals were used in this experiments. These materials are three types of austenitic stainless steel of SUS27, SUS27N which contained 0.2 % N_2 and 30Cr—16NiN which contained 0.2% N_2 and higher percentage of Cr and Ni, two carbon steels of semi-killed steel SM4l and killed steel S35C, two aluminum materials of 1200 aluminum and 5083 alloy which contained 4.5 % Mg and additionally TP28 titanium. All of these materials were rolled except the cast steel of 30Cr-16NiN, then prepared as a board shape of 200 to 300 mm in length, 100 mm in width and 30 mm in thickness.

2.2 Welding Equipment and Welding Conditions

150 KV-40 mA Hamilton type EB-welder was

Table 1. Chemical composition of material used.

Element / Material	C	Si	Mn	P	S	Fe	Ni	Cr	Cu	Al	Mg	Zn	Ti	H	O	N
												wt (%)			ppm	
SUS 27	0.08	0.79	0.99	0.025	0.010	Bal.	8.76	18.29						4.4	29	230
SUS 27N	0.02	0.45	1.53	0.006	0.009	Bal.	10.10	18.70						2.5	99	1950
30Cr-16NiN	0.25	0.83	1.30	0.020	0.010	Bal.	16.15	30.56						7.0	75	2325
SM 41	0.18	0.04	0.79	0.090	0.150	Bal.								0.0	316	35
S 35C	0.33	0.25	0.74	0.160	0.170	Bal.								0.1	6	35
1200		0.13	0.01			0.540			0.030	Bal.		0.010		0.4	24	16
5083		0.14	0.64			0.210			0.15	0.040	Bal.	4.60	0.010 0.020	0.4	6	16
TP 28	0.006					0.053							Bal.	21.0	700	50

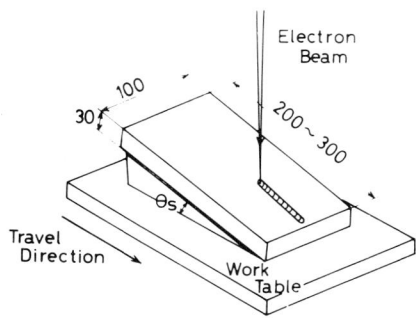

Fig. 1. Slope-welding employed in this experiment.

employed to make the welding bead on these specimens. The experimental method of slope welding was employed as shown in **Fig. 1.** The welding bead was made on the workpiece fixed on a moval table traveling horizontally, and it is set at a certain slope angle, $\pm\theta_s$, with respect to the horizontal plane, crossing at right angle with the beam axis. We call "upslope welding" for $+\theta_s$ slope and "downslope welding" for $-\theta_s$ slope, and it has been done at $\pm30°$ in this experiment.

Defining as given in **Fig. 2** various kinds of welding parameter the beam active parameter(active parameter), $a_b(\equiv D_0/D_F$, D_0: object distance, D_F: focal length), becomes the most important once, and its value is changed continuously by using slope welding mentioned above as show in **Fig. 2.** In this experiment, as the visual beam active parameter a_b^* was

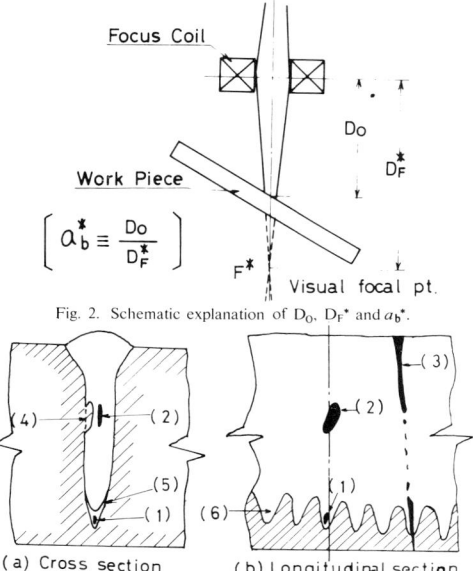

Fig. 2. Schematic explanation of D_0, D_F^* and a_b^*.

(a) Cross section   (b) Longitudinal section

Fig. 3. Schematic illustration of various macroscopical beam weld defects.
(1) R-Porosity, (2) A-porosity, (3) AR-Porosity, (4) Unmelted lump, (5) Cold Shut and (6) Spiking

used, where $a_b^*\equiv D_0/D_F^*$ (D_F^* is visual focal length and defined as the distance between the center of the focusing lens and the visual focal spot which is observed visually as the smallest brilliant point on the surface of workpiece, with so small current that unable to melt it), and its value was the range from 0.46 to 0.93 at $D_F^*=280$ mm.

The welding was mainly performed under following conditions: $V_b=150$ KV (beam voltage), $I_b=10\sim40$ mA (beam current), $v_b=30\sim240$ c. p. m. (welding speed), $P_{ch}=4\times10^{-4}$ Torr (pressure of work chamber), $D_F^*=280$ mm and $\theta_s=\pm30°$.

3. Experimental Results

3.1 Classification of Weld Defects except Cracks

The weld defects observed in longitudinal and transversed cross-section of the specimens are basically classified as shown in **Fig. 3.**

(1) R-Porosity (Root Porosity)

This porosity occurs at the tip of the penetration depth, that is, the vicinity of the root and tends to occur the more at the deeper zone of the penetration. **Photo. 1** and **Photo. 2** show the macro structure in the transverse section and the longitudinal section respectively. Especially in **Photo. 2,** the distribution of the R-porosity is observed at a glance.

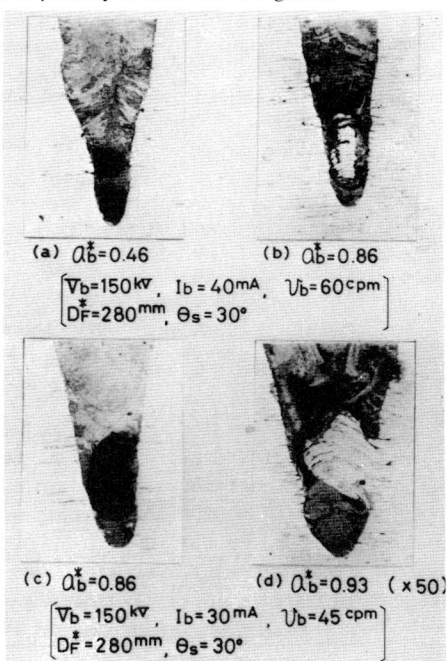

(a) $a_b^*=0.46$ (b) $a_b^*=0.86$
$\begin{bmatrix} V_b=150^{kv}, & I_b=40mA, & v_b=60^{cpm} \\ D_F^*=280^{mm}, & \theta_s=30° \end{bmatrix}$

(c) $a_b^*=0.86$ (d) $a_b^*=0.93$ ($\times50$)
$\begin{bmatrix} V_b=150^{kv}, & I_b=30mA, & v_b=45^{cpm} \\ D_F^*=280^{mm}, & \theta_s=30° \end{bmatrix}$

Photo. 1. R-porosity and unmelted lump in the cross-section of SUS27 beam weld.

343

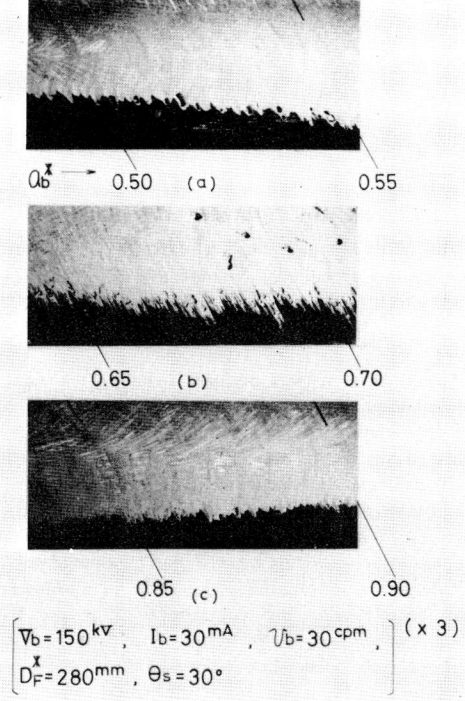

$a_b^x \to$ 0.50 (a) 0.55

0.65 (b) 0.70

0.85 (c) 0.90

$$\left[V_b = 150^{kv}, \quad I_b = 30^{mA}, \quad \upsilon_b = 30^{cpm}, \atop D_F^x = 280^{mm}, \quad \theta_s = 30° \right] \, (\times 3)$$

Photo. 2. Spiking, R and A-porosity in the longitudinal-section of SUS27 beam welds.

(2) A-Porosity (Active Zone Porosity)

This porosity occurs due to the violent action of the active zone of the beam and appears in the vicinity of the travelling path of the beam active zone which makes a bead width minimum or so. **Photo. 3** and **Photo. 2 (b)** show its macro-structures in the transverse section and the longitudinal section.

Especially **Photo. 2 (b)** is indicated precisely the distribution of the A-porosity. This porosity was recognized to occur in stainless steel and carbon steel, but not in aluminum and titanium under this experimental condition.

(3) AR-Porosity (Arcing Porosity)

As shown in **Photo. 4** authors recognized a long porosity which was composed of a chain porosity in the longitudinal section, which connects up usually to the bead surface

This porosity occurs due to the arcing in the electron gun. Such arcing occurs mainly due to the weld vapours generating violently during welding, and in general it tends to occur severely in aluminum and titanium compared with stainless steel and carbon steel. Moreover it is recognized that the AR-porosity usually brings about the formation of needle-like spiking with the R-porosity.

(4) Unmelted Lump

As shown in **Photo. 1 (b), (d)** and **Photo. 3 (b),** unmelted zone either connected or disconected with the parent metal, "unmelted lump", was observed

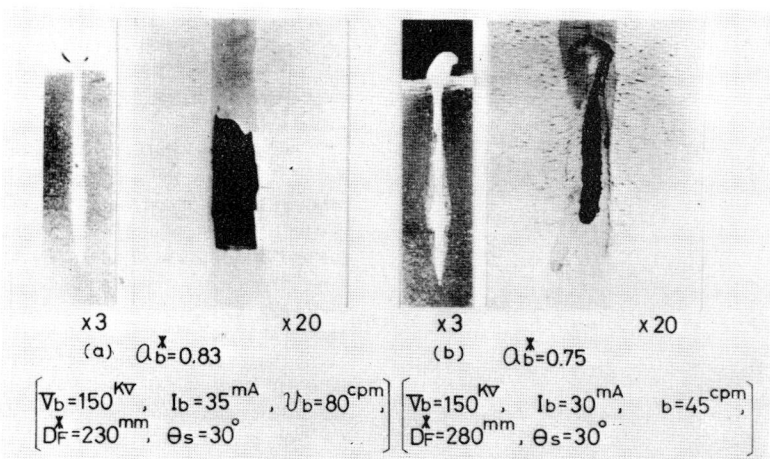

×3   ×20   ×3   ×20

(a) $a_b^x = 0.83$ (b) $a_b^x = 0.75$

$$\left[V_b = 150^{kv}, \quad I_b = 35^{mA}, \quad \upsilon_b = 80^{cpm}, \atop D_F^x = 230^{mm}, \quad \theta_s = 30° \right] \left[V_b = 150^{kv}, \quad I_b = 30^{mA}, \quad b = 45^{cpm} \atop D_F^x = 280^{mm}, \quad \theta_s = 30° \right]$$

Photo. 3. A-porosity and unmelted lump in the cross-section of SUS27 beam welds.

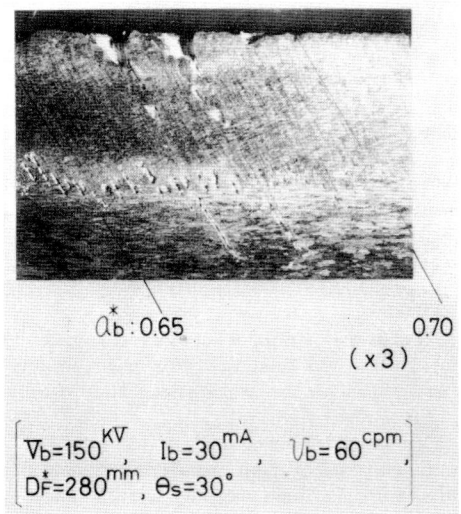

$$\overset{*}{a}_b : 0.65 \qquad 0.70$$
$$(\times 3)$$

$$\left[\begin{array}{l} V_b = 150^{KV}, \quad I_b = 30^{mA}, \quad U_b = 60^{cpm}, \\ D_F^* = 280^{mm}, \quad \theta_s = 30° \end{array} \right]$$

Poto. 4. R and AR-porosity in the longitudinal-section of 1200 beam welds.

inside the weld deposite. Usually it has a tendency to be co-existence with R or A-porosity.

(5) Cold Shut

It is well known that this defect is often found in the electron beam welds[2] and especially near the root zone as shown in **Photo. 5.**

(6) Spiking

The spiking-like penetration, "spiking", is indicated in **Photo. 2** adopted the longitudinal section of welds. As it is well known,[3] it is peculiar phenomenon generating in the high energy density beam welding such as the electron beam welding and is recognized

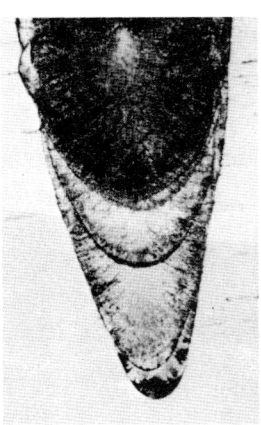

Photo. 5. Cold shut in the cross-section of SUS27 beam welds.

to occur in all materials used. Some of them present strong featurs, others weak ones, and also needle like spiking often appears, especially near the existence region of the former. We call strong spiking, weak spiking and needle spiking respectively. The strong spiking or needle spiking will be taken occasionally as a weld defect.

3.2 Characteristics of R and A-Porosities

The effect on the occurring of A and R-porosity of the welding procedure, materials and so on, was examined and discussed.

3.2.1 Effect of Welding Procedure

The experiment was performed under the same welding conditions as described in chapter **2.2** using the four materials (SUS27, SM41, 1200 and TP28). Typical results obtained are shown in **Fig. 4 (a)~(d).**

To examine the occurring zone of the porosity, weld bead appearance, radiograph and macro-structure of longitudinal section are observed, and its zone was indicated using beam active parameter, a_b^*.

One of these examples is shown in **Photo. 6.** From these results, the occurrence of A and R-porosity are recognized in SUS27 and SM41, while only R-porosity occurs but A-porosity doesn't ocuur in 1200 and TP28 under our experimental condition. The occurence of porosity is shown in **Fig. 5** used an indication, Δa_b^* (a_{b1}^*- a_{b2}^*) which expresses the existence range of the defects: "defect range" From this figure, it is understood that the defect range, Δa_b^* of A-porosity is fairly proportional to beam current, but that of R-porosity show different feature in each materials.

And using the EB-weldability based on the defect range of porosity, TP28 is the best, as the order of 1200, SUS27 and SM41. So, SM41 is worst among these materials.

The effect on the occurring zone of the focal distance, D_F^* and the slope ·angle, $'\theta_s$ in the slope-welding was examined for SUS27. Using D_F^* at 180, 280 and 380 mm, obtained results are shown in **Fig. 6 (a) and (b).** In case of upslope welding (θ_s=30°) and downslope (−30°) obtained results are shown in **Fig. 7** and **Fig. 8,** also the radiographs of both case are shown in **Photo. 7.** Then it is recognized that the occurring zone of R-porosity in the both case have a little difference but that of A-porosity have the distinguished difference.

This phenomena is also recognized in SM41.

3.2.2 Effect of Material

The porosity diagram as shown in **Fig. 4 (a), Fig. 9** and **Fig. 10,** was obtained about three types of austenitic stainless steel specimens. In this case, the

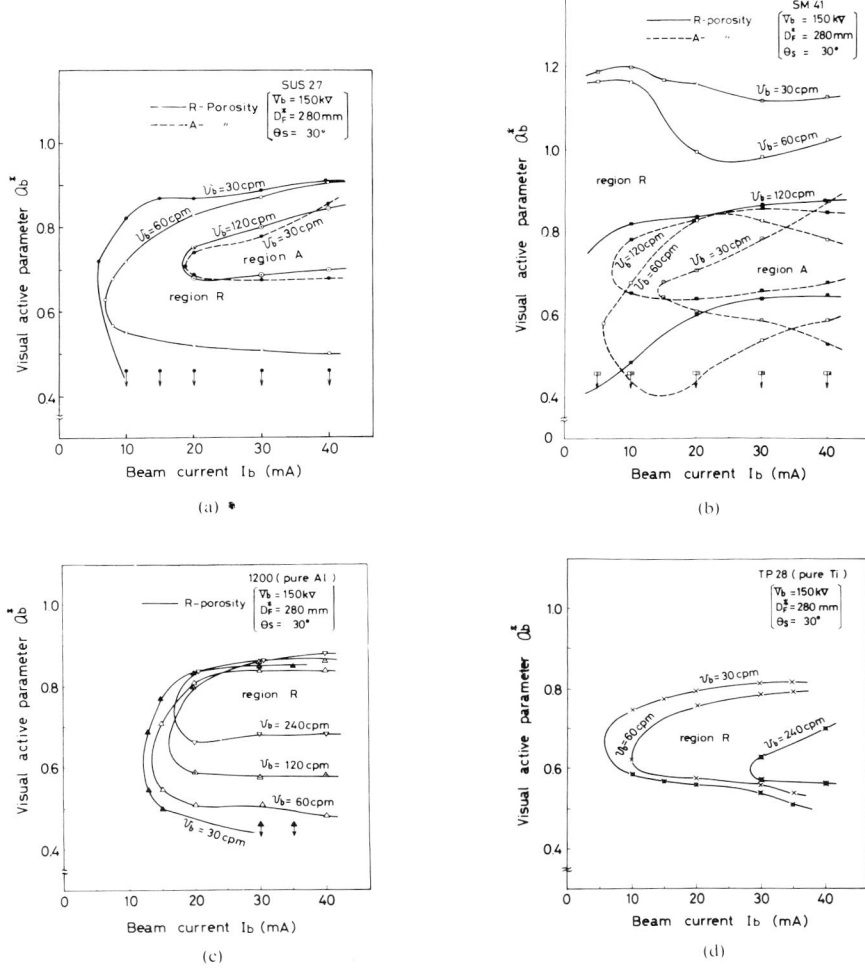

Fig. 4. Porosity diagrams of beam welds. (a) SUS27, (b) SM41, (c) 1200 and (d) TP28.

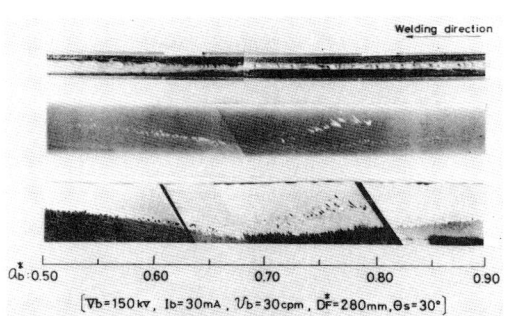

Photo. 6. Bead appearance (top side), radiograph (middle) and
longitudinal-section (bottom side) of SUS27 beam welds.

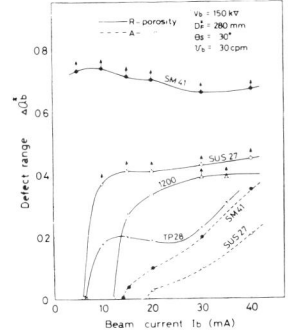

Fig. 5. Porosity diagram indicated relation between beam current and
defect range in typical used specimens.

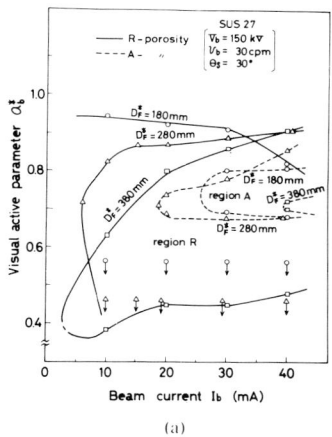

(a)

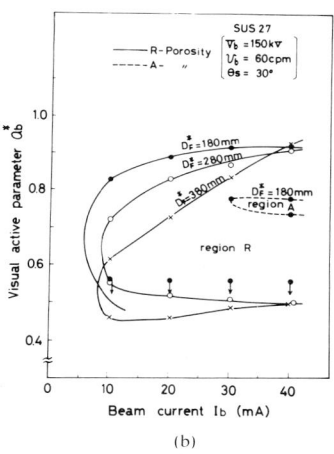

(b)

Fig. 6. Effect of focal length.
(a) $v_b=30$ c. p. m. and (b) $v_b=60$ c. p. m.

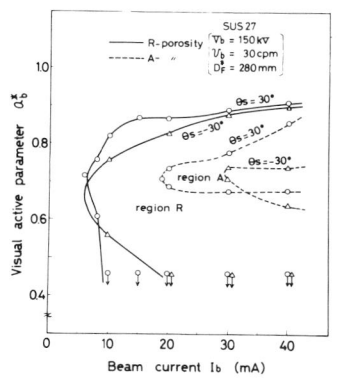

Fig. 7. Porosity diagram indicated effect of upslope and downslope-welding. (SUS27)

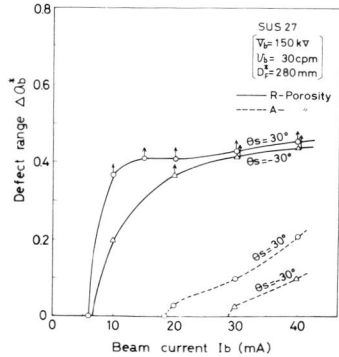

Fig. 8. Porosity diagram indicated effect on defect range of upslope and downslope-welding. (SUS27)

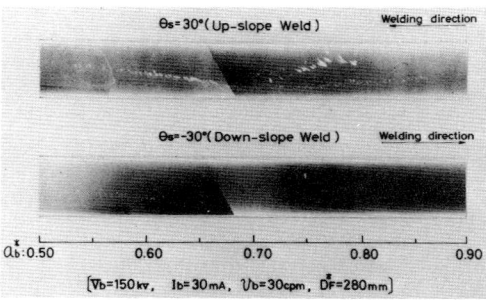

Photo. 7. Radiographs of upslope and downslope beam welds of SUS27.

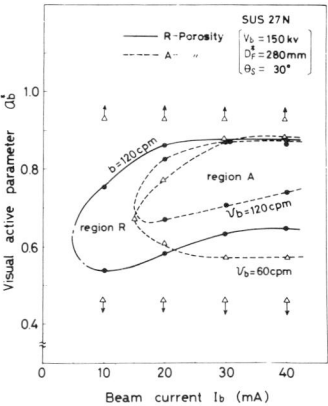

Fig. 9. Porosity diagram in SUS27N beam welds.

347

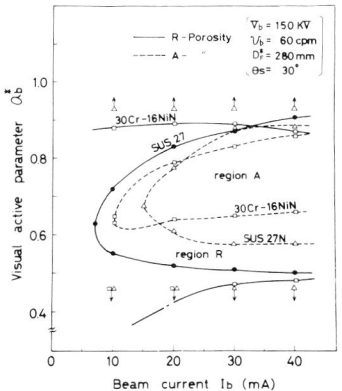

Fig. 10. Porosity diagram of austenitic stainless steels.

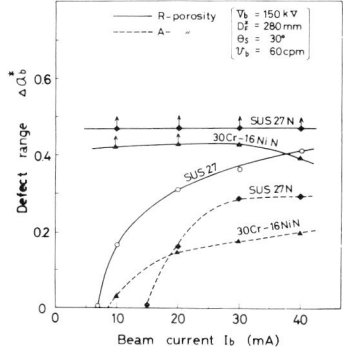

Fig. 11. Porosity diagram indicated relation between beam current and defect range in the austenitic stainless steel.

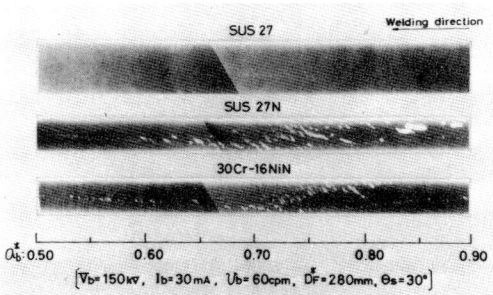

Photo. 8. Radiographs of upslope beam welds of SUS27, 27N and 30Cr-16NiN.

defect range, $\varDelta a_b^*$ and radiograph of the welds were shown in **Fig. 11** and **Photo. 8** respectively.

In these materials, it was proved that nitrogen gas contained fairly controlled the occurrence of porosity,

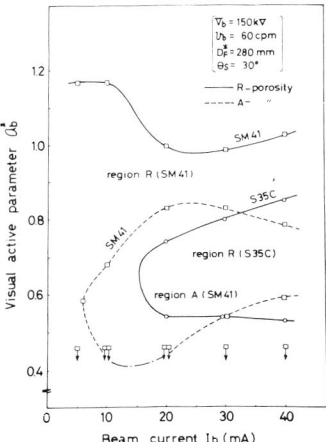

Fig. 12. Porosity diagram of SM41 and S35C.

and that its occurring range increased with increasing nitrogen content. The porosity diagram of SM41, S35C carbon steel is shown **Fig. 12** using 60 c. p. m. welding speed.

Thus, it was recognized that the porosity range in the carbon steel remarkably was increased by oxygen same as nitrogen in austenitic stainless steels.

In general, the penetration depth of 5083 alloy[5] is deeper considerably compared with that of 1200 aluminum and needle-like R-porosity occurs remarkably, but A-porosity doesn't occur as shown in **Photo. 10.** However, liny AR-porosity sometimes occurs through the location corresponding to R and A-porosity region considered as shown in **Photo. 9.**

With increasing the beam current, arcing tends to occur easily and AR-porosity is caused, and when the arcing continues generating of the beam becomes impossible. Compared with the oxygen gas content of 1200 aluminum, TP28 titanium contains extremely high,

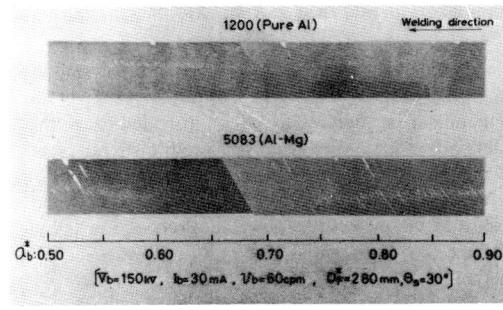

Photo. 9. Radiographs of upslope beam welds of 1200 and 5083.

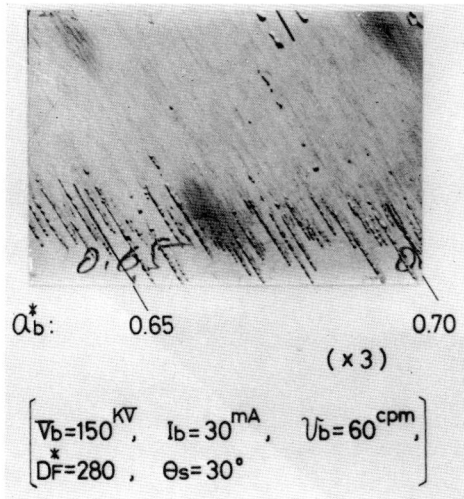

a_b^*:   0.65   0.70

(× 3)

$$\left[V_b = 150^{KV}, \quad I_b = 30^{mA}, \quad V_b = 60^{cpm}, \\ D_F^* = 280, \quad \theta_s = 30° \right]$$

Photo. 10. R-porosity in the longitudinal-section of 5083 beam welds.

nevertheless, R-porosity range is small, and A-porosity does not occur. It seems that titanium prevents the occurrence of porosity due to the oxygen.

3.2.3 Some Discussions

The relation between penetration depth and numbers of R-porosity and spike is shown in **Fig. 13 (a), (b)** and **(c)**. Moreover these figures contain the both results of upslope welding (θ_s=30°) and downslope welding (θ_s=−30°).

Then it seems that the occurrence of R-porosity has no connection directly with the depth of the penetration but is crosely connected with the spiking, because each maximum point of numbers of R-porosity, N_p, and numbers of spikes, N_s, agrees with each other around a_b^*=0.65 (θ_s=30°) or a_b^*=0.55 (θ_s=−30°) without regard to the materials and welding procedure, such as upslope or downslope but that of penetration depth, h_p, and N_p doesn't in case of downslope (θ_s=−30°).

Provided that symbol A_a represents sum of the area of A-porosity located at given a_b^*-point in the longitudinal section, **Fig. 14 (a), (b)** indicates the relation between A_a, h_p and bead width, d_B, and both of the maximum point of A_a and minimum point of d_B agree with each other around a_b^*=0.70～0.75 without regard to the materials and welding procedure such as upslope or downslope. Then it seems that the formation of A-porosity has no connection directly with maximum depth of the penetration, but is closely connected with the region of the minimum bead width.

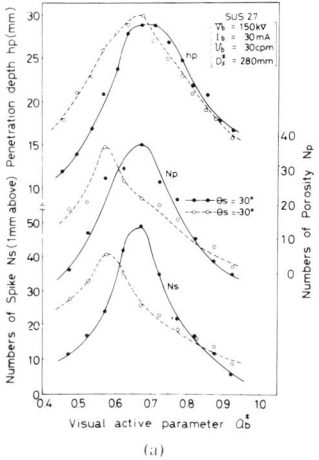

(a)

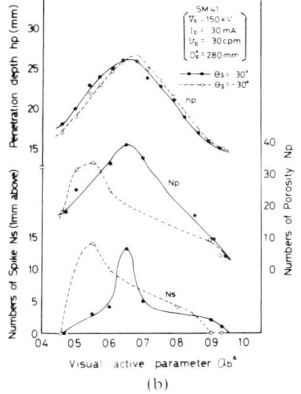

(b)

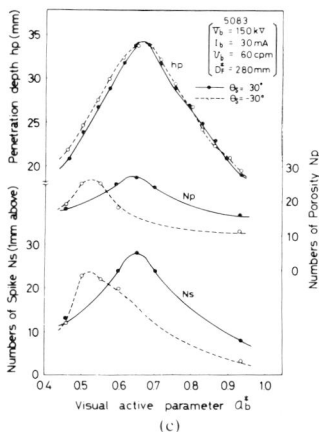

(c)

Fig. 13. Relation between penetration depth and both numbers of R-porosity and spike.
(a) SUS27, (b) SM41 and (c) 5083.

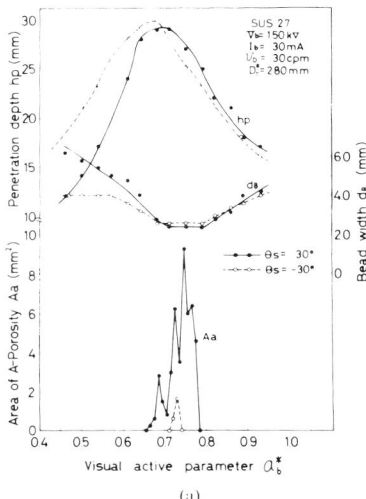

(a)

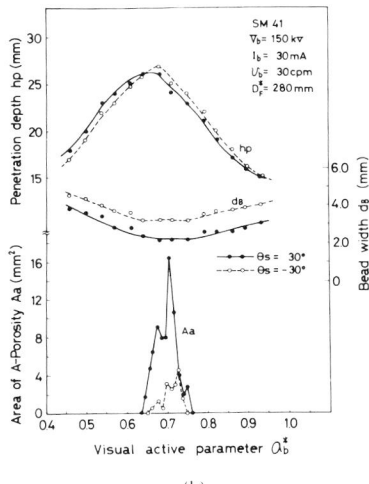

(b)

Fig. 14. Relation between area of A-porosity, bead width and penetration depth.
(a) SUS27 and (b) SM41

In the slope-welding in general, the mutual relation on situations between shape of the beam, specimen and shape of the penetration is given as shown in **Fig. 15.**[4]

The location of the porosity in this schematic explanation was obtained from **Fig. 15,** and moreover the shape of the bead obtained due to the upslope-welding was employed in stead of the shape of the actual beam, because both of them fairly with each other.

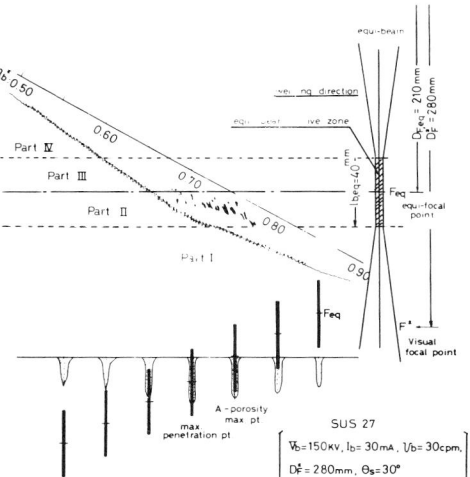

Fig. 15. Schematic explanation of mutual relation on situations between occurring range of porosity and equi-beam.

Then we call such slope-weld bead shape "equivalent beam" or "equi-beam" whose profile is obtained projecting such slope bead at the location of the actual beam.

The mutual relation on the situation between equibeam and specimen is considered in four parts, I～IV in **Fig. 15.** The difference of the characteristics of porosity occurrence in each section are recognized as follows:

(1) Part I: welding is performed using the upperfocus beam whose energy density is relatively low, and the beam active zone has no action there. In this region, only R-porosity tends to occur and its shape of penetration is well type as shown in **Photo. 11 (b).**

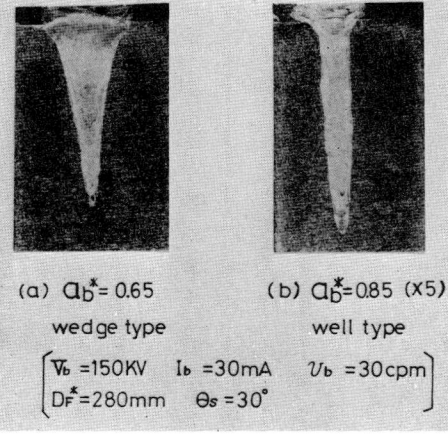

Photo. 11. Sectional profile the bead penetration of 1200.

(2) Part II: lower part, l_{b2}, of the beam active zone with the strong energy density. In this region, the occurrence of A-Porosity are the most. Especially around the just focus, $a_b=1$ or $a_b{}^*=0.75$, it occurs violently and much R-porosity occur also.
In this region, the penetration depth becomes the deepest around $a_b{}^*=0.70$.

(3) Part III: The surface of the specimen is exposed to the upper part, l_{b1}, of the beam active zone. In this region, R-porosity occurs so much, but A-porosity occurs very few, and the penetration depth gradually decreases with being apart from Part II and its shape tends to wedge type.

(4) Part IV: welding is performed using the under-focus beam whose energy density is as low as in Part I and its beam active zone is no concern there. In this region, only R-porosity tends to occur and the penetration depth decreases still more than in Part III.

4. Conclusion

Kind of macroscopic beam weld defects (except crack) and occurring characteristics of R, A and AR-porosity have been investigated using slope-welding.
Obtained results are as follows:

(1) R-porosity, A-porosity, AR-porosity, unmelted lump, cold shut and spiking as macroscopic defects in the beam welds have been observed.

(2) In electron beam weld of Austenitic stainless steel and carbon steel, occurring of porosity was remarkably affected by content of Nitrogen and Oxygen in these materials, and as these contents increase, occurring range of A and R-porosity increase.

(3) In electron beam welding of Aluminum alloys contained such high vapour pressure element as Magnesium, both of the occurring zone of R-porosity and the penetration depth remarkably increases in comparison to pure Aluminum and the porosity shape becomes like a needle. Although A-porosity does not occur, AR-porosity tends to occur there.

(4) Occurring of R-porosity directly is no concern with penetration depth but closely concern with spiking.

(5) Occurring of A-porosity is closely concern with both of the location and shape (length, diameter and profile) of the beam active zone, beam power and its density.

(6) Porosity diagram which is indicated using parameters of $a_b{}^*$, $\Delta a_b{}^*$, l_b and so on, can be sufficiently utilized in a electron beam welding and also it can be employed to search useful welding condition to protect porosity occurrence.

References

1) Meleka: "Electron-beam Welding", Published by Mcgraw-Hill (1971).

2) F. Matsuda, T. Hashimoto and Y. Arata: "Some Metallurgical Investigations on Electron-beam welds", Trans. Japan Welding Society, Vol. 1 No. 1 72-85 (1970).

3) R. E. Armstrong: "Control of Spiking in Partial Penetration Electron Beam Welds" Welding Journal, Vol. 50 No. 8 (1970).

4) Y. Arata: "Characteristics of the Electron Beam Heat Source and View of the Development on its Welding Technology", J. Japan. Welding Society, Vol. 41, No. 11 (1972).

5) M. Ohosumi: Lecture in EBW Comittee of Japan Welding Society, Society, (1971).

Fracture Path Transition Temperature

Mechanical Properties on Electron Beam Welds of Constructional High Tension Steels (1)
Y. Arata, F. Matsuda, Y. Shibata, S. Hozumi, Y. Ono and S. Fujihira
[Trans. JWRI **3** (1974), 185.]

Mechanical Properties on Electron Beam Welds of Constructional High Tension Steels (2)
Y. Arata, F. Matsuda, Y. Shibata, S. Hozumi, Y. Ono and S. Fujihira
[Trans. JWRI **4** (1975), 181.]

Mechanical Properties on Electron Beam Welds of Constructional High Tension Steels (1)

Abstract

Square-butt welding was performed on 25 mm thick plates of commercial high tension steels (HT-50, 60 and 80) with electron-beam welding. Two electron-beam welders (conventional high voltage and low voltage types) were used on this investigation and weld heat input was respectively varied with and without preheating.

Some results of the mechanical properties of the welds of HT—50, 60 and 80 steels such as hardness distribution, tensile and impact properties were investigated in this paper.

Most of welds have had no defects that were detectable by dye penetrant and X-ray inspections. It seemed that only Mn in welds was a little vaporized during electron-beam welding. The maximum hardness of welds was decreased with the adoption of preheating but even in the welds without preheating it was not so high to originate the cold cracks.

Most of tensile-tested specimens for the welded joint fractured in the base metal, and all of face and root bended showed the complete bend angle of 180 degrees without any defects.

Impact strength of welded joints showed adequate value in HT—60 and 80 steels, but inadequate value in some of weld metals of HT—50 in comparison with the criteria of JIS (Japanese Industrial Standard) and WES (Welding Engineering Standard) Specifications. However the results of these mechanical tests indicate that the electron-beam welds in these high tension steels will be anticipated in practical use for many fields of application.

1. Introduction

By use of electron-beam with high power density, the electron beam welding is performed for thicker plates with high speed. As the power density of electron-beam is higher than that of conventional arc heat source, extremely a narrow and deep penetrated bead and small quantity of distortion in welded joint is obtained. Furthermore, the additional benefit of electron-beam welding is that the molten weld metal is free from atmospheric gases of oxygen, nitrogen and moisture owing to vacuum. On the other hand, as compared with the conventional arc welding methods, the cooling rate of molten metal for electron-beam welding is much higher. Namely, time of chemical reaction in fusion zone during welding is extremely short because of a small weld heat input in electron-beam welding.

Then, the release of gases and impurities from the molten metal before solidification is not sufficiently accomplished, and therefore the weld defects occur often. Therefore, electron beam welding must be carefully conducted because the weldability of electron-beam welds is still unknown.

There are few data[1,2], so far, concerning the mechanical characteristics of the high tension steels which are used in constructions as bridge, building, tower and so on. Then, in this investigation, the authors aimed to make clear the mechanical properties on the electron-beam welds for commercial constructional high tension steels as HT—50, HT—60 and HT—80. The authors treated here about tensile, bending and impact properties of these welds as the first stage.

2. Experimental Procedure

2.1 Materials used

The materials used in this investigation are HT—50, HT—60 and HT—80 steels which are widely used in the constructional bridges and buildings. The chemical compositions of these three types of high tension steel are listed in **Table 1.** 25 mm thick plates of these steels were square butt-welded with electron-beam welding processes.

2.2 Welding condition

(1) Electron-beam welders

354

Table 1. Chemical composition of HT—50, 60 and 80 steels.

(wt%)

Composition / Steel	C	Si	Mn	P	S	Ni	Cr	Mo	V	Ceq*
HT—50	0.18	0.43	1.54	0.027	0.021	0.028	0.019	0.002	0.004	0.46
HT—60	0.13	0.32	1.32	0.015	0.013	0.025	0.012	tr	0.030	0.37
HT—80	0.13	0.29	0.85	0.016	0.008	0.98	0.48	0.39	0.023	0.50

$$*: \ Ceq = C + \frac{1}{6} Mn + \frac{1}{24} Si + \frac{1}{40} Ni + \frac{1}{5} Cr + \frac{1}{4} Mo + \frac{1}{14} V \qquad \text{(JISZ 3106)}$$

Table 2. Capacities of high and low voltage type electron-beam welders.

	High voltage type electron-beam welder	Low voltage type electron-beam welder
Maximum beam power	150—40 (KV-mA) =6 (KW)	30—500 (KV-mA) =15 (KW)
Chamber size	1.0 m (Width) ×1.0 m (Height) ×1.3 m (Length)	1.0 m (Width) ×1.0 m (Height) ×2.0 m (Length)
Vacuum of chamber during welding	10^{-4} (mm Hg)	10^{-4} (mm Hg)
Filament	Hair pin-type (Tungsten used)	plate-type (Tantalum used)

All of welding in this investigation were done by using two types of conventional high and low voltage electron-beam welders. **Table 2** shows the capacities of two types of electron-beam welder used in this investigation.

(2) Welding procedure

Welding for all materials to be treated was performed for one pass or two passes without and with preheating of 100°C under the joint profile of butt type as shown in **Fig. 1**. The size of the specimens is 200 mm in width, 500 mm in length and 25 mm in thickness, and each two specimens is welded in longitudinal seam. The welding conditions used on various materials are given in, **Table 3**. The beam active parameters[3] a_b, was respectively selected for 0.97 and 0.93 in high and low voltage electron-beam

welders. Every square butt joint had a good fitup and did not have any excessive gap and misalignment by use of restraint jig and tack welding. Root faces of joint for the electron-beam welding were completely machined to have no seam gap, and the scale on both plate surfaces was also eliminated along the welding direction with the width of 10 mm from the seam. All of the specimens used in this investigation were completely de-magnetized, and the root faces of joint were made clean by acetone in advance of electron-beam welding.

The following shows the additional notes in welding performance.

a) Welding conditions, A and C for the one pass-welding were selected from the result of preliminary test, the weld heat input of which was burnt through 20 mm thick steel plate, and welding conditions, B and D for the two passes-welding were also selected as the weld heal input to burn through 15 mm thick steel plate.

b) 100°C preheating was performed by driving the defocused electron-beam along the welding line by three times, the electron beam of which was about 20 mm in diam. Preheating temperature, 100°C was recognized by use of a chromel-alumel thermocouples.

c) In case of two passes-welding, the raised temperature in welded joint due to the first pass-welding with or without preheating was fully cooled to R·T

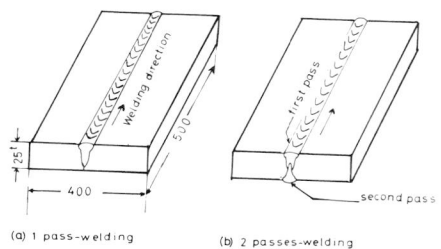

(a) 1 pass-welding (b) 2 passes-welding

Fig. 1. Dimension of specimen.

Table 3. Welding condition.

Condition / Series	Beam power (KV-mA)	Welding speed (cm/min)	Preheat temp. (°C)	Welding procedure
D5A, D6A, D8A	150—33 ≒(5 KW)	15	R·T	one pass-welding as Fig. 1 (a)
D5B, D6B, D8B	150—33 ≒(5KW)	24	R·T	two passes-welding as Fig. 1 (b)
D5C, D6C, D8C	150—33 ≒(5 kW)	15	100	one pass-welding as Fig. 1 (a)
D5D, D6D, D8D	150—33 ≒(5 KW)	24	100	two passes-welding as Fig. 1 (b)
N5A, N6A, N8A	30—250≒(7.5 KW)	30	R·T	one pass-welding as Fig. 1 (a)
N5B, N6B, N8B	30—250≒(7.5 KW)	60	R·T	two passes-welding as Fig. 1 (b)
N5C, N6C, N8C	30—250≒(7.5 KW)	30	100	one pass-welding as Fig. 1 (a)
N5D, N6D, N8D	30—250≒(7.5 KW)	60	100	two passes-welding as Fig. 1 (b)

Note: D; High voltage type E·B-welder
N; Low voltage type E·B-welder
5; HT50　6; HT60　8; HT80
A, B, C, D; Welding condition

before the second pass-welding. Then the second pass was performed without and with 100°C preheating. Both the first and the second passes were welded in the same direction.

3. Experimental Results

3.1 Non-destructive inspection

1) X-ray inspection

All of the electron-beam welds of high tension steels were X-ray inspected. Any defect such as porosity and crack could not be detected in the electron-beam welds of HT—50, 60 and 80 steels. Referring to JIS Specification (JIS Z 3104—1968), most of electron-beam welds were allowable in the first class for general grade.

2) Dye penetrant inspection

All of the weld surfaces of high tension steels were also checked by dye penetrant test. No surface defects were observed at all in all electron-beam welds.

3.2 Chemical analysis of weld metal

Table 4 shows the results of the quantitative

Table 4. Chemical analysis of weld metal.

(wt.%)

Steel		C	Si	Mn	P	S
HT—50	Base metal	0.18	0.43	1.54	0.027	0.021
	D5A— weld metal	0.19	0.42	1.23	0.028	0.021
	N5A— weld metal	0.18	0.45	1.29	0.026	0.021
HT—60	Base metal	0.13	0.32	1.32	0.015	0.013
	D6A— weld metal	0.14	0.36	1.06	0.016	0.014
	N6A— weld metal	0.15	0.36	1.14	0.016	0.015
HT—80	Base metal	0.13	0.29	0.85	0.016	0.008
	D8A— weld metal	0.14	0.32	0.69	0.015	0.008
	N8A— weld metal	0.13	0.28	0.70	0.014	0.005

analysis of C, Si, Mn, P and S for the one pass-welds of HT—50, 60 and 80 steels after welding. Regardless of the difference of the electron beam welder, only Mn in the welds is clearly decreased to that in the base metal. It seems that Mn in the welds vaporizes during electron-beam welding, as the partial pressure of Mn is high enough in calculation in molten metal during welding[4].

3.3 Metallographic and hardness examinations

1) Metallographic examinations

Macrophotographs and microphotographs of the entire welds for HT—60 steels are shown in **Photo. 1,** which were welded with one pass and two passes by high voltage type electron-beam welder. The welds have a typical cast structure with rapid solidification, and the cellular dendrites are extremely small. As welding speed decreases to 15 cm/min, the cellular structure parallel to the welding direction is observed in the center of weld bead as shown in upper Photo. 1. Preheating did not show any obvious effect upon variation of the microstructures.

2) Hardness examinations

Harkness distribution was measured in each cross-section of the welds using Vickers hardness tester with 1 kg load. Vickers hardness distributions of welds

for HT—50, 60 and 80 steels using low voltage-type welder are respectively shown in **Figs. 2, 3** and **4.** It is anticipated in general that the hardness must be maximum at the weld metal because the cooling rate is theoretically the highest in the temperature range to be quenched. However the weld metal was usually softer

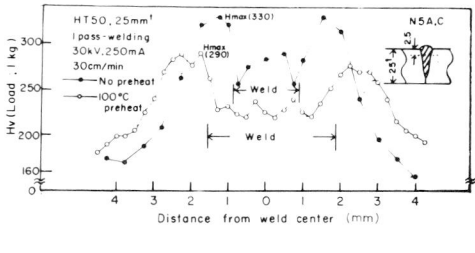

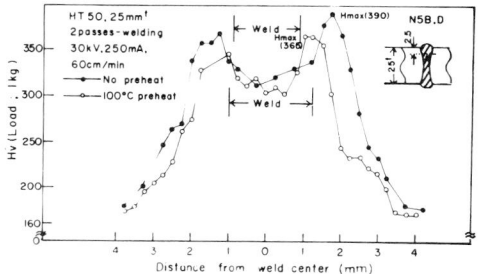

Fig. 2. Vickers hardness distribution of welds for HT—50 steel using low voltage type E·B welder.

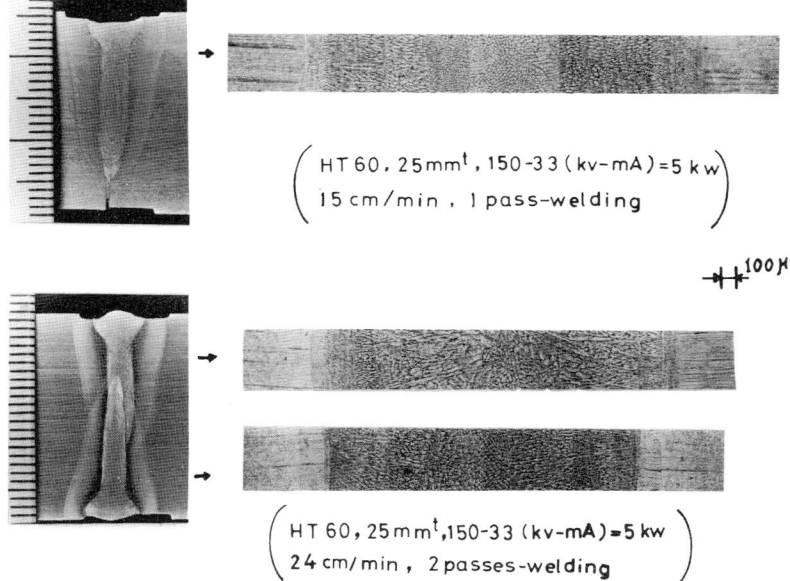

Photo. 1. Macro and microphotographs of welds without preheating for HT—60 steel by use of high voltage type E·B welder.

357

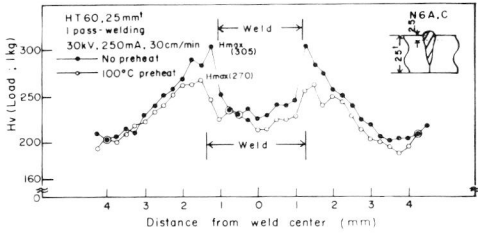

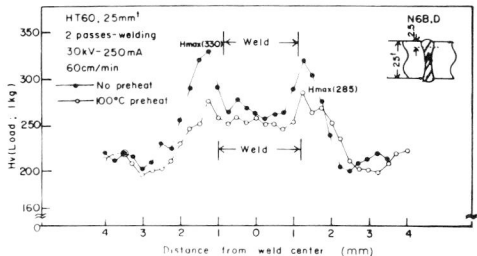

Fig. 3. Vickers hardness distribution of welds for HT—60 steel using low voltage type E·B welder.

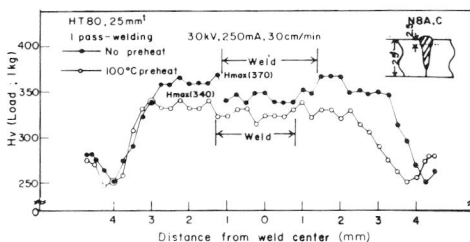

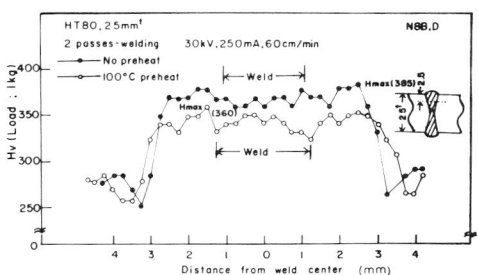

Fig. 4. Vickers hardness distribution of welds for HT—80 steel using low voltage type E·B welder.

than the HAZ near fusion boundary as shown in Figs. 2, 3 and 4. The authors think that it is due to the vaporization of Mn element which is analysed in Table 4. Furthermore, the hardnesses of welds with 100°C preheating usually showed lower value than that of the welds without preheating for both one pass and two passes-welding.

Figs. 5, 6 and **7** respectively show the relations between the weld heat input and hardness of welds for HT—50, 60 and 80 steels. Hw is the average hardness of weld metal and Hmax is the maximum hardness of welds which usually occurs in the HAZ. It is found that the hardness of welds decreases with an increase of weld heat input. This tendency shows irrespective of the difference of electron-beam welder.

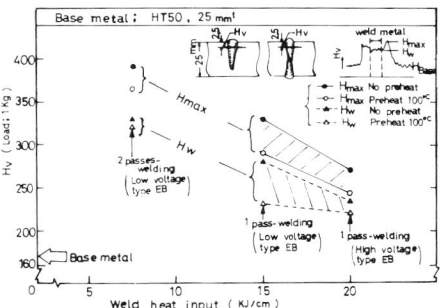

Fig. 5. Relation between weld heat input and hardness of welds for HT—50 steel.

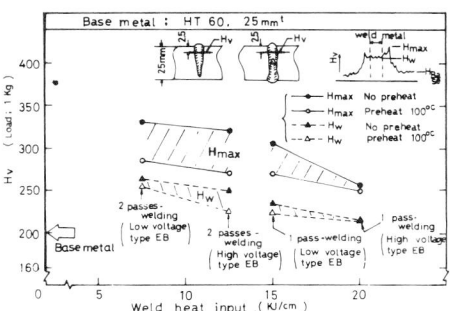

Fig. 6. Relation between weld heat input and hardness of welds for HT—60 steel.

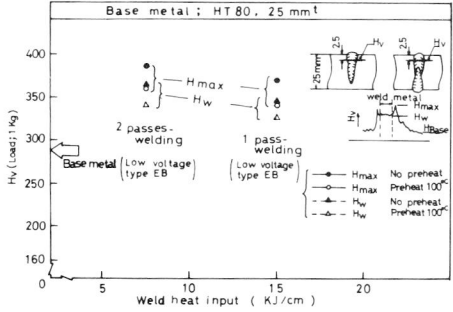

Fig. 7. Relation between weld heat input and hardness of welds for HT—80 steel.

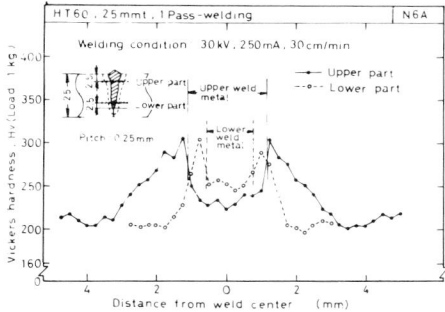

Fig. 8. Difference of Vickers hardness distribution between upper and lower part of weld metal for one pass-welding.

Namely it results that the hardness of welds is directly related to the cooling rate. Furthermore, as shown in **Fig. 8,** the hardnesses in the lower part of weld metal for one pass-welding usually show higher value than those of the upper part, which is due to the difference of cooling rate.

3.4 Tensile test results

Two types of uniaxial tensile test specimen were made from the welded high tension steels of HT—50, 60 and 80, the lengths of the parallel part of which were 220 mm and 12 mm, respectively as shown in **Fig. 9** (a) and (b). Two specimens were examined for each welding condition in the respective tensile test.

An example of the two types of tested specimens for HT—50 is shown in **Photo. 2.** In case of 220 mm parallel length, all of the tensile tested specimens were fractured in base metal. Meanwhile, in case of 12 mm

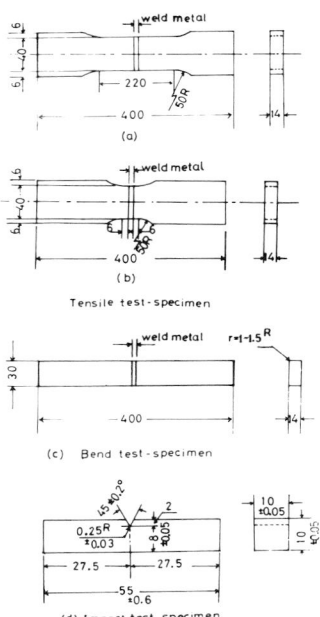

Fig. 9. Dimension of each test specimen.

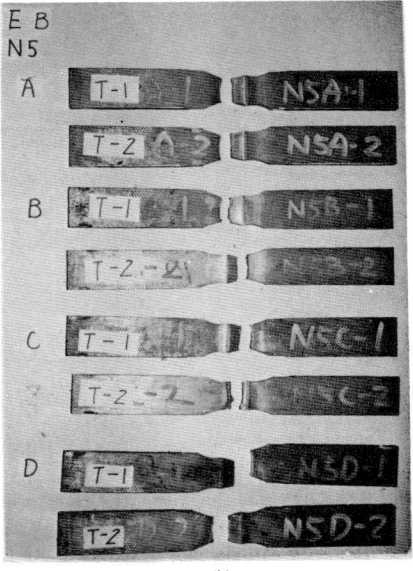

Photo. 2. An example of two types of tested specimens for HT—50 steel.
(a) Length of parallel part: 220 mm (b) Length of parallel part: 12 mm

359

parallel length, two specimens were fractured in the welds but the tensile strengths of them were equal to those of the other specimens which were fractured in base metal. The tensile test results are tabulated in **Tables 5** and **6.** Specimen number in those tables corresponds to the welding condition in Table 3.

Table 5. Tensile test results (Length of parallel part; 220 mm).

Steel	Specimen number	Tensile Strength, σ_T (kg/mm^2)	Elongation in GL : 100 mm (%)	Fractured in
HT—50	D5A—1	56.7	20.	Base metal (BM)
	A—2	56.7	24	"
	D5B—1	56.4	—	"
	B—2	56.9	—	"
	D5C—1	56.7	—	"
	C—2	56.3	—	"
	D5D—1	57.1	—	"
	D—2	56.7	—	"
	N5A—1	53.9	—	BM
	A—1	54.8	—	"
	N5B—1	55.0	—	"
	B—2	55.5	—	"
	N5C—1	55.1	—	"
	C—2	54.4	—	"
	N5D—1	54.5	—	"
	D—2	53.4	—	"
HT—60	D6A—1	60.3	—	BM
	A—2	58.5	—	"
	D6B—1	57.9	—	"
	B—2	58.0	—	"
	D6C—1	58.7	—	"
	C—2	58.1	—	"
	D6D—1	57.9	23	"
	D—2	58.3	—	"
	N6A—1	59.8	20	BM
	A—2	59.8	—	"
	N6B—1	59.7	—	"
	B—2	59.9	21	"
	N6C—1	60.0	—	"
	C—2	60.9	—	"
	N6D—1	59.8	20	"
	D—2	59.0	—	"
HT—80	D8A—1	89.5	13	BM
	A—2	89.8	—	"
	D8B—1	86.2	12	"
	B—2	85.7	—	"
	D8C—1	89.2	—	"
	C—2	88.7	12	"
	D8D—1	88.7	—	"
	N8A—1	86.7	13	BM
	N8B—1	86.9	12	"
	B—2	87.3	13	"
	N8C—1	87.3	13	"
	C—2	86.8	—	"
	N8D—1	86.6	—	"
	D—2	87.2	—	"

Table 6. Tensile test results (Length of parallel part; 12 mm).

Steel	Specimen number	Tensile strength, σ_T (kg/mm^2)	Fractured in
HT—50	D5A—1	64.4	Base metal (BM)
	A—2	65.0	"
	D5B—1	64.4	"
	B—2	64.6	"
	D5C—1	65.2	"
	C—2	64.3	"
	D5D—1	65.2	"
	D—2	64.1	"
	N5A—1	62.5	BM
	A—2	63.2	"
	N5B—1	61.8	"
	B—2	61.0	"
	N5C—1	62.8	"
	C—2	62.4	Weld metal
	N5D—1	61.7	BM
	D—2	61.5	"
HT—60	D6A—1	66.2	BM
	A—1	65.4	"
	D6B—1	65.0	"
	B—2	65.4	"
	D6C—1	64.4	"
	C—2	64.4	"
	D6D—1	64.5	"
	D—2	65.1	"
	N6A—1	67.3	BM
	A—2	66.7	"
	N6B—1	66.0	"
	B—2	65.4	"
	N6C—1	66.7	"
	C—2	65.7	"
	N6D—1	66.9	"
	D—2	66.1	"
HT—80	D8A—1	98.3	BM
	A—2	97.2	"
	D8B—1	92.6	"
	B—1	92.6	"
	D8C—1	96.4	"
	C—2	96.9	"
	D8D—1	95.0	"
	N8A—1	97.1	HAZ
	N8B—1	95.8	BM
	B—2	95.1	"
	N8C—1	98.0	"
	C—2	99.1	"
	N8D—1	96.4	"
	D—2	96.0	"

3.5 Bend test results

Roller bend test (bend radius; 24 mm) was made for the welded joints (HT—50, 60 and 80) with the bend test specimens as shown in Fig. 9 (c).' Two specimens were prepared for one welding condition, and one of them was face bend tested and the other was root bend tested. An example of bend tested-specimens is shown in **Photo. 3.** All of the face and the root bend tested specimens showed the complete bend angle of 180 degrees without any crack, but extremely small defect was observed on the surface of two root bend tested specimens.

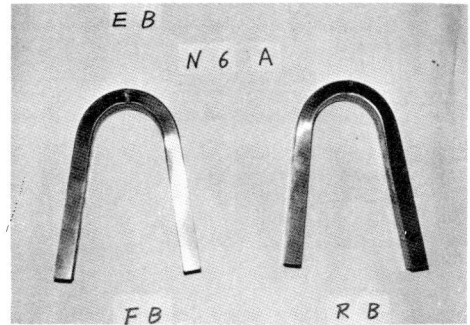

Photo. 3. An example of bend tested specimens.

3.6 Impact test results

1) Testing method

Impact test was performed with standard Charpy 2 mm V notch specimens as shown in Fig. 9 (d). Test specimens were machined from the center of the plate thickness for one pass-welding, and from the first bead for two passes-welding. The center of weld metal, the fusion boundary (we call here "bond") and middle of heat-affected zone ("HAZ") were selected as the notch location of impact test specimens. Six levels of testing temperature, -80, -50, -30, -10, 10 and room temperature (20 to 30°C) were selected in this impact test. Three test specimens were usually tested for each testing temperature.

2) Impact test properties

Transition temperature curves for the absorbed energy for the welds of HT—50, 60 and 80 steels are given in **Figs. 10** through **33.** Furthermore, **Tables 7, 8** and **9** show the transition temperatures of T_{rE} and T_{r15} for HT—50, 60 and 80 steels, respectively.

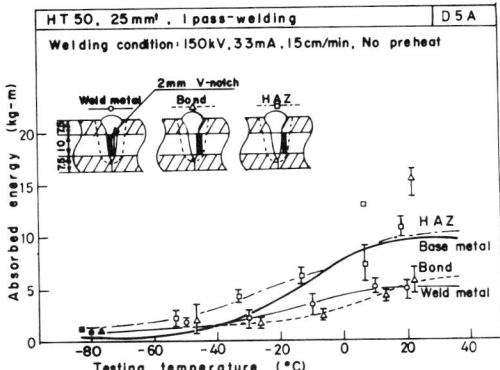

Fig. 10. Transition temperature curve of absorbed energy for welding condition D5A.

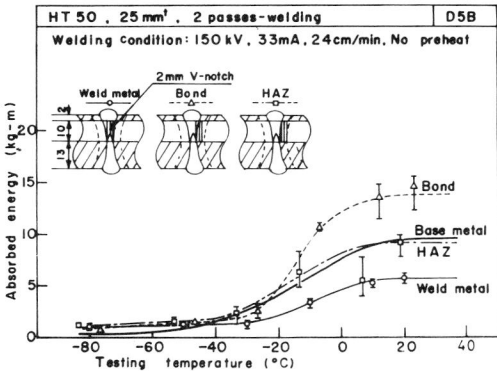

Fig. 11. Transition temperature curve of absorbed energy for welding condition D5B.

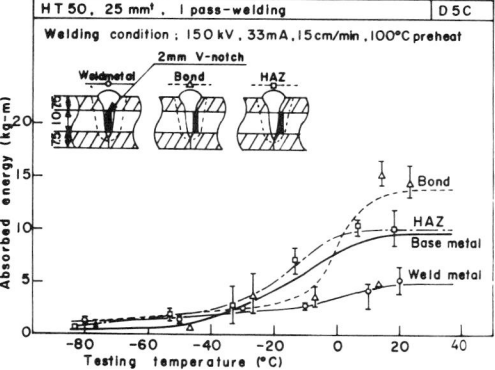

Fig. 12. Transition temperature curve of absorbed energy for welding condition D5C.

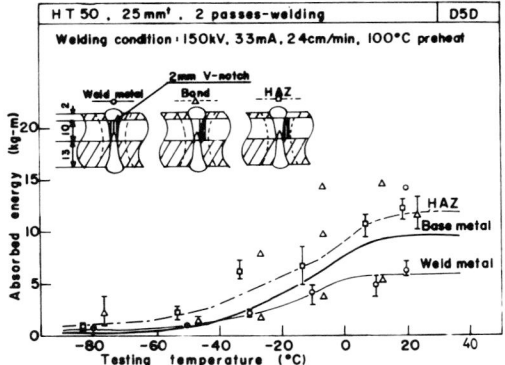

Fig. 13. Transition temperature curve of absorbed energy for welding condition D5D.

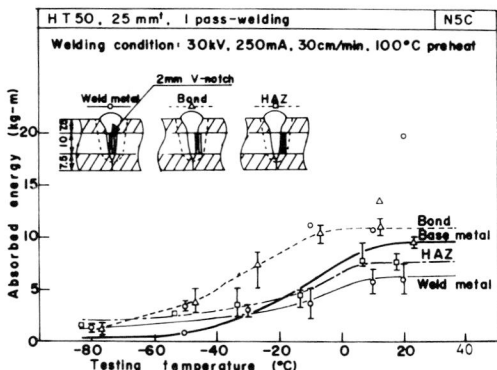

Fig. 16. Transition temperature curve of absorbed energy for welding condition N5C.

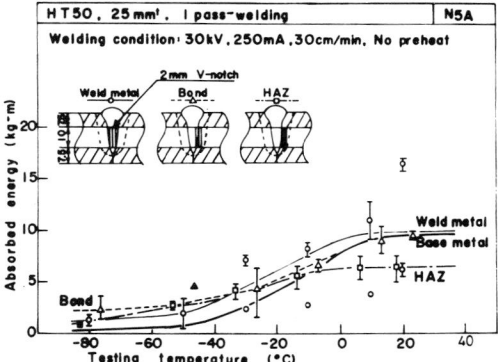

Fig. 14. Transition temperature curve of absorbed energy for welding condition N5A.

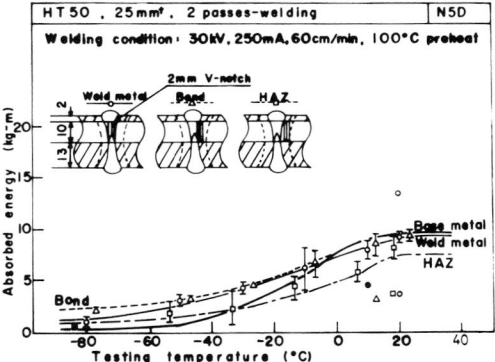

Fig. 17. Transition temperature curve of absorbed energy for welding condition N5D.

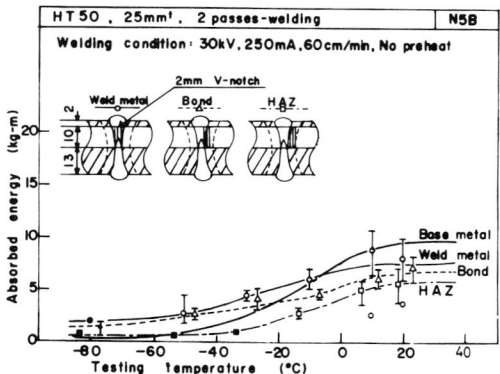

Fig. 15. Transition temperature curve of absorbed energy for welding condition N5B.

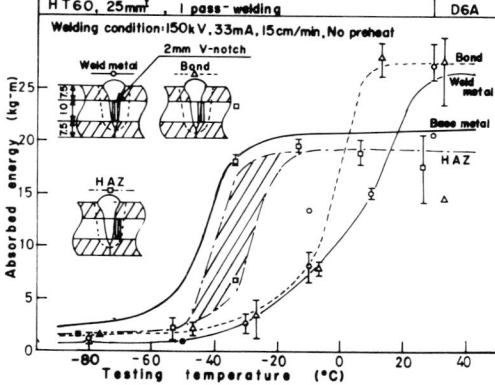

Fig. 18. Transition temperature curve of absorbed energy for welding condition D6A.

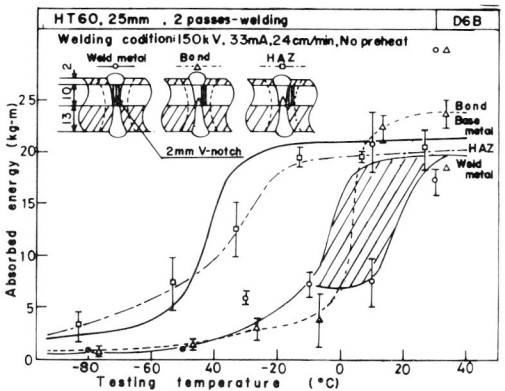

Fig. 19. Transition temperature curve of absorbed energy for welding condition D6B.

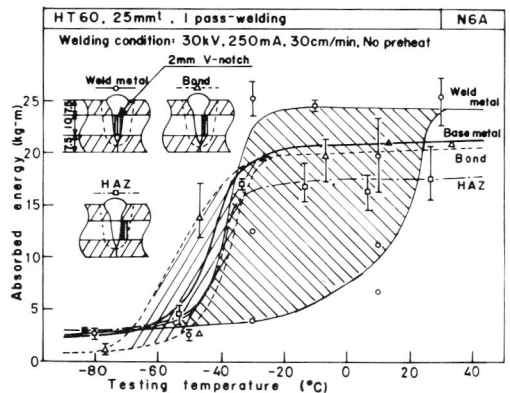

Fig. 22. Transition temperature curve of absorbed energy for welding condition N6A.

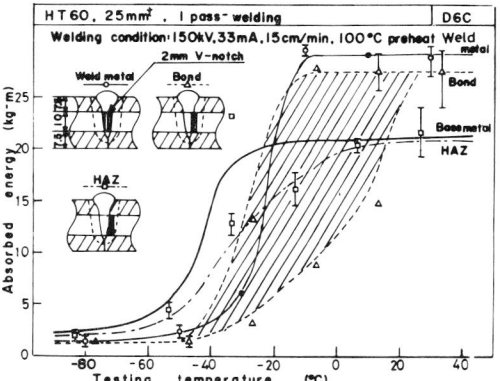

Fig. 20. Transition temperature curve of absorbed energy for welding condition D6C.

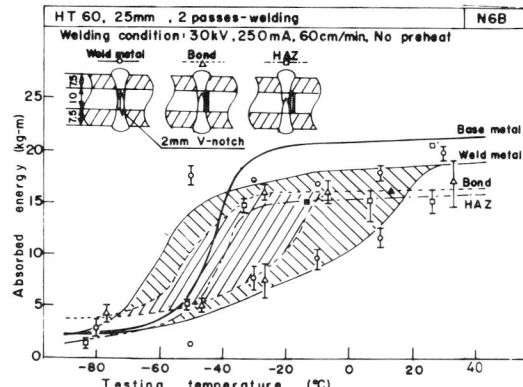

Fig. 23. Transition temperature curve of absorbed energy for welding condition N6B.

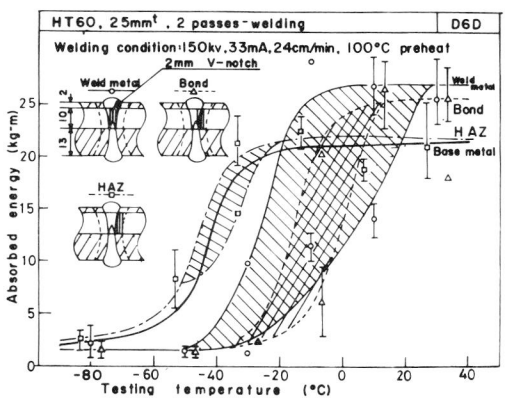

Fig. 21. Transition temperature curve of absorbed energy for welding condition D6D.

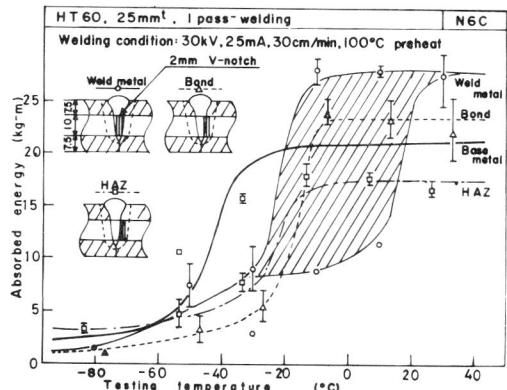

Fig. 24. Transition temperature curve of absorbed energy for welding condition N6C.

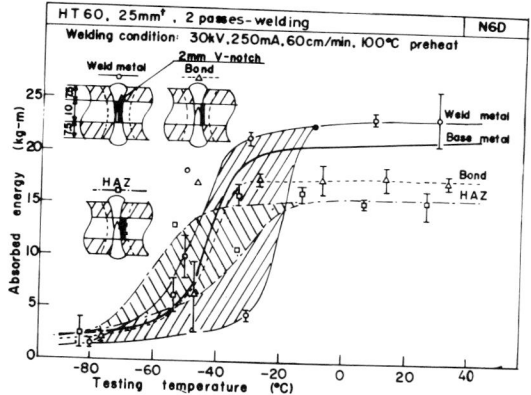

Fig. 25. Transition temperature curve of absorbed energy for welding condition N6D.

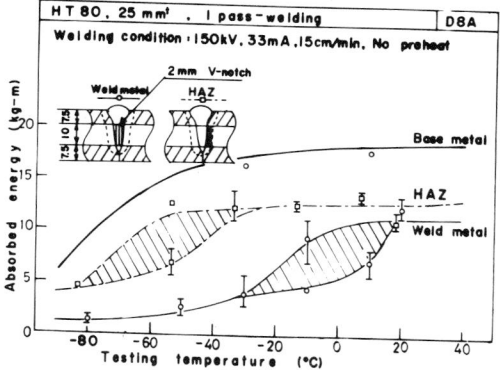

Fig. 26. Transition temperature curve of absorbed energy for welding condition D8A.

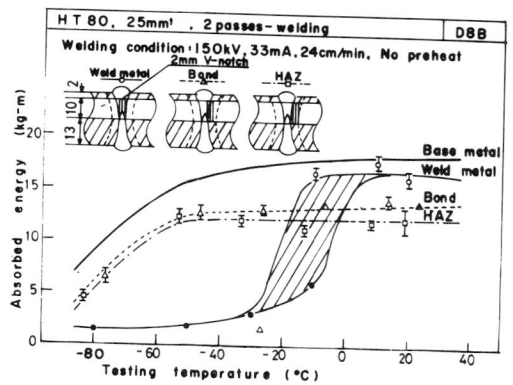

Fig. 27. Transition temperature curve of absorbed energy for welding condition D8B.

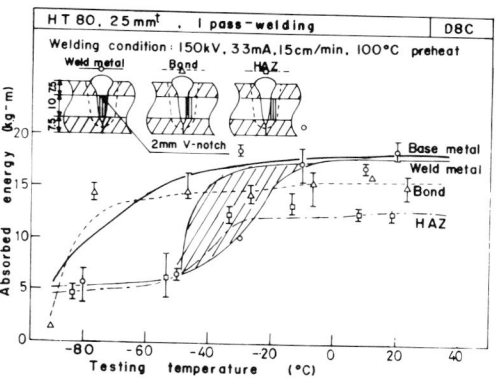

Fig. 28. Transition temperature curve of absorbed energy for welding condition D8C.

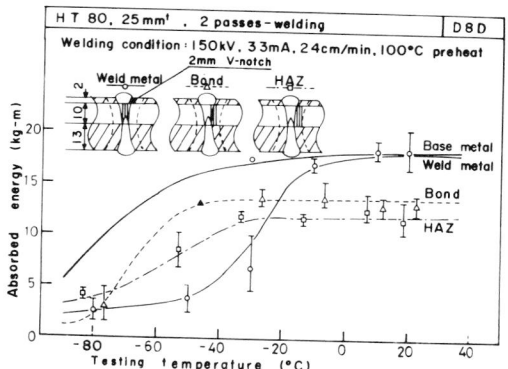

Fig. 29. Transition temperature curve of absorbed energy for welding condition D8D.

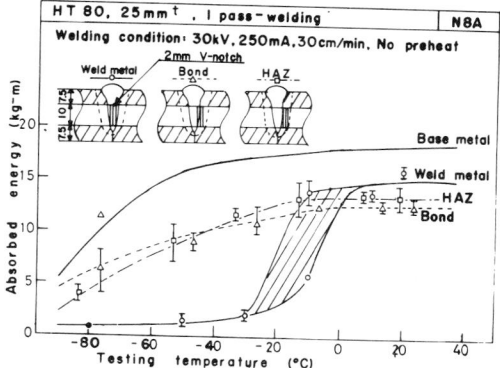

Fig. 30. Transition temperature curve of absorbed energy for welding condition N8A.

365

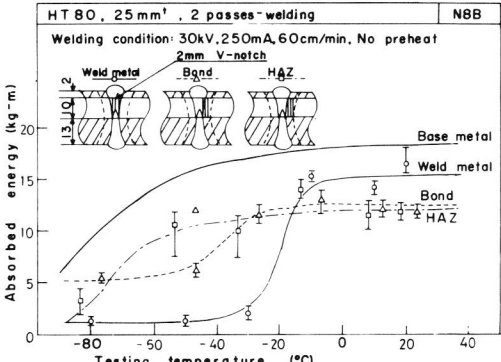

Fig. 31. Transition temperature curve of absorbed energy for welding condition N8B.

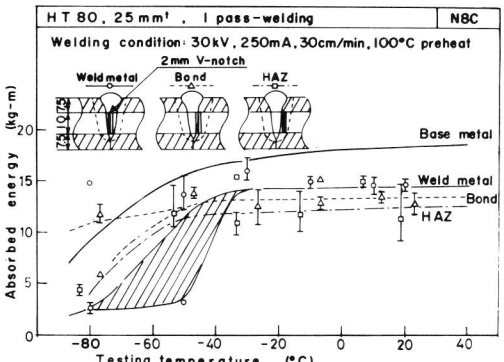

Fig. 32. Transition temperature curve of absorbed energy for welding condition N8C.

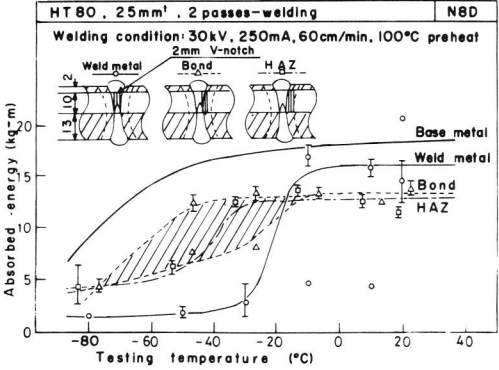

Fig. 33. Transition temperature curve of absorbed energy for welding condition N8D.

3) Study on impact test results

The absorbed energies at $R \cdot T$ for the base metal of HT—50, 60 and 80 steels were about 10, 20 and 18 kg·m, respectively. Impact strength at $R \cdot T$ generally decreased in order of weld metal, bond and HAZ for HT—60 and 80 steels, while it is clearly distinguished in case of HT—50.

In some figures for HT—60 and HT—80, the value of impact strength for weld metal, bond and HAZ at the same temperature frequently showed two levels, the temperature range of which is surrounded by oblique line. It was due to difference in the form of fracture pass in impact specimen as illustrated in **Fig. 34.** When the fracture had propagated straightly along the notched direction as shown in Fig. 34 (a), the absorbed energy showed the low value, however, when the fracture had propagated out the notched direction as shown in Fig. 34 (b), it showed high value.

According to Tables 7, 8 and 9, there are no special relations between weld heat input and T_{rE} and T_{r15}, however it seems that T_{rE} and T_{r15} rise with an increase of weld heat input in case of one pass-welding. Furthermore, T_{rE} and T_{r15} for the weld metal of HT—60 and 80 steels apparently rise in comparison with those of base metal, while in case of HT—50 steels, they have no difference. Nextly it seems that T_{rE} and T_{r15} are not clearly influenced by 100°C preheating, but T_{rE} for the weld metal tends to be improved with 100°C preheating in case of HT—80 steels.

According to JIS and WES Specifications, the minimum absorbed value required in the impact test is prescribed for base metal of HT—50, 60 and 80 steels. The minimum absorbed energies required in these Specifications are tabulated for each steel in **Table 10** In **Figs. 35** through **37** the relations between the absorbed energy at the specified temperature in JIS and WES and weld heat input are shown. In each figure the minimum absorbed value which prescribed in JIS and WES is shown by the broken line as "Lowest limit". In general it seems that the absorbed energy in the weld metal of HT—50 and 80 steels is gradually decreased with an increase of weld heat input, while

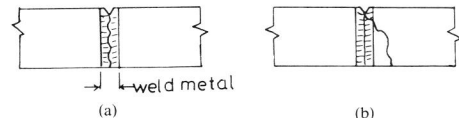

Fig. 34. Difference in form of fracture pass in impact specimen.

Table 7. Summary of transition temperature by V-Charpy impact test of HT—50 welded joint. (Base metal: T_{rE}: −10 (°C), T_{rls}: −35 (°C))

Welding Condition	D5—series												N5—series											
	1 pass-welding (Weld heat input, 20 KJ/cm)						2 passes-welding (12.5 KJ/cm)						1 pass-welding (15 KJ/cm)						2 passes-welding (7.5 KJ/cm)					
Preheating	No			100°C			No			100°C			No			100°C			No			100°C		
Notched Location*	W	B	H	W	B	H	W	B	H	W	B	H	W	B	H	W	B	H	W	B	H	W	B	H
T_{rE}^* (°C)	−15	5	−15	−10	0	−20	−10	−15	−20	−15		−5	−25	−15	−35	−15	−35	−10	−25	−20	−10	−20	−15	−10
T_{rls}^* (°C)	−35	−20	−55	−30	−30	−45	−20	−30	−40	−30		−50	−50	−80	−60	−50	−70	−70	−70	−55	−20	−60	−80	−35

* Notched location, W; weld metal B: Bond H; HAZ T_{rE}; Energy transition temp. T_{rls}; 15 f-lb transition temp.

Table 8. Summary of transition temperature by V-Chanpy impact test of HT—60 welded joint. (Base metal: T_{rE}; −45 (°C), T_{rls}; −90 (°C))

Welding Condition	D6-series												N6-series											
	1 pass-welding (Weld heat input, 20 KJ/cm)						2 passes-welding (12.5 KJ/cm)						1 pass-welding (15 KJ/cm)						2 passes-welding (7.5 KJ/cm)					
Preheating	No			100°C			No			100°C			No			100°C			No			100°C		
Notched Location*	W	B	H	W	B	H	W	B	H	W	B	H	W	B	H	W	B	H	W	B	H	W	B	H
T_{rE}^* (°C)	−5	5	−40 ~ −30	−25	−30 ~ 5	−30	−5 ~ 15	0	−35	−25 ~ 5	−15 ~ 0	−45 ~ −35	−40 ~ 20	−55 ~ −35	−40	−25 ~ 15	−15	−25	−45 ~ 0	−50 ~ −20	−40	−45 ~ −20	−45	−60 ~ −35
T_{rls}^* (°C)	−35	−45	−55 ~ −40	−55	−50 ~ −40	−75	−40 ~ −30	−40	** ~ −80	−45 ~ −35	−30	<−80	<−80	−75 ~ −55	−70	−60	<−80	−90	<−80	−70	−75 ~ −60	−80	<−80	

*: Notched location, W; Weld metal B; Bond H; HAZ T_{rE}; Energy transition temp. T_{rls}; 15 f-lb transition temp.
**: less than −80 (°C)

Table 9. Summary of transition temperature by V-Charpy impact test of HT—80 welded joint. (Base metal: T_{rE}; −80 (°C), T_{rls}; <−80 (°C))

Welding Condition	D8-series												N8-series											
	1 pass-welding (Weld heat input, 20 KJ/cm)						2 passes-welding (12.5 KJ/cm)						1 pass-welding (15 KJ/cm)						2 passes-welding (7.5 KJ/cm)					
Preheating	No			100°C			No			100°C			No			100°C			No			100°C		
Notched Location*	W	B	H	W	B	H	W	B	H	W	B	H	W	B	H	W	B	H	W	B	H	W	B	H
T_{rE} (°C)	−20	5	−75 ~ −50	−40	−85	−45	−20 ~ −5	−70	−70	−30	−65	−60	−20 ~ −5	−70	−65	−60 ~ −40	−95 ~ −75	−70	−20	−45	−70	−20	−65 ~ −45	−45
T_{rls}^* (°C)	−50		**<−80	<−80	<−80	<−80	−45	−90	−90	<−80	<−80	<−80	−30	<−80	<−80	−80	<−80	<−80	−35	<−80	<−80	−40	<−80	<−80

*: Notched location, W; Weld metal B; Bond H; HAZ T_{rE}; Energy transition temperature T_{rls}; 15 f-lb transition temperature
**: less than −80 (°C)

in the other parts it has no obvious relation. Furthermore, the impact strength in weld metals of these steels except that of HT—80 with preheating is lower than the others in general. The absorbed energies in HT—60 steel fluctuated widely due to the difference of form of the fracture pass. The value of impact strength for the welds (weld metal, bond and HAZ) of HT—60 and 80 steels is exceeding the lowest limit in JIS and WES Specifications, however in case of HT—50 steel it is not always enough to these Specifications, and some of welds show the value less than

WES' limit. The reason is thought that the absorbed energy in the base metal of HT—50 is rather low and is near the lowest limit.

4. Conclusion

(1) Elimination of the scale, cleaning of the contamination for the joints and de-magnetization must be completely performed before electron-beam welding in order to make the defect-free weld metals of HT—50, 60 and 80 steels. The defect-free weld metals

Table 10. The minimum absorbed energy required in V-notched impact test specimen in JIS and WES Specifications.

Materials \ Specification	Materials	HT—50	HT—60	HT—80
JIS* G3106—1973	Corresponding	SM50B	SM58	HW70
	T·S (kg/mm²)	50~62	58~73	80~95
	Temp. (°C)	0	−5	−15
	Absorbed energy (kg·m)	>2.8	>4.8	>3.6
WES* 135—1961	Corresponding material	HW36	HW45	HW80
	T·S (kg/mm²)	53~65	60~72	80~95
	Temp. (°C)	0	−5	−15
	Absorbed energy (kg·m)	>4.8	>4.8	>3.6

* JIS; Japan Industrial Standard
  WES; Welding Engineering Standard

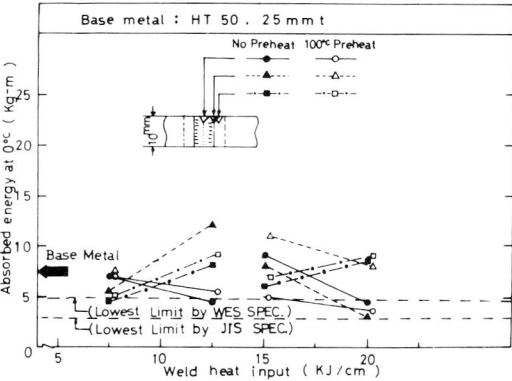

Fig. 35. Relation between absorbed energy of HT—50 at specified temperature in JIS and WES and weld heat input.

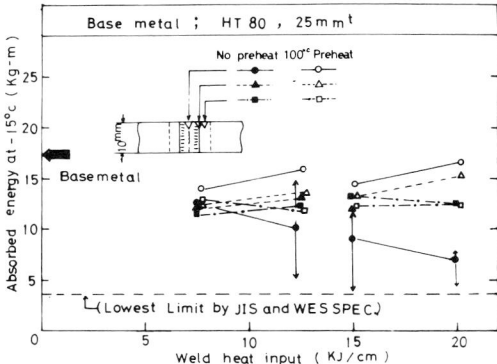

Fig. 37. Relation between absorbed energy of HT—80 at specified temperature in JIS and WES and weld heat input.

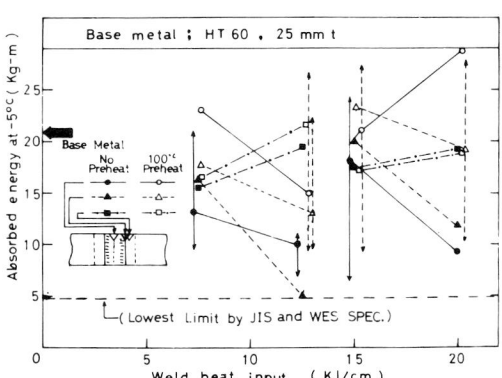

Fig. 36. Relation between absorbed energy of HT—60 at specified temperature in JIS and WES and weld heat input.

which were inspected by X-ray and dye penetrant inspections were obtained for these steels in this experiment.

(2) According to quantitative analysis for weld metal, it seems that there is a little vaporization of Mn element in weld metal.

(3) Judging from the microphotographs of the welds, 100°C preheating has no notable effect upon the variation of microstructures in the weld metal of HT—50, 60 and 80 steels. However, as the welding speed decreases, the cells which grow parallel to the welding direction are obviously observed.

(4) Microhardness of welds

(a) The hardness in weld metal is usually softer than that in the HAZ. This reason is considered due to reduction of Mn element in weld metal.

368

(b) The hardness of welds for both one pass and two passes-welding are reduced with 100°C pre-heating.

(c) The hardnesses of welds are reduced with an increase of weld heat input regardless of high and low voltage type electron-beam welders.

(d) The bottom part of a weld penetration for one pass-welding shows harder in hardness than the surface part.

(5) The tensile tested specimens of electron-beam welded joints of HT—50, 60 and 80 steels showed sound properties and were mostly fractured in base metal.

(6) All of the face and the root bend tested specimens of electron-beam welded joints of HT—50, 60 and 80 steels showed a complate bend angle of 180 degrees without serious defects.

(7) Impact tests by standard V-Charpy test specimen

(a) Absorbed energies for the base metal of HT—50, 60 and 80 steels are about 10, 20 and 18 kg·m at room temperature, respectively. T_{rE} and T_{r15} for the weld metal of HT—60 and 80 steels apparently rise in comparison with those of respective base metal, while in case of HT—50 steels, no obvious difference is observed between them.

(b) T_{rE} and T_{r15} for the welds of these steels had no obvious relations with weld heat input, however they tended to rise with an increase of weld heat input for one pass-welding, and then they were not clearly influenced with 100°C pre-heating.

(c) In case of the electron-beam welds of HT—60 and 80 steels, the absorbed energies of weld metal, bond and HAZ are usually exceeding the minimum absorbed energy required in JIS and WES Specifications, even though they showed lower value than that of base metal. In case of HT—50, however, the absorbed energies in weld metal, bond and HAZ are not always satisfied with criteria of WES.

Acknowledgement

The authors wish to acknowledge the co-operative support of Japan Electron Optics Industries and Osaka Transformer Co. Ltd.

References

1) H. I. Mchenry et al: "Electron Beam Welding of D6AC Steels", Welding Journal, Vol. 45, No. 9, Sep., 419S–425S, 1966.

2) A. J. Williams et al: "Properties of Electron-Beam Welds in Ultra-High-Strength Steels", 1st International Conference of Electron and Ion Beam Science and Technology, 1965, 674~712.

3) Y. Arata et al: "Study on Characteristics of Weld Defect and Its Prevention in Electron Beam Welding (Report I)", Trans. of JWRI, Vol. 2, No. 1, p. 103~112, 1973.

4) F. Matsuda et al: "Some Metallurgical Investigations on Electron-Beam Welds", Trans. of JWS, Vol. 1, No. 1, p. 72~85, 1970.

Mechanical Properties on Electron Beam Welds of Constructional High Tension Steels (2)

Abstract

In this investigation, the authors aimed to make clear the weldability of electron-beam welds for some commercial constructional high tension steels. Then, the influence of weld heat input on the weld defects, the hardness distributions and the impact properties of the electron-beam welds was investigated in this report. The materials used were three grades of high tension steel HT50, HT60 and HT80 with 25 mm thickness. The conventional low voltage type electron-beam welder, 30 KV-500 mA(15 KW) in maximum was employed in this experiment.

All of weld bead were performed with bead-on-plate type welding. The active beam parameter, a_b adopted was 1.0 through the experiment.

The remarkable conclusions are as follows:

(1) All of weld seam except unstable location of bead have had no defect according to dye penetrant and X-ray inspections.

(2) The hardness in weld metal was generally reduced with an increase of weld heat input. In Vickers hardness distributions of HT50 and 60 steels, the hardness in the HAZ near fusion boundary was harder than that in the weld metal. This is considered due to the vaporization of Mn element in weld metal.

(3) The value of impact strength for weld metal usually showed the different two levels, lower and higher, at particular testing temperature due to the difference in the fracture mode for 2 mm notched standard Charpy impact test specimen. Namely, when the fracture occurred straightly along the notched direction, the absorbed energy showed lower value, however, when the fracture occurred out of the notched direction, it showed higher value.

(4) The particular testing temperature, at which the fracture mode differs, tended to be raised with an increase of the weld heat input. This seemed to be related with the bead width of fusion zone and the hardness difference between base and weld metal.

(5) As far as the fracture propagates along the notched direction, there is little variation in the absorbed energies of weld metal for respective material within a limited change for the weld heat input from 10 KJ/cm through 40 KJ/cm.

(6) In case of the welds of HT50 and 80 steels, the absorbed energies of the weld metal, which show lower value than that of the base metal are usually exceeding the minimum absorbed energy required in JIS or WES specification, even though the fracture occurs within weld metal. In case of HT60, however, they are not always satisfied with the criteria of JIS specification.

(7) Furthermore, the impact strength of electron-beam welds showed higher value than that of submerged arc welds for HT50 and 80 steels.

1. Introduction

There are few data concerning the mechanical properties on electron-beam welds of the constructional high tension steels (HT50, 60 and 80 steels) which are used in large-sized constructions, and its weldability is still unknown. Then, in this investigation the authors aimed to make clear them.

Some results of the mechanical properties of electron-beam welds such as hardness distribution, tensile, bend, impact and fatigue properties were reported in the two previous reports.[1],[2] It was recognized that the results of these mechanical tests except some impact properties of HT50 steel would be satisfied with the practical use for many fields of application.

Then, in this report, the authors mainly treated about the more detailed behavior of impact properties for electron-beam welds, which were characterized by narrow bead width as compared with that of the welds by conventional arc welding method. Namely, the influence of the weld heat input on the impact properties of weld metal was investigated, and it was discussed that the behavior of impact properties of electron-beam welds is closely related to both the bead width and the hardness difference between base and weld metal.

2. Materials and Welding Conditions

The materials used in this investigation are HT50, 60 and 80 steels which are widely used in the constructional bridges and buildings. The chemical compositions of these three types of high tension steels are listed in Table 1. The carbon equivalent (C_{eq}) of HT 50, 60 and 80 steels is 0.37, 0.36 and 0.50, respectively. All of the plate thickness of these steels are 25 mm.

High vacuum type-EB welder, conventional low voltage type (30 KV-500 mA, 15 KW in maximum), was used in this investigation. The welding conditions used on various materials are tabulated in Table 2. The weld heat input is varied for the different four levels from 10

Table 1   Chemical composition of HT50, 60 and 80 steels used.

Composition / Steel	C	Si	Mn	P	S	Ni	Cr	Mo	V	Ceq*
HT50 (SM50)	0.14	0.46	1.27	0.023	0.020	–	–	–	–	0.37
HT60 (SM58)	0.12	0.35	1.16	0.018	0.010	0.02	0.10	–	0.030	0.36
HT80	0.13	0.29	0.85	0.016	0.008	0.98	0.48	0.39	0.023	0.50

*: Ceq = C + 1/6Mn + 1/24 Si + 1/40 Ni + 1/5 Cr + 1/4 Mo + 1/14 V

(JIS G 3106)

through 40 KJ/cm. the welding for all materials was performed with one pass bead-on-plate welding method. In case of the welding conditions, (c) and (d), in which 25 mm thick steel plates were burnt through, the welding was performed by means of the stacking of respective two plates.

In this experiment, the focal length of electron-beam was beforehand certified by the slope-welding method[3] and the active beam parameter, a_b was selected for 1.0. The focal length (D_F) of each welding condition was shown on the right part of Table 2. The size of each specimen is 150 mm in width and 500 mm in length. Any oxide and scaling on the plate surfaces was completely machined.

3. Experimental Results
3.1 Defect of welds

All of the electron-beam welds of high tension steels were X-ray inspected. The defects such as porosity and crack could not be particularly detected in the electron-beam welds of HT50, 60 and 80 steels. However, hot cracks were often observed in the crater points, or the places where the electron-beam irregularily stopped. Referring to Japan Industrial Standard Specification (JIS Z 3104-1968), most of the electron-beam welds except irregular points were allowable in the first class for synthetic grade. All of the weld surfaces of the high

tension steels were also checked by dye penetrant test. No surface defects were observed at all in them.

Table 2   Welding conditions used.

Designation of welding condition	Beam power (KV-mA)	Welding speed (cm/min)	Weld heat input (KJ/cm)	Focal length (mm)
(a)	30 200 (6KW)	36	10	255
(b)	30 300 (9KW)	36	15	234
(c)	30 400 (12KW)	36	20	218
(d)	30 400 (12KW)	18	40	218

3.2 Metallographic and hardness examinations

Photo. 1 shows the macro-and micro-photographs for HT80 welds in various weld heat inputs from 10

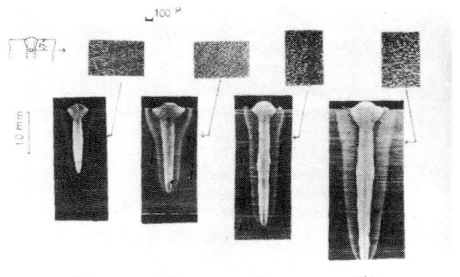

Photo 1   Macro-and microphotographs for HT80 welds in various weld heat inputs.
(a) 10 KJ/cm   (b) 15 KJ/cm   (c) 20 KJ/cm
(d) 40 KJ/cm

through 40KJ/cm. The micro-photographs indicate the micro-structures of weld metal at 7 mm downwards from the plate surface. The grain size of micro-structure of the weld metal tended to grow coarser with an increase of the weld heat input. This tendency shows irrespective of the difference of materials.

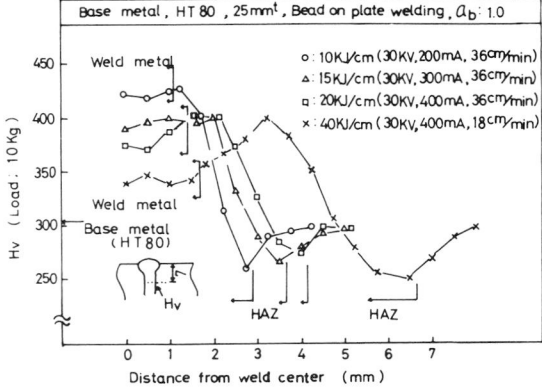

Fig.1   Vickers hardness distributions for HT80 welds in various weld heat inputs.

371

The hardness of welds was measured by using Vickers hardness tester with 10 Kg load. Vickers hardness distributions of HT80 welds in various weld heat inputs are shown in Fig. 1. The measured points are at 7 mm downwards from surface, which are accordant with the center of impact test specimens adopted. With an increase of the weld heat input, the cooling rate becomes faster and the hardness of weld metal are graduately decreased. Fig. 2 shows an example of the hardness distributions of HT50, 60 and 80 welds, the weld heat input of which is 15KJ/cm.

As mentioned in the previous report, the maximum hardness in HT50 and 60 welds is observed in the HAZ near fusion boundary, which is caused by the vaporization of Mn element in the weld metal during welding. In the welds of HT60 and 80, the softened zones are observed because they are quenched and tempered steels. Fig. 3 shows the relation between the weld heat input and the hardness of welds for HT50, 60 and 80 steels. Hw is the average hardness of weld metal.

H_w tends to be decreased with an increase of the weld heat input regardless of the difference of materials. Especially, the hardnesses in HT80 welds showed remarkably higher value than those in the welds of the other materials. Namely, the degree of hardness in the weld metal is closely related to the value of carbon equivalent in Table 1. Fig. 4 shows the hardness distributions of weld metals inward direction for HT50, 60 and 80 welds in case of 15KJ/cm weld heat input. In HT80 welds, there were no variations in the hardness of weld metal inward dire-

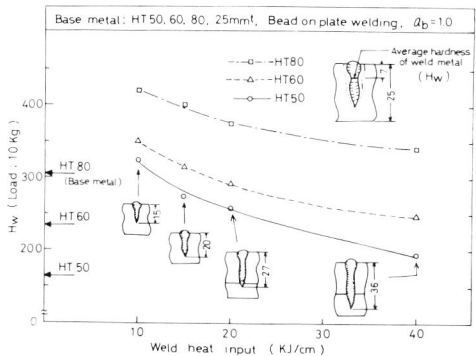

Fig. 3　Relation between weld heat input and average hardness of weld metal for HT50, 60 and 80 steels.

ction, while in HT50 and 60 welds, there are obvious differences in the hardness between the surface and bottom part of the bead, and the maximum hardnesses are observed in the tops of spiking in both weld metals. It is considered that these differences in hardness are caused by the difference of the cooling rate between the bead surface and bottom, which is related to the degree of hardness.

This difference in the degree of hardness due to the difference of materials should be considered by the combination with CCT diagram for respective material.

3.3 Impact properties

Impact test was performed with standard Charpy 2 mm V notch specimens. The capacity of impact test machine is 30 Kg-m in maximum. The test specimens were machined from 2 mm under the plate surface, and all specimens were notched at the center of weld metal. Six levels of testing temperature, −80, −60, −40, −15, 0°C and room temperature (15°C) were selected in this examination and the effects of weld heat input on the impact properties were investigated for HT50, 60 and 80 welds. Usually, two to four test specimens were tested for each testing temperature. As the common phenomenon in impact test for the welds of each materials, it was often observed that the value of impact strength usually showed the different levels even at the same testing temparature, which has already been reported previously. It was due to the difference in the three types of fracture mode for impact test specimens as illustrated in Fig. 5.

Namely, the different fracture mode concurrently occurs at the same testing temperature, and the value of impact strength is varied in response to respective fracture mode. In general, at the lower temperature side of that, the fracture mode ordinarily becomes (A) type, on the

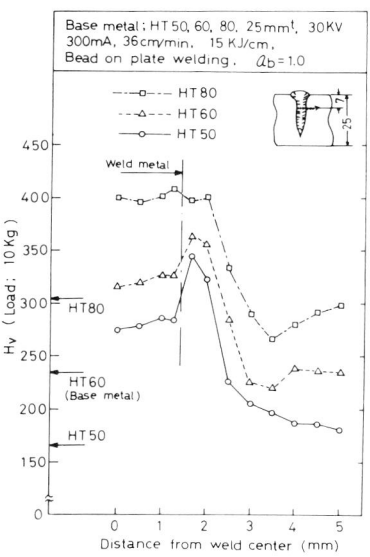

Fig. 2　Vickers hardness distributions of welds for HT50, 60 and 80 steels in 15 KJ/cm weld heat input.

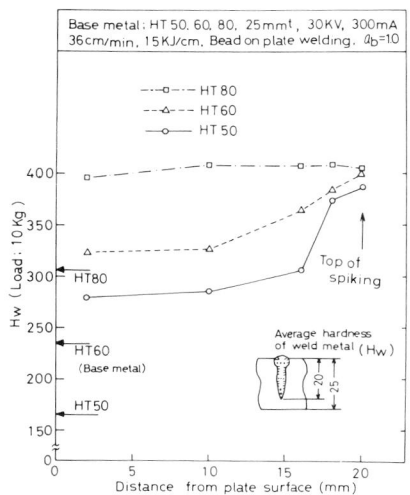

Fig. 4   Vickers hardness distributions of weld metals inward direction for HT50, 60 and 80 steels.

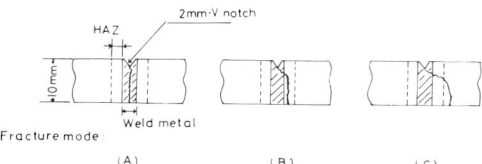

Fig. 5   Three types of fracture mode for impact test specimens.

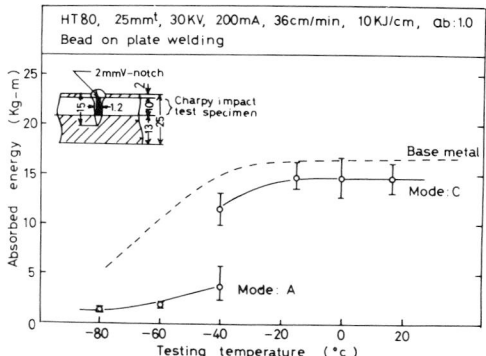

Fig. 6   Transition temperature curve for HT80 welds in 10KJ/cm weld heat input.

other hand, at the higher temperature side of that it ordinarily becomes (B) or (C) type. In case of (A) type the fracture occurs straightly along the notched direction, while in case of (B) and (C) types it deviates to HAZ and base metal, respectively. Then as the distance of fracture path is increased in order of type (A), (B) and (C), the absorbed energy of weld metal in case of type (A) shows the lowest value in all of types. Then all of transition curves in this report were shown with the distinction of the type of fracture mode in the tested specimens.

As an example, the transition curve for HT80 welds in 10KJ/cm weld heat input is shown in Fig. 6. It is found that the absorbed energy obviously showed two different values at −40°C testing temperature due to the difference of the fracture mode. And in the higher temperature than −40°C, the fracture path ordinarily deviated to the base metal and its energy showed remarkably high value, while in the lower temperature it occurs within the weld metal and its energy showed low value. It seems that there is little fluctuation in impact strength for respective fracture mode at the particular testing temperature. As mentioned in the above, it is found that the toughness of the weld metal in itself, which is made

clear only at lower testing temperature than −40°C, showed lower value than that of base metal in general. Figs. 7, 8 and 9 show the comparisons of transition curves in various weld heat inputs for HT50, 60 and 80 welds, respectively.

In case of HT50 (Fig. 7), the fracture mode respectively began to separate into two types at −60 and −15°C in 10 and 15KJ/cm weld heat inputs, however it didn't separate even at 20°C in 40KJ/cm. Namely, the transition temperature for the fracture mode tended to shift to higher temperature side with an increase of the weld heat input. However even in the two formers, the test specimens for (A) type fracture mode were observed at about room temperature, and then the impact strength of the weld metal in itself was able to be obtained. Judging from this result, it is found that there is little variation in the impact strength of the weld metal in itself within the changing range of the weld heat input in this investigation, and its value corresponds to that of base metal. The energy transition temperatures for weld metal are about −20°C.

On the other hand, in case of HT60 (Fig. 8), at about room temperature, (A) type-fractured specimens were not observed except in large weld heat input of 40KJ/cm. Furthermore, the transition temperature for the fracture mode has the tendency to shift to higher temperature side with an increase of the weld heat input, which is the same as that of HT50. According to mentioned in the above, the toughness of the weld metal in itself is not clear all temperatures range. However, with the analogy from the impact properties in 40KJ/cm in general its impact strength seems to show lower value than that of base metal. And also it seemed that there is no obvious difference in the impact strength of weld metal due to the variation of weld heat input.

Nextly the result of HT80 welds (Fig. 9) was almost similar to that of HT60. Fig. 10 shows the transition

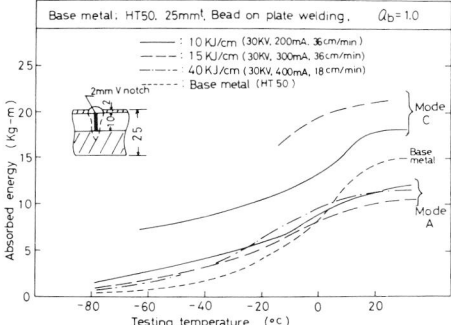

Fig. 7　Comparison of transition temperature curves for HT50 welds in various weld heat inputs.

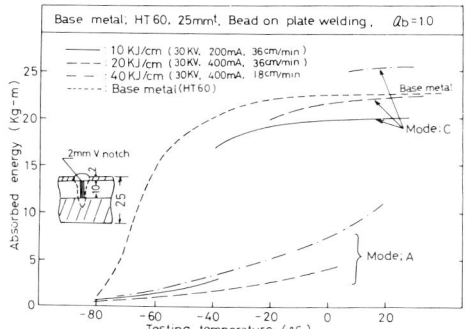

Fig. 8　Comparison of transition temperature curves for HT60 welds in various weld heat inputs.

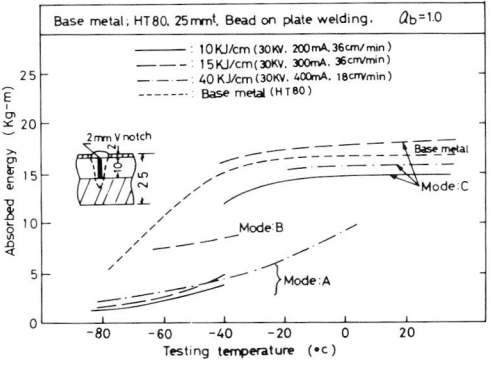

Fig. 9　Comparison of transition temperature curves for HT80 welds in various weld heat inputs.

curves for the submerged arc welds of HT50, 60 and 80 steels, which were asked in order to compare the impact properties of electron-beam welds with those of conventional arc welds. Each material is identical with that

used for the electron-beam welding, and the welding conditions used are given in the upper part of Fig. 10. The fracture modes for the tested specimens were all (A) types in Fig. 5. All of the impact strengths showed extremely low

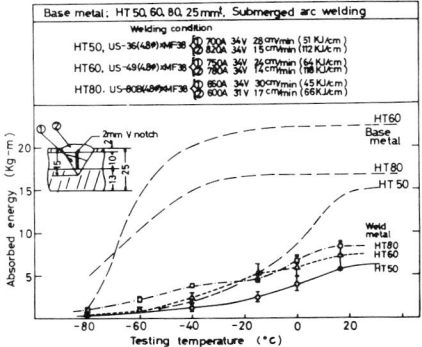

Fig. 10　Transition temperature curves for submerged arc welds of HT50, 60 and 80 steels.

values as compared with repective base metal, which is considered due to a little large weld heat input. And also it is found that there is no obvious difference in the value of the impact strength between the electron-beam and submerged arc welds.

Nextly, the relation existing between the weld heat input and the transition temperature for fracture mode in HT50, 60 and 80 welds is shown in Table 3. Moreover both the bead width (d_B) and the average hardness of weld metal (H_w) at 7 mm downwards from the surface which corresponds to each weld heat input are also listed in the table. Additionally, the bead width ratio, (d_B/h) and the hardness difference ratio, $H_w - H_s/H_s$ are calculated and given, too. h indicates the remained thickness in the bottom of notch for Charpy specimens, which is 8 mm for this case. H_s indicates the hardness in base metal for HT50 or in softened zone for HT60 and 80. In general, regardless of the difference of materials, the transision temperature for fracture mode has the tendency to rise with an increase of the bead width ratio, or with an decrease of the hardness difference ratio. That is, it is found that as the bead width ratio is decreased smaller, and as the hardness difference is increased larger, the fracture for Charpy specimens tends to deviate to the base metal side even at lower testing temperature.

Then, for instance, concerning to HT60 welds in 15KJ/cm weld heat input, the impact properties for the weld metal were examined by means of 5 mm V notched specimens. They were produced by way of trial to have 5 mm V notch to 10 mm square. Fig. 11 shows the impact test results of HT60 welds by using 5 mm V notch specimens. In this instance, as the bead width ratio is increased larger as compared with 2 mm V notch, even at 100°C testing temperature the fracture in this specimen did not deviate to the base metal side, that is, the fracture modes were all (A) types. From this result it is obvious that the transition temperature for fracture mode is influenced by the bead width ratio. And also the value of impact strength becomes constant at 80°C testing temperature, and the energy transition temperature for this curve is about 40°C, which is extremely high as compared with that (about −65°C) of the base metal.

As above-mentioned, it is found that the transition temperature for fracture mode are obviously affected by both the bead width ratio and the hardness difference ratio. However, it is still unknown which ratio between them has more influence on the fracture mechanism. Nextly, Fig. 12 shows the relation existing between the weld heat input and the absorbed energy of weld metal at specified temperature in JIS (Japan Industrial Standard)

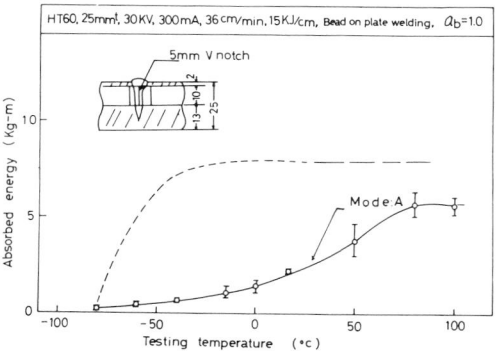

Fig. 11   Impact test results of HT60 welds by using 5mm-Vnotch specimens.

for HT50 and 60, or in WES (Japan Welding Standard) Specification for HT80. According to these specifications, the minimum absorbed energy required in impact test is prescribed to be 2.8 Kg-m at 0°C, 4.8 Kg-m at −5°C and 3.6 Kg-m at −15°C for the base metal of HT50, 60 80 steels, respectively.

In this figure each minimum value prescribed by JIS or WES is shown by the broken line as "Lowest limit". As compared with those prescribed values, the absorbed energy for some of HT60 welds in (A) type fracture mode showed the value less than JIS' limit, however the value of absorbed energy for the welds of HT50 and 80 steels is exceeding the lowest limit in JIS and WES Specifications even in case of (A) type fracture modes. On the other hand, it is natural that the absorbed energy for the welds of each material showed extremely higher value than respective lowest limit in case of (C) type.

Furthermore, in case of HT50 and 80 steels the absorbed energy for the electron-beam welds showed rather higher value than that for the submerged arc welds.

4. Conclusion

Nondestructive inspection, hardness distribution, and impact properties of weld metal were made clear on the electron-beam welds for three commercial high tension steels (HT50, 60 and 80). The results obtained were as follows:

(1) Defect of welds

According to X-ray inspection for the electron-beam welds, hot cracks were often observed in the crater points, or the plasces where the electron beam irregularily stopped due to arcing etc.. However, in accordance with JIS Specification most of electron-beam welds except irregular points were allowable in the first class

Table 3   Relation between weld heat input and transition temperature for fracture mode in various materials.

Steel	Weld heat input (KJ/cm)	Transition temperature for fracture mode (°C)	d_B (mm)	H_W (VHN)	d_B/h	H_W-H_S/H_S
HT50-weld metal	10	−60	1.2	323	0.15	0.96
	15	−15	2.6	280	0.33	0.70
	20	−15	2.6	259	0.33	0.57
	40	> 15	3.3	196	0.41	0.19
HT60-weld metal	10	−40	1.2	349	0.15	0.59
	20	−15	2.6	293	0.33	0.33
	40	15	3.3	247	0.41	0.12
HT80-weld metal	10	−40	1.2	420	0.15	0.58
	15	−40	2.6	402	0.33	0.52
	40	−15	3.3	342	0.41	0.29

d_B: Bead width at 7 mm depth from surface
h: Remaind length of Charpy 2 mm-Vnotch specimen, 8mm
H_W: Average hardness of weld metal, VHN
H_S: Average hardness of base metal or HAZ, VHN
     (HT50; H_S=165, HT60: H_S=220, HT80; H_S=265)

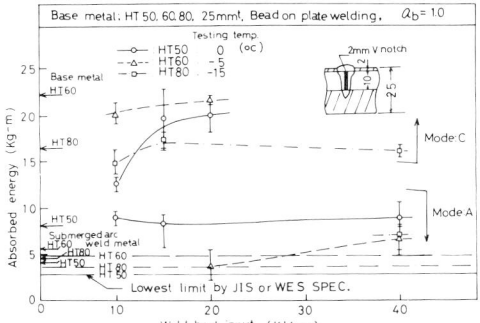

Fig. 12   Relation between weld heat input and absorbed energy of
weld metal at specified temperature in JIS or WES Specification for various materials.

for synthetic grade. And also there were no surface defects in dye penetrant test.

As a result it was found that particular cares should be taken for crater treatment and not to beam-stop irregularly due to the arcing for electron-beam welder in order to obtain perfect welds.

(2) Hardness examinations

(i) The hardness of weld metal are gradually decreased with an increase of the weld heat input.

(ii) In the hardness distributions of weld metals inward direction for HT50 and 60 welds, the tops of spiking showed the maximum value of hardness.

(iii) The softened zones were observed in the hardness distribution of HT60 and 80 welds.

(3) Impact properties

(i) There were not particular variations in the impact strengths of the weld metal in itself within the changing range of weld heat input in this investigation. And in case of HT50 welds the value of them corresponded to that of the base metal, but in case of HT60 and 80 welds it was remarkably lower than that.

(ii) In general, the value of impact strength for the weld metal showed the different levels at the particular testing temperature due to the difference in the fracture mode. And at lower temperature, the absorbed energy showed the low value, as the fracture occurred straightly along the notched direction, however at higher temperature it showed the high value as the fracture occurred out of the notched direction.

(iii) The transition temperature for fracture mode in general tends to shift to higher temperature with an in-

crease of the weld heat input.

(iv) It was found that the transition temperature for fracture mode was closely related to both the bead width and the hardnened degree of weld metal.

(v) As compared with the minimum abosrbed energy for respective base metal prescribed by JIS or WES Specification, in some of HT60 welds, the absorbed energy of the weld metal in itself showed the value less than JIS' limit. However in HT50, HT80 and the other HT60 welds it is exceeding the lowest limit in JIS and WES Specifications. And as compared with submerged arc welds, in case of HT50 and 80, the absorbed energy for electron-beam welds showed rather higher value than that for them.

Acknowledgement

Sincere appreciation is expressed to Mr. T. Kojima, Katayama Iron Works, Ltd. and Mr. S. Katayama the graduate student of the Welding Department, Osaka University, who kindly assisted us for carring out the various tests.

References

(1) Y. Arata, F. Matsuda, Y. Shibata, S. Hozumi, Y. Ono and S. Fujihira, "Mechanical Properties on Electron Beam Welds of Constructional High Tension Steels (Report I)", Trans. of JWRI, Vol 3, No. 2, 59~74, 1974.

(2) Y. Arata, F. Matsuda, Y. Shibata, S. Hozumi, Y. Ono and S. Fujihara. "Mechanical Properties on Electron Beam Welds of Constructional High Tension Steels((Report II)," Trans. of JWRI, Vol.4, No. 1, 65~69, 1975

(3) Y. Arata, K. Terai and S. Matsuda, "Study on Characteristics of Weld Defect and Its Prevention in Electron Beam Welding (Report I)", Trans. of JWRI, Vol. 2, No. 1, 103~112, 1973.

Dynamic Behavior of Welding and Cutting

A Study on Dynamic Behaviours of Electron Beam Welding—The Observation by a Fluoroscopic Method
Y. Arata, E. Abe and M. Fujisawa
[Trans. JWRI **5** (1976), 1. Proc. 2nd Int. Symp. of JWS (1975).]

Tandem Electron Beam Welding—Analysis of Front Wall of Beam Hole by Beam Hole X-ray Observation Method
Y. Arata, N. Abe and S. Yamamoto
[Trans. JWRI **9** (1980), 1.]

Dynamic Behavior of Laser Welding
Y. Arata, H. Maruo, I. Miyamoto and S. Takeuchi
[IIW Doc. IV−222−77.]

Beam Hole Behavior during Laser Beam Welding
Y. Arata, N. Abe and T. Oda
[ICALEO '83. IIW Doc. IV−339−83.]

Dynamic Behavior in Laser Gas Cutting of Mild Steel
Y. Arata, H. Maruo, I. Miyamoto and S. Takeuchi
[Trans. JWRI **8** (1979), 175.]

A Study on Dynamic Behaviours of Electron Beam Welding
—The Observation by a Fluoroscopic Method—

Abstract

An experimental technique utilizing a cine-fluoroscopy and a metal tracer was developed to continually monitor the electron beam welding process. This technique made it possible to observe the dynamic nature of the beam hole more directly and more clearly than some others.

Using aluminum alloys as the parent metal, some phenomena related to the beam-metal interactions such as the big wall-fluctuation at the back side of the beam hole ("wall-cave" and "wall-knob"), the formation of root porosity due to the arched curvature of the beam hole, the behaviours of spiking concerned with the hole shape, and in part the rapid movement of the molten metal were apparently revealed.

1. Introduction

It has been suggested that a vapor filled capillary is formed in the path of an electron beam[1 2] and is rapidly filled with molten metal as electron beam welding progresses. The term "beam hole" is here adopted to indicate such a capillary which has been called a (weld) cavity or (penetration) channel and so on.

Little is known, however, about the actual mechanism which determines the size and the shape of the beam hole and their time development during welding, because of many difficulties lying on the technique to observe directly those dynamic behaviours, while some works were attempted utilizing elavorate equipments. Tong and Giedt[1] first used radiography of pulsed X-ray for such investigation. However, since they could only take single pictures of which the definition and resolution were still unsatisfactory to reproduce, it was impossible to acquire continually an overall impression of the very rapid processes. Funk, McMaster et al.[2] employed a pinhole streak camera and recorded the X-ray intensity distribution which was emitted from the beam hole. While they could roughly reveal the time related position and intensity of electron beam impingement, dynamic behavior of beam hole and molten pool could not be observed on their high speed pictures. For this reason, a trial was initiated in which a fluoroscopic technique was used for continuous observation of the phenomena.

2. Experimental Method

2.1 System for observation and recording

The experimental arrangement of the equipment is shown in Fig. 1. The X-rays emitted from the X-ray source pass through a work piece during actual welding, and impinge onto an X-ray image intensifier in which they are collimated and amplified being converted into light-rays and electrons. Finally, the visible images are obtained on the second fluoresent screen and they are filmed with a movie camera.

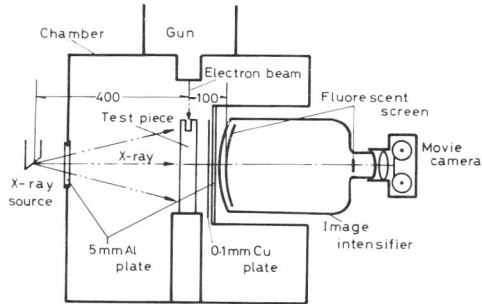

Fig. 1 Schematic figure of experimental technique

In electron beam welding, the welder itself works as an X-ray generator and X-ray is emitted from the wall of the beam hole produced by welding electron beam. So such X-ray is here called "beam hole X-ray." The resolution and contrast of the image on the fluorescent screen are reduced by the incidence of the beam hole X-ray and the other scattering X-ray. Thus a mask of copper plate 0.2 mm thick was provided upon the input window of the image intensifier to absorb these disturbing X-ray. The decrease of discernability caused by the copper plate was only 0.5% in the penetrameter sensitivity (that is the indication of radiographic resolution). In order to obtain a good image quality, it is important to set up suitable distances between the X-ray source, the welding zone to be detected and the image intensifier. The image intensifier should be set as close as possible to the welding zone for preventing the decrease of picture definition due to the enlargement. For the same reason, the smaller focal spot of the X-ray tube is, the better definition and resolution are gained on the picture. So, in this experiment, a d.c. X-ray equipment Müller MG 150 was employed as the X-ray source with its focal spot of 0.7 mm x 0,7 mm and Philips image intensifier tube MB 13 (diameter 6 inches) was located at the bored vacuum chamber as illustrated. Being aligned the center axis, a 16 mm cine camera was placed behind the viewing screen of the tube. The films used were Fuji type RKS-FG for fluoroscopic use. The input and output side windows of the vacuum chamber were both constructed of 5 mm thick aluminium plate instead of steel to decrease X-ray absorption.

For a given X-ray spectrum and intensity, the resolution and contrast of the image also depend to a great extent on the thickness ratio of X-ray penetration between the beam hole to be detected and the material to be welded. Hence a given void size can be most easily detected in thin specimens. But in thin specimens, the depth of weld penetration is severely limited since the specimens are apt to be over heated with a low power of electron beam, and show some behaviours different from the ordinary welding. Compromise specimen thickness of 12—20 mm was chosen in this experiment.

2.2 Experimental Procedure

Experiments were composed of two serieses of welding runs. In the first series, aluminum alloy 5083 (4.6Mg, 0.01Zn, 0.02Ti, 0.15Cr, 0.04Cu, 0.21Fe, 0.64Mn, 0.14Si, Bal, Al) was used as the parent metal and silver brazing alloy (75Ag-25Cu) was inserted as the tracer.

Without any tracers, the fluoroscopic images would yield no informations about the extent and the flow motion of the molten metal since the density difference between melt and solid is very slight. To discern the movement of the melt at least in part, a material which had higher absorption coefficient than the parent metal had to be employed as a tracer.

The size and shape of the specimens and the position of inserted tracers were illustrated in Fig. 2, and the conditions of welding and X-ray irradiation were as shown in Table 1. The lens current of 0.82 amperes made the

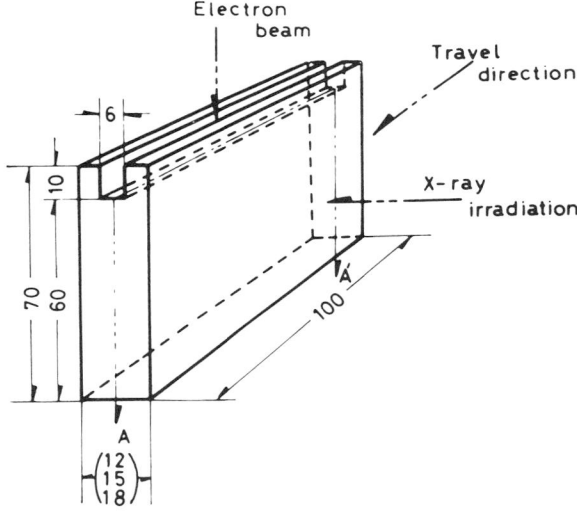

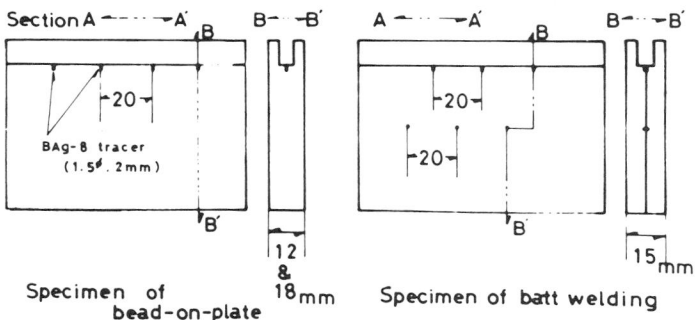

Fig. 2 Test specimen and position of inserted tracers (Series. 1)

focal point just on the surface of specimen (a_b value = 1). The weld penetrations reached into about 50 mm in these conditions.

Table 1   Experimental condition (Series 1)

Welding Condition			Width of plate	X-ray irradiation (Focal spot 0.7 × 0.7 mm)		
Fixed factors	Lens current (A)	Process		Tube volage	–	Tube current
30kV	0.82	Bead-on -plate	12mm	60kV	–	8mA
250mA	0.84					
75mm (W.D.)	0.82	Butt weling	15mm	63kV	–	8mA
300mm/min	0.84					
	0.82	Bead-on -plate	18mm	65kV	–	8mA
2 × 10⁻⁵ Torr	0.84					

* Note; W.D. = work distance

At the bottom part of the beam hole, the size of the void becomes smaller than at the upper part. The resolution of the film decreases to make it difficult to discern the shape of the hole.

In the second series, in order to improve the discernabity, copper foils were inserted at the lower part of the specimen. It was considered to be important to observe the bottom part more clearly since the spiking, which is one of the inherent defects in partial penetration electron beam welding, appeared at the part. By changing the inserted metal from pure copper to brass and also by changing its location, the effect of highly volatile element on the shape of beam hole and also on the formation of spiking was examined.

The location of inserted metal and the size of specimens were illustrated in Fig. 3, and in this series a grid made of stainless steel wire was located at the front of the first fluorescent screen to indicate the demensions. One spacing of this grid was about 8 millimeters on the

obtained films. The welding conditions were as shown in Table 2.

Table 2   Welding condition (Series 2)

Film	Material (width mm)	Welding speed cm/min	Work distance mm	Lens current⁴ Amp.	Inserted Metal (0.2t)
No. 4	5083 Alloy (20)	30	65	0.84	Copper
No. 5		50	40	0.95	Brass
No. 6	Aluminum A1100(20)	50	40	0.965	Copper
No. 7		50	40	0.965	Brass *

Location of inserted foils are at the lower part and (*) at the upper part. (Specimens are butt welded)

Experiments were conducted with a low voltage type welding unit which had the maximum capacity of 7.5 KW at 30 KV (and 15 KW at 60 KV).

The chamber was evacuated, welding parameters and X-ray irradiation parameters were set, and sequentially, the camera motor was started, and a little later, the welding operation was initiated. The film was moved at a framing rate of about 50 per second.

3.   Experimental Results and Discussions

3.1   Quality of obtained films.

The data was obtained in the form of exposed films which were several meters in length.

The contrast of the original film was rather low, but it could be risen to a certain extent by means of reprinting the original on a reversal film. As for the resolution, wires in the penetrameter which had a diameter of more than 0.8 mm were clearly visible on the viewing screen of image intensifier, but on the frames of the film, the minimum

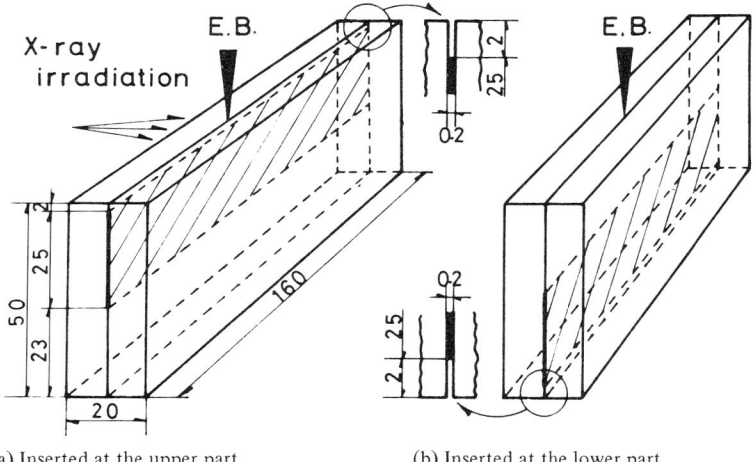

(a) Inserted at the upper part   (b) Inserted at the lower part

Fig. 3 Dimension of specimens and position of inserted foils (Series 2)

diameter of discernable beam hole was to be believed about 1 mm. A reasonable first approach to get a better resolution would be to select the material of the chamber windows such as beryllium which has a less X-ray absorption ability than aluminium employed in practice. It is not preferable to use thinner specimens for welding since the natural behaviours of the process may be distruved in return for decreasing the total thickness of the materials through which X-ray must penetrate.

In the second series of welding runs, the resolution and contrast of the film increased at the part where copper or brass foils was inserted.

3.2 Influence of inserted metals on the natural bahaviours of beam hole and molten metal

As for the selection of added tracers and inserted foils, each of these materials should have a larger atomic number than aluminium, the parent metal, from the view point of X-ray absorption. On the other hand, such a material has a higher specific gravity, and there is the possibility that they sink or sediment at the bottom of the moten pool.

The tracers which have a better affinity for the parent metal are expected to flow together with the molten metal giving less effect upon its natural movement. Therefore, a silver brazing alloy was considered to be preferable because its melting point is not so far from that of aluminium, and silver has a good solubility for aluminium, beside it was commonly available as a form of wire rod.

In spite of the high vapor pressure of silver, no remarkable change was recognized in the shape of beam hole even at the very place where the tracer was inserted. It might account for this that as concerned with the vapor

pressure silver is higher than aluminium but lower than magnesium which is involved in 5083 alloy. Consequently, it was supposed that the behaviours of the molten pool or beam hole were not affected so much by the addition of the small amount of such tracers. (Photo. 2)

As copper foils are concerned, the problem of the vapor pressure can be neglected since they have the lower one than aluminium. But the difference of the penetration depth or the spiking phenomenon between the weld with and without the copper foil was not yet examined, so the influence of inserted copper was remained as a problem for a further investigation.

3.3 Behaviours of the beam hole

Observing these films, it is immediately apparent that, as the welding is initiated, the formation of the beam hole begins with vaporizing the material along the path of electron beam. The beam hole increases its depth until an equilibrium is established, and then a nearly constant depth is continually produced for the duration of welding. While the beam hole varies its size and shape with the time, and shows a oscillatory nature.

In the first series, the cyclic wall-fluctuation of beam hole happened very rapidly and developed upward. In this case the strong wall-fluctuation gave rise to the big cave and knob at the wall. The terms "wall-cave" and "wall-knob" are here adopted to indicate such caved part in and swelled part on the wall respectively. Such wall-caves, which seemed the local expansion of gas bubbles due to the high pressure of the evaporating metal, were observed in the beam hole. These wall-caves grew up gradually moving upwards at the rate of 20~30 cm/sec. as shown in Photo. 1 where some selected frames of the film were reproduced.

E.B. moving direction

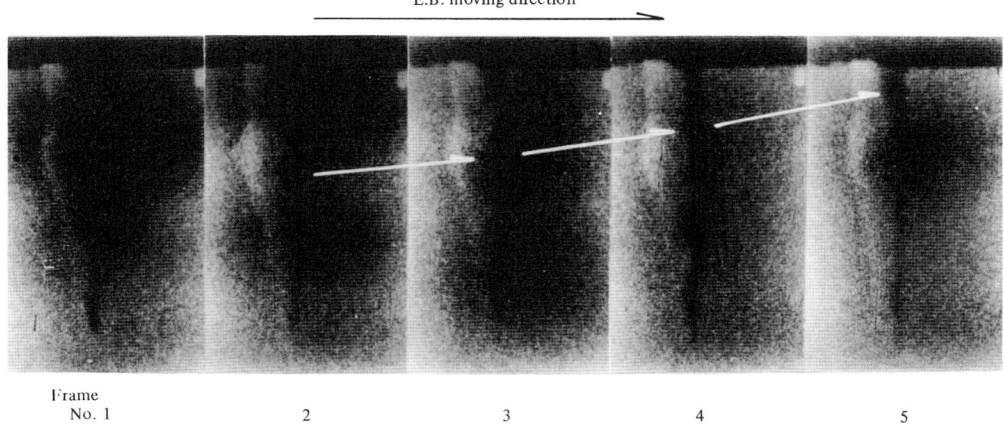

Frame
No. 1 2 3 4 5

Photo. 1 Upward movement of the "wall-caves" and "wall-knobs"

At the tip of the wall-knob, the beam hole was narrowed by the dinanric force of molten fluid on the back side. But the complete closure[5] of the beam hole could not be recognized upon all these films excepting what occured at the bottom of the hole. It was difficult to estimate the wall-oscillation frequency, because the mode of fluctuation was seemed different at each part of the beam hole and the sharpened bottom of the hole was less discernable. However, the frequency was considered to be about 5 cycles/sec. in view of the movement of the wall-caves. It is different from the "natural frequency of equilibrium state" mentioned by Funk et al.[2]. This

frequency was comparable to that of the spiking which was revealed by means of etching the longitudinal section of specimen after welding as illustrated in Photo. 2.

So far as this experiment was concerned, the basic fluctuation mode of the beam hole was the same for both butt-and bead-on-plate welding.

3.4 Arched curvature of the beam hole and formation of root porosity

In most cases an arched curvature deflected in a direction opposite to that of moving was observed at the bottom side of the beam hole as shown in Photo. 3. The reson for this event is yet unknown. But it is considered that this is due to inertia, and in all liklihood it is closely related to the "wall focusing" effect[3],[4] or the "gas focusing" effect[5],[6],[12] of electron beam which produces the local delay of energy assumption against the welding progression.

This event is also supposed to be concerned with the formation of root porosity. When the inflection point of the curvature is put upon a narrow part at the tip of the knob, a complete closure of the hole is easy to happen at the point. Near the bottom of beam hole, the cooling rate of molten metal is increased due to less input beam energy and becomes sufficiently rapid, and the molten metal may freeze before the trapped gas bubbles can be convected to the out side, thus giving rise to porosity in the root of weld.

3.5 Movement of the molten metal

The movement of molten fluid suggested by the inserted tracers of silver brazing alloy was as follows. The flow of molten metal had a very rapid rate of more than 9 m/min. in the vertical direction.

E.B. moving direction

Photo. 2 Macrograph of the spiking
(Longitudinal section after welding)

E.B. moving direction

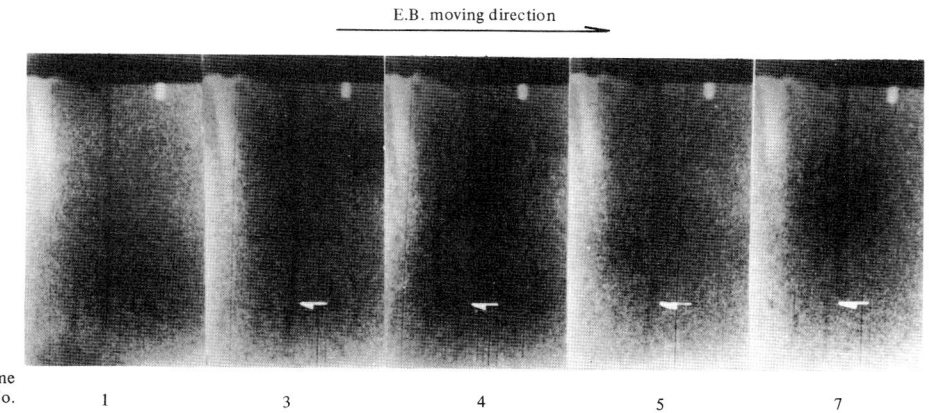

Frame No.   1   3   4   5   7

Photo. 3 Arched curvature of the beam hole and the formation of root porosity

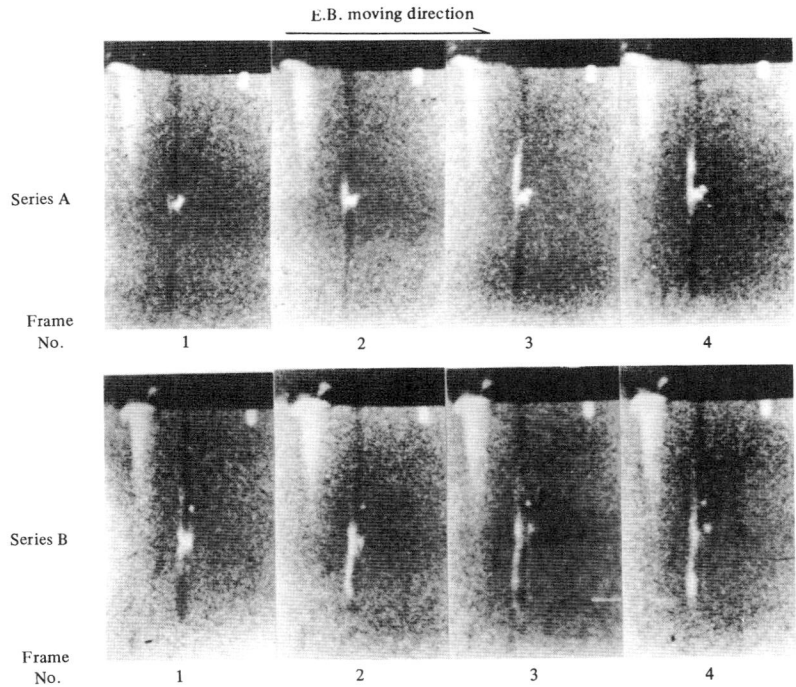

Series A

Frame
No.   1   2   3   4

Series B

Frame
No.   1   2   3   4

Photo. 4   Flow of the molten metal tracer (1) —middle part—

It was, of cource, recoginzed that the tracer was also spread to the rear wall through the horizontal direction[7] which was perpendicular for the X-ray passage. The details of this movement could not be clarified because the diameter of beam hole was very small.

At the middle part of the beam hole, there observed two ways of the movement. One was that the molten metal tracer contained in the fluid was at first risen up along the wall of beam hole, and about 0.1 second later, it

begun to fall down towards the bottom, and finally, it was distributed through out the beam hole to be rather a symmetric appearance. In this case, the beam hole had a sharp edge at the bottom as shown in Photo. 4 series A. The other was that the tracer spread up-and downwards at the beginning, but the upward motion was less than the former and ceased after about 0.06 sec. while the downward motion continued 0.12 sec.. The final distribution was inclined towards the obtuse bottom of the

E.B. moving direction

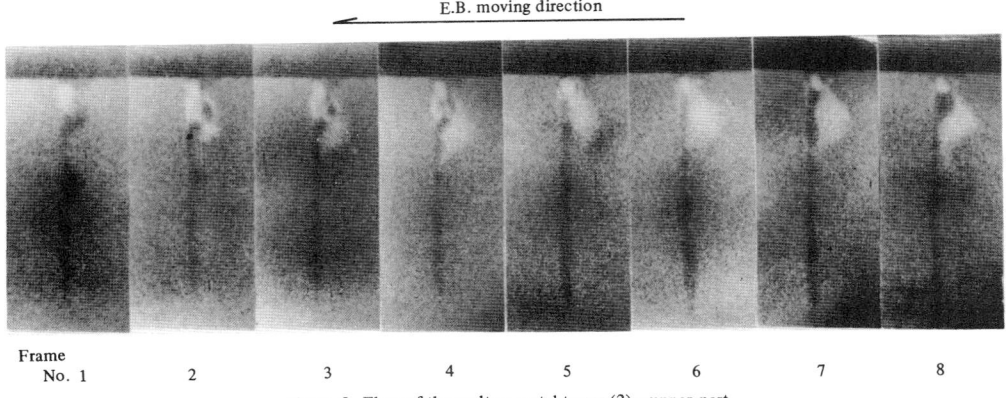

Frame
No. 1   2   3   4   5   6   7   8

Photo. 5   Flow of the molten metal tracer (2) —upper part—

383

hole as shown in Photo. 4 series B..

Near the surface, the fluid moved most likely in a whirling fashion, and this event was expected to have a relation with the wave of melt reflected at the weld pool. But the attempt to observe the wave simultaneously with the hole shape was unsuccessful because the heat deposition at the surface could not make a normal bead as a result of the limitation by the employed shape of specimen head. (Photo. 5)

Although it should be noted that the difference of the specific gravity between the tracer and the base metal might disturb the natural behaviours, those turbulance in the beam hole would be very efficient for mixing the melt and resulted in a homogeneous fusion zone.

3.6 Formation of spiking and the influence of highly volatile elements on it.

The spiking was revealed to have an intimate relation to the instantaneous shape of beam hole. The sequence of this phenomena was as illustrated in Photo. 6 and could be explained as follows.

When the beam hole is first constricted at the upper part, underlying wall-caves of gas bubbles increase their pressure. Then at the second stage, an explosion occurs which results in breaking the balance of surroundings.

Successively the back pressure of the explosion excludes the melt at the bottom of beam hole and the electron beam impingement on the dry bottom produces an additional penetration. Finally, the increase of hydrostatic head pressure makes the melt return downwards to the bottom, and the state of near equilibrium recovers again in which, for a brief instant, the back presssure of evaporation is in delicate balance with the forces of surface tension of the beam hole and the hydrostatic head of the molten pool, etc.. Metallurgical investigations of reference (10) which confirms the lack of molten metal at the root of spikes support this argument, although the movement of melt at the bottom of beam hole can not be observed in these films.

Tong and Giedt[1] proposed a model of the penetration mechanism which is a step in this direction and asserted that the formation of spiking is due to the periodic closure of beam hole by the molten metal. On the other hand, some investigators[6],[8],[9] hold that the fluctuation of the energy density of electron beam caused by the collision of electrons with ionized metal vapor results in the irregular penetration which involves the spiking.

The results of our observation provide grounds for maintaining that the spiking can be formed without

E.B. moving direction

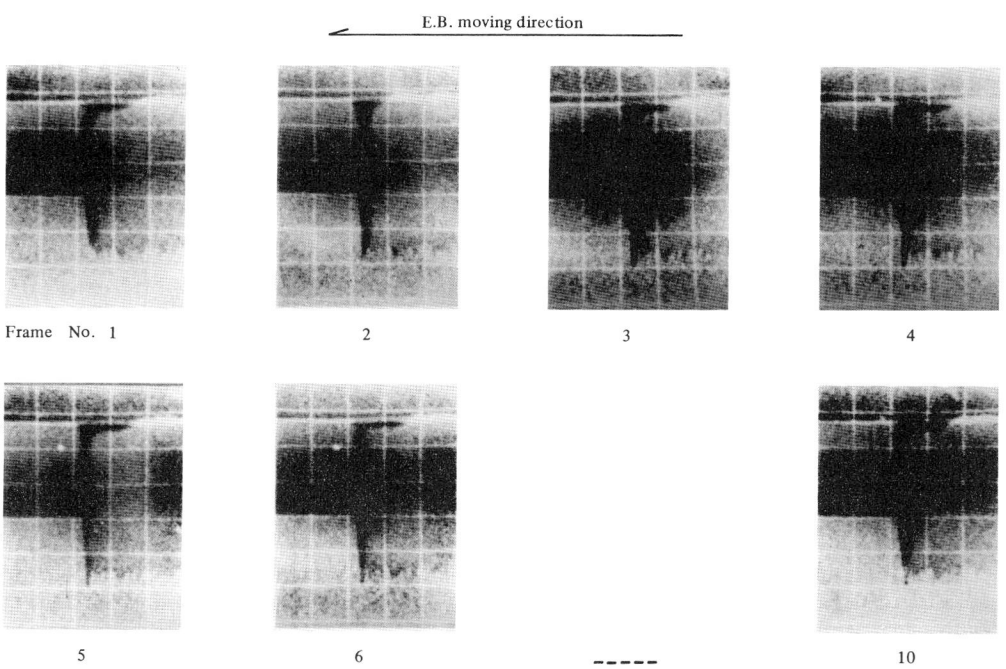

Frame No. 1   2   3   4

5   6   -----   10

Photo. 6 Formation of spiking (from welding run No. 4)
Note: Time interval of each frame is 0.02sec

384

assuming the existence of such complete closure events as proposed by Tong and Giedt. In accordance with what has been stated above, the main factors which have great influence on the spiking are the pressure of metal vapor and the overhanging behaviour of molten metal. Namely occurring of the wall-caves and wall-knobs in the beam hole gives rise to spiking. It can be esily considered that the breaking of the delicate balance between these factors is not necessary to have a precise periodicity because the fluctuation of energy density of electron beam is suspected to be not directly due to the ripples of power input which shows generally a higher frequency[3],[11] than that of the spiking macrographically observed. Consequently, if the beam hole has a sort of shape which is easy to keep the balance, there appears a possibility of suppressing the spiking.

In practice, when the foils of some alloys containing a highly volatile element were inserted at the upper part of the specimen as illustrated in Fig. 3-b the spiking decreased as shown in Photo. 7.

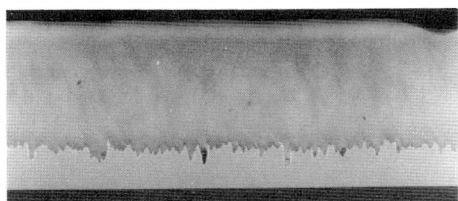

(a) Copper foil is inserted at the upper part

(b) Brass foil is inserted at the upper part

Photo. 7 Radiographs of aluminum specimen after welding

In this case, the beam hole became to swelled out being constiricted at the upper opening (Photo. 8) because of the increased vapor pressure, and the wall-fluctuation got to be small, while hydrostatic head increased with the growth of overhanging melt. Since the abrupt change of vapor pressure is difficult to happen under such condition, the beam hole is rather stable and the energy density in the bottom decreases because the electrons pass through the beam hole filled with dense metallic vapor. That is the most likely cause of the decrease of spiking.

E.B. moving direction

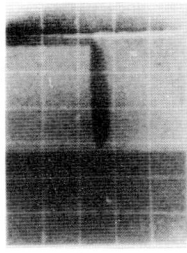 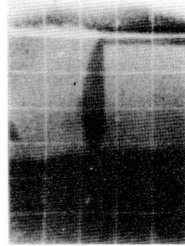

(a) A1100-(Brass)   (b) 5083-(Brass)

Photo. 8 Shape of beam hole −(I)−
When brass foils are inserted at the upper part of specimens

However, when the alloy foils were inserted at the bottom side of the specimen, it was found that the influence of these foils upon the hole shape and the spiking was different with the kind of parent metal employed as the specimen. The specimens of pure aluminum eliminate the effect of inserted foils and the spiking formed again, although the specimens of 5083 alloy remained the effect.

E.B. moving direction

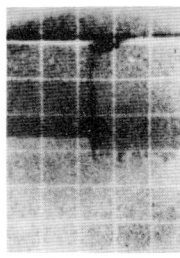 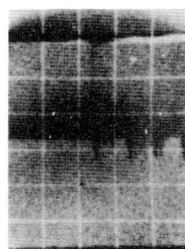

(a) A1100-(Copper)   (b) A1100-(Brass)

Photo. 9 Shape of beam hole −(II)−
When copper or brass foil is inserted at the bottom side of pure aluminum

In the case of pure aluminum the surface opening of beam hole was wide. The reason for the widening is suspected to be the larger molten pool due to the transfer of less heat in the layer close to the surface. As a result of this, the metal vapor vents itself easily through the opening and the inner expansion of bem hole does not occur. Consequently the .beam hole becomes narrow at the middle and lower part where the constriction of the hole and the explosion of gas babbles occur to form the spiking.

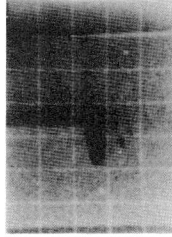 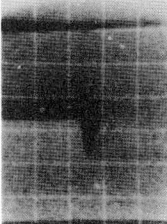

(a) 5083-(Brass)　　　(b) 5083-(Phosphor bronze)

Photo. 10　Shape of beam hole —(III)—
When brass or phosphor bronze foil is inserted at the bottom side of 5083 alloy

On the other hand, the passage in 5083 alloy was constricted and narrowed between the opening and the upper end of inserted foil as shown in Photo. 10, because of heat dissipation by the intense evaporation of magnesium contained in the alloy. As the expansion of beam hole still remained at the bottom side, the spike decreasing effect was not disappeared.

4.　Summary

A cine-fluoroscopic technique has been developed for studying the dynamic mechanism of electron beam welding using aluminum alloys. Although further experiments are necessary to clarify the mechanism more precisely, because a different phenomenon may be observed in the specimens of steels, etc., the following conclusions are possible from the evidence obtained in this experiment.

(1) By means of the technique combined with metal tracers, a direct observation of the beam hole and in part the molten metal during electron beam welding can be achieved with a fairly excellent resolution on the movie film.

(2) The beam hole generally does not hold a steady shape and oscillates in width and depth. The strong wall-fluctuation occurs so frequent and it gives rise to the wall-caves and wall-knobs. They move upward at the rapid rate of 20–30 cm/sec, for example.

(3) The higher welding speed becomes the more the beam hole bends like an arch convexed toward the direction of welding progression. This event has a considerable effect on the formation of root porosity.

(4) The formation of spiking is resulted from the breakdown of the balance between the vapor pressure and the hydrostatic head of the melt and in addition dynamic force of the molten fluid, and is revealed to have an intimate relation with occurring of the wall-caves and wall-knobs in the beam hole. It can be suppressed when the stable beam hole is produced by inserting the alloy foils which containes a highly volatile element.

(5) The molten metal flows up and down along the wall of beam hole at quite a high travel rate. That accounts for the favourable homogeneity of fusion zone caused by metal mixing.

References

1) H. Tong and W. H. Giedt: A Dynamic Interpretation of Electron Beam Welding, W. J. 49-6 (1970), 259s.

2) E. R. Funk, R.C. McMaster et al.: Penetration Mechanisms of Electron Beam Welding and the Spiking Phenomenon, W. J. 53-6 (1974), 246s.

3) Y. Arata et al. : Energy Distribution in the Hole of Electron Beam Welding, Report of EBW Research Committee of JWS (1975), EBW-123-75, (in Japanese)

4) Y. Arata and I. Miyamoto; Processing Mechanism of High Energy Density Beam (Rept. 1) —Mechanism of Drilling—, Trans. of TWRI, Vol. 2, No. 2 (1973) 19

5) H. Schwarz: Present Knowledge of the F oundamental Processes of Electron Beam As a Material Working Tools, The 3rd Inter. Conf. on Electron and Ion Beam Science and Technology (1968), 301

6) V. P. Ledovskoy et al.: Plasma-Beam Interaction during Electron Beam welding, IEE Conf. Publ. (Inst. Electr. Eng.) (1974), 563

7) N. A. Ol'shanskii et al.: Movement of Molten Metal during Electron Beam Weldong, Svar. Proiz. 21-9, (1974), 12

8) R. E. Armstrong: Spiking in Partial Penetration Electron Beam Welds, The 4th Intern. Conf. on Electron and Ion Beam Science and Technoloty (1970), 179.

9) G. K. Hicken and W. G. Booco: Penetration Variations in Electron Beam Welding, The 3rd Intern. Conf. on Electron and Ion Beam Science and Technology, (1968), 398.

10) Y. Arata, S. Matsuda et al. Study on Characteristics of Weld Defect and Its Prevention in Electron Beam Welding (Report III), Trans. of JWRI, 3-2, (1974), 81.

11) V. I. Leskov et al.: Mechanism in Deep Weld Pools During Electron-Beam Welding, Avt. Svarka. 28-1, (1975), 12

12) J. W. Meier: Electron Beam Welding Characteristics of Several Materials, Proc. 3rd Symp. on EB Tech., (1961), 145

Tandem Electron Beam Welding
— Analysis of Front Wall of Beam Hole by Beam Hole X-ray Observation Method —

Abstract

X-ray images emitted from the beam hole have been observed by a pin-hole camera method. The behaviour of the front wall of the beam hole during electron beam welding has been analysed by a high speed movie and a streak photograph. In case of single electron beam welding, the scraping process of the front wall, the formation process of spiking and the suppression mechanism of welding defects by the beam oscillation method have been revealed. It is found that the front wall of the beam hole is not always scraped uniformly but it is scraped by the locally melted region on the front wall. The suppression mechanism of defects by the Tandem Electron Beam Welding Method has been also analysed. It is compared with that of the beam oscillation method in ordinary single electron beam welding. It is suggested that the Tandem Electron Beam Welding Method is much more effective.

KEY WORDS: (Electron Beam Welding) (X-ray Analysis) (Beam Hole) (Mechanism) (Defects)

1. Introduction:

The electron beam welding can utilize an extremely higher power density beam than ordinary welding method. Therefore, it can be applied to a high speed welding of thin plates or a deep penetration welding of thick plates which can not be realized by ordinary methods. However, it is known that special defects such as humping in a high speed welding, spiking and porosity in a deep penetration welding are brought about. It is difficult to suppress these defects by using an ordinary single electron beam welding without disadvantages in welding speed and penetration depth.

In order to overcome these difficulties, the authors have proposed the Tandem Electron Beam Welding Method, which uses two electron beams at a time. This method was applied to a high speed welding with suppressing humping successfully. It has been found that the suppression of humping is caused by the control of the molten metal flow by the second electron beam[1]. This method was also applied to a deep penetration welding with suppressing spiking successfully without disadvantage in penetration depth[2].

In this report, in order to reveal the formation process of spiking in ordinary single electron beam welding and the suppression mechanism of spiking in the Tandem Electron Beam Welding Method, the image of so-called beam hole X-rays emitted from the region where the electron beam interacts with the metal is observed. The behaviour of the X-ray image in a beam oscillation method is also analysed, and then it is compared with that of the Tandem Electron Beam Welding.

2. Beam Hole X-ray Observation Method

2-1 Observation of beam hole during welding

On the electron beam welding, some experiments have been performed where the shape of the beam hole was observed during welding. For example, Tong and Giedt took transmission X-ray photographs of the beam hole during the electron beam welding by a flush X-ray source with finding that the shape of the beam hole changes with time[3]. Bryant took a series of photographs of the shapes of the beam hole under the welding by a high voltage ultra short pulsed X-ray source[4]. Weber *et al.* took streak photographs of the so-called "beam hole X-rays" which was caused by the interaction between the electron beam and the metal. They proposed the periodical closure-fallback mechanism for the formation of spiking[5].

In the previous report[2], the authors examined the method in which the edge welding phenomena was filmed by a high speed movie camera through a heat-proof glass, and also examined the method in which the image of the beam hole X-rays was visualized by a pin-hole camera and a X-ray image converter, and then it was also filmed by a high speed movie camera. In this report, in order to reveal the suppression mechanism of defects by the Tandem Electron Beam Welding Method, the images of beam hole X-rays were analysed by both a high speed movie camera method mentioned above and a streak photograph method.

2-2 Experimental apparatus for observation of beam hole X-ray images

2-2-1 High speed movie camera method

Figure 1 shows the layout of the filming apparatus. The image of beam hole X-rays, which is emitted from the

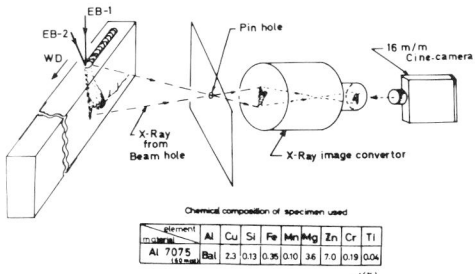

Fig. 1. : Layout of filming apparatus.

interacting region of the electron beam with the metal, is focused on the input panel of a X-ray image converter, and then it is converted to a visible image on the output panel. Then it is filmed by a 16 mm high speed movie camera. The pin-hole is made from a lead sheet. The thickness of the sheet depends on the X-ray intensity, that is, accelerating voltage and current of the electron beam. In this experiment, the lead plate of 1 mm-thickness is used for accelerating voltage of 58 kV and current of 100 mA. On the other hand, the diameter of the pin-hole depends on the sensitivity and resolution of the X-ray image converter. Since the brightness of the output panel of the converter is finite, there is minimum X-ray intensity for maximum brightness. Therefore, the diameter of the pin-hole is decided to be 0.6 mmφ in this experiment in order to satisfy above minimum X-ray intensity and to make the size of the half-shadow smaller than that of the spatial resolution of the converter itself.

The images for various shutter speed are shown in photo 1. It is seen that since the shape of the image

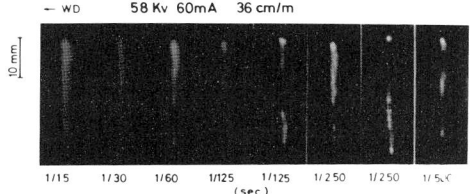

Photo 1.: X-ray images in various shutter speed.

changes with time very fast, it becomes wide with decreasing shutter speed. The time resolution depends on both the maximum brightness of the converter and the sensitivity of the movie film. In this experiment, the maximum time resolution is obtained in the condition that the shutter speed is 1/375 sec., the frame rate is 150 fps and the film of the sensitivity of ASA 400 is developed with 4 times higher sensitivity than that of standard.

2-2-2 Streak photograph method

Figure 2 shows the explanation drowing for a streak photograph method. If the image of the output panel of the converter moves from up to down and the film moves from left to right at the same time, the image on the film becomes the line inclined to the left. In photo 2, an example of the streak photograph is shown. In this experiment, the moving speed of the film is the constant value of 23 cm/sec..

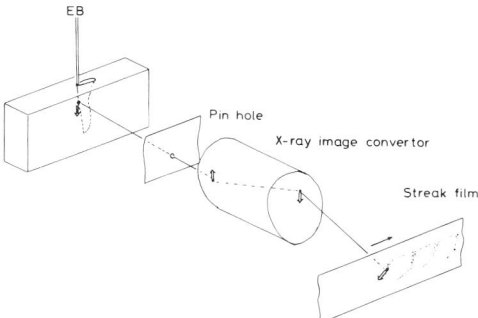

Fig. 2. : Explanation scheme for a streak photograph method.

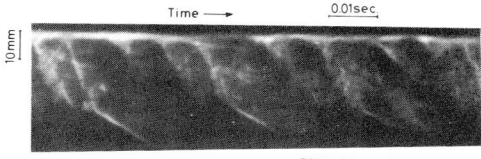

Photo 2.: An example of a streak photograph.

3. Analysis of the Formation Mechanism of Spiking

3-1 Analysis of the single electron beam welding

At first, these observation methods were applied to the analysis of the single electron beam welding. Photo 3 shows high speed photographs during the electron beam welding with the high power density beam. It is clearly

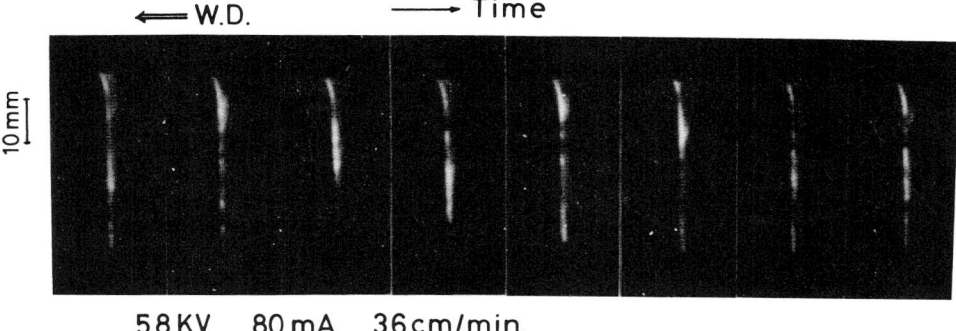

Photo 3.: Photographs in a high speed movie during electron beam welding with high power density beam.

seen that the image is not always stable but is changing with time. Because the electron beam reaches to the bright part, this means that the beam reaching depth is changing with time. It should be noticed that the bright point in the image seems to be moving downward. Furthermore, observing the images more carefully, it is found that the image is not simply perpendicular but it inclines with some degree against the normal direction.

From these facts, it is thought that the electron beam impinges to the front wall of the beam hole as illustrated in Fig. 3. The beam reaching depth (h_b), the angle of the

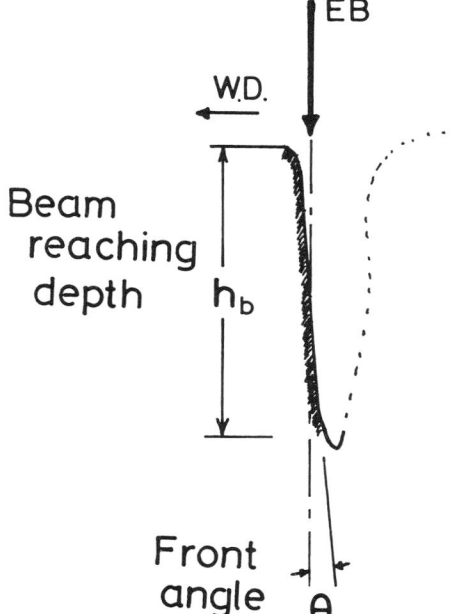

EB

W.D.

Beam
reaching
depth

h_b

Front
angle　θ

Fig. 3. : Beam reaching depth and front angle of beam hole.

front wall of the beam hole (θ) are analysed for various power density or welding speed.

3-1-1　Beam power density

Figure 4 shows the result of film analysis for h_b and θ in case of high power density (Photo 3). Upper part of

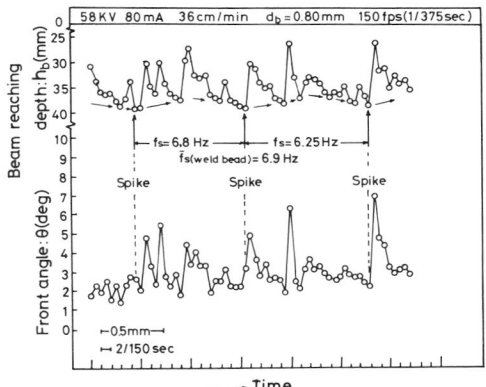

58 KV　80 mA　36 cm/min　d_b=0.80mm　150 fps(1/375sec)

Beam reaching depth: h_b (mm)

Front angle: θ (deg)

fs= 6.8 Hz　　fs= 6.25 Hz
$\bar{f}_{s(weld\ bead)}$= 6.9 Hz

Spike　　Spike　　Spike

←0.5mm→
←2/150 sec→

→ Time

Fig. 4. : Film analysis for h_b and θ in case of high power density shown in Photo 3. f_s and $\bar{f}_s$ are spiking frequency in film analysis and in bead analysis, respectively.

this figure represents the change of the beam reaching depth and lower part the front angle. It is seen that both of them change periodically and the changes of these two parameters correspond each other. In other words, just after the beam reaching depth becomes large, the front angle becomes large. These facts mean that the electron beam does not always uniformly interact with the metal but there is a strongly interacting region (such a region is recorded on the film as a bright point) which melts the

389

front wall of the beam hole locally. Therefore, the front wall is scraped not uniformly in time in case of high energy density beam. Film analyses in case of low energy density beam are shown in Fig. 5. Figures 4 and 5 are summarized to Fig. 6. It is seen that when the beam power density is small, the maximum beam reaching depth of cource decreases, while the fluctuations of the beam reaching depth and the mean front angle become also small. This change on fluctuation of the beam reaching depth corresponds with the occurance of the defects as shown in the longitudinal sections of the bead of Photo 4. Spiking phenomena grows up with increasing beam power density.

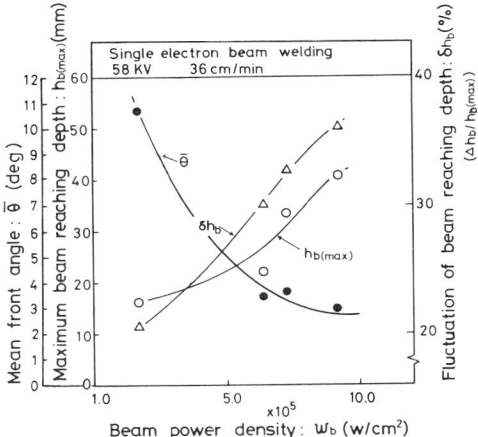

Fig. 6. : Dependency of h_b, δh_b (its fluctuation) and $\bar{\theta}$ to power density.

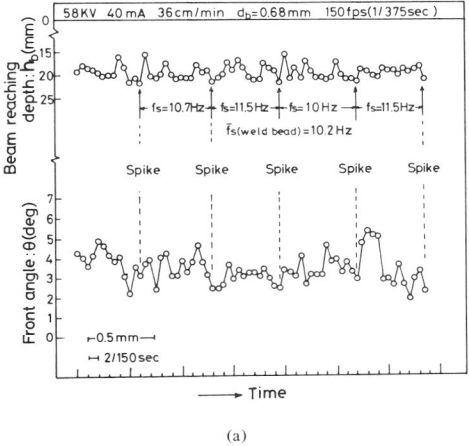

(a)

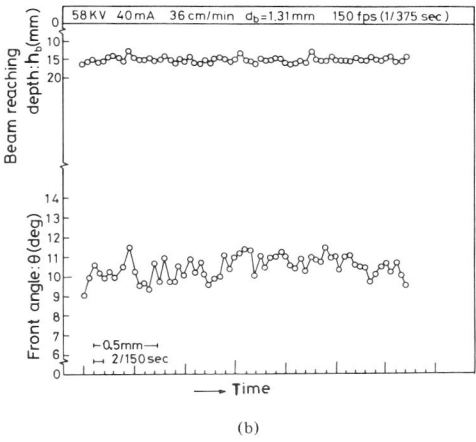

(b)

Fig. 5(a) and (b).: Film analysis for h_b and θ in case of (a) middle power density and (b) low power density.

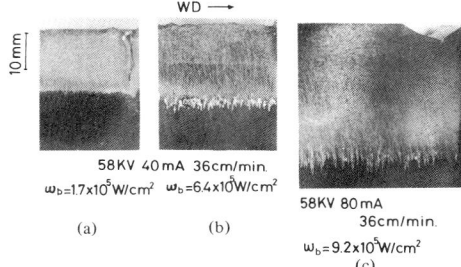

Photo 4(a), (b) and (c).: Londitudinal section of bead for various power density. (a): low, (b): middle and (c): high

3-1-2 Welding speed

Figure 7 shows the results of the film analyses in case of various welding speed. These figures are summalized in Fig. 8. As the welding speed becomes greater, the maximum beam reaching depth, the fluctuation of the beam reaching depth and the front angle also become smaller.

3-2 Formation process of spiking

These two results are summalized together as shown in Fig. 9 which represents the value of the beam reaching depth divided by the beam diameter and the front angle for various beam power density and welding speed. It is seen that the plots shift to the left corner with both increasing power density and decreasing welding speed. This means that the front wall becomes deep and steep. It is

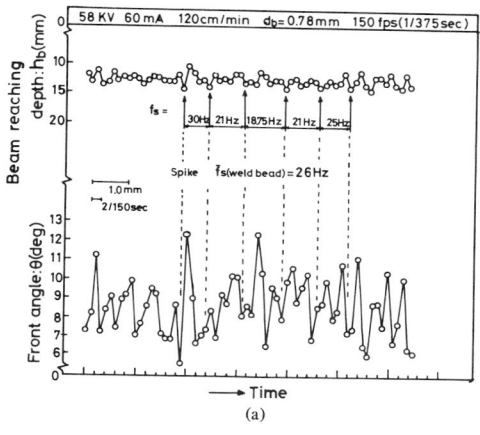

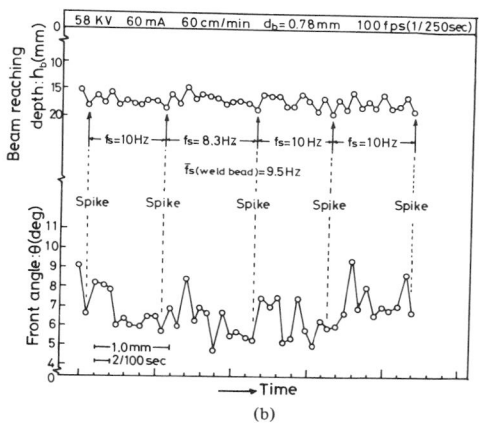

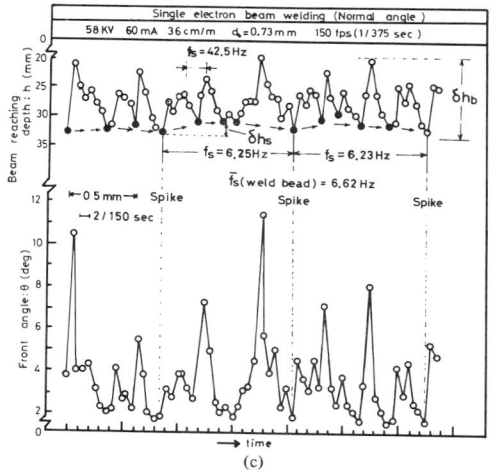

Fig. 7(a), (b) and (c).: Film analysis for h_b and θ in various welding speed. (a): high, (b): middle and (c): low.

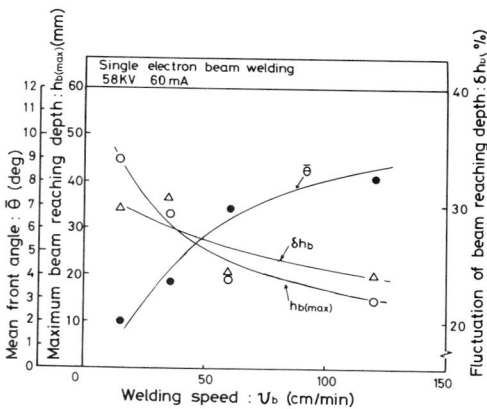

Fig. 8. : Dependency of h_b, δh_b and $\bar{\theta}$ to welding speed.

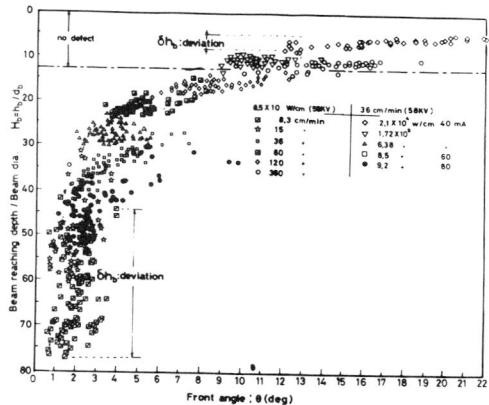

Fig. 9. : Dependency of h_b and θ to both power density and welding speed.

also seen that the fluctuation in same specimen becomes large with increasing power density and decreasing welding speed. These facts mean that the beam power density or the welding speed, in other words, the heat input which is impinged into the specimen per unit area and unit time plays an important role for the formation of the beam hole, whose change affects the beam reaching depth and the front angle with their fluctuations. That is, as the heat input to the specimen per unit area and time increases, the electron beam reaching depth becomes large, and the front angle becomes small. This causes the increase of their fluctuations and unstability of the beam hole.

Summalizing these facts together, following model will be proposed for the welding process of the metal by the electron beam welding. The local region where the electron beam interacts the metal strongly occurs on the edge of the beam hole. Since the angle to the normal direction of this region is greater than other part on the front wall, the heat input on this region per unit area and time becomes greater than other parts. Therefore, this region melts down the front wall to the root of the beam hole. As this region come close to the root of the beam hole, the angle of the front wall becomes smaller. Then a steep front wall is formed. During this process, the electron beam goes ahead. Then such a region mentioned above (called the "shoulder") occures again near the top of the beam hole, which begins to scrape the front wall again. It will be thought that with repeating this process, the front wall of the beam hole goes ahead. This process is illustrated in Fig. 10. If this process is supposed, when the beam power density is very high, the beam reaching depth is very large and the angle of the front wall becomes very small. Then, the shoulder mentioned above impinges to the root of the beam hole strongly. In such case, the beam hole becomes deeper than usual and the spiking phenomena are brought about.

Streak photographs prove this model more directly.

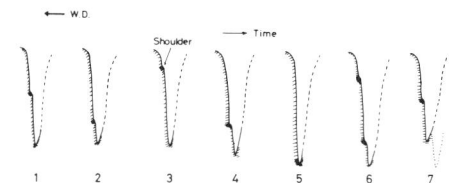

Fig. 10.: Scraping model for electron beam welding.

Photo 5(a) shows the streak photograph in case of high power density shown in Fig. 4. Discrete slant lines are clearly seen. If the interaction between the beam and the metal is uniform, the streak photograph should be seen as broad bright band shown in Photo 5(b). This fact indicates that the region where the electron beam interacts is a small region. Furthermore, the slope of these lines is classifyed into two kinds. This fact proves the stay of the shoulder at the root of the beam hole. In this second region where the beam stays, uneven penetration is brought about.

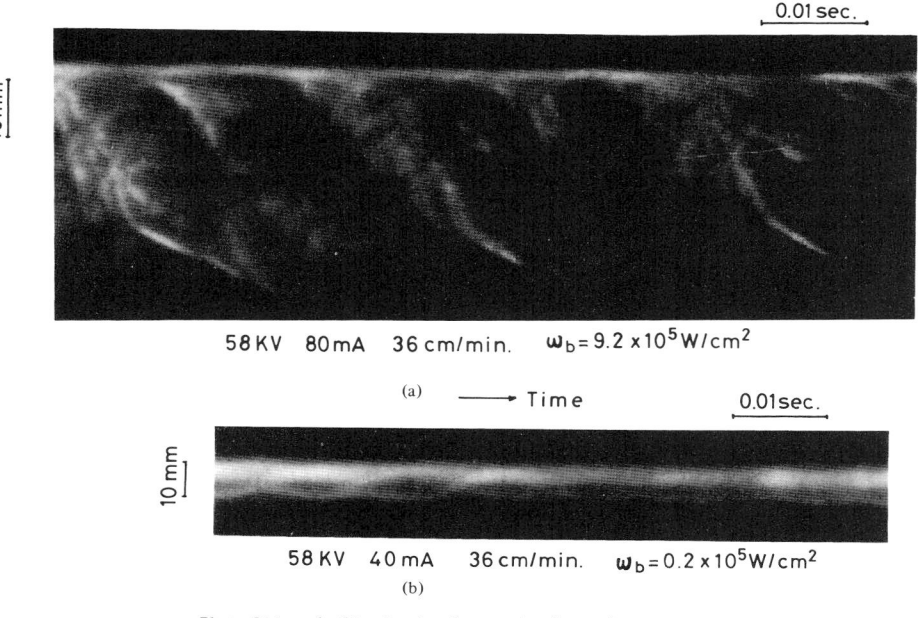

0.01 sec.

10 mm

58 KV 80 mA 36 cm/min. $\omega_b = 9.2 \times 10^5 \text{W/cm}^2$

(a) $\longrightarrow$ Time

0.01 sec.

10 mm

58 KV 40 mA 36 cm/min. $\omega_b = 0.2 \times 10^5 \text{W/cm}^2$

(b)

Photo 5(a) and (b).: Streak photographs for various power density. (a): high and (b):low.

392

4. Analysis of Suppression Mechanism of Defects

4-1   Tandem Electron Beam Welding

Photo 6 shows one frame of a high speed movie film of the image of beam hole X-rays during the Tandem Electron Beam Welding. The parameters of two electron

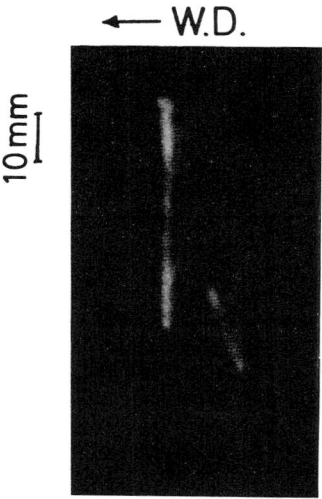

← **W.D.**

10mm

58KV EB-1   60mA
EB-2   40mA
36cm/min.

Photo 6.:   Photograph in a high speed movie film of the Tandem Electron Beam Welding. Beam parameters are in the region of suppression of defects.

beams are in the region previously reported[2] where defects are suppressed successfully. It is proved that the second electron beam penetrates slightly deeper than the first beam. The film analysis for these two beams are shown in Fig. 11. The change for the first beam is quite the same tendency as the single electron beam which brings about the defects. However, for the second beam, the change of the beam reaching depth and front angle are small. From these facts, in the Tandem Electron Beam Welding, the following mechanism will be thought. The first beam plays the role of only deep penetration by a high power density and the second beam plays the role of only effective suppression of defects by a low power density. In this succeeded case, because the second electron beam is impinged into the beam hole made by the first beam and the rear wall of the beam hole is preheated by the first beam, the second beam, in spite of the low power density, penetrates easily into the metal. It is found

that in the Tandem Electron Beam Welding, each beam displays fully their ability of deep penetration and repairment of defects.

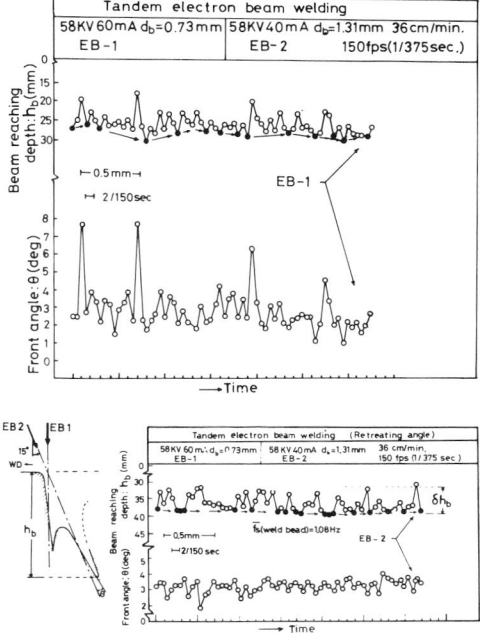

Fig. 11.:   Film analysis for h_b and θ in the Tandem Electron Beam Welding.

4-2   Single electron beam welding - beam oscillation method

The beam oscillation method has been used to suppress the defects in the single electron beam welding. However, there is a few research for its suppression mechanism. Therefore, the beam hole X-ray image during beam oscillation was analysed to compare with the Tandem Electron Beam Welding. It is well known that the direction of oscillation depends on the material and that the transverse oscillation is very effective for aluminum alloy. Therefore, the behaviour of the image of beam hole X-rays in the transverse oscillation was observed.

Figure 12 (a) shows the result of the film analysis for beam oscillation. It is seen that two parameters which were previously unstable become quite stable. Figures 12 (b) and (c) show the results of the film analyses in case of the different frequency. It is seen that the effect of beam oscillation dose not appear when the frequency is small. The results of the film analyses in case of various

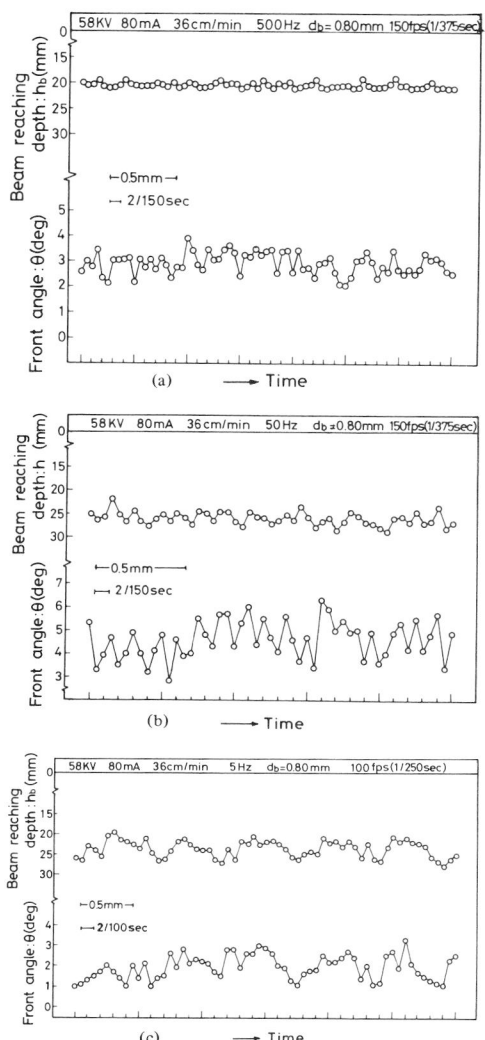

Fig. 12(a), (b) and (c).: Film analyses for h_b and θ in beam beam oscillation of various frequency. (a): high, (b): middle and (c): low.

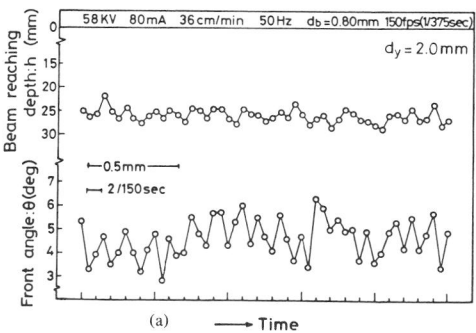

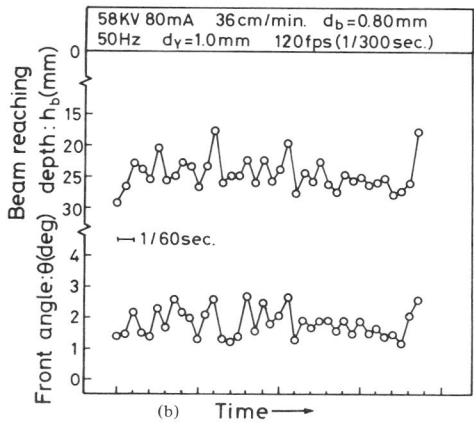

Fig. 13(a) and (b).: Film analyses for h_b and θ in beam oscillation of various amplitude. (a): 2.0 mm and (b): 1.0 mm.

amplitude are shown in Fig. 13. It is seen that the effect of amplitude also appears from large frequency.

By using the streak photograph and the section of beads, these situation will be understood more clearly. In case of low frequency, very clear lines are seen in streak photograph (Photo 7 (a)), and both longitudinal and cross sections of bead indicate clearly that there are two kinds of beam hole (Photo 7 (b)). That is, the beam behaves as a

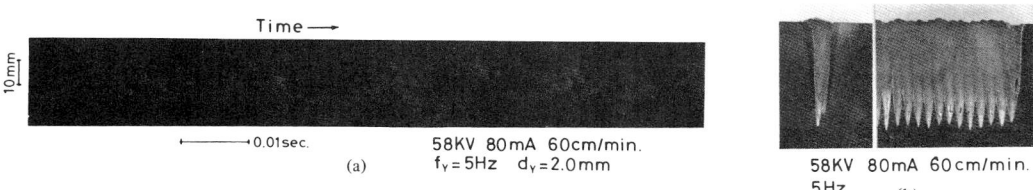

Photo 7.: (a): Streak photograph and (b): bead section for frequency of 5 Hz.

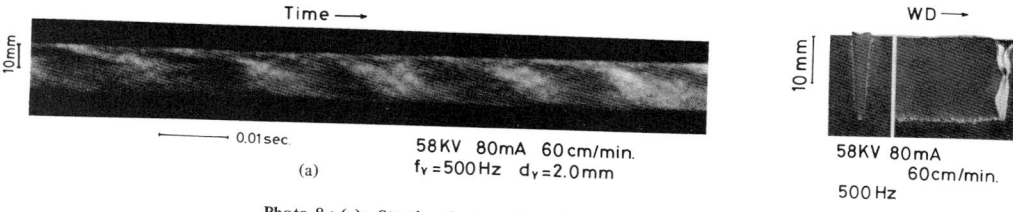

Photo 8.: (a): Streak photograph and (b): bead section for frequency of 500 Hz.

single electron beam moving diagonally. In case of high frequency, streak photograph seems to be rather uniform (Photo 8 (a)), and both loditudinal and cross sections of bead seems rather sound (Photo 8 (b)). This reason is thought that before the shoulder mentioned above reaches near the root, the electron beam is shifted by the beam oscillation of high frequency with interacting with the front wall uniformly.

4-3   Comparison of the Tandem Electron Beam and the single electron beam

When the suppression mechanism of the Tandem Electron Beam Welding compared with the single electron beam welding, the difference is clear. In case of the beam oscillation method, scraping characteristic of the beam itself is changed, because the beam is shifted before the shoulder reaches to the root. On the contrary, in case of the Tandem Electron Beam Welding, scraping characteristic is not changed. Therefore, in case of the beam oscillation method, much power is required for the same penetration depth. In case of the Tandem Electron Beam Welding, each beam displays their ability fully (the first beam is used only for the deep penetration and the second beam is used only for the repairement).

5. Conclusion

By using the pin-hole camera method with a X-ray image converter and a high speed movie camera, the image of beam hole X-rays emitted from the interacting region of the electron beam with the metal during welding is observed with time. It is also analysed with streak photographs and the section of the bead. From these analyses, following facts are found. The electron beam dose not always interact with the metal uniformly. There is the region where the electron beam interacts with the metal locally. It quickly melts the front wall downward periodically. It will thought that the spiking phenomena which appear frequently in a deep penetration welding occurs when the electron beam reaches the deeper part and the angle of the front wall of the beam hole becomes

small, that is, when the region where the beam interacts strongly ("shoulder") reaches to the root of the beam hole. In the previous paper[2], it has already shown that the Tandem Electron Beam Welding Method can suppress such defects. In this report, the reason why it can suppress the defects is revealed. It is found that the defects made by the first beam are repaired by the second beam with reheating and remelting. In spite of the low power density than the first beam, the second beam easily penetrates deeper than the first beam, because the second beam is impinged into the beam hole made by the first beam and the root of the beam hole is preheated by the first beam.

By using the same method, suppression mechanism of defects by the beam oscillation method is also analysed. The beam seems to be changed in its characteristic from a single beam to a widened beam with increasing the frequency and amplitude. Finally it becomes the same as low power density beam.

It is thought that the beam oscillation makes the discrete scraping process of the shoulder mentioned above uniform. Therefore, in order to utilize that method effectively, it must be required that the scraping characteristic is in the region between a single beam and a widened beam. The comparison between the Tandem Electron Beam Welding Method and the beam oscillation method are performed. It is suggested that the former is better than the latter for the deep penetration welding, because the latter is decreased in its ability of penetration for improvement of the scraping characteristic, while the former is not changed in ability of each beam.

395

References

1) : Y. Arata and E. Nabegata: Tandem Electron Beam Welding (Report I): Trans. JWRI, Vol. 7 (1978), No. 1, pp. 101-109.

2) : Y. Arata, E. Nabegata and N. Iwamoto: Tandem Electron Beam Welding (Report II): Trans. JWRI, Vol. 7 (1978), No. 2, pp. 233-243.

3) : H. Tong and W. H. Giedt: Radiographs of the Electron Beam Welding Cavity: Rev. Sci. Instrum., Vol. 40 (1969), No. 10, pp. 1283-1285.

4) : L.E. Bryant: Flash Radiography of Electron Beam Welding: Materials Evaluation, Vol. 29 (1971), No. 10, pp. 237-240.

5) : C. M. Weber, E. R. Funk and R. C. McMaster: Penetration Mechanism of Partial Penetration Electron Beam Welding: W. J., Vol. 51 (1972), No. 2, pp. 90s-94s.

Dynamic Behavior of Laser Welding

Abstract

A new experimental technique using transparent material was developed, which enables direct observation of the dynamic behavior occuring during the laser welding with deep penetration. Motion of the cavity and the melt flow in soda-lime glass were taken in the high speed color films. The films showed that the recoil force of evaporation at laser-heated region was responsible for maintaining the cavity in the liquid material and driving the melt to flow providing the vortex behind the cavity. The influence of the travel speed on the welding phenomena was discussed on the basis of the observation.

(1) Introduction

In recent years, the processing of materials by CO_2 lasers has rapidly advanced, and a deep penetration bead has also been achieved in laser welding as well as EB welding. Little is known, however, about the actual mechanism that occures during welding. This is mainly due to the difficulties of observation of the interaction between the laser beam and materials.

Some work has been done by using the X-rays technique [1-4] in order to observe the events occuring in the weld cavity, and the possible unsteady nature of this cavity has been suggested. But the contrast of the resulting radiographs was not high enough to understand the mechanism.

To overcome this problem, a new experimental technique using transparent material has been developed, which enables direct observation of the dynamic behavior occuring during the laser welding. Various kinds of transparent materials were tested, and it was found that the welding phenomenon in glasses having lower viscosity, soda-lime glass, was very similar to that in metals. This study was directed mainly to the direct observation of the laser-material interaction in the weld cavity. The welding phenomena were then taken on the high speed color films, and mechanism of the laser welding was discussed based on these films some of which will be exhibitied when this paper is presented.

(2) Experimental method

In this experiment, a conventional type of DC CO_2 laser with the maximum power output of about 500W was used. The beam with diameter of about 15mm from the laser was deflected downwards by a plane mirror to a focusing Ge lense with a focal length 100mm, and was then focused on to the workpiece surface. Radial distribution of power density at the focal point was measured, and it was approximated by a Gaussian curve with 1/e diameter of about 0.2mm as shown in Fig.1.

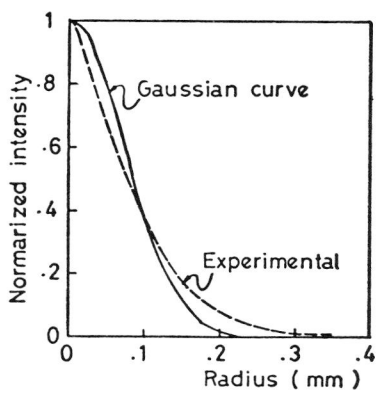

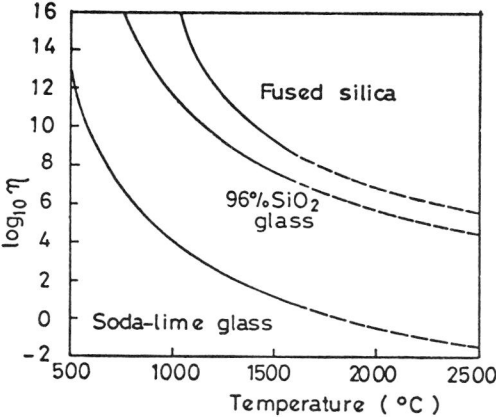

Fig.1 Radial distribution of
energy density in the focal plane.

Fig.2 Relation between viscosity of
glasses and temperature.[5]

Fig.2 shows relation between temperature and viscosity for various
glasses [5]. Dotted lines in this figure were expraporated using the
following fomula [6].

$$\log \eta = A + \frac{B}{T - To} \qquad ----- (1)$$

where η is viscosity, T temperature and A, B, To constants. At vaporization
temperature, about 2250°C, soda-lime glass has the viscosity of about 0.1
poises, which is comparable to that of molten metals, whereas 96% silicate
glass has the viscosity of about 100,000 poises, which is higher than that
of working point. In this experiment, soda-lime glass was mainly used, because
it has lower viscosity and is available easily. Table 1 shows chemical compositions and physical properties of glasses used.

Table 1

	soda-lime galss	96% SiO$_2$ glass
chemical composition	SiO$_2$ 72% Na$_2$ 15% CaO 9% MgO 3% Al$_2$O$_3$ 1%	SiO$_2$ > 96% B$_2$O$_3$ < 3%
density (g/cm^3)	2.47	2.18
heat conductivity (cal/cm·sec·°C)	0.0034 (50°C) 0.0035 (100°C)	0.0035 (25°C) 0.0038 (200°C) 0.0047 (800°C)
coefficient of thermal expansion (x10^7/°C)	80-90	8.0
sofening point (°C)	695	1500
working point (°C)	1000	————
boiling point (°C)	2250°C	
surface tension (dyne/cm)	320-330 (1200°C)	

398

It was observed that a cavity existed in the path of the laser beam, resulting in the large depth-to-width ratio of fusion zone. A bright orange plume of vapor of the glass was seen to be ejected from the cavity. This plume was associated with a cloud of very fine white powder and pungent acid oder. The laser beam is considered to decompose soda-lime glass to silica and compounds of sodium and calcium. Apparently, the vapor was ionized by the laser beam, and absorbed the incoming laser beam resulting in a reduced penetration depth. The compressed air was directed parallel to the workpiece and slightly off set from the surface in order to prevent the formation of the ionized gas. Extreme care was then exercised, because the head of the weld tended to be blown off due to dynamic pressure exerted on the molten glass by the compressed gas. Thus little material was lost during welding process.

High speed films were taken for the bead-on-plate weld pass in the glass at 1000 to 4000 frames per second with the camera perpendicular to the weld. No auxiliary lighting was used; the welding phenomena were photographed on color film by the self-luminosity of the molten glass. No preheating was made and cracking could be almost suppressed during welding in spite of the fracture characteristics of soda-lime glass.

Fine metallic powder was arranged on the surface in order to help to trace the direction of the melt. As the tracer, nickel powder was used. There was, however, some room for further improvement in that considerable amount of the metallic tracer near the bead center was vaporized away before it was involved in the melt. Almost all tracer near the bead edge where the beam power was not so high was involved in the melt. The tracer inserted in the workpiece was followed the movement of the melt without evaporation loss.

(3) Experimental results

In this section, the welding phenomena observed on the high speed films of soda-lime glass are described. Filming was carried out with the bead-on-plate welding in soda-lime glass for various travel speeds.

In general the weld depth in soda-lime glass was approximately in inverse proportion to the travel speed higher than about 15cm/min as shown in Fig.3. At speeds below about 10cm/min, however, decrease in the speed provided little increase in the weld depth, even some decrease below 5cm/min, and the cavity behaved apparently unsteadily. A lot of molten fluid was then formed around the cavity which inclined very little to the vertical, and had some waists.

The high speed films at 4 cm/min demonstrated that the cavity changed its shape near-periodically as schematically shown in Fig.4 [7]. As described in Ref.[7], there was a short period during which the cavity was apparently unstable in

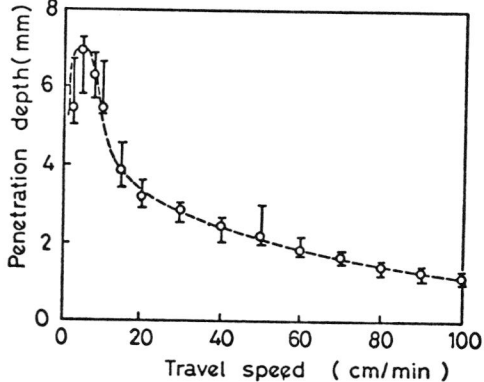

Fig.3   Relation between weld depth and travel speed (330W).

399

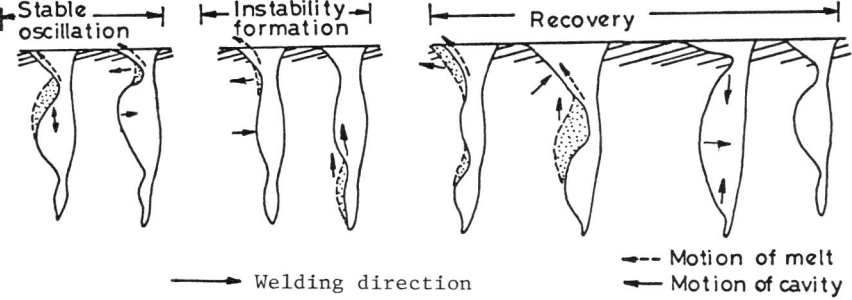

Fig.4   Schematic diagram showing the motion of the cavity.
Each cycle was divided into three periods. (4cm/min,150W)

each cycle, and after which it recovered to stable one. This unstable cavity is considered to be caused by the excess molten material surrounding it,
although it is difficult to provide complete explanation about these processes
at this stage.   There was little ordered motion of the melt around the cavity except for a very sluggish vortex behind the cavity near the surface in
the film.

    As the travel speed increased, the cycle of the cavity motion repeated
more rapidly with the smaller   amplitude, and the temperature of the molten
material decreased gradually.   Then the width of the cavity in a direction
of the weld tended to increase near the surface.   At speeds above 20cm/min,
unsteady nature of the cavity described earlier almost disappeared, and the
melt ablated from the front cavity face was observed to be transported behind more fluently.

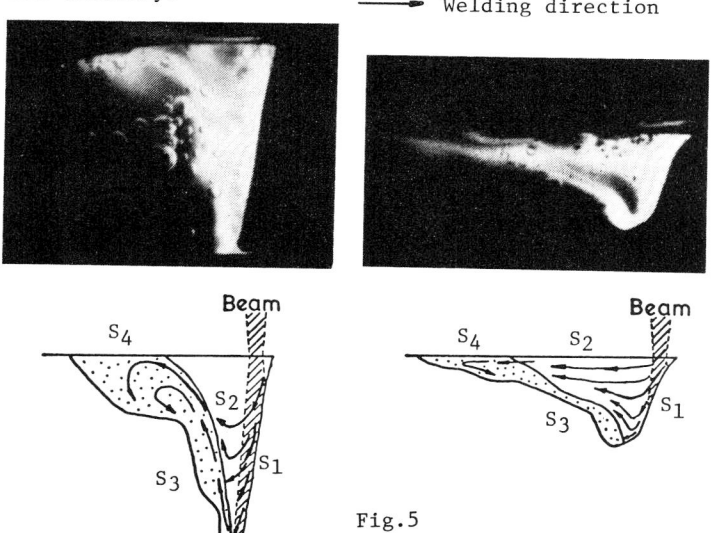

Fig.5

The weld cavity and movement of the melt (330W).
(a)   20cm/min   (b)   80cm/min

400

Fig.5 shows typical cavities and motions of molten glass at the travel speeds 20cm/min and 80cm/min. The direct beam from the lens irradiated the upper part of the front wall. As the travel speed increased, the lower part was out of range of the direct beam because of obvious bent and was considered to be heated by the beam reflected from the upper part. It is believed that this probably provides the cavity with a larger inclination angle in the laser welding than in the EB welding.

It was generally not difficult to distinguish movement of the melt, if not too fast or too slow, containing small bubbles or metallic tracer. However, rapid movement in an uniform thin layer was difficult to recognize. On the high speed films, several ordered streams of molten material could be observed around the cavity. These were considered to be divided into four streams as shown in Fig.5; S_1 is a downward stream on the front face of the cavity, S_2 a U-shaped stream from front face to the rear via side walls, S_3 a upward stream on the rear face of the cavity and S_4 a vortex behind the cavity. Each stream was significantly influenced by the shape of the cavity, which depended upon the travel velocity.

In the earlier stage of the experiment, high speed films suggested that there might be very rapid stream, S_1, at the beam-heated region where a violent evaporation took place. The flow velocity of the stream, if actually existed, had to be considered extremely rapid on the film taken at 1000 frames/sec, which was high enough to observe the other rather slower streams. At 4000 frames/sec, signs of the rapid movement could sometimes be observed, but these were so rapid that it was almost impossible to appreciate the flow velocity.

The existence of this rapid movement down of the melt was confirmed by the fact that on reaching the stream S_1, the nickel tracer disappeared suddenly, resulting in a slight color change in the downstream of S_1. In addition it should be pointed out that the upward stream S_3 observed on the rear surface of the cavity also supported the existence of S_1. Because of the high flow speed and very thin thickness of the stream S_1, Frude number should be large enough for S_1 to be supercritical.

Unlike the stream S_1, in the film taken at 2000 frames/sec S_2 could be clearly observed, because it contained a lot of small bubbles and the metallic tracer. The stream S_2 was observed to branch off from S_1, out of range of the beam, to pump the melt up to the rear edge by way of the side walls in a whirling fashion. Fair amount of nickel tracer arranged at the workpiece surface was observed to move in the stream S_2 at the travel speed below 80 cm/min. A part of S_2 struck the rear wall resulting in a smaller mass of melt just below the U-shaped stream. Films were also taken at 4000 frames/sec of the laser welding at 20 cm/min using a focusing lense with a somewhat shorter focal distance. Then the stream S_2 was obscure. This seems to be partially because the film was in ultra-slow speed (about 170:1 time ratio). At least, however, it is considered that the stream S_2 depended on the cavity shape especially near the surface. As the travel speed increased, these streams gradually inclined to the vertical, and S_2 became almost horizontal at 80 cm/min.

Along the rear face of the cavity, movement up, the stream S_3, of the melt with higher temperature than that for the stream S_2 was observed, which was associated with characteristic wave motions. At the bottom of the rear face, it was seen that from the front face the high temperature melt struck the stream S_3 with thicker depth, and that a discontinuous phenomenon considered to be a hydroulic jump occured.

The weld cavity had a large mass of melt behind it, in which a characteristic vortex motion was seen, resulting in remarked mixing of melt. A mode of the vortex depended apparently on a direction of streams flowing into the molten mass. As the travel speed increased, the inclination angle of the incoming stream to the vertical increased, resulting in a prolonged shallow molten pool.

At 80 cm/min, the melt flowed almost horizontally back into the pool, and this changed the vortex to approximately one dimensional flow associated with a long molten pool in consequence. In this long and shallow pool, however, movement forward of the melt along the bottom of the pool, which was not so fast, was still observed near the rear edge of the pool.

These evidences suggest the possibility that rapid movement back of molten metal on the workpiece surface reported by many workers might be associated with the movement forward in consequence of the vortex motion in the pool which could not be seen due to its opaque nature.

In order to confirm the existence of movement upward of the melt along the rear face of the cavity, S_3, the nickel powder was inserted at a depth of 2mm, and the melt run was made. As shown in Fig.6, the tracer distributed over whole depth of the cavity. It should be noted that the concentration of the tracer was richer near the surface. These facts indicated that material near the bottom moved upwards by way of the bottom and then solidified.

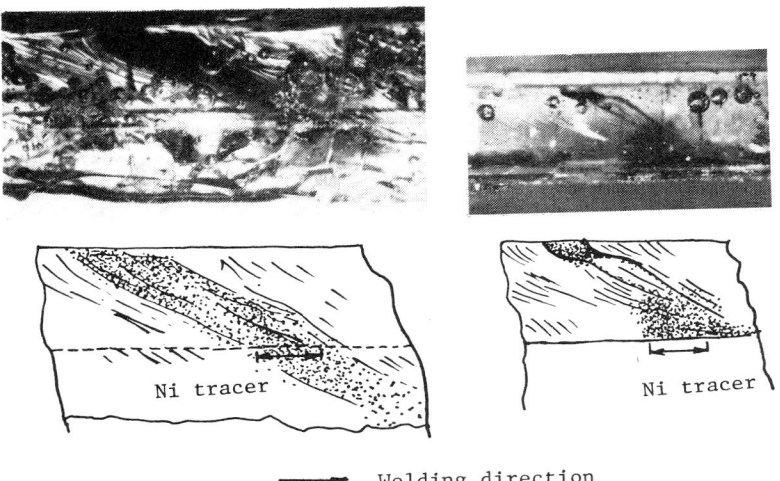

Welding direction

Fig.6 Redistribution of the tracer inserted at the depth of 2mm (330W). (a) 20cm/min (b) 50cm/min.

The melt run was also made in 96% silicate glass, of which viscosity is approximately 10^5 poises at 2250°C. The cavity was observed to be covered by a very thin film of molten material, and motions of the cavity and the melt were very slow due to the high viscosity even at the evaporation temperature. Unlike soda-lime glass, 96% silicate glass tended to produce some kinds of porosities at speed above a critical value. Therefore, this limited the travel velocity of sound weld to a lower value than 1.5 cm/min in this experiment.

At travel speeds above 2 cm/min, porosities were formed on the back side of the cavity, and then they grew in a direction perpendicular to the cavity wall there, resulting in the long porosity usually inclined away from the horizontal line. The authors have named this "cross porosity". The nickel powder was located at the both surfaces of the workpiece with the thickness of 5mm, and the melt run was made at 1 cm/min. Fig.7 shows redistribution patterns, and movement of molten material is illustrated in Fig.8. Both nickel tracers were found to move over almost full thickness in spite of the high viscosity. The movement, however, was very sluggish in compared with soda-lime glass.

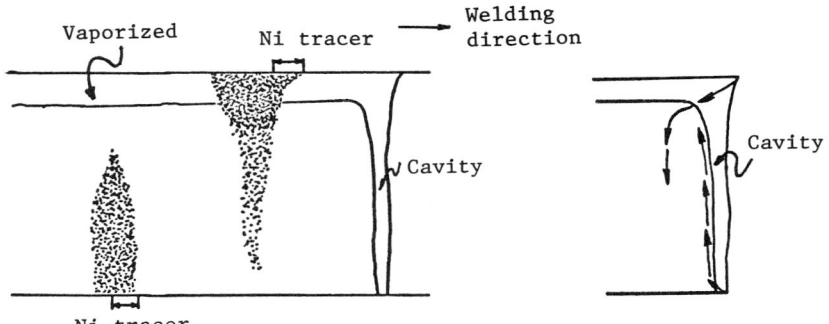

Fig.7 Redistribution of the tracer
from both surfaces. (1cm/min, 96%
SiO₂ glass, 330W).

Fig.8 Movement of melt
in 96% SiO₂ glass.

As shown in Fig.8, it was found that movement back of the melt along the side walls was considerably rapid on the bottom surface. It was then followed by movement upward along the back face of the cavity. Material from the top surface was observed to cross the cavity and then move vertically down apart from the rear wall of the cavity. These facts showed that these motions did not contradict with the vortex motion observed in soda-lime glass. At speed 2 cm/min, a local dark part with lower temperature tended to be formed just below the cavity waist [7] at the rear wall, and it grew into a cross porosity as shown in Fig.9. Then the form of redistribution of the nickel tracer from the lower surface showed that material from the lower surface could not move upwards so far away from the cross porosity as shown in Fig.9. These facts indicated that lack of mixing action due to high viscosity tended to produce defects such as a cross porosity.

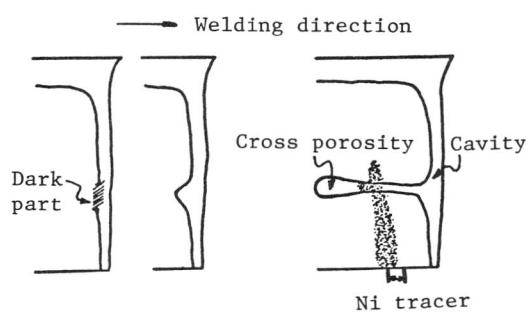

Fig.9 Formation process of cross
porosity in 96% SiO₂ glass
(330W, 2cm/min).

(5) Discussion

At low speeds, unsteady nature of the cavity in soda-lime glass is considered to be caused by an excess of molten material ablated at leading edge wich could not be carried away out of range of the laser beam. As the travel velocity increases, the inclination angle of the leading edge of the cavity increases, and then the movement back of the melt becomes smooth. In consequence of the smooth movement of the melt, the dwell time of the melt in range of the beam is considered to be shorten. These suggest that in order that the ablated material is rapidly swept away out of range of the beam, it may be necessary for the leading edge to incline to some extent to the vertical.

The movement of glass around the cavity are considered such that the stream S_1 at the beam-heated area causes the U-shaped stream S_2 and the upward stream S_3, and then the conflux of S_2 and S_3 results in the vortex S_4 in the pool behind the cavity. As the force to drive molten glass resulting in stream S_1 at front wall of the cavity, the following causes are considered:

(1) Surface tension forces.
(2) Expelling force of the advancing laser beam which behaves like a solid body in the channel filled with liquid.
(3) Some driving force caused by the laser beam-heating.

In view of the flow direction and faster flow velocity of S_1, it is reasonable to consider that the driving force is caused by (3), and is a recoil force of vaporization.

When material vaporizes at a rate of m g/sec, the recoil force F, which acts perpendicularly to the vaporizing surface is given by

$$F = mv, \qquad ----- (2)$$

where v is the speed at which molecule leaves the liquid surface, and is given by the following equation,

$$v = \frac{mRT}{PM}, \qquad ----- (3)$$

where M is molecular weight, P pressure in the vicinity of the vaporizing area, R gas constant and T temperature.

Assuming that the beam-heated area is completely smooth, no recoil force occures except for the vertical force F_v to the surface which is responsible for maintaining the cavity against hydrostatic head and surface tension in the fluid. On the other hand, if the laser beam heats the surface which is not smooth, it produces the force F_H tangential to the solid wall at point A as shown in Fig.10(b). This forces the melt to move down along the wall.

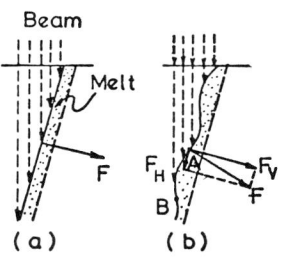

(a)   (b)

Fig.10 Recoil force of evaporation.

Assuming that the laser beam heats the uneven surface uniformly, the force acting tangentially to the wall at point B will compensate that for point A. In the deep penetration welding, however, the power density is apparently lower at B than at A, so that the melt is forced to flow down along the front wall. The driving mechnism of the melt is considered to be similar for the laser and EB beams.

Evidence in support of this theory was provided by a continuous record of the events occuring in the weld cavity [2,3] taken with an X-ray pinhole movie camera set perpendicularly to the weld. As shown in Fig.11, this technique reveals the location where electrons are colliding with metal atoms in the cavity, although this can not determine the cavity shape. It should be noted that several interaction points move downwards as the electron beam advances from left to right. This suggests that the front face of the cavity was uneven, and that the molten material was intermittently driven towards the cavity base.

The uneven surface of the molten layer is considered to be caused by an extremely violent evaporation due to the intensive beam. Besides the speed of thin sheet flow must be fast enough to be supercritical flow which tends to be associated with an uneven surface as well known. It is believed

Fig.11 Continuous record of the events occuring in the EB-weld cavity taken with an X-ray pinhole movie camera (7075 Al, 50cm/min) [3].

that even though the initial unevenness is minute, the recoil force tends to enlarge the unevenness.

At the bottom of the cavity where the stream S_1, which is supercritical, is followed by the stream S_3, which is subsritical, hydraulic jump is considered to occure there, resulting in characteristic wave motion mentioned earlier. A large measure of kinetic energy of the stream S_1 is considered to be expended.

In order to confirm the movement of molten material caused by the laser beam around the weld cavity as described earlier, a two dimensional "weld cavity" with a "molten pool"was formed with a metallic foil in water, and was impinged by the laser beam as shown in Fig.12. An anticlockwise vortex was observed to occure in the "molten pool" in a similar way as stream S_4, during the laser beam impingiment. It is believed that the driving mechanism of molten material is apparently very similar for electron and laser beams.

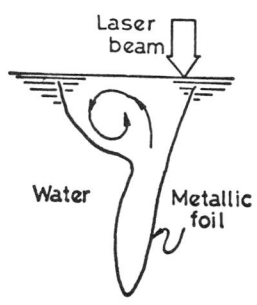

Energy contained in the molten glass heated and forced to move downwards by the laser beam is collected in the pool behind the cavity, resulting in a wider bead near the surface. As the travel velocity increased, the flow velocity of the molten material behind the cavity was observed to increase, providing longer molten pool. Hence the resulting temperature distribution and the cooling rate would deviate from the values calcurated on the basis of the heat conduction theory. Such a high flow velocity of the molten material would probably influence the weld defects such as undercutting and humping encountered in high speed EB welding.

Fig.12 A vortex motion formed by the laser beam.

(6) Conclusions

A new experimental technique using transparent material has been developed in order to directly observe the dynamic behavior occuring during laser welding. Results obtained are summerized as follows:

(1) In the deep penetration welding, the high intensity beam provides the recoil force of evaporation, which is responsible for maintaining the cavity in the liquid, and forcing the melt ablated from the leading edge to run down.

(2) At lower travel speeds, the ablated material is not transported back fluently and tends to be over-heated, because the front face of the cavity inclines very little to the vertical. In consequence, the cavity surrounded by a lot of molten material behaves unsteadily resulting in reduced weld depth.

(3) As the travel speed increases, the ablated material has a tendency to be transported fluently back due to increasing inclination angle of the front face and to be collected behind the cavity, so that the cavity covered with the thinner melt on side and front walls becomes stable.

(4) The ablated material forced to move downwards at the beam-heating region, which is probably supercritical, is transported to the pool behind the cavity near the surface by ways of the side walls and the rear wall, resulting in a characteristic vortex.

(5) The melt, which is heated excesively by the laser beam, releases excess heat in the vortex. This tends to produce a prolonged pool near the surface during welding, resulting in a wine-cup shaped weld bead.

The welding mechanism is considered to be very similar for both laser and EB-weldings, except that the reflection of the laser beam tends to provide more inclined front face of the cavity than for EB. It is considered that this leads to steadier cavity in laser welding than in EB-welding.

References

[1] H.Tong and W.H.Giedt, "Dynamic Interpretation of Electron Beam Welding", Welding Journal, Vol.47, p259s (1970).

[2] C.M.Weber, "Direct Observation of the Penetration Mechanism of Partial Penetration Electron Beam Welding", 5th International Conference on Electron and Ion Beam Science and Technology, p307s (1972).

[3] E.R.Funk and McMaster et al, "Penetration mechanisms of Electron Beam Welding and Spiking Phenomenon", Welding Journal, Vol.53, p246s. (1974).

[4] Y.Arata, E.Abe and M.Fujisawa, "Study on Dynamic Behaviours of Electron Beam Welding by a Fluoroscopic Observation", 7th International Conference on Electron and Ion Beam Science and Technology (1976).

[5] Corning Product Directory (1972).

[6] G.S.Fulcher, Journal of American Ceramic Society, Vol.8, p339s (1925).

[7] Y.Arata, H.Maruo, I.Miyamoto and S.Takeuchi, "Dynamic Behavior of Laser Welding and Cutting", 7th International Conference on Electron and Ion Beam Science and Technology (1976).

Beam Hole Behaviour during Laser Beam Welding

Abstract

Fundamental phenomena during laser beam welding in steel and glass, including beam hole shape, molten metal flow and peculiar plasma behaviour, were observed dynamically using a transmission X-ray system and high speed camera. The effect of altering the flow rate of helium assist gas were also studied. It was found that the gas flow rate had a strong effect on the beam hole shape, molten metal flow and plasma production. In order to avoid the interference effect of laser plasma, a new laser welding process, which we have named "Laser Spike Seam Welding", was developed. This process allows considerably deeper penetration than conventional continuous welding. The reasons for this superiority were analyzed by the above high-speed film method.

I. Introduction

Dynamic observation during laser beam welding is a useful method not only for analyzing the mechanism of laser welding, but also for developing new welding processes. Except for the laser welding of glass by the authors (1,2), however, very little research employing this method has been done.

The authors succeeded in observing the behaviour of the beam hole in steel during laser beam welding(3). In this report, the shape of the beam hole, molten metal flow and the peculiar motion of plasma are studied by direct observation with a transmission X-ray system and high-speed camera. By analyzing the films obtained with this high-speed imaging system, a clear basis is established for recommending the employment of a new laser welding process: "Laser Spike Seam Welding".

II. Experimental Condition and Apparatus

The main phenomena of laser beam welding can be classified into following four subcategories.
1) focusing of the laser beam
2) interaction of the laser beam and material
3) phenomena closely related to item 2)
4) molten metal flow and solidification

The characteristics of beam focusing are the most important parameter for laser heat processing. Conventional focusing systems are shown in Fig.1. System (a) in this figure, however, cannot be utilized for high power beams for extended periods of time. In system (b), on the other hand, sharp focusing is nearly impossible because of the difficulty of making angle θ small enough so that it is within a few degrees, as we have already reported(4).

In order to focus a high power beam, Professor Arata developed a system utilizing both a concave mirror and a plane mirror in 1967. This system is shown in Fig.2. The more the beam power increases, the more effectively this system works compared with other systems. This system was named the "Arata Laser Focus System" by AVCO (5) in 1980, and was recognized under that name at subsequent international conferences (6) and so on. It was adopted for the 15kW system of AVCO and the 100kW system of the U.S. Naval Research Laboratory(7). System (a) is suitable for a columnar beam such as a Gaussian beam, and it was applied for the 1.2kW and 5kW systems of Spectra Physics by the authors. Systems (b) and (c) utilized the same principle, which is suitable for hollow beams. In the case of system (b), the diameter of hole H on plane mirror A cannot be made effectively large enough with respect to the direction of the beam focused by concave mirror B. On the other hand, in the case of system (c), by adding another plane mirror "C", the effective diameter of hole H is enlarged and the flexibility of the focused beam is increased. The authors applied this system to the 15kW beam of AVCO.

In this report, we examine the laser beam welding phenomena subcategories outlined in 2), 3) and part of 4). In other words,
1) What kind of beam hole is formed as a result of the interaction of the laser beam and metal? How does the beam hole shape change when the assist gas flow rate is modified?
2) How does the plasma which is produced by the interaction of laser and metal behave?
3) What is the effect of removing this plasma?

407

4) How does the molten metal flow change if the gas flow rate is changed?

In order to answer the above four questions, observations were performed on the following four items:
1) the shape of the beam hole
2) molten metal flow
3) plasma
4) the phenomena in glass

The phenomena in items 1) and 2) were filmed by the transmission X-ray method (8,9,10). a schematic drawing of the experimental apparatus is shown in Fig.3. X-rays emitted from an X-ray tube (180 kVp, 3.6 mA) pass through the specimen during laser welding. They come into the input screen of an X-ray image converter where they are converted to visible images on the output screen. These visible images are filmed by a 16mm high speed movie camera at 300 frames per second. The phenomena in item 3) were filmed directly by a high speed camera at 300 frames per second and 6000 frames per second. The phenomena in item 4) were filmed at 6000 frames per second.

All of the experiments were performed using a 15kW CO_2 laser at our institute with an a_b value of 1.0. Helium gas was used as the assist gas, and was blown from the back of the laser beam at an angle of 60 degrees. The materials used were mild steel (SM41) and soda-lime glass. Their dimensions are 100mm long, 10mm wide and 50mm thick. For easier X-ray penetration, narrow specimens about 10mm wide were used in the experiments. Other experimental conditions are summarized in Table 1.

III. Results

3.1. Beam Hole Shape

Figure 4 shows the typical shape of the beam hole at various gas flow rates. Schematic drawings of each photograph are also shown in Fig.4. When there is no assist gas, the beam hole is very unstable and shallow. Because the behaviour of the beam hole is very violent, the beam hole cannot be seen clearly in this photograph even at a filming speed of 300 frames per second.

At an assist gas flow rate of 36 l/min, a narrow and deep wedge-type beam hole is formed. The beam hole still undergoes rapid fluctuation. When the gas flow rate is increased to 51 l/min, the upper part of the beam hole is enlarged by the assist gas, while the lower part of the beam hole is still wedge-type. The behaviour of the beam hole is still unstable and the shape of the beam hole undergoes large changes. This photograph is one typical example. At 81 l/min, the beam hole opening becomes much larger, and the beam hole is no longer wedge-type. The behaviour of the beam hole becomes more stable. Consequently, as the gas flow rate increases, there is a large increase in the length of the beam hole opening, but no increase in depth is seen as is the case with a gas flow rate of 120 l/min. At 120 l/min, the assist gas pushes the molten metal backwards to such a degree that humping phenomena is observed behind the beam hole. Thus it was recognized that the assist gas, which removes plasma, plays a very important role in high power laser welding.

When there is no assist gas, the behaviour of the beam hole is quite violent. As the gas flow rate increases, the behaviour of the beam hole becomes more stable. There is, however, an optimum gas flow rate for maximizing the depth of the beam hole. Beyond this optimum value, the beam hole opening lengthens but the depth of the beam hole does not increase. When the maximum depth has been reached, a wedge shape appears at the bottom of the beam hole.

For comparison purposes, examples of beam hole shapes in several kinds of steel during electron beam welding are shown in Fig.5. In the case of electron beam welding, the beam hole is narrower than in laser welding and the front wall is more steep.

3.2. Laser Plasma

When no assist gas is used, peculiar intense plasma is ejected perpendicularly from the beam hole at periodic intervals, as shown in the series of photographs in Fig.6(a). These photographs were taken by a high speed camera at 6000 frames per second horizontally from the side of specimen, as shown in Fig.6. Two kinds of plasma are seen. One is sky blue (A) and another is pink (B). The former stays near the specimen while the latter flies a further distance away from the surface. The pink plasma (B) is easily suppressed, even by a gas flow rate as low as 15 l/min as shown in Fig.6(b). There is still, however, some plasma remaining in the beam hole and around the opening of the beam hole as shown in Fig.6(c), even at flow rates of 36 l/min and 51 l/min. The photographs in Fig.6(c) were taken from an angle of 30 degrees above the specimen, as shown in the schematic drawing in Fig.6. This is the sky-blue plasma described before, which is never blown completely away by the assist gas.

3.3. Glass

In order to support the understanding of the phenomena in steel, the beam hole and plasma phenomena were observed during laser beam welding in soda-lime glass, because compared to steel the phenomena in glass are clearer and easier to observe at higher speeds. Figure 7(a) shows a side view of the glass during laser beam welding without assist gas. There is strong and periodic plasma production as in steel, and fast periodic horizontal flow is seen in conjunction with plasma production. However, an assist gas flow rate of only 7.5 l/min reduced the plasma sufficiently for the flow to become uniform. (Fig.7(b))

In order to observe the phenomena more precisely, close-ups were taken of the upper, middle and lower parts of the beam hole along the front wall without assist gas - these are shown in Fig.8. For the upper part of the beam hole, a series of pictures (Fig. 8(d)) was taken at high speed. They show the periodical movement of the irradiated zone of the laser, from the front wall to an area slightly away from the front wall. After strong impingement of the laser on the front wall, it happens that only a small amount of laser power can reach the front wall. These phenomena can be explained as follows (Fig.9):

1) When there is no plasma, the laser beam impinges directly on the front wall of the beam hole. (The front wall glows brightly.)
2) After strong impingement of the laser beam, intense laser plasma in produced and the laser beam is partially absorbed, reflected and/or scattered. (The front wall glows less brightly and the bright part moves a little inside the beam hole.)
3) This plasma is ejected upwards and the brightness of the beam hole is reduced.
4) When the plasma dissipates, the beam again intensely impinges on the front wall and this zone glows brightly.

Most of the drilling and melting occur only in the first stage, and strong flow is also associated with this stage. During this period, the laser beam rapidly melts the glass, producing a great deal of vapor and molten glass. In other stages, the plasma absorbs, reflects and/or scatters the laser beam although there is some difference in the degree to which these phenomena occur, thereby reducing the degree of vaporization and melting.

3.4. Molten Metal Flow

In order to observe molten metal flow in steel during welding, a silver tracer was inserted in the steel. When the flow rate of the assist gas is low, the flow speed of the molten metal is so high that it is not apparent on film. At 36 l/min, the molten metal flows horizontally and downwards as shown in Fig.10. This speed is much higher than the specimen speed.

IV. Laser Spike Seam Welding ("LSSW")

From the above analysis, it was found that the plasma produced during laser welding interrupts the laser beam and reduces its drilling ability. In order to remove or suppress the plasma, an assist gas is usually used. This assist gas, however, widens the beam hole, leading to failure of the "Wall Focusing Effect"(11). Furthermore, it has been found that two kinds of plasma are produced, and that one of them can be suppressed by assist gas, but the other cannot be completely removed.

In order to overcome these difficulties, the authors utilized a new welding process: "Laser Spike Seam Welding"("LSSW"), invented by Professor Arata(12). In this process, the laser beam is oscillated so as to follow the movement of the specimen. The laser beam stops relative to the specimen for a certain period. It drills the specimen as a pulsed beam, then it quickly returns to its original position. In Fig.11, the typical shapes of the beam hole in the continuous mode, pulsed mode and LSSW mode, respectively, are shown. Figure 11(a) shows the beam hole in the conventional continuous mode. Figure 11 (b) shows the drilling phenomena in the pulsed mode. The pulse duration is the same as for the LSSW mode. It can be seen that the beam hole is deeper than in the conventional continuous mode. Figure 11(c) shows the shape of the beam hole during LSSW. The beam hole is deep and stable. As can be seen from the illustration, the shape of the beam hole in LSSW is a combination of the continuous mode and pulsed mode. The upper part of the beam hole is bowl-typed, as in the continuous mode, and the lower part is wedge-shaped, as in the pulsed mode.

The mechanism and the reasons why the LSSW process is superior are shown in Fig.12. At the top of Fig.12 are high speed photographs of the beam hole taken by the transmission X-ray method. The photographs at the bottom were taken at the same speed (300 frames per second) by conventional photography from 30 degrees above the specimen. In the middle are schematic illustrations of these two kinds of photographs. The sequence of the LSSW process can be explained as follows;

409

a) When the laser beam is in its original position, the laser beam easily drills the specimen, because there is only a small amount of plasma.
b) The laser beam moves backwards along the specimen with the same speed as the worktable. It stops relative to the specimen, and quickly melts down the front wall.
c) As the laser beam drills the specimen, the amount of plasma produced gradually increases.
d) Just before a large amount of plasma is produced, the laser drills the specimen deeply.
a) Then the laser beam is quickly shifted forwards to its original position to avoid the plasma, and it again starts to drill the specimen.

In the top photographs, it can be seen that the gradient of the front wall changes during those stages. When the beam is in its original position (Stage (a)), the gradient of the front wall is gentle and the beam hole shape is bowl-typed. When the beam impinges perpendicularly (Stage (d)), however, the front wall becomes steep and the bottom of the beam hole becomes wedge-shaped. In the photographs on the bottom of Fig.12(c), which show the beam hole from above, a small wedge can be seen at the front of the round beam hole opening. It represents the shifting of the laser beam to its original position. With this periodic process, LSSW can avoid the plasma which reduces the penetration depth, thus allowing the laser beam to penetrate more deeply. Figure 13 shows an example of a bead crosssection welded by LSSW process. Figure 14 shows the comparison of bead appearance and crosssection between LSSW and conventional laser welding under the same welding conditions (power, welding speed, gas flow rate etc). It can be seen that LSSW process is superior in both the penetration depth and the bead appearance.

V. Conclusion

Fundamental phenomena during laser beam welding in steel were observed dynamically. It was found that the laser plasma weakens the drilling force of the laser and this mechanism was shown in detail. Two kinds of laser plasma appear, one pink and the other sky blue, each showing characteristic behaviour. Assist gas was effective in suppressing the pink plasma, but the sky-blue plasma could not be completely removed, even at high gas flow rates. The assist gas used for removing the laser plasma, however, also pushed away the molten metal and widened the beam hole. Therefore, too high a flow rate for the assist gas spoils the "Wall Focusing Effect". This also caused a reduction in the penetration depth.

In order to solve these problems, "Laser Spike Seam Welding" was developed. This process can penetrate considerably deeper than conventional continuous welding by suitably avoiding the laser plasma.

References

(1) Arata Y., Maruo H., Miyamoto I. and Takeuchi S. (1976). Dynamic Behaviour of Laser Welding and Cutting. Proc. 7th Int. Conf. on Electron and Ion Beam Science and Technology: 111-128
(2) Arata Y. (1980). What Happens in High Energy Density Beam Welding and Cutting?
(3) Arata Y., Abe N. and Oda T. (1983). Dynamic Observation of Beam Hole during Laser Beam Welding. IIW Doc. IV-339-83
(4) Arata Y., Miyamoto I. and Kubota M. (1969). Some Fundamental Properties of High Power CW Laser Beam as a Heat Source. IIW Doc. IV-4-69
(5) AVCO Tech-Note, No.6.
(6) Feinberg R. M. (1980). A 15kW High Energy Density Apparatus. Proc. Int. Conf. on Welding Research in the 1980's: 27-32
(7) Laser Welding at 100 kilowatts. Laser Focus. March (1977).
(8) Arata Y., Abe N. and Yamamoto S. (1980). Tandem Electron Beam Welding (Report III). Trans. of JWRI 9: 1-10.
(9) Arata Y., Abe N. and Abe E. (1982). Tandem Electron Beam Welding (Report IV). Trans. of JWRI 11(1): 1-5.
(10) Arata Y., Abe N., Wang H. and Abe E. (1982). Tandem Electron Beam Welding (Report V). Trans. of JWRI 11(2): 1-6.
(11) Arata Y. and Miyamoto I. (1972). Studies of High Power CO_2 Gas Laser as a Heat Source (Report V) -Heat Processing by CO_2 Laser-: J. Japan Welding Society 41: 81.; Arata Y. and Miyamoto I. (1973). Processing Mechanism of High Energy Density Beam (Report I)-Mechanism of Drilling-. Trans. of JWRI 2(2): 19-22; Arata Y. and Miyamoto I. (1975). Wall Focusing Effect of Laser Beam. Proc. 2nd Symp. Japan Welding Society on the Advance Welding Technology: 125.
(12) Arata Y. Patent No.57-27819.

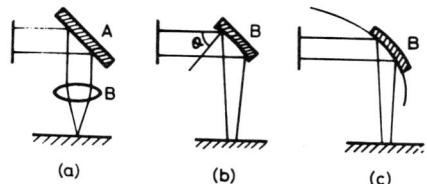

(a)   (b)   (c)

Fig. 1. Conventional focus system

		laser power (Kw)	gas flow rate (l/min)	filming rate (f.p.s)
beam hole shape		7.5	0, 36, 51, 81, 120	300
molten metal flow		9	0, 27, 375, 51, 81	300
plasma	side view	9	0, 15, 30, 36, 39, 51	6000
	bird's-eye view	9	0, 36, 51	300
glass	beam hole	10	0, 75	6000
	plasma	10	0, 75	6000
	close-up	10	0	6000

welding speed : 60 cm/min

Table 1 Experimental conditions

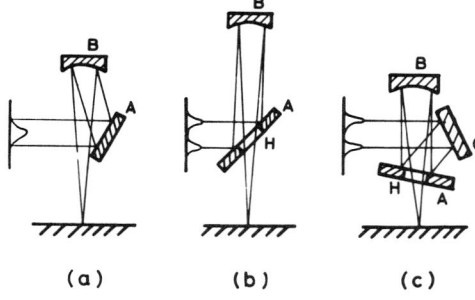

(a)   (b)   (c)

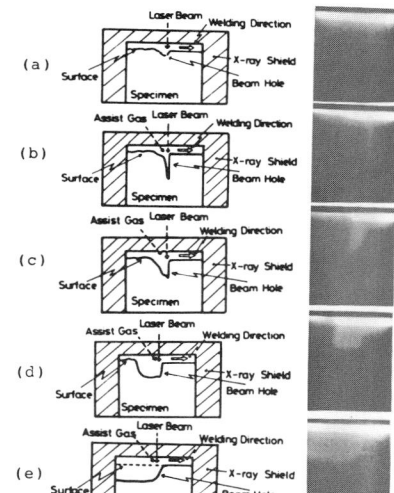

Fig. 4 Typical shapes of the beam hole at various gas flow rates
(a) 0 l/min, (b) 36 l/min, (c) 51 l/min,
(d) 81 l/min, (e) 120 l/min

Fig. 3. Experimental apparatus

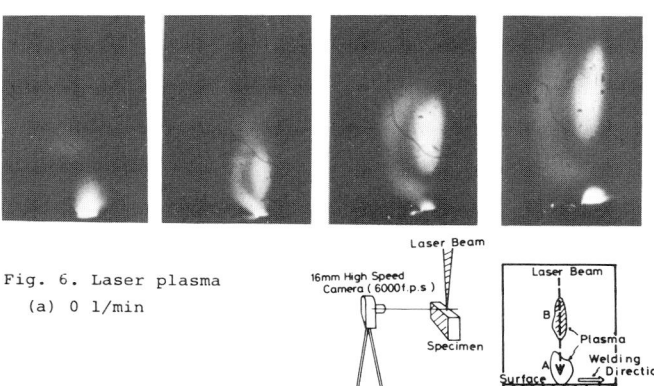

Fig. 6. Laser plasma
(a) 0 l/min

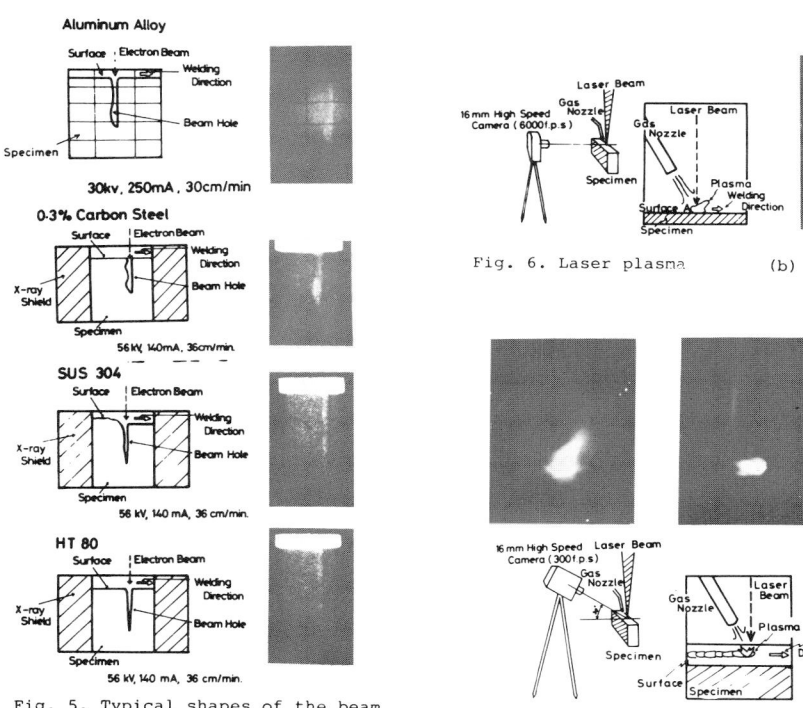

Aluminum Alloy

30kv, 250mA, 30cm/min

0.3% Carbon Steel

56 kV, 140mA, 36cm/min.

SUS 304

56 kV, 140 mA, 36 cm/min.

HT 80

56 kV, 140 mA, 36 cm/min.

Fig. 5. Typical shapes of the beam
hole in various materials during
electron beam welding

Fig. 6. Laser plasma   (b) 15 l/min

Fig. 6. Laser plasma
(c) left: 36 l/min, right: 51 l/min

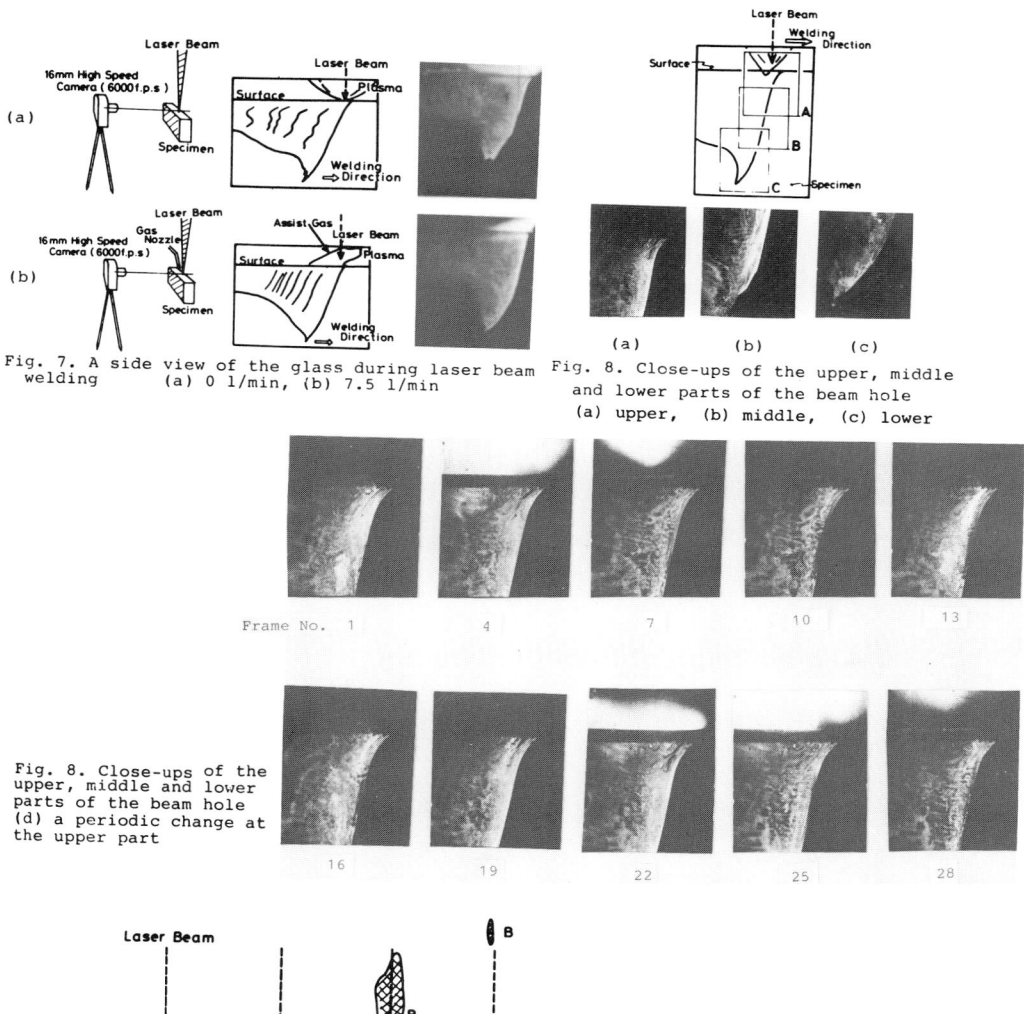

Fig. 7. A side view of the glass during laser beam welding (a) 0 1/min, (b) 7.5 1/min

Fig. 8. Close-ups of the upper, middle and lower parts of the beam hole
(a) upper, (b) middle, (c) lower

Frame No. 1   4   7   10   13

Fig. 8. Close-ups of the upper, middle and lower parts of the beam hole (d) a periodic change at the upper part

16   19   22   25   28

Fig. 9. Laser welding phenomena of glass

(a) ⟶ (b) ⟶ (c) ⟶ (d)

413

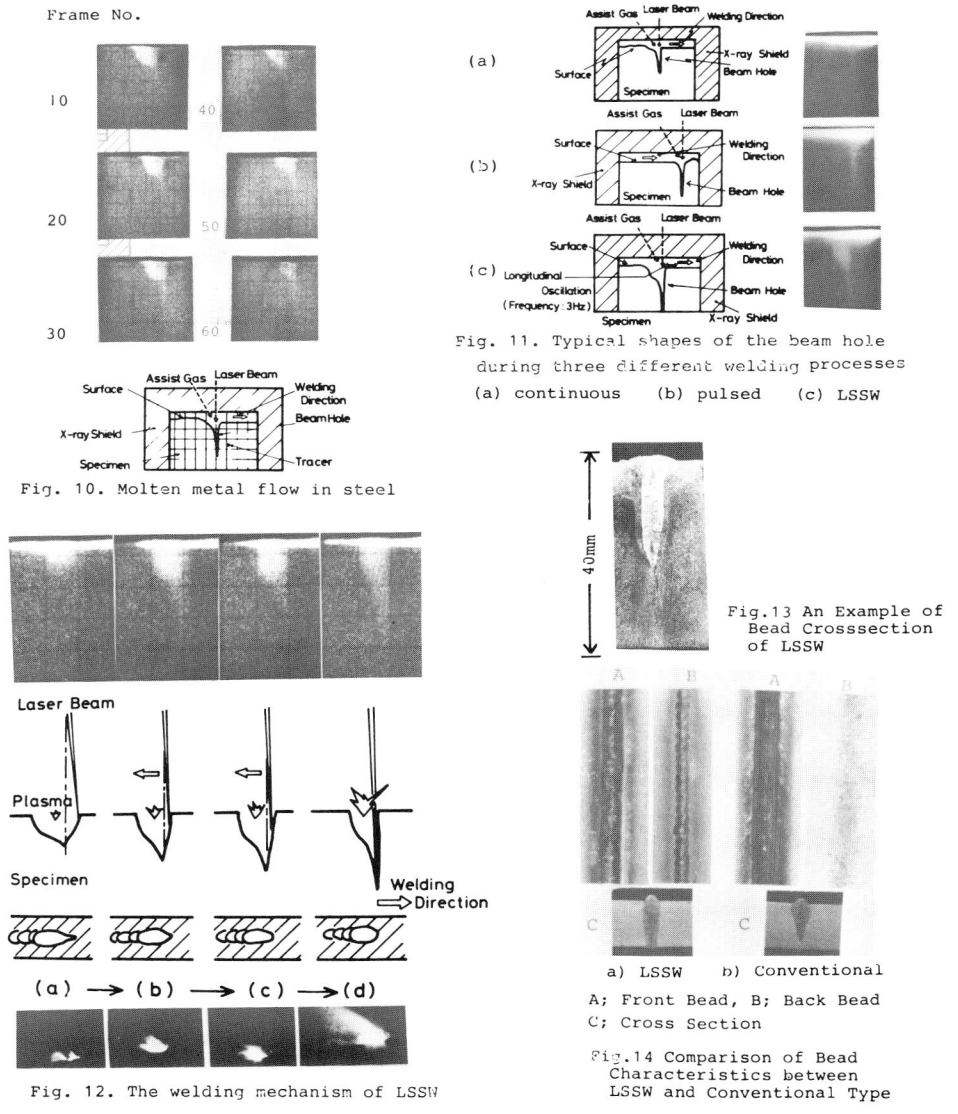

Frame No.

10 40

20 50

30 60

Fig. 10. Molten metal flow in steel

Fig. 11. Typical shapes of the beam hole
during three different welding processes
(a) continuous   (b) pulsed   (c) LSSW

Fig.13 An Example of
Bead Crosssection
of LSSW

Fig. 12. The welding mechanism of LSSW

a) LSSW   b) Conventional

A; Front Bead, B; Back Bead
C; Cross Section

Fig.14 Comparison of Bead
Characteristics between
LSSW and Conventional Type

Prof. Dr. Yoshiaki ARATA: Director of Welding Research Institute of Osaka University
Dr. Nobuyuki ABE: Research Instructor of W. R. I. of Osaka Univ.
Mr. Tatsuharu ODA: Researcher of W. R. I. of Osaka Univ.

Dynamic Behavior in Laser Gas Cutting of Mild Steel

Abstract

Mechanism for the laser gas cutting mild steel with oxygen jet is discussed based on information including the cut appearances, high speed color films of the cutting behavior, power density distribution of the focused laser beam and temperature distribution at the cutting front. At the cutting speed Vb below cirtical one, 2 m/min, which is independent of the laser power, the cutting phenomena are found to vary periodically because the reaction speed at the cutting front is faster than V_b by activation of the intense laser beam. At speeds higher than 2 m/min, the cutting phenomena become steady and the high temperature produced by high power density of the beam povides high speed cutting, which could not be achieved in the conventional oxy-acetylen gas cutting. The effect of such periodic cutting phenomena on the quality of the cut is described and conditions available for qualified cuts are recommended.

KEY WORDS: (Laser Gas Cutting) (Oxidation) (Mild Steel) (Mechanism) (High Speed Filming) (Temperature) (Beam Power Density) (Quality)

1. Introduction

In principle most materials can be cut by focused laser beam through melting or vaporization due to its high power density. In ferrous materials the addition of O_2 gas jet to the laser cutting has been shown to extremely improve the cutting performances because of the high beam absorptivity of the oxide film, the highly concentrated exothermic energy released in the reaction zone, and the low melting temperature and the high fluidity of oxide slag.[1] [2] Such a cutting method, which is referred to as "the laser gas cutting", has been much noticed because it provides so high quality and so high speed cuts that the conventional oxy-acetylen cutting cannot achieve.

A substantial amount of work has been performed on the laser gas cutting, demonstrating its feasibility. [3] ~ [12] However, in spite of potential capability of the laser gas cutting, few industrial applications have been reported. This is thought to be caused by the absence of a detailed, systematic investigation into the mechanism for the laser gas cutting which may reveal how the cuts are affected by the parameters including the beam power density, the cutting speed and the thickness of plate. As for the cutting mechanisms, the interaction between the focused laser beam and material to be cut has to be revealed. This includes the three-dimensional motion of the cutting front, behavior of molten material and the temperature of the reaction front as a function of the cutting speed.

Two-dimensional behavior of the cutting front[13] has been inferred based on high speed films taken with edge cutting which machines a thin slice from the workpiece, thereby periodic phenomena have been found to occur in the laser gas cutting.

In the present study, a new filming technique which enables one to observe the practical cutting has been introduced reproducing the three-dimensional motion of the cutting front precisely. The mechanism of the laser gas cutting has been discussed based on this information, the power density distribution in the focused laser beam spot and the temperature distribution in the cutting front. The effect of those phenomena on the quality of the cuts is also discussed.

2. Experimental Procedures

The data presented were obtained by using two continuous wave CO_2 lasers; one is a proto-type 200 W CO_2 laser and another CO_2 gas transport laser with the maximum output of 1.5 KW, GTE Sylvania, Inc. Model 971.

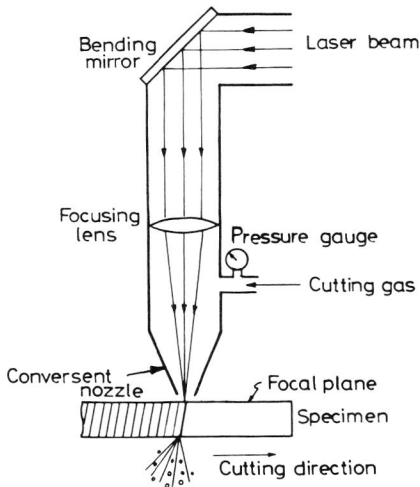

Fig. 1 Schematic diagram of laser gas cutting head.

The laser beam was focused onto the workpiece surface through a ZnSe lens with a focal length of 100 mm as shown in **Fig. 1**. The cutting experiments were carried out by directing the laser beam and oxygen gas coaxially through a convergent nozzle with 1.5 mm diameter, and by moving the workpiece, structural mild steel, SS41.

High speed color films were taken at 4000 frames/sec. The high speed movie camera was set so as to observe directly the leading face of the cut at an angle of about 30° downwards to the workpiece surface as illustrated in **Fig. 2(a)**. By this filming arrangement, suitable to observe

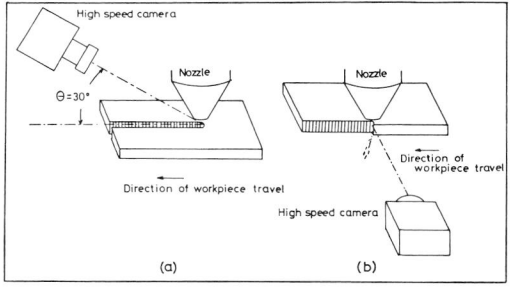

Fig. 2 Arrangement for high speed filming of laser gas cutting
(a) For normal cutting   (b) For edge cutting.

both vertical and lateral motions of the cutting front, the motion of the cutting front in the proceeding direction at the workpiece surface can also be observed. We also supplementally carried out filming with the camera

perpendicular to the cut and with the edge cutting, which machines a thin slice from the workpiece edge, as shown in **Fig. 2(b)**. Motion of the cutting front was analyzed primarily from the films taken by the former technique and supplementally by the later.

The temperature in the cutting front was measured by means of a radiation pyrometer, which was set in stead of the high speed camera shown in **Fig. 2(a)**. Since the width of the laser gas cuts was generally narrower than a minimum sensing spot diameter, 0.6 mm, of the radiation pyrometer, the temperature observed was calibrated as follows. The temperatures of a tungsten ribon lump were measured by the pyrometer with and without a slit having same width as the cut width to obtain temperature reading-ratio.

3. Experimental results

3.1 Beam power density distribution

A power meter was arranged just below a copper plate with a sharp edge which moved in the x direction as shown in **Fig. 3**, and the power was recorded as a function

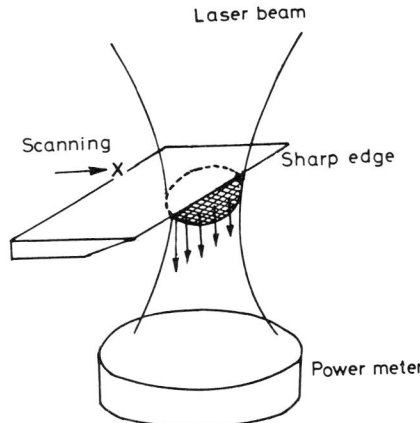

Fig. 3 Arrangement for measurment of power density distribution in focused beam.

of x. The beam power density distribution was determined by differentiating the power with respect to x, and by Abel inversion.

Figure 4 shows the power density profile in the focal plane of the laser beam of the 200 W laser. This profile can be approximated by a Gaussian curve with a radius 0.1 mm at which the intensity falls to $1/e$ of the center, and the maximum power density at the axis was about 5×10^5 W/cm^2 at 200 W power level.

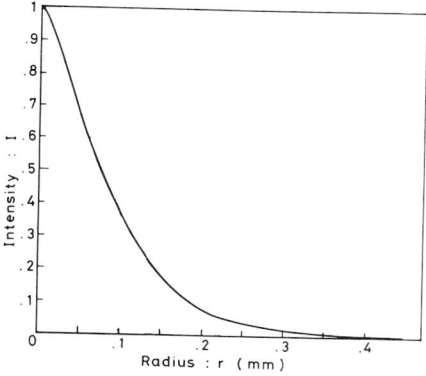

Fig. 4 Radial intensity distribution of laser beam on focal plane.

3.2 Cross-section and edge of the cuts

In the laser O_2 gas cutting of mild steel, the cross-sections are significantly affected by the cutting speed V_b, and are classified into five groups as shown in **Fig. 5**. At

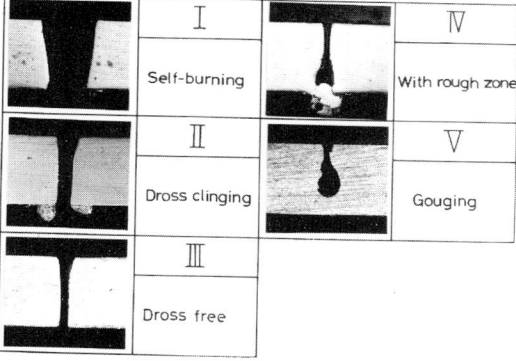

Fig. 5 Classification of laser gas cuts based on shape of cross section.

very low speeds, the cuts contain irregularly spaced holes of which diameter are much larger than the beam spot size and hence are of irregular width. These rough cuts, group I, are referred to as self-burning cuts. At higher speeds, the almost parallel sided cuts, group II and III, with smooth surfaces and narrow kerf width comparable to the beam spot size, are obtianed. Cuts III can be distinguished from the cuts II by no dross attached to the rear surface of the workpiece. More increase in the cutting speed produces group IV cuts of which kerf is wider and irregular at the lower part, and the upper retained as smooth as group III. The appearance at the lower part

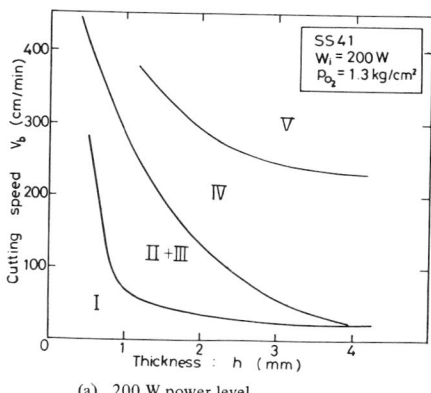

(a) 200 W power level

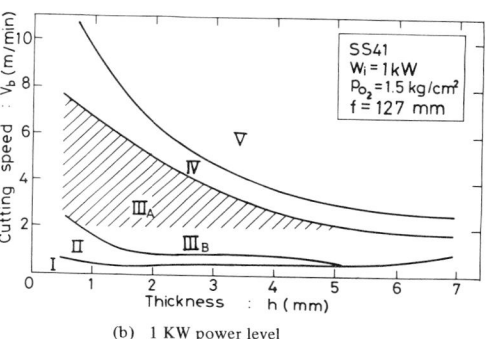

(b) 1 KW power level

Fig. 6 Effect of cutting speed and plate thickness on cut quality.

for the group IV is a sympton of a lack of laser power there.[11] At still higher speeds, cutting-off becomes impossible as in group V. Only the cuts, groups II and III, are available for practical cutting application. **Figure 6** shows the these quality regions in thickness-speed diagram obtained at 200 W and also 1 KW power levels. At higher power level, II and III regions spread to higher speed and larger thickness sides. At 1 KW power level, for example, 0.8 mm thick plate can be cut at 7 m/min, which is much faster than the conventional oxyacetylen cutting, about 2 m/min at most. **Figure 7** demonstrates

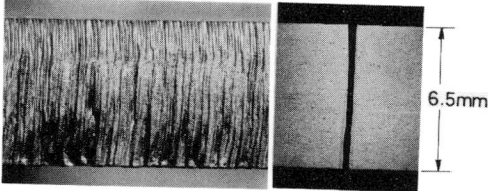

Fig. 7 Example of the laser gas cut. (thickness : 6.5 mm, laser power : 1 KW, cutting speed : 100 cm/min, focal length : 127 mm)

an example of high quality cut of 6.5 mm thickness with very narrow kerf, about 0.25 mm, and very thin heat affected zone.

In the regions II and III, the roughness of the cut surface is one of the important factors for evaluating the cut quality, and so the cut surface was examined in detial. **Figure 8** shows the cut surfaces of 1.2 mm thickness for

← Cutting direction

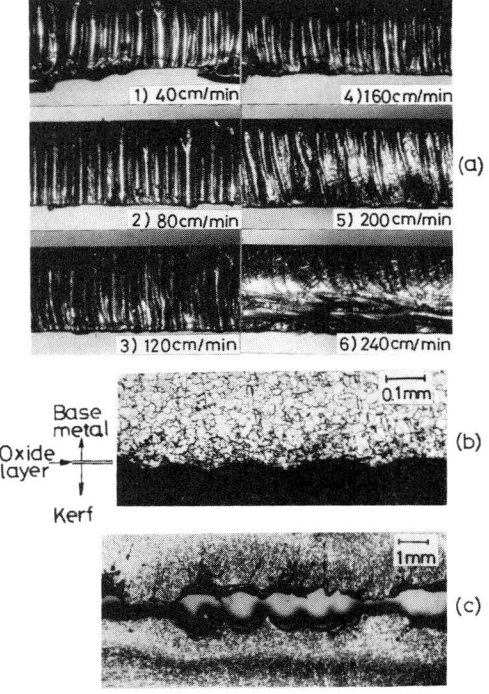

Fig. 8 Cut appearances at 200 W power level. (thickness : 1.2 mm, oxygen pressure ; 1.3 kg/cm^2)   (a)   Cut surface
  (b)   Contour of cut surface (cutting speed ; 80 cm/min)
  (c)   Example of self-burning cut (cutting speed : 10 cm/min)

various cutting speeds at 200 W power level. The cut surfaces at a speed range below 200 cm/min show a series of regularly spaced striations. In detail, the cut surface has a contour with circular arcs and raised ridges as shown in **Fig. 8(b)** and is covered with very thin oxide layer, less than 10μm in thickness, without any symptons of the melted metal layer.

On the other hand, the cut surfaces of 3 mm thickness obtained at 1 KW power level exhibit the regularly spaced striations only in the upper portion accompanied by rather irregular striations in the lower portion as shown in **Fig. 9**.

→ Cutting direction

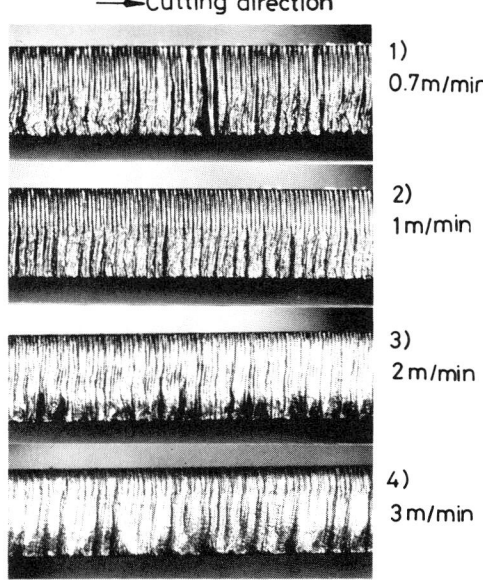

Fig. 9 Appearances of cut surface at 1 KW. (thickness : 3 mm, oxygen pressure : 1.5 kg/cm^2, focal length ; 127 mm)

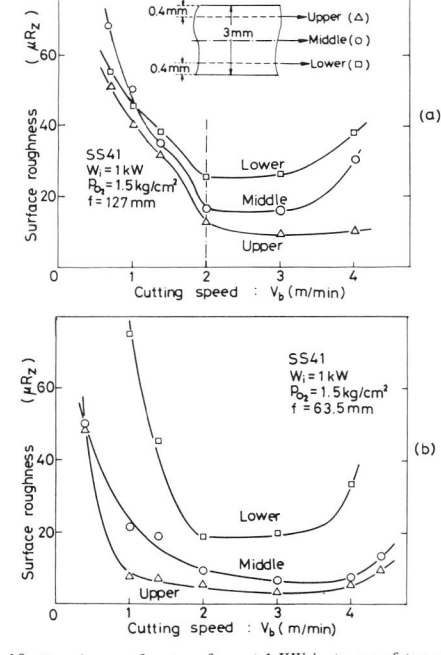

Fig. 10   Roughness of cut surface at 1 KW in terms of ten point height as a function of cutting speed.
   (a)   Focal length : 127 mm
   (b)   Focal length : 63.5 mm

418

Figure 10 shows the roughness of the cut surface in terms of ten points height measured at various locations of 3 mm thick plate. The roughness of the upper portion tended to decrease with increasing the cutting speed until it became constant value, about $10 \mu R_z$, at speeds of $V_b >$ 2 m/min. The roughness at the middle and lower parts varied with similar tendency to the upper one.

3.3 Dynamic behavior of the cutting front

In the early stage of the work, filming was carried out with the edge cutting shown in **Fig. 2(b)**. In the edge cutting, however, the cut appearances were found to be significantly dependent on the thickness of the slice machined from the workpiece. With the selected thickness of the machined slice, the cut appearances became similar to those of normal cuts except for the lower part of the cut which became less regular as shown in **Fig. 11**. Although the thickness of the slice was too large to observe the behavior of the cutting face in front of the beam axis, the upper part of the edge cut gave some auxiliary information on the mechanism for the laser gas cutting. The problem that occurred in the edge cutting was resolved by introducing the filming technique shown in **Fig. 2(a)** which enables the camera to observe the behavior of the normal cutting front.

High speed filming was carried out with the arrangement shown in Fig. 2(a) in a cutting speed range from 40

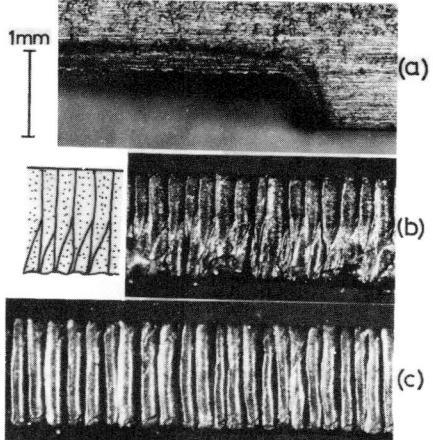

Fig. 11 Comparison between edge cut and normal cut. (thickness : 1.2 mm, cutting speed : 80 cm/min)
(a) Top view of edge cut
(b) Surface of edge cut
(c) Surface of normal cut

cm/min to 300 cm/min. The films showed that the cutting phenomena in the speeds range below 2 m/min changed periodically with a same frequency as that of the striations left on the cut surfaces shown in **Fig. 8**, and that the cutting phenomena became almost steady at the speeds above 2 m/min.

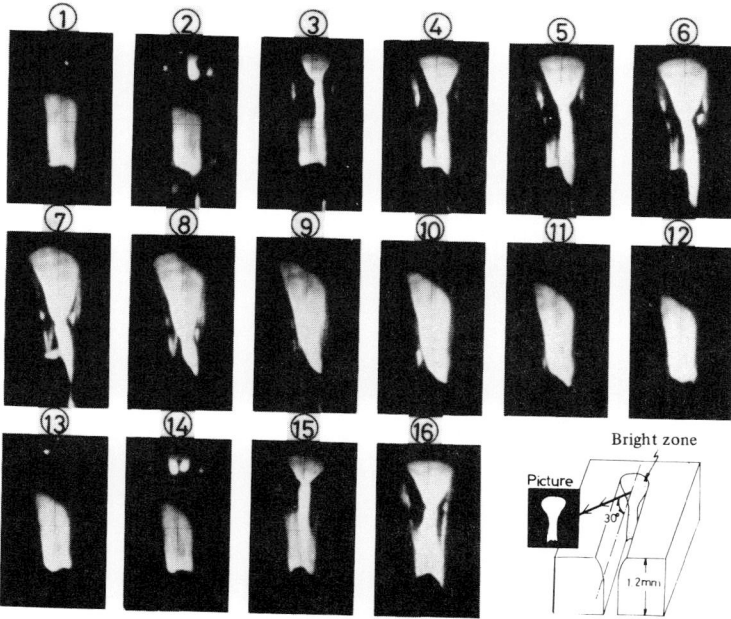

Fig. 12 High speed movie pictures of normal cutting at 80 cm/min (thickness : 1.2 mm, laser power : 200 W). Time interval of each picture is 1/1000 sec.

419

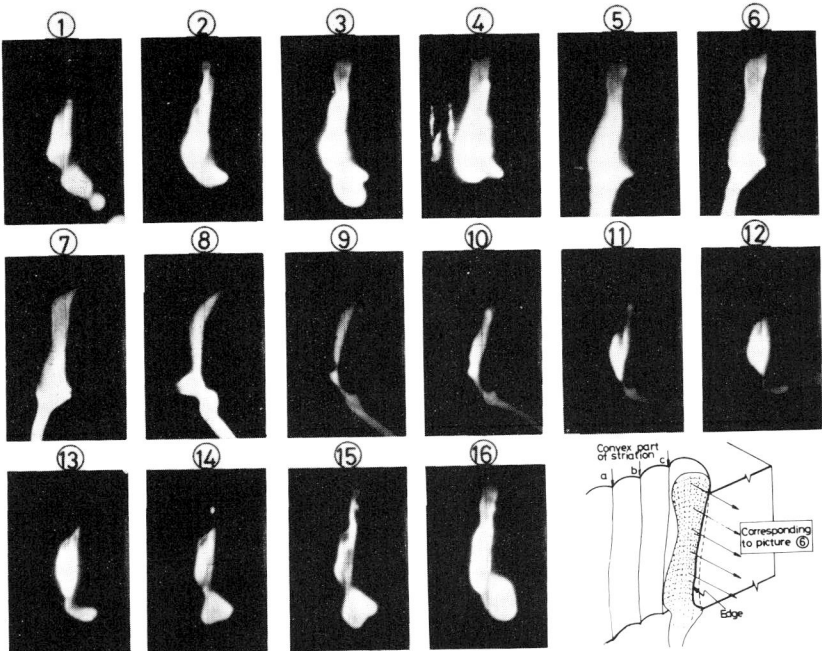

Fig. 13 High speed movie pictures of edge cutting at 80 cm/min (thickness : 1.2 mm, laser power : 200 W). Time interval of each picture is 1/1000 sec.

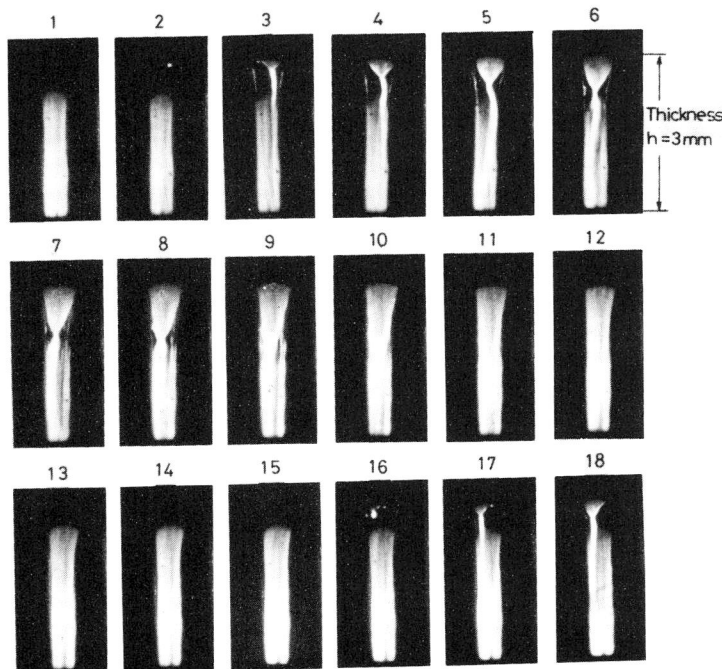

Fig. 14 High speed movie pictures of normal cutting at 100 cm/min (thickness : 3 mm, laser power : 1 KW). Time interval of each picture is 1/1000 sec.

Figure 12 shows a typical example of periodic cutting phenomena, at the laser power level of 200 W. It is seen in this figure that the cutting phenomena at the upper part are periodic whereas at the lower part almost steady. The phenomena observed in one cycle were as follows: an isolated bright spot appeared at the top of the cuutting front and was seen to spread laterally and downwards to grow into an inverted triangle. In the process of growth of the triangle the molten material poured down through the bottom corner. It was also found that the top of the cutting front spread upwards slightly in this stage. After growing up, the triangular zone moved downwards and eventually the upper part lost its brightness completely. Although the shape of the bright area at the lower part hardly changed, the brightness itself changed correspondingly to the periodic change at the upper part.

Figure 13 shows the corresponding behavior of the cutting front observed in the edge cutting. The growth and downward motion of the triangular reaction zone observed in **Fig. 12** corresponded to the change in the width of the melt flow in the direction of the cutting near the top surface of the plate observed in **Fig. 13**. The molten material is, however, seen to flow down unsmoothly causing less regular striations at the lower part.

At the power level of 1 KW, the cyclic phenomena at the upper part could be more clearly observed as shown in **Fig. 14**.

Such cyclic cutting phenomena were observed only at speeds $V_b < 2$ m/min. An increase in the speed V_b decreased the intensity of the cyclic change until almost steady cutting was established at $V_b = 2$ m/min; at 2 m/min a very little change in brightness only at the both sides of the cutting front was observed. At 3 m/min, the cutting was completely steady as shown in **Fig. 15**.

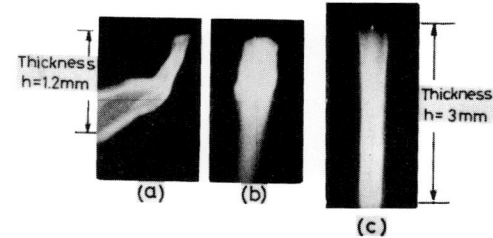

Fig. 15 Example of steady cutting.
  (a) Edge cutting at 300 cm/min (200 W)
  (b) Normal cutting at 300 cm/min (200 W)
  (c) Normal cutting at 250 cm/min (1 KW)

3.4 The reaction speed

Four values, l_r, l_s, l_n and l_w, shown in **Fig. 16**, which were important parameters characterizing three-dimensinal motion of the triangular reaction zone, were measured from the films. The origion of the Cartesian coordinate which moved at a constant velocity V_b was set at the point where a bright spot appeared at the top of the cutting front.

Although these values exhibited a scatter to some extent, they were found to be obviously affected by the cutting speed V_b. In **Fig. 16(b)** and **(c)**, the time variations of these values in typical cycles are plotted for $V_b = 80$ cm/min and $V_b = 140$ cm/min. The maximum value of l_r, which is the amplitude of the cyclic motion of the cutting front in the direction of cutting, was found to decrease with increasing the cutting speed V_b; 0.1 mm at 40 cm/min, 0.09 mm at 80 cm/min and 0.06 mm at 140 cm/min.

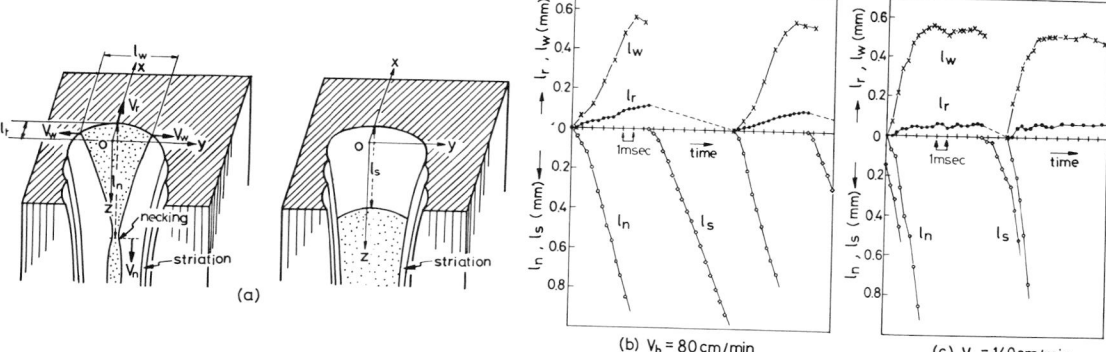

(b) $V_b = 80$ cm/min   (c) $V_b = 140$ cm/min

Fig. 16 Time variations of bright zone seen in upper part of cutting front in shape which are characterized by four dimensions and corresponding four velocities.
  (a) Each dimension
  (b) Data for 80 cm/min
  (c) Data for 140 cm/min

421

From **Fig. 16**, mean velocities V_w, V_r, V_n and V_s were obtained. The velocities V_n, V_s and V_w increased with increasing the cutting velocity V_b, whereas V_r remained almost at contant value, V_{r0}, about 2 m/min, at speed range $V_b < 2$ m/min as shown in **Fig. 17**. In a speed range $V_b \geq 2$ m/min, V_r coincided with V_b in consequence of steady cutting.

As shown in **Fig. 18(b)**, the oxidation reaction started at a point S was found to continue with the aid of the laser heating until a point F was reached, where S and F are the location at the distance D_S and D_F from the beam axis, and $D_F - D_S$ is equal to the maximum value of 1_r in **Fig. 16**. The distance D_F was measured during cutting by means of a telescope as shown in **Fig. 18(a)**, and the value $D_F - D_S$ was determined from **Fig. 16**. The values D_F and D_S thus obtained are plotted against the cutting speed at power level of 200 W in **Fig. 18(c)**.

Two beam power densities, ϵ_S and ϵ_F, corresponding to S and F respecively, were also determined from the beam power density profile in Fig. 4; ϵ_S and ϵ_F are the power density of the laser beam with which oxidation reaction begins and interrupts, respectively. As shown in **Fig. 18(d)**, at speeds $V_b < 2$ m/min, ϵ_S was almost kept constant at about 6×10^4 W/cm^2, whereas ϵ_F increased with increasing V_b until it coincides with ϵ_S at $V_b = 2$ m/min. At speeds $V_b > 2$ m/min, ϵ_S increased with V_b. It is believed that the cutting seed can be increased until ϵ_S reaches the maximum power density at the beam axis, 5×10^5 W/cm^2 at 200 W power level, as far as concerned with thin material.

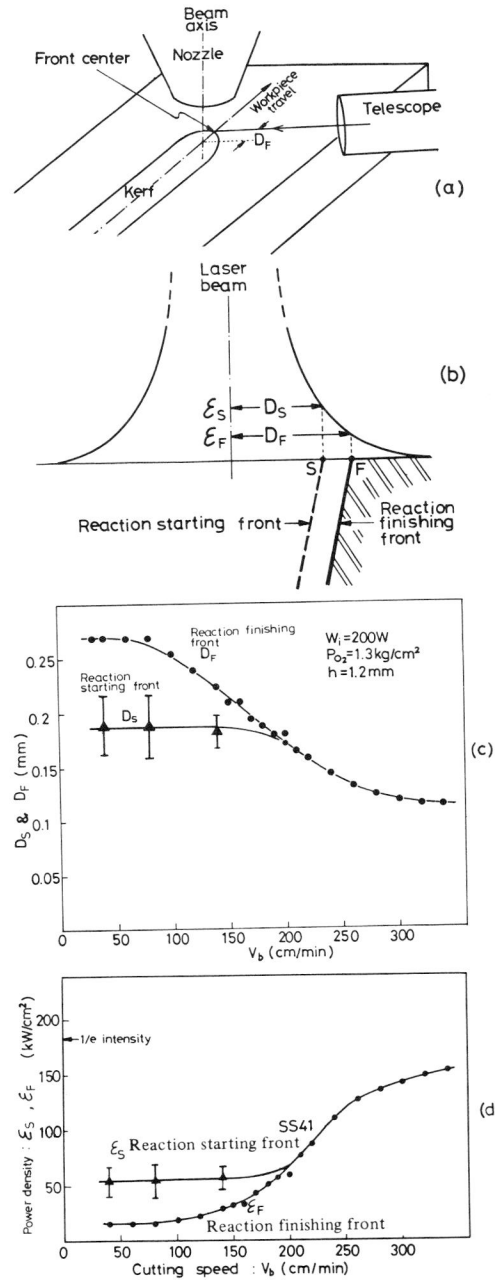

Fig. 18 Locations of reaction starting and finishing points S and F, and corresponding power densities, ϵ_S and ϵ_F.
(a) Arrangements for measuring location of F
(b) Reaction starting and finishing fronts, D_S and D_F distant from beam axis, respectively
(c) Effect of cutting speed on D_S and D_F
(d) Effect of cutting speed on beam power densities ϵ_S and ϵ_F which correspond to D_S and D_F, respectively

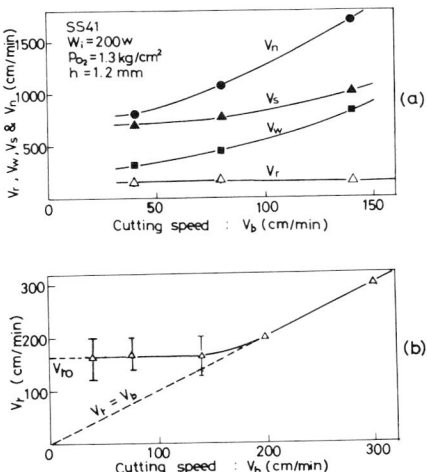

Fig. 17 Relationship between cutting speed and each speed.

3.5 Temperature at the cutting front

Figure 19 indicates the temperature distributions along the z axis. The temperature tended to increase with increasing the distance from the top surface z until the constant temperature distribution is established.

At 1 m/min the region where the temperature increased with z corresponded to the region where periodic cutting phenomena were observed, and the region showing constant temperature distribution corresponded to the part where the rather irregular striations were observed in the lower portion shown in **Fig. 9**. At 4 m/min, the region where the temperature decreased with z was seen to occur. In this region the cut appearances were extremely rough. This is caused by the fact that the laser energy supply is not enough to produce a clear cut [11] because the cutting speed is too high.

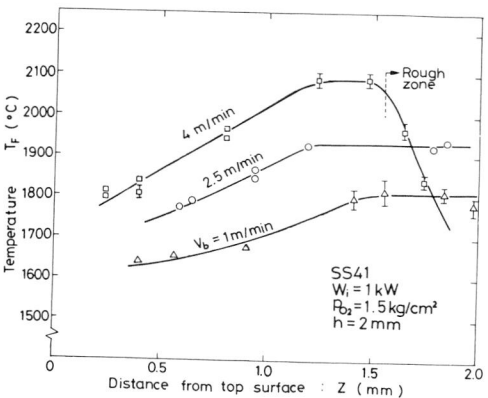

Fig. 19 Relationship between temperature of cutting front and distance from top surface for various cutting speeds.

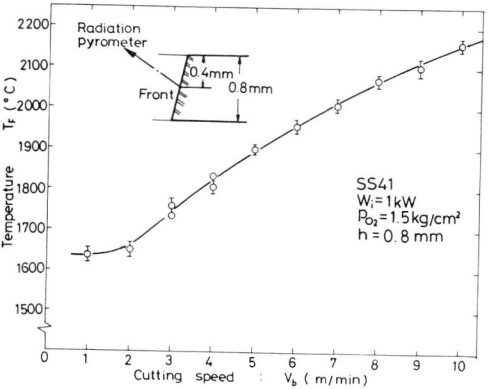

Fig. 20 Relationship between cutting speed and temperature.

Figure 20 exhibits the relationship between the temperature at the cutting front and the cutting speed. At speeds below 2 m/min, the temperature was almost constant, about 1650°C, which corresponds to that of oxy-acetylen cutting.[14]-[16] At speeds above 2 m/min, the temperature increased directly with the speed. The maximum temperature, 2150°C, was obtained in our experiment at V_b=10 m/min, which is the upper limit of the carriage used. It is believed that as the cutting speed increases further, the temperature at the cutting front will reach to the evaporation temperature.

4. Discussion

4.1 Mechanisms for laser gas cutting

The mechanisms for the laser gas cutting mild steel were discussed based on information including the cut apperances, the beam power density distribution, and the periodic motion and temperature obtained at the cutting front under various operating conditions. The prosposed interactions between the laser beam and the cutting front are illustrated in **Figs. 21, 22** and **23**.

Figure 21(a) exhibits the top view of model proposed to explain periodic cutting observed at the cutting speeds below 2 m/min. The circles O_S and O_F, of which centers are P_S and P_F, are equipower density circles with ϵ_S and

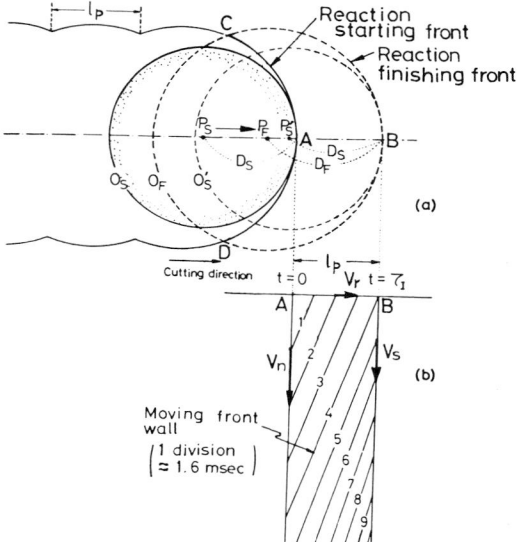

Fig. 21 Model showing periodically changing cutting.
(a) Top view
(b) Longitudinal cross section of cutting front of thin plate

ϵ_F, respectively. On coming in contact with the edge of the face CAD, the circle 0_S with the reaction starting power density ϵ_S, triggers off the oxidation reaction. The reaction front moves along the x-axis at V_r which is faster than V_b until it reaches the circle O_F with the reaction finishing power density, ϵ_F. In this process oxidation reaction spreads laterally so that the region CADB is removed. Then no action follws until the circle O's reaches the cutting front at B again. Thus a circular arc and a ridge at C are left. **Figure 21(b)** shows the corresponding behavior in the longitudinal cross section.

Figure 22 illustrates the time varing longitudinal cross section of the cutting front in the periodic cutting thick plate. When the circle O_S comes in contact with the cutting front, the isolated bright spot appears since the cutting front at the upper part is near vertical, whereas the

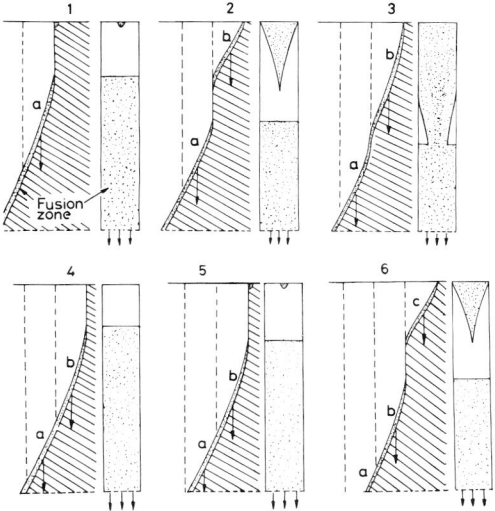

Fig. 22 Schematic illustration showing periodically changing cutting behavior of thick plate in longitudinal cross section.

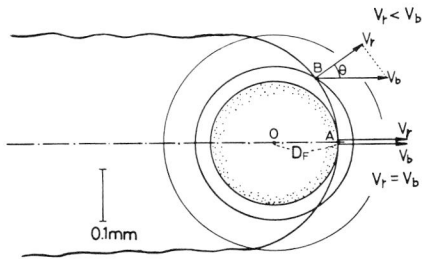

Fig. 23 Schematic illustration showing steady cutting.

lower part is constantly heated by the laser beam. As the spot grows into the triangle the upper part is removed so as to have an inclined face, and then the melt from the upper part pours into the lower part. Although the striations in the both parts are produced in a different way each other, the amount of the melt and hence energy included in the lower part is affected by the upper phenomena as shown in **Fig. 10**. Such an intermittent cutting is caused by the fact that the velocity of the reaction front, V_{r0}, activated by the laser beam with the power density ϵ_S is faster than the cutting speed V_b.

At the cutting speed $V_b=V_{r0}$, the cutting becomes steady leaving no obvious striations on the cut surfaces, since the circle of ϵ_S is always in contact with the cutting front. At the cutting speeds higher than V_{r0}, increase in the density ϵ_S has an effect to raise the temperature of the cutting front so as to enhance the reaction velocity up to V_b keeping the steady cutting. Such a high speed cutting which could not be achieved by the conventional oxy-acetylen cutting is due to the high power density which was realized by the laser beam.

At the very low cutting speeds, the cuts with the self-burning shown in **Fig. 8(c)** occurred which contains irregularly spaced holes with a diameter much larger than the beam spot size as described already. The oxidation reaction then will spread out beyond the circle of the reaction finishing power density ϵ_F, because the thermal conduction becomes predominant in the temperature distributions. This effect will be accelerated by the large amount of heat contained within the dross which easily deposits to the bottom surface of the thin plate in low speed cutting.

The kerf width at any cutting speeds should be also affected by the thermal conduction; in general moving heat sources provide isothermal lines of which maximum width is larger than the width at the heat source center. This means that the kerf width can be larger than $2D_F$. Strictly speaking, the thermal conduction effect should be taken into consideration even when the values D_F as well as D_S ahead of the beam axis are determined. In other words, the region in which the oxidation reaction can be maintained should be determined based not on the power density of the heat source but on the resultant temperature distribution. It should be noticed, however, that the kerf width is considerably close to $2D_F$ at a cutting speed range below 1 m/min. In the laser gas cutting in which materials are processed at high speeds with highly concentrated heat sources, the idea of the processing region limited by the power densities such as ϵ_S and ϵ_F is thought to be reasonable at the first approximation.

According to Ref. [13], it has been reported that the

cutting front activated by the laser beam moves forwards until it is out of range of the laser beam at the cutting speed 25 cm/min, and that at the cutting speeds above a critical value, 60 cm/min, the cutting becomes steady. It was found by the authors, however, that the distance over which the front progressed before the oxidation reaction ceases was limited within the range of the laser beam, and that the critical speed was about 200 cm/min, as aforementioned. Those discrepancies are thought to be caused by the facts that cutting speed, 25 cm/min, in Ref. [13] is extremely slow probably resulting in self-burning phenomenon as for the former, and that the cutting phenomena in the edge cutting are different from the normal cutting we adopted, as for the later.

4.2 Recommendation for smooth cutting

Since the contour of the cut surfaces is consisted of a series of circular arcs, the roughness of the cut surface is related to the pitch of the striations to some extent.

The pitch of the striations is proportional to the value $D_F - D_S$ and inversely proportional to the value $V_{r0} - V_b$ at cutting speeds higher than the self-burning limit. When the cutting speed increases, $(D_F - D_S)$ and $(V_{r0} - V_b)$ decrease, hence the pitch remains almost constant by mutual compensation. In detail, however, the value 1_r in **Fig. 16(c)** is seen to increase faster at the earlier period and then to become almost constant so that the mean value of V_r is equal to V_{r0}. This means that the reaction speed is faster than the cutting speed V_b during only short period just after the circle for ϵ_S comes in contact with the cutting front and then the circle itself moves at V_b leaving rather smooth surface there. Thus the cut surfaces become smooth with increase in the cutting speed as shown in **Fig. 10**, when the other cutting conditions are all the same. On the other hand, it is obvious that the well focused beam with smaller value of $D_F - D_S$ will bring more smooth cuts at given cutting speed. Therefore, the application of focused beam with smaller spot diameter is strongly recommended, which is practically available by the laser beam with a fundamental transversed mode, TEM_{00}, and/or by a lens having a short focal length. In **Fig. 10**, the effect of short focal length lens on the cut surface roughness can be seen; then lens having shorter focal length, f=63.5 mm, is seen to give the smooth cut surface in the upper part, but rather rough cut surface in the lower part due to the short focus depth. Lenses with short focal length are generally adequate for cutting thin plates.

Higher power laser beam is also recommended for smooth cutting. In other words, for given thickness of plate, higher speed cutting with the higher power is better than lower speed cutting with the lower power. In **Fig. 6(b)**, a recommended cutting region for 1 KW laser is shown with III_A.

5. Summary

High speed movie films for the laser gas cutting mild steel indicated that the cutting phenomena varied periodically resulting in regularly spaced striations on the cut edges until the steady cutting was established at the cutting speeds above a critical value V_{r0}. Based on the information obtained from the cut appearances, motion of the cutting front observed in the films, the power density distribution in the focused beam and the temperature distribution in the cutting front, the mechanisms for the laser gas cutting mild steel are discussed. The results obtained are summarized as follows:
(1) At cutting speeds below V_{r0}, about 2 m/min, which is independent of the laser power, the reaction front moves at a speed equal to V_{r0} within the range where the laser beam power is high enough to activate the reaction, resulting in periodically changing cutting.
(2) At speeds higher than V_{r0}, the cutting phenomena become steady and the high power beam elevates the temperature at the cutting front so that the high speed cutting is achieved which could not ever been achieved by the oxy-acetylen cutting.
(3) With well focused intense beam qualified cuts with smooth surface are obtainable at rather higher cutting speeds.

References

1) Y. Arata and I. Miyamoto; "Generation and Applications of CW High Power CO_2 Gas Laser", Technol. Repts. Osaka Univ., Vol. 17 (1967) No.285
2) A.B.J. Sullivan and R.T. Houldcroft: "Gas-jet Laser Cutting", British W. J., Vol. 14 (1967) No.8, 443-446
3) F.W. Lunau and E.W. Paine: "CO_2 Laser Cutting", Weld. & Metal Fab., Jan. (1969), 9-14
4) M.M. Scheartz, "Laser Welding and Cutting", W.R.C. Bulletin, Nov. (1971), No. 167, 1-34
5) P.T. Houldcroft: "The Importance of the Laser for Cutting and Welding", Weld. and Met. Fab., Feb. (1972), 42-46
6) G. Brandt, K.D. Kegel and J.V. Hulle: "Einige Ergebnisse von Schneid-und Schweißversuchen mit einem 900 W CO_2 Laser", Schweissen und Schneiden, Vol. 24 (1972), H7
7) I.J. Spalding, "Lasers - Their Applications and Operational Requirements", Opt. and Laser Tech., Dec. (1974), 263-272
8) J.D. Russell: "The Development of the Laser as a Welding and Cutting", British Weld. Inst. Res. Bulletin, Vol, 16 (1975), 245-248
9) S. Roy: "A Comparative Surface Integrity Study of Laser Cutting with Other Conventional Cutting Technique", Sheet Metal Industries, Oct. (1977), 994-1014

10) J. Clarke and M.M. Steen: Proceedings of Laser '78 Conference, London, March (1978)

11) Y. Arata, S. Takeuchi and I. Miyamoto: "Fundamental Research of Laser Gas Cutting - I", Journal of High Temperature Society (in Japanese), Vol. 4 (1978), No. 2, 122-134

12) V.S. Kovalenko, Y. Arata, H. Maruo and I. Miyamoto: "Experimental Study of Cutting Different Materials with a 1.5 KW CO_2 Laser", Trans. of J.W.R.I., Vo. 7 (1978) No. 2, 101-112

13) M.J. Adams: "Gas Jet Laser Cutting", Proceeding of the Conference on Advance of Welding Process, The British Weld.

Inst., April (1970), 140-146

14) K. Teske: "Contribution to the Explanation of Gas Cutting Process", Schweissen und Schneiden, Vol. 9 (1956), No. 8, 122-129

15) H. Hofe: "New Information on the Mechanism of Gas Cutting", Schweissen und Schneiden, Vol. 19 (1967), No. 5, 213-219

16) A.K. Ninburg: "Some Aspects of the Metal Combution Mechanism during Oxygen Cutting", Svar. Proiz., (1968), No. 2, 40-42

CHAPTER 5

APPLICATION OF ULTRA HIGH ENERGY DENSITY HEAT SOURCE

Electron Beam Welding

Laser Beam Welding

Laser Gas Cutting

Laser Surface Treatment

Application of Plasma Beam

Electron Beam Welding

Narrow Gap High Energy Density Beam Welding—Principle
Y. Arata
[Trans. JWRI **2** (1973), 119.]

100kW Class Electron Beam Welding Technology—Fundamental Research on Horizontal Electron Beam Welding
Y. Arata and M. Tomie
[Trans. JWRI **9** (1980), 157. IIW Doc. IV−308−81.]

Electron Beam Welding of High Strength Aluminum Alloy
Y. Arata, M. Ohsumi and Y. Hayakawa
[Trans. JWRI **5** (1976), 19. IIW Doc. IV−196−76.]

Insert-type Electron Beam Welding Technology—Characteristics of Insert-type Welding
Y. Arata, K. Terai, H. Nagai, T. Aota, I. Futami and S. Shimizu
[Trans. JWS **4** (1973), 63.]

Study on Local Vacuum Electron Beam Welding
K. Shinada, K. Kondo, S. Satoh, T. Shimoyama, G. Takano, M. Minami, T. Tanaka and Y. Arata
[Int. Conf. on Welding Research in 1980s, A-(10) (1980), 55.]

Fundamental Studies on Electron Beam Welding of Heat-resistant Superalloys for Nuclear Plants—Effect of Welding Conditions on Some Characteristics of Weld Bead
Y. Arata, K. Terai, H. Nagai, S. Shimizu and T. Aota
[Trans. JWRI **5** (1976), 219.]

Tandem Electron Beam Welding
Y. Arata and E. Nabegata
[Trans. JWRI **7** (1978), 101. IIW Doc. IV−221−77.]

Narrow Gap High Energy Density Beam Welding
—Principle—

It is considered that the narrow gap welding is an excellent welding process for the thick plate.

As a typical example, the narrow gap welding in which arc heat source is used to perform the multi-pass welding of the square groove zone with about 5 ~10 mm wide gap and over several to ten cm depth, has been developed recently and its research for the industrial application progresses in some countries. In this process, the core wire, shielding gas and their guides are inserted into the narrow gap zone and travel there, so the automatic control is necessary essentially because manual handling is very difficult.

The accurate automatic control of the arc behaviour and the operation of these tools in the narrow gap are complicate in general. Moreover, the narrower width of the gap, the more difficult it becomes.

Therefore, from now on the development of the new control technique is necessary in this process.

On the other hand, in the narrow gap welding proposed in this paper, the fine beam with the high energy density such as the electron beam or the laser beam is passed into the narrow gap zone using the method shown in **Fig. 1,** and due to their energy, the adequate filler or insert metal wire plate or powder is melted, and the narrow gap zone is welded[1].

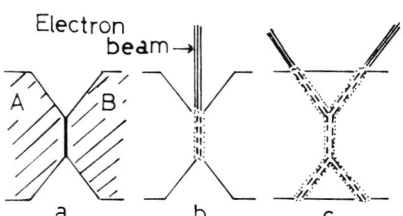

Fig. 2. Joining method given maximum thickness.

It is well known that the fundamental difference between the high energy density beam heat source and the arc heat source come out in their penetration depth. In the welding availing oneself of such beam to produce a deep penetration (the maximum penetration depth: h_p), the maximum plate thickness to be possibly jointed is smaller than $3 h_p$ as shown in **Fig. 2.** In the weld joint penetrated from both sides of it, the maximum plate thickness penetrated is smaller than $2 h_p$.

On the contrary, in the narrow gap welding process proposed in **Fig. 1,** we may suppose the possibility to be jointed even the ultra thick plate with several ten cm thickness. As shown in **Fig. 3,** it is well known that if the angle, θ_b with which the electron beam or the laser beam run against the material's wall is smaller than the certain angle, the beam is reflected efficiently on the wall surface, and the reflected beam energy becomes extremely larger than the absorbed beam energy on it. Therefore, adapting this principle to the narrow gap welding, the beam energy can be transported up to the very deep zone and distributed wide range. Namely, because the phenomenon such that the beam doesn't scatter but is focused with the wall, take place as mentioned above, it is to be called the "wall-focusing".

In the narrow gap welding proposed here, putting the wall-focusing to practical usage, most of the beam energy is concentrated upon the filler or the insert metal, and they are melted with the basemetal. In such case the multipass welding is usually used. The number of the pass-layers depend on the beam output, which can determine the adaptable one-pass maximum

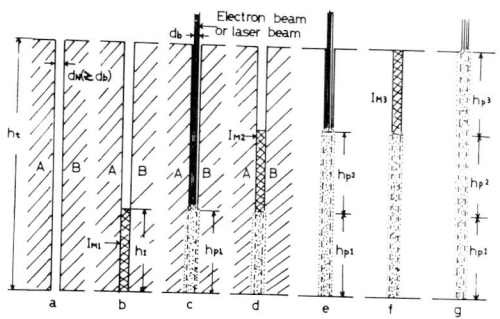

Fig. 1. Schematic illustration of narrow gap high energy density beam welding (NG-EBW, NG-LBW)
A, B: thick material to be joined, I_{M1}, I_{M2}, I_{M3} -----: insert metal, h_{p1}, h_{p2}, h_{p3}, ----: melted zone, d_N: narrow gap between A and B, d_b: electron or laser beam diameter.

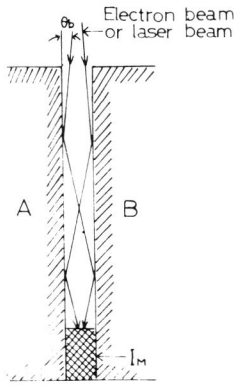

Fig. 3. Schematic illustration of "wall focusing".

Fig. 5. Another example of NG-EBW (thickness; 8 cm, materials; SUS304, number of multipasses; 4).

penetration depth (h_p). Moreover, such one-pass maximum penetration depth corresponding to hight of each layer or the number of it often, is decided due to the metallurgical characteristics of the weld zone. **Figure 4** is one example obtained experimentally, using this process. In this case, the welding depth of each layer is 2 cm and also the welding joint with 10 cm thickness has been welded with five layers. Because the electron beam has been used as the heat source, this welding process as shown in **Fig. 4** is called "Narrow gap electron beam welding" (NG-EBW). And in a similar manner, the laser beam can be used, and so it is called "Narrow gap-laser beam welding" (NG-LBW).

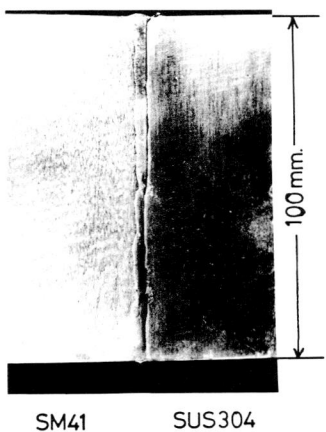

SM41   SUS304

Fig. 4. One example of NG-EBW (thickness; 10 cm, dissimilar metals; SM41 and SUS304, number of multipasses; 5).

SM41: plane carbon steel (0.15 % C)
SUS304: correspond to AISI304

The merits of these welding process are as follows.

1)   In case the ultra thick plate is welded with one-pass, the high energy density beam with the very large output is needed. But the narrow gap welding proposed here, becomes possible to joint the ultra thick plate without a very high power level of beam output as mentioned above. Moreover, in comparison with the narrow gap arc welding process, this process fully displaies the narrow gap welding process's real ability. For example, we may expect that the width of the narrow gap is usually about $2 \sim 3$ mm and its depth become possible even several ten cm.

2)   The welding of the dissimilar materials become easer than other process. **Fig, 4** and **Fig. 5** are examples of the NG-EBW. But in the welding of the very thick materials, the ploblems such as the cracking and the joining difficulties on the metallurgical properties, often may occur.

In such case, it may be needed that the properties of the welded joint is improved by unifying the deposited metal with not the one pass but the multipass narrow gap welding.

Acknowledgment

The author thanks the electron beam groop of JWRI, Kawasaki Heavy Industry Co. and Japan Electron Optics Industy Co. Ltd., for their effort in the experiment.

Reference

1) Y. Arata: A Special Lecture in a Symposium of Vacuum Metallurgy in the Iron and Steel Institute of Japan, and the Japan Institute of Metals, November (1972).

430

100 kW Class Electron Beam Welding Technology
— Fundamental Research on Horizontal Electron Beam Welding —

Abstract

A new type deflector was developed with which a high power vertical beam from the electron gun could be effectively converted to the horizontal one. To obtain a sound weld for a plate more than 100 mm in thickness, the influences of beam oscillation and material composition (especially O, N, Mn and Si) were tested in a transverse welding condition and the good welding conditions without any defects were decided. Furthermore the possibility was demonstrated to certify the full penetration welding of an ultra-thick plate more than 300 mm by the vertical upward welding with the horizontal beam.

KEY WORDS: (Electron Guns) (Electron Beam Welding) (Defects) (Hardness) (Weldability) (High Strength) (Horizontal E. B. Welding) (Vertical E. B. Welding) (All Position E. B. Welding) (Thick Plate E. B. Welding)

1. Introduction

In electron beam welding, it is essential that the electron gun should operate efficiently, and, since the gun is considerably heavy and a high degree of accuracy is required, it must in general be fixed in position or supported by a machine such as a robot. Therefore, in electron beam welding, it was heretofore said to be difficult to perform welding in various positions. The writers[1], however, have demonstrated that welding in all positions is possible, as shown in **Fig. 1**, by developing a beam deflector. Among these positions, horizontal welding[2] and vertical welding[3] by means of a horizontal beam are being studied by many researchers and are starting to be put into practical use.

A horizontal beam may be produced by deflecting[4] a vertical beam in 90 degrees with a deflector, or by using an electron gun fixed horizontally[5]. The above-mentioned horizontal electron beam welding has been found to be more effective in thick plate welding compared with ordinary flat position welding, and it is therefore frequently used in welding thick plates. We here intend to give further characteristics of horizontal electron beam welding on thick plates.

2. Evolution of Horizontal Electron Beam

An electron gun is composed of a beam acceleration chamber, a beam channel and an injection port. The electron beam emitted from the injection port reaches the part to be welded and generates plasma and thermal vapour jet as well as a beam hole there. Some of this plasma and thermal vapour jet flows back along the E.B. axis into the E.B. gun and contaminates the acceleration chamber, damaging the cathode. This effect becomes greater, the greater the increase in beam power.

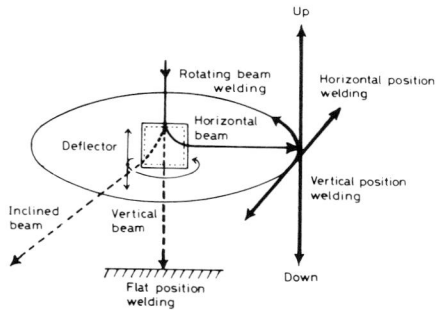

Fig. 1   All position electron beam welding method.

In the case of horizontal electron beam welding, a high power beam is usually used, and countermeasures must therefore be considered to deal with the situation. **Fig. 2**[6] shows the contamination rate in the E.B. gun according to the deflection angle of the beam. **Fig. 2** shows the advantages of the 90° deflecting beam in relation to contamination in the E.B. gun.

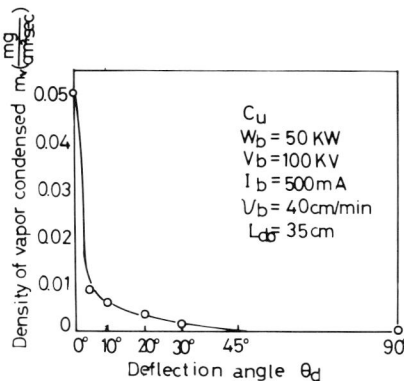

Fig. 2 Effect of beam deflection on metal vapor contaminating electron gun.

To simplify the following explanation, the above apparatus will be called Type A in this paper.

Two other methods of decreasing the contamination rate are to increase the distance (working distance) between the beam injection port and the welded part, and to alter the trajectory on the way along the beam axis. The former will be called Type B and the latter Type C.

Since these types A, B, and C all have almost indentical effects on welding, it follows that the type chosen should be the one most suitable for the particular welding conditions. Type A Welder (W_b = 100 kW, V_b = 100 kV, I_b = 1,000 mA) as shown in **Fig. 3** was used in this research, the beam condition was α_b ($\equiv D_O/D_F$) $\simeq 0.8 \sim$ 0.9, D_O = 150 mm in which D_O indicates objective distance and D_F focal Length.

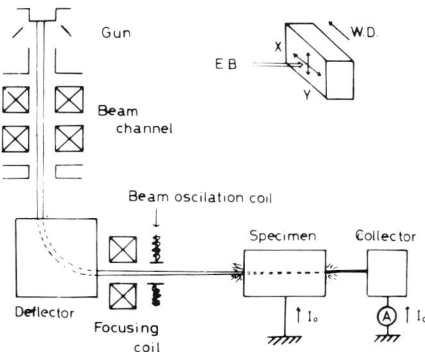

Fig. 3 Schematic diagram of beam oscillation at horizontal position. (A type welder)

3. Experimental Results and Discussions

3.1 Horizontal welding

There are basically two kinds of horizontal beam welding, i.e., welding in a horizontal position and welding in a vertical position. In the case of electron beam welding in general, shaking of the beam hole and the molten pool becomes more violent with increases in power.

In the case of flat position welding on thick plates in particular, the shaking becomes so violent as to make welding extremely difficult.

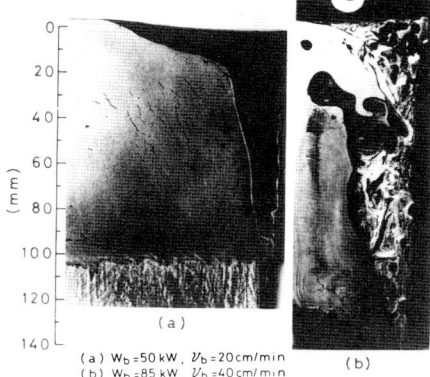

(a) W_b =50 kW, V_b =20cm/min
(b) W_b =85 kW, V_b =40cm/min

Fig. 4 Bead sections indicating condition of violently shaken molten pool during flat position welding. (θd=0)

This is clearly shown in **Fig. 4**[7] and motion picture analysis.

The fundamental conditions for achieving good welding are:
1) maintaining the beam-hole in a stable condition,
2) inhibiting shaking of the molten pool
These conditions will be called ConditionA.

It becomes increasingly difficult to maintain Condition A as the beam-hole grows deeper. In the case of flat position welding in particular, the influence of gravity on the molten pool must not be ignored. In the case of horizontal and vertical welding, this effect disappears, and it is possible to forecast a better result compared with flat position welding.

However, the drawback is that the molten pool is apt to flow outward and result in what is called "porosity" due to lack of molten metal. An example is shown in **Fig. 5** in the case of horizontal welding.

The materials used in this experiment were Cr - Mo steel (2¼Cr - 1Mo), stainless steel (SUS304), high tensile strength steel (HT50, 80) and centrifugally cast steel pipe for welded structures (SMK50) as shown in **Table 1** and thick plates of a thickness of 100 mm or more.

Table 1 Chemical composition of materials used.

Elements / Materials	C	Si	Mn	P	S	N	O
						Wt% +	+ ppm
2¼ Cr-1Mo (CM)	0.11	0.21	0.51	0.016	0.009	63	20
2¼ Cr-1Mo (Y)	0.12	0.15	0.49	0.011	0.010	109	30
SUS 304	0.050	0.74	1.74	0.030	0.010	360	75
HT 50	0.15	0.34	1.26	0.020	0.015	—	—
HT 80*	0.11	0.32	0.92	0.008	0.006	34	10
HT 80 (N)	0.12	0.29	1.56	0.010	0.001	75	69
HT 80 (A)	0.12	0.06	0.96	0.007	0.006	57	56
HT 80 (B)	0.12	0.10	0.94	0.008	0.005	102	47
HT 80 (C)	0.11	0.09	0.93	0.008	0.006	274	47
HT 80 (D)	0.12	0.03	0.98	0.008	0.007	46	61
HT 80 (E)	0.12	0.03	0.93	0.009	0.007	53	109
HT 80 (F)	0.12	0.03	0.90	0.007	0.006	118	271
HT 80 (G)	0.13	0.12	0.24	0.008	0.009	44	71
HT 80 (H)	0.11	0.11	1.06	0.010	0.007	57	82
HT 80 (I)	0.12	0.14	4.98	0.012	0.006	43	81
HT 80 (J)	0.11	0.03	0.92	0.010	0.010	79	63
HT 80 (K)	0.12	0.16	0.97	0.009	0.009	70	73
HT 80 (L)	0.13	1.02	0.93	0.010	0.009	66	77
SMK 50	0.15	0.30	1.03	0.016	0.023	123	28

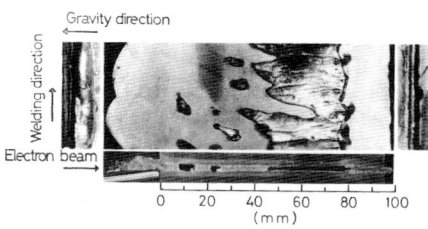

$V_b = 100$ kV, $W_b = 52$ kW, $I_b = 520$ mA, $\mathcal{V}_b = 40$ cm/min, $h_t = 95$ mm

Fig. 5 Fully penetrated cross sections and appearance of horizontal position welding.

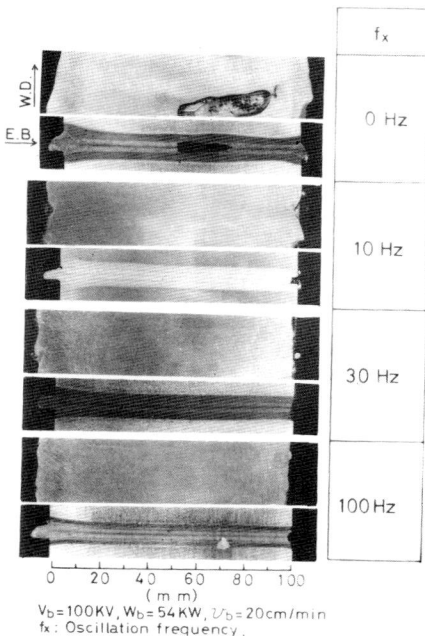

$V_b = 100$ KV, $W_b = 54$ KW, $\mathcal{V}_b = 20$ cm/min
f_x: Oscillation frequency,
Amplitude $d_x = 3$ mm, 2¼ Cr-1Mo

Fig. 6 Fully penetrated bead section of horizontal position welding with various beam oscillation frequency.

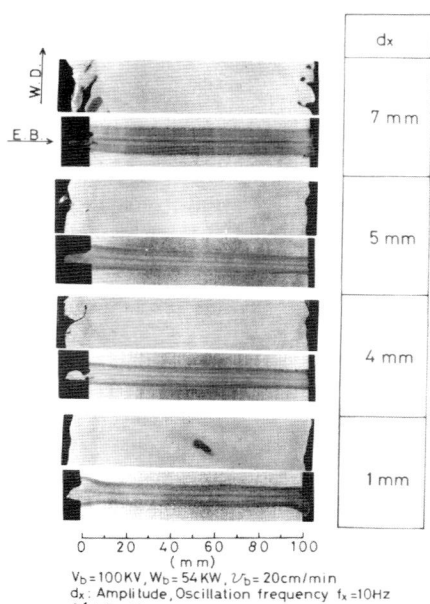

$V_b = 100$ KV, $W_b = 54$ KW, $\mathcal{V}_b = 20$ cm/min
d_x: Amplitude, Oscillation frequency $f_x = 10$ Hz
2¼ Cr-1Mo

Fig. 7 Fully penetrated bead section of horizontal position welding with various beam amplitude.

In the case of these thick plates, however, a number of defects such as porosity appear, if a suitable beam condition is not chosen. One means of inhibiting such phenomena is to produce beam oscillation, i.e., to cause the beam to oscillate along the transverse line of welding by applying X-oscillation, or to cause it to oscillate perpendicularly by applying Y-oscillation by means of a suitable magnetic field, as shown in **Fig. 3**. As a matter of course, these oscillations are indicated by frequency f_x, f_y (Hz) and amplitude d_x, d_y (mm).

Which oscillation should be chosen or whether a compound oscillation (circular or elliptical in shape) would be suitable depends on the material. **Fig. 6** and **Fig. 7** show the effect of X-oscillation (f_x and d_x) when 100mm–thick 2¼Cr - 1Mo steel was welded at a welding speed of $\mathcal{V}_b = 20$ cm/min., the beam condition being $d_x = 3$ mm, $W_b = 54$ kW and $V_b = 100$ kV.

The illustrations show the condition of a bead cross section, where large porosities appeared when there was

433

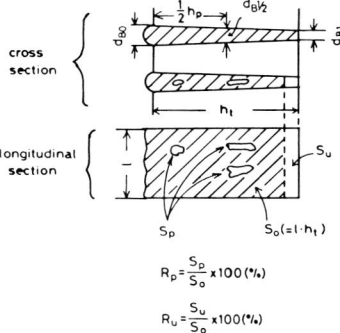

$$R_p = \frac{S_p}{S_o} \times 100 \, (\%)$$

$$R_u = \frac{S_u}{S_o} \times 100 \, (\%)$$

Fig. 8 Bead defect explanation.
　　Note: So indicates the cross section area of the bead.
　　(So = l · ht; l = width, ht = plate thickness) Sp
　　indicates the area of porosity in So.
　　Su indicates the area of under-fill in Su. l indicates
　　the width of the cross-section and in this case
　　l = 40 mm.

no oscillation ($f_X = 0$), but disappeared when $f_X = 5$.

To illustrate these bead conditions more specifically, the results shown in **Fig. 9** and **Fig. 10** are obtained by using the symbols shown in **Fig. 8** and by indicating the rate of porosities and under-fill by: - porosity rate R_p (= S_p/S_o), and under-fill rate R_u (= S_u/S_o). When the value of f_X and d_X are $f_X \simeq 10$, $d_X \simeq 2 \sim 5$, conditions are optimum, R_p and R_u being limited to almost 0.

Furthermore, a condition called parallel bead may be obtained at the welded part, where the bead width is almost uniform throughout. Neither Y-oscillation nor compound oscillation of X and Y were taken up this time because they produced rather poor welding results.

In full penetration welding, the important factor is the value of ϑ, the beam pass rate of the beam current. ($\vartheta = I_c/I_b$: I_b indicates the incident beam current, I_c indicates the collected beam current shown in **Fig. 3**. The influence

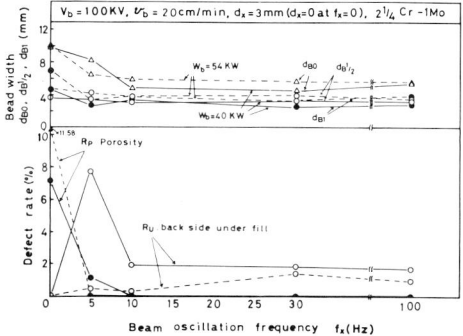

Fig. 9 Relation between bead width, defect rate and beam oscillation frequency at horizontal position welding.

of ϑ values is shown in **Fig. 11**, and its range of efficacy is $\vartheta \simeq 10 \sim 50$. When $\vartheta \simeq 10$ and $f_X \sim 10$, the porosity disappeared though the under-fill still remained; but when $f_X \simeq 30 \sim 100$, the under-fill also almost totally disappeared (except for $1 \sim 2\%$). The relationship of f_X, d_X and ϑ is most important.

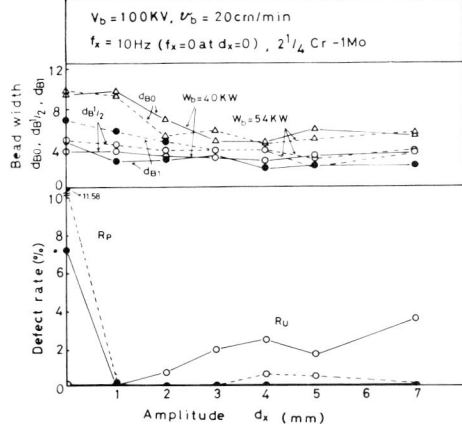

Fig. 10 Relation between bead width, defect rate and beam amplitude at horizontal position welding.

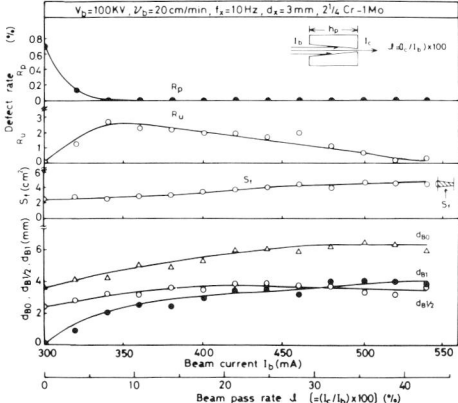

Fig. 11 Relation of Ru, R_p, S_f bead width and I_b, beam pass rate at horizontal position welding with x-oscillation beam.

The result mentioned above relates to the case where 100mm-thick plates were used. Extremely good results were obtained up to a thickness of $150 \sim 175$ mm as shown in **Fig. 12**, when welding was performed under a beam condition such as $f_X = 10$, $d_X = 3$, $\vartheta = 20 \sim 40$.

However, large porosities appeared when the plate thickness was increased to 200 mm. Therefore, the beam condition must be reconsidered when the plate thickness

h_t	υ_b
100mm	20 cm/min
150mm	15 cm/min
175mm	10 cm/min

$V_b = 100\,KV$, $W_b = 50\,KW$, $f_x = 10\,Hz$, $d_x = 3\,mm$, $2\frac{1}{4}Cr$-$1Mo$

Fig. 12 Fully penetrated bead section of horizontal position welding at various plate thicknesses.

is further increased. Not only 1¼Cr - 1Mo steel, but other steels exhibited a tendency similar to the above as shown in **Fig. 13**.

It was found from the above-mentioned results that it was possible to achieve a flawlessly welded part on fairly thick plates, if the beam condition was regulated properly. There were still, however, some points that could not be solved by merely regulating the beam condition, and some problems related to the materials themselves.

The influence of the gas constituent (especially O_2 and

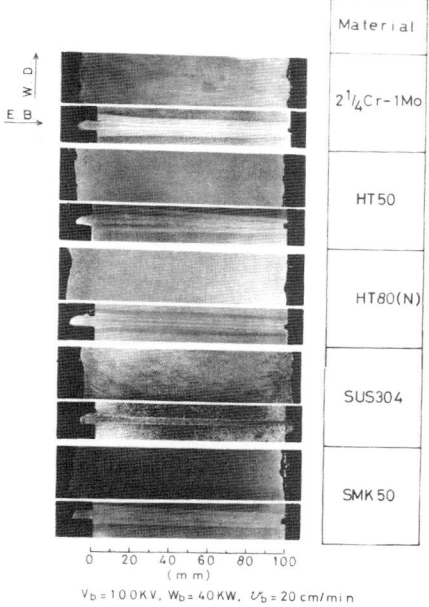

Material
$2\frac{1}{4}Cr$-$1Mo$
HT 50
HT 80(N)
SUS 304
SMK 50

$V_b = 100\,KV$, $W_b = 40\,KW$, $\upsilon_b = 20\,cm/min$
$f_x = 10\,Hz$, $d_x = 3\,mm$

Fig. 13 Fully penetrated bead section of horizontal position welding with various metals.

N_2) and elements having high vapour pressure such as Mn, Si were problems common to all these materials. High tensile strength steel HT80 (see **Table 1 (A)** ~ **(L)** and **(N)**) was used, and beam conditions such as $f_x = 10$, $d_x = 3$, $\vartheta = 20 \sim 30$ and $W_b = 50$ were adopted. **Fig. 14** and **Fig. 15** illustrate the results of when the nitrogen content varied from [N] = 57 ~ 273 ppm, while the oxygen content was held almost constant [O] = 47 ~ 63 ppm.

The bead shapes appeared normal and similar in shape up to [N] ≃ 100 ppm, but the bead width on the top side and the bottom side increased a little and defects appeared at around 300 ppm. This is thought to have resulted from the violent perturbation caused by the effusion of nitrogen gas.

This effect was heightened as plate thickness increased and the same defect appeared when the plate thickness was 175 mm, even though [N] = 100 ppm. The effect of

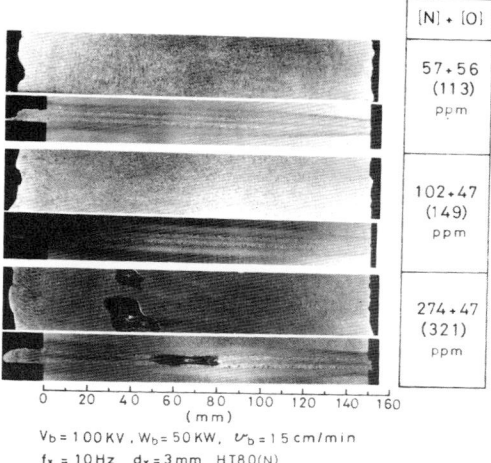

[N] + [O]
57+56 (113) ppm
102+47 (149) ppm
274+47 (321) ppm

$V_b = 100\,KV$, $W_b = 50\,KW$, $\upsilon_b = 15\,cm/min$
$f_x = 10\,Hz$, $d_x = 3\,mm$, HT80(N)

Fig. 14 Characteristics of fully penetrated bead section with variations in [N] + [O].

the gas was brought about not only by the amount of [N] but also by the total gas amount of [N] + [O].

In this case, similar results appeared as in the case of [N]. **Fig. 16** and **Fig. 17** show the effects of gas when [N] + [O] = 107 ~ 389 ppm and, they show clearly that the increase in gas content and the thickening of the plate result in defects. Thus defects resulted even when the plates were quite thin, if the gas content was high.

The effect of Mn is quite obvious from **Fig. 18** though that of Si in **Fig. 19** is not so apparent. However, the effects of both may be expressed as a whole by the parameter;

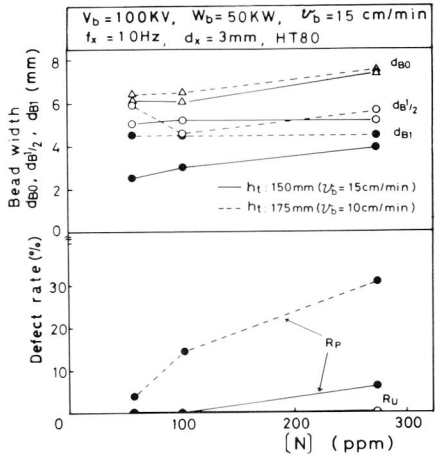

Fig. 15 Relation between bead width, defect rate and content of [N].

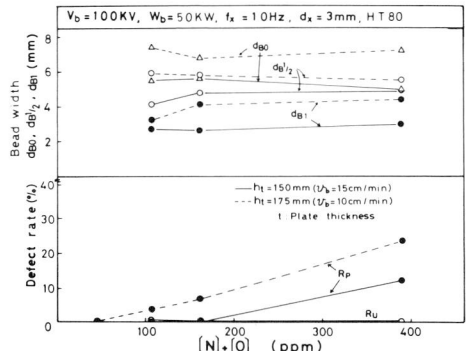

Fig. 16 Relation between bead width, defect rate and contents of [N] + [O].

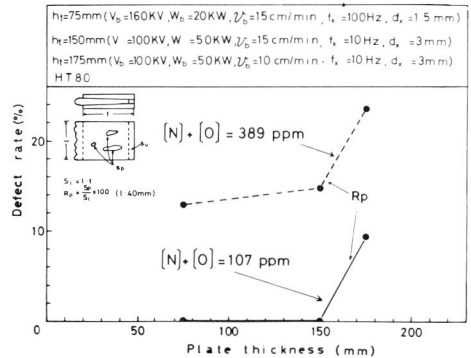

Fig. 17 Relation between defect rate and plate thickness with variation in [N] + [O].

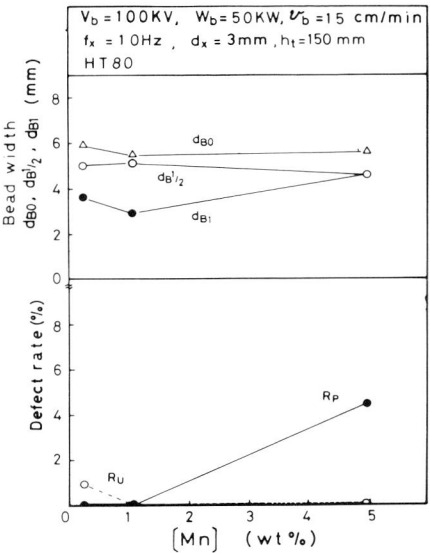

Fig. 18 Relation between bead width, defect rate and content of [Mn].

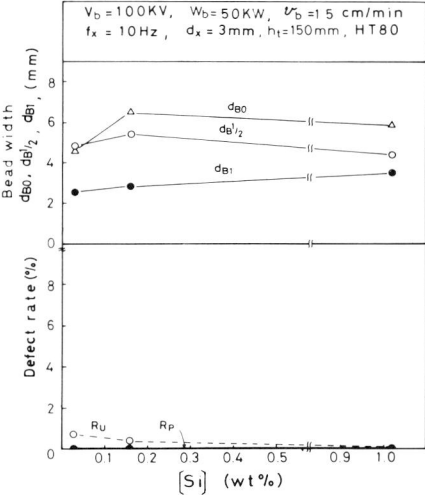

Fig. 19 Relation between bead width, defect rate and content of [Si].

436

$$C_R = \frac{[Si]}{[Mn][N + O]}$$

Fig. 20 shows that defects disappeared altogether at around $C_R \simeq 8$, for a plate thickness below 150 mm.

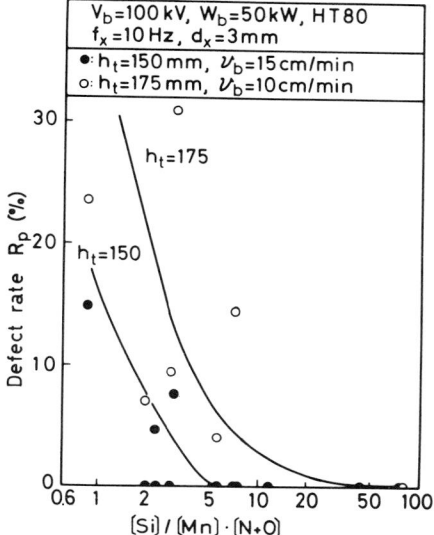

Fig. 20 Relation between defect rate [Si]/[Mn] [N+O] in steel.

The strength of welded joints is to be mentioned next. Centrifugally cast steel pipe for welded structures (SMK 50: cast while adding 50 ~ 100G) and high tensile strength steel (HT 50) show almost the same welding results, and this paper refers to the former here.

50mm-thick plates were welded under beam conditions where $W_b = 20$, $V_b = 20$ cm/min., $f_x = 30$, $d_x = 3$, $\vartheta = 30$. As a result, a fracture occurred on the base metal part. The strength of the welded part was 57.9 kg/mm² which slightly excelled the strength of the base metal's 56.3 kg/mm².

In the bending test, the welded part could be bent 180° in three directions and showed no defects. The impact test was performed with the notch position at $d_{B\frac{1}{2}}$ (in the center), and the charpy was decreased at the weld bond as shown in **Table 2**. This may be improved by preheating and by using a filler metal.

The hardness is very great in the vicinity of the weld bond with a maximum hardness of $H_{max} > 300$ when the plate thickness is about 50 mm as shown in **Fig. 21**; but the hardness falls below 300 when the plate thickness is more than 100 mm due to the increase in heat input.

Table 2 Results of V-notch charpy test.

	No.	Charpy (2mmV, 0°C)
Weld metal	1	5.6 kg-m
	2	22.1 kg-m
Bond	1	2.2 kg-m
	2	3.2 kg-m
HAZ.	1	10.2 kg-m
	2	9.6 kg-m

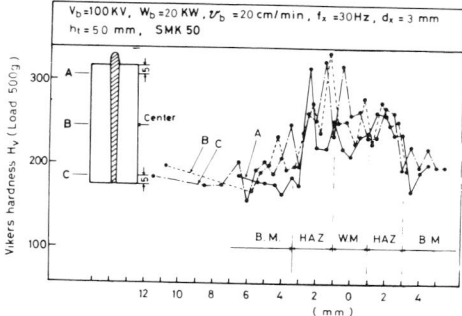

Fig. 21 Hardness curves from horizontal position welding. (50mm^t)

Even 50mm-thick plates showed hardness of $H_{max} \sim 280$, and there was hardly any dispersion of hardness when it was preheated at a temperature of around 150°C. The hardness may be lowered to that of the base metal by using a filler metal.

3.2 Vertical upward welding

Of all welding positions, vertical upward welding is most suitable for welding heavy thick plates because it is least influenced by the molten pool. Welding was performed according to this method in this instance.

Fig. 22 shows photographs of bead cross sections of vertival upward welding performed on plates of various thicknesses of high tensile steel HT80 (N) using an A type welder.

This welding method has the great advantage of producing high-grade beads whose penetration depth becomes greater in proportion to the power. **Fig. 23** shows the effect of beam currents on the bead width and

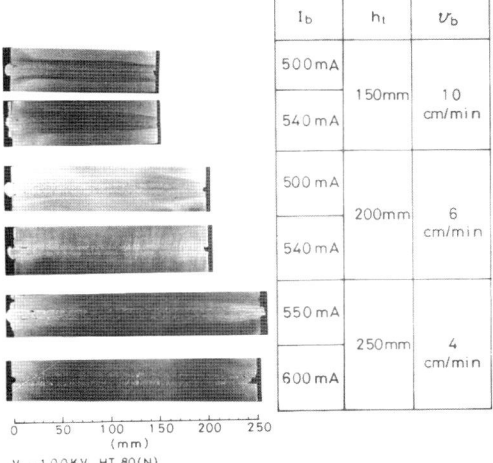

I_b	h_t	v_b
500 mA	150mm	10 cm/min
540 mA		
500 mA	200mm	6 cm/min
540 mA		
550 mA	250mm	4 cm/min
600 mA		

V_b = 1 0 0 KV, HT 80(N)

Fig. 22 Bead sections of vertical upward position welding with various plate thicknesses.

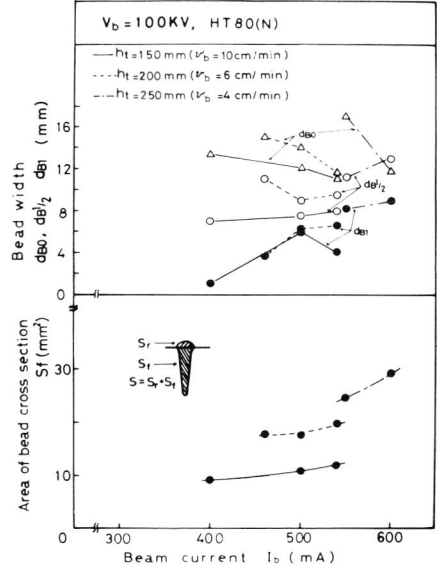

Fig. 23 Relation between bead width, melting area of weld bead and beam current at vertical upward position welding.

on the cross section area of beads produced on various plate thicknesses. It was found that a uniform crosssectional bead width (called a band bead) could be obtained all along the area even when using heavy thick plates,

when a suitable beam condition was provided.

Fig. 24 and **Fig. 25** show the effects a chemical composition has on these bead shapes, and they indicate that the effect is negligible.

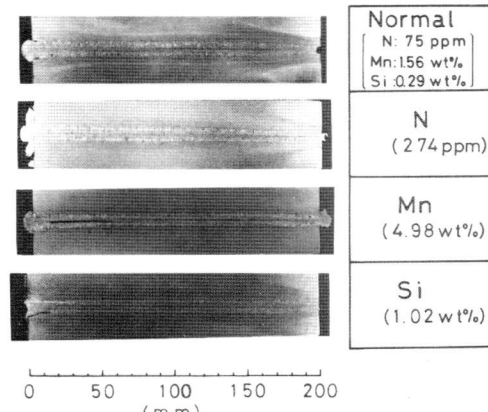

V_b = 1 0 0 KV, W_b = 5 0 KW v_b = 6 cm/min
h_t = 200 mm ,HT 80

Fig. 24 Bead sections of vertical upward position welding with verious element contents.

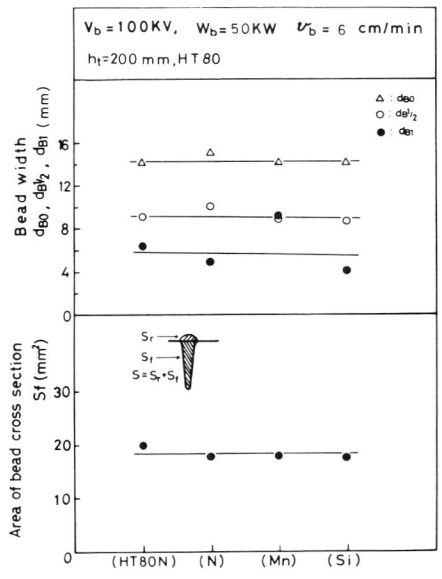

Fig. 25 Relation between bead width, melting area of weld bead and changing of elements in steels at vertical upward position welding.

438

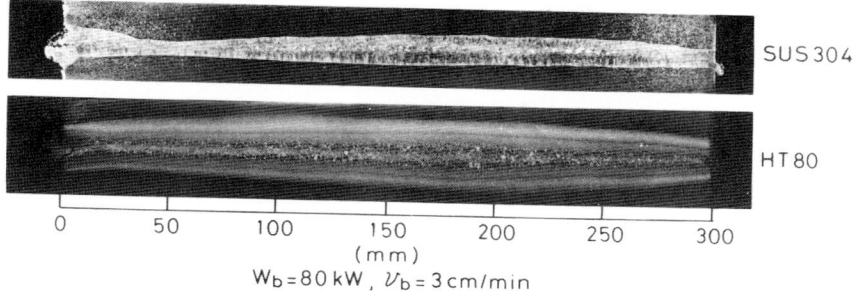

$W_b = 80\,kW$, $\mathcal{V}_b = 3\,cm/min$

Fig. 26 Bead sections of vertical upward position welding.

N, Mn, Si were chosen as the chemical components. HT80 (C), HT80 (I) and HT80 (L) in **Table 1**, which contain the largest amount of the respective components, and ordinary HT80 (N) were used and compared. It was proved that these chemical compositions had little effect on bead shape. Only HT80 (I), however, which contains Mn 4.98%, caused a slender shrinkage due to lack of liquid metal caused by the over-flowing of molten metal on account of high vapour pressure. HT80 (C), which contains a large quantity of N, showed no defects in vertical upward welding and no difference compared with ordinary HT80 (N) was detectable, but defects appeared in the case of horizontal welding.

Fig. 26 shows the results obtained by welding 300 mm-thick HT80 (N), SUS 304 plates at $W_b = 80kW$, $\mathcal{V}_b = 3cm/min$.

It may be induced from the above results that the fundamental condition for obtaining a high-grade bead with thick plates is to set up a beam condition which satisfies the above-mentioned A condition, thus devising a way to obtain a narrow band bead.

Vertical upward welding, compared with all other welding positions, not only satisfies this condition the best, but also shows a lower maximum hardness (Hmax) at the welded part as seen in **Fig. 27** than the others and little dispersion. Furthermore, these values are not changed by oscillation.

4. Conclusion

1. A horizontal beam was produced by 90°-deflection of a 100kW class vertical electron beam. By using this beam an effeicient horizontal and vertical welding was performed successfully for a plate thickness of more than 100 mm.

2. A horizontal welding without any defects was possible with a low frequency (10-100 Hz) beam oscillation method in X-direction. The proper beam pass rate ranged 10-50% (for a plate thickness of more than 100 mm).

3. When the gas compositions of [N] and [N] + [O] were both below 100 ppm, a sound weld was obtained. Some defects appeared when [Mn] was commingled more than several percentages.

4. For a plate thickness below 150 mm, no defect was observed at around 8 of the composition parameter C_R as show in **Fig. 12**.

5. In case of a vertical welding at $W_b = 80$ kW, a full penetration welding of a plate of 300 mm in thickness was possible with a penetration depth two times deeper than in the case of a vertical welding.

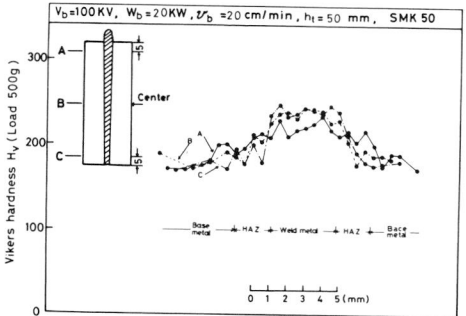

Fig. 27 Hardness curves of vertical upward position welding without X-oscillation.

439

Acknowledgement

The authors would like to express their appreciations to Nippon Steel Corporation, Sumitomo Metal Industries, LTD. and Kubota, LTD. for their supply various steel specimens for the experiment.

References

1) Y. Arata, M. Tomie: "Fundamental Features of 100 kW class Electron Beam Welding Technology", 2nd International Symposium J.W.S., (1975)., 7th International Conference on Electron and Ion Beam Science and Technology (at Washington D.C, U.S.A.) (1976).

2) A. Sanderson: "A 75 kW Electron Beam Installation for Thick-Section Welding", Metal Construction B.W.J., 6-1 (1974)., K. H. Steigerwald: "High Energy Density Beam Welding", 2nd International Symposium J.W.S., (1975)., T. Shida, H. Okamura, H. Kita and Y. Akutsu: "A Study on Occurrence and Prevention of Defects of Electron Beam Welding (Report 5)", J.W.S., 48-10 (1979).

3) H. Irie, T. Hashimoto and M. Inagaki: "Vertical Position Beam Welding", Research Committee for Welding J.W.S., No. EBW-163-76, (1976)., Y. Arata, M. Osumi, K. Higuchi and K. Noda: "Study on Electron Beam Welding of High Strength Aluminum Alloy", Preprints of the National Meeting of J.W.S., No. 16 (Spring 1975).

4) Y. Arata, M. Tomie: "Study of Ultra High Energy Density Heat Sourse of Electron Beam and its Application for Welding (Report 3)", J.W.S., 46-9, (1977).

5) References 2 and 3.
   Y. Arata, M. Tomie: "Study on Open Atmosphere and Low Vacuum Pressure Electron Beam Welding", Research Committee for Welding, Kansai Electric Power. (1978).,

6) References 1 and 4.

7) Y. Arata, M. Tomie and Y. Kato: "100kW Klasse-Electronenstrahlen - Schwe β technologie (Bericht II)", Trans. of J.W.R.I., 4-1, (1975)., Y. Arata, M. Tomie: "Study of Ultra High Power Heat Source of Electron Beam and its Application for Welding (Report 2)", J.W.S., 46-8 (1977).

8) Y. Arata, E. Abe, E. Nabegata and M. Fujisawa: "Dynamic Welding phenomena during E. B. Welding", Second Colloquium International Electron Beam Welding, Melting, AVIGNON, Sept. (1978).

440

Electron Beam Welding of High Strength Aluminum Alloy

Abstract

It has been a general consideration that super super dulalmin, the representative of which is 7075, can not be successfully welded with conventional fusion welding method. So we have been investigating the application of EB welding process to this sort of materials.

In this paper we report the penetration characteristics and the mechanical properties of 7075.

Considerable differences in welding penetration phenomenon were found between alloys depending on the sort and amount of alloying element when the penetrations do not reach to the bottom of the plate thickness (partial penetration). In particular, the penetration of 7075 which contains Zn and Mg of higher vaporization pressure as principal alloying elements is much deeper than 2xxx aluminum alloys whose main alloying element is Cu.

From the view point of mechanical properties, the properly EB welded 7075 exhibits tensile strength comparable to the yield strength of base metal, but the fracture mode seems to be rather brittle with small elongation. The fracture toughness of the weld, however, is superior to that of base metal.

1. Introduction

It has been a general consideration that super super dulalmin, the representative of which is 7075, can not be successfully welded with conventional fusion welding method. This is due to the fact that it has a severe inclination of weld cracking and extreme degradation in weld strength and quality including anti-corrosion characteristics.

We have been investigating the application of EB welding process to this sort of materials. EB welding was considered worth trying application to these materials because it is a new welding method in which heat influence can be limitted more locally than in traditional welding methods. We are reporting the penetration characteristics and the mechanical properties of 7075.

Considerable differences in welding penetration phenomenon were found between alloys depending on the sort and amount of alloying element, when the penetration do not reach to the bottom of the plate thickness (partial penetration). In particular, the penetration of 7075 which contains Zn and Mg of higher vapourization pressure as principal alloying elements is much deeper than 2xxx aluminum alloys whose main alloying element is Cu. This is probably due to the elevated pressure in beam hole caused by the vaporization of Zn and Mg during welding, and in this case, weld defects such as spikes and cold shuts were apt to be formed at the root of the weld bead.

On the other hand, from the view point of mechanical properties, the properly EB welded 7075 exhibits tensile strength comparable to the yield strength of base metal, but the fracture mode seems to be rather brittle with small elongation. The fracture toughness of the weld, however, is superior to that of the base metal, so it seems that from the standpoint of fracture toughness, this 7075 weld joint is useful if proper welding procedure is applied and sound weld is obtain.

2. Penetration Characteristics of Aluminum Alloys

2.1 Materials and Procedure of Experiment

The chemical compositions of aluminum alloys used in this experiment are listed in Table 1. Test pieces

Table 1 Chemical composition of materials used (%)

Material	Cu	Si	Fe	Mn	Mg	Zn	Ti	Sn	V	Cd	Zr	Cr
2014	3.9-5.0	0.5-1.2	<1.0	0.4-1.2	0.2-1.2	<0.25	<0.15	—	—	—	—	—
2021	5.8-6.8	<0.2	<0.3	0.2-0.4	<0.02	<0.1	0.02-0.1	0.03-0.08	0.05-0.25	0.02-0.2	0.1-0.25	—
2024	3.8-4.9	<0.5	<0.5	0.3-0.9	1.2-1.8	<0.25	—	—	—	—	—	<0.1
2219	5.8-6.8	<0.2	<0.3	0.2-0.4	<0.02	<0.1	0.2-0.1	—	0.05-0.15	—	0.1-0.25	—
5083	<0.1	<0.4	<0.4	0.3-1.0	3.8-4.8	<0.1	<0.2	—	—	—	—	<0.5
7075	1.2-2.0	<0.4	<0.5	<0.3	2.1-2.9	5.1-6.1	<0.2	—	—	—	—	0.18-0.35

shown in Fig. 1 were successively set in line and were bead-on-plate welded in one chamber. Welding condition of Table 2 was applied in this experiment.

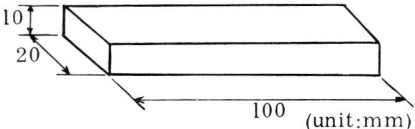

Fig. 1 Size of test piece

Table 2 Welding condition

Parameter	
Accelerating voltage	150 KV
Beam current	7 mA
Welding speed	1.5 m/min
Work distance	150 mm
Focus	a_b ; 0.92 at 7 mA current
Atmosphere	1×10^{-4} Torr

2.2 Results and Discussion of Experiment

Penetration geometries of aluminum alloys listed in Table 1 are shown in Photo. 1, from which it is evident that there are considerable differences among materials. As for bead appearance, 7075 and 5083 are irregular and rugged as if eruption has occured, 2024 and 2014 are fairly smooth and wide, and most beautiful, uniform and widest are 2021 and 2219. As for the depth of penetration, 7075 and 5083 are especially narrow and deep, approximately twice those of other materials such as 2014, 2024, 2021, 2219, and penetration becomes shallower and rounder in the order of 2024, 2014, 2021, 2219. Penetration depth and bead appearance are conversely related. Longitudinal cross-section of weld beads were taken to observe the ripple of penetration, which has made clear that the same influence of chemical composition as on spikes also exists. Alloys 7075 and 5083 were accompanied with root porosities (R-porosity) and cold shuts at almost all of the spike areas.

In the case of fully penetrated welding, 7075 still exhibits, not conspicuously, above phenomena, obtaining narrower penetration than 2219 etc. (Photo 2).

Closely related to the remarkably deep penetration of 7075 and 5083 are probably the evaporation of Zn and Mg during welding which are the principal alloying elements of these materials.

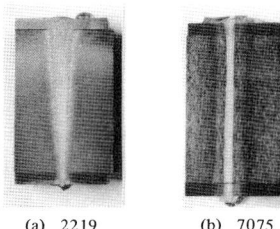

(a) 2219   (b) 7075

Photo. 2 Lateral cross-section of welded bead (full penetrated, 57mm^t)

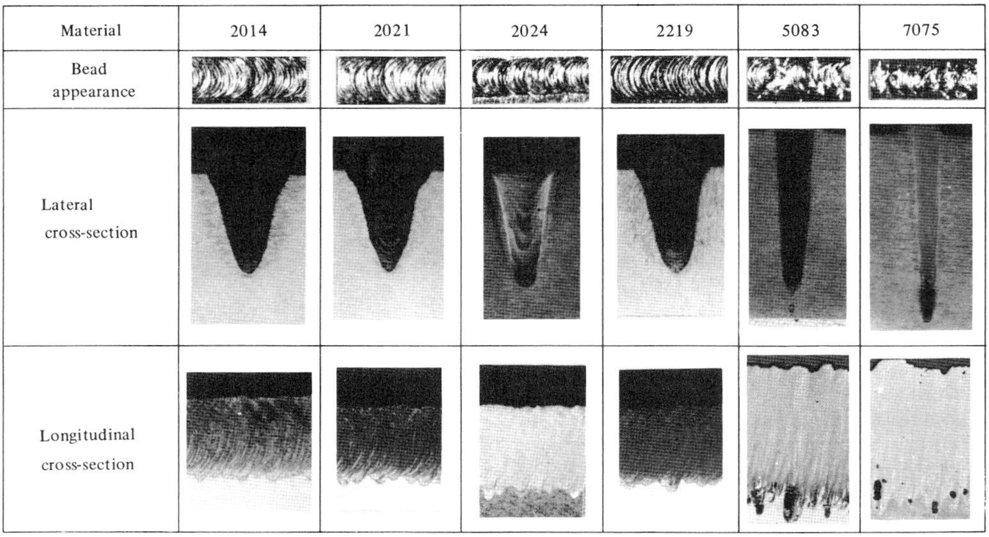

Material	2014	2021	2024	2219	5083	7075
Bead appearance						
Lateral cross-section						
Longitudinal cross-section						

Photo. 1 Penetration geometries of aluminum alloys

442

A vapor pressure vs. temperature curves for individual alloying elements are shown in Fig. 2. In Fig. 2,

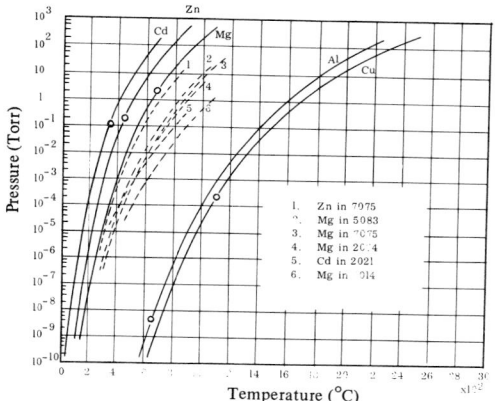

Fig. 2 Vapor pressure vs. temperature diagram

solid lines are for pure metals, while dotted lines are for
alloy corrected by Raoult's law according to the
following equation;

$$P_{AS} = \gamma_A N_A P_A \qquad (1)$$

where

P_{AS} = the vapor pressure of alloying element
γ_A = activity coefficient assuming 1 ($\gamma_A=1$)
N_A = mole fraction of alloying element
P_A = vapor pressure of alloying element
    as pure metal

$$N_A = \frac{Wt\%_A/M_A}{Wt\%_A/M_A + Wt\%_B/M_B}$$

    A: alloying element
    B: base element
    M_A: the atomic weight of alloying
         element
    M_B: the atomic weight of base element

In Fig. 2, it is found that the vapor pressures of Zn
and Mg are approximately 10^2 and 10 Torr respectively
at 1000°C, which are extraordinally higher than
4×10^{-5} Torr of Cu, the principal alloying element of
2xxx grade aluminum alloys.

The effect of this vaporization phenomenon is proved
by spectrum analysis on the longitudinal cross-section of
weld fusion zones of 7075. The results are summerized in
Table 3, in which vaporization rate is greater for Zn
than for Mg, and increases as welding speed is reduced,
Zinc content becoming below the specification requirement at a welding speed of 1250mm/min.

Table 3 Chemical composition of weld fusion zone (%)

Welding speed (mm/min)		Si	Fe	Cu	Mn	Mg	Cr	Zn	Ti	Al
Specification		<0.50	<0.70	1.2-2.0	<0.30	2.1-2.9	0.18-0.40	5.1-6.1	<0.20	Bal
Base metal		0.10	0.21	1.50	0.03	2.50	0.20	5.38	0.01	
Weld fusion zone	125	0.10	0.21	1.48	0.03	2.14	0.20	3.70	0.01	
	500	0.10	0.21	1.48	0.03	2.35	0.20	4.50	0.01	
	1250	0.10	0.21	1.48	0.03	2.45	0.20	4.28	0.01	
	1750	0.10	0.21	1.48	0.03	2.50	0.20	5.10	0.01	

Observation of weld bead formation during welding
and the resulted bead appearance also implies that
alloying elements vaporize at considerable rate in 7075.
Distinct differences in the colors of electron beam
(spectra) also exist.

3. Mechanical Properties of EB Welded 7075-T6

3.1 Preliminary test for set up of welding condition.

Plates of 7075-T6 57mm thick, were EB full
penetration welded at 1 pass, by the welding condition
shown in Table 4-1, at several welding speed. Internal
and external quality were tested and shown briefly in
Table 5. The tensile test specimens across the weld seam

Table 4-1 Welding conditions for 57 mm thick 7075-T6 plates

Accelerating Voltage	Welding Speed	Beam Current*
55 kv	127 mm/min (5 ipm)	240 mA
''	254 '' (10 '')	290 mA
''	635 '' (25 '')	360 mA
''	1,016 '' (40 '')	470 mA
''	1,524 mm'' (60 '')	600 mA
''	2,032 '' (80 '')	600 mA**
''	2,540 '' (100 '')	600 mA**

Welder   : 60 kv, 500 mA
Work distance : 127 mm (5")
Focus   : 25 mm (1") above the specimen surface at
                 small current
a_b : 1.2 at small current; 1.0 at 250 mA current
Atmosphere   : 1×10^{-4} torr

* Beam current was set at 20% over the current value with which
   the 57 mm thick plate can be penetrated thru by one pass.
** No higher current due to the welder capacity limitation.

Table 5 Weld quality at various welding speed

Welding Speed (ipm)	127mm/ min(5)	254mm/ min(10)	635mm/ r.in(25)	1.016 mm/ min(40)	1,524mm/ min(60)	2,032mm/ min(80)	2,540mm/ min(100)
Appearance & consistency of weld bead	◐	◐	○	○	○	○	◐
Crack in deposit metal	○	○	○	○	◐	●	●
Crack in HAZ	○	○	○	◐	◐	●	●
Cold shut	●	●	○	◐	◐	◐	●

○   ●
good   bad

443

(Fig. 3) were machined from this weld plates and subjected to test. The test results were as shown in Fig. 4. According to these experimental results, weld-

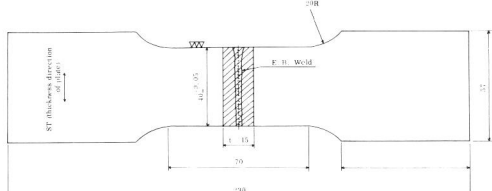

Fig. 3-1　Tensile test specimen

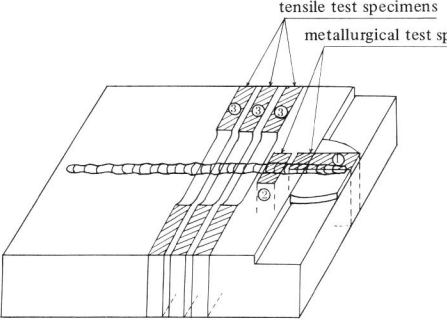

tensile test specimens
metallurgical test specimens

Fig. 3-2　Cutting plan for 57mm thick plate

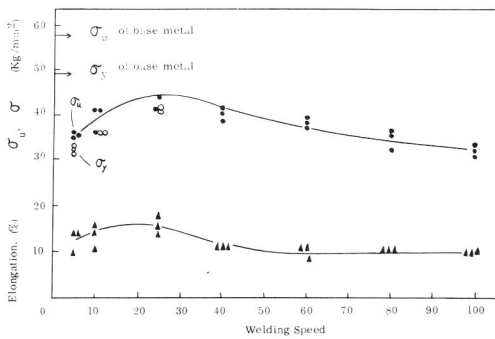

Welding Speed

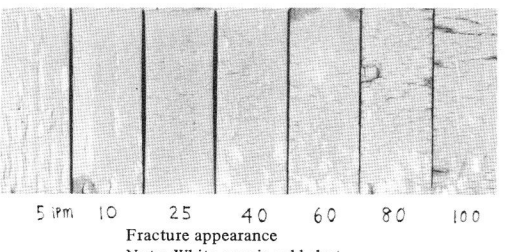

5 ipm　10　25　40　60　80　100
Fracture appearance
Note: White area is cold shut
Fig. 4　Tensile test results of various welding speed on 57mm thick plate

ing speed of 25 ipm seems to give the best quality and strength in the range of welding condition tested. As the speed becomes higher, there appeared the intergranular cracking of heat affected zone along the flow of the plate and the deposit metal cracking (Photo. 3). At the

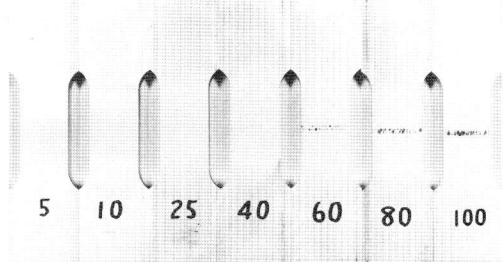

Penetrant inspection on intergranular fusion

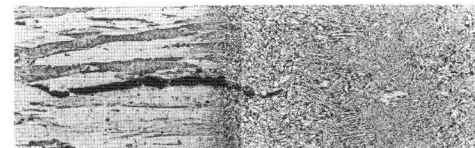

Micro photo of above intergranular fusion　×100

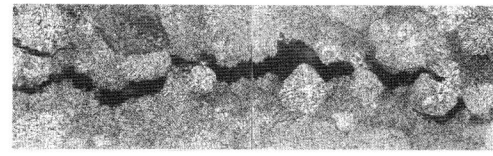

Crack in deposit metal　×100

Photo. 3　Defect in weld

slower speed region, the formation of weld bead was not uniform any more (in this welding position) and the resultant cold shut prevails very much. Cold shut trouble also increased as the welding speed became higher.

3.2　Static tensile test

Another tensile test was accomplished on 6.0 mm thick 7075-T651 plates. The welding speed of 25 ipm was chosen by the experimental results above-mentioned, and the whole welding condition was as shown in Table 4-2.

The strength under tension both along and across the weld seam direction was taken this time. For tensile test along the direction of weld seam, 4 different kind of specimen with the width of 2 mm (almost all deposit metal specimen), 15 mm (deposit metal and heat

444

Table 4-2 Welding condition for 6 mm thick and 30 mm thick plate

Plate thickness	6.0 mm	30 mm
Accelerating voltage	40 kV	50 kV
Welding speed	635 mm/min(25 ipm)	635 mm/min(25 ipm)
Beam current	55 mA	225 mA
Focus	1.3	1.2 at small current 1.0 at 250 ~ 50 mA range

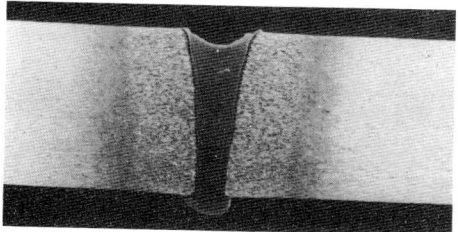

cross section of 6mm thick plate

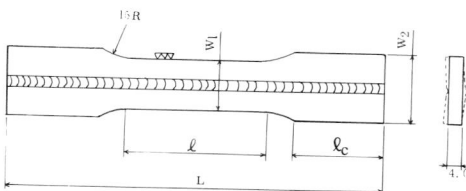

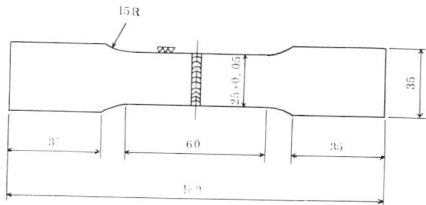

Dimension (mm)

T.P. No.	W_1	ℓ	W_2	ℓ_c	L
1	2 ± 0.05	60	5	60	210
2	15 ± 0.05	60	25	60	210
3	40 ± 0.1	80	55	80	270
4	80 ± 0.2	160	100	80	350

Fig. 5 Tensile test specimen of 6mm thick   7075-T651

affected zone contained), 40 mm and 80 mm was pulled. The cutting plan details and the configuration of the specimen are shown in Fig. 5.

These test results (Fig. 6) across the seam show that the ultimate joint strength as high as 45 kg/mm²

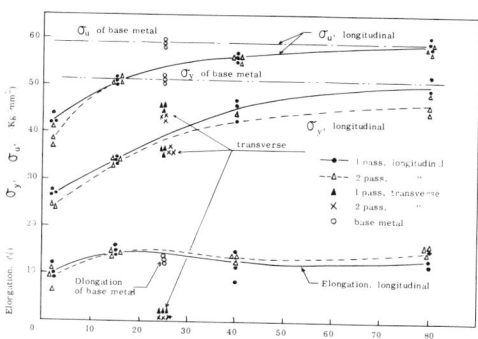

Fig. 6 Tensile test results

can be get? But the elongation was very small (2% at gage length 25 mm) and the fracture mode does not appear to be ductile enough (the same as Fig. 4., 25 ipm specimen, the fracture plane was flat and perpendicular to the tensile direction and the plate surface). However, the elongation is restricted to the narrow soft zone of weld deposit and HAZ. So this lack of elongation does not mean that the weld joint is very brittle and useless, we will discuss this matter from the fracture toughness standpoint afterwards.

The test results along the weld seam reaches very near to the base metal strength at the specimen width as short as 40 mm (only 5 kg/mm² less than base metal) due to the narrow weld characteristic of EB welding.

In this test, no defect including cold shut was found in the specimen and fracture face. But the strength of joint was about the data of 25 ipm specimen in Fig. 4. The effect of cold shut defect on the joint static tensile strength seems to be negligible if the extent is less than those found in the 25 ipm specimen here.

3.3 Fracture toughness test

Plates of 7075-T6, 30 mm thick, were butt welded according to the welding condition shown in Table 4-2.

Non destructive and metallurgical inspections of welded zone did not indicate weld defects at all. CKS type fracture toughness test specimens (Fig. 7) of 25 mm thickness were machined from welded plates above mentioned. Tests were made both in the as-welded and postweld T6 treated conditions. Notch was located at the center of weld fusion zone or weld bond. Test pieces were precracked on a bend testing machine rated 1690 cps, and then tensile tested using a universal testing machine in LT direction.

Test results are listed in Table 6. Fracture mode is shown in Photo. 4.

Table 6 Fracture toughness test result

Specimen No.	Condition	Location of notch	Location of crack point	Fractured at	Kc $Kg \sqrt{mm}/mm^2$
WD-1	as welded	bead center	bead center	center center/bond	139
-2	"	"	"	" "	127
-3	"	"	"	bead center	127
-4	"	"	bond	bond	110
-5	"	"	bead center	bead center	135
WB-1	"	bond	HAZ	base metal	108
-2	"	"	"	"	122
-3	"	"	"	"	126
-4	"	"	"	"	123
HD-1	T6 after weld	bead center	bead center	bond	130
-2	"	"	"	bead center	102
HB-1	"	bond	bond	HAZ	112
-2	"	"	"	"	103
-3	"	"	"	"	105
Base metal-1	T6	center			84
-2	"	"			84
-3	"	"			83

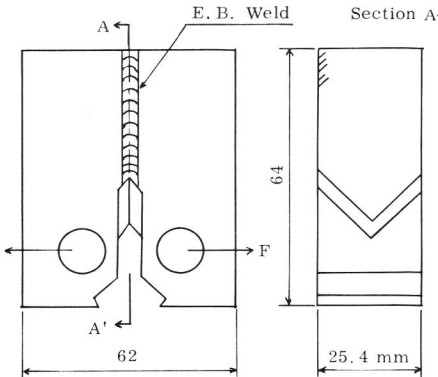

Section A–A'

E. B. Weld

64

62

25.4 mm

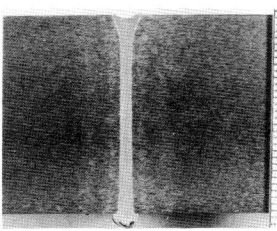

Fig. 7 Fracture toughness test specimen

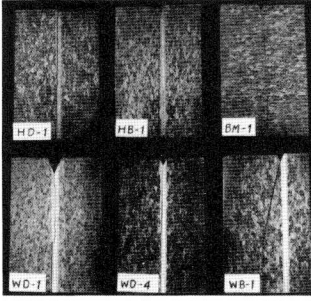

Photo. 4 Fracture mode (fracture toughness test)

Test results are as follows;

1) The fracture toughness of welded zone exceeds that of base metal ($84kg\sqrt{mm}/mm^2$), independently of the conditions after welding and the location of notch. The strength in the as-welded condition, in particular, reaches 124 kg $\sqrt{mm}/mm^2$, which is 1.5 times as high as that of base metal. Individual toughness values are considerably uniform.

2) The fracture toughness value of postweld heat treated specimen also exceed 100 kg $\sqrt{mm}/mm^2$. In this case hardness distribution is uniform and there exist no soft zone of weld.

3) Some of the fracture surfaces shift from weld fusion zone to weld bond. Imcomplete fusion is suspected at the shift point, but no decrease in fracture toughness is accompanied.

446

4) The fracture of test pieces notched at weld bond
wholly propagates from heat affected zone to base
metal.
Test pieces recovered in hardness through postweld
T6 treatment has shown the same inclination,
proving the superior toughness of weld fusion zone.
5) All of the fractured surfaces exhibit normal plain-
strain condition.

4. Summery

1) Partial penetration phenomenon of EB welded
aluminum alloys is considerably affected by vaporizing alloying elements. In the welding of 7075 and
5083 which contain high-vaporization-pressure
elements of Zn and/or Mg, the bead surface is
irregular and rugged, and narrow penetration about
twice as deep as 2xxx aluminum alloys is obtained
with high inclination of occurring the root defects
such as cold shuts or spikes.

2) Tensile strength of EB welded 7075 T6 reaches as
high as yield strength (0.2% offset) of base metal if
proper welding condition is employed, but the
fracture appears to be brittle with very small
elongation and bend test develops the same result.
Fracture toughness of welded zone, however,
exceeds that of base metal, proving superior weld
toughness.

Reference

1) M. Ohsumi, et al,: MITSUBISHI JUKO GIHO,
(1970-5)
2) A. H. Meleka: Electron Beam Welding: Published by Mcgraw-
Hill (1971)
3) Y. Arata et al.: Study on Characteristics of Weld Defect and
its Prevention in Electron Beam Welding, (Report I, II, III),
Trans. of JWRI Vol 2, No.1 (1973); Vol 3, No.1, No.2 (1974)

Insert-type Electron Beam Welding Technology
—Characteristics of Insert-type Welding—

Abstract

It is well known that electron beam welding process has many advantages over conventional welding processes and that it can be effectively applied to weld various kinds of metals. In electron beam welding, however, some complications appear, such as: mechanically and metallurgically poor properties of weld metal, poor weld geometry and difficulty of addition of filler material.

This paper deals with need to solve abovementioned problems by introducing insert metal at 6 mm-thick and 12 mm-thick joints, carbon steel to carbon steel and 2.25 Cr-1 Mo steel to austenitic stainless steel.

obtained results may be summarized as follows:

(i) Proper joint configuration and uranami welding conditions are found which produce excellent weld geometry and X-ray properties. Desirable "Active Beam Parameter" a_b, is 1.05 to 1.20 (where $a_b \equiv D_0/D_F$, D_0: object distance, D_F: focal length). Further, by adoption of joint configuration with both insert metal and root space stable uranami bead can be easily formed and range of proper uranami welding conditions widens.

(ii) Metallurgical and mechanical properties at room and elevated temperatures can be remarkably improved by introduction of insert metal.

1. Introduction

Electron beam welding process has certain important characteristics as follows: easy to control penetration depth precisely, possible to weld precisely because of resultant low distortion and narrow welds, easy to weld reactive metals and possible to weld at high speed. And further, it can be effectively applied to weld various kinds of metals.

In electron beam welding, however, following complicated problems appear: mechanically and metallurgically poor properties of weld metal, poor weld geometry and difficulty of addition of filler material. Such problems seem to be improved by introducing insert metal designed specially. Some papers[1-4] discuss need of insert metal in electron beam welding. However, each of them is either fragmentary or suggestive.

Herein authors discuss effect of insert metal on properties of electron beam welds with regard to 6 mm-thick and 12 mm-thick joints, carbon steel (SM41A) to carbon steel and 2.25 Cr-1 Mo steel (ASTMF22) to austenitic stainless steel (AISI316).

2. Experimental instrument

Welding experiment was performed on 150 kV-40 mA type electron beam welder shown in Photo. 1.

3. Basic study to find proper welding conditions

Characteristics of about 6 mm-deep and 12 mm-

Photo. 1. Electron beam welder used

deep welds were investigated to find proper welding conditions for 6 mm-thick and 12 mm-thick joints.

3.1. Proper welding conditions at maximum penetration part

3.1.1. Welding procedures

23 mm-thick 2.25 Cr-1 Mo steel and austenitic stainless steel plates were used. These plates were set on work table as shown in Fig. 1, where slope angle, θ_s, equals 30°. Thereafter work table was positioned so that object-distance-to-focal-length ratio, that is, '"Active beam parameter", $a_b \equiv D_0 D_F$ (where, D_0: object Distance, D_F: focal length) might be around 0.90*** at center line C-C of steel plate, where max. penetration depth is obtained as indicat-

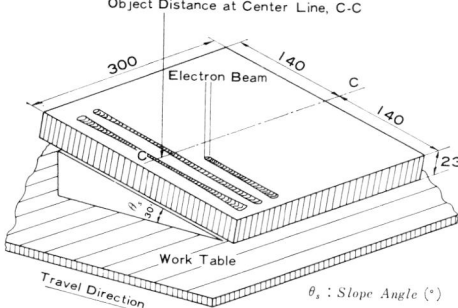

Fig. 1. Bead-on-plate welding method

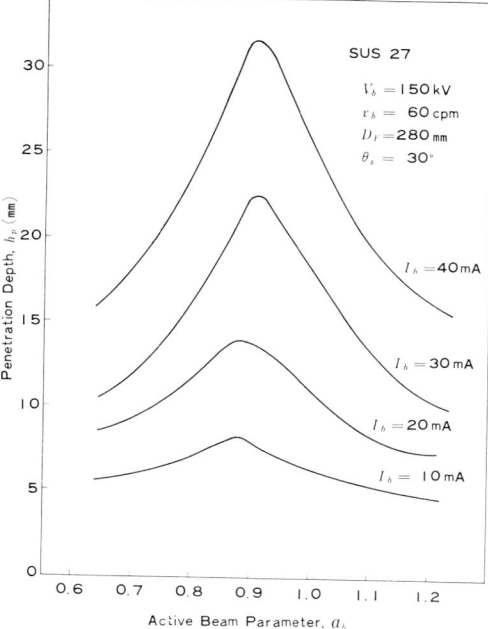

Fig. 2. Relationship between penetration depth and active
beam parameter

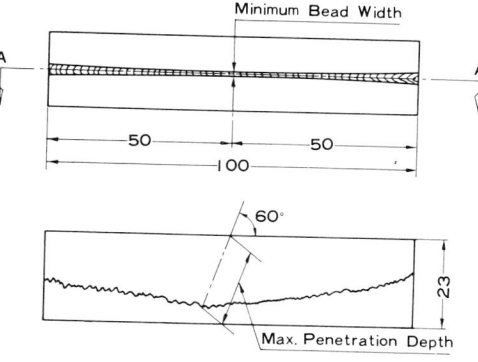

A-A Cross Section

Fig. 3. Survey method of bead appearance and longitudinal
profile section at max. penetration part

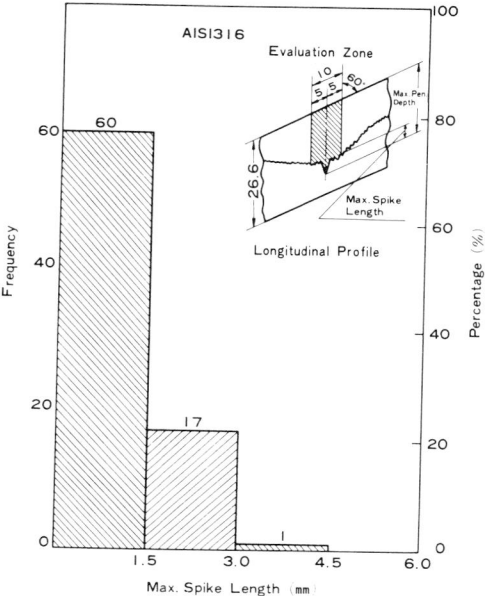

Fig. 4. Relationship between max. spike length and its frequency

ed in Fig. 2.   Bead-on-plate welding was carried out
at various heat inputs with electron gun fixed, moving
work table in such a direction as arrow shows.   Accelerating voltage, beam current and welding speed
were decided in order that max. penetration depth
might be either 6 mm or 12 mm at each heat input.

Test specimen was cut out from every bead-on-plate
weld as shown in Fig. 3.   Thereafter penetration depth,
defects associated with narrow and deep welds and
bead appearance were respectively surveyed.

3.1.2. Characteristics of welds at maximum penetration part
(1) Spiking

It is recognized that there appear three types of
spikes, that is, needle spike, sharp spike and obtuse
spike.   Sharp or obtuse spikes occurred at every longitudinal profile section as shown in Photo. 2.

Spikes are very sharp and longest (needle and sharp
spikes) in particular at max. penetration part as it is
well known[5)6].   It can be considered that irrespective
of materials (2.25 Cr-1 Mo steel and austenitic stainless
steel), frequency of spikes less than 1.5 mm in length
forms about 80% of all specimens and that exceeding
1.5 mm in length decreases drastically.   Fig. 4 shows
relation between max. spike length and its frequency
in case of austenitic stainless steel welds.   Based on

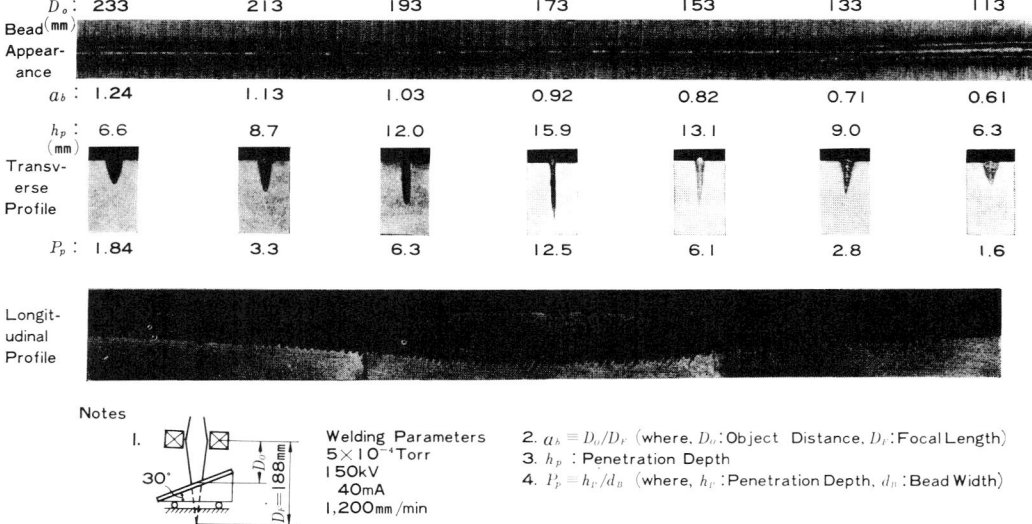

D_o :	233	213	193	173	153	133	113
a_b :	1.24	1.13	1.03	0.92	0.82	0.71	0.61
h_p : (mm)	6.6	8.7	12.0	15.9	13.1	9.0	6.3
P_p :	1.84	3.3	6.3	12.5	6.1	2.8	1.6

Bead (mm) Appearance

Transverse Profile

Longitudinal Profile

Notes

1.

Welding Parameters
5×10^{-4} Torr
150kV
40mA
1,200 mm/min
30° $D_f = 188$ mm

2. $a_b \equiv D_o/D_F$ (where, D_o : Object Distance, D_F : Focal Length)
3. h_p : Penetration Depth
4. $P_p = h_P/d_B$ (where, h_P : Penetration Depth, d_B : Bead Width)

Photo. 2. Bead appearance, transverse and longitudinal profile of bead-on-plate welds

Table 1. Criteria for evaluation of spikes, porosities and cold shuts

Evaluation Zone	Item	Criteria
Max. Penetration Depth / Evaluation Zone 60° / 26.6 / Max. Spike Length / Spike / Porosity / Longitudinal Profile	Spikes	○ : Max. Spike Length $\leqq$ 1.5mm ✕ : Max. Spike Length 1.5mm
	Porosities and Cold Shuts	○ : Sum 4 : Sum 5to8 ✕ : Sum 9

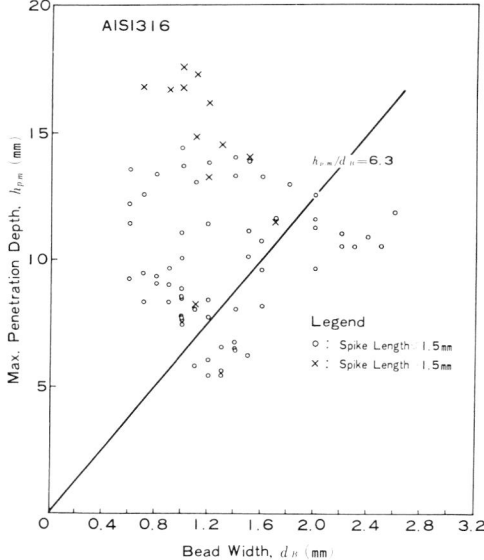

Fig. 5. Effect of max. penetration depth and bead width on spiking

AISI316

Max. Penetration Depth, h_{pm} (mm)

$h_{pm}/d_B = 6.3$

Legend
○ : Spike Length 1.5mm
✕ : Spike Length 1.5mm

Bead Width, d_B (mm)

such a result, criteria for evaluation of spikes was established in Table 1. It can be fully noticed that irrespective of materials (2.25 Cr-1 Mo steel and austenitic stainless steel), spikes larger than 1.5 mm in length are apt to occur when penetration-depth-to-bead-width ratio, that is, "Penetroparameter", $P_P \equiv h_P/d_B$ (where, h_P: penetration depth and d_B: bead width), exceeds 6.3 as shown in Fig. 5.

(2) Porosities and cold shuts
Porosities and cold shuts were found at longitudinal profole sections. These defects were also evaluated in accordance with criteria shown in Table 1. In this

basic study these defects occurred without distinction of max. penetration depth and heat input because P_P value becomes very large at max. penetration part irrespective of heat input and resultant weld geometry is apt to induce them. And further, they occurred more often in 2.25 Cr-1 Mo steel than austenitic stainless steel welds.

(3) Weld bead appearance

Humping, which consists of a regular series of swellings in weld bead, occurred at max. penetration part. Fig. 6 shows effect of max. penetration depth and bead width on it ocncerning austenitic stainless steel welds. From this figure it is certainly apt to occur at deeper and narrower welds, that is, at higher energy density and welding speed. Although it did far less occur in 2.25 Cr-1 Mo steel than austenitic stainless steel welds, this phenomenon was sometimes noticed at narrower welds.

(4) Transverse profile

Photo. 3 shows transverse profile sections at max. penetration part. Every section has extremely sharp edge at bottom as shown in this photo.

3.1.3. Proper welding conditions selected

In consideration of resultant characteristics of welds

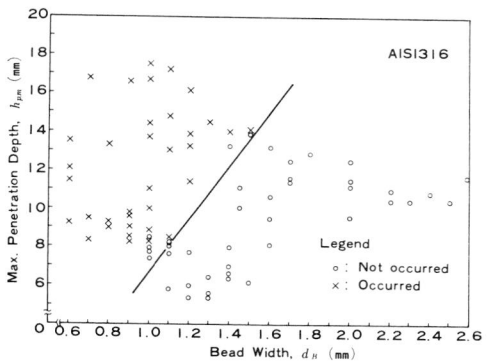

Fig. 6. Relationship between max. penetration depth, bead width and humping

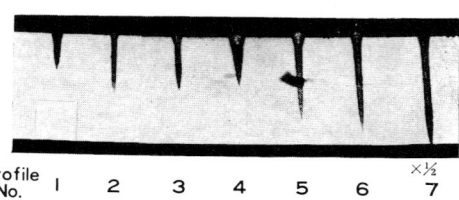

| Profile No. | 1 | 2 | 3 | 4 | 5 | 6 | 7 | ×½ |

Vacuum : 5×10^{-4} Torr

Profile No.	Accelerating Voltage (kV)	Beam Current (mA)	Welding Speed (mm/min)
1	100	20	500
2	150	40	14,00
3	150	20	600
4	100	20	250
5	150	40	750
6	150	40	650
7	150	40	500

Photo. 3. Transverse profile sections at $a_b \fallingdotseq 0.90$

at max. penetration part proper welding conditions, which produce either 6 mm-deep or 12 mm-deep penetration concerning 2.25 Cr-1 Mo steel and austenitic stainless steel, were found and are tabulated in Table 2.

3.2. Proper welding conditions at other parts except for maximum penetration part

3.2.1. Welding procedures

Using 23 mm-thick 2.25 Cr-1 Mo steel plates, bead-on-plate welding was carried out at various heat inputs

Table 2. Proper welding conditions selected

Material	Penetration Depth (mm)	Vacuum (Torr)	Accelerating Voltage (kV)	Beam Current (mA)	Welding Speed (mm/min)	Object Distance, D_o (mm)	Focal Length, D_f (mm)	a_b $(=D_o/D_f)$
ASTM F22	6	5×10^{-4}	100	20	1,200	240	268	0.90
				30	1,600	238	268	0.89
			150	20	1,800	258	298	0.86
				30	3,000	173	188	0.92
				40	4,000	273	338	0.81
	12			20	600	208	238	0.87
				40	2,000	233'	268	0.87
AISI 316	6	5×10^{-4}	100	20	1,000	288	338	0.85
					1,100	243	268	0.91
						288	338	0.85
					1,200	243	268	0.91
				30	1,600	238	268	0.89
						293	338	0.87
	12		150	20	600	223	268	0.83

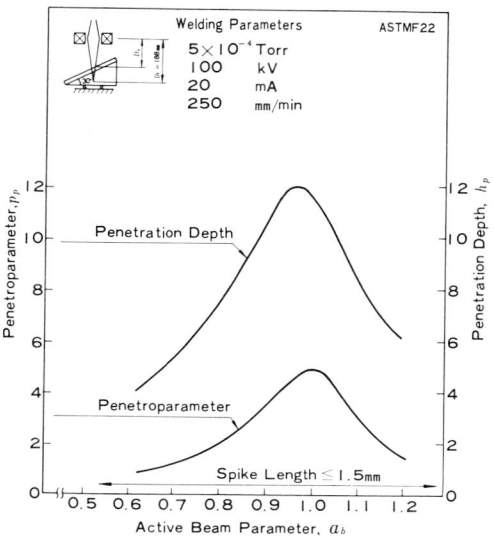

Welding Parameters   ASTMF22
5×10^{-4} Torr
100 kV
20 mA
250 mm/min

Fig. 7. Effect of penetroparameter and active beam parameter on spiking

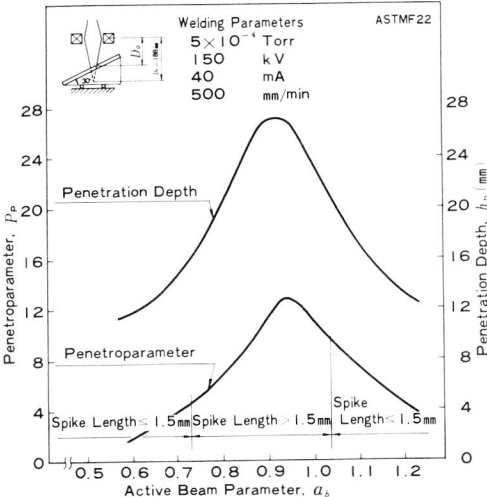

Welding Parameters ASTMF22
5×10⁻⁴ Torr
150 kV
40 mA
500 mm/min

Fig. 8. Effe Effect of penetroparameter and active beam parameter on spiking

Table 3. Proper welding conditions selected

Penetration Depth (mm)	Welding Conditions						
	Vacuum (Torr)	Accelerating Voltage (kV)	Beam Current (mA)	Welding Speed (mm/min)	Object Distance D_o (mm)	Focal Length D_f (mm)	a_b $(=D_o/D_f)$
6	5×10⁻⁴	100	20	250	223	188	1.20
	″	100	20	500	248	268	0.92
	″	150	40	1,200	223	188	1.20
	″	150	40	1,400	223	188	1.20
12	″	150	20	600	183	188	0.97
	″	150	40	500	223	188	1.20
	″	150	40	650	223	201	1.10
	″	150	40	750	223	213	1.05

by same welding procedures as shown in item 3.1.1. to find proper welding conditions, producing 6 mm-deep and 12 mm-deep welds at other parts except for max. penetration part. Thereafter, characteristics of welds were investigated by same method as shown in item 3.1.1.

3.2.2. Characteristics of welds at arbitrary part

(1) Spiking

Figs. 7 and 8 show typical examples of effect of P_P value and ab value on spike length. It was noticed from Fig. 5 that spikes, larger than 1.5 mm in length at max. penetration part, were apt to occur when P_P value exceeds 6.3. In Fig. 7 P_P value at max. penetration part is 4.9. Therefore spikes are less than 1.5 mm in length at arbitrary part including max. penetration part. In Fig. 8, however, P_P value at max. penetration part is 12.8 and exceeds 6.3. Therefore spikes, larger than 1.5 mm in length, occur continuously herein. Critical P_P value at which spikes exceeding 1.5 mm in length are apt to occur is 9.4 for a_b value larger than 0.90 and 4.6 for a_b value smaller than 0.90 respectively. From these results it may be summarized that critical P_P value as defined above is larger than 6.3 for a_b value larger than 0.90, 6.3 for a_b value nearly equal to 0.90 and smaller than 6.3 for a_b value smaller

than 0.90. In other words, as a_b value decreases, above-defined critical P_P value does either.

(2) Weld bead appearance

At every heat input, penetration depth at certain a_b value smaller than 0.90 becomes equal to that at another a_b value larger than 0.90, which can be noticed in Fig. 8. However, transverse weld profiles at these different a_b values quite differ from each other. As shown in transverse weld profiles in Photo. 2, weld geometry at smaller a_b value is like wedge-type bead as shown in Fig. 9, while that at larger a_b value is like well-type bead in this figure. This remarkable difference in weld geometry due to a_b value seems to have resulted mainly from electron beam geometry through welding joints.

3.2.3. Proper welding conditions selected

At a_b value smaller than 0.90 two problems resulted. One is that weld geometry is apt to become wedge-type bead and the other is that critical P_p value is smaller than that at a_b value larger than 0.90. Therefore it is undesirable to select welding conditions at a_b value smaller than 0.90 as proper welding conditions. Table 3 shows proper welding conditions selected at which a_b value exceeds 0.90.

4. Investigation on proper uranami welding conditions by introduction of insert metal

In this experiment, 6 mm-thick and 12 mm-thick carbon steel (SM41A), 2.25 Cr-1 Mo steel (ASTMF22) and austenitic stainless steel (AISI316) plates as shown in Table 4 were used. Table 5 shows chemical com

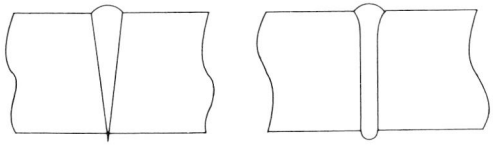

(a) Wedge-type Bead   (b) Well-type Bead

Fig. 9. Typical electron beam weld profile

452

Table 4. Chemical composition and mechanical properties of steel plates used

Base Metal	Thickness (mm)	Mechanical Properties			Chemical Composition (%)									
		T.S. (kg/mm²)	Y.P. (kg/mm²)	EL. (%)	C	Si	Mn	P	S	Al	Ni	Cr	Mo	Fe
Austenitic Stainless Steel (AISI 316)	12	55.3	—	63.0	0.06	0.75	1.60	0.034	0.006	—	11.2	17.8	2.36	Bal.
	6	56.4	—	62.5	0.05	0.62	1.66	0.032	0.007	—	11.3	17.1	2.50	Bal.
2.25Cr-1Mo Steel (ASTMF22)	12	45.1	30.1	45.1	0.06	0.25	0.47	0.020	0.028	—	—	2.04	1.01	Bal.
	6	46.8	31.2	42.3	0.13	0.21	0.43	0.015	0.011	—	—	1.95	0.94	Bal.
Carbon Steel (SM41A) A	12	45.2	28.2	26.3	0.17	0.01	1.11	0.009	0.027	Trace	—	—	—	Bal.
B	"	44.9	29.5	27.6	0.17	0.02	0.84	0.022	0.018	Trace	—	—	—	Bal.
C	6	44.8	27.9	26.9	0.21	0.01	0.83	0.019	0.019	Trace	—	—	—	Bal.
D	"	48.2	29.3	25.4	0.19	Less than 0.01	0.75	0.009	0.029	Trace	—	—	—	Bal.

Table 5. Chemical composition and mechanical properties of insert metal used

Material	Width (mm)	Mechanical Properties			Chemical Composition (%)									
		T.S. (kg/mm²)	Y.P. (kg/mm²)	EL. (%)	C	Si	Mn	P	S	Al	Ni	Cr	Cu	Fe
Austenitic Stainless Steel (AISI 309)	0.4 0.8 1.2	60.0	—	20.5	0.07	0.64	1.63	0.037	0.004	—	14.05	22.8	—	Bal.
Inconel 600	0.4 0.8 1.2	110.1	—	56.1	0.04	0.20	0.29	0.008	0.003	—	Bal.	15.4	0.02	8.34
Al-killed Carbon Steel (SM41A)	0.4 0.8	44.5	27.1	26.9	0.09	0.23	1.24	0.012	0.007	0.01	—	—	—	Bal.

Table 6. Joint configuration adopted

Type	Joint Configuration
A : Joint Configuration without Insert Metal	
B : Joint Configuration with Insert Metal but not Root Space	H_{Iu}, H_I, H_{II}, d_I
C : Joint Configuration with both Insert Metal and Root Space	H_{Iu}, d_I, H_I, H_{II}, H_R Root Space

position and mechanical properties of insert metal-used.

Insert metal was selected from following viewpoints; Insert metal of Al-killed carbon steel (SM41A) is suitable to welding of semi-killed carbon steel which is not fully deoxidized. Because spattering in welding of semi-killed carbon steel will disappear by introduction of this kind of insert metal including more deoxidizer. While, insert metal used for dissimilar metal joints, 2.25 Cr-1 Mo steel (ASTMF22) to austenitic stainless steel (AISI316), must meet such requirements that weld metal doesn't form any hardened structure by dilution, its thermal-expansion coefficient lies between two dissimilar metals, it can significantly influence prevention of σ phase formation and carbon migration resulting from long-time heating at high temperature and it possesses high corrosion resistance, acid resistance and further high strength and ductility at low and high temperatures. Therefore Inconel 600 insert metal was used for dissimilar metal joints, 2.25 Cr-1 Mo steel (ASTMF22) to austenitic stainless steel (AISI316). And further, insert metal of austenitic stainless steel (AISI309) was also used for dissimilar metal joints though it leaves some problems in thermal-expansion coefficient and carbon migration. Because it is generally considered that it is effective because of high weld cracking resistance in case service condition is not thermally strict.

Three types of joint configurations were adopted herein as shown in Table 6.

Hereby effect of joint configuration and a_b value on weld geometry, X-ray properties and so forth was investigated systematically, based on result obtained in item 3. Result obtained in this item is explained in details as follows.

4.1. Selection of proper uranami welding conditions

In general, narrowest and deepest welds can be produced in electron beam welding at a_b value nearly equal to 0.90 irrespective of beam parameter, welding speed and object distance as shown in Fig. 2.

In this case characteristics of electron beam welding appear intensively. Weld geometry, however, becomes wedge-type bead at this a_b value and following troubles resulted from sharp apex angle of molten metal represented in Fig. 9(a);

(i) Even small local variations in heat input may produce appreciable differences in penetration depth. As a result, incomplete penetration or undersirable uranami bead arises.

(ii) Lack of fusion is apt to result from beam-waver and "D-M deflection" which occurs at dissimilar metal joints.

Accordingly, it is desirable in uranami welding by

introduction of insert metal to select welding conditions producing weld geometry of well-type bead at which spiking less occurs, bead widthat top and that at bottom are larger than insert metal width and nearly equal to each other. In other words a_b value around 1.20 should be selected.

Further, sound welds could be produced with nearly same proper uranami welding conditions irrespective of material of joints.

4.2. *Selection of proper joint configuration*

In 6 mm-thick joints desirable uranami bead could be easily formed just by adjusting a_b value around 1.20 without distinction of joint configuration, while in 12 mm-thick joints it is next to impossible to obtain satisfactory weld geometry even at a_b value, 1.20 or so. However, by adoption of joint configuration with insert metal and root space*, satisfactory uranami bead becomes possible to be formed with considerable ease even at a_b value, 1.05.

Further, formation mechanism of uranami bead can be fully regarded as key-hole type** in 6 mm-thick joints but not in 12 mm-thick joints from uranami bead at crater and luminant condition of fluorescent paint under joints.

4.3. *Proper uranami welding conditions selected*

Table 7 shows proper uranami welding conditions selected for 6 mm-thick and 12 mm-thick joints.

Table 7.   Proper uranami welding conditions selected

Steel Plate		Welding Conditions				Joint Configuration
Material	Thickness (mm)	Vacuum (Torr) Accelearing Voltage (kV)	Beam Current (mA) Welding Speed (mm/min)	Object Distance D_s	Focal Length D_F	
ASTMF 22 + AISI316	6		40 1,400	223 $a_b=$1.20	188	ASTMF22 AISI316
	12	5×10⁻⁴ 150	40 650	223 $a_b=$1.10	201	
			40 750	223 $a_b=$1.05	213	
SM41A + SM41A	6		40 1,200	223 $a_b=$1.20	188	SM41A SM41A
	12		40 650	223 $a_b=$1.10	201	
			40 750	223 $a_b=$1.05	213	

5.   Effect of insert metal size and its position on weld geometry

In this experiment, 6 mm-thick and 12 mm-thick steel plates shown in Table 4 and insert metal shown in Table 5 were used. Effect of insert metal size and its position on weld geometry was investigated syste-

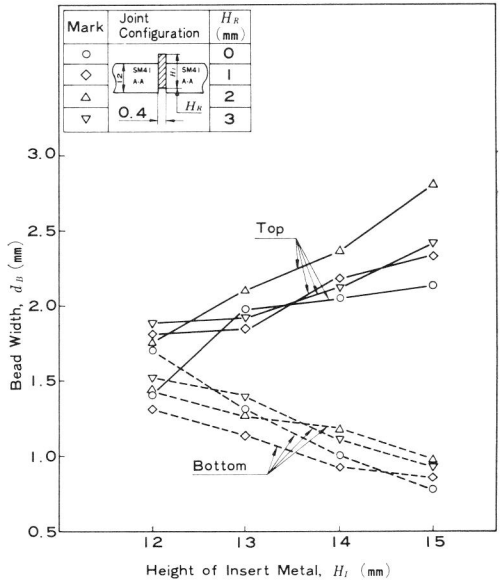

Fig. 10.   Effect of insert metal on bead width

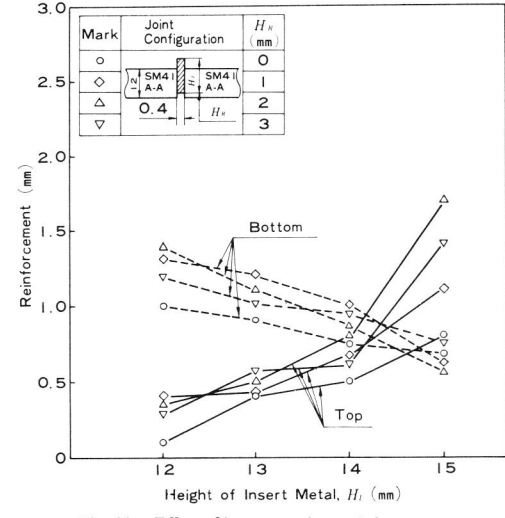

Fig. 11.   Effect of insert metal on reinforcement

---

* Root space is defined as opening space at root of joint as indicated in Table 6(C).

** Key-hole type uranami bead is defined as uranami bead formed under key-hole formation by electron beam.

Tz Table 8. Weld bead appearance and its profile

Base Metal		Joint Configuration	Bead Appearance ×1 (reduced 50% on reproduction)		Bead Profile
Material	Thickness (mm)				
ASTMF 22 + AISI316	6	Inconel 600 Insert Metal — ASTMF22 / AISI316 — 0.4	Top Bead / Bottom Bead	Top Surface Y3 / Bottom Surface Y3	
	12	Inconel 600 Insert Metal — ASTM F22 / AISI 316 — 0.4	Top Bead / Bottom Bead	F 52	

matically.

Fig. 10 shows a typical example of effect of insert metal size on bead width at top and bottom. As insert metal height, H_I, increases, bead width at top gradually increases irrespective of insert metal width, d_I, while that at bottom decreases. In other words, weld bead comes near to wedge-type bead with increase of H_I.

Fig. 11 shows a typical example of effect of insert metal size on reinforcement at top and bottom. Re-

inforcement also shows a similar tendency. Reinforcement increases gradually at top with increase of H_I and decreases at bottom.

In 6 mm-thick joints weld bead was not so much influenced by root space height, H_R. In 12 mm-thick joints, however, weld bead was significantly influenced by H_R. As a result very stable and uniform weld bead could be formed in case of 2 mm-high root space. Table 8 shows a typical example of weld bead appearance and profile in this case.

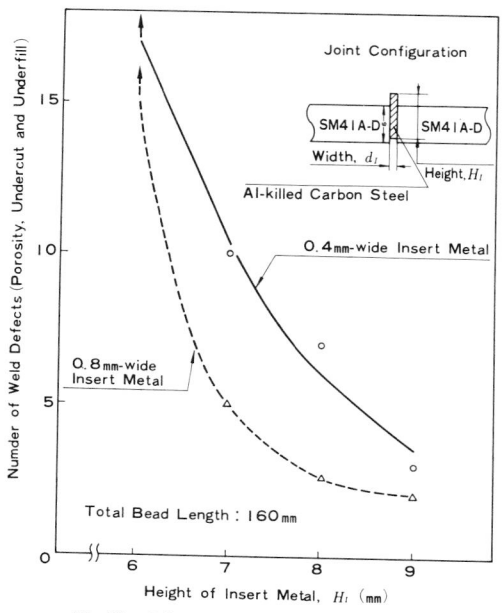

Fig. 12. Effect of insert metal on weld defects

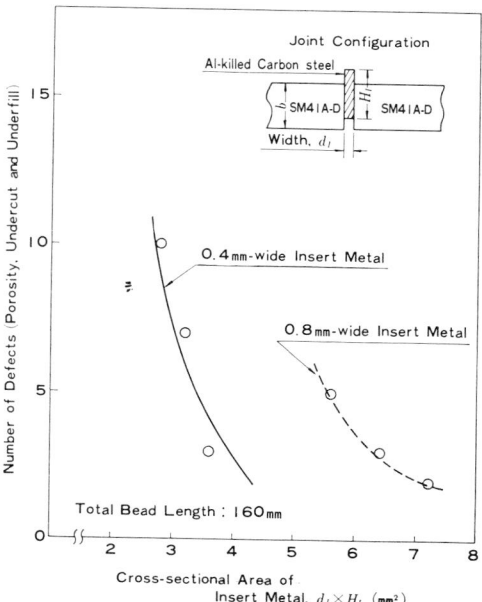

Fig. 13. Effect of insert metal on weld defects

455

Table 9. Proper joint configuration and uranami welding conditions selected

Base Metal Material	Thickness (mm)	Insert Metal Material	Width (mm)	Joint Configuration	Welding Conditions
ASTM F22 + AISI316	6	Nil			5×10^{-4} Torr 150 kV 40 mA 1,400 mm/min $D_O=223$ mm $D_F=188$ mm $a_b=1.20$
		AISI 309 Inconel 600	0.4	0.4 ... 0.4 ... 0.4 ... 0.4 (6-9)	
			0.8	0.8 ... 0.8 ... 0.8 ... 0.8 (6-9 / 6-7)	
			1.2	1.2 ... 1.2 ... 1.2 ... 1.2 (6-9 / 6-7)	
	12	AISI 309 Inconel 600	0.4	0.4 (12)	5×10^{-4} Torr 150 kV 40 mA 650 mm/min $D_O=223$ mm $D_F=201$ mm $a_b=1.10$
			0.8	0.8 (12)	
			1.2	1.2 (12)	
SM41A-C + SM41A-C	6	Al-killd Carbon Steel	0.4	0.4 ... 0.4 ... 0.4 (6-9)	
			0.8	0.8 ... 0.8 ... 0.8 (6-9 / 6-7)	5×10^{-4} Torr 150 kV 40 mA 1,200 mm/min $D_O=223$ mm $D_F=188$ mm $a_b=1.20$
SM41A-D + SM41A-D	6	Al-killed Carbon Steel	0.4	0.4 ... 0.4 ... 0.4 ... 0.4 (9)	
			0.8	0.8 ... 0.8 ... 0.8 ... 0.8 (7-9 / 8-9 / 8)	
SM41A-A + SM41A-A SM41A-B + SM41A-B	12	Al-killed Carbon Steel	0.4	0.4 (12)	5×10^{-4} Torr 150 kV 40 mA 650 mm/min $D_O=223$ mm $D_F=201$ mm $a_b=1.10$
			0.8	0.8 (12)	

By introduction of Al-killed carbon steel insert metal at 6 mm-thick and 12 mm-thick joints, carbon steel to carbon steel including less deoxidizer (i.e. SM41A-B and SM41A-D in Table 4), effect of this insert metal on weld defects was investigated. Every weld was radiographically inspected after welding. Figs. 12 and 13 show effect of insert metal size on the number of weld defects, that is, porosities, undercuts and under-

Table 10. Proper uranami welding conditions selected

Base Metal		Insert Metal	Joint Configuration	Vacuum (Torr)	Accelerating Voltage (kV)	Beam Current (mA)	Welding Speed (mm/min)	Object Distance D_0 (mm)	Focal Length D_F (mm)	a_b ($\equiv D_0/D_F$)
Material	Thickness (mm)									
ASTMF22 + AISI316	6	Nil	(ASTMF22 / AISI316)	5×10^{-4}	150	40	1,400	223	188	1.20
		AISI309	0.4 / 0.8 / 1.2							
		Inconel 600	0.4 / 0.8 / 1.2							
	12	Nil	(ASTMF22 / AISI316)	5×10^{-4}	150	40	650	223	201	1.10
		AISI309	0.4 / 0.8 / 1.2							
		Inconel 600	0.4 / 0.8 / 1.2							
SM41A-C + SM41A-C	6	Al-killed Carbon Steel	SM41A / SM41A 0.4 / 0.8	5×10^{-4}	150	40	1,200	223	188	1.20
SM41A-A + SM41A-A	12	Al-killed Carbon Steel	SM41A / SM41A 0.4 / 0.8	5×10^{-4}	150	40	650	223	201	1.10

fills. From these figures it can be considered that these defects decrease drastically with increase of H_I and cross-sectional area. Further, 0.4 mm-wide insert metal has more remarkable effect than 0.8 mm-wide one even if cross-sectional area equals. In other words, insert metal height influences weld defects

457

more significantly than insert metal width. Similar tendency was also shown in 12 mm-thick joints.

As a result of systematical experiments, proper joint configuration was decided as shown in Table 9.

6. Effectiveness of root-spaced joint configuration in uranami welding

It became clear after investigation on proper uranami welding conditions by introduction of insert metal that satisfactory weld bead can be easily formed by adoption of root-spaced joint configuration especially in 12 mm-thick joints.

Herein effect of root-spaced joint configuration on proper uranami welding conditions was investigated furthermore to examine its effectiveness in 6 mm-thick, 9 mm-thick and 12 mm-thick dissimilar metal joints respectively.

Fig. 14 shows effect of root space on range of proper uranami welding conditions. From this figure it has been clarified that range of proper uranami welding conditions widens by adoption of root-spaced joint

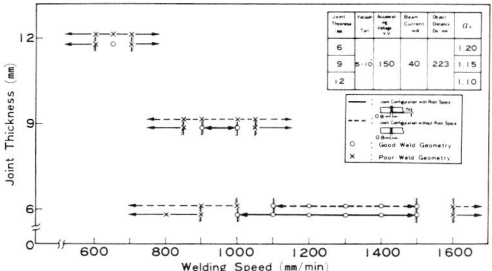

Fig. 14. Effect of root space on range of proper uranami welding conditions

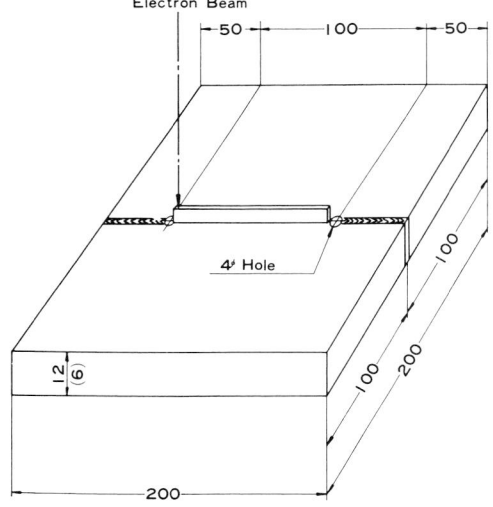

Fig. 15. Weld cracking test specimen

configuration and also root-spaced joint configuration becomes more effective in uranami welding with increase of joint thickness, as compared with joint configuration without root space.

7. Effect of insert metal on metallurgical and mechanical properties of welds

7.1. Material used and welding procedures

6 mm-thick and 12 mm-thick 2.25 Cr-1 Mo steel, austenitic stainless steel and carbon steel plates including more deoxidizer (i.e. SM41A-1 and SM41A-C) shown in Table 4 were used. Insert metal of austenitic stainless steel, Inconel 600 and Al-killed carbon steel shown in Table 5 were also used. Welding was carried out with proper uranami welding conditions shown in Table 10 which was selected from in Table 9.

7.2. Effect of insert metal on weld cracking

Lehigh restraint cracking test was carried out with regard to 6 mm-thick and 12 mm-thick dissimilar metal joints, 2.25 Cr-1 Mo steel to austenitic stainless steel. Test specimen was shown in Fig. 15.

Crater cracking occurred at every crater irrespective of insert metal width and its material.

However other cracking, that is, hot cracking and quench cracking[3)7)] did'nt occur. And further, transverse cracking and so forth[3)7)] were not recognized.

7.3. Effect of insert metal on metallurgical properties of welds

7.3.1. Microstructure

(1) Dissimilar metal joints, 2.25 Cr-1 Mo steel to austenitic stainless steel

When insert metal was not introduced, martensite was produced in weld metal zone irrespective of joint thickness. In heat-treated condition carburized and decarburized band, due to carbon migration during stress relief annealing, were observed at weld metal zone adjacent to 2.25 Cr-1 Mo steel base metal and at 2.25 Cr-1 Mo steel heat affected zone adjacent to weld metal zone respectively.

By introduction of austenitic stainless steel insert metal, martensite was produced in weld metal zone as shown by photo. 4(a) in case of 0.4 mm-wide insert metal, while austenite was produced in weld metal zone in case of 0.8 mm-wide and 1.2 mm-wide insert

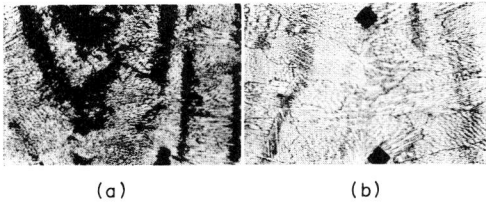

(a)   (b)

Photo. 4. Microstructure of weld metal of 12mm-thick dissimilar metal joint, 2.25 Cr-1 Mo steel to AISI 316
×100 (reduced 50% on reproduction)

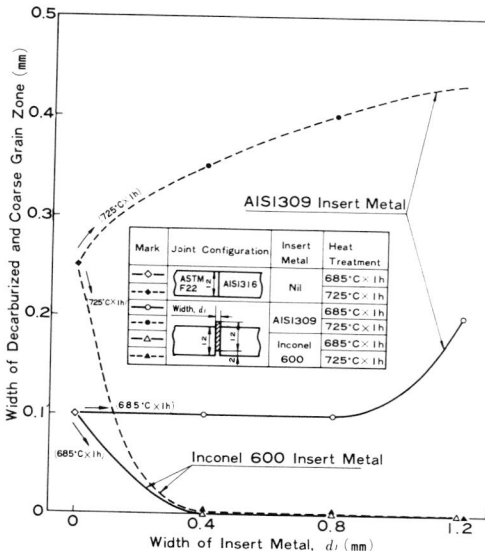

Fig. 16. Effect of insert metal on width of decarburized and coarse grain zone

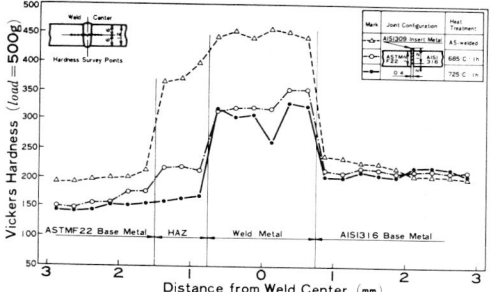

Fig. 17. Hardness distribution across electron beam weld

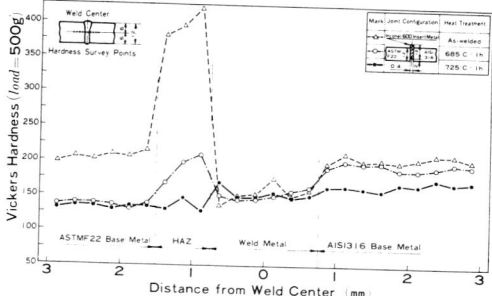

Fig. 18. Hardness distribution across electron beam weld

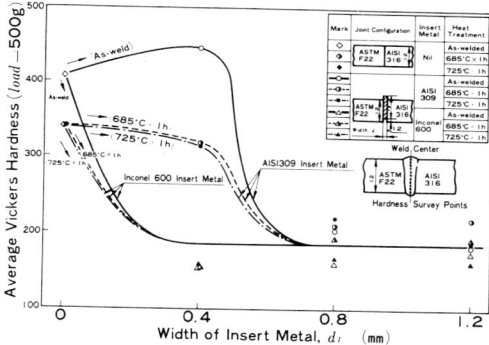

Fig. 19. Effect of insert metal on average hardness of weld metal

metal. In heat-trdeated condition carburized and decarburized band were observed at boundary layer adjacent to 2.25 Cr-1 Mo steel base metal either.

By introduction of Inconel 600 insert metal, austenite was porduced in weld metal zone as shown by Photo. 4(b) even in case of 0.4 mm-wide insert metal. In heat-treated condition very narrow carburized and decarburized band were observed at boundary layer adjacent to 2.25 Cr-1 Mo steel base metal in case of 0.4 mm-wide and 0.8 mm-wide insert metal.

Microstructure in weld metal zone almost agreed with structure of weld metal estimated from average value of XMA point scanning result, using Schaeffler's diagram.

Fig. 16 shows relation between width of decarburized and coarse grain zone and insert metal width, d_I in case of 12 mm-thick joints. When austenitic stainless steel insert metal is introduced, width of decarburized and coarse grain zone increases with increase of d_I irrespective of stress relief annealing temperature and joint thickness. This tendency is more conspicuous in case of stress relief annealing at 725°C for 1h than at 685°C for 1h. By introduction of Inconel 600 insert metal, however, width of decarburized and coarse grain zone decreases with increase of d_I. In case of stress relief annealing at 685°C for 1h carbon migration doesn't occur, while in case of stress relief annealing at 725°C for 1h this carbon migration hardly occur when insert metal is 0.8 mm-wide or 1.2 mm-wide. Quite a similar tendency was shown in 6 mm-thick joints.

(2) Carbon steel to carbon steel joints

In as-welded condition hardened structure was observed in weld metal zone. In heat-treated condition, however, ferrite and pearlite were observed.

7.3.2. Hardness distribution

(1) Dissimilar metal joints, 2.25 Cr-1 Mo steel to austenitic stainless steel

When insert metal was not introduced, weld metal zone was hardened up to about Hv 400 in as-welded condition and about Hv 320 even in heat-treated condition.

By introduction of austenitic stainless steel insert metal, weld metal zone was hardened up to about Hv 450 in as-welded condition in case of 0.4 mm-wide insert metal as shown in Fig. 17. In case of 0.8 mm-wide and 1.2 mm-wide insert metal, however, weld metal zone was not hardened even in as-welded condi-

459

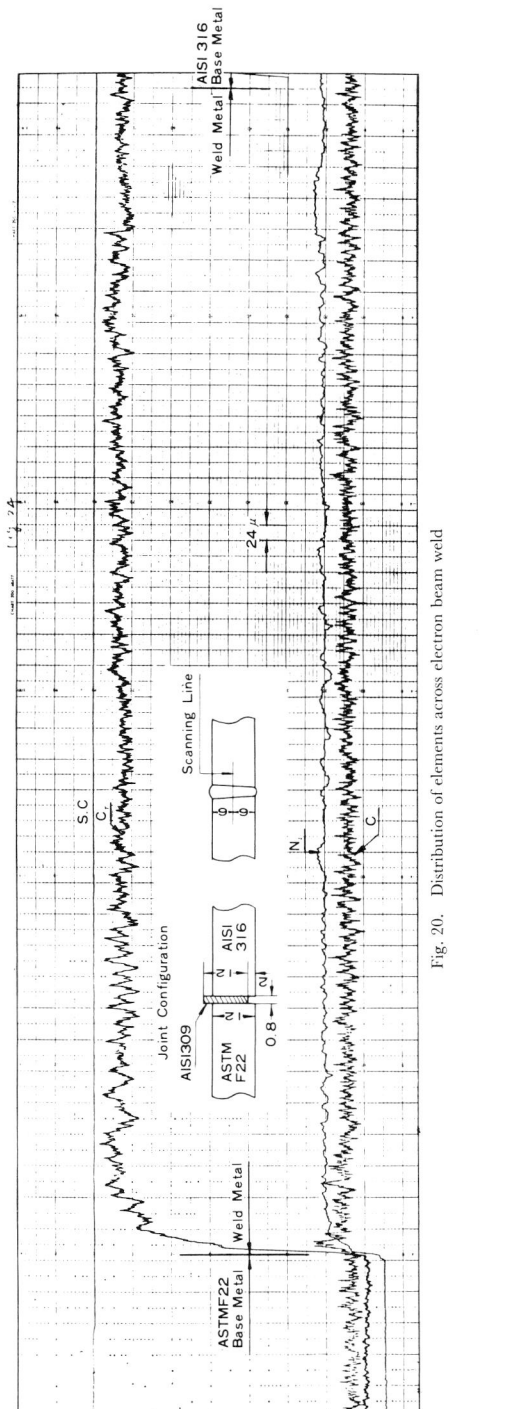

Fig. 20. Distribution of elements across electron beam weld

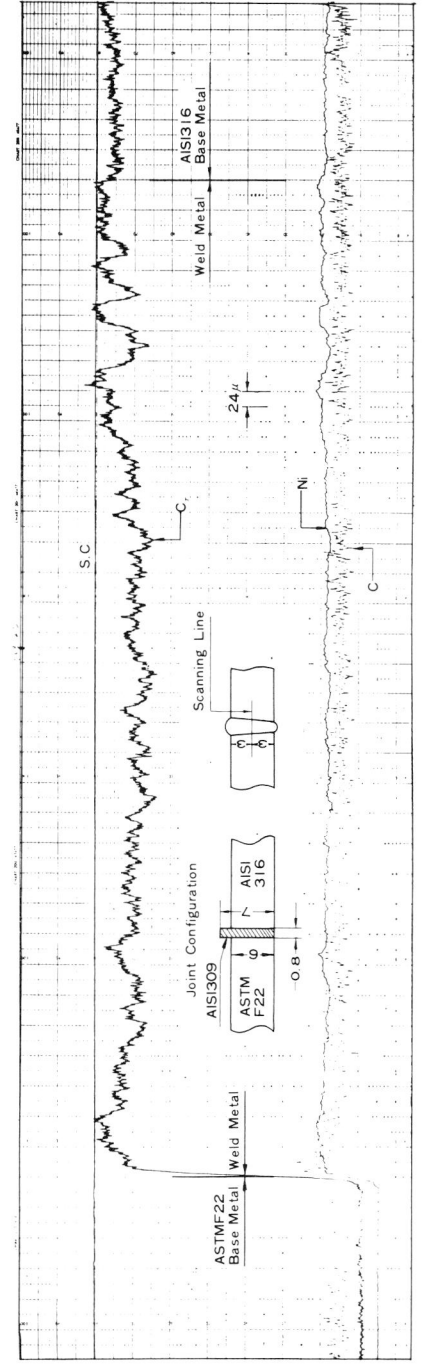

Fig. 21. Distribution of elements across electron beam weld

460

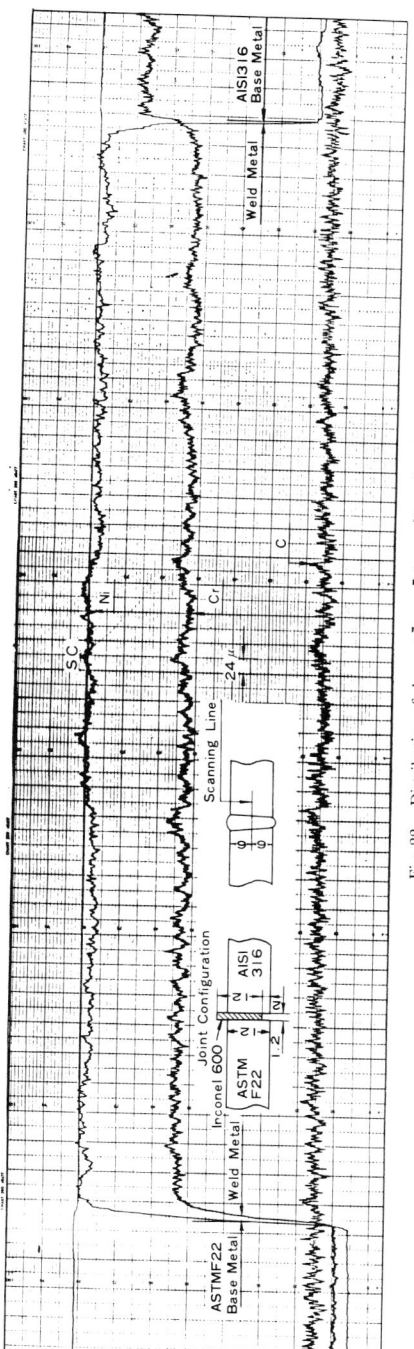

Fig. 22. Distribution of elements across electron beam weld

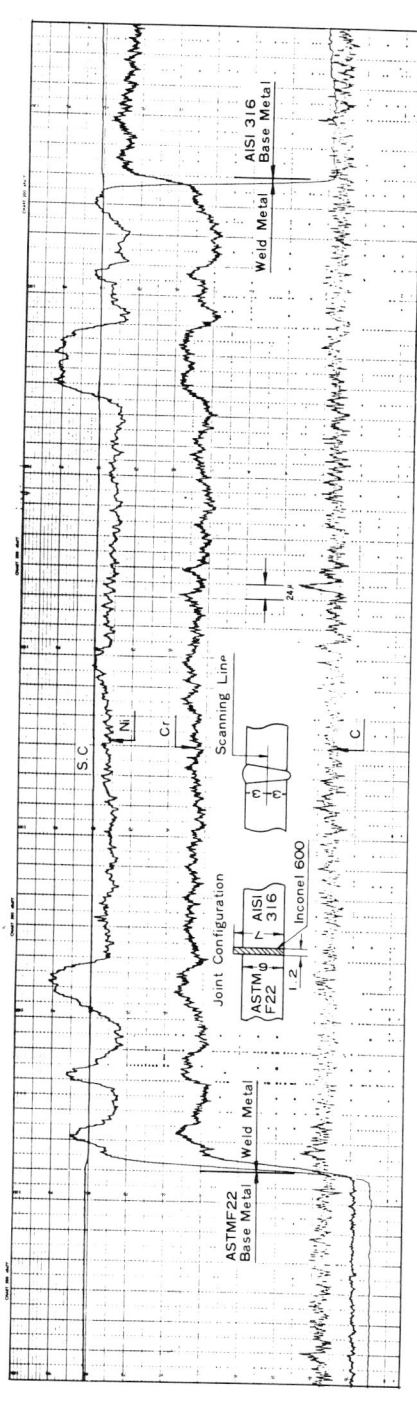

Fig. 23. Distribution of elements across electron beam weld

461

tion.

By introduction of Inconel 600 insert metal, weld metal zone was not hardened irrespective of insert metal width. Fig. 18 shows a typical example of hardness distribution.

On the other hand, hardness distribution along weld center was considerably uniform without distinction of insert metal width and its material.

Fig. 19 shows relation between average hardness of weld metal zone and insert metal width in case of 12 mm-thick joints.

When austenitic stainless steel insert metal is introduced, weld metal zone is hardened in case of 0.4 mm-wide insert metal but not in case of 0.8 mm-wide and 1.2 mm-wide insert metal. While, by introduction of Inocnel 600 insert metal, weld metal zone is not hardened even in case of 0.4 mm-wide insert metal. Quite a similar tendency was shown in 6 mm-thick joints.

(2) Carbon steel to carbon steel joints

In any case weld metal zone was hardened up to about Hv 300 in as-welded condition but not in heat-treated condition.

7.3.3. Distribution of elements in welds

Line scanning across welds and point scanning along weld center were performed respectively. Result is explained in details as follows.

(1) Dissimilar metal joints, 2.25 Cr-1 Mo steel to austenitic stainless steel

(a) Line scanning across welds

When insert metal was not introduced, chromium fluctuated a little in weld metal zone in case of 6 mm-thick joint. Carbon-rich part which seems to be a carburized band was observed at weld metal zone adjacent to 2.25 Cr-1 Mo steel base metal irrespective of joint thickness.

By introduction of austenitic stainless steel insert metal, chromium and manganese fluctuated a little in weld metal zone. This tendency was more conspicuous in case of 6 mm-thick joints than 12 mm-thick joints.

Carbon-rich part which seems to be carburized band could be also observed at weld metal zone adjacent to 2.25 Cr-1 Mo steel base metal. Figs. 20 and 21 show typical examples of distribution of nickel, chromium and carbon across welds.

Figs. 22 and 23 show typical examples of distribution of Nickel, chromium and carbon when Inconel 600 insert metal was introduced. As shown in Fig. 23, nickel and chromium fluctuate in weld metal zone adjacent to two boundary layers in 6 mm-thick joints.

(b) Point scanning along weld center

Figs. 24 to 29 show distribution of elements along weld center. From these figures it can be recognized that nickel and iron show fluctuation only when Inconel 600 insert metal was introduced.

In general, fluctuation of elements in welds does'nt seem to be so remarkable in this analysis as expected.

(2) Carbon steel to carbon steel joints

In any case, fluctuation of elements was hardly

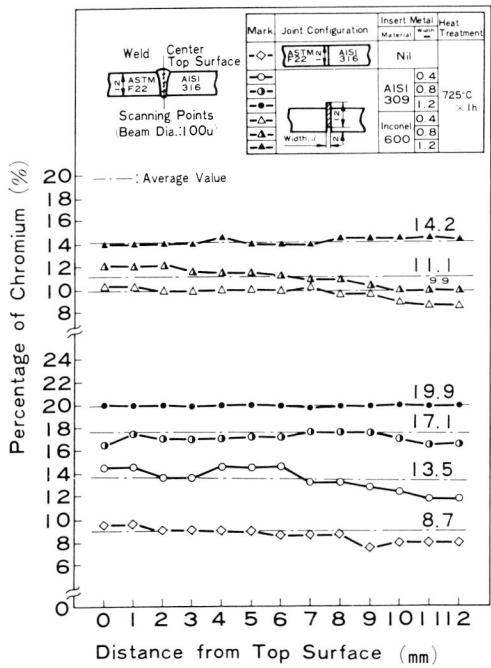

Fig. 24.   Fluctuation of chromium along weld center

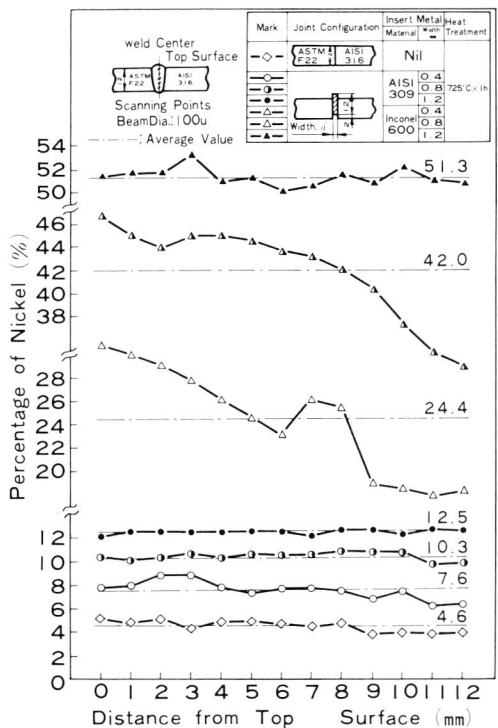

Fig. 25.   Fluctuation of nickel along weld center

462

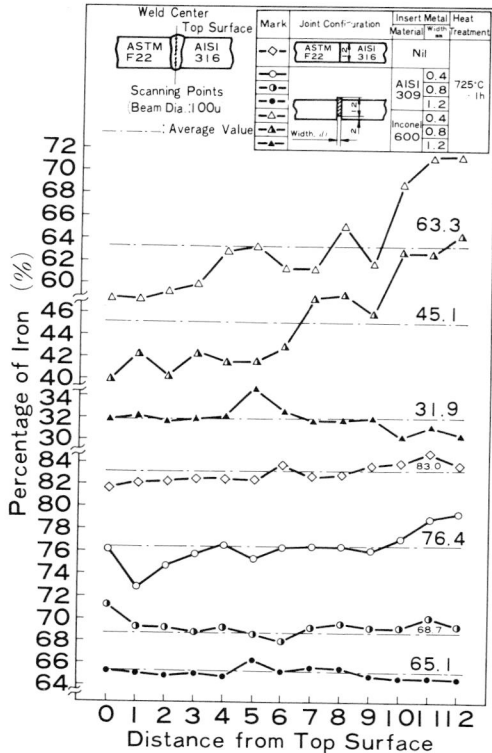

Fig. 26. Fluctuation of iron along weld center

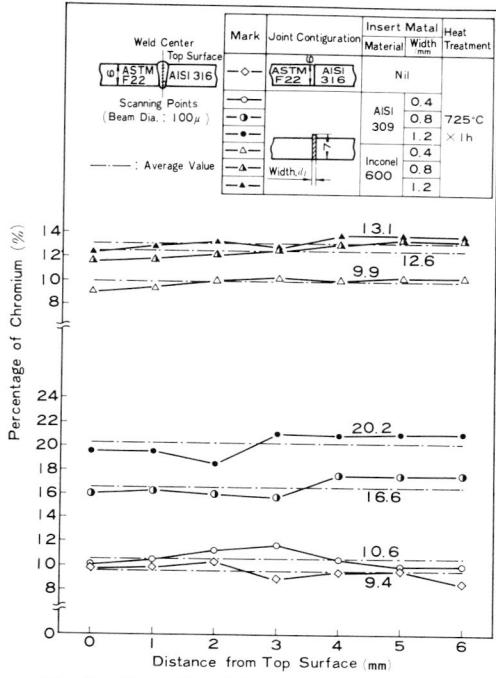

Fig. 27. Fluctuation of chromium along weld center

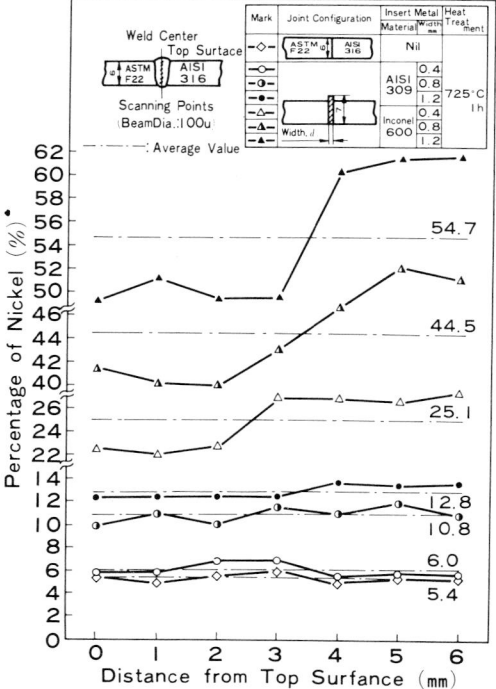

Fig. 28. Fluctuation of nickel along weld center

recognized in weld metal zone.

7.4. *Effect of insert metal on mechanical properties of welds*

7.4.1. Tensile properties

Tensile test was carried out at room and elevated temperatures, that is, 450°C, 500°C, 550°C and 600°C. Every test specimen ruptured satisfactorily at base metal in carbon steel to carbon steel joints and at 2.25 Cr-1 Mo steel base metal in dissimilar metal joints.

7.4.2. Bending properties

Face bend, root bend and side bend test were conducted. Every bend test specimen was successfully bent by 180°C. Photo. 5 shows typical bend test specimens.

7.4.3. Impact properties

Impact properties of weld metal and boundary layer adjacent to 2.25 Cr-1 Mo steel base metal in dissimilar metal joints were significantly improved by introduction of insert metal at low temperatures (−20°C, 0°C and 20°C) and elevated temperatures (450°C, 500°C, 550°C and 600°C). However, effect of insert metal width and its material on impact value was not clear.

7.4.4. Creep rupture properties

463

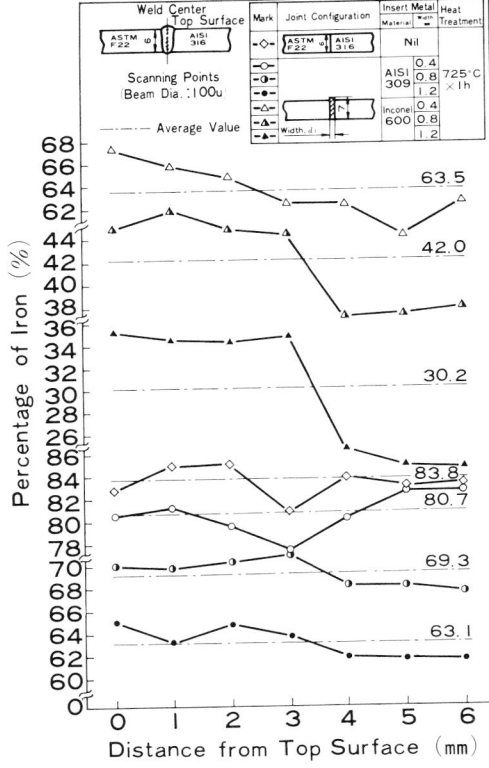

Fig. 29. Fluctuation of iron along weld center

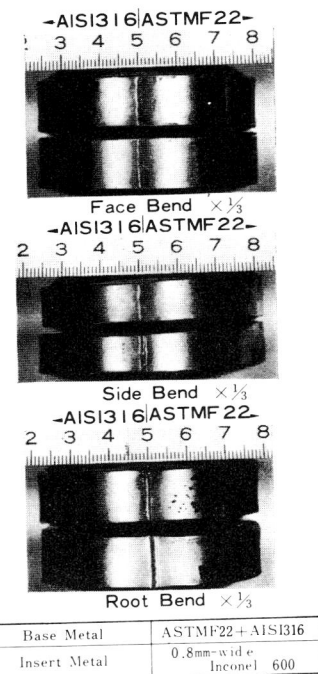

Base Metal	ASTMF22 + AISI316
Insert Metal	0.8mm-wide Inconel 600

Photo. 5. Bend test specimens

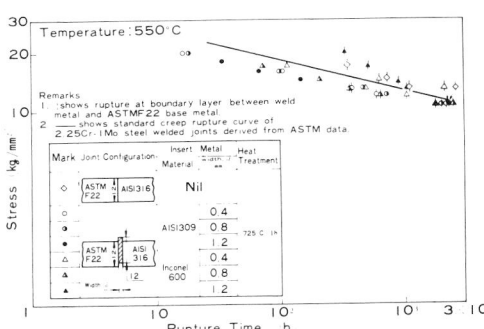

Fig. 30. Creep rupture properties of electron beam welded joints

Figs. 30 and 31 show creep rupture test results at 550°C and 600°C respectively. It might be considered from these figures that there is hardly appreciable difference in creep rupture strength resulting from insert metal width and its material at 550°C and 600°C. It should be also added that creep rupture strength at 550°C approaches creep rupture strength of 2.25 Cr-1 Mo steel welded joints derived from ASTM data irrespective of insert metal width and its material.

At 550°C failure occurred at 2.25 Cr-1 Mo steel base metal at shorter rupture time than 300h and at boundary layer adjacent to 2.25 Cr-1 Mo steel base metal at longer rupture time. At 600°C every specimen ruptured at boundary layer adjacent to 2.25 Cr-1 Mo steel base metal except for two specimens at short rupture time. From microstructure of fractured part, failure at base metal and that at boundary layer are transcrystalline fracture and intercrystalline one respectively.

Photo. 6 shows typical microstructures at vicinities of fractured part. In Photo. 6(a) failure occurs at 2.25 Cr-1 Mo steel base metal. In this case, however, intercrystalline cracking can be also observed at boundary layer adjacent to 2.25 Cr-1 Mo steel base metal. This cracking seems to have propagated from small intercrystalline one which occurred before rupture at

boundary layer where metallurgical notch induces stress concentration. In Photo. 6(b) failure occurs at boundary layer adjacent to 2.25 Cr-1 Mo steel base metal. Herein two interfaces at fractured part differ from each other in diameter and a part of 2.25 Cr-1 Mo steel base metal remains attached to center of fractured part. From above-mentioned feature it might be assumed that a small intercrystalline cracking at boundary layer propagated along it with elapse of time and failure occurred. Further necking occurred at 2.25 Cr-1 Mo steel base metal adjacent to fractured part. It is also assumed that this necking occurred with propagation of small intercrystalline cracking

464

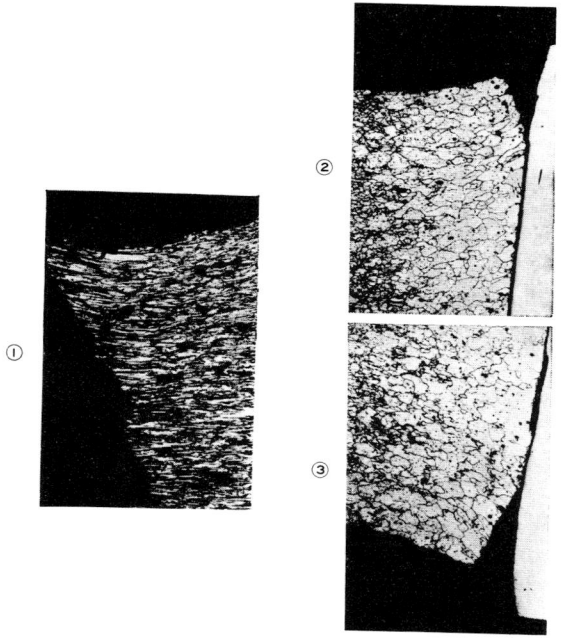

Base Metal	ASTMF22+AISI316
Insert Metal	0.4mm—wide Inconel 600
Temperature (°C)	600
Stress (kg/mm²)	12.5
Rupture Time (h)	60

(a) Rupture at 2.25 Cr-1 Mo Steel Base Metal

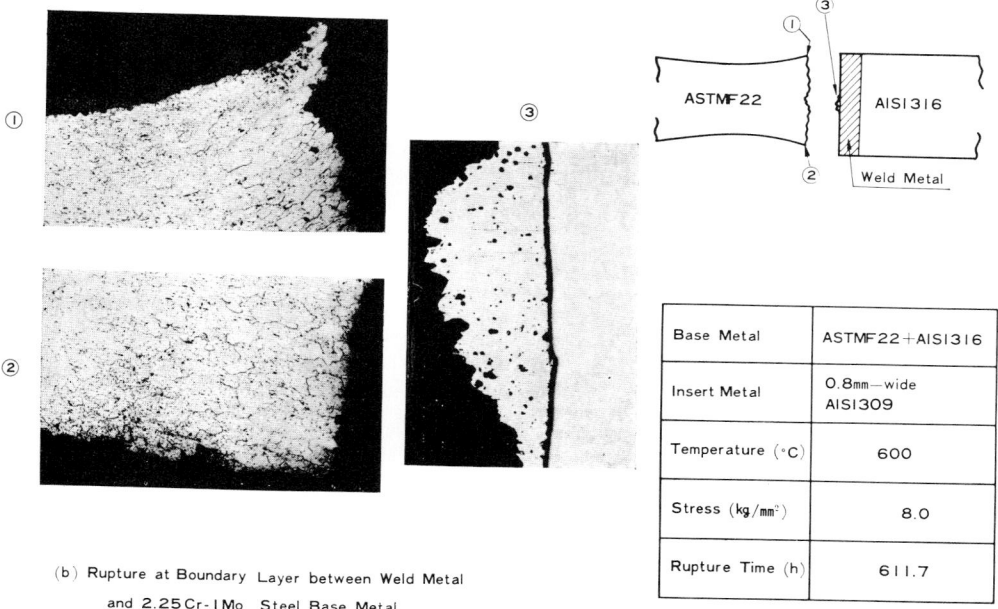

Base Metal	ASTMF22+AISI316
Insert Metal	0.8mm—wide AISI309
Temperature (°C)	600
Stress (kg/mm²)	8.0
Rupture Time (h)	611.7

(b) Rupture at Boundary Layer between Weld Metal
and 2.25 Cr-1 Mo Steel Base Metal

Photo. 6.　Microstructure of creep rupture test specimens ×100　　(reduced 50% on reproduction)

Table 11 Effect of insert metal on mechanical and metallurgical properties of dissimilar metal joint, 2.25 Cr-1 Mo steel to AISI316

Base Plate		Joint Config-uration	Inset Metal		Metallurgical Properties				Sensitivity to Cracking	Mechanical Properties			Remarks
Material	Thick-ness (mm)		Material	Width d (mm)	Tensile Strength	Bending	Impact Value	Creep Rup-ture Strength		Hardness	Microstruc-ture	Segrega-tion	
ASTMF22 + AISI316	6	ASTMF22|AISI316	Nil		Good	Good			Good	Bad	Bad	Good	
			AISI309	0.4	〃	〃			〃	〃	〃	〃	
				0.8	〃	〃			〃	Good	Good	〃	
		ASTM F22 / AISI 316 Width d		1.2	〃	〃			〃	〃	〃	〃	Lack of fusion sometimes occurs.
			Inconel 600	0.4	〃	〃			〃	〃	〃	〃	
				0.8	〃	〃			〃	〃	〃	〃	
				1.2	〃	〃			〃	〃	〃	〃	Lack of fusion sometimes occurs.
	12	ASTMF22|AISI316	Nil		〃	〃	Bad	Good	〃	Bad	Bad	〃	Stable uranami bead cann't be formed.
			AISI309	0.4	〃	〃	Bad	〃	〃	〃	〃	〃	
				0.8	〃	〃	Bad	〃	〃	Good	Good	〃	
		ASTM F22 / AISI 316 Width d		1.2	〃	〃	Good	〃	〃	〃	〃	〃	Lack of fusion sometimes occurs.
			Inconel 600	0.4	〃	〃	〃	〃	〃	〃	〃	〃	
				0.8	〃	〃	〃	〃	〃	〃	〃	〃	
				1.2	〃	〃	〃	〃	〃	〃	〃	〃	Lack of fusion sometimes occurs.

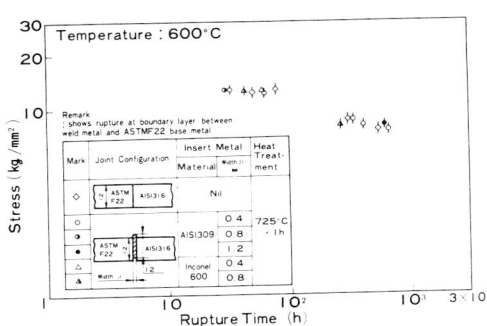

Fig. 31. Creep rupture properties of electron beam welded joints

along boundary layer as described above. After all it probably depends upon test temperature and rupture time where test specimen fractures.

Test specimens, fractured at boundary layer adjacent to 2.25 Cr-1 Mo steel base metal where decarburized and coarse grain zone occurred by introduction of austenitic stainless steel insert metal or no insert metal, hardly differ in creep rupture strength and feature of fracture from other specimens in which such a zone scarcely occurred by introduction of Inconel 600 insert metal. Therefore it can be considered that there is hardly any effect of decarburized and coarse grain zone on creep rupture strength and feature of fracture if rupture time is shorter than 2000h at 550°C and 1000h at 600°C respectively.

7. Summary

Based on results discussed in this paper, obtained conclusions may be summarized as follows:

(i) By introduction of insert metal 6 mm-deep and 12 mm-deep welds can be satisfactorily produced with proper uranami welding conditions as in Table 10 from viewpoint of weld geometry and X-ray properties. And further, by adoption of joint configuration with both insert metal and root space stable uranami bead can be easily formed and range of proper uranami welding conditions widens.

(ii) By introduction of insert metal metallurgical and mechanical properties of electron beam welds can be remarkably improved as shown in Table 11 in summary.

References

1) L.N. Sayer, B.Sc., A.I.M., A.M. Inst. W.: "Quality in electron beam welding" British Welding Journal April 1967
2) T. Boniszewski Ph. D., and D.M. Kenyon, A.M., Ph. D.: "Examination of electron beam welds in 18% Ni, Co, Mo maraging steel sheet" British Welding Journal July 1966
3) Meleka: "Electron-beam Welding" P171, Mcgraw-Hill for the welding Institute
4) Stanley T. Walter: "Welding Handbook-welding, cutting and related process (6th Edition Section 3A)" 47.40, American Welding Society
5) R.E. Armstrong: "Control of spiking in partial penetration electron beam welds" Welding Journal August 1970
6) E.H. Bradburn, R.A. Hubber and P.W. Turner: "Multi-pass electron beam welding for controlled penetration" Welding Journal April 1970
7) K. Terai, T. Toyooka and Y. Nagai: "Electron beam welding of heat-resistant steels" Committee of Electron Beam Welding E.B.W.-34-71 Japan Welding Society

Study on Local Vacuum Electron Beam Welding

I.   Introduction

     Recently, high power electron gun is developed and then electron beam welding is able to apply heavy thickness structures. But in order to apply electron beam welding to heavy components, it is necessary to develop a local vacuum electron beam welding machine.

     The authors have already developed local vacuum electron beam welding machine (1) with slide seal mechanism for butt-welding of flat plates. This mechanism is useful for seam welding but useless for circular seam welding.

     This paper describes electron beam welding conditions under a vacuum of 6.7 Pa (5×10^{-2} Torr) and development of local vacuum electron beam welding machine for circular seam welding of cylindrical structures. And it's performance was tested with 60 kgf/mm^2 grade high tensile strength steel pipe (diameter: 1.4 m, thickness: 50 mm).

II.   Welding conditions under low vacuum

     Usually, electron beam welding is done under a high vacuum 0.013 pa (1×10^{-4} Torr). But penetration depth is not changed between 0.013 pa (1×10^{-4} Torr) and 6.7 Pa (5×10^{-2} Torr). Then local vacuum electron been welding is done under a low vacuum 6.7 Pa (5×10^{-2} Torr).

     The authors investigated welding conditions and its performance under low vacuum.   The used HT60 steel is 50 mm in plate thicknesss.   Table 1 shows the chemical compositions and mechanical properties.

Table 1   Chemical compositions and mechanical properties of material used

Steel	Chemical compositions (%)								Tensile properties			Impact properties
	C	Si	Mn	P	S	Ni	Mo	V	0.2% P.S. (kgf/mm^2)	T.S. (kgf/mm^2)	El. (%)	vE $_{-10}$ (kgf.m)
HT60	0.13	0.24	1.29	0.017	0.004	0.26	0.12	0.03	52	63	27	30.2

P.S. ; Proof Stress
T.S. ; Tensile Strength

2.1   Relation between welding

     Low vacuum type high power electron gun, the maximum beam power of which is 120 kW (100 kV x 1.2 A) is used in this investigation. Welding position is horizontal.

Fig.1 shows the relation between welding conditions and bead appearance. In good conditions there is no faults (crack of porosity) and fair welding conditions are same between 0.067 Pa (5x10⁻⁴ Torr) and 6.7 Pa (5x10⁻⁴ Torr).

Used heat input is not considered penetration beam power. High heat input shows bad penetration bead and low heat input shows lack of penetration.

Table 2 shows the chemical compositions of weld metal. Gas contents and chemical compositions are almost same between 0.067 Pa (5x10⁻⁴ Torr) and 6.7 Pa (5x10⁻² Torr).

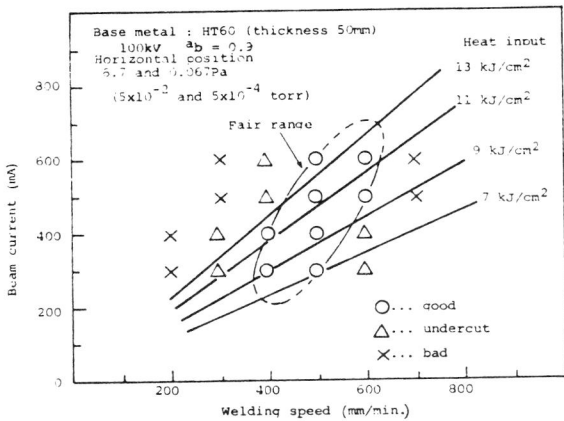

Fig.1 Relation between welding conditions and bead appearance

Table 2 Chemical analysis of weld metal and base metal

	Vacuum (Pa)	C	Si	Mn	P	S	Ni	Cr	Mo	V	Cu	ppm O	N	H
Weld Metal	6.7	0.12	0.23	1.24	0.016	0.005	0.27	0.03	0.12	0.032	<0.01	15	36	0.6
	0.067	0.12	0.23	1.22	0.016	0.005	0.28	0.04	0.12	0.032	<0.01	12	33	0.6
Base Metal	–	0.12	0.23	1.31	0.016	0.005	0.27	0.03	0.12	0.032	<0.01	21	41	0.7

2.2 Gap of groove face

In actual practice, small gap of groove face always exsists. But electron beam diameter is so small that it is impossible to weld soundly at large gap.

Fig.2 shows the feect of gap of groove face. Accelerating voltage is 100 kV, welding position is horizontal, and oscillation is not used. Sound weld is possible under 0.5 mm gap of groove face but over 1 mm gap the surface of weld bead hollows largely. The effect of welding conditions (beam current or welding speed) on shape of bead is not clear in fair range.

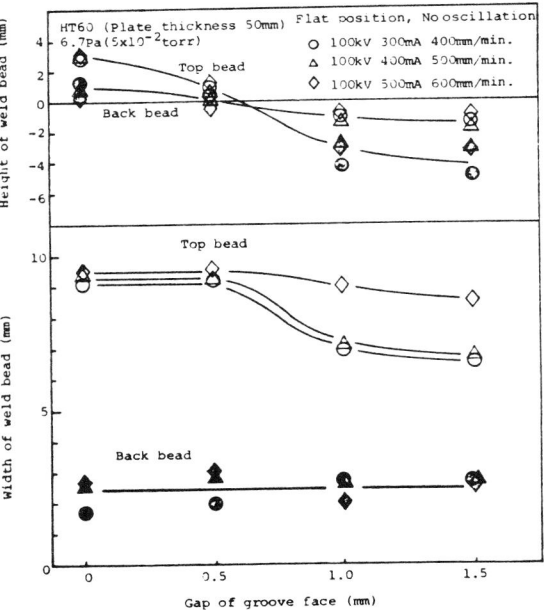

Fig.2 Effect of gap of groove face on shape of bead

III. Development of local vacuum electron beam welding machine for circular seam welding of cylindrical structures

The authors developed local vacuum electron beam welding machine for circular seam welding of cylindrical structures which diameter is 1.4 m and thickness is 50 mm.

A schematic diagram of this machine is shown in Fig.3. The vacuum seal equipment consists of two parts. One is inner fixed chamber and the other is outer fixed chamber which has a slide seal. Low vacuum type 120 kW electron gun is fixed at outer chamber. Welded pipe and fixed chambers move during welding. Seal equipment is double. Intermediate room of chamber is vacuumed by rotary pump and welding room is vacuumed by rotary pump and mechanical booster. A vacuum is more than 6.7 Pa (5×10^{-2} Torr). Fig.4 shows appearance of this welding machine. Fig.5 shows an example of a vacuum exhaust feature. It was ascertained that welding room's vacuum reaches to 6.7 Pa (5×10^{-2} Torr) in 2 min. and a vacuum during rotation is not differed from that during not rotation.

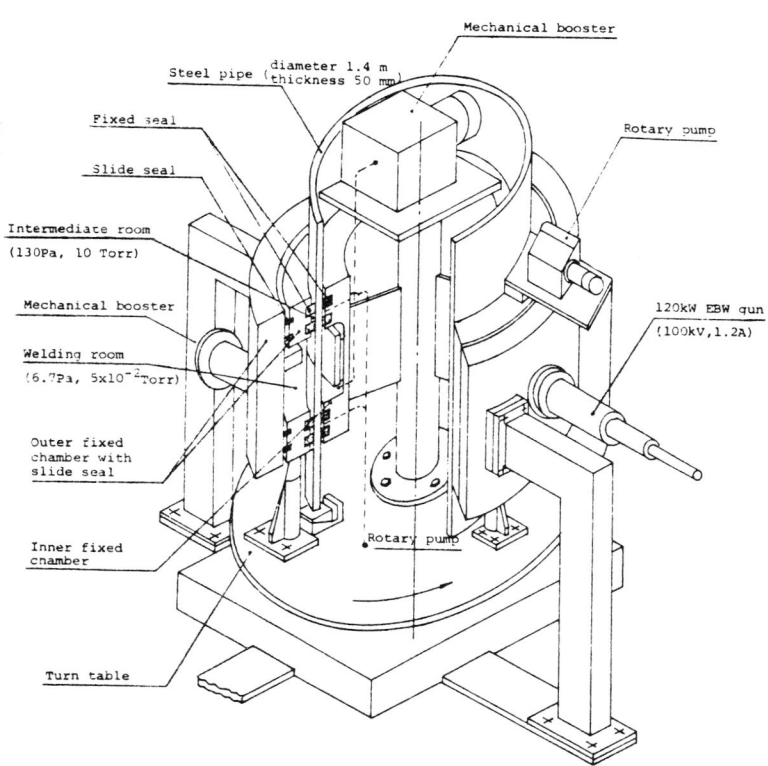

Fig.3 Schematic diagram of local vacuum electron beam welding machine

Fig.4 Local vacuum electron beam welding machine

469

Results of vacuum test are shown in Table 3. A good result satisfying a vacuum of 6.7 Pa (5×10^{-2} Torr) needed for good welding could be obtained in 10 min.

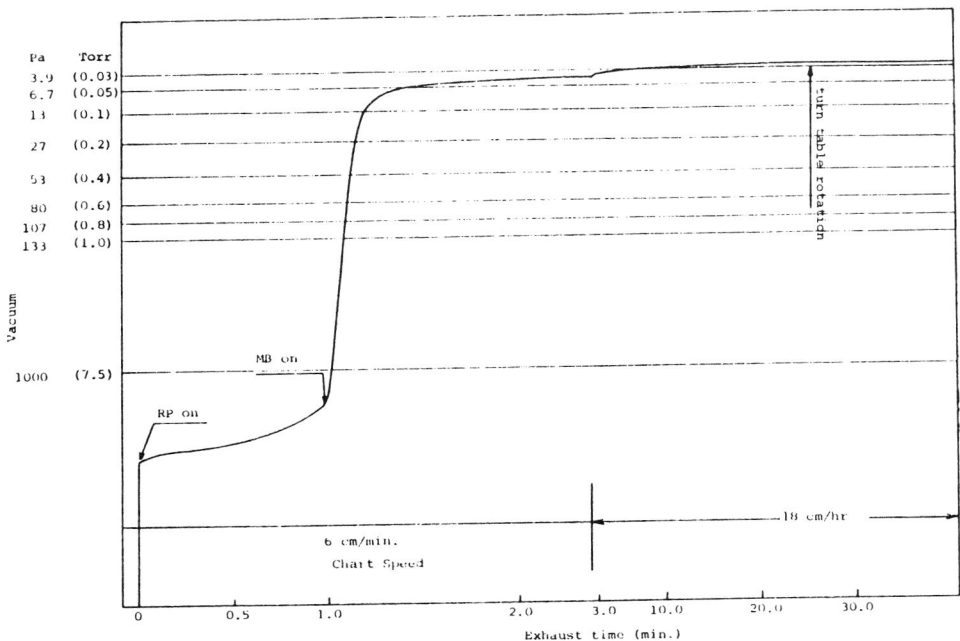

Fig.5   Vacuum pump-down curve (Welding room)

Table 3   Results of vacuum test

Position		Plan	After 10 min.	After 1 hr.
Outer fixed Chamber with slide seal	Intermediate room of fixed seal	under 1300 Pa (10 Torr)	39 Pa (0.3 Torr)	-
	Intermediate room of slide seal	under 1300 Pa (10 Torr)	13 Pa (0.1 Torr)	9 Pa (7×10^{-2} Torr)
	Welding room	under 6.7 Pa (5×10^{-2} Torr)	3.9 Pa (3×10^{-2} Torr)	0.9 Pa (7×10^{-3} Torr)
Inner fixed chamber	Intermediate room of fixed seal	under 1300 Pa (10 Torr)	33 Pa (0.25 Torr)	-
	Welding room	under 6.7 Pa (5×10^{-2} Torr)	6.7 Pa (5×10^{-2} Torr)	1.0 Pa (8×10^{-3} Torr)

470

IV. Performance test

Performance of local vacuum electron beam welding machine was tested with HT60 steel pipe (diameter: 1.4 m, plate thickness: 50 mm). Table 4 shows welding conditions. Gap of groove face is under 0.5 mm. Fig.6 and Fig.7 show good appearance of welded HT60 steel pipe and weld bead. Welding was started at a vacuum 4.0 Pa (3×10^{-2} Torr) and the same vacuum level was maintained during welding.

Table 4. Welding Conditions

Accelerating Voltage	Beam Current	Welding Speed	Oscillation	$A_b = \dfrac{D_c}{D_f}$	Work Distance	Down Slope Time
100 kv	300 mA	400 mm/min.	None	0.9	340 mm	30 sec.

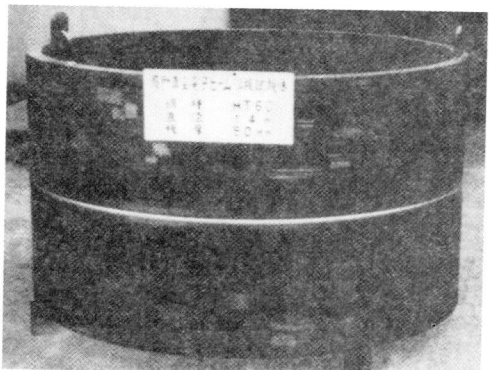

Fig. 6 Appearance of welded HT60 steel pipe
(diameter 1.4m, thickness 50mm)

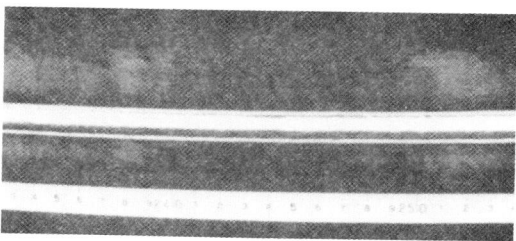

Top bead

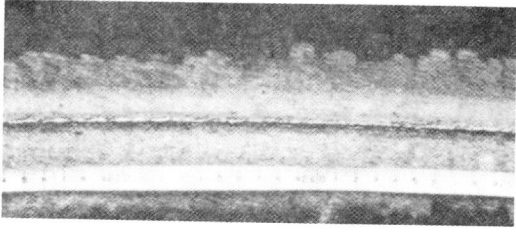

Back bead

Fig. 7 Appearance of weld bead

4.1 Non destructive test

In order to evaluate the soundness of weld joint, X-ray and ultrasonic test were practiced all the weld bead of the model.

There was no faults such as crack, porosity and so on.

4.2 Mechanical test

Table 5 shows the results of tensile test. Fracture location was on the base metal.

Table 6 shows the results of side bend test. There was no faults on the surface of bend specimens.

Table 7 shows the results of charpy impact test. Weld metal and heat affected zone showed good value.

Table 5 Results of tensile test

No.	Tensile strength (kgf/mm^2)	Fracture location	Specimen size
1	68.1	Base metal	* JIS Z 2201 No.5
2	68.3	Base metal	

* Japan Industrial Standard

Table 6 Results of side bend test

No.	Bending Condition	Judgement	Specimen size
1	Specimen thickness: 9 mm	Good	JIS Z 3122 No.3
2	Bend radius :18 mm 180 deg. bend	Good	

471

Table 7   Results of charpy impact test

Test temperature	Weld metal		Heat affected zone		Specimen size
	Absorbed energy (kgf.m)	Brittle fracture (%)	Absorbed energy (kgf.m)	Brittle fracture (%)	
RT (21 °C)	20.8	43	10.2	27	
	25.8	24	10.2	29	
	29.8	0	9.3	30	
	(25.5)	(22)	(9.9)	(29)	JIS Z 2202 No.4
-10°C	25.4	40	7.2	46	
	12.4	73	9.9	43	
	8.4	81	7.7	55	
	(15.4)	(65)	(8.3)	(48)	Radius of notch 0.25 mm Angle of notch 45°

Notes: ( ) shows average value.

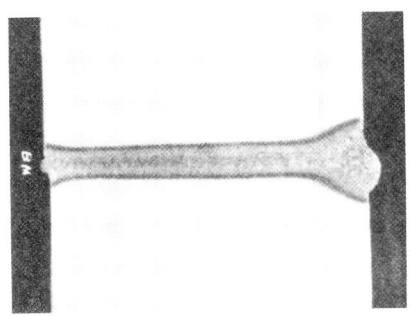

Fig.8   Macro structure of weld joint

4.3   Macro structure test

Fig.8 shows a macro structure of weld joint.   Any faults were not observed.

V.   Conclusion

The authors investigated the welding conditions under low vacuum and developed local vacuum electron beam welding machine for circular seam welding of cylindrical structures.   HT60 steel pipe (diameter: 1.4 m, thickness: 50 mm) was used for performance test of this machine.
Results obtained are summarized as follows.

(1)   Electron beam welding conditions at horizontal position are same between 0.067 Pa (5×10^{-4} Torr) and 6.7 Pa (5×10^{-2} Torr).   Permissible gap of groove face is under 0.5 mm without oscillation at 50 mm plate thickness.

(2)   The slide seal mechanism always maintains a good vacuum (4.0 Pa) higher than expected object of 6.7 Pa.   A circular seam weld joint which has good appearance without faults is obtained.

(3)   X-ray test, ultrasonic test, mechanical tests and metallurgical test were examined.   The results of them show good performance and a prospect of local vacuum electron beam welding application to circular seam welding of heavy components has been obtained.

This electron gun's maximum power is 120 kW (100 kV x 1.2 A) and maximum penetration at horizontal position is 300 mm.   As a next step of this study, the application of local vacuum electron beam weling for heavy components is programmed.

REFERENCE

(1)   A Ujiie, S. Satoh, T. Shimoyama, H. Kono, N. Sakamoto, and K. Enami: Development of Local Vacuum Electron Beam Welder, IIW-IV-224-77.

Fundamental Studies on Electron Beam Welding of Heat-resistant Superalloys for Nuclear Plants

— Effect of Welding Conditions on Some Characteristics of Weld Bead —

Abstract

In this paper, the effect of the welding conditions on the characteristics of the weld geometry and the weld defects was made clear, concerning the heat-resistant superalloys for the nuclear plants. Obtained conclusion may be summarized as follows, using technical symbols which are given meanings in this report.

1) The weld defects were R-porosity and microcrack.

2) S_p and Δa_b are considered to be the important criteria for the evaluation of the susceptibility to the R-porosity.

3) Superalloys could be evaluated in the sensitivity to the microcrack in terms of the critical heat input to avoid microcrack q_{cr}. This q_{cr} is considered to be one of the proper criteria for evaluating the superalloys in the susceptibility to the microcrack.

4) Most microcracks were apt to occur when h_C/h_N came near to 1.0. These microcracks came to occur easily with the increase of $d_{B \cdot N}/d_B$.

1. Introduction

In recent years, such superalloys as Hastelloy type, Inconel type, Incoloy type, etc. have been taken into consideration as one of the materials for very high temperature gas-cooled reactor, where dimensional tolerance is strictly limited from the structural viewpoint for the safety assurance peculiar to the nuclear plants and the strength at very high temperatures is also required. In this sense, welding of these superalloys involves many problematical points.

As it is well recognized that the energy density in the electron beam welding process is extremely high by nature, the large penetration depth can be easily obtained with small heat input. Furthermore, very precise welding can be performed at high quality in the vacuum chamber. For these reasons, electron beam welding seems to be most suitable for welding of the heat-resistant superalloys for the nuclear plants. However, very rapid melting of the limited area with the heat source of high energy density tends to cause such individual weld defects as porosity, cold shut, spiking and cracking. There are some fundamental reports[1]~[5] on these weld defects. However, there is no report on these weld defects of the superalloys.

As described above, weld defects in the electron beam welds of these superalloys have not been clarified yet. In this report, authors herein determined the effect of the welding conditions on some characteristics of weld bead, that is, the weld geometry and the weld defects by the slope welding method to establish the proper welding conditions for these superalloys.

2. Material and Experimental Instrument used

Such superalloys as Hastelloy type, Inconel type and Incoloy type and austenitic stainless steel for compararison were used which generally seem to be fit for very high temperature gas-cooled reactor. Chemical composition and mechanical properties of these superalloys are shown in Table 1. Brief marks shown in Table 1 are below employed to distinguish Hastelloy X of different heat. Welding of these materials was conducted by 150kV-40mA type electron beam welder of hard vacuum.

3. Experimental Procedure

20mm-thick materials shown in Table 1 were set on work table as in Fig. 1, where slope angle θs equals $30°$. Thereafter, upslope welding was performed with the electron gun fixed, moving work table in such a direction as arrow shows. In this case, a_b parameter[6] defined by D_O/D_F (where, D_O: object distance, D_F: focal length)

473

Table 1 Chemical Composition and Mechanical Properties of Superalloys used

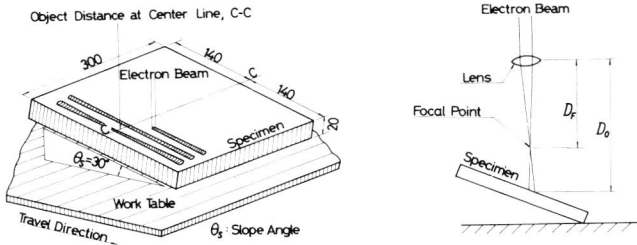

Material	Mark	Thickness (mm)	Melting Process	Final Heat Treatment	Grain Size (ASTM)	0.2%PS (kg/mm²)	T.S (kg/mm²)	El. (%)	R of A (%)	C	Si	Mn	P	S	Ni	Cr	Co	Mo	W	Nb+Ta	Al	Ti	B	Zr	Ce	Fe	O	N Sol	N Insol	N Total
Hastelloy X	HAEN	2.0	AE	1120°C·18Min W.Q.	1~4	30.4	71.1	59.0	65.0	0.066	0.48	0.88	0.016	<0.002	Bal.	21.36	1.77	9.05	0.45	—	0.07	0.15	0.002	—	—	19.09	3	405	42	447
	HAEM		AE	1150°C·50Min W.Q.	5~6	38.7	75.0	47.6	—	0.068	0.37	0.59	0.012	<0.002		20.74	1.03	8.70	0.50	—	0.21	0.02	0.001	—	—	18.23	2	248	20	268
	HVEN		VE	1120°C·18Min W.Q.	1~6	34.3	73.8	49.3	56.9	0.084	0.17	0.84	0.001	<0.002		20.70	1.59	9.20	0.55	—	0.02	0.19	0.002	—	—	18.53	5	306	16	322
	HVERN		VE	1170°C·30Min A.C.	1~4	32.4	74.0	54.3	52.2	0.065	0.35	0.72	0.001	<0.002		21.40	1.45	8.93	0.51	0.05	0.05	0.06	0.002	0.009	<0.005	19.08	4	121	22	143
Inconel 625	Ini 625AE		AE	1000°C·1Hr W.Q.	6	45.5	90.1	46.8	—	0.053	0.28	0.24	0.003	<0.002		22.09	0.06	8.81	0.69	3.53	0.24	0.13	—	—	—	2.54	3	44	211	255
Inconel 617	Inl 617V		V	≈1177°C W.Q.	3~4	30.0	74.6	70.0	57.0	0.056	0.17	0.02	0.004	<0.002		21.24	12.60	9.00	—	—	0.93	0.52	—	—	—	1.45	4	244	180	424
Incoloy 800	Iny 800V		V	1100°C·15Hr W.Q.	2.5	22.1	58.2	52.0	72.1	0.056	0.37	0.77	0.010	0.002	32.13	21.21	0.50	0.18	—		0.51	0.59	—	—	—	Bal.	6	32	93	125
Incoloy 807	Iny 807A		A	1230°C·3Hr W.Q.	1	25.7	64.1	52.2	60.3	0.057	0.50	0.70	0.002	0.002	40.10	20.58	8.28	0.20	4.85	0.99	0.47	0.24	—	—		7	68	144	212
SUS 316	S 316		A	1100°C·13Min W.Q.	3~5	—	60.3	61.4	—	0.045	0.79	1.26	0.028	0.004	11.51	17.52	—	2.58	—	—	—	—	—	—	—		13	257	32	289

※ AE : Air Melting followed by Electroslag Remelting.   VE : Vacuum Induction Melting followed by Electroslag Remelting.
V : Vacuum Induction Melting .   A : Air Melting

is called "beam active parameter" or "active parameter" which is one of the important variables in the electron beam welding. As max. penetration depth is experimentally clarified to be obtained around the a_b parameter of 0.9, materials were set so that the a_b parameter might agree precisely with this value at the center line C-C in Fig. 1. This enables a_b parameter to vary over considerably wide range from 0.6 to 1.2.

After the bead-on-plate welding, X-ray inspection was carried out by such a method as shown in Fig. 2 to investigate the weld defects. Thereafter, every bead-on-plate weld was machined so that transverse profile might be microscopically inspected after polishing and etching. Thereby, penetration depth and bead width were respectively measured. Microcrack was also investigated by the microscope.

The diameter of the electron beam was measured by means of the AB-test (Arata Beam Test)[7].

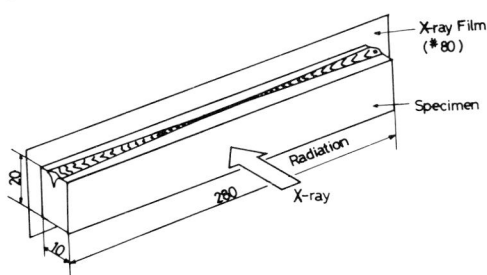

Fig.1   Bead-on-plate Welding Method by Slope Welding

Notes   1. Specimen is machined from bead-on-plate welds.
         2. *80 means film sensitivity.
         3. X-ray film is apart from X-ray Source by 600mm approximately.

Fig.2   X-ray Inspection Method of Bead-on-plate Welds

4. Experimental Result and Discussion

It was recognized through the X-ray inspection and microscopical survey that porosity and microcrack were observed in almost all the materials. These results are described in details below.

4.1 Porosity

Porosity observed in X-ray inspection was root porosity (R-porosity) to occur at the root of the penetration. Typical R-porosity is shown in Photo. 1.

It is reported that occurrence of the R-porosity depends upon the heat input[1]. In this experiment, occurrence of the R-porosity is arranged on w_B $(= \frac{I_b V_b}{2 r_b v_b})$ – a_b diagram (where, I_b: beam current, V_b: accelerating voltage, r_b: radius of the electron beam, v_b: welding speed). Above-mentioned w_B is herein defined as the "bead energy density" of the electron beam, that is, the heat input divided by the actual beam diameter of the electron beam, which means the energy density of the electron beam supplied on the surface of the material.

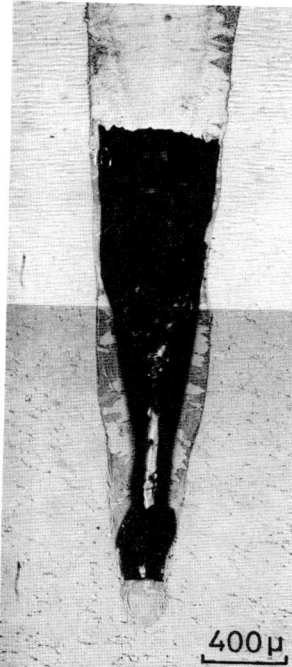

V_b =150kV
I_b =40mA
v_b' =160cm/min
a_b =0.87

Photo. 1 Typical R-porosity (SUS 316)

As far as it concerns with the electron beam welding, the diameter of the electron beam is very dependent both upon the welding conditions and upon the type itself of the electron beam welder. Therefore, estimation of the proper welding conditions and comparison of the suscepti-

bility to the porosity are considered to be done in any case by use of the diagram proposed above.

Figs. 3 and 4 show the examples of w_B–a_b diagram concerning the occurrence of the R-porosity. It was noted on the curve of the constant heat input whether R-porosity did occur or not. As a result, R-porosity was recognized to occur within the certain range of a_b parameter. When the penetration depth and heat input are limited less than 15mm and 5000 joule/cm (I_b : 40mA, V_b : 125kV, v_b': 60 cm/min) respectively, the area where R-porosity occurs can be shown as in the figures by plotting these critical a_b parameter on the curve of the constant heat input. In these figures, the penetration depth is constant on the certain curve. In other expression, h_p=15mm means that penetration depth is constantly 15mm on this curve.

Susceptibility to the R-porosity might be evaluated by above-mentioned individual area S_p, max. bead energy density $w_{B.m}$ and the difference of the critical a_b parameters, that is, Δa_b ($\equiv a_{b.2} - a_{b.1}$) as shown in Fig. 5. Herein, smaller S_p and Δa_b and larger $w_{B.m}$ means less susceptibility to the R-porosity.

Table 2 evaluates the susceptibility of the superalloys to the R-porosity in terms of these criteria, S_p, $w_{B.m}$ and Δa_b. As shown in this table, there is appreciable difference among the superalloys in terms of S_p and Δa_b. There is also clear correlation between S_p and Δa_b. It may be safely said from these results that the susceptibility of the superalloys to the R-porosity is evaluated in terms of S_p and Δa_b. Generally speaking, Fe-base superalloys such as Incoloy 800 and austenitic stainless steel (SUS316) are more susceptible to the R-porosity as compared with Ni-base superalloys such as Hastelloy X, Inconel 625.

Table 2 Criteria obtained for Evaluation of Susceptibility to R-porosity

Material	$S_p/S_{p.m}$	$w_{B.m}$ (joule/cm²)	$a_{b.1}$	$a_{b.2}$	Δa_b
Hastelloy X (HAEN)	0.75	2.35	0.83	1.08	0.25
Hastelloy X (HAEM)	0.66	2.35	0.89	1.09	0.20
Hastelloy X (HVEN)	0.60	2.35	0.87	1.07	0.20
Hastelloy X (HVERN)	0.73	2.95	0.85	1.09	0.24
Inconel 625 (Inl 625AE)	0.75	2.35	0.78	1.13	0.35
Inconel 617 (Inl 617V)	1.00	1.26	0.71	1.16	0.45
Incoloy 800 (Iny 800V)	0.91	2.95	0.69	1.12	0.43
Incoloy 807 (Iny 807)	0.55	2.95	0.81	1.09	0.28
SUS 316 (S 316)	0.91	1.89	0.71	1.16	0.45

S_p means the area schematically shown in Fig. 5 where R-porosity occurs.
$S_{p.m}$ means max. S_p in this experiment.

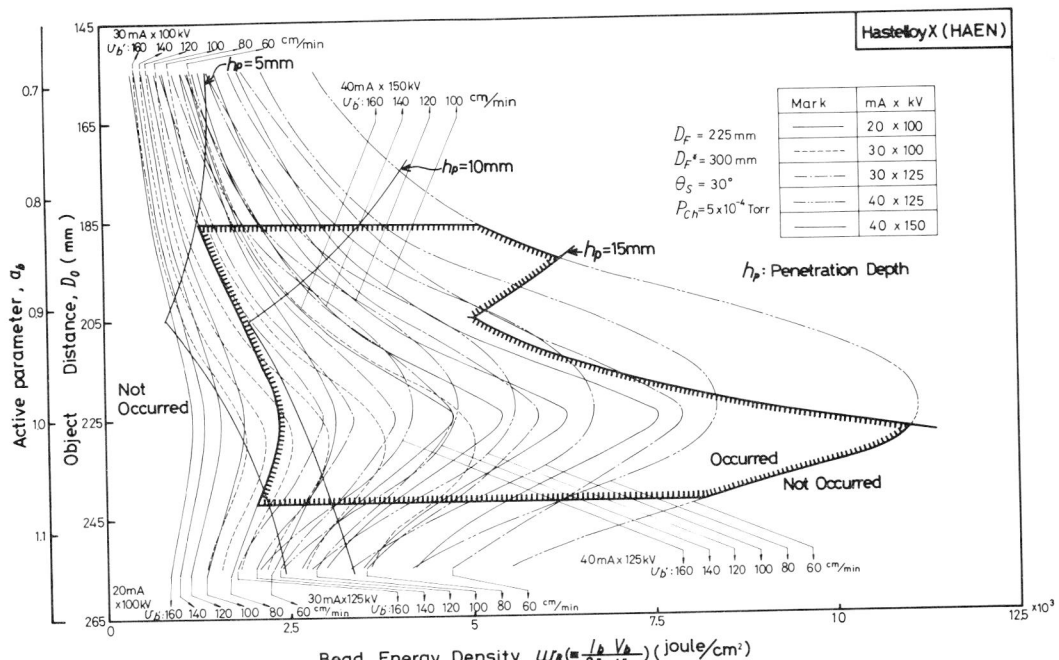

Fig.3   Effect of Bead Energy Density and Active Parameter on R-porosity

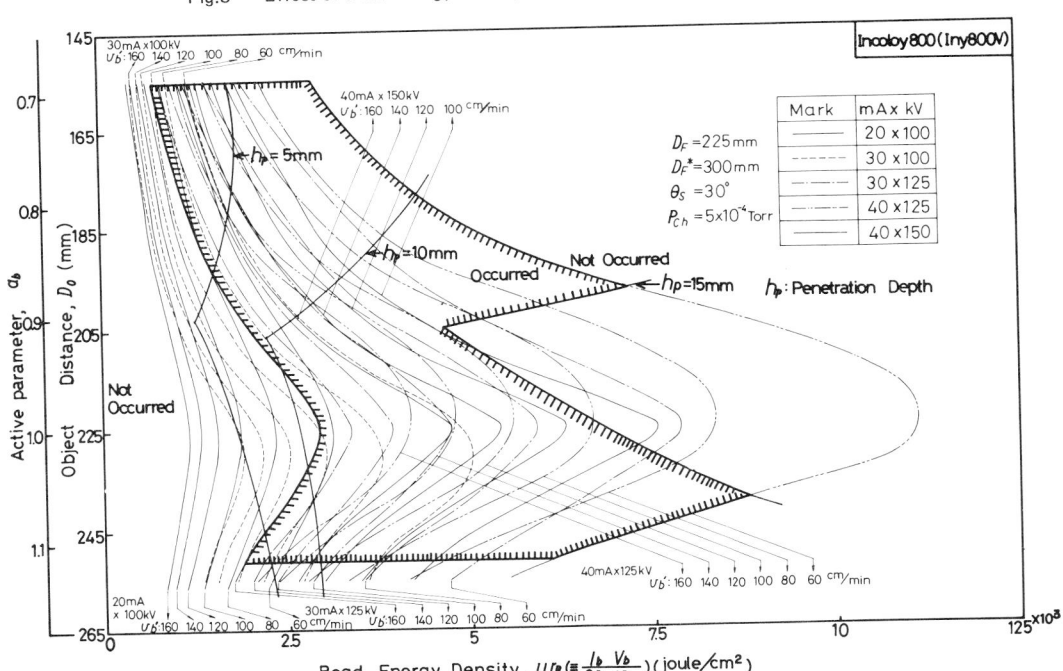

Fig. 4   Effect of Bead Energy Density and Active Parameter on R-porosity

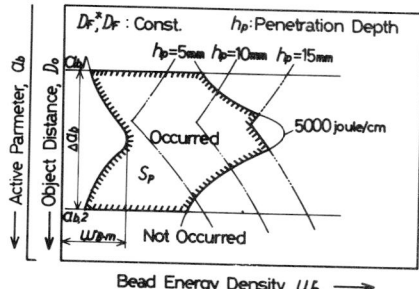

Fig. 5 Evaluating Criteria for R-porosity

Figure axes: Active Parameter, a_b / Object Distance, D_b (vertical); Bead Energy Density, u_b (horizontal). Labels: D_F^*, D_F : Const. h_p: Penetration Depth; $h_p=5mm$ $h_p=10mm$ $h_p=15mm$; 5000 joule/cm; Occurred; S_p; Not Occurred; a_{b1}, Δa_b, a_{b2}, $u_{b m}$.

4.2 Microcracking

Every transverse cross section was electrolitically polished with the mixed solution (1.5% perchloric acid, 5.7% sodium thiocyanate, 7.1% citric acid, 75.5% ethyl alcohol, 9.4% N-propyl alcohol and 0.8% oxine in weight percent) and also etched with 10% oxalic acid solution. Thereafter, microcrack in the electron beam welds was investigated through the microscopic survey.

Microcrack was observed in the electron beam welds of all the superalloys except for Inconel 625 and SUS316. In most cases, microcrack was continuously seen in the interdendritic boundary of weld metal and the grain boundary of heat affected zone and also nearly perpendicular to the fusion boundary. Typical microcrack is shown in Photo. 2.

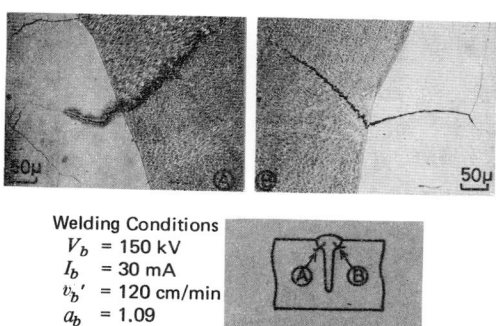

Welding Conditions
V_b = 150 kV
I_b = 30 mA
v_b' = 120 cm/min
a_b = 1.09

Photo. 2 Typical Microcrack (Hastelloy X, HAEN)

(1) Effect of Power of Electron Beam and Welding Speed on Microcrack

Fig. 6 shows the effect of the power of electron beam $W_b(\equiv I_b V_b)$ and the welding speed v_b' on the occurrence of the microcrack in the superalloys. On this $w_b - v_b'$ diagram, there is drawn the critical line on the disappearance of the microcrack. It can be said from this figure

that microcrack is completely prevented in this experiment by limiting the welding conditions in consideration of this critical line. Furthermore, these critical lines are nearly parallel to one another. From this result, it can be considered that the superalloys are not so susceptible to the microcrack in which microcrack never occur in a larger region. The superalloys are below placed in terms of this region in the order of less susceptibility to the microcrack.

(SUS316, Inconel 625), HAEM, HAEN, Incoloy 800, HVEN, (HVERN, Inconel 617)

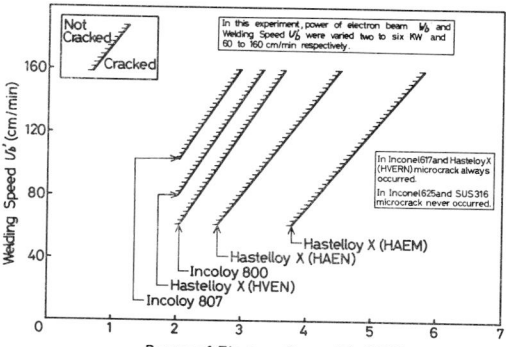

Fig. 6 Effect of Power of Electron Beam and Welding Speed on Microcrack

(2) Effect of Heat Input on Microcrack

Fig. 7 shows the example of the relation between the heat input and microcracking percentage (cracked specimens' ratio of all the specimens at the constant heat input). From this figure, microcracking percentage tends to increase with the increase of the heat input and the microcrak disappear at the heat input less than the certain value. This value of every superalloy, under which microcrack never occurs, can be regarded as the critical heat input to avoid microcrack q_{cr} which is tabulated in Table 3. That is to say, q_{cr} is one of the important criteria to evaluate the superalloys in the susceptibility to the microcrack. In this table, max. heat input in this experiment are used as the critical heat input to avoid microcrack for Inconel 625 and SUS316 in which microcrack never occurred. On the other hand, minimum one is also used as that for Inconel 617 and HVERN in which microcrack invariably occurred. In terms of q_{cr}, the superalloys are below placed in the order of less susceptibility to the microcrack.

(SUS316, Inconel 625), HAEM, HAEN, (HVEN, Incoloy 800) Incoloy 807, (HVERN, Inconel 617)

This order agrees nearly with the above-mentioned one.

Microcrack was widely scattered with considerable irregularity on the $w_B - a_b$ diagram. Therefore, it was difficult to clarify the effect of W_B and a_b parameter on the microcrack. Generally speaking, however, microcrack was apt to disappear both around a_b parameter of 0.9 and with the decrease of bead energy density W_B.

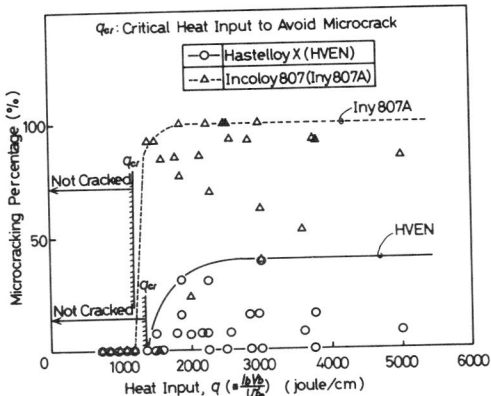

Fig. 7   Effect of Heat Input on Microcracking

Table 3   Critical Heat Input to Avoid Microcrack

Material	q_{cr} (joule/cm)
Hastelloy X (HAEN)	1406
Hastelloy X (HAEM)	2143
Hastelloy X (HVEN)	1286
Hastelloy X (HVERN)	< 750
Inconel 625 (Inl 625AE)	> 5080
Inconel 617 (Inl 617V)	< 750
Incoloy 800 (Iny 800V)	1286
Incoloy 807 (Iny 807A)	1200
SUS 316 (S 316)	> 5080

q_{cr}: Critical Heat Input to Avoid Microcrack

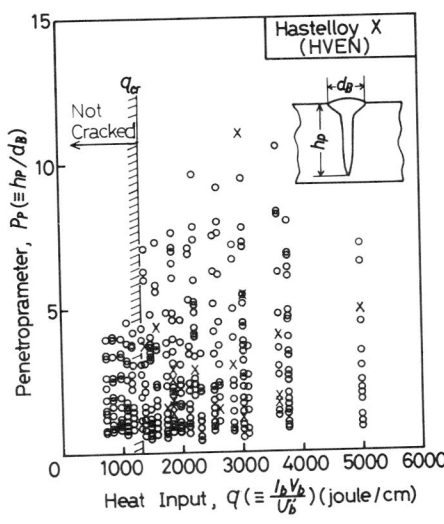

Fig. 8   Effect of Heat Input and Penetroparameter on Microcrack

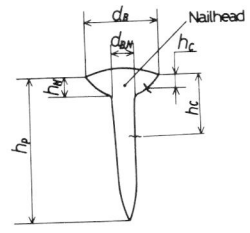

d_B :   Bead Width at Top Surface
d_{BN} :   Minimum Width of Nailhead part
h_P :   Penetration Depth
h_N :   Depth of Nailhead
h_c :   Distance between Top Surface and Microcrack

Fig. 9   Electron Beam Weld Geometry defined

(3) Correlation between Microcrack and Penetroparameter

Fig. 8 shows the example of the effect of heat input and penetroparameter P_P ($\equiv h_p/d_B$, h_P: penetration depth, d_B: bead width) on the microcrack. In every cracked superalloy, there was no definite effect of P_P value on the microcrack. As described above, there was clear correlation between the microcrack and the heat input. In this figure, there is also recognized the critical heat input to avoid microcrack. This corresponds to q_{cr} described above.

(4) Position of Microcrack

Typical geometry of the electron beam welds is herein defined as shown in Fig. 9 to make the microcrack clearer in its position. Every microcrack was carefully examined in its position by the microscope.

Figs. 10 to 12 show typical examples in which the microcrack is plotted on h_N/h_P–h_c/h_N diagram (where, h_N: depth of the nailhead, h_P: penetration depth, h_c: distance between the top surface and the position of microcrack). Inconel 617, HVERN and Incoloy 807, very sensitive to the microcrack, had quite similar features in its position. In case of Inconel 617, most microcracks are

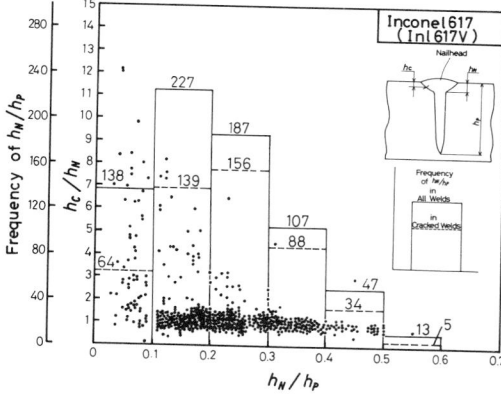

Fig. 10   Effect of h_N/h_P and h_c/h_N on Microcrack

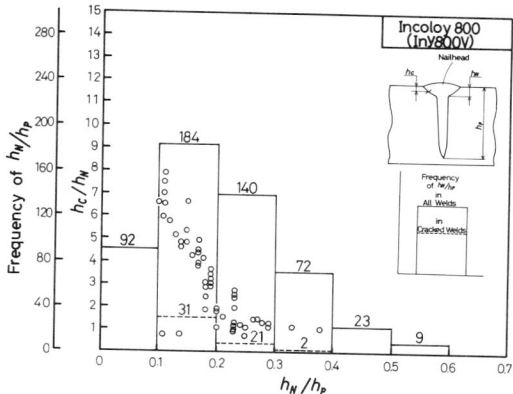

Fig. 12   Effect of h_N/h_P and h_c/h_N on Microcrack

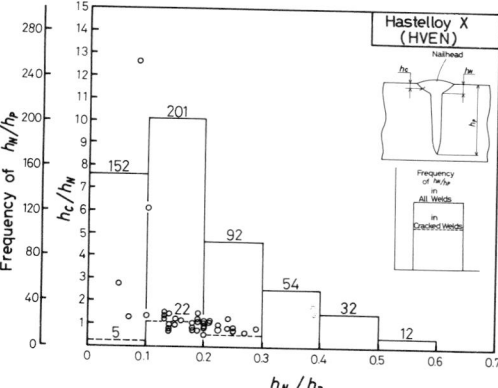

Fig. 11   Effect of h_N/h_P and h_c/h_N on Microcrack

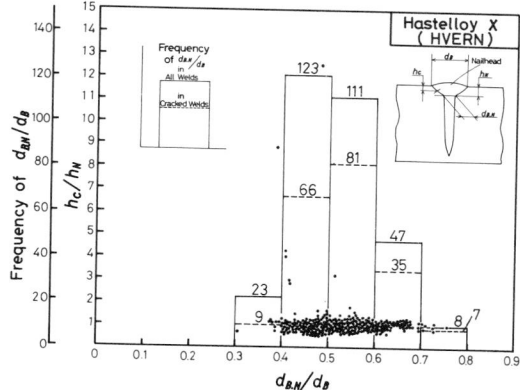

Fig. 13   Effect of $d_{B.N}/d_B$ and h_c/h_N on Microcrack

scattered around h_c/h_N of 1.0 as in **Fig. 10** and called nailhead crack. In this figure, the frequency of h_N/h_P in all the welds and that in the cracked welds are also noted with the solid line and dotted line respectively. However, there is no clear correlation between this frequency and h_N/h_P.

In HAEN, HAEM and HVEN, not so sensitive to the microcrack as the above-mentioned three superalloys, similar tendency was also seen as shown in **Fig. 11**.

In Incoloy 800, microcracks are irregularly distributed regardless of h_c/h_N as shown in **Fig. 12**.

Fig. 13 shows the typical example in which microcrack is arranged on $d_{B.N}/d_B - h_c/h_N$ diagram (where, $d_{B.N}$: min. width of nailhead part, d_B: bead width at the top surface). In this figure, the frequency of $d_{B.N}/d_B$ in all the welds and that in the cracked welds are also noted with the solid line and dotted line respectively.

From this figure, it may be recognized that the microcrack comes to occur easily with the increase of $d_{B.N}/d_B$. Similar tendency was also shown in case of Inconel 617, Incoloy 800 and Incoloy 807.

5. Conclusion

The effect of the welding conditions on some characteristics of electron beam weld bead of the heat-resistant superalloys for the nuclear plants was herein investigated systematically. The obtained conclusion may be summarized as follows.

1) The weld defects were R-porosity at the root of the penetration and microcrack observed continuously in the interdendritic boundary of weld metal and the grain boundary of heat affected zone, which was nearly perpendicular to the fusion boundary.

479

2) R-porosity occurred within the individual areas on w_B-a_b diagram. Superalloys could be evaluated in the susceptibility to the R-porosity in terms of $\dot{S}_p$ and Δa_b. Fe-base superalloy was generally more susceptible to the R-porosity than Ni-base superalloy.

3) Superalloys could be evaluated in the sensitivity to the microcrack in terms of the critical line on w_b-v_b' diagram. Microcracks did never occur by limiting the welding conditions.

4) It was recognized from the effect of the heat input on the microcracking percentage that microcrack was apt to occur at the heat input larger than the critical value which was herein defined by q_{cr}. In terms of this critical heat input to avoid microcrack q_{cr}, superalloys are below placed in the order of less susceptibility to the microcrack.

   (SUS316, Inconel 625), HAEM, HAEN, (HVEN, Inoloy 800), Incoloy 807, (HVERN, Inconel 617) In Inconel 625 and SUS316, no microcrack did occur.

5) There was no clear correlation between the microcrack and penetroparameter P_p. Most microcracks were apt to occur when h_c/h_N came near to 1.0. These microcracks came to occur easily with the increase of $d_{B.N}/d_B$.

References

1) Y. Arata, K. Terai, S. Matsuda et al, "Study on Characteristics of Weld Defect and its Prevention in Electron Beam Welding (Report I) – Characteristics of Weld Porosities –", IIW Doc. IV-112-73 (1973)

2) Y. Arata, K. Terai, S. Matsuda et al, "Study on Characteristics of Weld Defect and Its Prevention in Electron Beam Welding (Report II) –Some Metallurgical Features of Weld Porosities –" IIW Doc. IV-147-74 (1974)

3) Y. Arata, K. Terai, S. Matsuda et al, "Study on Characteristics of Weld Defect and Its Prevention in Electron Beam Welding (Report III) – Characteristics of Cold Shuts –" Trans. of JWRI, Vol. 3 (1974) No. 2, p.81~88

4) J. D. Russel, A. J. Rodgers and R. J. Stearn, "Electron-beam welding of structural steels", Metal Construction & British Welding Journal, Vol. 6 (1974), No. 10, p.307~312

5) M. J. Bibby, J. A. Goldak and G. Burbidge, "Cracking in Restrained EB Welds in Carbon and Low Alloy Steels", Welding Journal, Vol. 54 (1975), No. 8, p.2535~2585

6) Y. Arata, "Terms and Definitions proposed from Japan" IIW, 1972

7) Y. Arata, M. Tomie, K. Terai et al, "Shape Dicision of High Energy Density Beam" Trans. of JWRI, Vol. 2 (1973), Nol 2, p.131~146

Tandem Electron Beam Welding

Abstract

In case of high speed welding, by using the conventional single electron beam welding method it was seen that the formation of irregular bead such as undercutting and humping is occurred under the condition that the flow of molten metal is restricted in narrow channels formed in central region of the pool cavity walls because the solidified wall had been produced on the top and bottom parts of lateral walls in the pool cavity during welding, simultaneously the pool cavity length becomes very long.

By the development and use of the TEB-welding method, we have proved the possibility of preventing the irregular bead formation, and it was clarified that the pool cavity length becomes short, while the flow of molten metal behind the pool cavity is dammed up by a new beam hole, and this flow direction of molten metal is changed toward the lateral walls associated with the solidified wall, there the molten metal is deposited on this solidified wall, and consequently the flow path of molten metal is broaden out and sound bead can be obtained even at high welding speed of 10 m/min where irregular beads appear by using the single electron beam welding method.

1. Introduction

Electron beam welding method, as is well known, has many superior characteristics such as producing a high quality and high efficiency in the welded part. And in recent years, it has been desired, still more, to increase the welding speed to reduce the welding time in production industry. High welding speed, however, introduces a serious problem of the occurrence of irregular weld bead formation such as humping and undercutting. This formation of irregular beads limits the welding speed to an economically undesirable value.

In arc welding, it has been seen that the appearance of these irregular beads is associated with the flow patterns of molten metal behind the arc[1],[2]. In electron beam welding, little investigation[3],[4] has been done with respect to the formation of these irregular beads at high welding speed, so that there is no reliable technique to prevent the occurrence of these irregular beads.

The aim of this work is to observe the formation phenomena of these irregular weld beads in detail by investigating the configuration of irregular bead formed by the electron beam in high welding speed, and to prove the possiblity of preventing the occurrence of these irregular beads by the use of the TANDEM ELECTRON BEAM WELDING method. There, the term of the "TANDEM ELECTRON BEAM WELDING" is named after that the dual electron beam is utilized in tandem and we call this, "TEB-welding"[5] as a short name.

2. Experimental apparatus and procedure

A schematic diagram of the TEB-WELDER is shown in **Fig. 1.** Two streams of electron beam are produced by Gun-1 and Gun-2, both of them are settled compactly and EB-1 is perpendicularly applied to the surface of a specimen, while EB-2 is controlled to applied on the proper position of a specimen surface using the beam deflector.

EB-1 is used as a heat source to melt the specimen with full-penetration, and EB-2 is used to control the flow condition of molten metal. Here, the distance L_b as shown in figure, between the location of EB-1 and EB-2 on the surface of a specimen is named-for "TANDEM gap", and EB-1 is called the "leading" electron beam and EB-2 is called the "trailing" electron beam with respect to the welding direction respectively.

Each power of these beams is 6 KW max. (60 KV, 100 mA max.), and total power is 12 KW max.. Besides, its beam power is arbiturarily selected by

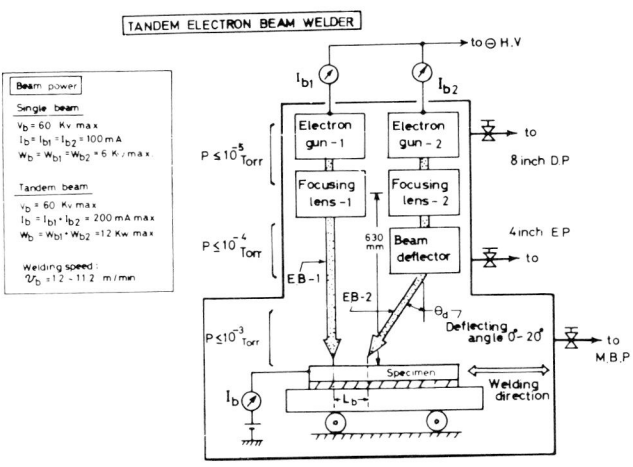

Fig. 1 Schematic diagram of TANDEM ELECTRON BEAM
WELDING apparatus.

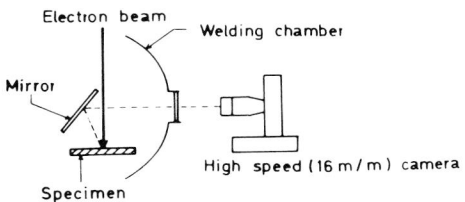

Fig. 2 Observation of flow condtion of molten metal during
welding by use of High-Speed Camera with turnning
mirror.

changing the beam current of each gun independently
under the constant beam voltage.

Welding speed of this apparatus is in the range of
1.2～11.2 m/min, and all of experiments in this work
were carried out in bead-on-plate welding. Specimen
used is austenitic stainless steel, and chemical compositions of specimen are shown in **Table 1,** and size
of specimen is 300 mm in length, 80 mm in width, and
1.2 and 2.0 mm in thickness.

The flow condition of molten metal was observed
using by the high-speed camera with turnning mirror
arranged as shown in **Fig. 2.**

3. Single electron beam welding
—Characteristics of irregular bead formation—

The crater-cavities near the molten puddle in the
case of the conventional single electron beam welding
are shown in **Photo. 1.** Both typical high-speed
photographs and its schematic figures during welding
are shown in **Photo. 2** and **Fig. 3** respectively. As
shown in them, here we named a hole behind the
impinging electron beam, "pool cavity" and its length,
"pool cavity length L_c" and we call the region of
molten metal behind this pool cavity, "main molten
pool".

At a low welding speed of 3 m/min the shape of
weld bead is sound, at a high welding speed of 6 m/
min humping bead appears and this is a irregular
weld bead.   Moreover it is clear that the cavity length

Table 1 Chemical compositions of specimen used

SUS 304 Wt(%)	C	Si	Mn	P	S	Fe	Ni	Cr
1.2 mmt.	0.078	0.66	0.99	0.025	0.007	Bal.	8.66	18.18
2.0 mmt.	0.049	0.55	1.03	0.032	0.009	Bal.	9.19	18.30

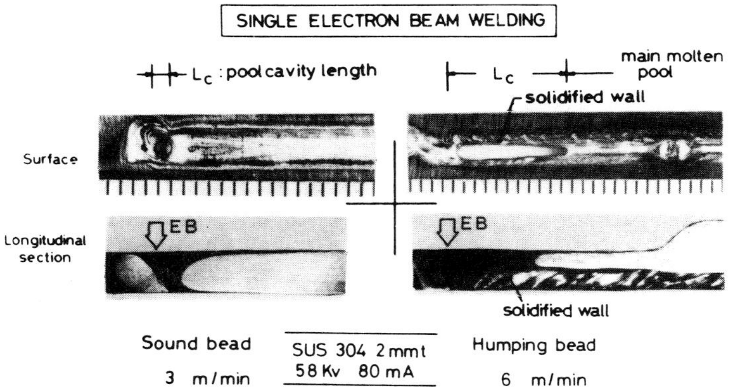

SINGLE ELECTRON BEAM WELDING

L_c : pool cavity length

main molten pool

L_c

solidified wall

Surface

Longitudinal section

EB

EB

solidified wall

Sound bead

3 m/min

SUS 304 2mmt
58 Kv 80 mA

Humping bead

6 m/min

Photo. 1 Shapes of weld beads near the molten puddle using by convention single electron beam welding method with low and high welding speed.

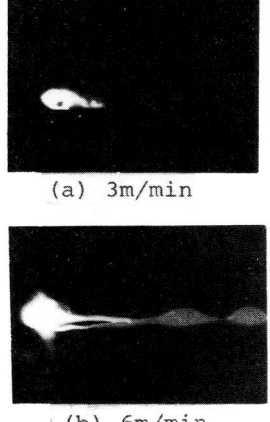

(a) 3m/min

(b) 6m/min

Photo. 2 (a) (b) Typical High-speed photograph during conventional single electron beam welding method. 58 kV-80 mA SUS 304 2 mmt (a) 3m/min, (b) 6 m/min.

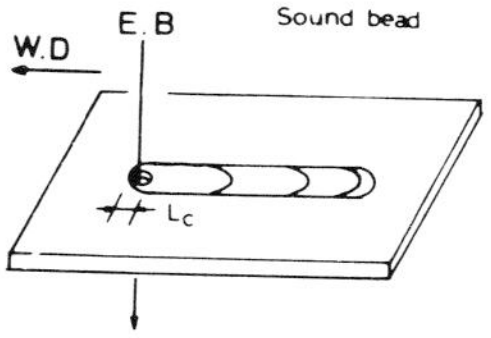

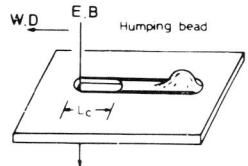

W.D   E.B   Sound bead

L_c

(a) Low welding speed

W.D   E.B   Humping bead

L_c

(b) High welding speed

Fig. 3 (a), (b) Schematic figure of conventional single electron beam welding during welding.

L_c in the case of humping bead is longer compared with the case of sound bead since the main molten pool of humping bead goes backwards remarkably with respect to the welding direction.

While, on the photograph of the longitudinal section of humping bead it is seen that a broad solidified wall* as shown in **Photo. 2,** had been existed during welding, on the top and bottom parts of lateral walls in the pool cavity but we can find no solidified wall in sound

bead. The formation process of this solidified wall along both lateral walls on each cross-section of humping weld bead is shown in **Photo. 3,** and its schematic figure is shown in **Fig. 4,** easily to realize the situation of **Photo. 3.**

Molten metal which flows along the wall surface inside the pool cavity increases gradually as molten metal advanced from (a–1) to (a–3) as shown in **Photo. 3.** On the other hand, the scale of solidified

---

* Here, "solidified wall" is defined to be the region where the thin layer of molten metal solidified on the top and bottom parts of both lateral walls in the pool cavity.

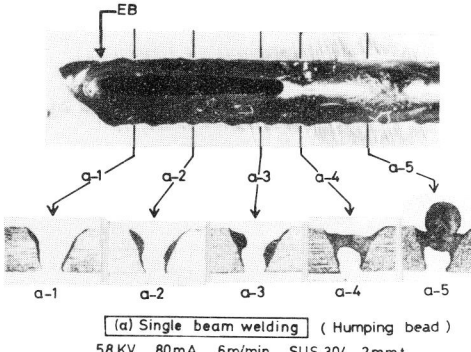

(a) Single beam welding (Humping bead)
58 KV 80 mA 6 m/min SUS 304 2 mm t.

Photo. 3 Flow condition of molten metal and growing process of solidified wall along lateral walls in pool cavity of humping bead.

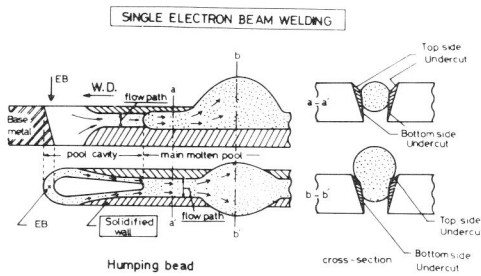

Fig. 4 Configuration of humping bead associated with broad solidified wall, narrow flow path of molten metal.

wall grows rapidly because the more molten metal goes away from the location of electron beam as a heat source and the faster molten metal is cooled. And moreover, the flow of molten metal is gathered together on the central region of pool cavity wall by the force of surface-tension of molten metal, at last these flows which exists on both lateral walls are bridged each other, there this pool cavity becomes the main molten pool.

According to these observations, it is seen that the solidified wall as shown by shadowed portion in **Fig. 4,** restricts the flow path of molten metal in narrow channels which are formed in the central region of pool cavity walls, and therefore it brings about finally undercutting in the weld bead as shown in (a–4) and a–a' cross-section of **Phto. 3** and **Fig. 4**, and its inner pressure[1] caused by the surface configuration of molten metal is higher than that of the sound bead, then as shown in (a–5) of **Photo. 3** and b–b' of **Fig. 4** the

shape of weld bead consequently becomes humping.

4. Tandem electron beam welding

(a) Principle of preventing irregular weld bead formation

By the use of the TEB, we can prevent the occurrence of irregular bead formation as schematically described in **Fig. 5**. EB-1 is used as a heat source, as was stated previously, to melt the specimen with full-penetration in much the same manner as the case of the conventional single electron beam welding. Of course, using by the single electron beam welding of only EB-1 at high welding speed the irregular bead formation appears.

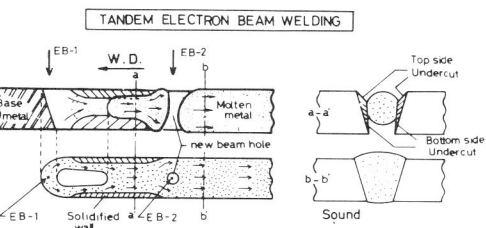

Fig. 5 Principle of preventing occurrence of irregular bead formation using by TEB-Welding method.

Then, when EB-2 of the TEB is impinged in the molten pool produced by EB-1, EB-2 generates a new beam hole as shown in **Fig. 5**. This new beam hole has a function of changing the flow direction of molten metal toward both lateral walls where the solidified wall formed by EB-1 exists, and the following phenomena arise in there; this molten metal is deposited on the surface of solidified wall which causes the irregular weld bead to occur, therefore the flow path of molten metal in the molten pool is consequently broaden out and its inner pressure is reduced compared with the case of single beam welding. In this way, the irregular bead formation can be prevented and sound bead is obtained. This is a principle of preventing the irregular bead formation at high welding speed by the use of the TEB.

(b) Important welding parameters in TANDEM ELECTRON BEAM WELDING

In the TEB-welding process, the following two welding parameters should be selected carefully in order to prevent the irregular bead formation for the purpose of controlling the flow condition of molten metal effectively in high welding speed.

(1) Location of applied trailing electron beam (EB-2) with respect to the molten pool produced by leading electron beam (EB-1).

(2) Ratio of beam power of trailing electron beam (EB-2) to that of leading electron beam (EB-1).

(1) Location of impinging EB-2

In the TEB-welding process, as shown in **Fig. 6** the location of impinging EB-2 with respect to that of EB-1 can be classified into A, B, C three regions according to the configuration of humping bead formed by the single beam welding of only EB-1.

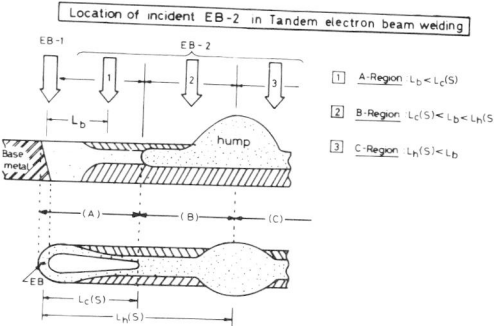

Fig. 6 Location of impinging EB-2 with respect to molten pool caused by single electron beam of only EB-1.

Here, in **Fig. 6** the location of incident EB-2 is given as the TANDEM gap L_b as was stated previously, and $L_c(S)$ is defined to be the cavity length made by the single beam welding of EB-1, $L_h(S)$ is the distance between the location of EB-1 and its hump in the single electron beam welding.

Further, $L_c(T)$ is defined to be the cavity length formed by the TEB-welding, and $L_h(T)$ is the distance between the location of impinging EB-1 of the TEB and its hump.

[1] A-Region: $L_b < L_c(S)$

When EB-2 is applied in the hole of pool cavity formed by the single electron beam welding of only EB-1, the occurrence of irregular bead formation is not prevented and the humping phenomena only be promoted in this region as shown in **Photo. 4** and **Fig. 7** of typical high-speed photograph and its schematic figure near the molten pool, respectively. In this case the force of evaporating recoil pressure

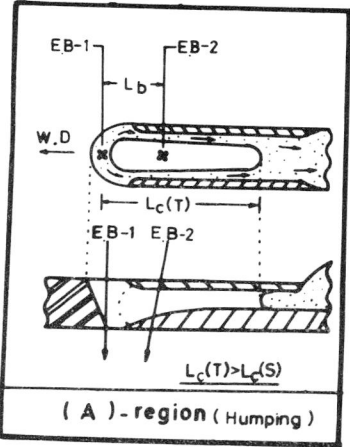

Fig. 7 Schematic figure of A-Region during TEB-Welding.

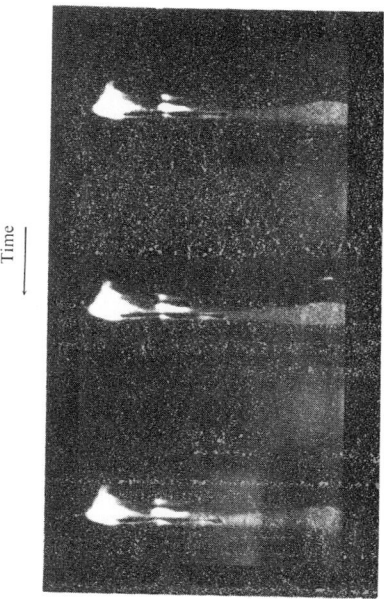

Photo. 4 Bead formation in A-Region during TEB-Welding. 58 kV-80 mA (EB-1), 58 kV-20 mA (EB-2) Lb=4mm 6 m/min SUS304 2 mmt (1800 pfs).

generated by EB-2 accelerates the flow rate of molten metal in the pool cavity, so EB-2 would rather make the bridging point of molten metal move toward backwards remarkably with respect to the welding direction, therefore the cavity length $L_c(T)$ is far longer than $L_c(S)$.

485

A typical high-speed photograph and schematic figure near the molten pool in B-Region are shown in **Photo. 5** and **Fig. 8** respectively.

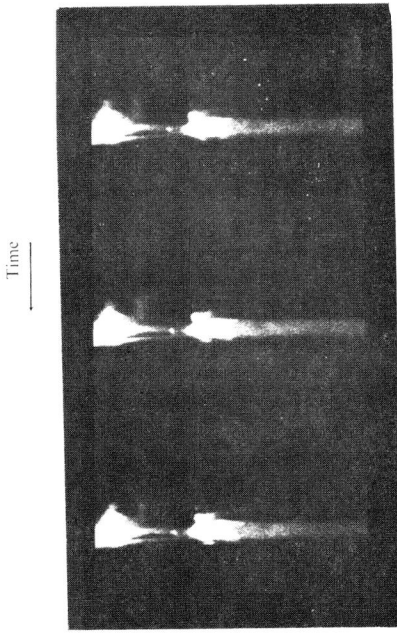

Photo. 5 Bead formation of successful preventing process in B-Region during TEB-Welding. 58 kV-80 mA (EB-1), 58 kV-20 mA (EB-2) Lb=7 mm 6 m/min SUS304 2 mmt (1800 fps).

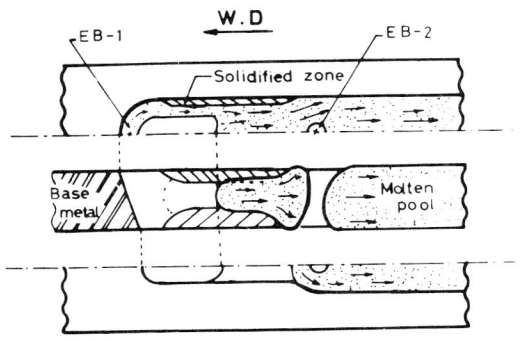

Fig. 8 Process of successful preventing iregular bead in B-Region during TEB-Welding.

When EB-2 is impinged on the surface of main molten pool between the end of pool cavity and its hump produced by the single electron beam welding of only EB-1, the shape of weld bead is sound.   In this region, as the flow of molten metal caused by EB-1 is dammed up effectively by the newly generated beam hole of EB-2, the flow direction of molten metal is changed toward both lateral walls where the solidified wall produced by EB-1 exists, besides the backward movement of molten metal is restricted and this pool cavity $L_c(T)$ becomes shorter than $L_c(S)$, while the perturbation of pool cavity is decreased, and humping phenomena and any irregularity no occur in there, as shown in **Photo. 5.**

An example of sound bead produced by the proper controlling the flow of molten metal by EB-2 is shown in **Photo. 6** with each various cross sections of crater-

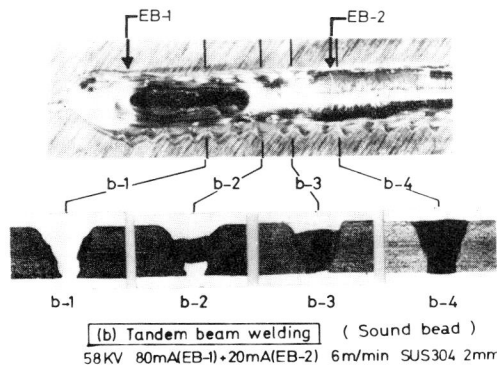

Photo. 6 Example of sound bead produced by proper controlling flow of molten metal by EB-2 in B-Region using by TEB-Welding method. Lb=7 mm.

cavity during the TEB-welding.   In the lateral walls of crater cavity as shown in **Photo. 6,** the solidified wall on the cross section (b-1) and (b-2) appears with same formation process as (a-1)~(a-3) shown in **Photo. 2** of the single beam welding.

But on the cross section (b-3) in front of the location of impinging EB-2 the molten metal flows in the bottom side of bead, and behind of that of inpinging EB-2 the top side of bead is smoothly filled up with the molten metal as shown in (b-4), there a sound bead is completed without humping and undercutting. On the other hand the flow path of molten metal on (b-4) is broader compared with that on (b-2).

[3]   C-Region: $L_b > L_h(S)$

A typical high-speed photograph and schematic illustration near the molten pool in C-Region are shown in **Photo. 7** and **Fig. 9.** In this case, when EB-2 is applied at a little farther than $L_h(S)$, where the molten metal formed by the single beam welding of only EB-1 is not solidified yet.

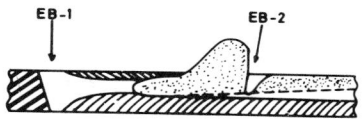

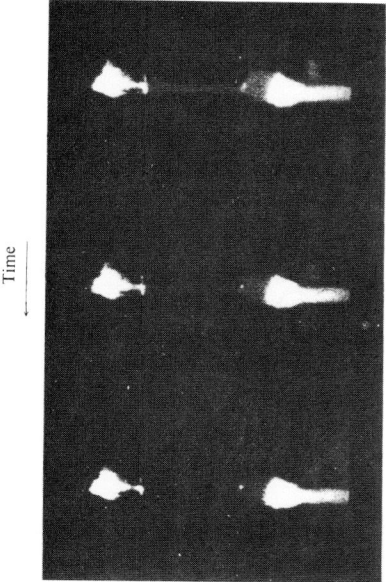

Fig. 9 Explanation of "Swelling" phenomenon peculiar to TEB-Welding in C-Region.

The solidified wall is only partially filled with the molten metal, and the cavity length $L_c(T)$ is nearly equal to $L_c(S)$ because the backward movement of main molten pool can not be obstructed.

Besides, in this region the "Swelling" phenomena appear, which causes an irregular weld bead peculiar to the TEB-welding. This "Swelling" phenomena occur as follows; the hump made by EB-1 meets the beam hole of EB-2 during this humping is moving backwards with respect to the welding direction with almost same velocity as the moving speed of specimen, and this beam hole of EB-2 restricts the movement of this hump in front of EB-2 also this hump remains with EB-2 during welding. While, when the molten metal of this hump increases gradually with time until the beam hole of EB-2 can not hold the mass motion of this hump, the hump is destroyed and simultaneously breaks away from the restriction of EB-2. This process is repeated periodically as described above and is named-for "Swelling" phenomena.

In summarization, **Photo. 8** shows typical shapes

Photo. 7 "Swelling" phenomenon in C-Region during TEB-Welding. 58 kV-80 mA (EB-1), 58 kV-20 mA (EB-2) Lb=16 mm 6 m/min SUS304 2 mmt (1800 fps).

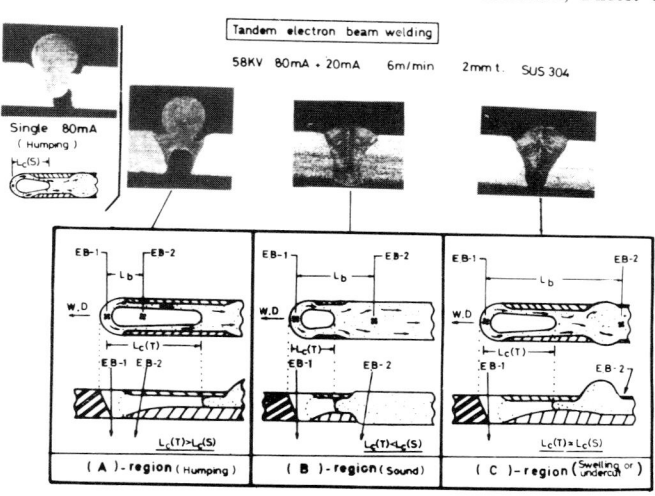

Photo. 8 Bead shapes in A, B, C-Region using by TEB-Welding.

487

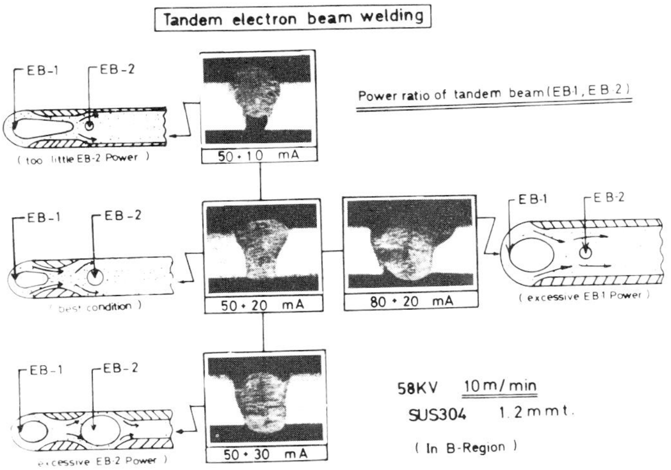

Photo. 9 Influence of power ratio of TEB on bead formation in each B-Region.

of weld beads in A, B, C-Regions by the use of the TEB-welding method.

(2) Beam power ratio of TANDEM ELECTRON BEAM

In this section, another important welding parameter, namely the beam power ratio of the TEB will be described here. In this experiment, welding speed is selected at a very high speed of 10 m/min using thinner metal plate of 1.2 mm thickness.

Their experimental results are shown in **Photo. 9** with schematic figure near the molten pool during the TEB-welding in each B-Region. In this situation, the conventional single electron beam welding, of course, only make a humping bead formation, but when the TEB is used the occurrence of irregular bead formation is prevented and sound bead can be obtained.

In this case, the best condition of beam power ratio to obtain a sound bead is 50 mA 58 KV and 20 mA 58 KV for EB-1 and EB-2 as shown in **Photo. 9.** When EB-2 power is a little low of 10 mA 58 KV than the best condition under a constant EB-1 power of 50 mA 58 KV, the solidified walls on the top and bottom parts of lateral walls produced by EB-1 are not filled up with the molten metal completely, therefore the broad undercutting remains, particularly on the bottom part of weld bead. While, when EB-2 power is a little large of 30 mA 58 KV over the best condition under a constant EB-1 power of 50 mA 58 KV, the solidified wall appears newly again on the lateral walls

of bead near the location of impinging EB-2 since the diameter of beam hole of EB-2 enlarges more of the distance between both lateral walls containing the solidified wall caused by EB-1, then the shape of weld bead is irregular associating with undercutting on the top part of specimen.

Next, we change the power of EB-1 under a constant EB-2 power of 20 mA 58 KV. When EB-1 power is larger of 80 mA 58 KV than the best value of 50 mA 58 KV. Indeed, the beam hole of EB-2 can change the flow direction of molten metal formed by EB-1, but they are not enough to fill the solidified wall because the width of bead produced by EB-1 power is larger compared with the beam hole diameter of EB-2, this welding situation is similar to the case of the beam power ratio of 10 to 50 mA (58 KV) with respect to the surface of specimen, and the undercutting formation can not be prevented. As described above, the power ratio of the TEB is a very important welding parameter to prevent these irregular bead formation.

5. **Conclusion**

In this study, the formation phenomena of the irregular weld beads are observed by the investigating the configuration of molten pool near the pool cavity, and the new welding method for preventing the irregular bead has been developed and named, "TANDEM ELECTRON BEAM WELDING"

488

method. Moreover, the process of preventing the occurrence of the irregular bead is investigated.

Results obtained are stated as follows:

(1) When the conventional single electron beam welding method is used for the high speed welding of a thin metal plate, the irregular beads such as humping and undercutting appear.

The irregular bead formation is occurred under the condition that the flow of molten metal along the lateral walls in the pool cavity is restricted in narrow channels which are formed in central region of pool cavity walls because the occurrence of the solidification of molten metal had existed at the top and bottom parts of these walls during welding, simultaneously the length of pool cavity becomes very longer compared with that of sound bead and the pool cavity is perturbed.

(2) When the TANDEM ELECTRON BEAM WELDING method is used, the occurrence of irregular bead formation can be prevented and sound bead is obtained without humping and undercutting even at high welding speed of 10 m/min where irregular bead formation appears by the use of the conventional single electron beam welding method.

The process of preventing the occurrence of these irregular weld beads is as follows; the flow of molten metal behind the pool cavity formed by the leading electron beam (EB-1) is dammed up by a generated beam hole of the trailing electron beam (EB-2), and the flow direction of molten metal is changed toward the both lateral walls where the solidified wall exists, there this solidified wall is filled up with the molten metal,

and consequently the flow path of molten metal is broaden out and sound bead is obtained. While the length and perturbation of the pool cavity is reduced.

In practice, the important welding parameters of the TEB-welding are the location of the trailing electron beam (EB-2) impinged with respect to the configuration of molten pool caused by the leading electron beam (EB-1), and the ratio of beam power of trailing electron beam (EB-2) to that of the leading electron beam (EB-1).

Acknowledgement

We are pleased to acknowledge the considerable assistance of Professor Dr. F. MATSUDA, Associate Professor Dr. M. USHIO, and Professor Dr. N. IWAMOTO. Thanks are due to Associate Professor Dr. S. MIYAKE, Mr. K. KATADA and Mr. M. OHARA with whom we have discussed this problem.

References

1) B.J. Bradstreet: "Effect of surface tension and metal flow on weld bead formation", B.W.J. vol. 47, 1968.
2) G.K. Hicken, C.E. Jackson: "Effects of applied magnetic field on welding arcs", J.A.W.S. vol. 45, 1966.
3) Y. Arata, E. Nabegara, F. Matsuda, M. Ushio: "Study on High Speed Welding Using by Electron Beam Welding Method", Comittee of Electron Beam Welding of J.W.S. No. EBW-157-76, 1976 (in Japanese).
4) Y. Arata, E. Nabegata, F. Matsuda, M. Ushio: "Study on Tandem Electron Beam Welding Method", Preprints of the National Meeting of J.W.S. No. 17, 1976 (in Japanese).
5) Y. Arata: "Terms and Definition for Electron Beam Welding, Laser welding and Laser Cutting Used in Japan", IIW Doc. IV-229-77.

Laser Beam Welding

Mechanism of Bead-Transition in Laser Welding
I. Miyamoto, H. Maruo and Y. Arata
[Int. Conf. on Welding Research in 1980s.]

The Role of Assist Gas in CO_2 Laser Welding
I. Miyamoto, H. Maruo and Y. Arata
[ICALEO '84. Int. Beam Tech. Conf. in Essen (1980).]

CO_2 Laser Welding of Ceramics
H. Maruo, I. Miyamoto and Y. Arata
[1st Int. Laser Proc. Conf. (1981).]

Fundamental Phenomena during Vacuum Laser Welding
Y. Arata, N. Abe, T. Oda and N. Tsujii
[ICALEO '84.]

Mechanism of Bead-Transition in Laser Welding

A "bead-transition" has been shown to occur in laser welding; the bead changes substantially in depth with a slight change in lens-specimen distance at a critical lens-specimen distance. At the critical distance the specimen surface has been shown to reach the boiling point of the material, at which a cavity is formed by the recoil force of evaporation. The shape of the cavity has been approximated by a wedge, and the cavity depth has been analized from multiple reflections of the incident laser beam in the cavity for clarifying the bead-transition mechanism. The calculated values have agreed well with experimental data.

I.   Introduction

There has been a great deal of interest in the industrial application of laser beam welding, since lasers can be used as a practical heat source for deep penetration welding [1][2] even in the atmosphere. The mechanism of laser beam welding, however, is not yet well understood; the laser welding is expected to have many different features from electron beam welding, because the laser beam has very high reflectivity to metal and very shallow depth of focus in comparison with the electron beam. This report describes the mechanism of bead-transition, which gives insight into the mechanism of laser welding, on the basis of beam-material interaction and the thermal conduction theory.

II.   Experimental procedures

In this experiment, 1 KW class CO_2 laser (Sylvania Inc., Model 971) was used. The output beam was focused downwards using a plane mirror and a lens (focal length f: 38, 64, 127 and 254mm). The beam diameter at the lens was about 16 mm. An acrylic plastic specimen was moved quickly across the beam perpendicularly to the beam axis, and then the beam power density distribution at any distance from the lense was approximated by contour of the cross-section of the groove vaparized in the specimen. The resuting power densities are approximated by a Gaussian curve:

$$w(r) = \frac{W}{\pi r^2} \cdot \exp(-r^2/a^2) \ \ldots\ldots (1)$$

where r is the distance from the beam axis and W the laser power. The beam diameter is defined by 2a according to this equation.
Figure 1 represents a plot of the

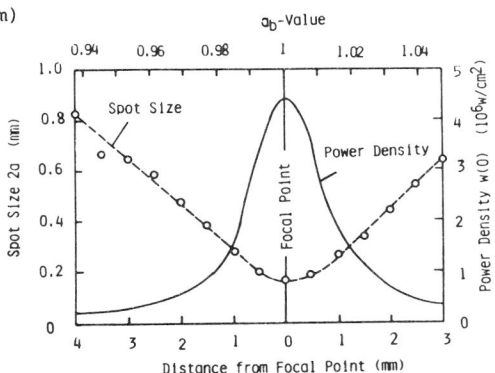

Fig. 1   Distribution of spot size and power density at beam axis at 1 KW power level.

beam diameter 2a and the power density on the axis w(0) versus the distance from the focal point for f=64 mm at 1 KW power level.

In this experiment, bead-on-plate run test was carried out at 1 KW power level measured on the workpiece of 18-8 stainless steel, which provides reproducible welding response. A helium gas jet was used to remove plasma produced at the beam impinging point, which tends to reduce the penetration depth.

III. Bead-transition

In Fig. 2, the dimensions of the welded bead at 0.5 m/min are plotted as a function of focal position of various focal length lenses. The focal position with respect to the workpiece surface is presented by an a_b-value, a ratio of the lens-specimen distance to the focal length. As the specimen surface approaches the focal point, it is seen that the bead changes rapidly from semicircular to a typical deep penetration bead, unlike the bead of electron beam welding; the depth increases approximately by five times, the width two times and the cross-sectional area ten times, in this case. The authors named this phenomenon that a slight change in the a_b-value substantially changes the shape and dimensions of the bead, a "bead-transition".

The substantial increase in the bead cross-sectional area, about a factor of ten in this case, associated with the bead-transition, suggests the corresponding increase in the beam coupling coefficient of the specimen. It is also noticed in Fig. 2 that the beam convergent side ($a_b < 1$) provides sharper bead-transition curves and

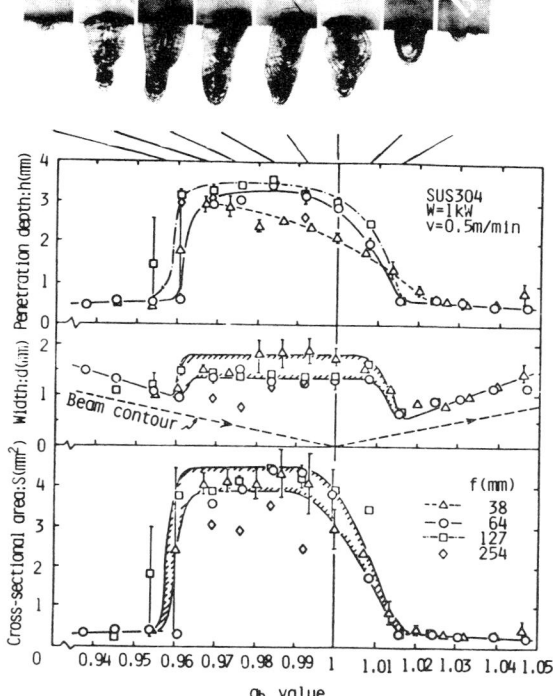

Fig. 2 Effect of a_b-value on penetration depth, width and cross-sectional area at 0.5 m/min.

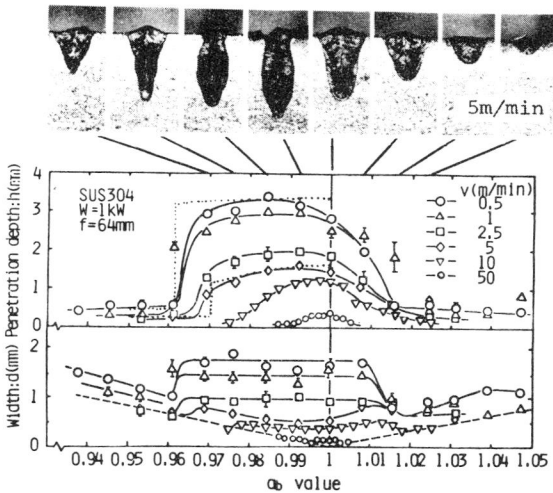

Fig. 3 Effect of a_b-value on penetration depth and width for various v.

493

larger spot size for the bead transition point than the beam convergent size ($a_b > 1$). The bead widths between two transition points are almost indipendent of the focal position, and are much larger than the beam spot size on the specimen. Whereas the bead width on the outside of the transition region appears to vary depending on the beam spot size.

As the travelling speed increases, the transition curves of the penetration becomes less sharp, with the inceased power density for the transition point, whereas the bead width decreases, which remains almost constant between two transition points at a given speed, as shown in Fig. 3. (In this figure, the solid lines through the data points are for convenience in viewing and do not represent a theoretical curve.) The spot sizes at the bead-transition points are plotted against the travelling speed in Fig. 4.

At a given speed, the maximum penetration depth h_m occurs at the center of these two points. The range of the focal position within which an appreciable fraction of h_m is obtained, tends to decrease with increasing speed. It should be noticed that very precise focal positioning in the laser welding is required when compared with EB welding, especially at high welding speeds.

IV. Mechanism of bead-transition

(1) Surface temperature

In the previous paper[3], the authors analyzed the temperature distributions in semi-infinite solids due to a Gaussian heat source moving at a constant speed v. Based on this analysis, the maximum temperature attained in the surface is approximated by

$$\theta_s = \frac{AW}{16\sqrt{\pi}ka} \cdot \frac{7.7}{\sqrt{va/2\alpha}+0.95} \quad \ldots \ldots \ (2)$$

where A is the absorptivity at the fusion point, k the thermal conductivity and α the thermal diffusivity of the material.

The relationship between the spot size and the travelling speed is plotted with solid lines for various values of θ_s in Fig. 4, where k= .061 cal/sec·cm·°C. $\alpha=0.045$ cm^2/sec [4] and A= 0.13[5] were adopted. It is seen that on the

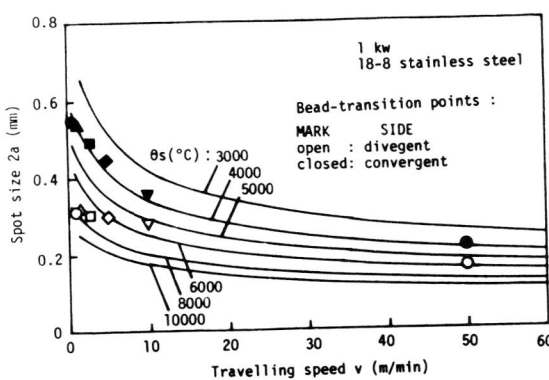

Fig. 4 Travelling speed vs spot size at bead-transition points. Solid lines show calculated values.

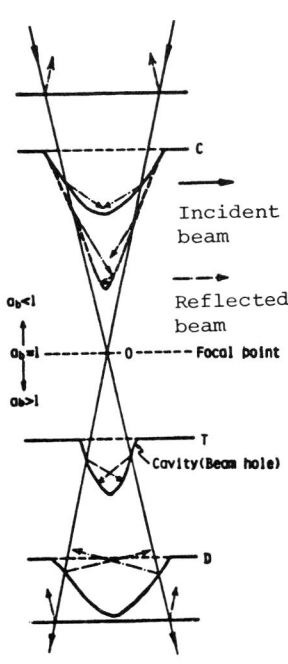

Fig. 5 Schematic illustration showing beam-material interaction.

convergent side of the transition, the temperature at the bead-transition point corresponds to the calculated value of about 4000°C. Since a discrepancy in the temperature of about 1000 °C between the boiling point of the material used, about 3000°C, and this calculated value may be regarded to be acceptable when it is considered that vaporization should occur not only at a given point but over an area as large as the cavity entrance, the bead-transition can be assumed to occur when the surface is violently vaporized. Much higher temperatures, on the other hand, were calculated from the data on the divergent side of the transition.

(2) Interaction between beam and material

Based on the results mentioned above, the authors have developed a model for material-beam interaction as shown schematically in Fig. 5, which qualitatively explains the bead-transition in laser welding. C and D represent the focal positions where 4000 °C is reached at the surface. On the outside of the region C-D, the bead is very shallow and semi-circular, since the most incident beam is reflected away owing to no cavity formation and high reflectivity of the metal. The liquid surface at C and D is depressed by the recoil force of evaporation, and then a shallow cavity is produced. On the divergent side, the depression of the liquid surface at D moves the cavity bottom away from the focal points to decrease the power density at the bottom, so that the shallow cavity cannot grow any longer. To the contrary, however, the depression of the liquid surface on the convergent side at C allows the growth of the cavity, because the bottom approaches the focal point. Increase in the power density at the bottom is enhanced by a wall-focusing effect[6], which is caused by refocusing the beam reflected from the cavity wall, and is available only on the convergent side. Besides, at the same time the increase in the cavity depth leads to an increased number of reflections, thereby increasing the beam power deposition in the cavity.

Thus, once a shallow cavity is produced at C, the cavity is allowed to deepen until the deposited power in the cavity is balanced by the heat conduction power required for producing the corresponding bead. On the divergent side, the bead transition does not occur until the beam power density becomes high enough to produce a slender, deep cavity at T where the wall-focusing effect on this side becomes available.

(3) Fundamental equations for heat balance

In this section, the bead-transition mechanism is quantitatively analyzed based on the heat balance between the deposited beam power in the cavity and the thermal conduction power required to maintain the corresponding bead.

For the sake of simplicity, it was assumed that the cavity is of wedge shape with depth h and width s as shown in Fig. 6, and that the laser-produced plasma is successfully removed to give no influence on the beam path. The former assumption is considered to be available only on the convergent side where the reflected beam concentrates to the bottom of the cavity to make the bottom sharp in shape. Although transition behavior on the convergent side alon is dealt with here, the phenomenon on the divergent side is considered to be qualitatively the same as that on the convergent side.

Consider the reflection behavior of the incident beam to the wedge-shaped cavity. The incident beam proceeds to the bottom of the cavity via multiple reflections.

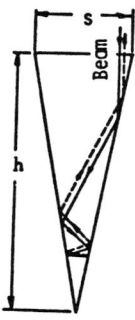

Fig. 6 Beam reflection in wedge-shaped cavity.

The incident angle of the beam to the cavity wall decreases by $\tan^{-1}$ (s/h) for each reflection. At the n-th reflection, the incident angle of the beam to the cavity wall is given by

$$\alpha = \frac{\pi}{2} - \frac{2n-1}{2} \tan^{-1} \frac{s}{h} - \theta \quad \cdots\cdots\cdots\cdots (3)$$

where θ is the convergent angle of the incident beam. When α becomes less than zero, the beam begins to recede from the bottom. Then the total number of the reflections in the cavity is approximately

$$n = \pi/\tan^{-1}(s/h) \quad \cdots\cdots\cdots\cdots\cdots\cdots (4)$$

where θ was neglected in comparison to $\pi/2$ for usual optical systems. Thus, the beam power deposited in the cavity for the incident laser power W is given by

$$W_{ab} = W[\ 1-(1-A)^{\pi/\tan^{-1}(s/h)}] \quad \cdots\cdots (5)$$

On the other hand, the power required to produce the bead having width d and depth h is expressed is the approximated equation of Wells[7]:

$$W_{cond} = 8k\theta_f\ (\ 0.2 + vd/4\alpha\ )h \cdots\cdots\cdots (6)$$

where θ_f is the melting point of the material. Figure 7 shows a plot of W_{ab} and W_{cond} versus cavity depth h for v=0.5 m/min and 5 m/min. In this computation, it was assumed that s is equal to 2a, d is given by the experimental data of the constant bead width between the transition points in Fig. 3, and the bead depth is equal to the cavity depth.

The value W_{cond} for large cavity depth is given by Eq. 6. For comparatively small cavity depth, W_{cond} was assumed to approach the value given by Eq. 6 progressively from the value of h=0, which is equal to the power required to heat the surface up to the boiling point of the material, along the curves appropriately drawn in Fig. 7. When the beam spot size 2a is larger than the critical values, about 0.6 and 0.4 mm for 0.5 m/min and 5 m/min, respectively, it is seen that the cavity cannot grow because $W_{ab} < W_{cond}$ is valid. When the spot sizes become a little smaller than the critical value, $W_{ab} > W_{cond}$ becomes valid at the surface, and therefore the cavity can grow up to the depth where W_{ab} is balanced by W_{cond}, so that the bead transition occurs.

The bead depth thus calculated is plotted with dotted curves in Fig. 3, and is seen to agree well with experimental data both qualitatively and quantitatively. The sharp bead-transition observed at 0.5 m/min is caused from the high coupling coefficient of the beam at the transition point so that further increase in the number of reflections produces almost no increase in the coupling coefficient, therefore almost no increase in the cavity depth. At a high speed, 5 m/min, on the other hand, since the bead-transition occurs at

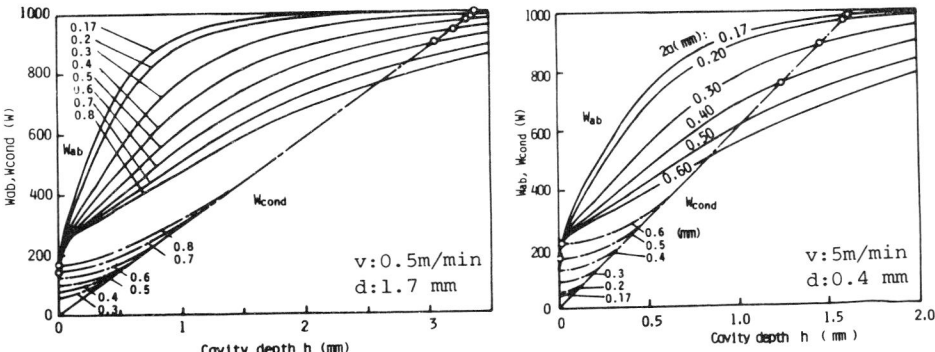

Fig. 7 W_{ab} and W_{cond} vs vavity depth at (a) 0.5 m/min and (b) 5m/min

comparatively low coupling coefficient, further decrease in spot size can increase the coupling coefficient, therefore the cavity depth, providing less sharp transition curves.

V. Concluding remarks

It has been shown that the beam coupling coefficient of the specimen very much depends on whether or not a cavity is produced, and how this phenomenon leads to the bead-transition.

As a welding heat source, laser beam is characterized by a small depth of focus and high reflectivity to metal when compared with EB welding. The former characteristics are due to LB's longer equivalent wavelength, and limit the penetration depth to smaller value. The latter has been shown to require high critical power densities for deep penetration bead by around order of unity compared with EB welding, and the combined former and latter effects require very precise focal positioning. The small depth of focus suggests that better performance can be obtained at higher welding speeds as far as precise positioning is done rather than at lower speeds. It should be noticed that penetration depth of about 0.5 mm with an aspect ratio of 3 was obtained at as fast as 40 m/min in this experiment.

It is seen in Fig. 2 that the shorter focal length lens , f=38 mm, tends to give smaller penetration depth at low welding speed. An analysis based on the wedge cavity model can provide insight into dependence of the bead shape on the focal length of the lens used the beam with a large convergent angle gives small attainable depth where proceeding direction of reflecting beam is reversed in the cavity, leading to smaller penetration depth. At higher welding speeds, on the other hand, higher power densities, which can be obtained by using shorter focal length lens as far as the astigmatism of the lens is negligible, are required to obtain deep penetration beads. If the focal positioning is optimized for the largest value of h according to Fig.3, it has been found that f=127 mm and 64 mm give the deepest penetration at v < 1 m/min and v > 1 m/min in this experiment, respectively.

It has also been found that the penetration data in this experiment are deeper than open literature data of up to 3 KW power level[8] at speed higher than 8 m/min and that the maximum penetration depth at the maximum output of the laser apparatus used, 1.2 KW, is about 5.5 mm which is comparable to these literature data. These are considered to be due to probably higher focusibility of the laser beam used and optimization of focal positioning carried out depending on the welding speed.

References

[1] E.v.Locke, E.D.Hoag and R.A.Hella, "Deep Penetration Welding with High-Power CO_2 Lasers", IEEE J. Quantum Wlectronics, Vol.QE-8, No.2, 1972
[2] E.B.Breinan and C.M.Banas, "Evaluation of Basic Laser Capabilities", UTRC Technical Paper, R75-911989-4, 1975
[3] Y.Arata, H.Maruo and I.Miyamoto, "Heat Flow in Laser Hardening", I.I.W. Doc IV-241-78, Doc 212-436-78, 1978
[4] Y.S.Touloukian, "Thermophysical Properties of Matter", The TPRC Data Series IFI/Plenum, 1970
[5] Y.Arata and I.Miyamoto, "CO_2Laser Beam Absorption Characteristics of Metal", Trans. J.W.S.,Vol.3, No.1, 1972
[6] Y.Arata and I.Miyamoto, "Processing Mechanism of High Energy Density Beam", Trans. J.W.R.I., Vol.2, NO.2, 1973
[7] A.A.Wells, "Heat Flow in Welding", Weld. J.,Vol.22,1952
[8] R.C.Crafer,"IMproved Welding Performance From a 2 KW Laser Welding Machine" Advances in Welding Process, Harrogate, May, 1978

The Role of Assist Gas in CO_2 Laser Welding

Abstract

Optimum gas-assisting parameters and their tolerable setting errors in controlling plasma in CO_2 laser welding at 1 and 10 kW power levels were determined by using a system by which height and angle of the nozzle, assist gas-workpiece interaction position and pressure of the assist gas can be precisely adjusted for various gas species and nozzle diameters. The role of the assist gas in controlling plasma was clarified on the basis of pressure measurements of assist gas and vapor in the cavity, high speed motion pictures and measurement of plasma brightness. It was found that the assist gas suppresses the plasma at pressures slightly higher than the vapor pressure by forcing the vapor to flow away from the focussed laser beam along the cavity rear wall, providing deep weld bead without weld defects. A method of monitoring plasma control by using phototransistor is also proposed.

1. Introduction

CO_2 lasers provide excellent welding characteristics at atmospheric pressure especially at high welding speeds [1,2], and is finding new industrial applications [3,4]. The disadvantage is, however, to produce plasma which absorbs and/or scatters the incoming laser beam [5-7] especially at low welding speeds, resulting in much shallower penetration depth than electron beam welding.

Assist gas directed to the weld-cavity can control the plasma [5], thereby increasing the penetration depth, whereas it also can produce weld defects by exerting excess pressure on the weld cavity (key hole or beam hole) and molten pool, depending on the gas-assist conditions [8,9]. Thus, in order to obtain deep weld bead without defects, it is essential to understand the interaction among the assist gas, plasma, cavity and molten pool. Little systematic work, however, has been done about the role of the assist gas so far.

One of the serious problems in employing the assist gas is poor reproducibility of welding [9]. So this research begins with constructing an experimental set-up by which the assist gas parameters can be precisely determined to see each exact effects. The aims of the present research are to optimise the gas-assisting parameters, and to reveal the role of the assist gas in controlling the plasma.

2. Experimentals

In this experiment, welding was performed with two CO_2 lasers, 1 kW (SP Model 971) and 15 kW (AVCO HPL10), which were focused by a ZnSe lens (f/8) and metallic mirrors (f/7), respectively. Assist gas was directed to the weld-cavity by using a straight nozzle directly behind the cavity. The gas-assisting parameters tested here include gas species, gas pressure, nozzle angle, nozzle diameter, nozzle height and gas-workpiece interaction position (in x-y plane). As the assist gases, He, N_2, Ar and CO_2 were examined in bead-on-plate welding of type 304 steel (0.06%C, 0.65%Si, 1.17%Mn, 0.03%P, 0.012%S, 8.5%Ni and 18.33%Cr) in a shielded chamber.

The authors directed their efforts to measuring the assist gas parameters (**Fig. 1**); Nozzle was attached to an xyz-microscope stage, and the movement of the nozzle (x,y,z) was precisely measured by a dial-gauges. Assist gas pressure was measured by three different ways depending on purpose: (1) For determining the exact operating pressure prior to each welding run, the assist gas pressure was routinely measured on a flat plate by a pressure sensor calibrated by a manometer through a perforated small hole (0.2 mm diam). The hole was set coaxially with the beam axis, as shown in Fig. 1(a). The pressure thus measured was referred to as

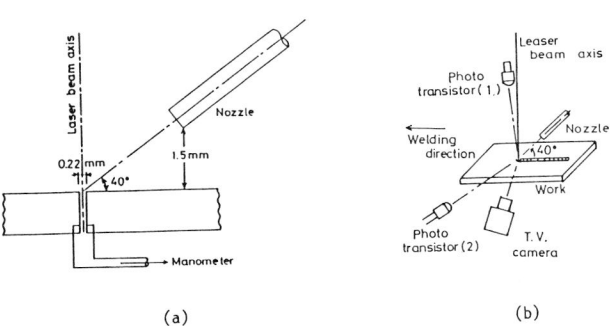

Fig. 1 Experimental setup. (a) Measurement of operation pressure and (b) arrangement of phototransistors and TV camera.

"operation pressure", while the pressure on the cavity during welding is affected by the existence of the weld cavity and uneven molten pool. The operation pressure is expressed with a superscript "'''". (2) In order to find the value with physical meaning, the pressure on the weld crater instead of the small hole was measured; the crater which was nearly the same size as the real cavity was produced by directing rather strong assist gas to the weld zone. The pressure thus obtained was referred to as " **mean cavity pressure**". The mean cavity pressure gives time averaged pressure exerting on the cavity during welding. (3) The real-time pressure in the cavity during welding was measured by the pressure sensor connected to the cavity bottom (see Fig. 14a). This is referred to as "**cavity pressure**". This measurement system responds up to frequency of 200-300 Hz. The welding was taken with 16 mm high speed camera, 35 mm camera and TV camera. Two photo-transistors of which peak sensitivity is at around 8000 angstroms were set above and on the side of the workpiece to detect the light emitted from the plasma and the molten pool as shown in Fig. 1(b).

3. Effects of gas-assisting parameters

A series of experiments were performed to find the effects of the assist gas parameters mainly with He assist gas at 1 kW power level at 50 cm/min. Experiments were also done using various gases and at 10 kW.

3-1 Angle

When the nozzle is parallel to the workpiece, little effect of the assist gas was seen. In cross-asisting, the penetration depth and cross-sectional area of the bead increase with increasing nozzle angle, and saturated at around 30 deg as shown in **Fig. 2**. Thus the angle was fixed to 40 deg hereafter.

3-2 Pressure

Figure 3 illustrates an example of pressure distribution profile of assist gas on a flat plate (θ=40 deg). The origin of the gas-workpiece interaction position for given height and angle was determined at the location of a peak pressure measured through the small hole set coaxially with the laser beam by adjusting the nozzle in x-y plane. Very small setting errors, less than 50 microns and 5%, were attained for the nozzle position and operating pressure, respectively, and thus satisfactory reproducibility of welding was realized.

The effects of assist gas pressure on penetration depth were examined for various heights and diameters of nozzle with the interaction point at the origin, and typical result is shown in **Fig. 4**. Light intensity was also measured by phototransistors from the side and above (Fig. 19 : examples of the wave form). **Figure 5** shows the time-averaged intensity of the emitted light. It was found that there are three pressure regions in each height and diameter in terms of bead shape as shown in Fig. 4:

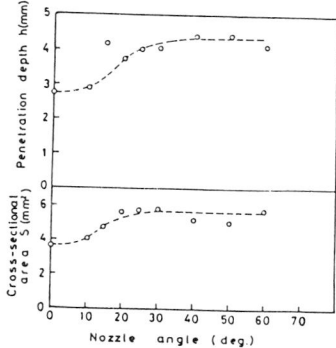

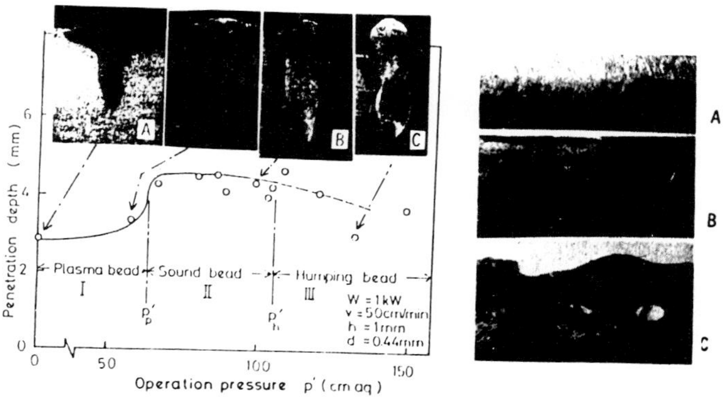

Fig. 2 Nozzle angle vs cross-sectional area and penetration depth.

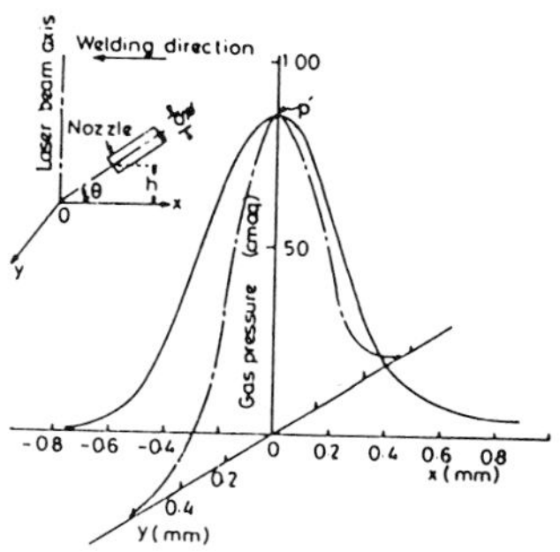

Fig. 3 Pressure distribution of assist gas on plate d=0.4mm, h=1.5mm, 40 deg).

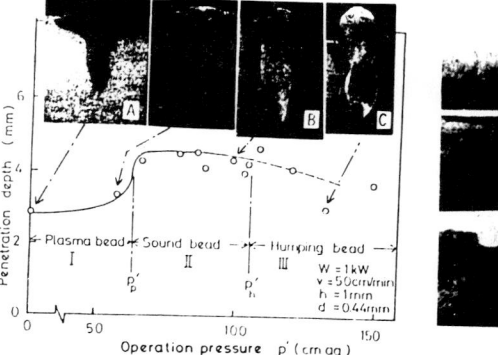

Fig. 4 Effects of assist gas pressure on penetration depth and longitudinal cross-section (1 kW, 50 cm/min).

[I] **Plasma bead** ($p'<p'_p$): This is characterized by shallow bead (2.8 mm) with wine-cup shape. Although increase in assist gas pressure decreases the average light intensity of the plasma (Fig. 5) and size of plasma bowl (see Fig. 17a,b), it gives little increase in penetration depth.
[II] **Sound bead** ($p'_p<p'<p'_h$): Penetration depth increases drastically at p'_p up to about 4.3 mm, which corresponds to about 50% increase of the plasma bead. With increasing operating pressure, the mean penetration depth remains almost constant, although change in penetration increases as the longitudinal cross-

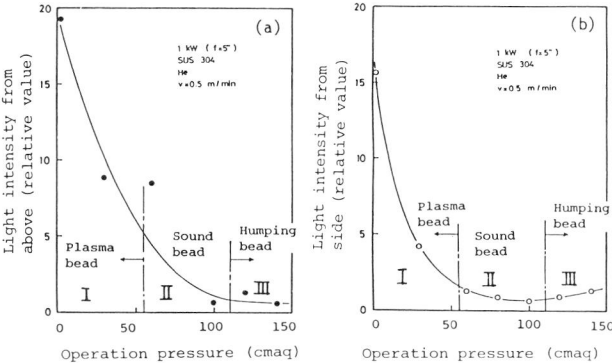

Fig. 5　Time-averaged light intensity measured by phototransistor from (a) above and (b) side (1kW, 50 cm/min, f=5 inch).

Fig. 6　Effect of beam impinging angle on penetration depth and cross-sectional area (1 kW, 50cm/min).

sections of the bead are shown in Fig. 4(b). Little change in the brightness is seen region II. No weld defects were seen except for fine porosities (<0.2 mm in diam).

[III] **Humping bead** ($p'>p_h'$): This shows irregular humping at the surface, and uneven penetration with large porosities (see Fig. 4 and Fig. 17d). The light intensity from above remains almost constant, whereas the intensity from the side tends to increase with increasing operation pressure.

Table 1 summarizes p_p' and p_h' for various heights and diameters of nozzle; p_p' and p_h' decrease with increasing nozzle height and diameter while the penetration depth attained remains constant. It is seen that the smaller height and diameter provide wider setting allowance of the operating pressure in region II.

Table 1　Critical operation pressures, p_p' and p_h' (1 kW).

Welding speed (cm/min)	Nozzle diam. (mm)	Nozzle height (mm)	p_p' (cm aq)	p_h' (cm aq)	Penetration depth (mm)
50	0.44	1	63	105	4.2 - 4.5
		1.5	59	92	
		2	58	86	
		2.5	51	72	
	0.9	1	33	64	
	1.5	1.5	32	59	
100	1	1	72	100	3.3

3-3 Cavity and molten pool behavior

High speed pictures were taken of cavity and molten pool behavior. Filming was made through a microscope (work distance 140 mm) vertically to the workpiece and laser beam impinged on the workpiece with inclined angle to overcome the shallow depth of focus of the microscope. Since the inclined beam impingement was found to produce little change in the penetration depth and cross-sectional area up to at least 15 deg as shown in **Fig. 6**, filming was made at a forward angle of 15 deg. High speed motion pictures were taken at 3000 frames/sec in each pressure region. **Figures 7 and 8** show the pictures taken with 35 mm still camera; the illuminated part seen behind the cavity is the reflection of halogen lump used for auxiliary lighting.

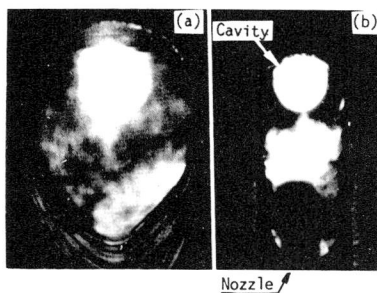

Fig. 7　Cavity and molten pool at (a) p'=0 (region I) and (b) p'=75 cmaq (region II) (1 kW, 50 cm/min).

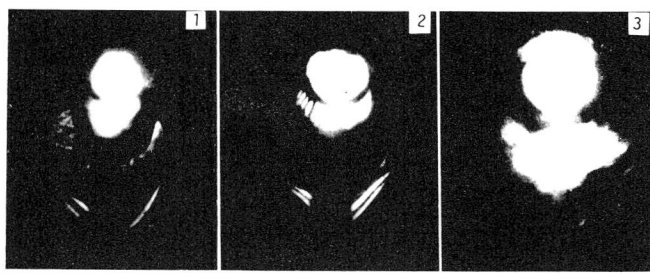

Fig. 8　Cavity and molten pool at operation pressure p'=130 cmaq (region III) (1 kW, 50 cm/min).

500

At pressure p=0, the contour of the cavity is seen through the bright plasma. The welding is rather stable except for small-scale fluctuation of the molten pool. The plasma bowl reported to decouple periodically from the workpiece at high power levels [6,7] was not observed at 1 kW power level. The molten pool with the diameter about 3 mm is seen to surround the cavity (approximately 1 mm in diam), and no macroscopic metal flow possible to develop such a large molten pool was not observed. This fact implies that such a large molten pool was produced by radiation and/or conduction from the hot plasma [10].

In region II (p'=75 cmaq), the cavity is seen to locate at the leading edge of the molten pool with drastically reduced brightness and size of plasma. The cavity changes its diameter semi-periodically in a small range 0.8-1 mm and correspondingly molten metal is pushed out of the cavity producing eddy flow behind the cavity. The bead width was about 1.6 mm with less fluctuation.

In region III, periodical change becomes more violent producing change in bead width; (1) small cavity is ahead of the molten pool with comparatively bright plasma, (2) increase in cavity diameter accompanies the molten metal pushed out by which molten pool front moves faster than the laser beam, and (3) the molten pool goes ahead of the cavity. The forward movement of the molten metal is partially by molten metal pushed out of the cavity and partially by plasma produced periodically.

3-4 Location of gas-workpiece interaction

Figure 9 shows the effects of the location of the gas-workpiece interaction on bead shape and depth. Sound beads are obtained in a region -0.5 mm <x< 0.7 mm, and -0.5mm <y< 0.5mm, which is nearly equal to the the cavity size. The maximum penetration depth about 4.7 mm was obtained when the assist gas aims at the leading edge of the cavity. When the interaction point is x<-0.5 mm, wine-cup bead is produced, which is the same as the bead in region I. This implies that the plasma adjacent to the cavity is very absobing. When the interaction point is far behind the cavity (x>0.7 mm), bright plasma fluctuating perpendicularly to the welding direction at very low frequency was seen on the cavity; the inclined plasma was seen to widen the bead width asymmetrically, producing snake like bead (Fig. 9b) with asymmetrical cross-section.

In the both regions of wine-cup beads, x<-0.5 mm and x>0.7 mm, time averaged brightness is seen to become stronger than that of sound bead region as shown in **Fig.10**. From Figs. 5 and 10, it is seen that whether or not the plasma bead is produced, at least, can be monitored by time-averaged light intensity detected by phototransistor. Monitoring of gas assisting is detailed in **section 5**.

3-5 Gas species

The penetration depth was shown to increase with ionization potential of shielding gas, [11] when no dynamic pressure of the gas is exerted. So in order to find the effect of ionization potential on the penetration depth when the dynamic pressure is exerted, He, N_2, CO_2 and Ar were compared at various operation pressures. As shown in **Fig. 11**, little difference in penetration depth in region II is seen. This means that the pressure is important rather ionization potential in gas-assisting, and that one can use cheaper gas like CO_2, N_2 or Ar for controlling plasma under optimized gas-assisting conditions, unless there is metallurgical problem.

3-6 Full penetration welding

In full-penetration welding, p_h' becomes very high, whereas p_p' remains unchanged; in 3 mm thick plate, for instance, p_h' increased up to about 200 cmaq. Such a high operation pressure of p_h' is due to decrease in dynamic pressure in the cavity where the assist gas passes through the plate.

3-7 High power laser

A series of experiments were also performed at 10 kW power level using nozzle of 1.5 mm diam at height 4mm. The penetration depth is plotted against the operation pressure in **Fig. 12**, which shows the same tendency as 1

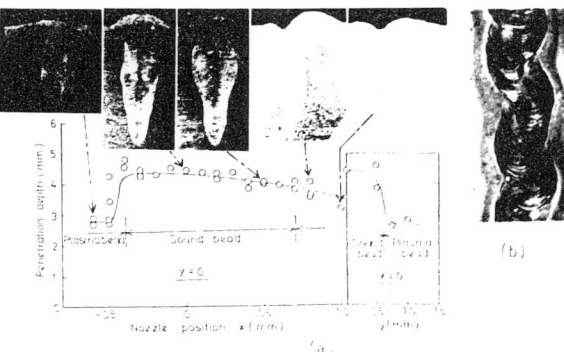

Fig. 9 Effects of gas-workpiece interaction position on penetration depth and bead shape (1kW, 50cm/min).

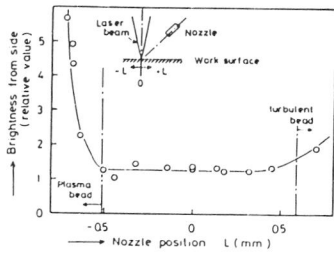

Fig. 10 Time-averaged light intensity from the side vs interaction position.

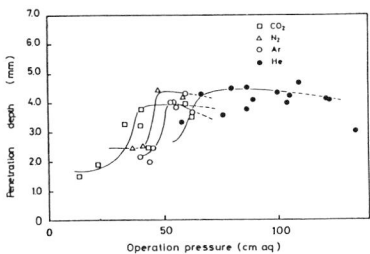

Fig. 11 Operation pressure vs penetration depth
for various gas species (1kW, 50cm/min).

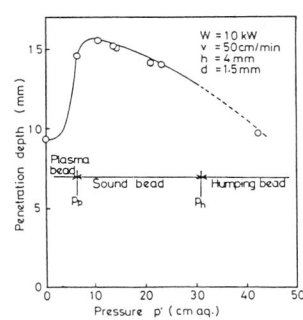

Fig. 12 Operation pressure vs penetration depth
at 10 kW power level (50 cm/min, f/7).

kW (Fig. 4). At 10 kW, the lower critical pressures of region II is 7-9 cmaq, which is at least 3 times lower than the lowest value of 1 kW (Table 1), but is 10 times higher than the values reported by Blake et al [8].

It was noticed that at 10 kW the humping bead tends to be suppressed unlike 1 kW, even if excess gas pressure is exerted to the cavity, producing humped molten pool or a cavity expanded along the welding line. This is because the cooling rate at higher power level is slow enough for the molten pool to be flattened before solidification. Excess gas pressure to some extent also tends to produce almost no porosities with round bead edge, as also reported by Estill at al [9]. This is thought to be due to prolonged cooling time in such an expanded cavity that enables the porosities to escape. The penetration depth, however, decreases with increasing operation pressure in region II, since wall-focussing of the laser beam [12] becomes less effective in such an expanded cavity.

4. Role of assist gas

4-1 Mean cavity pressure
The mean cavity pressures $\bar{p}_p$ and $\bar{p}_h$ were measured at p_p' and p_h', and are plotted in **Fig. 13**. It is seen that $\bar{p}_p$ and $\bar{p}_h$ are almost constant, 30-40 cmaq and 60-80 cmaq, respectively, independent of height and diameter of nozzle, and are, more or less, lower than p_p' and p_h' especially for smaller diameters and lower heights. This is because the non-uniform pressure as shown in Fig. 3 is averaged over the crater entrance. While the value $\bar{p}_p$ and $\bar{p}_h$ at 10 kW (Fig. 12) are almost the same as p_p' and p_h' respectively, since larger diameter and height of the nozzle used provide rather uniform pressure distribution over the cavity.

4-2 Cavity pressure and vapor pressure
Figure 14 shows the cavity pressure recorded during welding. Negative pressure seen prior to welding is due to temperature change of the workpiece during setting the workpiece. At p'=0, the cavity pressure, which represents the vapor pressure in the cavity, is seen to reach an equilibrium value of 20-25 cmaq in 1 sec. It was found that the lower critical value of $\bar{p}_p$ is somewhat higher than the vapor pressure.

For Fe, **2170-2220 C** is estimated from the vapor pressure of 25-30 cmaq [13]. It is noticed that thus estimated value is very close to the temperature measured by Guidt et al [14] in electron beam welding of type 304 steel. Assuming that the cavity is a cylinder of diameter D, the surface tension pressure $\mathbf{p_s}$ is given by

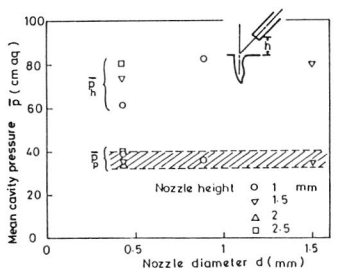

Fig. 13 Mean cavity pressure $\bar{p}_p$ and $\bar{p}_h$ at
various nozzle heights and diameters.

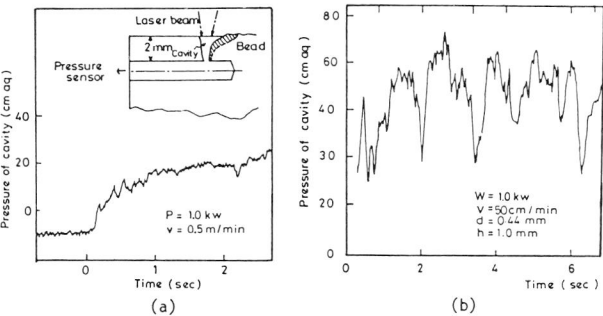

(a)   (b)

Fig. 14 Cavity pressure measured by pressure sensor at (a) p'=0
(I) and (b) p'=85 cmaq (II) (1 kW, 50 cm/min).

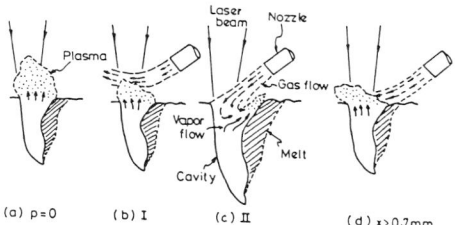

Fig. 15 Schematic illustration showing the role of assist gas in regions I (a)(b) and II (c), and at position x>0.7 mm (d).

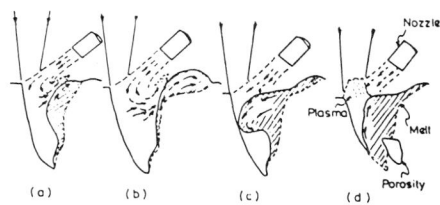

Fig. 16 Schematic illustration showing interaction among assist gas, plasma, cavity and molten metal in region III.

$$P_s = 2s/D \qquad \ldots\ldots\ldots\ldots\ldots\ldots\ldots \qquad (1)$$

where s is the surface tension of the metal. The surface tension of Fe at 2220 C is estimated to be about 1000 dyne/cm [15] by taking the effect of 0.012 wt%S and 0.52 wt%Si into consideration. Using this value and D=0.9 mm, which is approximated by the width of finger part of the bead at p'=0, p_s is estimated about 22 cmaq. This value is in satisfactory agreement with the measured vapor pressure, 20-25 cmaq. This means that an equilibrium between the surface tension pressure and the vapor pressure is established in region I. This also is confirmed by the rather smooth curve of cavity pressure at p'=0 in Fig. 14(a).

Figure 14(b) shows the cavity pressure at an operation pressure p'= 85 cmaq (region II). Unlike the case of p'=0, the pressure in the cavity varies semi-periodically ranging from the vapor pressure to about p'_b. The resultant bead appearance was also seen to change correspondingly to the pressure change, implying that the liquid metal is periodically pushed out of the cavity by the excess pressure. The experimental setup (Fig. 14), however, seemed to produce somewhat larger fluctuations of the pressure and bead appearance than real welding.

4-3 Mechanism of plasma control

On the basis of the aforementioned results, the plasma controlling by assist gas is schematically shown in Figs. 15 and 16. Figure 17 shows the pictures seen on TV monitor at various assist gas pressures.

In region I, since the vapor pressure of metal in the cavity, which is weekly ionized, is higher than the assist gas pressure, the metal vapor is ejected vertically along the beam axis, and hence is heated by the laser beam to become absorbing plasma. Although the assist gas can remove only upper part of the plasma, lower part of the plasma still absorbs the incident laser beam. Welding itself is stable due to an equilibrium established between the vapor pressure and the surface tension pressure, unless LSC [16] wave is produced. When the gas-workpiece interaction point is far behind the cavity, the assist gas forces the vapor to flow up along the front wall of the cavity where the vapor is exposed to the laser beam to become absorbing plasma.

In region II where the assist gas pressure is somewhat higher than the vapor pressure, the weekly ionized metal vapor is forced to flow out along the rear wall of the cavity (see Fig. 17c), without being heated by the laser beam above the cavity, providing deep weld bead due to little plasma absorption loss (Fig. 15c). At the same time, however, the assist gas pressure which exceeds the vapor pressure tends to, more or less, make the welding unstable. As the result, the pressure fluctuation (Fig. 14b) and cyclic change in the cavity diameter (see 3-3) occur, producing uneven penetration depth (see Fig. 4b). The metal vapor in the cavity is thought not to be heated up to high temperature because of the cooling effect of the cavity wall. The absorbing plasma is produced only above the cavity because the plasma is cooled only by diffusion.

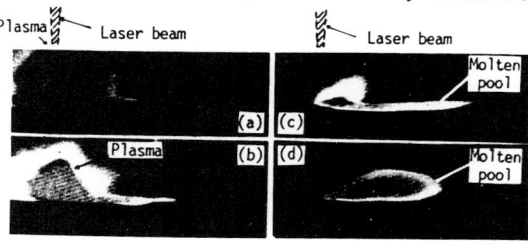

Fig. 17 Photograph of laser welding taken by TV camera. (a)(b):I, (c):II and (d):III (1 kW, 50 cm/min).

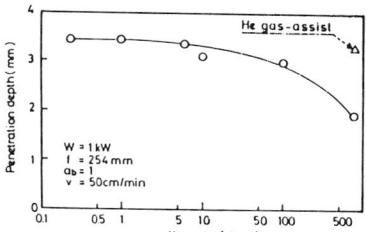

Fig. 18 Penetration depth at various vacuum pressures (1 m/min, 1 kW, focal length=10 in.).

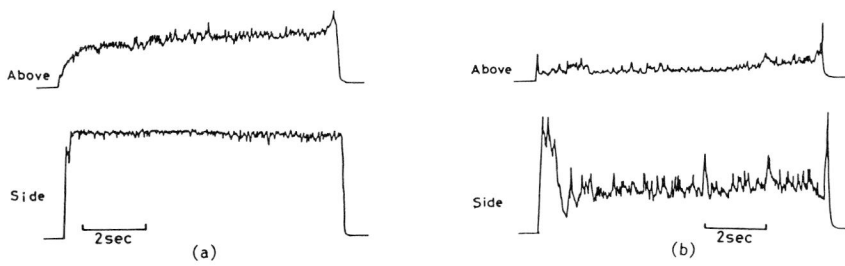

Fig. 19 Light intensity (arbitrary unit) measured by phototransistor from above and side at (a) p'=0 (I) and (b) p'=70 cmaq (II) (1 kW, 50 cm/min).

At 10 kW, the lower critical pressure of region II was approximately 7 cmaq, which is nearly 3 times as low as p'_h of 1 kW. This difference is due to low vapor pressure in the cavity at 10 kW; the cavity diameter D at 10 kW is approximately 3 times as large as that of 1 kW so that the surface tension pressure is 3 times lower than 1 kW according to Eq(1), assuming that the surface temperature of the cavity is the same at both of the power levels. The cavity diameter at higher power levels tends to be larger so that the vapor may become higher temperature due to less effective cooling by the cavity wall, becoming more strongly ionized plasma in the cavity.

In region III where the assist gas pressure exceeds upper critical pressure p_h, the fluctuation becomes too large to form sound bead; the excess gas pressure pushes out most molten metal surrounding the cavity (b), and then the inertia of the melt flowing in closes the upper part of the cavity (c), producing porosities (d) as shown in Fig. 16. Figure 17 (d) corresponds to Fig. 16(c). As shown in Fig. 16(d), plasma is produced intermittently in region III, since large incident beam energy per unit cavity depth overheats the melt. Similar plasma is transiently detected at the beginning of the welding in region II, as shown in Fig. 19.

4-4 Welding in vacuum

As mentioned above, controlling plasma by using assist gas potentially tends to produce unstable welding even in region II. Especially at very low welding speeds, the welding itself becomes so unstable that the cavity easily collapses even at very low assist pressure.

The authors tried to control plasma by evacuating instead of exerting pressure to the cavity. The laser beam was focussed by a lens with rather long focal length, 10 inches (f/14), and was directed to the workpiece through a NaCl window. **Figure 18** shows the result at 1 kW power level at 1 m/min. The penetration depth increases with decreasing pressure in the chamber, and at 20-30 torr it reaches the value somewhat larger than that of assist gas. The welding was seen to be more stable than the case of assist gas. The effect of evacuation is considered to be more remarkable at lower welding speeds where the cavity easily collapses.

5. Monitoring of gas-assisting

Since plasma control requires considerably precise adjustment of the gas-assisting parameters as mentioned above, it is desirable to monitor the plasma and molten pool behaviors from the industrial view points. For this purpose, two photo-transistors from above and the side were set with a TV camera as shown in Fig. 1(b).

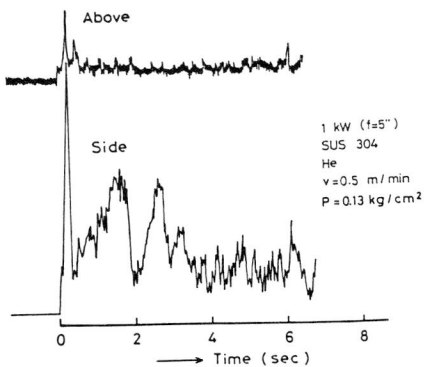

1 kW (f=5")
SUS 304
He
v=0.5 m/min
P = 0.13 kg/cm²

(a)

(b)

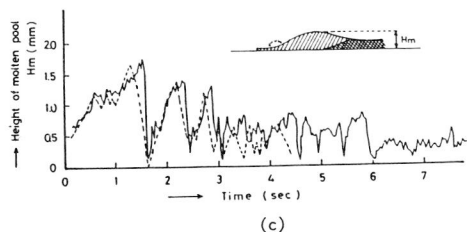

(c)

Fig. 20 Light intensity from (a) above and (b) side, and (c) heights of molten pool (solid line) and bead reinforcement (dotted line), at p'= 115 cmaq (region III).

Figure 19 shows the output of phototransistors in regions I and II. Region I provides very smooth wave form with high averaged intensity (Fig. 5). In region II, the curve is still not so uneven with lower signal level except for the beginning of the welding. During this cavity formation period, heat input per unit cavity depth is very high so that molten metal surrounding the cavity is over-heated to become very bright plasma. In sections 3-2 and 3-4, it was shown that in steady welding one can recognize whether or not the assist gas pressure is strong enough to remove plasma from the time-averaged output of the phototransistor.

In Fig. 20, the light intensity (a), height of molten pool behind the cavity (b) and bead reinforcement in resulted bead (c) are shown in region III (operation pressure is 130 cmaq). The height of molten pool onTV agrees well with resultant bead height. This indicates that the cooling rate at 1 kW power level is considerably fast due to narrow bead width. The light intensity detected from the side also agrees well with molten pool height except for the beginning of welding, whereas little agreement is seen between light intensity from above and molten pool height. This means that the light intensity detected by the side photo-transistor in region III is almost from the humped molten pool, and corresponds to the area of the molten pool above the workpiece surface. Photo-transistors having a peek sensitivity at around 8000 **A** were used in this experiment. By using phototransistors with peak sensitivity at 1.3 microns corresponding to peak thermal radiation of the molten metal, detectivity of molten pool height will be much improved.

6. Conclusions

Gas-assisting system using a straight nozzle for controlling plasma in CO_2 laser welding was constructed, by which gas pressure and gas-workpiece interaction position can be precisely adjusted, and the optimum gas-assisting conditions and their tolerable setting errors were determined. The role of the assist gas in CO_2 laser welding was discussed on the basis of measurement of assist gas pressure and vapor pressure in the cavity (beam hole or key hole) and high speed filming. Results obtained are summarized as follows:

(1) The plasma formation is suppressed at assist gas pressures higher than a critical pressure p_p. p_p is slightly higher than vapor pressure (region II). This is because the assist gas pressure forces the vapor in the cavity to flow away from the focussed beam. Under optimized gas-assisting conditions, He, Ar, N_2 and CO_2 as the assist gas provided no appreciable difference in penetration depth.

(2) In region II, since the assist gas pressure exerting to the cavity is higher than the surface tension pressure which keeps the cavity stable, the change in penetration depth increases with increasing assist gas pressure. Humping beads are produced eventually, as the assist gas pressure increases.

(3) At very low welding speeds, the cavity easily collapses by the assist gas pressure. Then evacuation can control plasma without exerting force to the cavity.

(4) The phototransistor set at the side of welding can monitor the behavior of plasma removal and humping bead formation during gas-assisted welding in real-time.

References

[1] Crafer, R. C. (1978). Improvement Welding Performance from a 2 kW Laser Welding Machine. Advance in Welding Process, Harrogate
[2] Miyamoto, I., H. Maruo and Y. Arata (Oct. 1980). Mechanism of Bead Transition in Laser Welding. Proceedings of International Conference on Welding Research in 1980's (Osaka): 103-108
[3] Mazumder,J.and W. M. Steen (1981). The Laser Welding of Steels Used in Can Making. Weld. J. 60:19-25
[4] Eckersley, J. S. (1982) CO_2 Laser Welding of Aluminum Air Spacers for Insulated Windows. Proceedings of ICALEO '82: 61-64
[5] Locke, E. V.,E. D. Hoag and R. A. Hella (1972). Deep Penetration Welding with High-Power CO_2 Lasers. IEEE Journal of Quantum Electronics. QE-8:132-135
[6] Donati,V.et al (Sept. 1983). On the Development of Absorption Waves During the Laser-Material Interaction. Proceedings of 3rd CISFFEL (Lyon): 71-80
[7] Arata. Y.,N. Abe and T. Oda (1983) Beam Hole Behavior during Laser Beam Welding. Proceedings of ICALEO '82: 59-66
[8] Blake,A.and J. Mazumder (1982). Control of Composition during Laser Welding of Aluminum-Magnesium Alloy Using a Plasma Suppression Technique. Proceedings of ICALEO '82: 33-46
[9] Estill, W. B. and B. D. Formisano (1982). Porosity Decrease in Laser Welds of Stainless Steel Using Plasma Control. Proceedings of ICALEO '83: 67-72
[10] McKay, J. A. and J. T. Schriempf (1978). The Spatial Distribution of Heating of Aluminum Targets by Laser-Ignited Air Plasma. Applied Physics Letters. 33: 877-878
[11] Seaman, F. D. (1977). The Role of Shielding Gas in High Power CO_2 (CW) Laser Welding. SME Technical Paper No. MR77-982
[12] Arata, Y. and I. Miyamoto (1973). Processing Mechanism of High Energy Density Beam. -Mechanism of Drilling-. Transaction of JWRI: 19-22
[13] Honig, R. E. (Dec.1962) S. Vapor Pressure Data for the Solid and Liquid Elements. RCA Review. 23: 567-586
[14] Schawer, D. A., W. H. Giedt and S. M. Shintaku (May 1978). Electron Beam Welding Cavity Temperature Distribution in Pure Metals and Alloys. Welding Journal. 127s-133s
[15] Allen, B. C. (1972). Liquid Metals. Edited by S. Z. Beer. Marcel Dekker, Inc., New York. 162-212
[16] Dixon, R. D. and G. K. Lewis (1983). The Influence of a Plasma during Laser Welding. Proceedings of ICALEO '83: 44-50

CO_2 LASER WELDING OF CERAMICS

A CO_2 laser was applied for welding Al_2O_3-SiO_2ceramic plates up to 4 mm in thickness with 48-99.5 wt% alumina content. The effects of welding parameters on the penetration depth and microstructure of the fusion zone are discussed. Prevention of weld deffects including crack and porosity was also dealt with. Strength of the weld joints is evaluated by a 3-point bend test. 100 % of weld joint efficiency is obtained, and the resitance of the weld joint to thermal shock and thermal cycle is the same as that of the base material. A variety of welding joint geometries including I, L and T are demonstrated.

I. Introduction

Ceramics are of considerable interest as engineering material that can be used under very severe conditions in various industrial fields because of their exellent resistance to heating, wearing and corroding. The primary requirement is that the desirable character-istics and properties of these materials not be lost or compromised by the joining oper-ation. However, the existing joining techniques, such as mechanical joining, brazing and organic adhesive bonding, cannot provide vacuum-tight, structurally strong, heat resistant joint. Considerable interest has developed in fusion welding of ceramics, which is thought to be the most reliable and efficient joining process.

Commonly available heat sources include electron beam, laser beam and arc plasma, for localized fusion welding ceramics. Welding of ceramics has been accomplished by means of electron beam [1], but reliable weld joints have not been achieved by it. In this study, a CO_2 laser is used for fusion welding of Al_2O_3-SiO_2 ceramics, which are most widely used in industry. The potential advantages of the CO_2 laser beam for fusion welding ceramics stem from the fact that it can be used in air without causing any beam scattering problems in heating high vapour pressure material, and can heat dielectric materials without any electric charge-up problems, unlike the electron beam.

In this paper, the effects of the laser beam factors on the penetration depth and bead shape are discussed. Microstructure of fusion zone, and prevention and control of weld defects including porosities and cracks are also dealt with. For estimating the practical availability of the welds, the strength of the weld joint and the resistance of the weld joints to thermal shock and thermal cycles are evaluated.

2. Experimentals and materials

In this experiment, 1 kW class CO_2 laser with nearly Gaussian intensity profile with a diameter of 16 mm at 1/e-power point was used. In order to prevent weld-cracking, ceramic specimens were preheated in an electric furnace with a maximum heating temperature of about 1500 °C, and then they were welded in air by focused laser beam through an opening on the top of the furnace, as schematically shown in Fig. 1. This arrangement required a rather long distance between the workpiece and the lens, and thus a comparatively long focal length, 10 inches, was used. The weld specimens were sectioned with a diamond wheel, and etched with hydroflouric acid for microscopical observation.

As the weld specimen, five Al_2O_3-SiO_2 ceramics having various alumina contents from 48 to 99.5 wt% were used. Table 1 shows their chemical compositions and physical properties. The three materials having Al_2O_3 contents from 48 to 59 wt% consist of mullite crystal and high silica glass matrix, and 95 and 99.5 wt% Al_2O_3 consist of alumina crystal and silica glass at grain boundary.

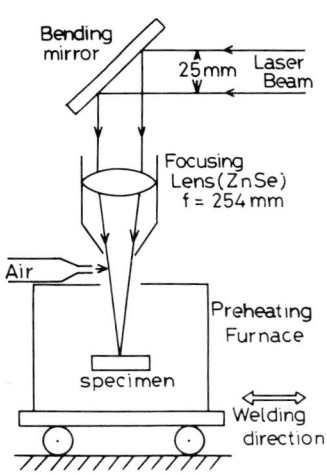

Fig. 1
Schematic diagram of focussing system.

Table 1　Chemical compositions and physical properties of ceramics used.

		Mullite			Alumina		
Chemical compositions	Al_2O_3(wt%)	48	55	59	91	95	995
	SiO_2(wt%)	47	41	38	7	4	0.1
Heat conductivity (400°C)cal/cm·sec°C		6.4 ×10⁻³	6.9 ×10⁻³	7.5 ×10⁻³	1.6 ×10²	1.7 ×10²	2.0 ×10²
Coefficient of thermal expansion(20~1000°C)×10⁻⁶		4.5	4.9	5.6	7.3	7.8	8.1
Apparent density g/cm³		2.5	2.6	2.8	3.6	3.7	3.9

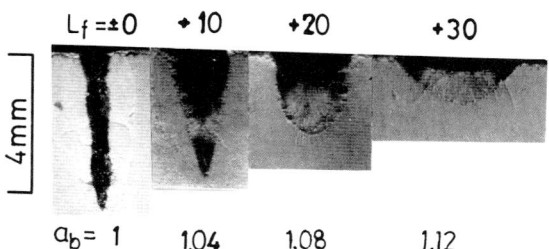

Fig. 2
Effect of focal position on bead shape at cm/min at 200 W power level (48 wt% Al_2O_3, no preheating).

3. Depth and shape of the bead

3-1 Vaporization loss and bead shape

A bead-on-plate test was carried out with no preheating at a fixed travelling speed for the various work distances for 48 wt% Al_2O_3. Figure 2. shows the cross-section of the weld bead. In this figure, the relative position of the focus with respect to the specimen is expressed by a_b-value, a ratio of objective distance to the lens focal length. As the workpiece surface approaches the focal point, the vapor plume produced at the beam impinging portion became brilliant so that the depression formed at the resultant bead surface increased. The maximum depression was reached at a_b=1.1 where the bead aspect ratio is approximately one. With still increase in power density, the bead became deeper and narrower, and the depression at the bead surface decreased until the maximum penetration depth is reached at a_b=1. If the specimen surface can be kept flat during welding, the amount of vaporization will increase with increasing power density of the impinging laser beam. However, in deep penetration welding, where a cavity is formed into the heated material, the vaporization rate is limited by the area of the cavity opening through which the vapor can pass. Thus it is reasonable that the mass vaporized away out of the cavity becomes smallest at a_b=1, where the opening of the cavity is smallest.

Heat loss rates due to vaporization were evaluated by calorimetrical measurement of the heat absorbed in the specimen, and is expressed by the ratio of heat loss due to evaporation to the incident laser heat in Fig. 3. The vaporization loss is seen to be much larger than the value for metal welding; the highest heat loss is about 75 %. Although the smallest vaporization loss in terms of thrmal quantity is as high as 25 %, the vaporization caused little decrease in the bead cross-sectional area.

507

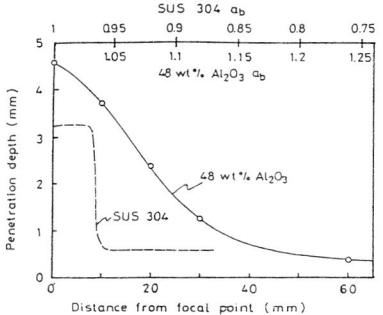

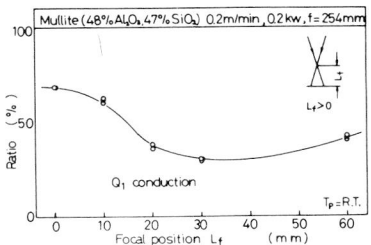

Fig. 3   Effect of focal position on absorbed power ration and penetration depth at 20 cm/min at 200 W power level (48 wt% Al$_2$O$_3$, no preheating).

3-2 Penetration depth

The penetration depth of mullite ceramic is plotted against the focal position in Fig. 3. In this figure, the penetration data for laser welding of SUS304 stainless steel are also plotted for comparison. As the power density increases, the penetration depth in mullite ceramic gradually increases, whereas that in SUS304 increases sharply at a certain a_b-value providing the bead-transition from a shallow to a deep penetration bead.

In a previous paper [2], it was declared that the bead-transition is caused by the multiple reflections of the incident laser beam to the cavity; the multiple reflections increase the absorbed beam power, and the reflected beam is concentrated to the bottom of the cavity. The absence of bead-transition in laser-welding ceramics is accounted for by their high absorptivity at 10.6 μm radiation, almost 100 %, which allows no multiple reflections of the laser beam in the cavity. The very low thermal conducticity in addition to the high absorptivity provided much lower critical power density for deep penetration welding, about 3x10^4 W/cm^2, which is about an order smaller than that of SUS304.

Figure 4 indicates the effect of welding speed on the penetration depth for various materials at a fixed focal position. Mullite ceramics gave narrow and straight sided bead, whereas as shown in Fig. 5 high alumina ceramics provided wine-cup shaped beads, which are indicative of plasma absorption observed in laser welding metals [3][4]. The penetration depth is seen to decrease with increasing alumina content because of increase

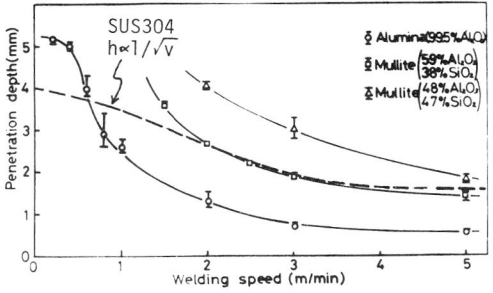

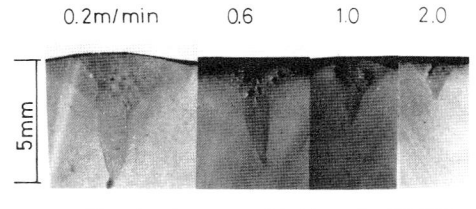

Alumina(99.5%Al$_2$O$_3$),1 kw, T$_r$=1200°C

Fig. 4
Relationship between travelling speed and penetration depth at 1 kW power level (preheating temperature 1200 °C).

Fig. 5
Cross-section of laser-welded 99.5 wt% Al$_2$O$_3$.

(a)

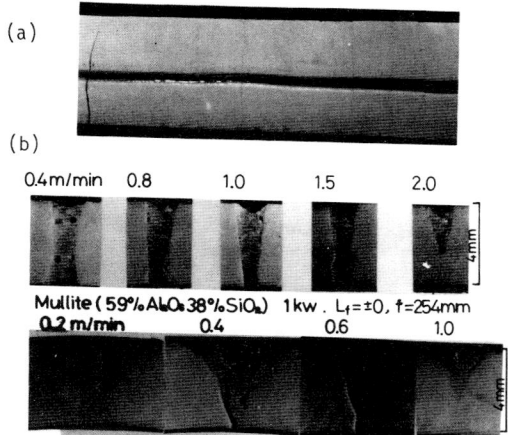

(b)

0.4m/min   0.8   1.0   1.5   2.0

4mm

Mullite (59%Al₂O₃,38%SiO₂) 1kw. L_f=±0, f=254mm
0.2 m/min   0.4   0.6   1.0

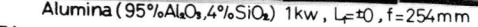

4mm

Alumina (95%Al₂O₃,4%SiO₂) 1kw , L_f=±0 , f=254mm

Fig. 6
Cracks in laser welding of ceramics.
(a) Transverse crack
(b) Longitudinal crack

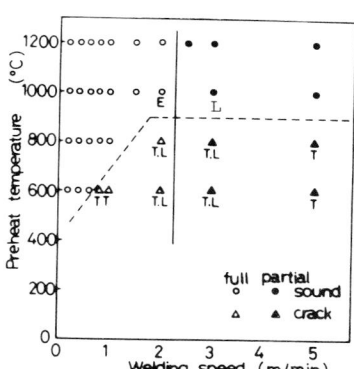

Fig. 7
Weld-cracking at various welding speeds
and preheating temperature.

in the thermal conductivity and melting temperature, and strong temdency to plasma absorption. Penetration depth of about 7 mm was obtained in 48 wt% Al_2O_3 at 50 cm/min.

When compared with SUS304 plotted in this figure, the penetration depth of ceramic is deeper than SUS304 at low welding speeds. At high welding speeds, penetration depth in ceramics decreased faster than SUS304 with welding speed; The penetration depth h in ceramics is proportional to $1/v$, whereas h in SUS304 is proportional to $1/\sqrt{v}$ (v=welding speed).

The effect of preheating temperature on the penetration depth was examined and it was found that the preheating ceramics in a range of 600-1200 °C provided little increase in penetration depth, although it gave rise to an appreciable increase in the bead cross-sectional area.

4. Prevention of weld crack

When the preheating temperature is not high enough, two different types of weld cracks appeared. One is a transverse crack, which initiates at the top surface of the bead and propagates perpendicularly to the bead, and the other is a longitudinal crack, which initiates from the bottom surface and propagates parallelly to the bead along the fusion boundary, as shown in Fig. 6.

The fact that in ceramics the tensile strength is much lower than the compressive strength suggests that those cracks result from the tensile stress that occurs during or following welding. In general, the welding process produces the tensile stress both in the unmelted area beneath the bead during heating, and in the srinking bead during cooling. When the penetration depth is comparable to the plate thickness resulting in very small load-bearing area, the former tensile stress is expected to cause the longitudinal crack. The transverse crack is thought to be caused by the latter tensile stress.

In order to find out crack avoidance conditions for both types of cracks, 48 wt% Al_2O_3 plates of 40 mm length, 10 mm width and 4 mm thickness were laser-welded at various travelling speeds, v, and preheating temperatures, θ,ranging 0.2-5 m/min and 600-1200 °C, respectively. Figure 7 shows whether each crack occurs or not in the v-θ plane; T and L in this figure indicate the occurrence of transverse and longitudinal crack, respectively. In this figure, the critical welding speed v_c for full penetration welding is shown in a solid line. As mentioned before, v_c is almost independent of preheating temperature.

As was expected, the occurrence of the longitudinal crack was limited to a narrow speed range around v_c. In other words, the longitudinal crack can be prevented in parallel sided, full penetration welding or in partial penetration welding in which the penetration depth is considerably smaller than the plate thickness. It should be noted that preheating is not so effective to prevent the longitudinal crack; A rather high temperature, 1200 °C is required for prevention of the longitudinal crack at speeds near v_c. The transverse crack is seen to be prevented at preheating tempratures higher than 1000 °C at speeds faster than v_c. At speeds lower than v_c, the prevention temperature against transverse crack decreased linearly down to 600 °C with decreasing welding speed.

It was found that 55 and 59 wt% Al_2O_3 have the same critical prevention temperature against both transverse crack and longitudinal crack as that of 48 wt% Al_2O_3, except for the tendency of transverse crack at speeds lower than v_c; The transverse crack prevention temperature in 59 wt% Al_2O_3 is constant, 1000 °C, and in 55wt% Al_2O_3 the temperature is between that of 48 and 59 wt% Al_2O_3, at speeds lower than v_c. In 95 and 99.5 wt% Al_2O_3, transverse crack was prevented at 1200 °C and 1400°C, respectively. On the other hand, the longitudinal crack occurred only at speeds near v_c similarly to mullite as shown in Fig.6, and was prevented at temperature of 1400 °C. When 2 mm thic plate were used in 99.5 wt% Al_2O_3, the longitudinal crack was prevented at temperature of 1400 °C.

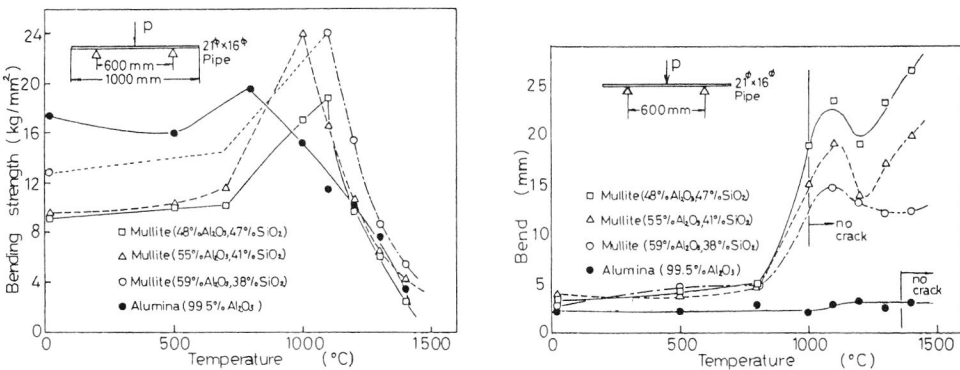

Fig. 8   Effect of temperature on fracture strength and bend in a 3-point bending test.

The occurrence of the transverse crack is thought to depend on whether or not the plasticity of the material around the bead is large enough to relax the tensile stress due to shrinkage of the bead. Therefore plasticity and fracture strength of the base materials used were determined on a 3-point bend test with a load span of 600 mm at various temperatures. The bend at the center and fracture strength are shown in Fig. 8. The bend of every mullite increases around 1000 °C, at which high silica glass of the matrix begins to have plasticity, whereas the fracture strength decreases at temperatures higher than 1000 °C.

This temperature, 1000 °C, agrees well with the prevention temperature of the transverse crack at welding speed higher than v_c in every mullite as shown in Fig. 7. This indicates that even though a large temperature gradient is formed near fusion boundary at high welding speeds, plastic deformation of the material around the bead can relax the shrinkage stress of the bead above 1000 °C. Since plasticity of 48 wt% Al$_2$O$_3$ is considerably large above 1000 °C, wide heat affected zone with a modest temperature gradient developed around the bead at low welding speeds can relax the shrinkage stress even at preheating temperatures lower than 1000 °C. With increasing Al$_2$O$_3$ content, however, the plasticity at high temperatures tends to decrease as seen in Fig. 8. and hence even at low welding speeds, prevention of the transverse crack requires as high preheating temperature as 1000 °C. In 99.5 wt% Al$_2$O$_3$, only a small amount of plastic deformation is seen at temperatures higher than 1200-1400 °C. This leads to the necessity of higher preheating temperature than mullite.

(a)   (b)

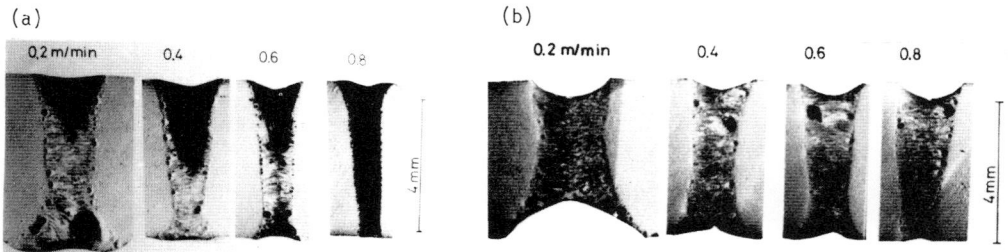

Fig. 9   Cross sections of weld bead (4 mm thickness).
          (a) 48 wt% Al$_2$O$_3$
          (b) 59 wt% Al$_2$O$_3$

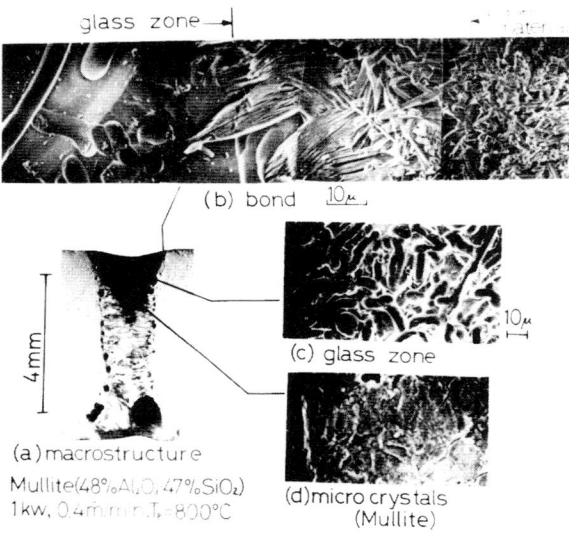

Fig. 10   Scanning electron micrograph of fusion zone (48 wt% Al$_2$O$_3$)

5. Mechanical strength of weld joint

5-1 Micro-structure and porosity in the weld bead

The 48 wt% and 59 wt% Al_2O_3 specemens of 4 mm thickness were held at 800-1000 °C for about 10 min in the furnace, and then full-penetration welding was performed at various welding speeds at 1 kW powerlevel. Figure 9 shows the examples of cross-sections of the weld bead in 48 wt% Al_2O_3 observed in an optical microscope. At low welding speeds, most part of the weld bead is seen to be covered with columnar crystal structure, and near the bead surface a V-shaped black zone appears. Figure 10. shows the SEM photograph in the crystallized zone; The mullite crystal of about 1 μm in diameter stretches from about 10 μm at the base material to 100-200 μm at the center of the bead. The black zone is phase-separated glass as was observed in rapidly quenched Al_2O_3-SiO_2 system by MacDowell et al [5], and shows an elliptical warm-like pattern of the phase-separation. X-ray diffraction of this part showed that some of the droplets crystallized to mullite, and that crystallization progressed with increasing the preheating temperature.

With increasing the cooling rate by increasing the welding speed, the phase-separated glass region became wider, and at speed 0.8 m/min it covered the entire area of the bead. As alumina content increased, glass stability decreased markedly; In 59 wt% Al_2O_3 even at the surface no glass structure was observed and mullite crystal covered the entire area of the bead, as shown in Fig. 9. The mullite crystal became gradually longer from 3 μm of the base material to about 10 μm in the weld bead of 59 wt% Al_2O_3.

Since ceramic material is porous, porosities tend to be produced in the weld bead. It was found that shape of porosity varies depending on Al_2O_3 content. In the weld bead of mullite, a lot of round small porosities which are not connected each other were found along fusion boundary. No cracks were seen in the weld specimens subjected to a penetrating dye under pressure, showing vacuum tightness. These porosities are thught to be produced because the high viscosity due to low temperature near fusion boundary prevents the bubbles in the melt from going up to the surface during welding. As the SiO_2 content increases, the number of porosities near fusion line tends to increase because the viscosity becomes larger. In the bead off the fusion line, on the other hand, few porosities were seen. This suggests that higher temperature there produces lower viscosity so that the most porosities could move up to the surface. In the weld bead of high Al_2O_3, the round porosities were not seen near the fusion line unlike the case of mullite specimens, but large elliptical porosities were seen to extend in the direction of temperature gradient only in the nail head part of the bead as shown in Fig. 5. The porosities were not produced in the finger part of the bead.

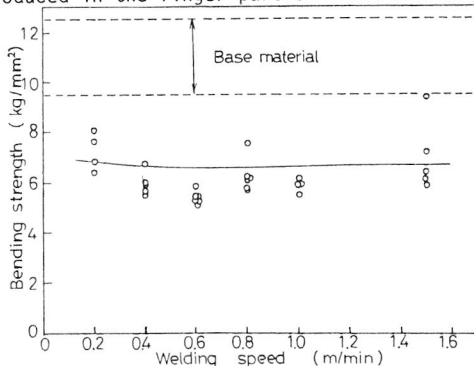

Fig. 11 Relationship between welding speed and bending strength in 48 wt% Al_2O_3 weld joint.

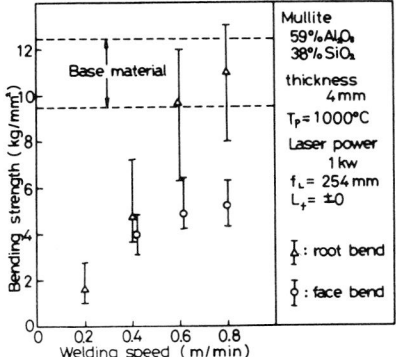

Fig. 12 Relationship between welding speed and bending strength in 59 wt% Al_2O_3 weld joint.

5-2 Bending strength

Strength of the weld joint were determined at room temperature in a 3-point bend test with a bend span of 20 mm. Since no significant difference in bending strength was seen between butt-weld joints and bead-on-plate specimens, all specimens used for the bending test were prepared in bead-on-plate welding. 4 mm and 2 mm thickness plates were used for welding mullite and high alumina ceramics. The weld specimens of 30 mm width and 50 mm length were sectioned by a diamond wheel to 7 mm width transverse to the welding derection, and then were subjected to the bending test in both face and root bending modes.

Figure 11 shows the effect of welding speed on the bending strength for 48 wt% Al_2O_3. The bending strength of the base material was also measured and it ranged between 9.5 and 12.5 kg/cm². The bending strength of the weld specimens was almost independent of the welding speed tested, about 7 kg/cm², that corresponds to a joint efficiency of about 60 wt%. No significant difference in the strength was found between face and root bending. All specimens fractured at the fusion boundary, and a lot of round porosities were observed on the most fracture surfaces. It is thought that the reduction in joint efficiency resulted from reduction in load-bearing area at fusion boundary due to a lot of porosities existing there. The phase-separated glass region in the bead caused no strength degradation.

Figure 12 shows the effect of welding speed on the bending strength in 59 wt% Al_2O_3. With decreasing welding speed, strengths both in face and in root bending modes are seen to decrease rapidly. Very low strength at low welding speeds is accounted for by significant depression on the both sides of the bead due to evaporation loss unlike the case of 48 wt% Al_2O_3 as shown in Fig. 9. It is also seen in Fig. 12. that the strength in face bending is much lower than that in root bending. At speed 0.8 m/min, the root bending gave the joint dfficiency as high as 100 %, whereas the face bending provided only 50 % of joint efficiency.

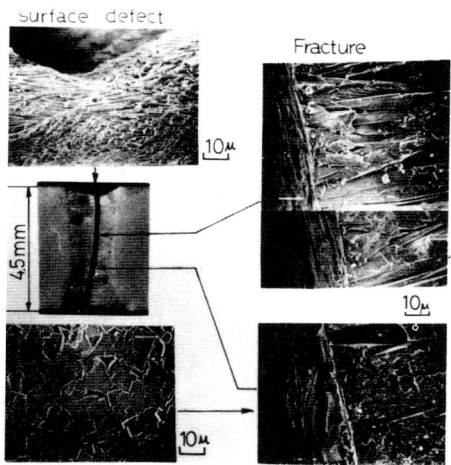

Fig. 13
Scanning electron micrograph of 59 wt% Al_2O_3 weldjoint.

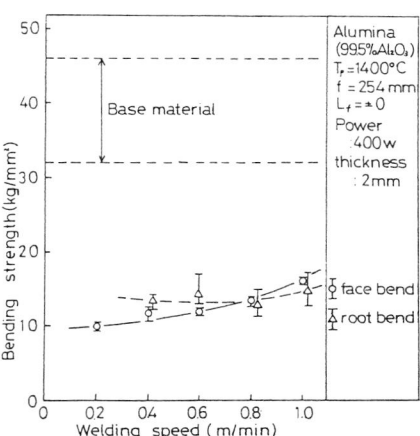

Fig. 14
Relationship between welding speed and bending strength in 99.5 wt% Al_2O_3.

As shown in Fig. 9, within the conditions tested the top surface of the 59 wt% Al_2O_3 bead does not contain any glassy structure, but consists of long columnar mullite crystals, which strech from the opposite sides, meet at the center of the bead surface and form the V-shaped depression unlike the round smooth depression in 48 wt% Al_2O_3. In face bending, fracture initiated at the root of the V-shaped depression, and no porosities were seen on the fracture surface as shown in Fig. 13. The strength degradation is accounted for by the notch effect of the V-shaped underfill .

Figure 14 shows the effect of welding speed on the bending strength for 99.5 wt% Al_2O_3. The bending strength of these material had as low joint efficiency as 30 %, but are higer than that of mullite. The low joint efficiency is thought to be caused from large blowhole in the nail head of the bead as mentioned before (see Fig. 5) If the plasma plume over the cavity can be removed successfully, the bending strength will increase markedly.

5-3 Improvement of weld strength

In the last section, laser welding Al_2O_3-SiO_2 ceramics is shown to be encountered with weld deffects that degraded the joint strength significantly in different ways depending on alumina content of of material. The authors tried to enhance the joint strength by improving the welding process; In 48 wt% Al_2O_3, porosities seen along the fusion boundary cause the strength degradation as mentioned above. In order to lower the viscosity of melt, thereby to make the movement of the porosities easy, a filler rod of alumina was added during welding.   By putting a thin alumina rod of around 2 mm in diamiter on the specimen along the welding line, the strength was enhanced up to 9.3 kg/cm^2.

In welding 59 wt% Al_2O_3, filler material was added to improve the joint strength. In this case, filler material of the same composition as the base material was used. Addition of filler material provided raised bead surface on both sides where no V-shaped depression was seen. Then the strength of face bending was enhanced up to 9 kg/cm^2, which is approximately twice as high as the value for no-filler material as shown in Fig. 15.

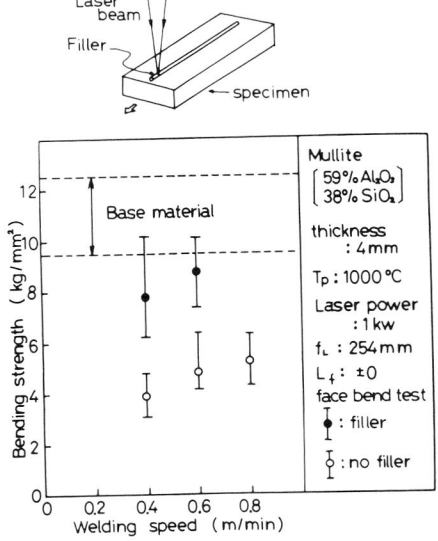

Fig. 15 Enhancement of joint by adding filler material in 59 wt% Al_2O_3.

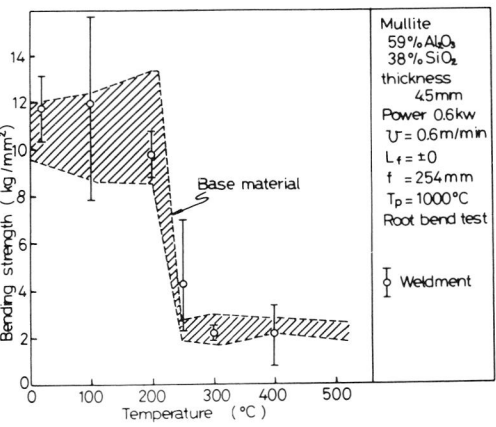

Fig. 16
Effect of thermal-shock temperature on bending strength in 59 wt% Al_2O_3 weld joint.

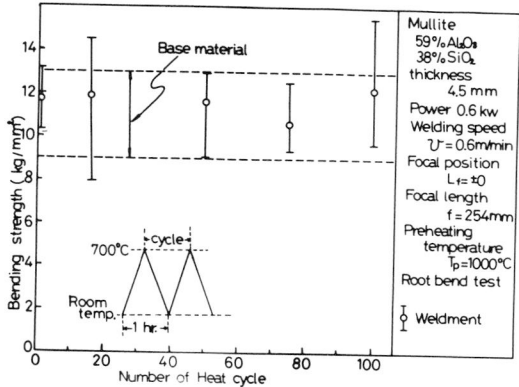

Fig. 17

Effect of thermal cycle on bending
strength in 59 wt% Al_2O_3 weld joint.

5-4 Resistance to thermal shock and thermal cycle

Since crystal structure of the bead is different from that of base material, information about resistance of the weld joint to thermal shock and thermal cycle is thought to
be important for practical applications. 59 wt% Al_2O_3 specimens welded with the filler
material were used for these resistance tests. The weld specimens held at a predetermined
temperature for 10 min were thrown into ice water, and then were subjected to the 3-point
bend test. The severity of the thermal shock was changed by adjusting the holding temperature. The strength data for the thermally shocked base material and weld joints are
given in Fig. 16. The data for the base material shown by a dashed band exhibited the
similar behavior as was predicted in Ref[6]; On quenching the strength remains nearly constant up to a critical temperature about 200 °C, at whichthe strength decreases discontinu-
ously to about one-third of the initial value. With a further increase in quenching temperature, the strength remains constant for a substantial temperature range. The weld-
specimens also showed the same behavior as the base material, and fractured in the base
material about 2 mm apart from the fusion boundary. These facts show that the weld joint
itself has the thermal shock resistance higher than or equal to the base material. The
weld specimen of 48 wt% Al_2O_3 with the filler material showed the similar resistance to
thermal shock as the base material. The weld specimens were also subjected to the thermal
cycle between room temperature and 700 °C, and then the bending strengths were determined
by the 3-point bending test as a function of number of the thermal cycle. As the strength
data are shown in Fig. 17, no degradation in strength is seen at least up to 100 cycles.

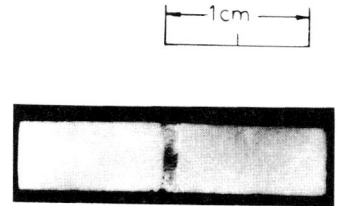
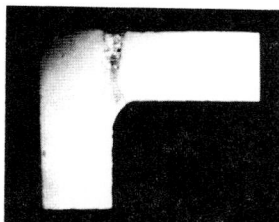
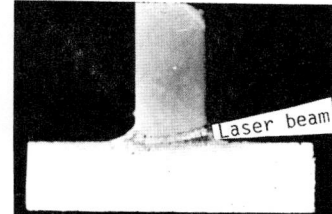

Fig. 18   Examples of laser welded ceramic joints.

515

6. Various weld joints

In Fig. 18, the example of various joints, I, L and T, are demonstrated. In L and T joint welding, addition of filler material resulted in smooth reinforcement of the bead at their corners. It should be noted in this figure that the T-joint was obtained in a single pass welding by irradiating the beam at small angle, about 15°, to the joint face as seen in Fig. 18; The incident laser beam proceeds with multiple reflections in the cavity, which is deepened through material via the route of the least laser energy thus resulting in bent bead. In the L joint, similar binding of the bead is seen.

7. Summary

A CO_2 laser was applied to welding Al_2O_3-SiO_2 ceramics with 48-99.5 wt% alumina contents, and was found to produce the deep penetration bead with excellent mechanical properties.

Effects of beam factors on the penetration depth and microstructure in the bead were demonstrated in comparison with laser welding metal. The prevention and control of the weld defects including underfill, crack and porosity are discussed, and the strength of the butt-weld joint with thickness up to 4 mm was evaluated by a 3-point bending test. Results obtained are summerized as follows:

(1) Two types of weld cracks occur in laser welding ceramics; a transverse crack is prevented by preheating the material up to the temperatures at which the materials show plasticity, and a longitudinal crack is prevented at welding speeds except for the critical speed of full-penetration welding.

(2) The joint efficiency is decreased by the porosities near the fusion line, the V-shaped underfill at the bead surface and elliptical porosities in the bead nail-head produced in low Al_2O_3 mullite, high Al_2O_3 mullite and high Al_2O_3 ceramics, respectively.

(3) The strength of the weld joints are enhanced by adding filler mateiral during welding, and are not deteriorated by being subjected to thermal cycles between room temperature and 700 °C. Their resistance to thermal shocking is as high as the base materials.

(4) A variety of welding joint geometries including I, L and T were performed in a single pass welding with the excellent bead shape.

References

[1] H. A. Hokanson, S. L. Rogers and W. I. Kern, "Electron Welding of Alumina", ceramic industry, August, 1963.
[2] I.Miyamoto, H.Maruo and Y.Arata, "Mechanism of Bead-Transition in Laser Welding", Proceedings of International Conference on Welding Reserch in the 1980's Osaka, 1980.
[3] E.V. Locke, E. D. Hoag and R. A. Hella, "Deep Penetration Welding with High-Power CO_2 Lasers", IEEE Journal of Quantum Electronics, vol. QE-8, No. 2. (1972) 132-135
[4] Y. Arata, H. Maruo and I.Miyamoto, "Characteristics of High Power CO_2 Laser Welding", International Beam Technology Conferenve, Essen, 1980.
[5] J. E. MacDowell and G.H. Beall, "Immiscibility and Crystallization in Al_2O_3-SiO_2 Glasses", Journal of The American Ceramic Society, vol. 52, No. 1. (1969) 17-25
[6] T. K. Gupta, " Strength Degradation and Crack Propagation in Thermally Shocked Al_2O_3" Journal of The American Ceramic Society, vol. 55, No. 5 (1972) 249-253

Fundamental Phenomena during Vacuum Laser Welding

Abstract

Laser welding under vacuum conditions (760 Torr - 10^{-5} Torr) was performed and the fundamental phenomena during Vacuum Laser Welding were dynamically observed, including the behavior of laser plasma, the molten pool and the beam hole. Observation was performed with a high speed camera and by the transmission X-ray method. It was found that Vacuum Laser Welding can almost completely suppress laser plasma and that this allows deep penetration at a very slow welding speed. Under these conditions, the shape and behavior of the beam hole during welding were very similar to electron beam welding. The Fundamental characteristics of Vacuum Laser Welding were also studied, including the effect of gas pressure and the welding speed on the penetration depth. The penetration depth increased with decreasing pressure and also as the welding speed decreased. A penetration depth of over 40 mm was achieved at a power of 11 kW, a pressure of 10^{-3} Torr, and a speed of 10 cm/min. Vacuum laser welding using an aerodynamic window was proposed for practical applications, and a penetration of over 25 mm was subsequently achieved, even at a pressure of 50 Torr.

I. Introduction

Laser beams, electron beams, ion beams and special plasma beams are generally classified as "High Energy Density Beam (HEDB)". The most important feature of HEDB welding is the deep penetration depth. This is due to the presence of a deep beam hole, which results from the violent melting and evaporation phenomena caused by the high energy density[1].

However, compared with the considerably deep penetration of vacuum electron beam welding, it was difficult to achieve such deep penetration in atmospheric laser welding. This is due to the fact that laser welding in an atmosphere produces a great amount of laser plasma, which strongly interferes with a CO_2 laser beam[2]. An assist gas easily suppresses such laser plasma and thus increases the penetration depth. However, an assist gas also affects the shape and behavior of the beam hole. When the flow rate of the assist gas is low, it has a very beneficial effect, but too high a flow rate enlarges the upper part of the beam hole. This causes failure of the wall focusing effect and the beam hole becomes shallow.

In order to solve this problem, one of the authors developed a new welding method, "Laser Spike Seam Welding (LSSW)". In this method, the laser beam stops relative to a specimen for a short period of time in order to drill more deeply. Just before an overabundance of plasma is produced, the laser beam is quickly shifted forwards to keep it away from the strong plasma. Thus, LSSW can penetrate more deeply than conventional laser welding at the same power and welding speed. However, the penetration depth is still shallower than can be achieved with an electron beam of the same power[2].

Another solution to this laser plasma problem, Vacuum Laser Welding, is described in this report. In order to completely suppress the laser plasma, laser welding was performed inside a vacuum chamber such as is usually used for electron beam welding. The fundamental phenomena during Vacuum Laser Welding were dynamically observed, and the fundamental characteristics were also studied.

II. Experimental Apparatus

A sketch of the experimental apparatus is shown in Fig. 1. It consists of a vacuum chamber, a vacuum pump system, a beam transport system and a work table. A 6" diffusion pump system can make a pressure of from one atmosphere to 10^{-5} Torr inside the vacuum chamber. A 15 kW CO_2 laser apparatus was used at an α_b value of 0.998. The laser beam enters through a ZnSe window and a transport system which protects the ZnSe window from vapor and/or laser plasma. Mild steel, stainless steel and high tension steel were used as the specimen.

III. Results and Discussion

III-1 Laser plasma

The laser plasma produced at a constant welding speed of 30 cm/min and at various pressures of from 200 Torr to 10^{-4} Torr was filmed with a high speed camera at a film speed of 3000 frames per second. The camera was placed at a side observation window of the vacuum chamber horizontally to the specimen as shown in Fig. 2. Several typical examples of the laser plasma produced at different pressures are shown in Fig. 3. When the pressure is 200 Torr, although there is no strong upward emission of plasma as under atmospheric conditions[2], a large amount of plasma is nevertheless produced. As shown in Fig. 3, the amount of laser plasma decreases with pressure down to a pressure of 5 Torr. Below 5 Torr, the laser plasma is almost completely suppressed up to a laser power of 11 kW, and only a minute amount of plasma can occasionally be seen. In this pressure range there is only a slight change in the amount of laser plasma as the pressure decreases further.

To observe the phenomena inside the beam hole and on the surface of the molten pool in more detail, the phenomena were filmed from the upper observation window with the camera at position B, as shown in Fig. 2. Typical examples observed at a constant welding speed of 30 cm/min are shown in Fig. 4. At a pressure of 50 Torr, some laser plasma still remains. The molten pool width is very wide, and the beam hole is positioned a little apart from the front wall. With decreasing pressure, the laser plasma inside the beam hole is suppressed. The molten pool width becomes narrower, and the beam hole moves near the front wall. The molten center line of the front wall and the solid wall on each side of the center line can sometimes be seen, as shown in Fig. 4. Below a pressure of 5 Torr, there is very little change in the molten pool width and in the beam hole position.

III-2 Beam Hole

In order to observe the behavior of the beam hole inside the specimen, a dynamic X-ray observation method[3] was employed, as shown in Fig. 5. An X-ray tube was set at the side window of the vacuum chamber, and the glass in the side window was replaced with a thin sheet of aluminum. The X-rays emitted from the X-ray tube enter through the aluminum window of the vacuum chamber and irradiate the specimen. The X-ray image of the beam hole is converted to a visible image by an X-ray image converter placed opposite the X-ray tube. The image is then filmed by a high speed camera at a film speed of 300 frames per second. Typical photographs of the beam hole at pressures of 5, 10^{-1} and 10^{-4} Torr and at a constant welding speed of 30 cm/min are shown with explanatory drawings in Fig. 6. It can be clearly seen that below a pressure of 5 Torr, at which point laser plasma is seldom apparent, there is little change in the beam hole shape and size. Also, the shape and behavior of the beam hole is similar to that in electron beam welding[2]. As shown in Fig. 7, a bulge first appears at the bottom of the beam hole, then moves smoothly upwards, with a strong effect on the motion of the molten pool.

III-3 Penetration depth

Figure 8 shows the pressure dependence of the penetration depth at a constant welding speed of 30 cm/min and at a constant laser power of 9 kW. It is clear that the penetration depth increases with decreasing pressure down to the level of a few Torr, below which little increase in depth is seen. This pressure dependence of the penetration depth shows great similarity to that already reported for electron beam welding[4].

Figure 9 shows the dependence of the penetration depth on the welding speed, at a constant pressure of 10^{-3} Torr and at a constant power of 9 kW. The penetration depth also increases with decreasing welding speed and reaches a maximum value at about 10 cm/min. It also increases with power, as shown in Fig. 10. Even at a power of 11 kW and a speed of 10 cm/min, a penetration depth of 40 mm was achieved. The bead cross section welded under these conditions is shown in Fig. 11.

Figure 12 - 14 show the phenomena when the pressure is kept at 10^{-3} Torr and the welding speed is reduced from 100 to 15 cm/min. Figure 12 shows a side view of the laser plasma. Even at a welding speed of 15 cm/min, there is no increase in laser plasma production. Figure 13 shows a top view of the specimen's surface from the upper observation window with the high speed camera at position B. When the welding speed is high, the molten pool is very narrow and the beam hole opening is at the very front of the molten pool. At a very slow speed, the molten pool becomes much wider and the beam hole opening is positioned a little apart from the front of the molten pool. Figure 14 shows the beam hole as imaged by the transmission X-ray method. At a speed of 100 cm/min, the beam hole is shallow and inclined, and appears as a sock-like shape with a large bottom. In contrast, at a low speed of 30 cm/min, the beam hole does not yet have a sock-like shape, and is instead similar to that produced by electron beam welding.

It is found that in Vacuum Laser Welding, laser plasma, which is a fundamental problem in atmospheric laser welding, is almost totally suppressed. This enables very low welding speed and deep penetration can thus be achieved. Under these conditions, the beam holes in both laser beam welding and electron beam welding display a very similar shape and behavior. Even at a laser power of 11 kW, deep penetration of over 40 mm was achieved at a welding speed of 10 cm/min.

IV. Vacuum Laser Welding with Aero dynamic Window

Compared to an electron beam, which is essentially well-suited for welding in a high vacuum, a high power laser beam is not easy to handle under high vacuum conditions. The authors therefore would like to propose a new method of vacuum laser welding for practical applications. Figure 15 shows a prototype incorporating a low vacuum welding chamber with no ZnSe or KCl window for laser beam transmission. Instead, it has only a simple hole which functions as an aerodynamic window, thus avoiding the problem of laser power limitation and vapor damage to the window or mirrors. Furthermore, by varying the pressure of the welding chamber and upper vacuum chamber, a dynamic pressure can be generated from the upper chamber to the welding chamber. This dynamic pressure serves the same function as an assist gas in atmospheric laser welding. Typical photographs of laser plasma at a constant welding pressure of 50 Torr and at different dynamic pressures are shown in Fig. 16. Above a dynamic pressure of 8 Torr, the laser plasma is visibly suppressed. However, as the dynamic pressure increases, the upper part of the beam hole enlarges and the depth of beam hole is reduced, as shown in Fig. 17. It is thus confirmed that the gas flow caused by this dynamic pressure has the same effect on the beam hole as an assist gas in atmospheric laser welding. There is also an optimum value for deep penetration. At the optimum dynamic pressure, even under a pressure of 50 Torr, deep penetration of over 25 mm was achieved at a power of 11 kW and a speed of 25 cm/min, as shown is Fig. 18.

V. Conclusion

The laser plasma problem, which is the most severe problem in atmospheric laser welding, was completely solved by Vacuum Laser Welding. The fundamental phenomena during Vacuum Laser Welding were studied, and dynamic observation of the laser plasma with a high speed camera showed that laser plasma production decreased with pressure. Below a few Torr, laser plasma was almost completely suppressed, even at a welding speed of 15 cm/min. Dynamic observation of the beam hole with a transmission X-ray method showed that the beam hole at this low pressure is very similar to that of electron beam welding. The fundamental characteristics of Vacuum Laser Welding were also studied, and it was found that the penetration depth increased with decreasing pressure at a constant welding speed and power. It also increased as the welding speed decreased, down to a speed of 10 cm/min under constant pressure and power.

It can be concluded that in Vacuum Laser Welding, laser plasma can be almost completely suppressed. This enables very low welding speed which provides deep penetration. Under optimum conditions, a penetration depth of over 40 mm was achieved even at a power of 11 kW.

Furthermore, Vacuum Laser Welding with aerodynamic window is proposed for practical applications of Vacuum Laser Welding, and it has the potential to allow much higher laser powers to be employed compared with other method under relatively low vacuum conditions of a few Torr to a few tens of Torr. Even at a pressure of 50 Torr, penetration depth of over 25 mm was acheived at a laser power of 11 kW.

References

(1) Y. Arata: What Happens in High Energy Density Beam Welding and Cutting?, 1980.
(2) Y. Arata, N. Abe and T. Oda: Beam Hole Behaviour during Laser Beam Welding: Proc. of ICALEO'83, 1983.
(3) Y. Arata, N. Abe and E. Abe: Tandem Electron Beam Welding (Report IV) -Analysis of Beam Hole Behaviour by Transmission X-ray Method-: Trans. JWRI, 11 (1) 1982, pp. 1-5.
(4) M. Tomie: Fundamental Study on High Power Electron Beam Welding: Doctor Thesis of Osaka University, 1978.

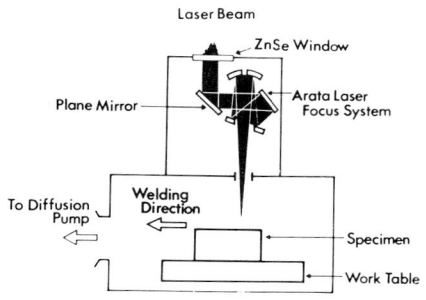

Fig.1 Experimental apparatus.

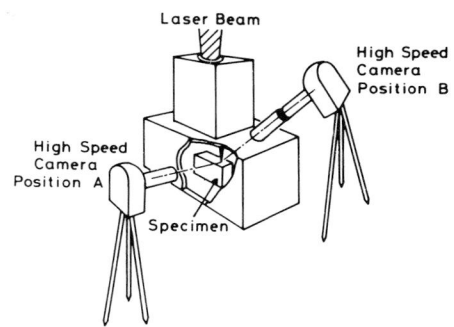

Fig.2 Camera positions.

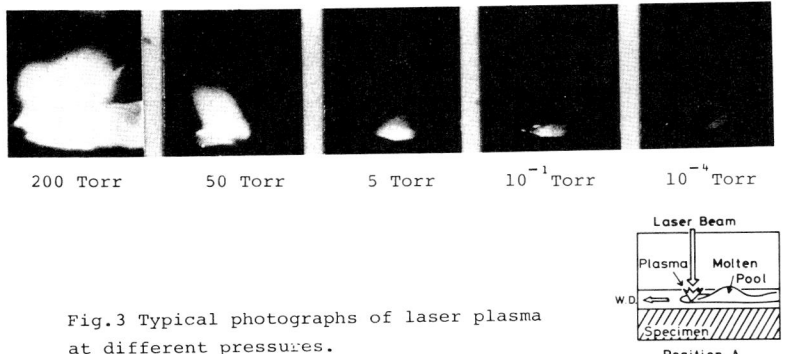

200 Torr   50 Torr   5 Torr 10^{-1} Torr 10^{-4} Torr

Fig.3 Typical photographs of laser plasma at different pressures.

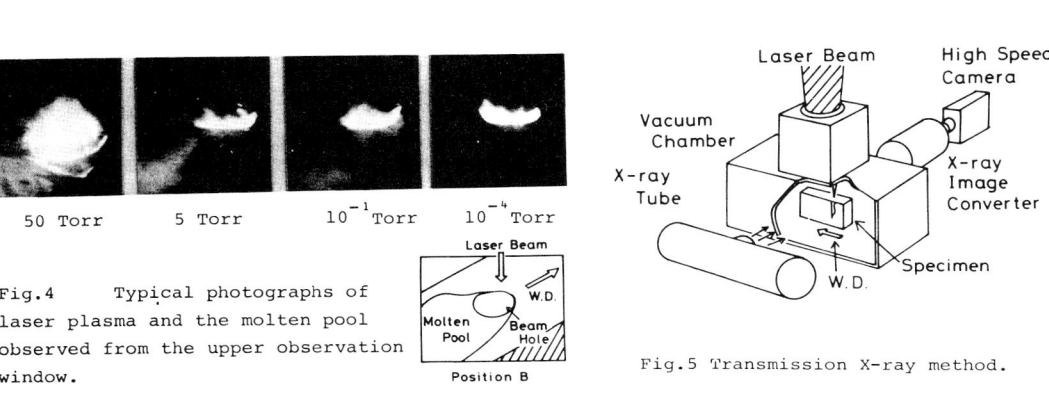

50 Torr   5 Torr 10^{-1} Torr 10^{-4} Torr

Fig.4 Typical photographs of laser plasma and the molten pool observed from the upper observation window.

Fig.5 Transmission X-ray method.

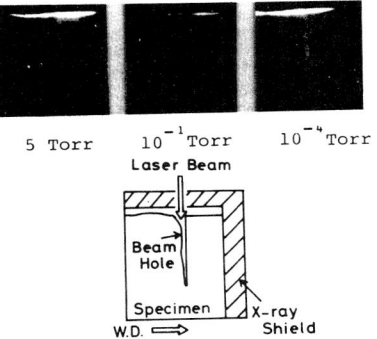

5 Torr 10^{-1} Torr 10^{-4} Torr

Fig.6 Typical photographs of the beam hole at different pressures.

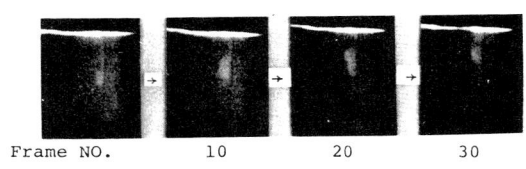

Frame NO.   10   20   30

Fig.7 Typical sequence of the motion of the beam hole. Numerical numbers represent frame numbers.

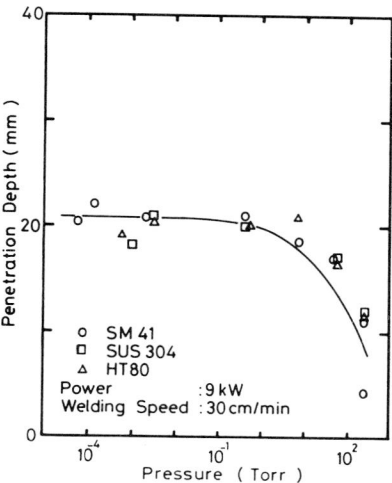

Fig.8 Pressure dependence of the penetration depth.

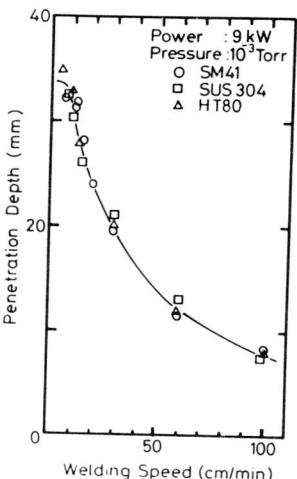

Fig.9 Welding speed dependence of the penetration depth.

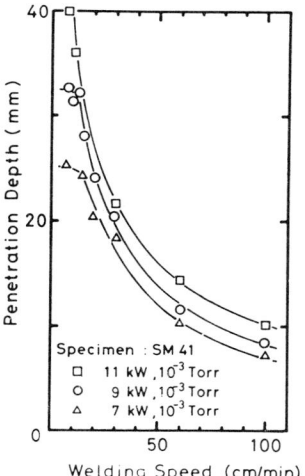

Fig.10 Power and welding speed dependence of the penetration depth.

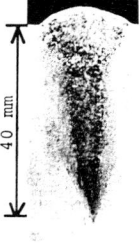

Fig.11 Typical example of bead cross section.

521

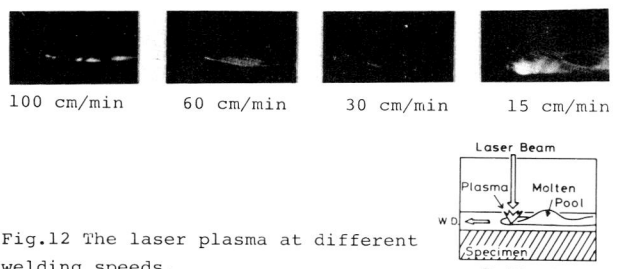

100 cm/min   60 cm/min   30 cm/min   15 cm/min

Fig.12 The laser plasma at different welding speeds.

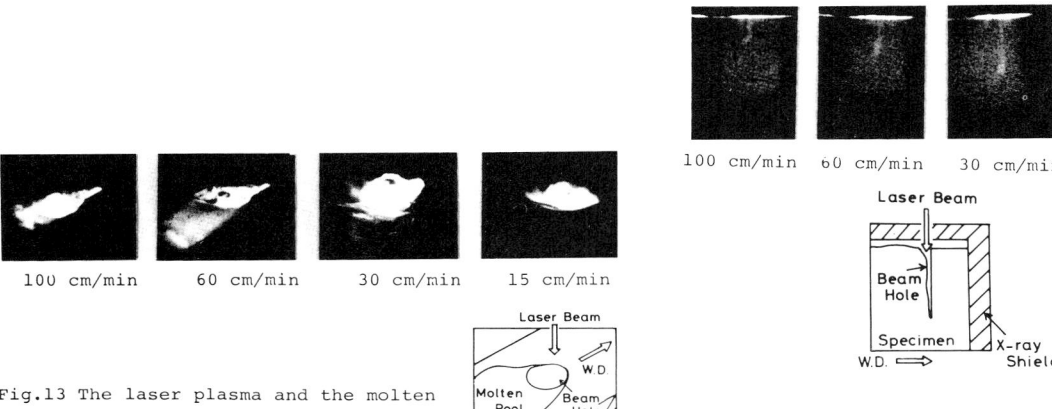

100 cm/min   60 cm/min   30 cm/min   15 cm/min

Fig.13 The laser plasma and the molten pool observed from the upper observation window at different welding speeds.

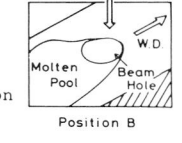

100 cm/min   60 cm/min   30 cm/min

Fig.14 The beam hole at different welding speeds.

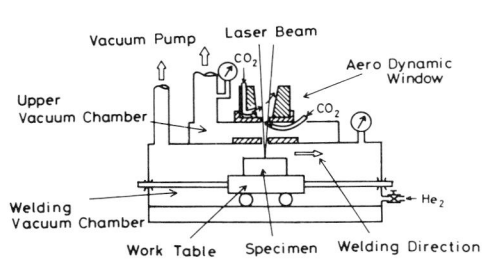

Fig.15 Experimental apparatus of vacuum laser welding with aerodynamic window.

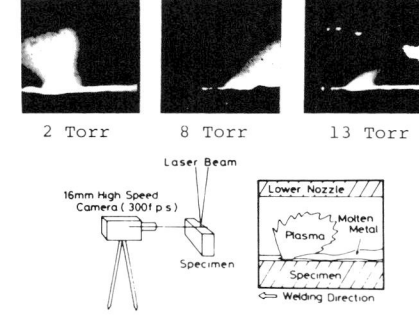

2 Torr   8 Torr   13 Torr

Fig.16 Laser plasma at different dynamic pressures.

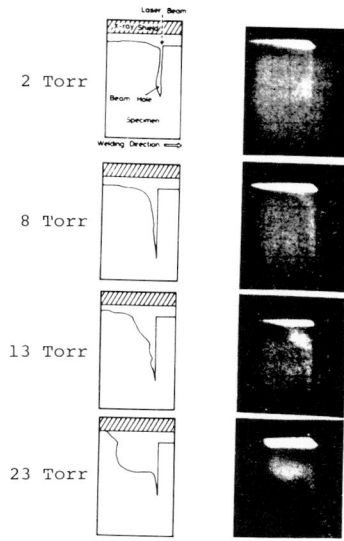

Fig.17 The beam hole at different dynamic pressures.

Fig.18 Typical example of bead cross section of vacuum laser welding with aerodynamic window.

Prof. Dr. Yoshiaki ARATA : Director of Welding Research Institute of Osaka University
Dr. Nobuyuki ABE   : Research Instructor of W. R. I. of Osaka Univ.
Mr. Tatsuharu ODA   : Researcher of W. R. I. of Osaka Univ.
Mr. Nobuhiro TSUJII: Graduate Student of Osaka Univ.

Laser Gas Cutting

New Laser-Gas-Cutting Technique for Stainless Steel
Y. Arata, H. Maruo, I. Miyamoto and S. Takeuchi
[IIW Doc. IV- 82. 1st Int. Laser Proc. Conf. (1981).]

New Laser-Gas-Cutting Technique for Stainless Steel

Abstract

The effects of cutting parameters in laser-gas-cutting on the qualities of cuts have been studied in stainless steel plate of 0.5 to 4 mm thickness using a 1 kW CO_2 laser. In the conventional laser-gas-cutting method, in which oxygen flow is coaxially arranged with the focused laser beam, the quality of stainless steel cut is undesirable because an oxide dross tends to cling to the bottom edge of the cut due to poor fluidity of the oxide dross. For this improvement, two new cutting techniques, "Pile Cutting" and "Tandem Nozzle Cutting", have been developed. Pile cutting, in which a thinner mild steel is piled on stainless steel, makes it possible to get completely dross-free cut, since the molten iron oxide supplied from mild steel enhances the fluidity of the cutting region of stainless steel. Tandem nozzle cutting, in which an off-axial nozzle is used with the coaxial nozzle to enhance the dross removal, also provides little or no dross cuts. These two techniques are also markedly effective to improve surface roughness and out-of-flatness of the cut.

1. Introduction

In recent years Laser-Gas-Cutting method [1, 2] is widely used in automobile industry, electrical engineering and so on. In spite of its wide acceptance [3-10] much potenial capability has not been sufficiently brought out.

The conventional laser-gas-cutting technique, in which a gas jet is used coaxially with focused laser beam, has made it possible for various metals to precisely cut, and especially using oxygen jet for oxidizable materials can highly enhance the cutting performance and the quality owing to addtion of oxidation energy to beam energy [11]. For instance, in cutting mild steel plate a dross-free, fine cut can be easily obtained at high speed, for example, 7m/min for 1mm thick plate at 1kW power level [12]. However, for stainless steel plate, which has been widely used in industry, the conventional laser-gas-cutting cannot provide a performance high enough to apply to various actual presision processings, because the oxide dross has a tendency to cling to the bottom edge of the cut. For this reson the laser-gas-cutting techniques for stainless steel are required by an exact study of cutting characteristics, and cutting mechanism so as to bring its potential performance to full play.

In the present study, the relationship between cut qualities and cutting parameters in conventional laser-gas-cutting stainless steels has been described in detail and the dross clinging mechanism has been revealed through observation by high speed filming of cutting region. On the bases of these results, the authors have developed two kinds of new cutting techniques, Pile Cutting Method and Tandem Nozzle Cutting Method, to improve the cut quality.

2. Experimental Procedures

The data presented here were obtained by using continuous wave CO_2 laser with the maximum output of 1.2 kW, GTE Sylvania, Inc. Model 1971. The laser beam was focused onto the workpiece surface through a ZnSe lens, and beam and oxygen gas flow were coaxially directed through a convergent nozzle with 1.5mmϕ orifice, as shown in **Fig. 1**. The three pieces of lenses with focal length f = 63.5, 127 and 254 mm were used. Austenitic stainless steel plate, SUS304 (18%Cr 8%Ni), was mainly tested here.

It is required for strict estimation of laser-gas-cutting performance to synthetically evaluate the quality of the cut surface. The cut quality can be successfully evaluated in terms of clinging dross, out-of-flatness and surface roughness of the cut, as far as we refer to the existing evaluation standards of thermal cut, WES 118 (Japan), DIN 2310 (West German) and IIW's tentative plan. Then the following evaluation method suitable to the laser-gas-cuts of stainless steel as shown in **Fig. 2** was used in the present study. Height of clinging dross H_D was measured with micrometer, out-of-flatness of the cut F was measured with reading micro-scope and surface roughness of the cut R (μR_z) was measured with talysurf meter and expressed as ten points roughness. The talysurf measurement was carried out only on flat surfaces of which roughness variation of less than $150 \mu R_z$ and the maximum value R_{max} in the three trace lines was regarded as the representative roughness.

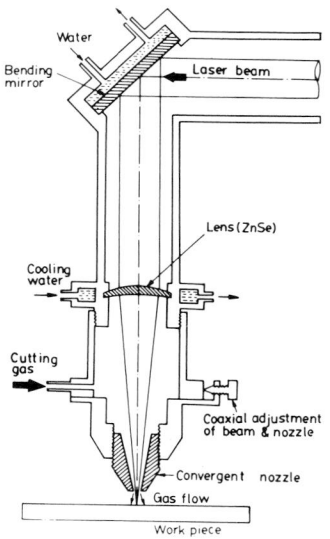

Fig. 1   Schematic diagram of laser-gas-cutting head.

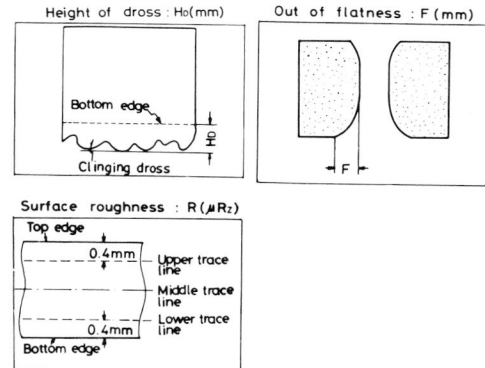

Fig. 2   Evaluation factors of cut quality and those method.

It is sufficiently useful to classify the cuts into several types by those kerf shapes, as a convenient indication of the cut surface characteristics. The five types of the cuts, which was modified from authors' method [12] for mild steel cuts, are shown in **Fig. 3**. These are generated as increase in cutting speed and/or plate thickness. IV_L type is accompanied with a rough zone in the bottom part at the low speed, II type is with a parallel kerf and some dross, III type is dross-free and fine, IV type is with a rough zone at the higher speed and V type is with a gouge at the still higher speed.

Dross clinging phenomena and flowing phenomena of molten material on the cutting front were analyzed by high speed movie pictures as shown in **Fig. 4** and **Fig. 5**. In the former case, the camera was set so as to observe the bottom edge of the cut at 4000 frames/sec and shutter speed 10^{-4} sec, as shown in Fig. 4. In the latter case it was set so as to observe directly the leading face of the cut at 4000 frames/sec and shutter speed of 10^{-5} sec, as illustrated in **Fig. 5** (refer to [12]).

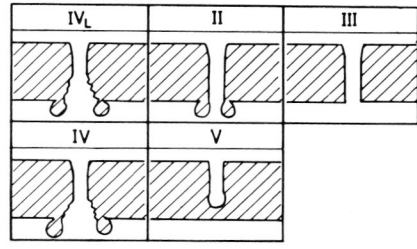

Fig. 3   Classification of laser-gas-cuts based on shape of cross section.

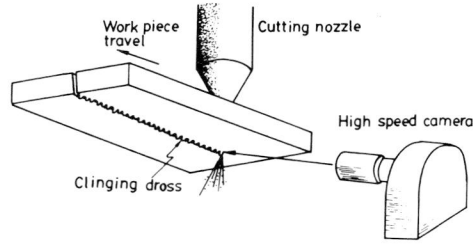

Fig. 4   Arrangement for high speed filming of dross clinging phenomena.

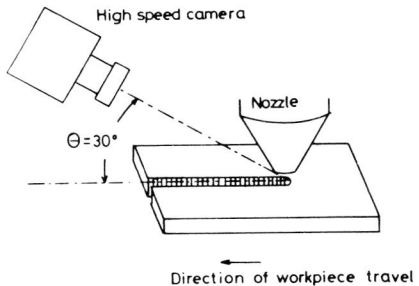

Fig. 5 Arrangement for high speed filming of melting phenomena on cutting front.

3. Performance of Conventional Cutting and its Limitation

From various cutting parameters which influence the cut qualities, cutting speed, laser power, oxygen pressure and focal length were chosen as the most important parameters, and their effects on the cut quality were tested in detail. **Figure 6** shows the result of the three evaluation factors in various incident laser powers W_i and cutting speed v_b, at the thickness of 2 mm. This result indicates that the cut qualities are best synthetically at the medium cutting speed in each power condition.

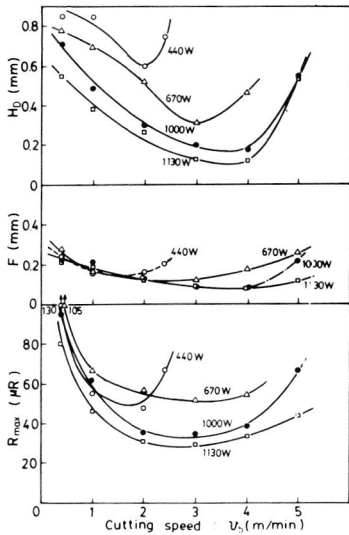

Fig. 6 Effect of laser power and speed on cut qualities. (thickness: 2 cm, oxygen pressure: 1.5 kg/cm², focal length: 127 mm)

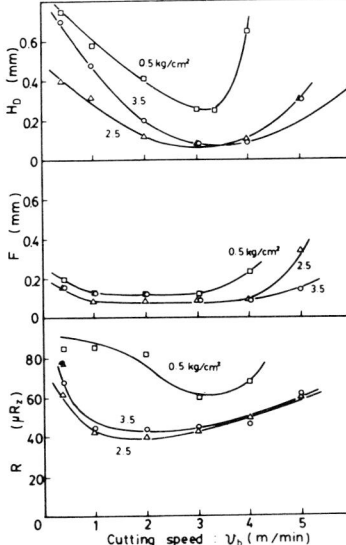

Fig. 7 Effect of oxygen pressure and speed on cut qualities. (thickness: 2 mm, laser power: 1.1 kW, focal length: 127 mm)

Furthermore with increase in laser power the optimum speed increases gradually and the corresponding qualities also are relatively improved. The similar inclination is noticed in case of various cutting oxygen pressures P_{O_2}, as shown in **Fig. 7**. However the effect of oxygen pressure, unlike laser power, has a limitation, because it has a saturation at the pressure above 2.5 kg/cm² as shown in the figure. **Figure 8** shows the effect of the focal length of lens on the cut qualities. The focal length also is a very influential factor and here f = 127 mm gives the best condition to the synthetic cut quality. These results indicate that stainless steel, unlike mild steel [12], is difficult to be cut without any clinging dross in conventional cutting.

The results obtained above have been further studied by examining a correlation between the three evaluation factors. **Figure 9** shows the correlation diagram which involves the data in Fig. 6 and Fig. 7, provided that the plots in IV and IV_L types, having an undoubtedly rough kerf zone, have been already eliminated. The height of clinging dross is in plus correlation with the other factors and the amount of those scatters becomes the smallest in the small H_D region. This is caused by that the cutting state is dominated mainly by the situation of melting flow on the cutting front and therefore out-of-flatness and surface roughness are finally accomplished as its traces, as well as clinging dross. In view point of actual applications of laser processing, those results are much important for the following reason. In case that the cutting parameters should be determined to obtain the best performance and/ or the higher cut quality, those optimum conditions can be almost found out by a visual judgement of the situation of clinging dross and/or by its amount.

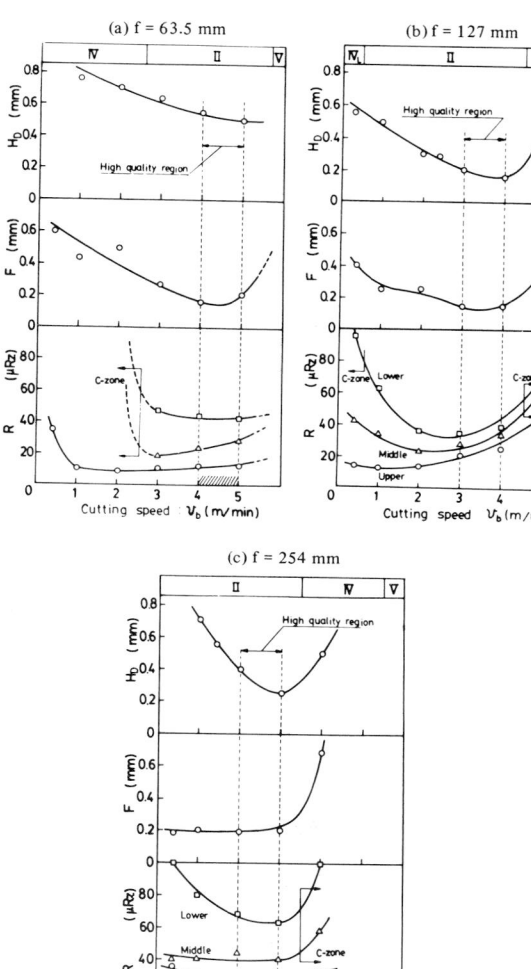

Fig. 8 Effect of focal length and speed on cut qualities. (thickness: 2 mm, oxygen pressure: 1.5 kg/cm², laser power: 1 kW)

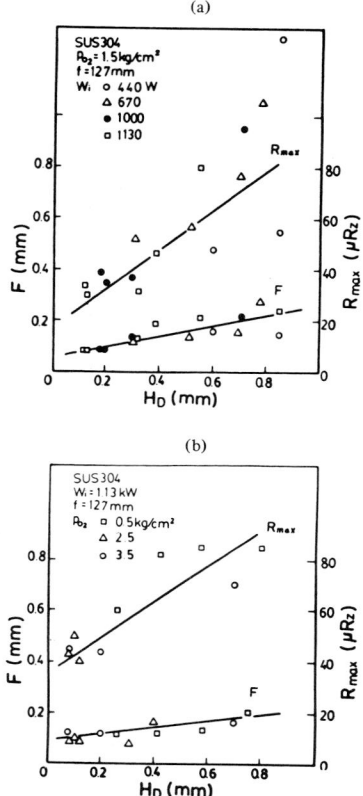

Fig. 9 Correlation between evaluation factors for various laser powers and oxygen pressures.
(a) Various laser powers
(b) Various oxygen pressures

4. Observation of Dross Clinging Phenomena

In the previous section, the cutting characteristics by the conventional laser-gas-cutting with a coaxial cutting nozzle were described, and the optimum controle of cutting parameters proved to be effective, but has a limitation to obtain the high cut quality with no dross. Furthermore, the correlation between the evaluation factors implicated a possibility that a further reduction of the clinging dross, if it is possible by a certain new technique, might lead to a synthetic improvement of the cut. For this reason, the study of the dross clinging mechanism is helpful to acquire an idea for improvement technique.

High speed films were taken to observe the dross clinging phenomena, with stainless steel plates of the thickness 2 mm. The results are shown in **Fig. 10** to **12**. Figure 10 shows the typical phenomena at the low speed, in which the cut is accompanied with a large amount of clinging dross. The melting flow makes a stagnated region at the bottom edge. A part of melting flow is ejected out into a large amount of small spherical drosses by the dynamic force of the gas flow. Simultaniously the other part flows backwards through the bottom edge or stays there, and is finally solidified into clinging dross. This phenomena changed remarkably with cutting speed. The dross clinging phenomena in which the dross is the minimum are shown in Fig. 11. The backward melting flow as shown in Fig. 10 is weak in this case, and a long pendent part (1 to 2 mm length) is formated at the ejecting region of the flow, so that the most of spherical drosses are scattered through this pendent part. In this case there exists a small disturbance at the ejecting region (the neck of the pendent part) and it develops into the clinging dross. Figure 12 shows the typical phenomena in the IV type cutting, in which the amount of clinging dross increases again. There exists a greatly disturbed flow, which ejects out along the inclined cutting front as making a rough zone.

Dross clinging process was deduced also through microscopic observation of the cut kerf. **Figure 13** shows the appearances of the cut near the bottom edge after lasergas-cutting. The thick solidified metal zone is left over the base metal and is connected with the clinging dross. This fact explains the following cutting process. In cutting a part of molten metal on the cutting front only is beeing oxidized by oxygen and another part in the sublayer remains non-reacted, so that a cohesive force is generated between both layers just when the oxide layer (or dross) is ejected out. This force and the less fluidity of the molten oxide make the flow disturb and suppresses the smooth separation of the molten dross from the bottom edge because there is a great disturbance produced by an intensive burning in the lower part of the kerf as illustrated in Fig. 12.

It is concluded from those results that an inclination of dross clinging is greatly influenced by the melting/flowing conditions of material on the cutting front. Therefore, a basic need in an improvement technique is to enhance relatively the fluidity of molten flow on the cutting front or to further increase the momentum of gas flow, or the gas dynamic force, so as to suppress the stagnation and/or disturbance of the melting flow. From this view point the two kinds of new improvement techniques have been developed as later described.

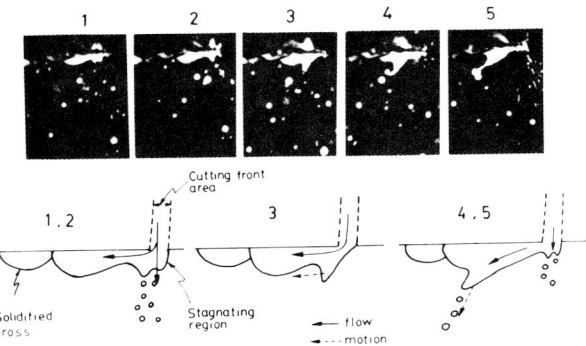

Fig. 10 High speed movie pictures of cutting accompanied with a large amount of clinging dross. (cutting speed: 1.4 m/ min, thickness: 2 mm, laser power: 1 kW, oxygen pressure: 1.5 kg/cm², time interval: 0.3 msec)

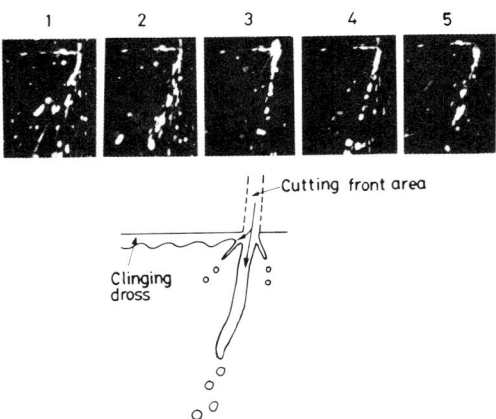

Fig. 11 High speed movie pictures of cutting accompanied with a little amount of clinging dross. (cutting speed: 3 m/min, other conditions: the same as in Fig. 10)

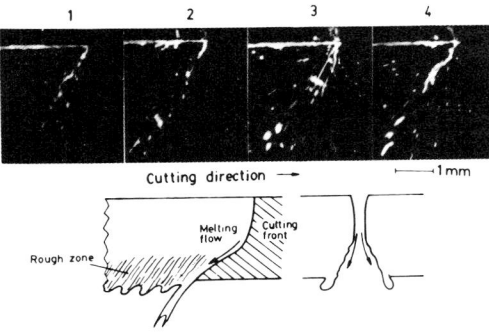

Fig. 12 High speed movie pictures of cutting accompanied with rough kerf zone. (cutting speed: 5 m/min, other conditions: the same as in Fig. 10)

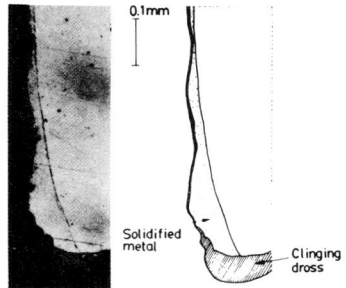

Fig. 13 Appearances of molten metal and clinging dross near bottom edge. (speed: 3 m/min, thickness: 2 mm)

5. Improvement of Cut Quality by Pile Cutting Method

5.1. Pile Cutting Method and its Characteristics

Pile cutting method developed by authors to improve the cut quality is illustrated in **Fig. 14**. This method is carried on through the simple procedure that the subject metal is laser-gas-cut together with a thin mild steel plate (supplement plate) piled on it by a small constraint force. The pile cutting characteristics are described here and the cutting mechanism is discussed later.

Figure 15 shows the typical cuts of stainless steel (SUS 304) 2 mm in thickness by this method. Perfectly dross-free cuts can be obtained between 2 to 4 m/min. A comparison of conventional laser-gas-cutting and pile cutting characteristics is made in **Fig. 16**. In the conventional cutting any dross-free cuts are not obtained in each thickness, as known from the height of dross H_D in Fig. 16 (a). Compared with this, pile cutting makes the stainless steel plate up to 4 mm in thickness perfectly dross-free, as shown in Fig. 16 (b). Such an excellent performance has been further confirmed by the three evaluation factors of the cut quality described in the previous section, and the result is shown in **Fig. 17**. Using this method, not only height of dross H_D but out-of-flatness F and roughness R are improved quitely well (refer to Fig. 8 (b)).

It was also confirmed that this method was surfficiently effective to the 18% Cr stainless steel and 13% Cr stainless steel. Thus pile cutting method is considered to be much applicable to most of stainless steels.

This technique was tested also for nickel plate (1 mm thick) and aluminum plate (1 to 2 mm), and proved to be much effective. For the nickel, some clinging dross was inevitable in conventional laser-gas-cutting, as well as stainless steel, but the pile cutting method not only made it perfectly dross-free but also enhanced the maximum limit of cutting speed up to 130%. For the aluminum, the dross-free cut could not be obtained also by pile cutting,

but the maximum limit of cutting speed was enhanced by it up to 250 to 350%. Thus, this method has a great possibility to apply to various materials and is expected to be further examined.

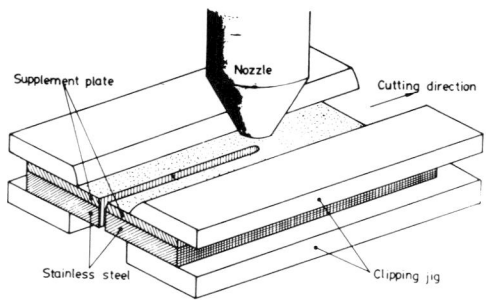

Fig. 14 Schematic diagram of pile cutting method.

(a)

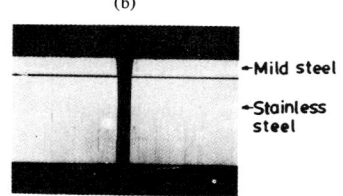

(b)

Fig. 15 Cut appearances in pile cutting stainless steels.
(a) dependence on cutting speed (2 to 4 m/min: dross free)
(b) typical kerf shape of dross-free cut (speed: 4 m/min)

531

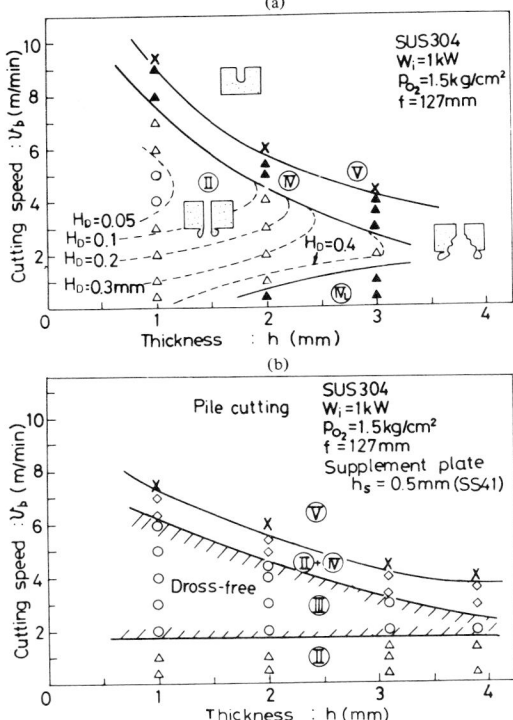

Fig. 16 Comparison between conventional cutting and pile cutting.
(a) conventional cutting characteristics
(b) pile cutting characteristics

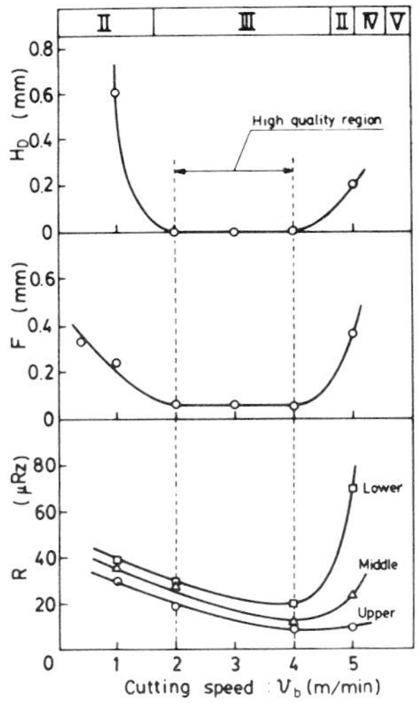

Fig. 17 Evaluation of pile cut qualitities of stainless steel plate. (thickness: 2 mm)

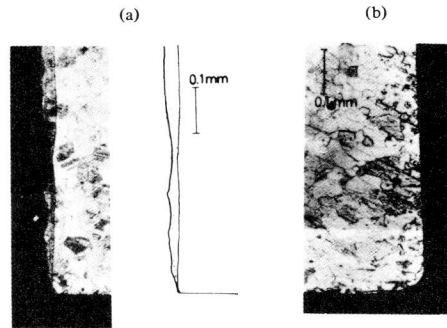

Fig. 18 Appearances of pile cut near bottom edge.
(a) pile cutting of stainless steel
(b) conventional cutting of mild steel

5.2. Discussion of Pile Cutting Mechanism

Detailed observation of the cut kerf and the melting state in pile cutting stainless steel is made and on the bases of those results the improvement mechanism is discussed here.

The kerf shapes of mild steel and stainless steel have a good continuity as shown in Fig. 15 (b). This means that the melting state of mild steel is smoothly kept on to stainless steel side. **Figure 18 (a)** shows the molten zone near the bottom edge after pile cutting. Compared with the cut by conventional cutting method (Fig. 13), this molten layer is quitely thin in the bottom edge, and is, therefore, comparatively close to the dross-free cut of mild steel **(Fig. 18 (b))**. This means that the molten material on the cutting front is much fluid in pile cutting, as well as cutting mild steel. The melting phenomena were actually confirmed by the high speed movie pictures of the cutting front. **Figure 19** shows the pictures filmed, in which **(a)** and **(b)** correspond to the front zones of mild steel (dross-free) and stainless steel (dross clinging) by the conventional cutting method respectively, and **(c)** corresponds to the stainless steel's one by the pile cutting method (dross-free). In the conventional cutting of stainless steel (b) the flow is in great disturbance, while in the conventional cutting of mild steel **(a)** and in the pile cutting of stainless steel **(c)** it is in no disturbance and almost in steady state. Furthermore in the pile cutting (c) the molten oxide from the supplement mild steel proves to flow smoothly into the stainless steel side. These results show the state in pile cutting stainless steel is very close to the state in the dross-free cutting of mild steel.

To deduce the pile cutting process, the scattered dross ejected from the bottom edge in cutting were collected and analyzed by X-ray diffractometer. It was proved from this that the main oxidized product was FeO in cutting only mild steel by conventional method, while it was FeO and Cr oxides (Cr_2O_3 and $FeCr_2O_4$) in pile cutting stainless steel. The FeO-Cr_2O_3 phase diagram, **Fig. 20**, means that such Cr oxides have high melting points in comparison of iron oxide. Cr oxides act as a restrainer for oxygen diffusion through the molten layer of cutting front, because they tend to make minute solid regions in the molten layer of the cutting front.

Based on those results, the effect of the supplement mild steel on the improvement is explained by the following cutting process. In pile cutting, the oxide supplied from mild steel has an effect to dillute the Cr oxides which are restrainers of oxygen diffusion, so that the molten zone of the cutting front becomes more oxidizable according to Fig. 20, that is, reaches the single liquid phase region. A raized oxidation reaction through such a process leads to reduce the thickness of molten metal layer. In this case the effect of viscosity of molten oxide and metal should be also considered. **Figure 21** shows the relationship between the viscosities of molten FeO [13] and stainless steel [14]. This means an increase in the amount of FeO in pile cutting makes it possible to enhance the fluidity in sufficiently high temperature region. Thus the melting flow whose fluidity has been raised through such a whole process is easily removed from the bottom edge by the dynamic force of gas jet.

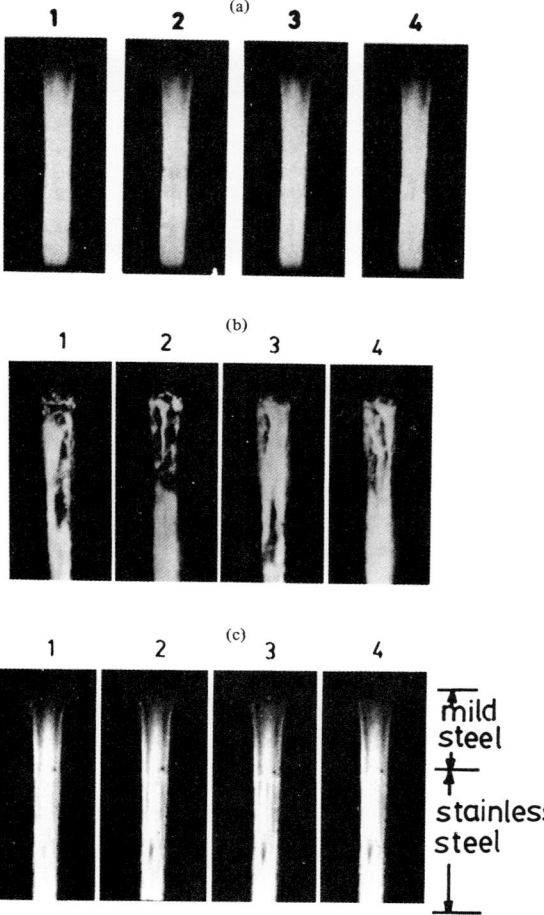

Fig. 19 High speed movie pictures of cutting front in cutting (time interval of pictures: 0.3 msec).
(a) conventional cutting of mild steel
(b) conventional cutting of stainless steel
(c) pile cutting of stainless steel

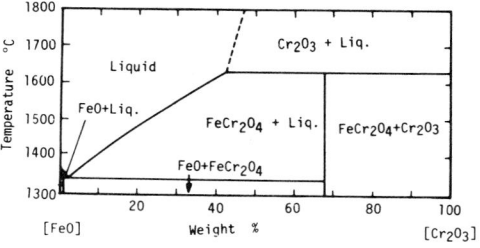

Fig. 20 FeO-Cr_2O_3 phase diagram.

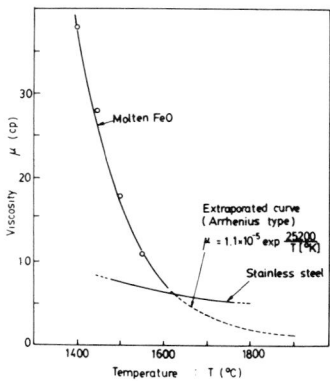

Fig. 21 Viscosity of molten FeO and stainless steel depending on temperature [13, 14].

6. Improvement of Cut Quality by Tandem Nozzle Cutting Method

It was shown in the previous section that in coventional laser-gas-cutting with a coaxial nozzle the increase in oxygen pressure of nozzle contributed to reduce the amount of clinging dross (Fig. 7), but its effect was saturated above 2.5 kg/cm², so that a perfectly dross-free cut could not be obtained finally. The purpose of tandem nozzle cutting shown in the present section is to overcome such a limitation of gas dyramic force in conventional laser-gas-cutting.

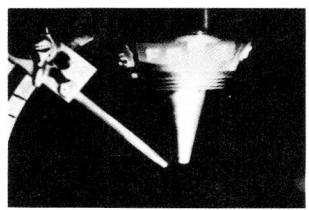

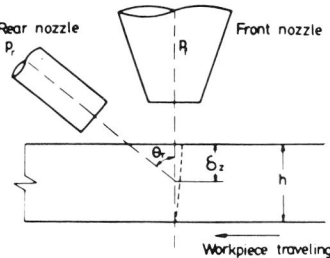

Fig. 22 Schematic diagram of tandem nozzle cutting method.

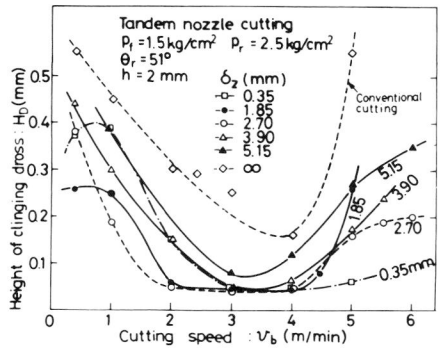

Fig. 23 Relationship between height of clinging dross and aiming position of rear nozzle in various speeds.

The tandem nozzle cutting method is illustrated in **Fig. 22**. The off-axial nozzle (rear nozzle) is arranged behind the coaxial nozzle (front nozzle) and the gas jets from these nozzles are used cooperatively. The rear nozzle conditions are much influential factors. These optima were experimentally determined by checking the height of clinging dross H_D. In case of the rear nozzle pressure P_r, the height of dross H_D decreased gradually with increase in P_r and was saturated at $P_r \gtrsim 2.5$ kg/cm² at the fixed front nozzle pressure $P_f = 1.5$ kg/cm². In case of the aiming angle θ_r, the optimum proved to be about 50°. This allowance was considerably limited because the H_D increased twice at $\pm 10°$ of the optimum. In case of the aiming position δ_z, when it was set near the bottom edge of the cutting front, the effect of rear nozzle became the best as shown in **Fig. 23**. This means that the rear nozzle is most effective when its jet is blown off so as to suppress the stagnation or the backward melting flow along the bottom edge. **Figure 24** shows a comparison between the appearances of the tandem nozzle cutting and the conventional cutting. The region of scattered drosses proves to be successfully removed with a small spreading angle by the tandem nozzle. **Figure 25** shows

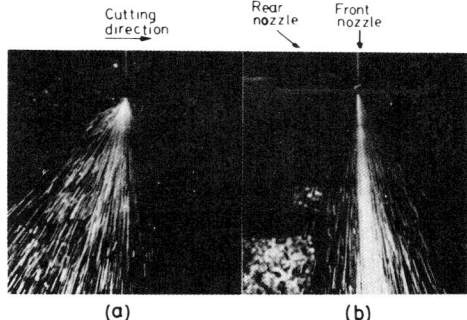

Fig. 24 Appearances of ejecting drosses.
(a) conventional cutting
(b) tandem nozzle cutting

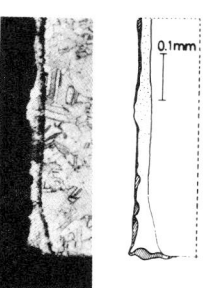

Fig. 25 Appearances of tandem nozzle cut near bottom edge.

the molten zone near bottom edge after tandem nozzle cutting. It presents an appearance that the molten layer is teared off by a strong force, and the amount of the clinging oxide dross is less than it of the cut by conventional cutting method (Fig. 13). This means that the dynamic force of gas jet heightened by the rear nozzle contributes effectively to remove the melting flow along the cutting front.

Figure 26 shows the cuts by this method in the optimum conditions; $P_r = 2.5$ kg/cm² ($P_f = 1.5$ kg/cm²), $\theta_r = 51°$ and $\delta_z = 2.7$ mm. It is clearly shown in this figure that the fine cuts with little or no dross can be obtained at speed of 2 to 4 m/min. **Figure 27** shows the cut qualities of the tandem nozzle cutting. Compared with the conventional cutting (Fig. 8 (b)), this method can reduce the height of dross up to 1/3, but cannot get the perfectly dross-free cut. Therefore this is, so to say, imperfect in comparison of pile cutting method. However, in another view point, this method is much convenient and effective to various actual applications, because a simple nozzle only is set up additionally to the conventional cutting method.

It was also tested whether this method was effective to other materials. As well known, laser-gas-cutting of titanium is inevitable with a rough cut kerf when oxygen gas jet is used, because it is too oxidizable and with an explosive burning. For this reason, an inert gas jet is desirable, though a large amount of clinging dross is inevitable. **Figure 28** shows the effect of tandem nozzle cutting on this material. It proves to be much effective not only to reduce the clinging dross but to improve the kerf shape. Thus, this method has a great possibility to apply to various materials.

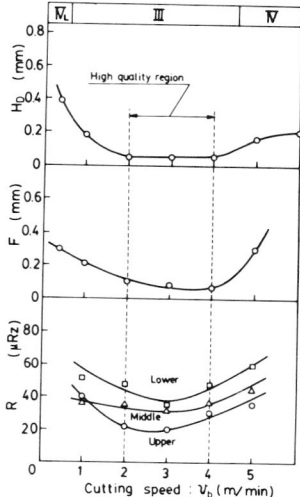

Fig. 27 Evaluation of tandem nozzle cut qualities. (thickness: 2 mm)

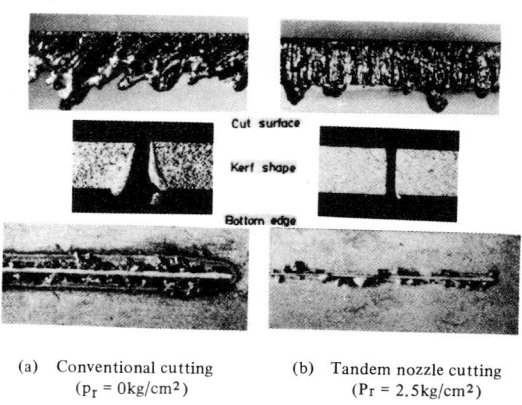

(a)　Conventional cutting
　　(p_r = 0kg/cm²)

(b)　Tandem nozzle cutting
　　(P_r = 2.5kg/cm²)

Fig. 28 Effect of tandem nozzle cutting on titanium plate. (Argon assist, front nozzle pressure: 1.5 kg/cm², thickness: 2 mm, cutting speed: 1 m/min)

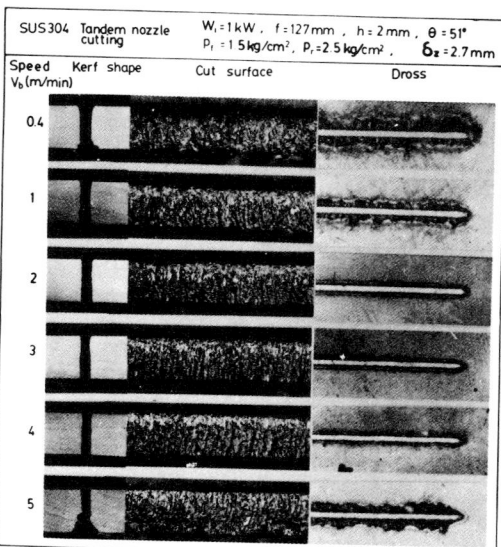

Fig. 26 Cut appearances by tandem nozzle cutting of stainless steels.

7. Conclusion

Cut quality of stainless steel by the conventional laser-gas-cutting with 1 kW class CO_2 laser and oxygen jet, in which an oxygen nozzle is coaxially arranged with the focused laser beam has been detailed through the evaluation of height of clinging dross, out-of-flatness and cut surface roughness, and dross clinging has been shown to deteriorate the cut quality. The dross clinging mechanism is discussed through the observation of the cut by high speed movie pictures. On the bases of these results two new laser-gas-cutting techniques are developed to overcome a limitation of the cut quality in the conventional laser-gas-cutting technique. Conclusion obtained from this study is summarized as follows.

1. Laser power, oxygen pressure and focal length of lens are important infulential factos to laser-gas-cut quality, and the cut quality is improved with increase in laser power and oxygen pressure. However, the conventional laser-gas-cutting cannot produce dross-free cut in cutting stainless steel.

2. The amount of dross clinging to the bottom of the plate is greatly influenced by the flowing conditions of the molten materials at the cutting front; clinging of the oxide dross is caused by the stagnation and disturbance of the melt flow at the bottom edge of the cutting front.

3. Pile cutting, in which stainless steel is cut with a thin mild steel plate piled on it, makes it possible to get completely dross-free cut. In this cutting, a molten oxide provided by mild steel enhances the fluidity of the melting flow on the cutting front of stainless steel, and promotes the separation of the molten oxide dross from the bottom of the cut.

4. Tandem nozzle cutting also produces fine cut having little or no dross. In this cutting, the off-axial gas jet nozzle used with the coaxial nozzle enlarge the momentum of gas jet, and promotes the removement of the molten metal and oxide.

5. These two techniques can also imporve the out-of-flatness and surface roughness of the cut, and these new cutting methods are effective for improving cutting quality and performance of other various materials.

References

1) Y. Arata and I. Miyamoto; "Generation and Applications of CW High Power CO_2 Gas Laser", Technol. Repts. of Osaka Univ. Vol. 17 (1967), No. 285.

2) A.B.J. Sullivan and R.T. Houldcroft; "Gas-jet Laser Cutting", British W.J., Vol. 14 (1967), No. 8, p.p. 443-446.

3) G. Brandt. K.D. Kegel and J.V. hulle: "Einige Ergebnisse von Schneidund Schweißversuchen mit einem 900W CO_2 Laser", Schweissen und Schneiden, Vo. 24 (1972), H7.

4) I. J. Spalding; "Lasers–Their Applications and Operational Requirements", Opt. and Laser Tech., Dec. (1974), p.p. 263-272.

5) J.D. Russell; "The Development of the Laser as a Welding and Cutting", British Weld. Inst. Res. Bulletin, Vol. 16 (1975), p.p. 245-248.

6) H. Herbrich; "The Economic and Idustrial application of CO_2 Lasers", Maschinenmarkt, Vol. 45 (1976).

7) J. Clarke and M.M. Steen; Proceedings of Laser '78 Conference, London, March (1978).

8) V.S. Kovalenko, Y. Arata, H. Maruo and I. Miyamoto; "Experimental Study of Cutting Different Materials with a 1.5 kW CO_2 Laser", Trans. of J.W.R.I., Vol. 7 (1978), No. 2, p.p. 101-112.

9) M.M. Schwartz; "Laser Welding and Cutting", W.R.C. Bulletin, Nov. (1971) No. 167, p.p. 1-34.

10) M. Hoffmann; "The Laser as an Industrial Cutting Tool", Metal Const., Vol. 11 (1979), p.p. 33-34.

11) F.W. Lunau and E.W. Paine; "CO_2 Laser Cutting", Weld. & Metal Fab., Vol. 37 (1969), p.p. 9-14.

12) Y. Arata, H. Maruo, I. Miyamoto and S. Takeuchi; "Dynamic Behabior in Laser Gas Cutting of Mild Steel", Transaction of JWRI, Vol. 8, No. 2 (1979) p.p. 15-26.

13) A. Adachi; "Measurement of Density and Viscosity of Liquid Metal and Slag", (in Japanese) Proceeding of the Research Group on Liquid Steel and Slag, Joint-Research Society for the Study of Steel, (1970) P.P. 97-115.

14) P.P. Arsent'ev and B.G. Vinogradov; "Viscosity Characteristics of Iron-Chrominum-Nickel Melts", Steel In The USSR, Vol. 5 (1975), No. 3, p.p. 145-146.

Laser Surface Treatment

Surface Hardening of Titanium by Laser Nitriding
S. Katayama, A. Matsunawa, A. Morimoto, S. Ishimoto and Y. Arata
[The Metallurgical Soc. of AIME (1984).]

Condition Setting Method Utilizing Data Base System in CO_2 Laser Surface Hardening—Fundamental Concept.
Y. Arata, K. Inoue and S. Matsumura
[Trans. JWRI **8** (1979), 21.]

Application of Laser for Material Processing—Heat Flow in Laser Hardening
Y. Arata, K. Inoue, H. Maruo and I. Miyamoto
[IIW Doc. IV–241–78.]

Effect of Heating Condition in Laser Hardening Carbon Steel
H. Maruo, I. Miyamoto, T. Ishida and Y. Arata
[1st Int. Laser Proc. Conf. (1981).]

SURFACE HARDENING OF TITANIUM BY LASER NITRIDING

Summary

Nitriding and related hardening of titanium (Ti) and its alloys were investigated as a new method of laser surface treatment. The treatment was carried out in nitrogen (N_2) atmosphere by utilizing a pulsed Nd:YAG laser with 3.6 ms pulse width at an energy of 14 to 36 J/p or a continuous–wave (CW) CO_2 laser at a power of 2 to 4 kW. In the case of the YAG laser, it was found that smooth, flat surfaces covered with TiN nitride layer could be produced by selecting a good combination of laser irradiation conditions. Compared with Ti base metal having a Vickers hardness (H_v) of about 200, the hardness of the YAG laser-nitrided surfaces could exceed 700 H_v and the fusion zones (near the surface) below TiN layer showed a hardness of 300 to 700 H_v. On the other hand, it was rather difficult to obtain a smooth, flat surface in the case of CW CO_2 laser treatment. However, CO_2 laser-melted zones consisted almost entirely of relatively larger TiN dendrites and α– Ti phase containing a large content of N, and hence the zones, which could exceed 700 H_v almost all over, were by far deeper than the YAG laser ones. Nitriding and hardening phenomena were also observed in titanium alloys such as Ti-6Al-4V and Ti-6Al-6V-2Sn. Furthermore, a feasibility of TiN-nitriding and hardening of other materials was examined by irradiating a pulsed YAG laser in N_2 atmosphere on each material sticking Ti powders. The results revealed that TiN nitride formed on the treated surfaces and consequently the surface hardnesses increased from about 50 - 200 H_v to 650 - 850 H_v depending on the conditions of the laser irradiation and the thickness of Ti powders.

Introduction

Titanium (Ti) and its alloys are generally regarded as excellent materials exhibiting good corrosion resistance and high strength-to-weight ratio. Moreover, nitriding surface treatment of these alloys, which makes Ti-N compounds and/or N-enriched zones form on the surface, may be carried out advantageously to harden them and hence improve wear resistance or to improve corrosion resistance to some acids more excellently.

Conventional nitriding processes such as ion-nitriding, nitrogen-gaseous nitriding, salt-bath nitriding, and so on are used in practice, but all of them have a disadvantage of requiring several hours and more of processing time. On the other hand, the treatment by a laser beam with high power density is expected to realize short time treatment and to provide greater selectivity and precision of the area treated. Therefore, in this study a new feasibility of laser nitriding and hardening surface treatment of Ti and its alloys was investigated by employing mainly a pulsed Nd:YAG laser with 3.6 ms pulse duration and partly a CW CO_2 laser. TiN-nitriding and hardening of other materials were also examined by YAG laser treatment of the specimens covered with Ti powders.

Experimental Procedure

Materials investigated were commercially available pure Ti (0.009%N, 0.21%O) in most cases and Ti-6Al-4V (6.39%Al, 4.25%V, 0.010%N, 0.178%O) and Ti-6Al-6V-2Sn (5.54%Al, 5.76%V, 2.09%Sn, 0.010%N, 0.16%O) titanium alloys in part. Besides, materials used were mild steel, austenitic stainless steel AISI 304, nickel, copper and aluminum on which Ti powders were stuck with acrylic resins. The thickness of all sheets was about 3 mm except for copper sheets of 1 mm in thickness.

Lasers used were chiefly a pulsed Nd:YAG laser (Control Laser: Model 428, delivering an average power of up to 200 W and a maximum energy of 40 J/p) and partly a CW CO_2 laser (GTE Sylvania (Spectra-Physics):Model 975 with a maximum power of 5.5 kW). A 1.06 µm wavelength YAG laser beam or a 10.6 µm wavelength CO_2 laser beam was focused by 127 or 254 mm focal length lens and was irradiated on the material polished with No.400 emery paper and cleansed with aceton. Laser irradiation was principally carried out in N_2 atmosphere and partly in air or argon (Ar) environment. For the purpose of obtaining a better understanding of features of laser-treated surface geometry and degrees of nitriding and hardening, the YAG laser output energy (E_o) and the defocused distance (f_d) were varied in the range of 14 to 36 J/p and -30 to 60 mm, respectively. and the CO_2 laser power (P_o), traverse speed (v) and f_d were varied between 1 and 4 kW, 0.5 and 2 m/min and 0 and 30 mm, respectively.

The morphology and microstructures of laser fusion zones were observed by scanning electron microscope (SEM) and the formed phases were identified by X-ray microanalyser (XMA) and X-ray diffractometer (XD).

Results and Discussion

Nitriding by Laser Irradiation

Pulsed YAG laser.   Figure 1 shows SEM microstructures of Ti fusion zones after pulsed Nd:YAG laser irradiation at E_o=20 J/p and f_d=30 mm in nitrogen (a) and argon (b) atmosphere. In the case of N_2 (a), a white phase layer consisting of a thin film of about 2 μm thickness and a dendritic zone of about 10 μm depth is observed on the surface of the melted zone. The white layer surface is actually golden with the naked eye. In Ar (b), on the other hand, martensitic structure is only seen but no white layer is formed, and the surface is silver.

Figure 2 represents examples of XMA line analysis results, showing the concentration distributions of Ti and N elements on the surface (a) and cross section (b) of Ti sheet melted by a pulsed laser beam in N_2 atmosphere. The surface layer is obviously enriched in N element and a slight enrichment in N is observed below it. Figure 3 exhibits X-ray diffractometer result of the laser-treated surface of Ti sheet. High peaks of TiN type nitride are recognized besides peaks of α-Ti but Ti_2N type nitride peaks cannot be detected. Consequently the layer formed in N_2 atmosphere was identified as TiN nitride. In air atmosphere, TiO_2 oxide thin film formed on the top surface of TiN nitride layer (1),(2).

CW CO_2 laser. Figure 4 shows SEM microstructures of Ti fusion zones made by CW CO_2 laser beam under the conditions of P_0 =4 kW (a) or P_0 =2 kW (b) , f_d =10 mm and v=1 m/min in N_2 atmosphere. The CO_2 laser-treated surface is rather rougher, while the melted zone is occupied by larger white dendrites almost all over.

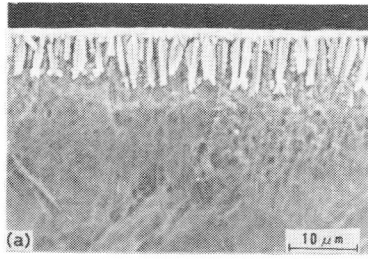

Figure 1 - SEM microstructures of Ti fusion zones produced by single YAG laser shot in (a) nitrogen and (b) argon atmosphere, showing formation of white layer in nitrogen atmosphere.

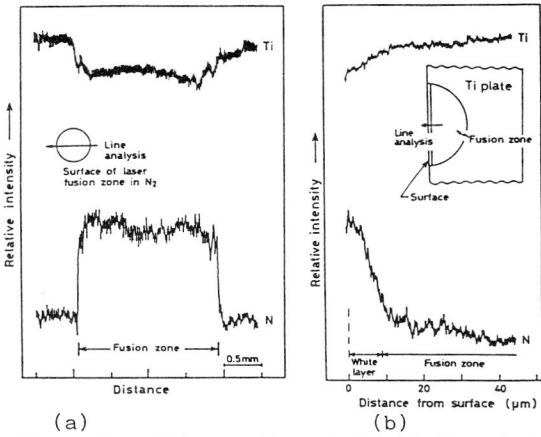

Figure 2 - XMA results of Ti and N on (a) surface and (b) cross section of laser spot weld, showing enrichment in N near surface.

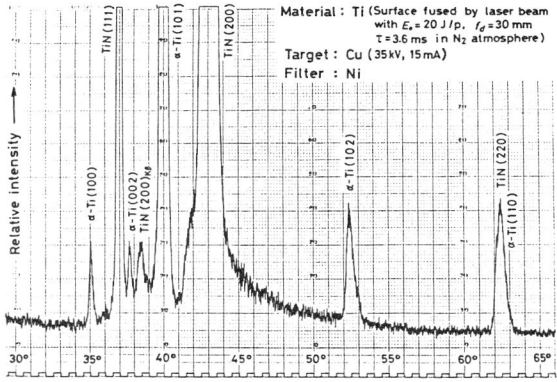

Figure 3 - XD result of Ti surface treated by 50% overlap of YAG laser spot welds in nitrogen, showing formation of TiN.

It was revealed from XMA and XD results that the zone consisted of dendritic TiN nitride and α-Ti phase containing a considerable amount of N.

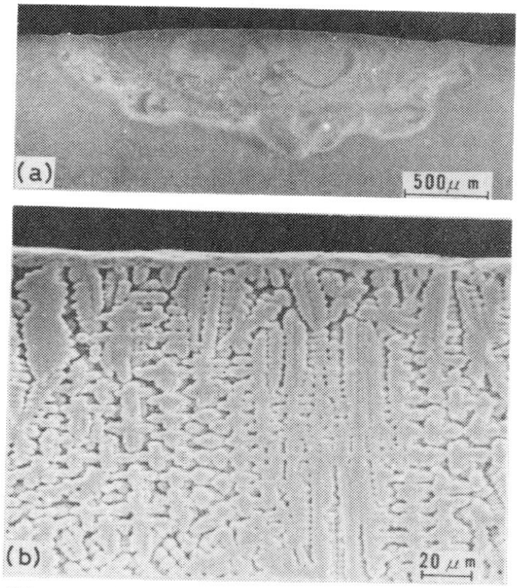

The above results confirmed that nitriding by YAG and CO_2 lasers was feasible, and the detail of nitriding will be further discussed in the following sections.

Fusion Zone Geometry and Hardness Distribution

Pulsed YAG laser. When the specimen was placed near the lens focal point, spattering occured together with plume due to high energy density, while in the case of larger defocused distance only plume took place. Correspondingly, various shapes of laser fusion zones were observed depending upon the degree of the defocussed distance (1),(2). Based on the surface features, appearances of fusion zones are classified into five types: Type H with a hole, rough Type R, Type C with convexity in a flat surface, smooth, flat Type S and irregular Type I. When laser energies are the same, these types are symmetrically observed in either side of a focal point in the order of Types H, R, C, S and I with an increase in the distance from the focal point. Laser-nitrided surface became fairly flat all over the specimen by laying down Type C or S spots with more than 65 or 35 % overlap, respectively. Figure 5 shows the formation regions for Type C and S surface in the relationship between the laser energy and the defocused distance on the positive side and Vickers hardnesses on the Type C and S surfaces made by a single shot. It is understood that in order to produce a flat and smooth surface it is necessary to select proper combination of the laser energy and the defocused distance. Besides, it is apparently seen that the laser-nitrided surfaces are extremely hardened up to 700 to 1250 H_v, which is 3.5 to 6 times larger than 200 H_v of base metal.

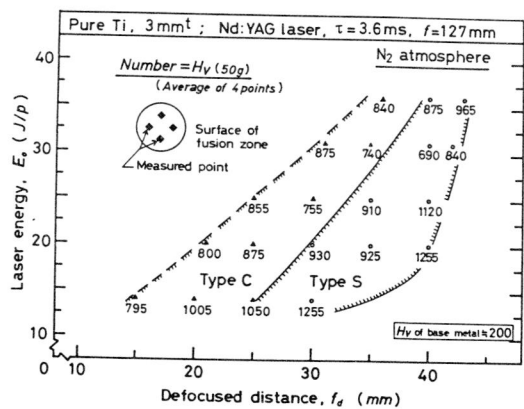

Figure 4 – SEM microstructures of cross sections of Ti fusion zones produced by CW CO_2 laser at $P_0 = 4$ kW (a) and $P_0 = 2$ kW (b) in N_2, showing formation of dendritic TiN in melted zones.

Figure 5 – Conditions of laser energy and defocused distance for formation of Type C and S nitrided-surfaces, and Vickers hardnesses on surfaces of Types C and S.

Figure 6 shows the hardness distribution on the cross sections of Ti, Ti-6Al-4V and Ti-6Al-6V-2Sn sheets after a single shot, together with a datum of ion-nitriding at 850°C for 5 hrs. Ti and Ti alloy fusion zones beneath TiN layer within 0.15 mm in depth were hardened to more than 400 H_v. The depth hardened by a single shot of laser with 3.6 ms pulse duration was much deeper than that hardened by ion-nitriding process for 5 hrs.

The effects of shooting times and the pressure of nitrogen atmosphere were investigated. In either case, according to the increase in the shooting number (2) or in the nitrogen pressure, TiN layer was observed to become deeper and consequently the surface hardness increased more.

CW CO_2 laser. The CO_2 laser melted zone width and penetration depth were observed to increase with increasing laser power and/or decreasing traverse speed. CW CO_2 laser-treated surface is rather rough, and thus special technique and strict selection of proper laser conditions were required to obtain a flat surface. Figure 7 indicates the hardness profiles on the cross sections of Ti fusion zones after CO_2 laser irradiation of P_0=2 & 4 kW at v= 1 m/min. The overall zones were hardened above 500 H_v, and thus the hardened zones expanded with an increase in laser power. The hardened zone by the multikilowatt CW CO_2 laser is much deeper than that by pulsed YAG laser due to longer interaction time between a laser beam and the material and a consequent enrichment in absorbed nitrogen content.

Hardening Mechanism

Hardness were measured on the surface of YAG laser spot welds made in nitrogen or air was remarkably hardened up to more than 900 or 750 H_v, but the hardness of the fusion zone surface produced in argon was still equall to that (200 H_v) of the base metal. From these results the surface hardening in nitrogen or air primarily attributed to the formation of hard TiN layer, and

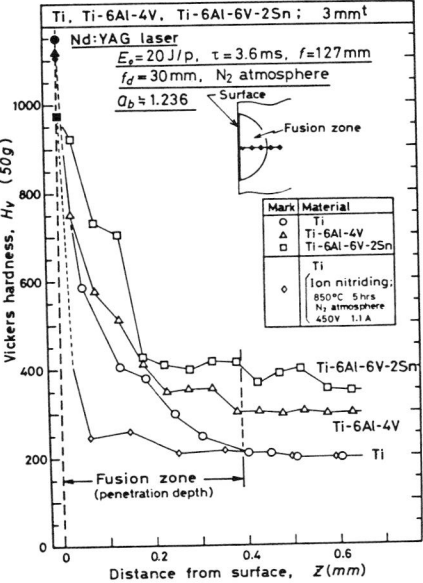

Figure 6 - Hardness profiles on cross sections of fusion zones and base metals of Ti, Ti-6Al-4V and Ti-6Al-6V-2Sn after laser- and ion nitriding treatment in nitrogen.

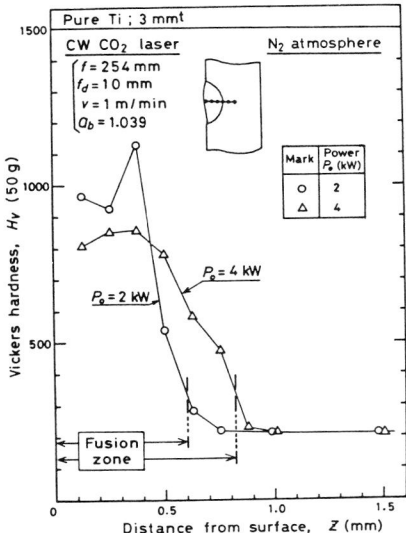

Figure 7 - Hardness distributions on cross sections of Ti fusion zones and base metals after CO_2 laser treatment at P_0 =2 & 4 kW.

the martensitic α'-Ti phase (alpha-prime) does not contribute to hardening of Ti. This differs the hardening of steels.(1), (2)

To provide a good understanding of an increase in hardness at the upper part of a fusion zone beneath TiN layer, the surface of the laser-nitrided specimen was polished and then the polished surface was examined by X-ray diffraction study. Examples of the results are shown in Figure 8, where the peaks of TiN decrease and dffraction angles of Ti peaks from (002), (101) and (102) lattice planes increase with increasing the polished degree. The measured lattice constants of c-axis are tabulated in Table I, together with N and O content in base metal and the fusion zone after removing the surface layer of TiN and TiO_2. The expansion of c-axis corresponds to enrichment in N and/or O content (3), and thus the results in Table I show clearly that N content and N and O contents were enriched in fusion zones produced in nitrogen and air atmosphere, respectively. It is therefore concluded that the hardening of fusion zones in nitrogen or air is the solid-solution hardening due to an increase in N or N and O.

A summary correlation among microstructure, N distribution and hardness profile observed in fusion zone by pulsed YAG or CW CO_2 laser in nitrogen atmosphere is schematically shown in Figure 9 (a) and (b). The microstructure of YAG laser fusion zone from the surface to the bottom is observed in the order of TiN nitride, α-Ti (alpha-case) and martensitic α'-Ti (alpha-prime), depending on N content. Accordingly, the hardness decreases gradually with an increase in depth. The CO_2 laser fusion zone consists almost entirely of TiN nitride and α-Ti (alpha-case) containing a large amount of N, and thus it results in an extreme hardening.

In the result the hardening of Ti and its alloy fusion zones produced by YAG and CO_2 lasers is essentially attributed to the formation of hard TiN nitride and the solid-solution hardening due to nitrogen-enrichment.

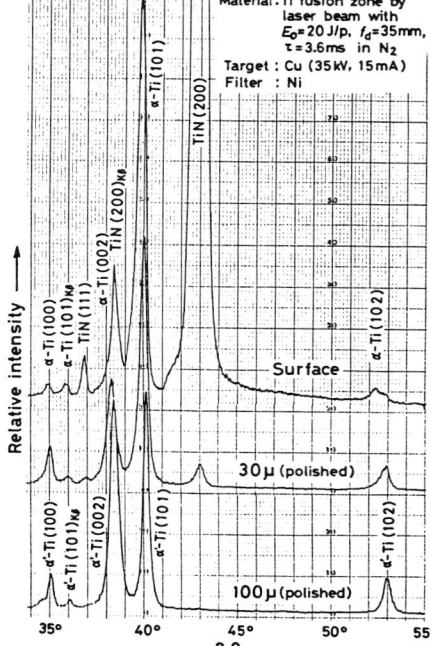

Figure 8 - XD results of laser-nitriding surface and polished surfaces of Ti fusion zone.

Table I. Variation in c-Axis (Obtained by XD Study) of Ti Fusion Zones Produced by YAG Laser in N_2 and Air and N and O Content in Base Metal and Fusion Zone in Air.

Object		Distance from surface (μm)	Lattice constant c (Å)	Content (wt%)	
				N	O
Base metal (alpha-Ti)		——	4.688	0.009	0.21
Fusion zone	in nitrogen	0	4.766	–	–
		30–60	4.700	–	–
		100–130	4.694	–	–
	in air	0	4.760	1.37	0.76
		40–70	4.706		
		100–130	4.694		
alpha-Ti according to ASTM card		——	4.686	–	–

543

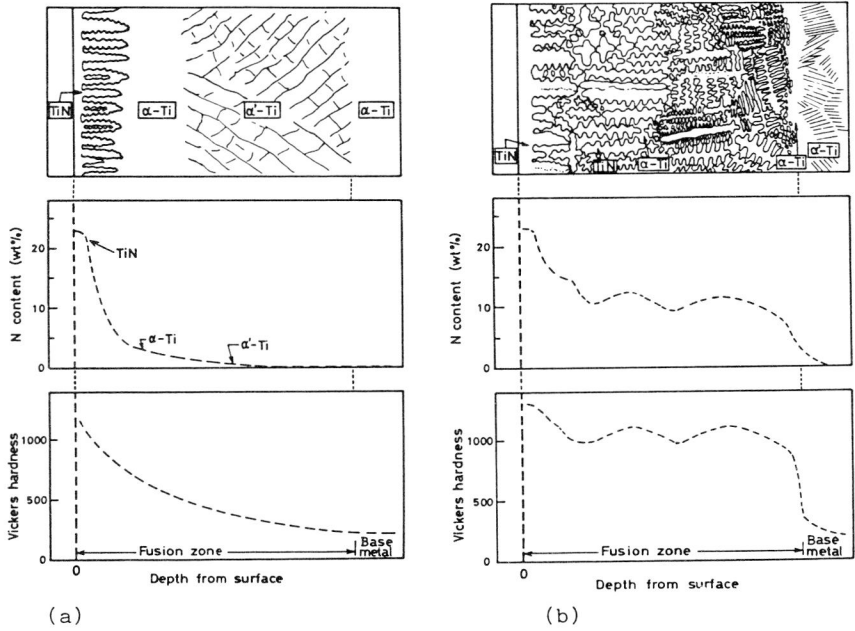

Figure 9 – Schematic representation of correlation among microstructure, N distribution and hardness profile in fusion zone and base metal of Ti in case of YAG laser (a) and CO_2 laser (b) nitriding treatment.

Laser TiN-Nitriding on Other Metals

A feasibility of TiN-nitriding and related surface hardening of other materials such as mild steel, austenitic stainless steel, nickel, copper and aluminum was investigated by the method that in N_2 atmosphere a pulsed YAG laser beam   was shot on the surface of each material covered with Ti powders of 350 mesh or less. Figure 10 shows examples of SEM micro-structres of laser spot welds of AISI 304 and copper produced at E_0=35 J/p and f_d=35 mm. It was found that the formation of golden TiN nitride on the fusion zone surface was easier for austenitic stainless steel, mild steel and nickel, but it was rather difficult for copper and aluminum to obtain a flat and smooth overall nitrided-surface. These tendencies were confirmed

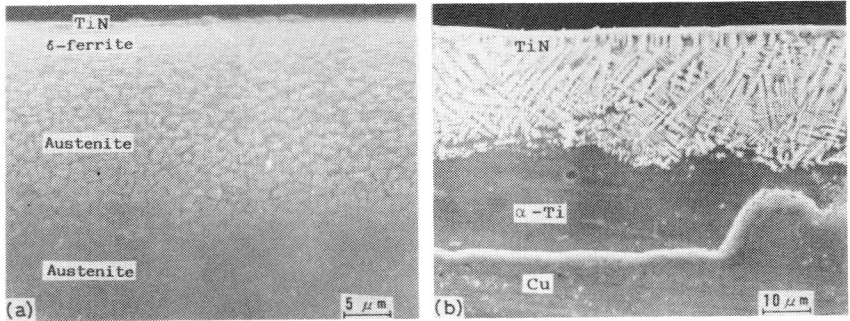

Figure 10 – SEM microstructures of YAG laser fusion zones of Ti-powder-spread AISI 304 (a) and copper (b) treated in N , showing formation of TiN nitride on surfaces.

by utilizing various alloys containing several levels of Ti contents. For example, nitriding was recognized for mild steel and AISI 304 containing about 10 % Ti but was impossible for alloys consisting of 60 % Ti and 40 % Al. Table II indicates a summary of the surface hardnesses of some materials and their laser-treated fusion zones. It was noted that the surface hardening of each material was feasible probably due to hard TiN-nitride layer and compounds between Ti and other alloying elements. In such application, unfortunately some cracks were seen in laser melted zones, and thus more procedures are required to prevent cracking.

Table II. Surface Hardnesses of Various Materials before and after YAG Laser TiN-Nitriding Treatment in Nitrogen

Materials	Vickers hardness, Hv (50g)		
	$(Hv)_s^*$	$(Hv)_b^{**}$	$(Hv)_s/(Hv)_b$
Mild steel	800 650–930	130	6.2
AISI 304	830 800–870	200	4.2
Nickel	650 580–760	110	5.9
Copper	740 630–810	85	8.7
Aluminum	580 500–650	45	13

* : Hardness of treated surface
**: Hardness of base metal

Conclusions

From laser nitriding and hardening surface treatment of Ti and its alloys in nitrogen atmosphere, the following conclusions can be drawn:
1. A laser nitriding process of Ti and its alloys has been developed, which produce a surface TiN-layer with an apparent Vickers hardness of about 700 or more.
2. YAG laser fusion zones consisted of TiN nitride, α(alpha-case)-Ti and α'(alpha-prime)-Ti, and the upper parts of the zones showed Vickers hardnesses of about 300 to 700 depending on N contents. On the other hand, CO_2 laser melted-zones consisted predominantly of TiN nitride and α(alpha-case)-Ti and therefore were hardened above 750 H_v almost all over.
3. The hardenings of the surface and fusion zone were interpreted in terms of the formation of TiN nitride and α(alpha-case)-Ti containing a large amount of N.
4. From the standpoint of the application of laser TiN-nitriding, Ti-powder-spread various materials were treated as the first step by a pulsed YAG laser in nitrogen atmosphere. The result revealed a feasibility of TiN-nitriding and hardening of laser-treated materials.

Acknowledgments

The authors wish to acknowledge Professors F. Matsuda and K. Inoue of JWRI, Osaka University, for their cooperation of ion-nitriding treatment and CW CO_2 laser treatment, respectively.

References

(1) S. Katayama, A. Matsunawa, A. Morimoto, S. Ishimoto and Y. Arata, "Laser Nitriding of Titanium and Its Alloys ," paper presented at 3rd CISFFEL, Lyon, France, Sept., 1983 (pp.219-226).
(2) S. Katayama, A. Matsunawa, A. Morimoto, S. Ishimoto and Y. Arata, "Surface Hardening of Titanium by Laser Nitriding," paper presented at ICALEO '83, LIA, Los Angeles, Nov., 1983.
(3) H. T. Clark, Trans. AIME, 185 (1949) pp.588-589.

Condition Setting Method Utilizing Data Base System in CO_2 Laser Surface Hardening
— Fundamental Concept —

Abstract

One of the welding condition setting method is investigated. Decision of the welding condition is performed by the heuristic method utilizing data base system. In this study, the authors avoid to fix the relations among parameters by using the data on welding in the form of data base where any parameters or any relations among them can be easily managed.

The study is restricted within the condition setting of "CO_2 Laser Surface Hardening", as the primary approach for welding condition setting.

KEY WORDS: (Computation) (Data) (Process Conditions) (Optimisation) (CO_2) (Lasers) (Hardening)

1. Introduction

The welding condition setting method has been studied by many researchers in the numerical method, on the basis of experimental data, using computer.[1, 2, 3]

However, it is well known that the actual welding is affected by the very various factors and it has many problems in practical ·use to set the welding condition under the fixed relations among parameters as usually done in the method which is widely used at present, because it is comparatively easy to construct the algorithm.

In this study, the authors avoid to fix the relations among parameters by using the data on welding in the form of data base where any parameters or any relations among them can be easily managed. Therefore, the flexible method can be used and we determine the optimum welding conditions or get the very powerful tool for the welding condition setting by using such data base.

It is important to utilize the data effectively in the laser welding that has only little data, partly because of the shortness of its history. Then, as the first experiment of "Condition Setting Method Utilizing Data Base", this study is restricted within the condition setting of "CO_2 Laser Surface Hardening".

The following is discussed on the condition setting method of CO_2 laser surface hardening utilizing data base, especially on the partial surface hardening method of the material of complicated shape that has high necessity for CO_2 laser.[4, 5, 6]

2. Organization of Software System

As this study is utilizing the data base, it is important not only to use it easily, but also to put it independent of the application system. That is, we use the data base free from care of its all maintenance.

The software system developed in this study is composed of two parts, the data base management part and the processing part that decides the surface hardening conditions on the basis of data. The block diagram of the system is shown in **Fig. 1**. The core of this system is condition calculation program, and the main subject in this study is how to construct it. The condition setting of surface hardening is made in this program. Therefore, it can be said that other program modules are prepared for servicing the necessary data to this program.

In a series of modules, a part of data base manager and semantic formatter is the resident part of the system, and other modules are activated at need. The comunication between data base management part and processing part is unified in the form of supervisor call that is called data base comunication line

The following is described the outline of each module.

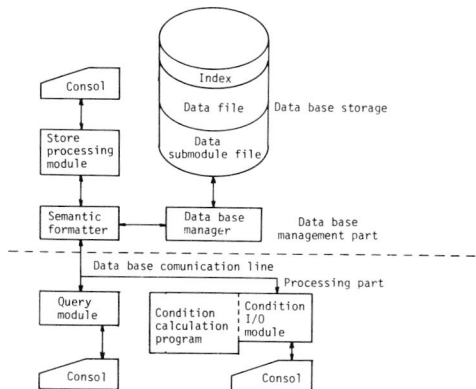

Fig. 1 Block diagram of software system

2.1. Semantic formatter

This module generates the structure of data base on the basis of the schema or subschema, and organizes and modifies all data submodels through data base manager.[7,8]

That is to say, it interfaces between two logical data structures, the first is defined and used at the processing part, the second is managed by data base manager. The integrated data base can be utilized and maintained and its independence can be assured with this method. Using this module, the data submodel of the only data that is necessary in any studies is organized in free format, and it can realize the short access pass and free operation.

2.2. Data base manager

This module converts between the structures of physical memory and logical memory, that is, it inputs and outputs logical data on request from the semantic formatter.

The physical memory structure is based on the concept of modified B-tree and the data structure of LISP.[9,10]

By this structure, the physical access time is shortened, and the organization of the data submodel based on a subschema is simplified. Usually all the necessary data can be included in the submodel. Therefore, it is possible to speed up the data submodel organization and to shorten the access path to the model.

2.3. Store processing module

This module is activated at the time the data are fed to the data base. By using this module in the query-answer form data are modified and saved to the data base in the format of the schema that was determined at the start of the data construction. This module requires the necessary data on the results of experiments with their conditions. Because few experimental data generally have all parameters known, the module fits the input data for the data base structure by editing their defecs and redundancies. This edition is important for the data base to have flexibility.

2.4. Query module

The data base is directly accessible through the consol by this module, and the data can be retrieved by the set of key words and, or the appointment of the data structure. In addition to the conventional commands[11,12] this module has some commands to examine the subschema and to construct the data submodel which are important functions for this study. As this module can run not only interpreter mode, but also direct mode, the definition of subschema and the data submodel construction based on the subschema are usually carried out with this module when each of condition calculation modules is required.

3. Condition calculation program

Surface hardening conditions are decided utilizing the data base prepared as mentioned above. Because the object of this study is the data processing rather by the form of expression or graph than by the numerical method. LISP is used as the program language in this program. It has the specific interface module for comunication to the data base.[10]

The block diagram of this program is shown in **Fig. 2**, which consists of LISP interpreter, interface module, and each routine for condition calculation written by LISP. The condition calculation routines consist of resident main routines and each utility routine that is rolled in and rolled out by the main routines.

The condition setting method is constructed of two phases.

———In the first phase, the program for condition setting is made from some experimental results. The data base has already been constructed on the basis of the data obtained from the surface hardening experiment of plane plates performed as pre-experiment and the data on the special shaped base metal obtained through the learning which will be discussed after. The processing of the first phase is made by condition evaluation main routine. The program of condition setting is constructed by this routine using any utility routines based on the heuristic method.[13] (Differentiation and integration of the primary functions, and evaluation of equations and dynamic examination of equations.)

As parameters for saving the experimental data to data base,

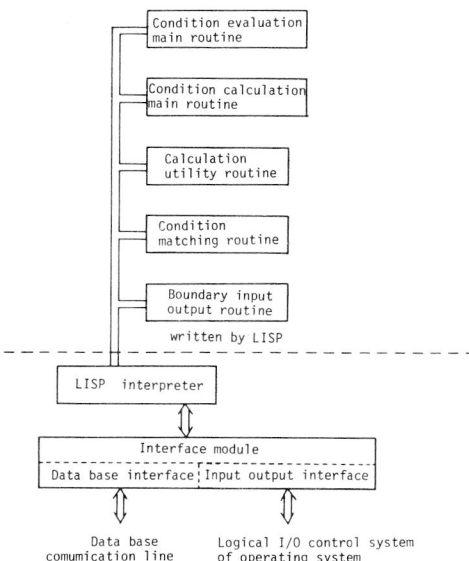

Fig. 2 Block diagram of condition calculation program

each condition parameter of

—Dimension and physical constants of base metal
(heat conductivity, specific heat, CCT curve, density,
hardness, absorptivity of laser beam, width, length,
thickness, shape)

—Characteristics of laser beam

(output power, spot size, distribution, scan speed)
and each result parameter of

—Surface hardness and three-dimensional distribution

—Metal structure
are all or partly used.
and as the base of processing, the basic equation of
point heat source equation (1) is used.

$$T = \frac{q}{c\rho} \frac{e^{-\frac{r^2}{4kt}}}{(2\sqrt{\pi kt})^3} \qquad (1)$$

where r^2 : $x^2 + y^2 + z^2$

q : impulse heat input by laser beam at origin (0,0,0) and time 0

c : specific heat of base metal

ρ : density of base metal

k : thermal diffusivity of base metal

t : time

T : temperature at position (x, y, z)
and the equation that integrates
the laser beam distribution and time are:

$$T = \frac{Ak}{4c\rho(\sqrt{\pi k})^3} \int_0^\infty \frac{dt}{\sqrt{t^3}} \int_{-\infty}^\infty dx' \int_{-\infty}^\infty w(x', y')$$

$$exp\ [-\frac{(x - x' + vt)^2 + (y - y')^2 + z^2}{4kt}]\ dy \qquad (2)$$

where w (x, y) : heat input density at position (x, y)

A : absorptivity of beam

v : velocity for x direction

T : temperature at (x, y, z) when the heat source moves on a semi-infinite plane from infinite distance on the x axis by the velocity v

These equations are evaluated by symbolic manipulation in the condition evaluation routine.

Equation (3) is used if the cooling rate is required.

$$\frac{\partial T}{\partial t} = k \nabla^2 T \qquad (3)$$

The program examines the symmetry and equivalence of each axis, and automatically analizes by the symbolic manipulation using differentiation and so on.

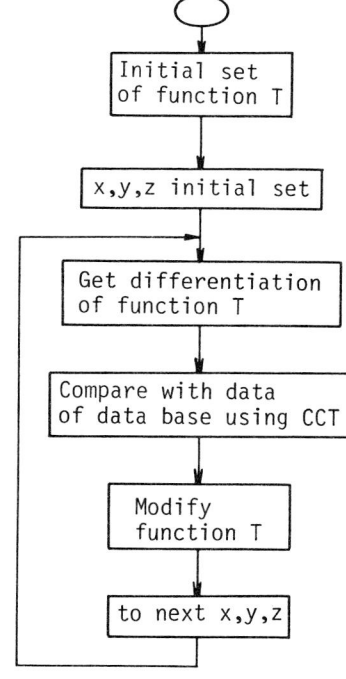

Fig. 3 Concept of function T investigation

And, as the data base has CCT curve data, the program examines three-dimensional distribution of temperature, and its cooling rate, refers them to CCT curves, automatically analizes the approximate expression of T or $\partial T/\partial t$ etc., and evaluates it as correct expression by fixing the parameters. The approximate expression can not be always proved theoretically. The concept of this algorithm is shown in **Fig. 3**.

——In the second phase, the practical conditions of surface hardening are decided. This processing is carried out at the condition calculation main routine. To satisfy the surface hardening requirements that were fed by boundary I/0 routine, this routine tries to find out the conditions by the heuristic method using all expressions that are obtained in phase 1. The expressions in this place are not only the equations, but also the procedures in general means. The routine outputs the most reliable conditions, or outputs that the requirements are impossible. Then we actually examine by the conditions of output, and feed the result to this routine. If the result is suited to the requirements, the process is finished, and if not, as the program tries to find out the conditions using the result with the priority, this process continues until the suitable result is got. It is regarded that the system learnes in the broader means on feeding the result to the store processing module.

As already stated, each processing of two phases is done by the condition evaluation main routine and the condition calculation main routine. These two routines utilize calculation utility routine for the analysis of expressions condition matching routine for the comparision with data in data base, and boundary input output routine for the input, output through the consol. Each of those routines is cut off from the main routine and treated as the garbages, when it becomes unnecessery.

The results of the evaluation are stored as the utility routine with the restricted conditions by manual, thus the program can learn. Once the utility routine was stored, until it is deleted by any reason, it is activated as one utility routine by the main routine.

4. Conclusion

The above mentioned method is to analize and decide the conditions by the heuristic method based on the thought that the data base is a concurrence of information and knowledge against the condition setting algorithms that is ordinarily made by simply using several equations or loops.

Although this method is not always better than the manual analysis by cut and try method at present, the authours are confident that such research method will take a good position of laboratory automation when it is taken into consideration that the computer science will be developed at increasingly rapid speed.

References

1) O. G. KASATKIN and V. F. MUSIYACHENKO "The Development of an Information-Calculation System for Welding" Avt. Svarka No. 11 pp. 27-30 (1977)
2) Isao MASUMOTO et al. "Program for CO_2 Welding parameters of I-Batt One-Pass Joint" Journal of the Japan Welding Society Vol. 48 No. 1 pp. 17-21 (1979) (in Japanese)
3) "The Collected Literatures of Algorithm of Welding Enginering The First Series" the Japan Welding Society, the Japan Welding Association 1974. 8.
4) Hiromichi KAWASUMI "Metal Surface Hardening CO_2 Laser" TECHNOCRAT Vol. 11 No. 6 pp. 11-20 (1978)
5) Edward V. LOCKE and Richard A. HELLA "Metal Processing with a High-Power CO_2 Laser" IEEE J. Quantum Electronics. Vol. QE-10 No. 2 pp. 179-185 (1974)
6) Fred D. SEAMAN and Daniel S. GNANAMUTHU "Using the Industrial Laser to Surface Harden and Alloy" Metal Progress pp. 67-74 (1975)
7) Donald D. CHAMBERLIN "Relational Data-Base Management Systems" ACM Computing Surveys Vol. 8 No. 1 pp. 43-66 (1976)
8) KAMIJO "Data Base System" Sangyo Toshyo 7. 10 (1972) (in Japanese)
9) R. BAYER and E. McCREIGHT "Organization and Maintenance of Large Ordered Indexes" Acta Informatica Vol. 1 No. 3 pp. 173-189 (1972)
10) C. WEISSMAN "LISP 1.5 Primer" Dichenson Pub. Company (1967)
11) H. UENO. M. SAITO and S. KAIHARA "POD: Patient-Oriented Data Base for Medical Research" Proc. MEDINFO 77 pp. 675-678 (1977)
12) FURUKAWA et al. Nikkei Electronics No. 183 pp. 118-143 4-3 (1978) (in Japanese)
13) James R. Slagle "Artificial Intelligence: The Heuristic Programing Approach" McGraw-Hill Inc. (1971)

Application of Laser for Material Processing
-- Heat Flow in Laser Hardening --

Abstract

Present paper describes a study conducted to investigate the basics of the laser hardening process primarily through heat conduction theory. New mathematical formulations are proposed by means of which the temperature distributions and the shape of the isotherms can be calculated as a function of heating conditions at the surface. These are shown to agree well with microscopical examinations of the laser hardened material. These results are represented diagrammatically and are expressed in simple approximated equations as well, thus providing immediately the best working for any case depth and width.

[1]   Introduction

High power CW CO_2 lasers developed for industrial application are powerful enough to be used for deep penetration welding and high speed cutting of metals, if they are focussed to a small spot. They are also found to excel as a case hardening tool. When the diameter of moving laser spot is adjusted to produce a desired power density to transform steels to austenite, self-quenching of the heated surface layer by heat conduction into the underlying mass of cold metal can transform the austenite to martensite. Thus, the laser is able to harden selective portions of a part at the surface. The CO_2 lasers are being actively considered for surface heat treating, as an alternative to other surface heat treating techniques, such as induction or flame hardening, since it causes less distortion than do other methods.

Although a few reports [1]-[4] about the laser hardening have been published, the experimental data and the theoretical analysis available are insufficient for the prediction of optimum laser parameters for the laser hardening. In the laser hardening process, defocused spots or focused spots oscillated at a high frequency are used to heat material with surface coatings which are usually employed to promote the absorptivity of the far-infrared laser energy. Some of the inpinging beam energy to the coated metals is reflected from the surface depending upon the hardening parameters employed. In order to predict and evaluate the surface hardening, therefore, it is essential to study the beam absorption characteristics of the coated metals subjected to the CO_2 laser beam and the heat flow within the metals due to moving surface heat source as well. Although some efforts of investigating the laser beam absorption characteristics of coated metals [5] and heat conduction due to surface heat source [6] have been published in the field of laser welding, there are little data available for the surface hardening.

Present paper describes a study conducted to investigate the basics of the laser hardening process primarily through the heat conduction theory. For this purpose new mathematical equations, which describe the temperature distributions due to moving surface heat source, were developed in order to evaluate and to predict the metallurgical response. As shown in Ref.5, the absorptivity of coated metallic surface varies with the heating parameters, and this suggests that the estimation of the laser hardening cannot be obtained only through analysis of heat flow. To overcome this difficulty, the authors also tried to predict the absorptivity of coated metallic surfaces based on the experimental data obtained and heat conduction analysis.

For the present study, a single lens was used to produce a defocused spots on the material to be hardened, and SK-4 tool steel with about 1.0% carbon coated with phosphates was employed.

[2] Experimental procedures

The data presented were generated by using GTE Sylvania, Inc. Model 971 continuous CO_2 Gas Transport Laser, with the maximum output of 1.5KW. The output beam from the laser head was reflected down and then focused by a ZnSe lens with a focal length of 10 inches as shown in Fig.1. The laser beam power which had passed through the lens was measured by using the powermeter [5] which had been designed to capture all incident beam power.

The power density of the beam spots was obtained by measuring the cross sectional dimensions of the grooves produced by scanning the beam spots along the two axes, x-and y-axes, on the material with a low thermal conductivity such as an aclylic resin. The scanning speed of the beam sopt was controlled depending upon the spot size to avoid the influence of a wall-focussing effect [7], which is caused from the reflection of the beam in the groove, on the shape of the groove. Figure 2 illustrates the spot sizes at the 1/e and $1/e^2$ power points of the center plotted as a function of the distance from the focal point. D_x and D_y represent the sopt size measured along the x- and y-axis, respectively. The laser cavity was adjusted so that the power density distribution along the y-axis became flat to obtain the constant case depth when the spot moved in the x-direction. An example of the trace pattern is also shown in Fig.2, which had nearly trapezoidal distribution in the y-axis, and a Gaussian distribution in the x-axis.

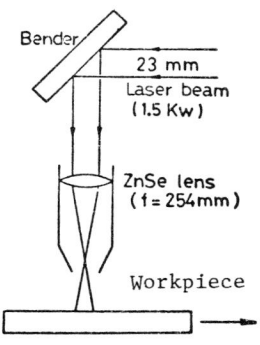

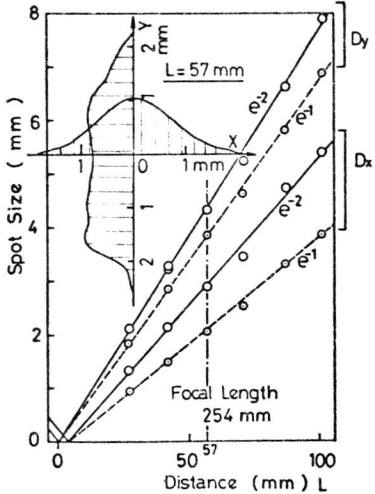

Fig. 1 Schematic diagram of optical system.

Fig.2   Relationship between spot size and distance from focal point (focal length=254 mm).

D_y is represented by the 1/e diameter in this study, throughout.

The ratio D_y/D_x selected was about 2.

In the present study, spot sizes in the y-axis were less than 9mm, although larger spots could be obtained. The material used was SK-4 tool steel with the dimension of 40mm × 80mm × 4mm. The chemical composition of SK-4 is shown in Table 1. D_y is represented by 1/e diameter in this study, throughout.

[3] Heat conduction theory

If heat is liberated at the rate w per unit time per unit area at the point (x,y) in the surface of the semi-infinite body, z=0, and the surrounding medium moves across it with velocity v in the direction of the x-axis as shown in Fig.3, the temperature rise in the steady state at the point (x,y, z) with no loss of heat from the place z=0 is given by

Table 1 Chemical composition of SK-4 tool steel.

C	Si	Mn	P	S	Cu	Cr	Ni
0.9~1.0	0.35≥	0.5≥	0.03≥	0.03≥	0.25≥	0.2≥	0.25≥

$$\theta = \frac{k_D}{4k(\pi k_D)^{3/2}} \int_0^\infty \frac{dt}{\sqrt{t^3}} \int_{-\infty}^\infty dx' \int_{-\infty}^\infty w(x',y') \exp\left[-\frac{(x-x'+vt)^2+(y-y')^2+z^2}{4k_D t}\right] dy' \quad --- (1)$$

where k is the thermal conductivity of the material, k_D the thermal diffusivity of the material. If the power density distribution on the workpiece is mathematically tractable so that the integrations in x and y can be performed independently, simpler expressions of the temperature rise can be derived from Eq(1). The power density distributions of the moving heat source delt with herein are a Gaussian, a rectangular and a rectangular-Gaussian source.

The Gaussian source is defined by the following distribution of the flux intensity,

$$w(x,y) = \frac{AW}{\pi ab} \exp[-x^2/a^2 - y^2/b^2] \quad ----(2)$$

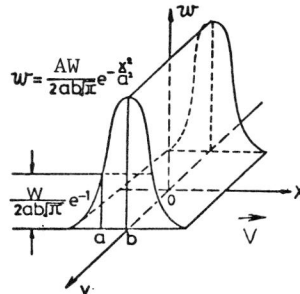

Fig. 3 Cartesian co-ordinates and moving direction.

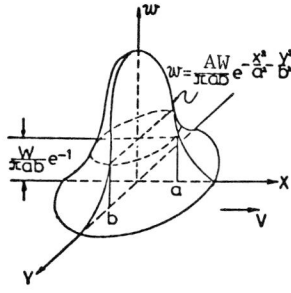

Fig.4 Gaussian source.

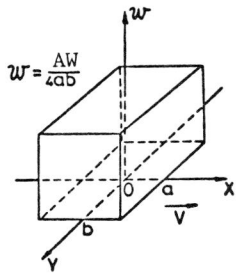

Fig.5 Rectangular source.

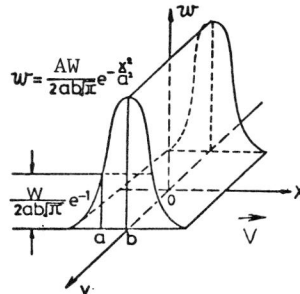

Fig.6 Rectangular-Gaussian source.

where W is the laser power, A the absorptivity of the workpiece to the CO_2 laser energy, 2a and 2b the diameters at the 1/e power point on x-and y-axes, respectively. As indicated in Fig.4, the Gaussian source expressed by Eq(2) provides elliptical spots unless a is equal to b. In the rectangular source, heat is liberated at the rate $W/4ab$ per unit time per unit area over a rectangle of sides 2a and 2b, parallel to the x- and y-axes, as is shown in Fig.5. The rectangular-Gaussian source is the combination of the rectangular and Gaussian sources, and is written by

$$w(x,y) = \frac{W A}{2\sqrt{\pi}ab} \exp(-x^2/a^2), \quad (-b \le y \le b) \quad --- (3)$$

The power density distribution shown in Fig.2, which was used in this study through-out , may be approximated by this equation.

To find the temperature distributions for these sources which are moving with velocity v for infinite time, each flux density $w(x,y)$ is substituted into Eq(1), and then Eq(1) is integrated with respect to x and y. In order to derive the universal equations, following dimensionless quantities are introduced.

$$\theta* = 16\sqrt{\pi}kr\,\theta/WA,$$

$$v* = vr/2k_D,$$

$$x* = \frac{x}{r}, \quad y* = \frac{y}{r}, \quad z* = \frac{z}{r},$$

$$a* = \frac{a}{r}, \quad b* = \frac{b}{r},$$

$$r^2 = ab. \qquad\qquad --- (4)$$

Then the following equations which describe the temperature distributions for the sources mentioned above are derived.

(a) Gaussian source

$$\theta* = \frac{16}{\pi} \int_0^\infty \frac{1}{\sqrt{(a*^2+t*^2)(b*^2+t*^2)}} \exp\left[-\frac{(2x*+v*t*)^2}{4(a*^2+t*^2)} - \frac{y*^2}{b*^2+t*^2} - \frac{z*^2}{t*^2}\right]dt* \qquad --- (5)$$

(b) Rectangular source

$$\theta* = \int_0^\infty (\mathrm{erf}\,\frac{2x*+v*t*^2+2a*}{2t*} - \mathrm{erf}\,\frac{2x*+v*t*^2-2a*}{2t*}) \times (\mathrm{erf}\,\frac{y*+b*}{t*} - \mathrm{erf}\,\frac{y*-b*}{t*})\, e^{-\frac{z*^2}{t*^2}} dt* \quad ---(6)$$

$$\text{where}\quad \mathrm{erf}\,\xi = \frac{2}{\sqrt{\pi}} \int_0^\xi \exp(-u^2)du$$

(c) Rectngular-Gaussian source

$$\theta* = \frac{4}{\sqrt{\pi}} \int_0^\infty \frac{1}{\sqrt{a*^2+t*2}} \exp\left[-\frac{(2x*+v*^2)^2}{4(a*^2+t*^2)} - \frac{z*^2}{t*^2}\right] (\mathrm{erf}\,\frac{y*+b*}{t*} - \mathrm{erf}\,\frac{y*-b*}{t*})\frac{dt*}{b*} ---(7)$$

For the sake of simplification, one-dimentional heat flow was also applied to calculate the temperature distribution where heat is liberated uniformly at the rate W/4ab per unit time per unit area over the surface of semi-infinite solid for laser-material interaction time given by 2a/v. The temperature rise at z at time t due to the heat delivered from 0 to T is given by

$$\theta^* = 4\sqrt{\pi}\left[\sqrt{2(x^*+1)/v^*}\ \text{ierfc}\ \frac{z^*}{\sqrt{2(x^*+1)/v^*}} - K\sqrt{2(x^*-1)/v^*}\ \text{ierfc}\ \frac{z^*}{\sqrt{2(x^*-1)/v^*}}\right] \quad ---(8)$$

where

$$\text{ierfc}\ X = \left(1/\sqrt{\pi}\right)e^{-X^2} - X(1-\text{erf}\ X),$$

$K = 1$, for $t \leqq T$ and

$K = 0$, for $t > T$.

At the surface, z=0, a simple expression of the temperature rise can be obtained;

$$\theta^* = 4\left[\sqrt{2(x^*+1)/v^*} - K\sqrt{2(x^*-1)/v^*}\right]. \quad ---(9)$$

The maximum surface temperature occures at x*=1, and is written in a very simple form,

$$\theta^* = 8/\sqrt{v^*} \quad ---(10)$$

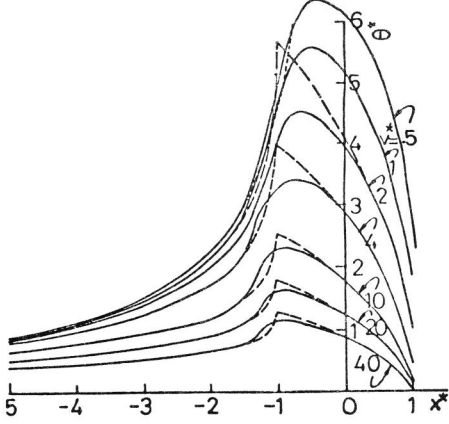
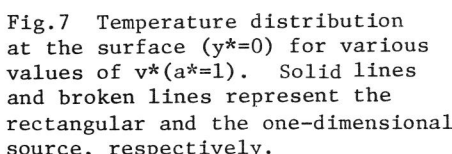

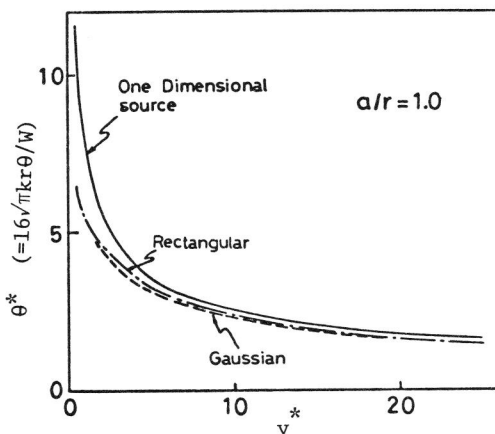

Fig.7 Temperature distribution at the surface (y*=0) for various values of v*(a*=1). Solid lines and broken lines represent the rectangular and the one-dimensional source, respectively.

Fig.8 Relationship between v* and θ*(maximum value) at the surface for the one-dimensional source, rectangular source and the Gaussian source.

554

The temperatures on the y-axis for a square source (a*=1) are compared with those obtained from the one-dimensional approximation in Fig.7, where $16\sqrt{\pi}kr\theta/AW$ is plotted against x* for verious values of v*. As is shown in Fig.7, the temperature obtained from one-dimensional approximation is higher than that for the square source around the rear edge of the source, x*=-1, but the cooling curves are almost same after x*=-1.5.

Figure 8 indicates the relationship between v* and the maximum values of θ* attained during heating for the square source, the Gaussian source and the one-dimensional source. At large values of v*, the surface temperature calculated from one-dimensional approximation is considerably close to that for the rectangular source, but this does not hold at smaller values of v*. It should be noticed that the one-dimensional approximation is acceptable only in the vicinity of the surface at high travelling velocity. It can be also said that the maximum surface temperature at the surface is in inverse proportion to the square root of the travelling velocity for large v*, and that the temperature gradient becomes sharper as v* increases.

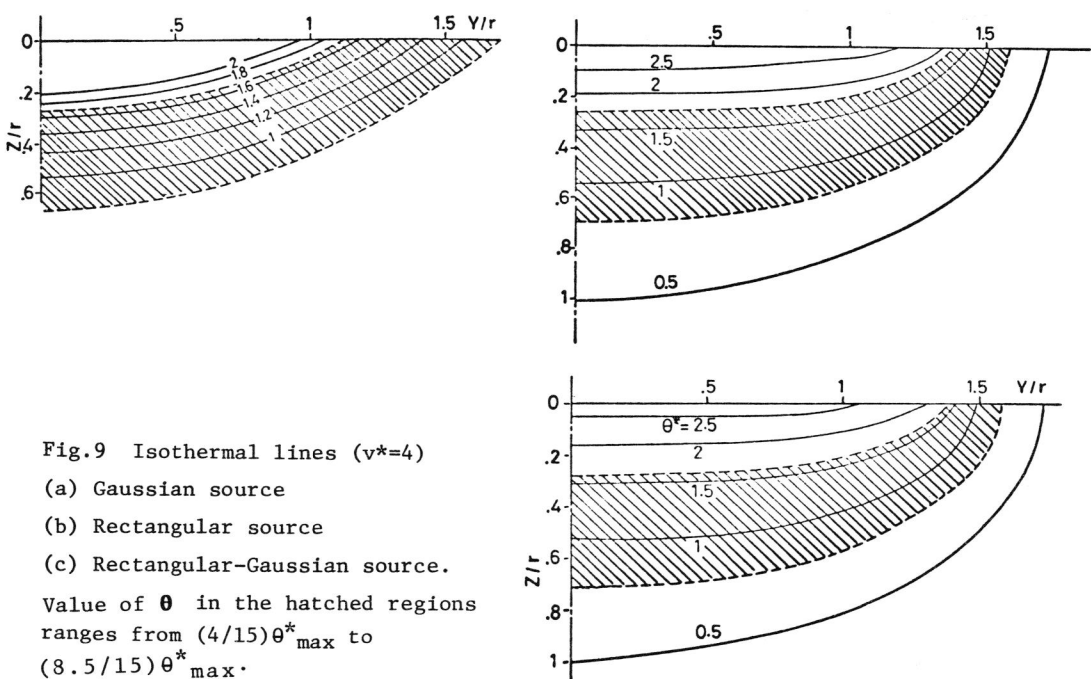

Fig.9   Isothermal lines (v*=4)

(a) Gaussian source

(b) Rectangular source

(c) Rectangular-Gaussian source.

Value of **θ** in the hatched regions ranges from $(4/15)\theta^{*}_{max}$ to $(8.5/15)\theta^{*}_{max}$.

The isothermal lines of the maximum temperature attained are indicated for selected value of a* and v* in Fig.9. As is clear in this figure, the Gaussian source gives rise to apparently non-uniform depth of the isothermal lines which will result in less and less exposed hardened structure remained as surface wear progresses. On the other hand, the rectangular and the rectangular-Gaussian sources provided the almost same isothermal lines which result in considerably uniform depth, although the temperature for the rectangular source is somewhat

higher than that for the rectangular-Gaussian
source. The isothermal lines corresponding
to $(1/2)\theta*$max are drawn at various v* for the
square source (a*=1) in Fig.10. These lines
are considered to give rough estimation of
the deepest hardened pattern obtained for
steels without causing melting since the
transformation temperatures are approximately
the half of the melting points. As v*
increases, the isothermal lines inclined to
be more uniform but shallower. At lower
values of v*, on the other hand, the isothermal lines become deeper near the center
with larger tapers in the depth. The width
of $(1/2)\theta*$max isothermal line at larger v*
provides little decrease with increasing v*.
As the value a* decreases, wider hardening
patterns with more uniform depth are obtained,
although the depth decreases as shown in Fig.11.

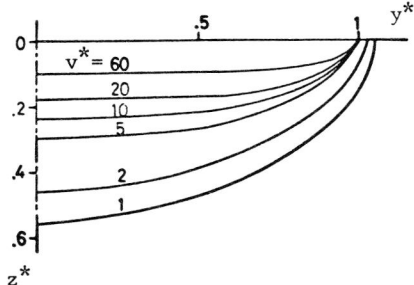

Fig.10 Isothermal lines corresponding to $(1/2)\theta*$max for various
v* (Rectangular source, a*=1).

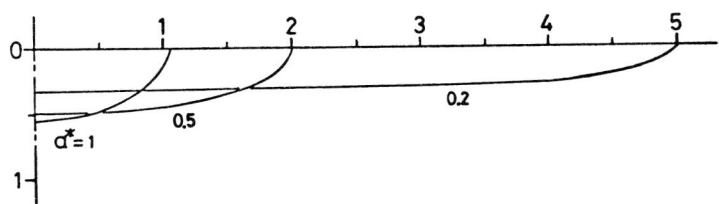

Fig.11 Isothermal lines corresponding to $(1/2)\theta*$max for various
a* (Rectangular source, v*=1). $\theta*$max is 5.56 for a*=1,
4.31 for a*=0.5 and 2.34 for a*=0.2.

[4]   Absorptivity of laser beam

The laser power actually absorbed by the specimen has to be determined in
order to evaluate and to predict the laser hardened pattern. As shown in Ref.5,
the absorptivity of polished metallic surfaces to the 10.6μ wavelength of CO_2
laser radiation at both room and fusion temperatures is proportional to the
square root of the resistivity of metal and generally provideds very low values,
abour 15% at most.
Accordingly various coating materials were used to promote the absorptivity of
the laser energy up to the melting point of SK-4 tool steel. The absorptivity of
the coated workpiece at 10.6μ wavelength was measured in the same way as used in
Ref.5. As the result, phosphate coatings, which are widely available in the metal
processing industry, were found to have the highest absorptivity due to their
excellent heat resistivity, adherence property to the workpiece surface and uniformity in thickness.
Two kinds of phosphates, manganese-and zinc-phosphates, were examined, and it
was found that the difference of the absorptivity between two phosphates, was
negligible. The absorptivity was measured for the workpieces heated in the air
and argon atmosphere. As the result it was found that the absorptivity was independent of the atmosphere applied and hence there was no need to shield the
specimen to be laser-hardened from the air. In this study, the zinc-phosphate

was employed throughout since it was easier to be used , and the laser hardening
tests were carried out in the atmosphere for the sake of simplification.

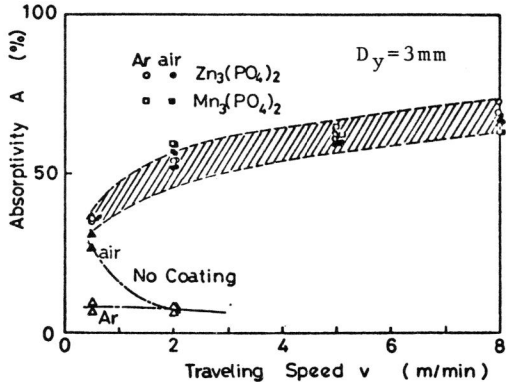

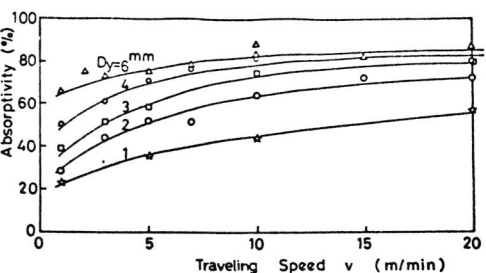

Fig.12 Absorptivity of specimens
coated with zinc-and manganese-
phosphates measured in the air and
argon atmosphere.

Fig.13 Effect of travelling speed
and spot size on the absorptivity
of steel coated with zinc-phosphate.

The absorptivity of zinc-phosphate coated specimens thus obtained is plotted
as a function of spot size of the beam and the travelling velocity in Fig.13. The
absorptivity, which varied widely ranging from 20% to 90% within the condition
tested, apparently tended to increase with increasing the spot size and the
travelling velocity. Since the surface temperature of base metal coated with thin
films was qualitatively known to influence the absorptivity of the laser beam [5],
in order to provide the quantitative estimation of its influence the maximum temperature reached at the surface was computed by substituting relevant processing
parameters and each absorptivity indicated in Fig.13, into Eq(7). Figure 14
shows the relationship between the absorptivity and the maximum surface temperature
thus calculated. It is seen in this figure that the absorptivity of the specimen
coated with zinc-phosphate is also strongly influenced by the surface temperature
of the base metal; that is, the absorptivity tends to decrease with increasing the
surface temperature. If the surface temperature of the metal used is given, the
absorptivity can be obtained from this figure. At the melting point of the metal,
for example, the specimen coated with zinc-phosphate provided the absorptivity of
about 55%, which is much higher than that for the uncoated polished metal.

But the absoptivity can not be obtained from the laser-hardening parameters
such as the spot size, the laser power and the travelling velocity, because the
surface temperature is dependent not only on the laser-hardening parameters, but
also on the absorptivity. The ultimate object of this section is to derive a
simple corelation between the absorptivity and the laser-hardening parameters and
thereby to make experimenters possible to easily predict the absorptivity. As the
temperature in Fig.14 had to be calculated from the complicated equation containing
an integration Eq(7), one-dimensional heat flow equation, Eq(10), which provides
mathematically much more tractable equation, was attempted to be used instead of
Eq(7). The results obtained are plotted in Fig.15. Although the temperature
calculated from the one-dimensional approximation is higher than the actual

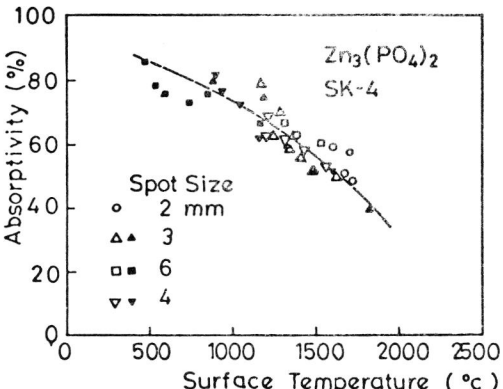

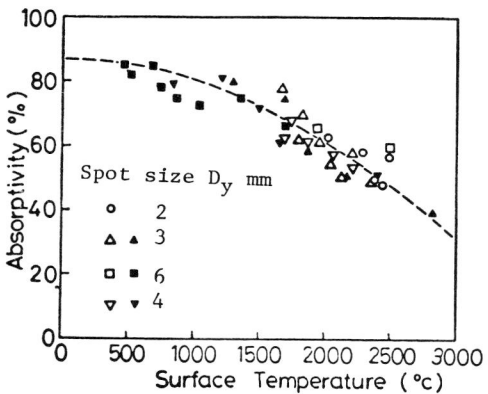

Fig.14 Relationship between absorptivity and surface temperature calculated from Eq(7).

Fig.15 Relationship between absorptivity and surface temperature calculated from one-dimentional approximation equation Eq(10).

temperature, especially at smaller v* as described above, the amount of the scattering in the resultant data was seen to be acceptable. As is obvious in this figure, the absorptivity can be expressed as a function of the surface temperature thus calculated as follows;

$$A = A_o - (\theta/\theta_o)^2 \qquad ----(11)$$

where A is the absoptivity of the workpiece, θ the maximum surface temperature of the base metal calculated from Eq(10), and A_o, θ_o constants. These constants are considered to depend upon the thermal properties of the coating material and base meatl used. For steel coated with phosphates, A_o and θ_o were about 0.87 and 4000°C, respectively. The theoretical values are plotted with a broken line in Fig.15.

The absorptivity can easily be obtained by solving two equations, Eq(10) and Eq(11). The solution which provides the absorptivity is obtained by eliminating θ from two equations;

$$A^2 + \frac{2\ k^2 r^4 v \theta_o^2}{a k_D W2} A - \frac{2\ k^2 r^4 v \theta_o^2}{a k_D W2} A_o = 0$$

$$A = \frac{2K}{Tw2} \left[(1 + \frac{Tw^2 A_o}{K})^{1/2} - 1 \right] \qquad --- (12)$$

where K is $\pi k^2 \theta_o^2 / 16 k_D$, T interaction time 2a/v defined as the time requred for the laser beam to traverse the length 2a, and w is the average power density given by W/4ab. In Fig.16, a solid curve indicates the absorptivity calculated from this equation as a function of Tw^2 and the experimental data in Fig.13 are also replotted. The experimental values are seen to agree well with the theoretical results except for those obtained from D_y=1mm which provides significantly higher temperature than that obtained from Eq(7) due to very small values of v*. As the result of the analysis described above, it was found that the absorptivity of steel coated with phosphates to the CO_2 laser beam can be predicted using Eq(12) or Fig.16, if the interaction time T and the power density w are specified.

558

Fig.16 Absorptivity plotted as a function of Tw^2 (T = interaction time, w = power density). The absorptivity can be predicted from this figure independently of beam spot shape, laser power and travelling velocity.

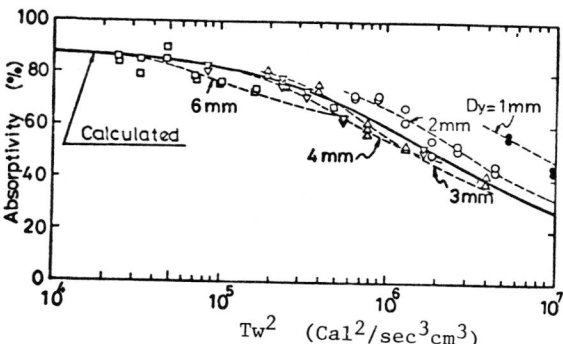

[5] Hardening results and discussion.

(1) Comparison with the experimental results

Zinc-phosphate coated SK-4 tool steel was laser-hardened in the air at the power level of 1.25KW with the beam spots of a*=0.7 which corresponds to the aspect ratio b/a of about 2. Figure 17 shows examples of the cross section of laser-hardened pattern. When the melting occurred together with hardening, the surface was roughened and cracks were produced within the melted zone. Then pronounced surface depressions were also observed. When the surface melting did not occure, the surface remained smooth enough to be used without any other surface finishing. Shallow and uniform case depth can be produced at higher velocities. Deeper case depth without surface fusion could be obtained for larger spot size and smaller travelling speed, but the hardened pattern was not uniform due to the smaller value of v*.

The average cooling rates calculated from 800°C to 400°C ranged from 10^5°C/sec to 200°C/sec depending upon the hardening parameters employed, and are always faster than the critical cooling rate for the martensite transformation in SK-4 tool steel. This fact demonstrates that in this case the hardened region is dependent only upon the mximum temperature attained during the process, and independent of the cooling rate.

The temperature raise was calculated from the equation of rectangular-Gaussian source using a*=0.7 (b/a=2) and the absorptivity measured as a function of the laser hardening parameters shown in Fig.13. The averaged values of the thermal conductivity k and thermal diffusivity k_D as shown in Eq(13) were employed in the calculation.

$$k = (1/\theta_m)\int_0^{\theta_m} k(\theta)d\theta$$

$$\qquad\qquad\qquad\qquad\qquad ---(13)$$

$$k_D = (1/\theta_m)\int_0^{\theta_m} k_D(\theta)d\theta$$

where θ_m is the fusion temperature of the metal. As the result, k=0.086 cal/sec·cm·°C and k_D=0.074 cm^2/sec were obtained.

Unlike the traditional hardening methods, the laser hardening process contains the especially high heating rates, and therefore the increased α-γ transformation temperature is expected to occure in this process. A variety of α-γ transformation temperatures higher than the values of equilibrium conditions was assumed in order

559

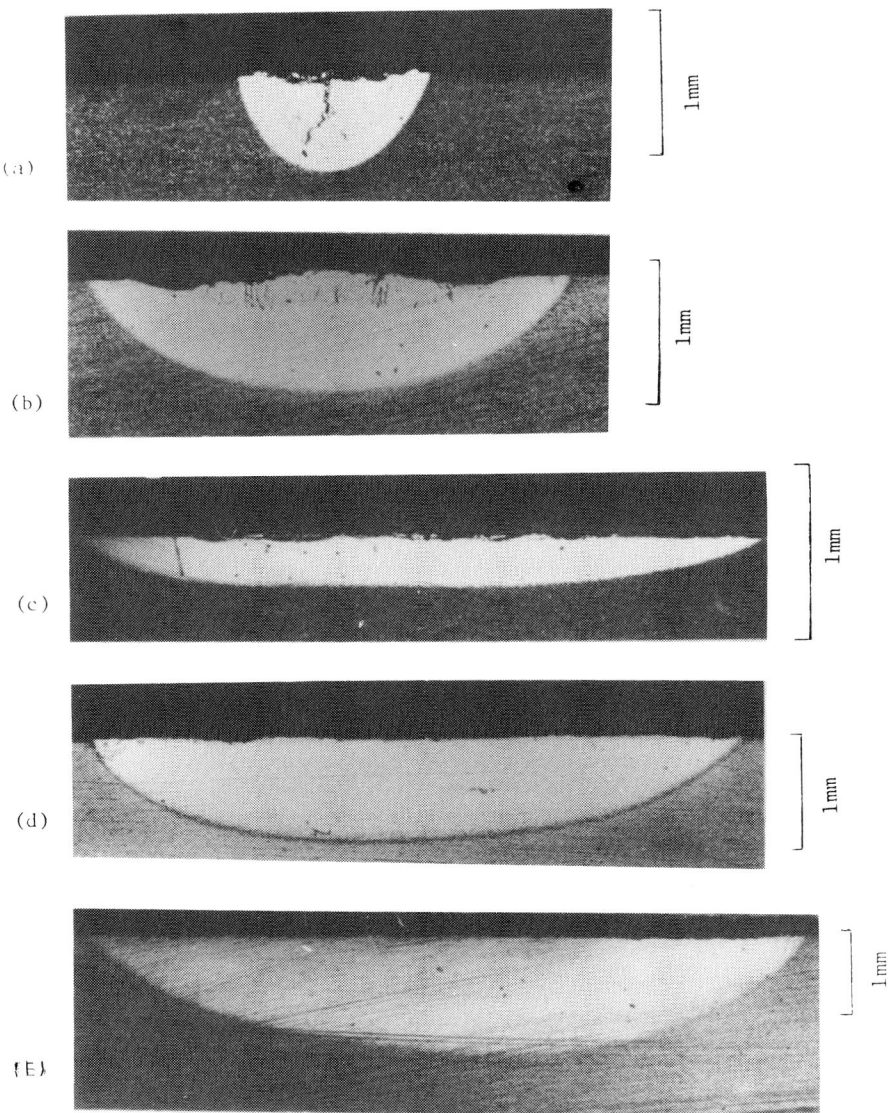

Fig. 17 Examples of laser-hardened pttern (zinc-phosphate coating, 1.25 KW).
(a) 100cm/min (b)50cm/min (c)250cm/min (d)40cm/min (e)20cm/min

$D_y=1mm$ $D_y=2mm$ $D_y=4mm$ $D_y=6mm$ $D_y=9mm$

560

to calculate the depth and width of the hardened zone for a series of hardening conditions applied, and these results were compared with the dimensions obtained from the etched cross sections. The comparison between the theoretical results and the experimental results indicated that the closest agreement between both values was obtained over a wide range of the hardening parameters when the transformation temperature was assumed to be 850°C as shown in Fig.18.

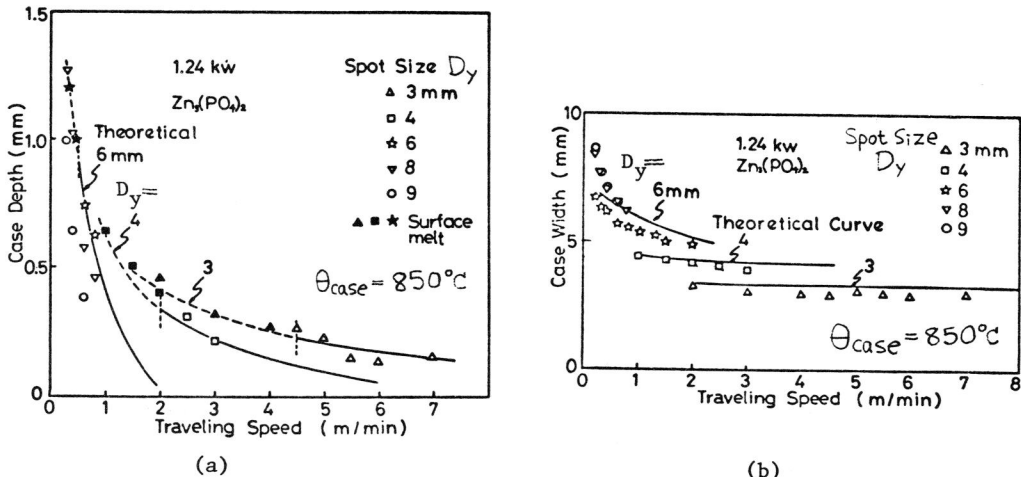

Fig.18   Comparison between theoretical results and experimental values.
(a) Hardened depth.   (b) Hardened width.   Each curve indicates theoretical value. Filled symbols and broken lines represent that the surface melting occurs. In the calculation the measured absorptivity was used.

In this figure each curve represents the theoretical values, and the experimental data points are also plotted; the filled symbols indicate the occurance of the surface melting, corresponding to the broken curves theoretically obtained. The value, 850°C, coincides well with the temperature at which the transformation in the eutectoid steel was reported to finish at the heating rates higher than 200°C/sec [8].

Whereas the width calculated using 850°C was somewhat larger than the measured one on the whole. This discrepancy in the width is not considered to be so important, because this would be improved by using the exact value of a* without influencing the hardened depth. But the width had a tendency to decrease more rapidly with increasing the travelling velocity for the measured than for the calculated. This is apparently caused by the assumption that the power density distribution is constant in the y-direction. It should be rather noticed that the difference is not so large in spite of apparent taper of the power density at the edges. The maximum depth and width of hardened zone obtained in the present experiment without causing surface fusion were 1.4mm and 8.5mm, respectively.

Now it is seen that the demensions of the laser hardened patterns calculated are in good agreement with the experimental results for a wide variety of the hardening parameters, if the absorptivity of the laser beam obtained experimentally is employed. However, it is also possible to predict the dimensions of the hardened zone only through the hardening parpmeters such as the beam spot size, the laser power and the travelling speed, if the absorptivity is determined by using Eq(12) instead of the experimental values. Figure 19 shows the theoretical values obtained using Eq(12), and they are also seen to be in good agreement with the experimental values. As the result above, it was found that the dimensions of the hardened pattern can be predicted if the hardening parameters are specified.

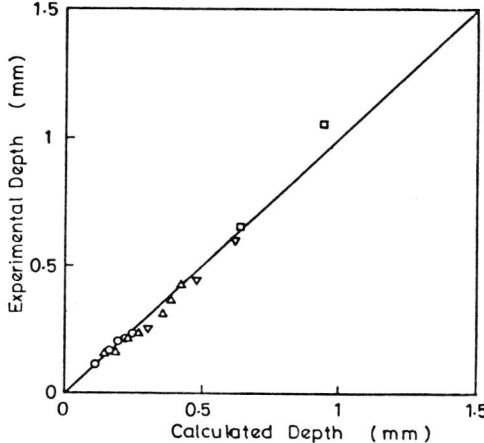

Fig.19 Comparison between calculated depth and experimental depth. Eq(12) was used in the calculation.

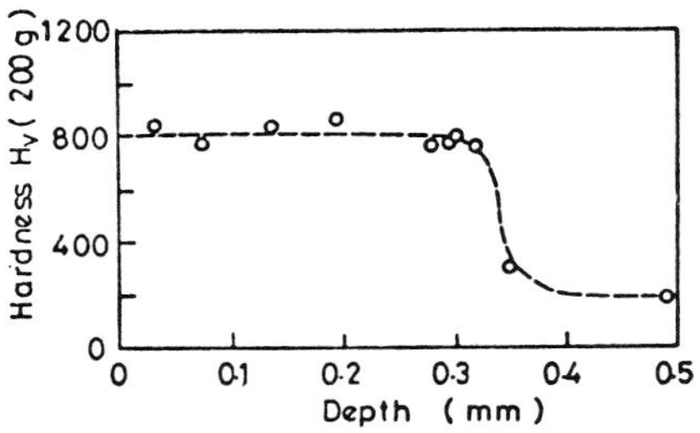

Fig.20 Distribution of hardness. (Dy=4mm, v=2m/min)

(2) Estimation of softened width

Figure 20 illustrates a typical hardness distribution of laser-hardened SK-4 tool steel. Hardness within the hardened region was fairly uniform showing about 850Hv, and varied sharply within 0.5mm to about 200Hv of the base metal.

In order to increase the width of the hardened pattern, multiple overlapping passes are necessary. This method causes soft streaks in the overlap region. So the specimen heated at 850°C for 10 minites in a furnace was water-quenched and then it was laser-hardened after coated with zinc-phosphate to estimate the width of the soft streaks. Here it was assumed that there was no difference about the softening characteristics between the laser-hardened and water-quenched regions. Figure 21 illustrates the distributions of hardness thus obtained which were measured prependicularly to the surface and parallelly to the surface at the depth about 100 microns. As shown in Fig.21 there was little difference in hardness between the water-quenched and laser-harnened regions. The minimum hardness measured in the resultant soft streaks was about 400Hv which was still higher than the hardness of the base metal. In order to give a brief estimation of softened width accompanied by the multiple overlapping passes, the softened width was defined by the one with the hardness lower than 700Hv which corresponded to the value of about 85% of the hardened zone, for convenience. As shown in Fig.21, the theoretical temperature corresponded to the hardness 700Hv resulted from the laser heating was about 400°C. The hardness obtained at the point where 400°C was reached for different spot sizes is plotted against the travelling velocity in Fig.22 and is seen to be almost constant, about 700Hv-750Hv, independent of the hardening parameters.

Thus the width of soft streaks at the surface can easily be calculated, assuming that it corresponds to the width adjoining to the hardened zone where the temperature ranges from 400°C to 800°C, as descrived above. In Fig.23 the depth

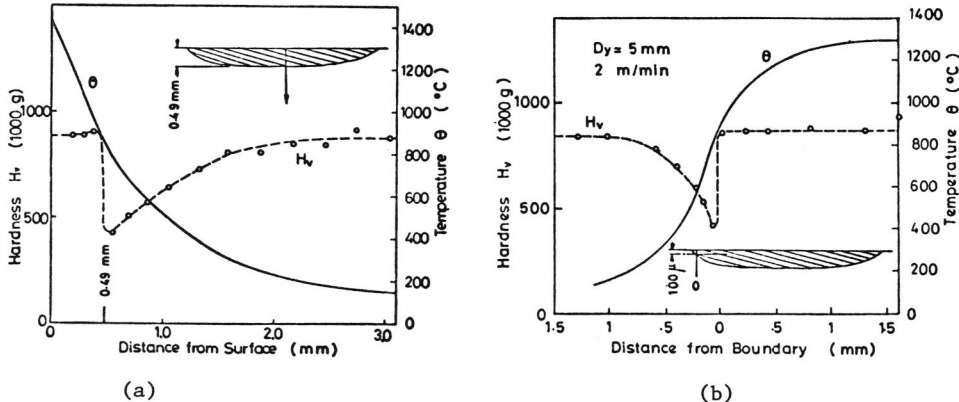

(a)   (b)

Fig.21 Distribution of hardness and calculated temperature. The
workpiece was coated with zinc-phosphate and then laser-hardened
after it had been water-quenched from 850°C (10minutes). (a)
Distribution parpendicular to the surface. (b) Distribution parallel to the surface at a depth about 100 microns.

and width of the hardened zone, the softened width and the ratio of the softened
to the hardened width, calculated for the rectangular-Gaussian source with $a*=0.7$
are plotted against $vr/2k_D$. It should be noticed that the non-dimensional values
are used to make this figure available for a wide range of hardening parameters,
and that the ratio of the temperature at the edge of the hardened zone and the
maximum tmperature at the surface is equal to be 850/1500 which provides the
maximum hardening depth without causing melting. As is seen in this figure, the
ratio $h*/H*$ decreases with increasing the value of $vr/2k_D$ showing a similar tendency with the value $z*$, whereas the width $H*$ remains almost constant. In Fig.9,
the hatched zone illustrates the softened zone for the Gaussian, rectangular and
rectangular-Gaussian sources when the surface temperature is 1500°C. It is seen
that the Gaussian source provides the significantly wide softened zone at the surface.

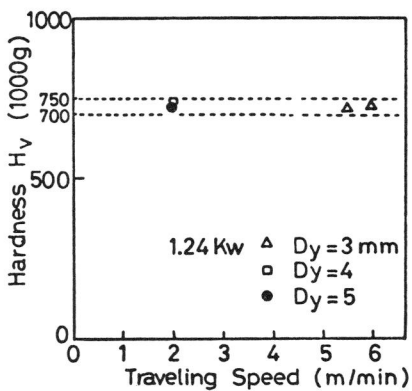

Fig.22 Hardness corresponding
to temperature rise of 400°C
for different spot sizes and
travelling velocities.

(3) Calculation for given laser power

In Fig.24 the depth and width of the hardened zone, and the softened width calculated from the rectangular-Gaussian source with a*=0.7 are plotted against the travelling velocity when the laser power output of 1.5KW is used, for example. In this calculation, the surface temperature is assumed to be the melting temperature which provides the maximum hardened depth with infinitesimal melted layer, and hence the beam absorptivity of 55% is also assumed. A broken line drawn in Fig.24 corresponds to the depth calculated from the rectangular source with a*=0.7, and gives somewhat larger depth than that for the rectangular-Gaussian source.

In this figure a hardening efficiency η is also plotted, which is difined as the ratio of the energy required for the specimen to be heated adiabatically up to 850°C, to the input energy, and is written by

$$\eta = 850 \ c\rho vS/W \quad ---(14)$$

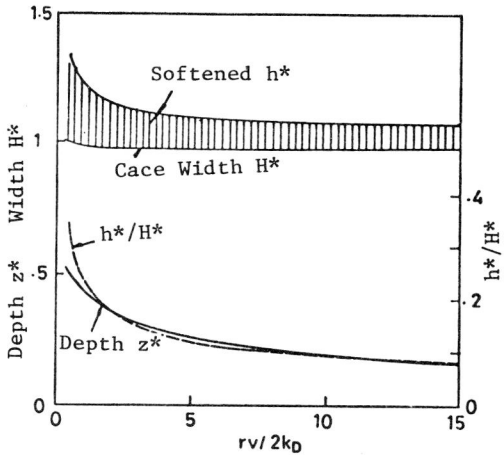

Fig.23 Hardened depth z*, width H* softened width h* and ratio h*/H* plotted as a function of v*. Surface temperature was assumed to be 1500°C (rectangular-Gaussian source, a*=0.7). H*, h* and z* are normarized with r ($=D_y/1.4$).

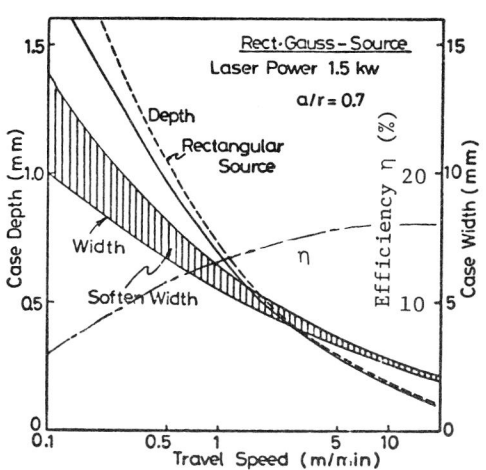

Fig.24 Hardened depth, hardened width and softened width (Laser power 1.5KW, rectangular-Gaussian source of a*=0.7, surface temperature 1500°C).

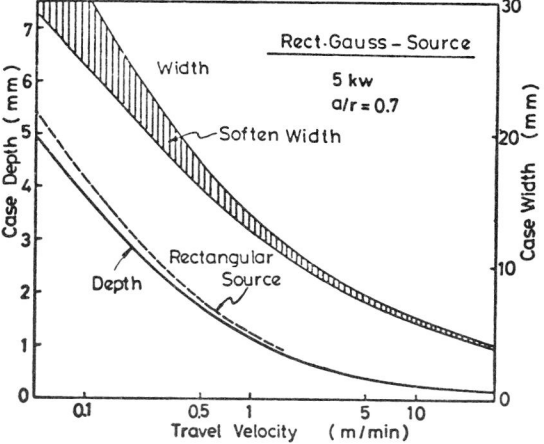

Fig.25 Hardened depth, hardened width and softened width (Laser power 5 KW, rectangular-Gaussian source of a*=0.7, surface temperature 1500°C).

where c is the specific heat of the meterial, ρ the density of the material, v the travelling velocity, W the input power and S the cross sectional area of the hardened zone. For the sake of simplification, the cross sectional area was calculated assuming that the hardened pattern is elliptical. It is seen that the hardening efficiency increases with increasing the travelling velocity at lower velocities, and approaches to about 15%. This very low value, 15%, is caused partially by the reflection loss of about 45% at the fusion temperature of the base metal, and partially by the assumption of the elliptical hardened pattern. Assuming that the specimen to be laser-hardened is a black body, the hardening efficiency η becomes about 30%. The hardening efficiency become about 38% by adding the assumption of the uniform case depth furthermore.

The theoretical results calculated from 5KW laser output are also given in Fig.25. When 5KW laser is used, it is seen that the case depth of about 3.5mm with the width of about 25mm will be obtainable at the travelling velocity of 10cm/min. However, it should be noticed that the cooling rate is assumed to be fast enough for the martensite critical cooling rate of the metal used to be reached in these calculations.

(4) Approximate equation

It is possible to predict and evaluate the surface hardening by using the results represented diagrammatically such as Figs.7-11, if the laser-hardening parameters are specified. It will, however, be much more helpful to derive simplified equations which enable the experimenters to easily predict and evaluate the laser-hardening. Here the exact temperature solutions which require the rather complicated calculations are approximated by elementary equations. The hardened depth alone is delt with herein, since the hardened width is almost equal to the spot size Dy except for small values of v* as described above.

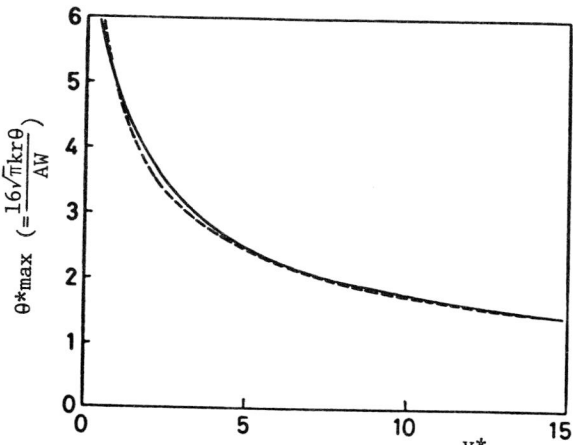

Fig.26 Relationship between θ^*_{max} and v* for rectangular-Gaussian source (a*=0.7). A solid line represents the exact solution Eq(7) and a broken line is the approximate equation Eq(15).

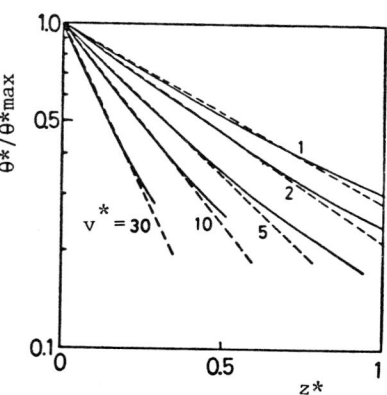

Fig.27 Relationship between θ^*/θ^*max and z* for rectangular Gaussian source (a*=0.7) at y*=0. Solid lines correspond the exact solution and broken lines are represented by Eq(16).

A solid line in Fig.26 shows the relationship between v* and θ^*_{max} which is the maximum value attained at the surface on the x-axis for the rectangular-Gaussian source of a*=0.7. As shown with a broken line, this can be closely approximated by a equation,

$$\theta^*max \cong 5.6/\sqrt{v^*+0.3} \quad --- (15)$$

The numarator of this equation is seen to coincide with 8 of Eq(10), if the value a*=0.7 is taken into consideration. The discrepancy of θ^*max between the precise value from Eq(7) and the value from one-dimensional approximation at smaller values of v* which was described in the section [3] becomes neglegible small by adding 0.3 to v* in the denominator of Eq(10).

Solid lines drawn in Fig.27 indicate the relationship between z* and θ^* which is normalized by θ^*max. As is clear in this figure, following equation approximately holds for any values of v* in a region $1 > \theta^*/\theta^*max > 0.5$

$$\frac{\theta^*}{\theta^*max} \cong \exp(-p \cdot z^*) \quad --- (16)$$

As shown in Fig.27, p is a function of v* and is given by

$$p = 0.886\sqrt{v^*+1} \quad --- (17)$$

As the result, we get a very simple equation instead of Eq(7);

$$\theta^* \cong \frac{5.6}{\sqrt{v^*+0.3}} \exp(-0.886z^*\sqrt{v^*+1}) \quad --- (18)$$

$$\theta \cong 1.15 \frac{AW\theta^*}{D_y} \quad --- (19)$$

where

$z^* = z/r = 2.8z/D_y$

$v^* = vr/2k_D = 2.42D_y \cdot v, (k_D = 0.074cm^2/sec)$

D_y = spot size on the y-axis

A = absorptivity : Eq(12)

W = laser power (cal/sec)

The case depth for Gaussian source with a*=0.7 is obtained substituting θ=850°C, and is written by

$$z_{case} \cong \frac{0.26D_y}{\sqrt{vD_y+0.413}} \ln\frac{0.0049AW}{D_y\sqrt{vD_y+0.124}} \quad --- (20)$$

The critical travelling velocity at which surface melting occurs is given by

$$v_c^* \cong 31.4/\theta^{*2} - 0.3 \quad ---(21)$$

$$v_c \cong 2.3 \times 10^{-6} W^2/D_y^3 - 0.124/D_y \quad ---(22)$$

For the rectangular source, a*=1, the temperature rise at y*=0 is written by

$$\theta^* \cong (8/\sqrt{v^*+1.1})\exp(-1.1z^*\sqrt{v^*+0.15}) \quad --- (23)$$

From this equation, the case depth for given hardening parameters and the critical travelling speed at which surface melting occurs are computed easily for the square source (a*=1). For example, the case depth (θ=850°C, 2r=D_y) is written by

566

$$z_{case} \cong \left(0.247D_y/\sqrt{vD_y}+0.044\right)\ell n\ 0.0042AW/D_y\sqrt{vD_y}+0.326 \qquad --- \ (24)$$

The absorptivity A is given by Eq(12) or Fig.16.

[6]   Conclusions

Laser-hardening characteristics are discussed primarily through heat conduction theory of surface heat source. Conclusions obtained from this study are summerized as follows;

1. New mathematical formulations were derived by means of which the temperature distributions and the shape of isotherms for moving surface heat source, Gaussian, rectangular and rectangular-Gaussian sources, can be calculated. These agreed well with microscopical examinations of the laser hardened material.

2. The absorptivity of phosphate-coated steel to the CO_2 laser beam decreases with square of the surface temperature, and can be predicted if the laser hardening parameters are given.

3. Approximation equations with simple form are derived by means of which hardened depth, hardened width softened width and critical travelling speed for surface melting can easily be predicted.

Acknowledgement

The authors wish to thank K.Sohno, who helped the authors with carrying out the experiments, and performing the tedious numerical calculation.

References

1) E.V.Locke, D.Gnanamuthu and R.A.Hella, "High Power Lasers for Metalworking", Research Report of AVCO Everett Research Laboratory, March 1974.

2) G.H.Harth, W.C.Leslie, V.G.Gregson and B.A.Sanders, "Laser Heat Treating of Steels", Journal of Metals, No.4 (1976).

3) C.Wick, "Laser Hardening", Manufacturing Engineering, June (1976).

4) S.L.Engel, "Basics of Laser Heat Treating", Technical Report of GTE Sylvania, Inc., March 1976.

5) Y.Arata and I.Miyamoto, "Some Fundamental Properties of High Power Laser Beam as a Heat Source (Report 2) — CO_2 Laser Absorption Characteristics of Metals —", Trans. Japan Welding Society, Vol.3, No.1 (1972).

6) Y.Arata and I.Miyamoto, "Some Fundamental Properties of High Power Laser Beam as a Heat Source (Report 3) — Metal Heating by Laser Beam —", Trans. Japan Welding Society Vol.3 No.1 (1972).

7) Y.Arata and I.Miyamoto, "Processing Mechanism of High Power Density Beam — Mechanism of Drilling —", Trans. JWRI, Vol.2, No.2 (1973).

8) K.J.Albutt and S.Garber, "Effect of Heating Rate on the Elevation of the Critical Temperatures of Low-carbon Mild Steel", Journal of the Iron and Steel Institute, Vol.204, No.8 (1966).

EFFECT OF HEATING CONDITION IN LASER HARDENING CARBON STEEL

Abstract

Effects of temperature history in terms of maximum cycling temperature, heating and cooling rate and holding time above a particular temperature in single truck laser hardening on the hardening process are evaluated for carbon steels with a wide range of carbon content, 0.1-0.84 C%. The hardeness in given materials is shown to depend only on the maximum cycling temperature under conditions that the cooling rate is faster than the critical values of martensite transformation. The effects of carbon content and carbide distribution on the temperature, θ_h and θ_s, at which the maximum hardness is attained and hardening begins, respectively, are demonstrated. Simple approximated formulae are derived which predict the case depth from given θ_h and hardening parameters.

1. Introduction

There has been a great deal of interest in the use of lasers for transformation hardening, since lasers can offer localized hardening with minimal thermal distortion of components in a very short operation time. It is essentially important in practical applications of laser hardening to predict the accurate case depth from the hardening parameters, and to find out the optimum hardening parameters.

In laser hardening process, heating is from the surface inwards in a very short time period so that wide range of temperature and heating and cooling rate which are generally much faster than conventional hardening processes result depending on the depth from the surface. The short thermal cycle times accompanied by laser hardening result in the lag in the α-γ transition [1][2] and non-uniform microstructure [3] due to limited diffusion distance of carbon. Such complicated metallurgical responses make it difficult to correlate between the thermal cycle and hardening. A substantial amount of work [3-5] has been done to demonstrate the effects of process parameters on the case depth for given material. However, few detailed, systematic investigations have been made into the effect of thermal cycle on hardening.

In a previous paper [6], the authors analized the temperature history in the laser hardening process for various intensity distributions in the laser beam spot with taking the beam absorption characteristics into consideration. In this paper, on the basis of this analysis the effects of the temperature history in terms of the maximum cycling temperature, heating and cooling rates and holding time above a given temperature, on the hardness and microstructure are dealt with for carbon steels with 0.1-0.84 % carbon contents at a wide range of heating conditions. Simple mathematical formulae are also derived for predicting the case depth from the heating parameters.

2. Experimental procedures and materials

A GTE Sylvania Model 971, gas-transport CO_2 laser of 1.2 kW output power was used for the work described in this paper. The laser beam was passed vertically through a ZnSe lens of 10 in. focal length. A single hardened track was produced by moving the workpiece beneath the beam at a given speed. The spot size on the workpiece surface was changed by changing the distance between the lens and workpiece which was placed below the focus.

This gas-transport laser can generate a rather elliptical beam spot of which longer axis is parallel to the direction of discharge for exciting the lasing medium. By adjusting the laser resonant cavity, the ellipticity was decreased down to about 0.6 with a reasonably uniform intensity distribution along the longer axis.

The intensity profile in the beam spot was determimined from the groove formed by scanning the beam spot along the two axes, x- and y-axis, on aclylic plastics. Figure 1 shows an example of the beam profile obtained. The intensity profile can be approximated by a rectangurar-Gaussian distribution:

$$w(x,y) = \frac{W}{2\sqrt{\pi}\,ab} \exp\left(-\frac{x^2}{a^2}\right) ; (-b<y<b) \qquad \text{---------------(1)}$$

where W is laser power, and $2a$ and $2b$ are the spot diameters along the x- and y-axis at which the power levels fall to 1/e of the center value. In order to obtain a hardened pattern with uniform depth, the beam spot was moved parallelly to the x-axis as shown in Fig. 1. When the laser power is fixed, the beam intensity distribution is characterized by a spot size $2b$, and a parameter $a*$:

$$a* = \sqrt{a/b} \qquad \text{--------------(2)}$$

In this experiment, $2b$ and $a*$ were varied from 2.6 to 4.3 mm, and from 0.76 to 0.89, respectively.

Table 1 shows the compositions of the commercial carbon steels used in this experperiment. Two types of carbon distribution, spherodized carbide and lamella pearlite, were provided in SK-5 tool steel and the 4 other materials, respectively. The specimens were coated with zinc-phosphte for enhancement of the laser beam absorptivity. The laser beam absorptivity of zinc-phosphate coated specimens is formulated by the authors as a function of their surface temperature [6]. The specimens were hardened in open atmosphere at various travelling speeds and spot sizes at a fixed laser power level, 1 kW.

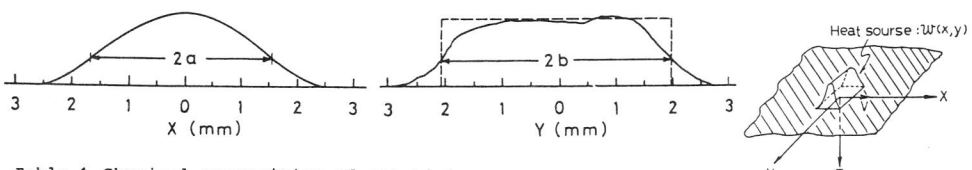

Table 1 Chemical composition of matetials used (wt-%)

Material	.C	Si	Mn	P	S	Cu	Ni	Cr
S10C	0.11	0.15	0.53	0.013	0.017	—	—	—
S20C	0.23	0.22	0.48	0.017	0.013	0.01	0.03	0.01
S30C	0.33	0.26	0.70	0.020	0.021	0.02	0.02	0.10
S58C	0.58	0.21	0.75	0.027	0.013	—	—	—
SK-5	0.84	0.19	0.38	0.017	0.012	—	—	—

Fig. 1

Power density distribution in laser spot.

3. Results and discussion

3-1 Hardness distribution

A series of laser-hardening experiment was carried out for each material and Fig. 2 shows an example of the laser-hardened pattern in SK-5. It is seen that fairly uniform hardened depth was obtained.

Figures 3-5 show the data of microhardness distributions. In every material, an approximately constants, maximum hardness has been observed at a region adjacent to the surface of specimen. With increasing the depth, the hardness decreased rapidly down to the matrix hardness in S58C and SK-5. The thickness of the transient hardness zone appeared to be larger in SK-5 than that in S58C.

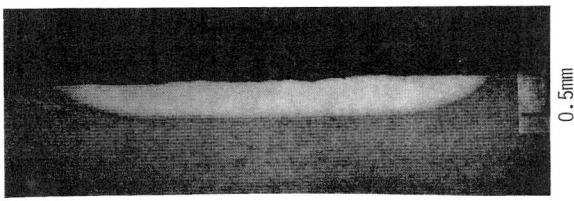

Fig. 2　Cross-section of laser hardened zone (SK-5, *v=2 m/min, 2b=3.2 mm, W=1 kW*).

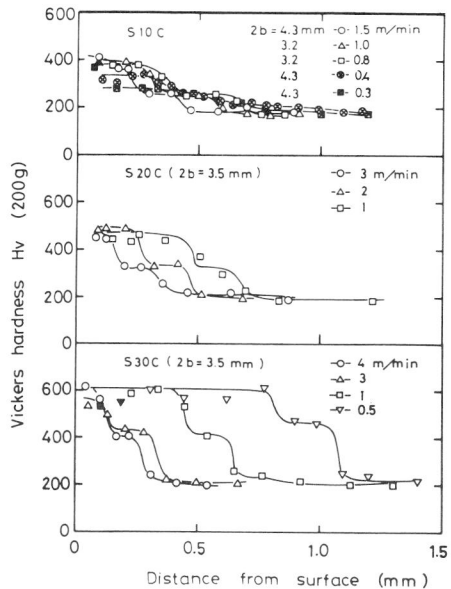

Fig. 3　Hardness distribution in S10C, S20C and S30C.

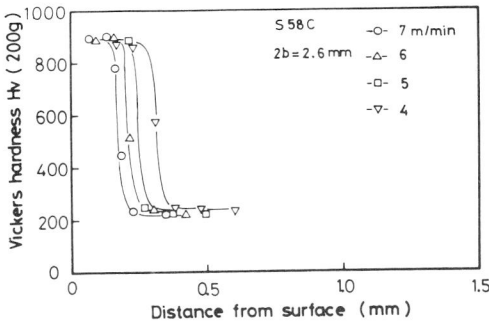

Fig. 4　Hardness distribution in S58C.

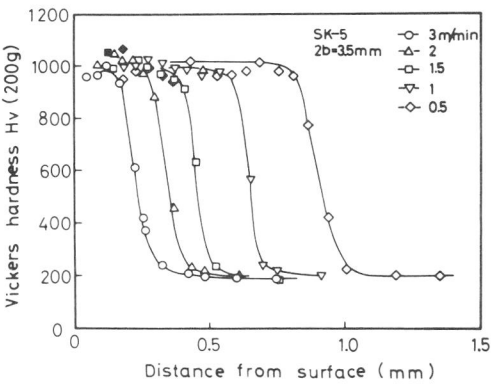

Fig. 5　Hardness distribution in SK-5.

In S10C, S20C and S30C, two-step hardness curves are observed; There appears an additional region showing rather flat hardness between the surface part and the matrix. At given carbon contents, the hardness of the mid-step was independent of travelling speed and spot size, except only for S10C hardened at speeds lower than 40 cm/min; with increasing the carbon content, the hardness of the mid-step decreased. In the laser hardening of S10C at speeds lower than 40 cm/min, the hardness both near the surface and in the mid-step tended to decrease with decreasing the travelling speed.

The maximum hardness in materials other than S10C was independent of travelling speed within the conditions tested. The maximum hardness obtained in the laser-hardening was higher than that in the conventional hardening methods by Hv=50-100 in S58C and SK-5.

3-2 Thermal conduction theory

If the heat is liberated at the rate $w(x,y)$ given in Eq(1) per unit time per unit area at the point (x,y) on the surface of the semi-infinite body, $z=0$, and surrounding medium moves across it with a constant speed v in the direction along the x-axis as shown in Fig. 1, the temperature rise in the steady state at the point (x,y,z) with no loss of heat from the surface [6] is given by

$$\theta = \frac{\sqrt{\bar{\alpha}}\,WA}{4\pi kb}\int_0^\infty \frac{1}{\sqrt{t(a^{*4}b^2+4\alpha t)}}\exp\left[-\frac{(x+vt)^2}{a^{*4}b^2+4\alpha t}-\frac{z^2}{4\alpha t}\right]\left(\operatorname{erf}\frac{y+b}{\sqrt{4\alpha t}}-\operatorname{erf}\frac{y-b}{\sqrt{4\alpha t}}\right)dt \,, \quad \text{------------(3)}$$

where $\operatorname{erf} X = \frac{2}{\sqrt{\pi}}\int_0^x e^{-t^2}d\xi$,

k=thermal conductivity, α=thermal diffusivity, v=velocity, W=laser power, A=laser beam absorptivity, $a^*=\sqrt{a/b}$ and $2b$=beam spot size. In this equation, the value A of specimens coated with zinc-phosphate (see Appendix 2) is given by

$$\theta_{MB}(0) = \frac{W}{16\sqrt{\pi}\,kb}\frac{\gamma}{\sqrt{\frac{bv}{2k}}\,a^* + \delta} \qquad \text{-----------(4)}$$

where $\theta_{MB}(0)$ is the maximum cycling temperature attained at the surface of the hypothetical solid with no reflection loss, and is obtained by substituting $A=1$ and $y=z=0$ into Eq(3), A_0 and θ_0 are constants, 0.9 and 2600 °C, respectively. In the computation of Eq(3), the average values of k and α between room temperature and melting point [9], k=0.086 cal/sec·cm·°C and α=0.074 cm²/sec were used, respectively.

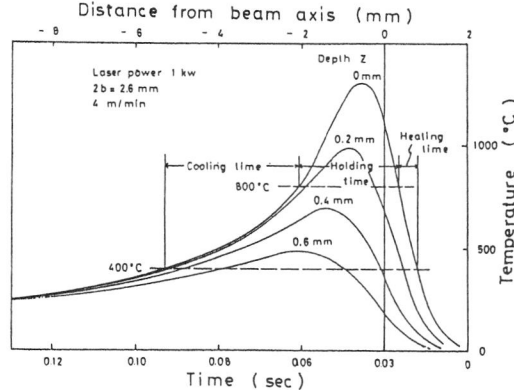

Fig. 6 Temperature distribution calculated in the x-y plane (v=4 m/min, $2b$=2.6 mm, a^*=0.86)

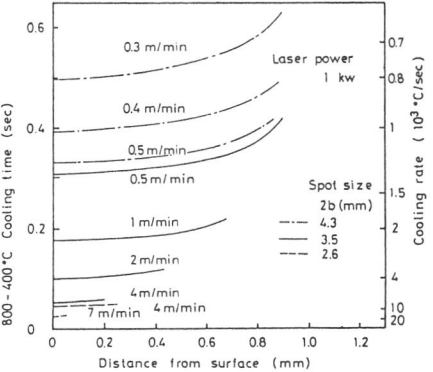

Fig. 7 Cooling time calculated between 800 °C and 400 °C vs. distance from surface.

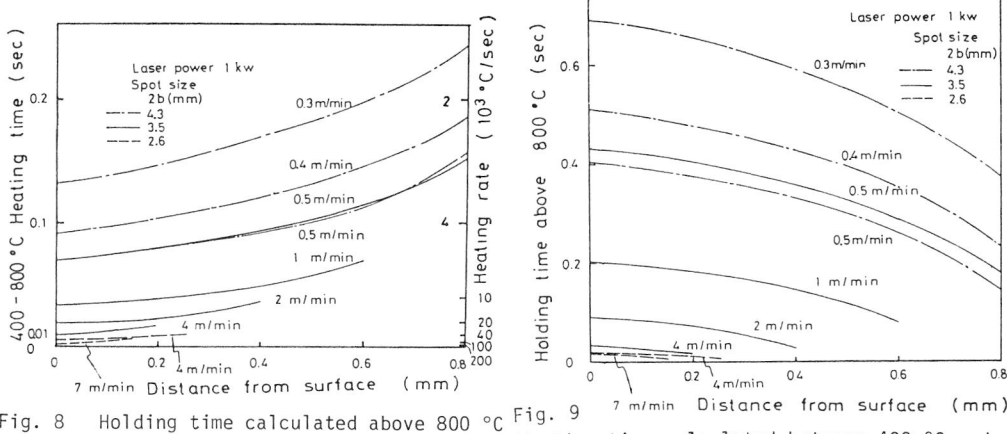

Fig. 8 Holding time calculated above 800 °C vs. distance from surface.

Fig. 9 Heating time calculated between 400 °C and 800 °C vs. distance from surface.

Figure 6 illustrates an example of the temperature distribution in the x-z plane calculated from Eq(3). The maximum temperature, heating and cooling time between 800 °C and 400 °C, and holding time above 800 °C can be determined from the temperature distribution curves in the x-z plane.

3-3 Thermal cycle

In Figs. 7-9, the calculated heating and cooling times between 400 °C and 800 °C, and holding time above 800 °C are plotted against z for various travelling speeds and spot sizes. It is seen that the holding time above 800 °C in this experiment ranges between about 0.01 sec and 0.6 sec, and is extremely shorter than that of the conventional hardening methods. The heating time from 400 °C to 800 °C ranges between 0.002 sec and 0.2 sec, corresponding to the average heating rate between 2×10^3 °C/sec and 2×10^5 °C/sec.

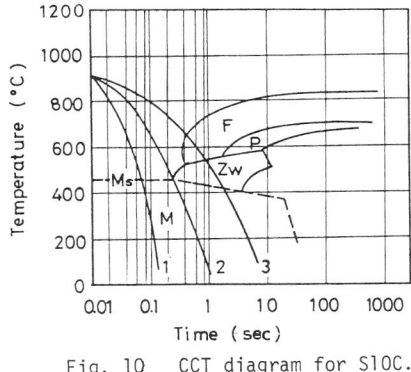

Fig. 10 CCT diagram for S10C.

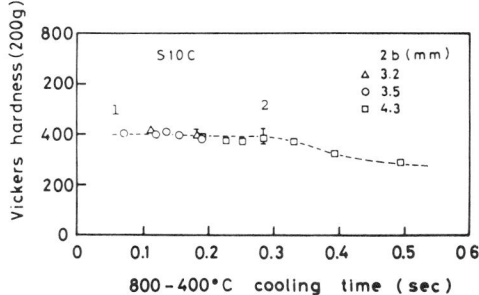

Fig. 11 Maximum hardness measured vs. cooling time calculated between 800 °C and 400 °C in S10C.

The cooling time from 800 °C to 400 °C ranges between 0.6 sec and 0.02 sec, corresponding to the average cooling rate between 700 °C/sec and 2×10^4 °C/sec. It is noticed that the cooling time changes only slightly with change in the spot size in this experiment. The calculated cooling time from 800 °C to 400 °C was compared with the critical cooling time from 800 °C to 400 °C, T_M, for the martensite transformation available in the CCT diagram [7], and was found to be shorter than T_M except for the value of S10C, $T_M=0.3$ sec, as seen in Fig. 10.

In Fig. 11, the maximum hardness in S10C is plotted against the cooling time; the maximum hardness decreases in a cooling time region longer than 0.3 sec. This fact proves that the calculations are accurate enough to estimate the temperature history. The maximum hardness in the materials other than S10C was found to be constant within the conditions tested. The hardening conditions under which the cooling time is shorter than T_M can be determined from Fig. 7, when T_M of given material is known.

3-4 Relationship between maximum cycling temperature and hardness

Figure 12 shows the maximum cycling temperature calculated for various travelling speeds. The hardness data plotted in Figs. 3-5 were correlated with the corresponding maximum cycling temperatures calculated, as shown in Fig. 13.

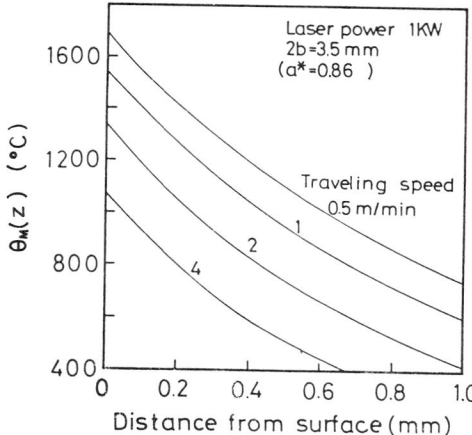

Fig. 12

Maximum cycling temperature calculated vs. distance from surface.

It is noticiable that the hardness data fall within a narrow band around a curve, independently of the heating rate and the holding time for materials other than S10C. In S10C, too, the hardness is seen to be a function of the maximum cycling temperature except for the travelling speeds slower than 0.4 m/min, which corresponds to the critical cooling rate for the martensite transformation.

The temperature at which hardening begins is about 750 °C without regard to carbon content of material. Low carbon steels exhibited two-step curve again in θ_M-H_v plane; the width of transient hardness part accompanied by the mid-step decreased as the carbon content increased. In S58C, the transient region having intermediate hardness became narrowest without showing the mid-step. In the material with higher carbon content, SK-5, however, the transient region became wider than S58C.

In Fig. 14, the temperatures θ_h and θ_s, at which hardening begins and the maximum hardness is reached respectively, are superimposed on the C-Fe phase-diagram. It is seen that θ_s does not depend on carbon content and its distribution in materials, and is about 750 °C, which is very close to Ac_1. On the other hand, θ_h depends on carbon content and carbide form ; θ_h of ferrite-perlite steel decrease with increasing carbon content and approaches about 900 °C near eutectoid carbon content. In SK-5 with spherodized carbide, θ_h is approximately 150 °C higher than that of ferrite-pearlite steel.

3-5 Austenite and stress retained in the hardened zone

Sub-zero treatment was performed in liquid N_2 for about 60 min just after laser-hardening in SK-5. Hardness data are plotted against the calculated maximum cycling temperature with closed marks in Fig. 13. There is no difference in hardness between with and without sub-zero treatment, showing no appreciable austenite retained in the laser hardening process within the conditions tested, in contrast to around 20 % austenite retained in the conventional water quenching from 1000 °C in this carbon content.

The authors measured the residual stress at the surface of the laser-hardened zone by X-ray technique and found that the large compressive stress, several tens kg/mm^2, was retained in the directions parallel and perpendicular to the beam traverse. Hagiwara et al found that the compressive stress that applys just below the M_s point in the cooling process reduces the residual austenite [8]. Possible accounts for less tendency to retain austenite in laser-hardening are the compressive stress applying during martensite transformation, and short cooling time during which austenite is hardly stabilized.

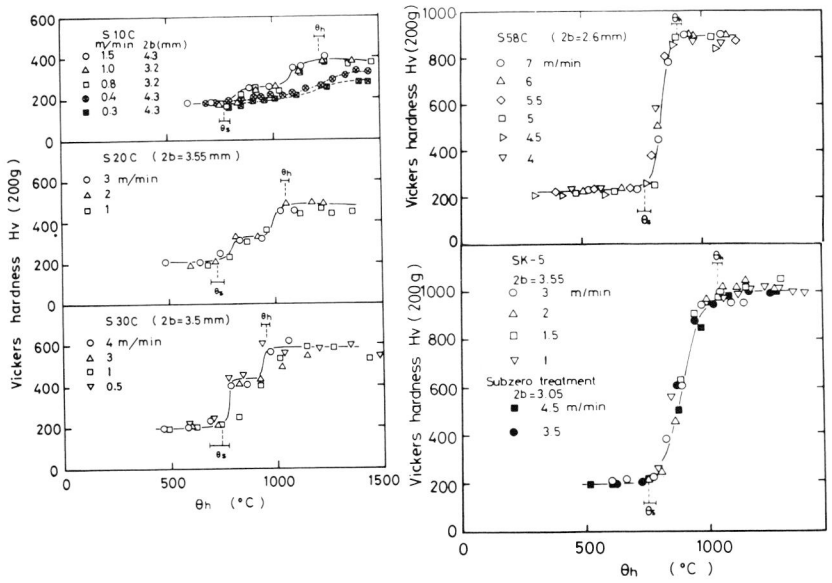

Fig. 13 Relationship between hardness and maximum cycling temperature.

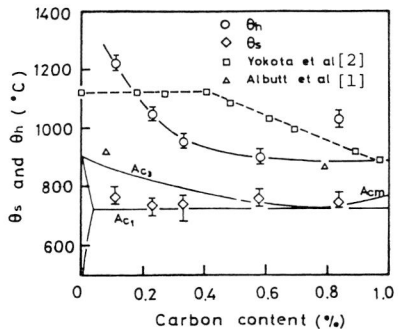

Fig. 14
θ_h and θ_s vs. carbon content.

3-6 Microstructure in hardened zone

In S10C, S20C and S30C, a two-phase region consisting of proeutectoid ferrite and martensite is seen near the border of the hardened zone as shown in Fig. 15. It appears that the material in this region is heated above the eutectoid temperature but below the A_{C3} of the ferrite for a short time period, in this case for about 70 msec. This time duration available above the eutectoid temperature was insufficient to allow the carbon in the original pearlite to diffuse out into the ferrite, but sufficient to allow the original pearlite to transform to austenite by a function of the large number of nucleation site available in the form of cementite lamellae.

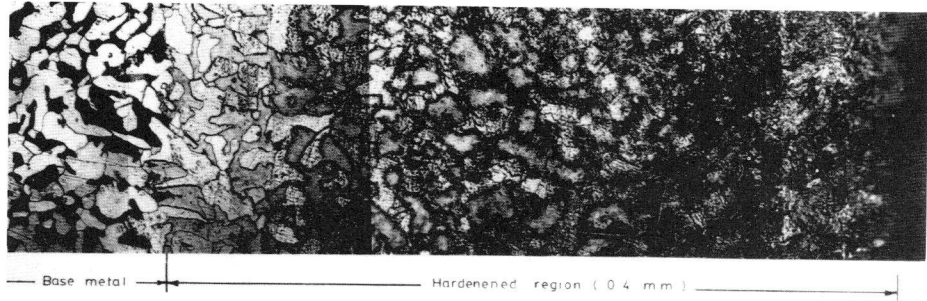

Base metal ——— Hardenened region (0 4 mm)

Fig. 15   Microstructure of laser-hardened S30C ($2b=3.5$ mm, $v=3$ m/min).

No acicular structure is seen in this martensite region in SEM (Fig. 16). This two-phase region corresponds to the mid-step region of the hardness curve. As the surface of the specimen is approached, ferrite is seen to diminish due to the diffusion of carbon.

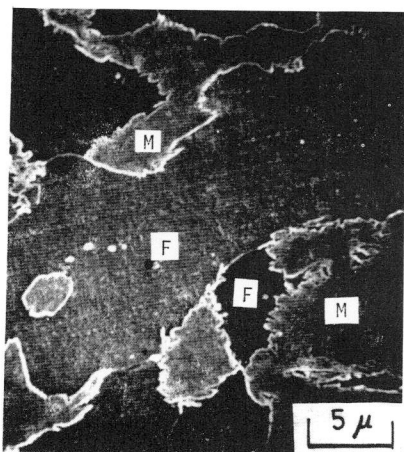

Fig. 16   Two-phase structure (S30C).

The hardness remains almost constant until glowing martensite colonies are connected each other. The maximum hardness is attained when a rigid network structure of growing martensite is formed.

In SK-5, white regions appeared adjacent to the spherodized carbide at 750 °C, and spread with increasing maximum cycling temperature as seen in Fig. 17. The fact that SK-5 gives higher θ_h than that of S58C is accounted for by the longer diffusion distance required to form connected martensite structure. This means that steels with finer carbide structre are desirable for deeper case depth.

In Fig. 14, the literature data of the A_{C3} temperature for ferrite-pearlite steels during rapid heating measured dilatometrily are plotted for comparison. In this figure, the values for the high heating rate limit in the data are adopted. Large discrepancy between θ_h and the dilatometory data is seen except for the values of eutectoid steel. Taking the possibility of diffusionless transformation of ferrite to austenite [1] into consideration at very high heating rates, it is conceivable that the α-γ transformation does not necessarily cause the

575

increase in the average hardness.

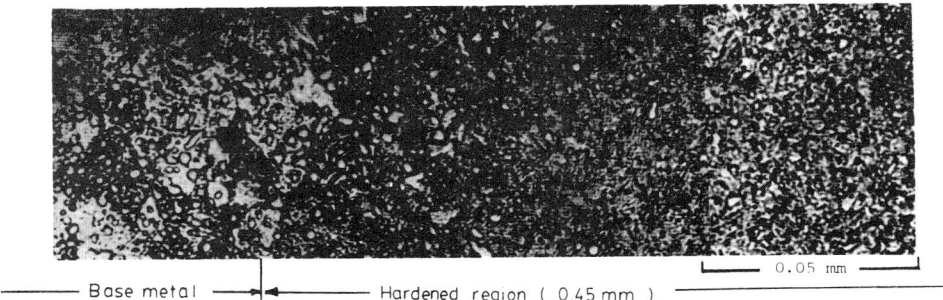

Base metal ─────►│◄───── Hardened region (0.45 mm) ──────

0.05 mm

Fig. 17 Microstructure of laser-hardened SK-5 ($2b=3.5mm$, $v=2$ m/min).

3-7 Approximated formula for predicting the case depth

In this section, the effects of θ_h on the case depth are discussed under conditions that the cooling rate is high enough for austenite to transform to martensite. The authors have derived simple approximated equations which give the maximum cycling temperatures as a function of beam parameters as shown in Appendix 1. From Eq (A2) in Appendix 1, the travelling speed which provides the maximum cycling temperature $\theta_M(0)$ at the surface of semi-infinite solid at $y=0$ is given by

$$v = \frac{2}{a^* b}\frac{\alpha}{}\left[\left(\frac{\gamma AW}{16\sqrt{\pi}kb\theta_M(0)}\right)^2 - \delta\right] \qquad \text{---------(5)}$$

In this equation, the absorptivity A for the surface temperature $\theta_M(0)$ is given in Eq (A7).

Eq (A6) gives the depth corresponding to the maximum cycling temperature θ_h as follows:

$$z_h = \frac{a^* b}{\left(\frac{mbv}{2k}+1\right)^n} ln[\theta_M(0)/\theta_h] \qquad \text{---------(6)}$$

In Eqs (5)(6), γ, δ, m and n are constants determined by the power distribution in the spot and value a^* as shown in Table A1. Under fixed heating conditions including beam factors, a^*, $2b$, and the travelling speed v, z_h is directly proportinal to ln $\theta_M(0)/\theta_h$. As shown in Fig. 18, z_h is seen to increase linearly with decreasing θ_h, and to decrease with increasing surface temperature $\theta_M(0)$.

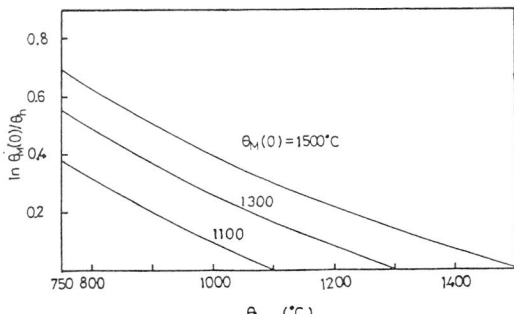

Fig. 18

θ_h vs. $ln\theta_M(0)/\theta_h$ for various $\theta_M(0)$.

576

When the surface temperature is limited by the melting point of the material θ_{melt}, the maximum case depth is proportional to $ln(\theta_{melt}/\theta_h)$. Deeper case depth can be obtained with material having higher melting point and finer carbide structure which gives low value of θ_h. It is noted that high carbon materials with coarse carbide distribution, such as cast iron, are typically difficult material to laser-harden.

4. Summary

In single track laser hardening of carbon steels with carbon contents between 0.1-0.84 C%, the effects of maximum cycling temperature, heating and cooling rate and holding time above a particular temperature on hardening process are studied on the basis of thermal conduction theory, microstructural observation and hardness distribution. Results obtained are summarized as follows:

(1) When the cooling rate is faster than the critical rate for Ar" transition, the hardness depends only on the maximum cycling temperature within the conditions tested.

(2) Hardening begins at the maximum cycling temperature of about 750 °C, independently of the carbon content and carbide distribution. In ferrite-pearlite steels, θ_h, at which the maximum hardness is attained, decreases with increasing carbon content, approaching progressively about 880 °C of eutectoid steel. At given carbide content, material with spherodized carbide has higher θ_h than material with lamellar pearlite.

(3) Two-step hardness curves are obtained in hypoeutectoid steel, and two-phase structure consisting of martensite and proeutectoid ferrite observed corresponds to the mid-step hardness region. The maximum hardness is reached when growing martensite colonies are connected each other.

(4) The effect of residual austenite in laser hardening on the hardness was negligible within the conditions tested.

(5) Simple approximated equations are derived which predict the case depth from given heating parameters and θ_h.

Appendix 1 : Maximum cycling temperature

In the previous paper [6], the authors derived non-dimentional equations for steady state temperature rise in a semi-infinite solid due to a Gaussian, a rectangular and a rectangular-Gaussian source moving at a constant speed as shown in Fig. A1. The dimen-

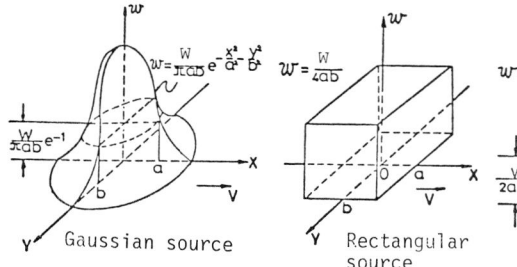

Fig. A1
Gaussian source, rectangular source and rectangular-Gaussian source.

Gaussian source   Rectangular source   Rectangular-Gaussian source

sional form for the rectangular-Gaussian source is given in Eq (3). In Fig. A3, the maximum cycling temperature $\theta_M*(0)$ at fixed position on the surface of the solid at $y*=0$ calculated for Gaussian source having $a*=1.0$ is plotted against $v*$ in a solid line. The value $\theta_M^*(0)$ is approximated by

$$\theta_M^*(0) = \frac{7.7a*}{\sqrt{a*v*+0.95}}, \qquad\qquad --------(A1)$$

where $\theta* = \frac{16\sqrt{\pi}kb\theta^*}{AW}$, $v = \frac{vb}{2\alpha} a*$ and $a* = \sqrt{a/b}$.

Table A1 Value of γ, δ, m and n

Source	m	n	a^*	δ	γ
Gaussian	1.4	0.4		$0.95a^{*2}$	7.7
Rectangular-Gaussian	0.65	0.5	$a^*<0.5$	0.1	7.1
			$a^*>0.5$	$2.7(a^*-0.5)^2+0.12$	
Rectangular	0.7	0.5	$a^*<0.5$	0.25	3.1
			$a^*>0.5$	$3.5(a^*-0.5)^2+0.28$	

In general, the maximum cycling temperature on the surface at $y^*=0$ for these three sources are given in a simple formula within accuracy of ±2 %;

$$\theta_M{}^*(0) = \frac{\gamma a^*}{\sqrt{a^*v^*+\delta}} , \qquad \text{---------- (A2)}$$

where γ and δ are constants which depend on the intensity distribution and a^*, and are given in Table A1. In Fig. A2, the value $\theta_M{}^*(z^*)/\theta_M{}^*(0)$ is plotted against the depth for the rectangular-Gaussian source ($\theta_M^*(z^*)$=maximum cycling temperature at depth z^* at $y^*=0$); $\theta_M{}^*(z^*)/\theta_M{}^*(0)$ is seen to decrease exponentially with increaing z^* in a range $0<\theta_M^*(z^*)/\theta_M^*(0)<1/2$, and is expressed in a form

$$\frac{\theta_M{}^*(z^*)}{\theta_M{}^*(0)} = exp(-pz^*) , \qquad \text{---------- (A3)}$$

where p is given by

$$p = \left(0.65\frac{v^*}{a^*} + 1\right)^{1/2} \qquad \text{---------- (A4)}$$

In general, the maximum cycling temperatures at depth z^* at $y^*=0$ for these three sources treated herein are givn by

$$\frac{\theta_M{}^*(z^*)}{\theta_M{}^*(0)} = exp\left[-\left(\frac{mv^*}{a^*} + 1\right)^n z^*\right], \qquad \text{---------- (A5)}$$

where m and n are given in Table A1. Thus the maximum cycling temperature at depth z is given in a dimensional form;

$$\theta_M(z) = \frac{AW}{16\sqrt{\pi}kb} \frac{\gamma}{\sqrt{\frac{vb}{2\alpha}a^*+\delta}} exp\left\{-\left(\frac{vb}{2\alpha}m+1\right)^n \cdot \frac{z}{ba^*}\right\} \qquad \text{---------- (A6)}$$

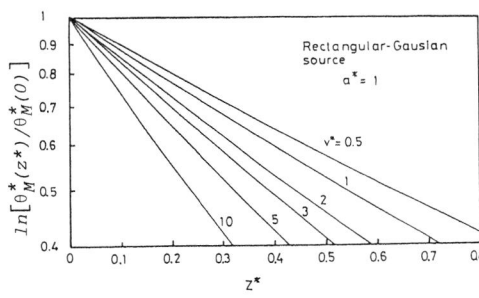

Fig. A2
$ln[\theta_M^*(z^*)/\theta_M^*(0)]$ vs. z^* for various v^* in rectangular-Gaussian source ($a^*=1$)

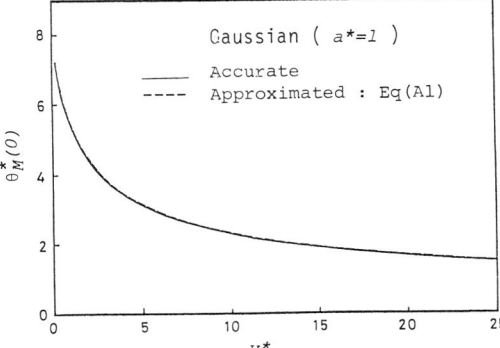

Fig. A3 $\theta_M^*(0)$ vs. v^* in Gaussian source ($a^*=1$)

Appendix 2 : Laser beam absorptivity

The authors measured calorimetrically the CO_2 laser beam absorptivity of steel specimens coated with zinc- and manganese-phosphate in wide ranges of travelling velocities and spot sizes, and found that A is given as a function of $\theta_M(0)$ in a range of $\theta_M(0) < 2000$ °C as follows;

$$A = A_0 \left(1 - \frac{\theta_M^2(0)}{\theta_0^2} \right) . \qquad \text{---------- (A7)}$$

Assuming that $\theta_{MB}(0)$ is the value of $\theta_M(0)$ of the solid with the absorptivity $A=1$,

$$\theta_M(0) = A \cdot \theta_{MB}(0) . \qquad \text{---------- (A8)}$$

Substituting Eq (A8) into Eq (A7),

$$\theta_{MB}^2(0) \cdot A^2 + \frac{\theta_0^2}{A0} A - \theta_0^2 = 0 \qquad \text{---------- (A9)}$$

The solution of this equation is

$$A = \frac{1}{2\theta_{MB}^2(0)} \left[\sqrt{\frac{\theta_0^4}{A0} + 4\theta_0^2 \theta_{MB}^2(0)} - \frac{\theta_0^2}{A0} \right] , \qquad \text{---------- (A10)}$$

where

$$\theta_{MB}(0) = \frac{W}{16\sqrt{\pi} kb} \frac{\gamma}{\sqrt{\frac{bv}{2k} a^{2*} + \delta}} \qquad \text{------------ (A11)}$$

Substituting (A10) into (A6), the maximum cycling temperature at depth z is obtained.

References

[1] K. I. Albutt and S. Garber; "Effect of Heating Rate on the Elevation of the Critical Temperatures of Low-carbon Mild Steel", JISI, Vol.204 (1966) 1217-1222.

[2] K. Miwa, N. Inokuchi & K. Yokota, "On the Transformation in Eutectoid Steel by Super-Rapid Heating", Journal of Japan Metal Society, Vol. 39, No.1 (1975) 24-28 (in Japanese).

[3] W. M. Steen & C. Courtney; "Surface Heat Treatment of En8 Steel Using a 2 kW Continuous-Wave CO_2 Laser", Metal Technology, No. 12 (1979) 456-462.

[4] F. C. Seaman and D. S. Gnanamuthu, "Using an Industrial laser to Surface Harden and Alloy", Metal Progress, Vol. 108, No. 3 (1975) 67-70, 72, 74.

[5] G. H. Harth, W. C. Leslie, V. G. Gregson & A. Sanders, "Laser Heat Treating of Steels", JOM, No. 4 (1976) 5-11.

[6] Y. Arata, H. Maruo & I. Miyamoto, "Application of Laser for Material Processing--Heat Flow in Laser Hardening--", International Institute of Welding, Doc. IV-241-78 & Doc. 212-436-78 (1978).

[7] S. Owaku et al, "Hand Book of Metal Heat Treatment Technique", Nikkan-Kogyo Shinbun, 1977 (in Japaneae)

[8] I. Hagiwara, S. Kanazawa & U. Kumada, "The Effects of Stress Induced by Preventing the Thermal Contraction on the Amount of Retained Austenite", Journal of Japan Metal Society, Vol. 23, No. 5 (1958) 299-302 (in Japanese)

[9] Y. S. Touloukian, "Thermophysical Properties of Matter", The TPRC Data Series IFI/ Plenum (1970).

Application of Plasma Beam

Hydrogen Content in Arc Atmosphere of Water Curtain Type Underwater Argon Arc Welding
Y. Arata, M. Hamasaki and J. Sakakibara
[Trans. JWRI **10** (1981), 19.]

Corrosion Behavior of Plasma-Sprayed Ceramic Coated Stainless Steel at High Humidity
Y. Arata, A. Ohmori, J. Morimoto, T. Kudoh and M. Kishida
[Proc. 10th Int. Thermal Spraying Conf. (1983), 197.]

The Present State of Ceramic Spraying in Japan
Y. Arata, A. Ohmori, J. Morimoto, K. Kishida and T. Kudoh
[7th Int. Conf. on Vacuum Metallurgy (1982).]

Hydrogen Content in Arc Atmosphere of Water Curtain Type Underwater Argon Arc Welding

Abstract

The hydrogen content in arc atmosphere in underwater welding was investigated by a direct measurement of the gas just under the arc or by indirect methods such as transfer phenomenon of droplets, generation of blowholes, diffusible hydrogen content and so on. The hydrogen content in arc atmosphere of water curtain type underwater MIG arc welding was estimated about 0.1%.

KEY WORDS: (Water curtain type underwater welding) (Hydrogen in arc at mosphere) (Underwater welding Blowhole) (Transfered droplets) (Diffusible by drogen)

1. Introduction

There are some reports on a diffusible hydrogen content in a "wet" underwater welding: manual metal arcs with covered stick electrodes indicate 25-60cc/100g[1], which are different from the types of coated materials and a usual CO_2 arc welding with a conventional nozzle indicates 20cc/100g[2].

In a water curtain type underwater MIG/CO_2 arc welding, it is thought that the surface of the plate near the arc is shielded perfectly and further the remained water is evaporated by the preheat effect of arc. These mean the partial pressure of hydrogen in arc atmosphere is low, but it is important to declare the partial pressure of hydrogen and the diffusible hydrogen content which affects on the strength of welds.

This report shows the hydrogen content in arc atmosphere by the direct measurement of gas just under the arc or by the indirect measurements such as transfer phenomenon of droplets, generation of blowholes, diffusible hydrogen content and so on.

2. Experimental procedure

2.1 Observation of preheat effect

It is difficult to observe directly the evaporation of water due to the small bubbles formed with the curtain water in an actual underwater welding. The observation was performed in surface, that is, the TIG arc was generated on the plate which had a 2 mm water film as shown in Fig. 2 and heat input was changed. The shield-ing gas was pure argon and its flow rate was 20 l/min. A nozzle was 15 mm outer diam. and 10 mm inner diam. and nozzle-base metal distance was 5 mm.

2.2 Hydrogen content in arc atmosphere

The gas just under the underwater TIG arc was collect-ed and its hydrogen content was measured. Fig. 1 shows a schematic drawing of gas collecting apparatus used in

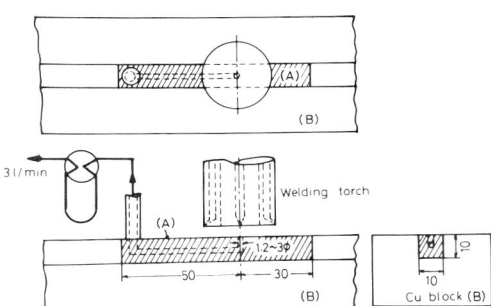

Fig. 1 Schematic drawing of gas sampling apparatus

this experiment, which consists of copper block(A) which has a gas collecting hole of 1.2-3.0 mm ϕ. A tungsten ele-ctrode was set just above this hole. This copper block (A) was moved in a groove of 350 mm long copper block(B) together with the torch. The gas in the arc atmosphere was sucked by a vacuum pump at a rate of 3 l/min through this hole and collected in a glass collecter in this pass. The hydrogen in this gas was analyzed with a gaschromatograph.

2.3 Critical hydrogen content on blowhole generation

The relation between blowhole generation and hydrogen content in argon as a shielding gas was investigated in surface with the MIG arc welding under spray arc condition. The dual shielding nozzle which had 65 mm ϕ O.D. was adopted and attention was paid to a laminar flow of gas with gas lens to prevent the air contamination. An arc voltage was selected to maintain the constant arc length. The results were arranged based on JIS Z 3104.

2.4 Numbers of transfered droplets

The MIG arc welding was performed using a motor generator as a welder. The numbers of transfered droplets in underwater welding were compared with those in surface welding in which the hydrogen was added to argon. The numbers of droplets were searched from the peaks of arc voltage.

2.5 Diffusible hydrogen

The diffusible hydrogen in underwater welding was compared with that in surface welding in which the hydrogen was added to argon. It was measured by the glycerine method (JIS Z 3113). The chemical compositions of the plate were shown in Table 1.

Table 1   Chemical compositions of base metal

C	Si	Mn	P	S
0.11	0.19	0.52	0.015	0.016

3. Results and discussion

3.1 Preheat effect of arc

In the water curtain type MIG/CO_2 arc welding, in spite of a "wet" welding method, both arc phenomena and bead shapes were similar to those in surface welding[3~5]. But sometimes the blowholes were generated in but welding especially in low heat input because it was

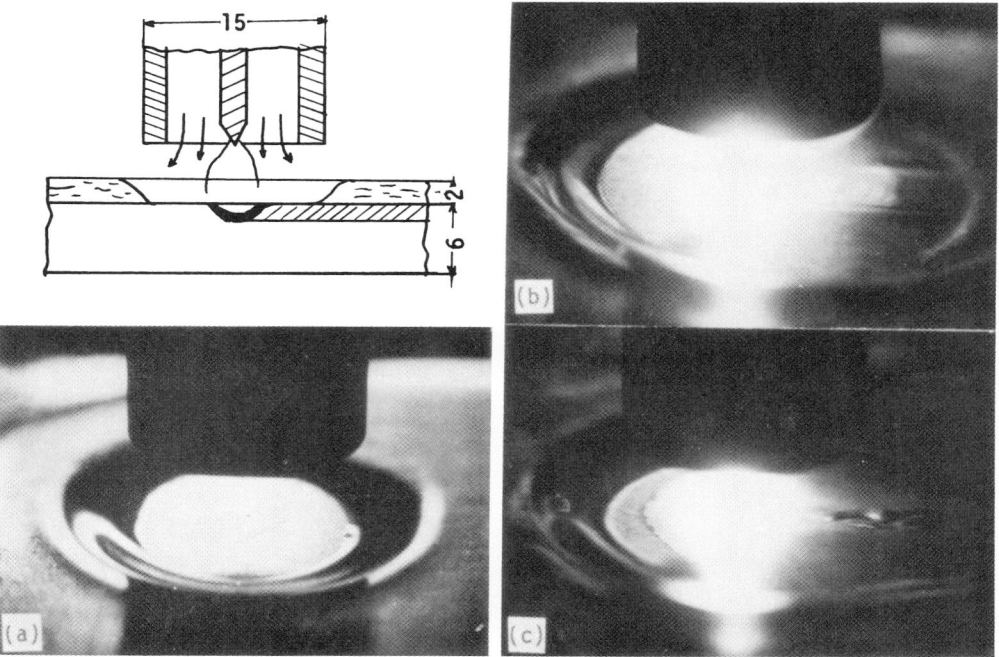

Fig. 2   Sham underwater TIG welding for observation of preheat effect of arc

583

difficult to remove the water in a root gap and joint phase between works and backing plate. In the same shielding condition, on the other hand, the blowholes were not generated in high heat input[6]. This may be based on the following consideration: even if the dry region was not obtained and the very thin water film was remained only by the shielding gas, the dry region could be obtained by the preheat effect of arc near the arc spot where the water in front of the arc was evaporated. This dry region may be larger in high heat input.

Fig. 2 (a) shows the condition that the torch is at a standstill and the shielding gas of 20 l/min is flowed out. The water is removed by the shielding gas and a circular region of 13 mm diam. becomes dry condition after the lapse of sufficient time. But when the torch is moved, the thin water film less than 0.5 mm is remained in this region because the fresh water is supplied from the moving direction. Even in this shielding condition, when the arc is generated, the water around the arc is evaporated and dried. Fig. 2 (b) and (c) show the observation results in case the welding current of 120A under welding speed of 25 and 100 cm/min respectively. The region from which the water film of 2 mm is removed becomes small with increasing of welding speed, moreover the very thin water film can be observed in (c). The water film keeps at a distance of 9 mm and 5 mm in front of the arc spot for (b) and (c) respectively, which means that the lower the heat input is, the nearer the water remains and becomes the source of the water vapour.

The evaporation of the water must be contributed to not only the rise of temperature by the heat conduction but also the increase of apparent gas flow rate and heat radiation from the arc column.

The generation of the vapour was notable on the surface of the formed bead which was cooled with the water, but the area was apart from the arc and almost all of the generated vapour was removed with the shielding gas flow to exterior of the arc atmosphere, therefore the arc phenomenon did not affected with this vapour.

3.2 Hydrogen content in arc atmosphere

In order to make sure of the decomposition of the water vapour to H_2 and O_2, the hydrogen content was measured when the arc was generated in the mixed shielding gas of argon and water vapour as shown in Fig. 3. A dew point of this gas was $-5°C$, which corresponded to hydrogen content of 0.4%. As the temperature of the arc increases with the current, the decomposition reaction, $2H_2O \rightarrow 2H_2 + O_2$, proceeds with increasing of the current. The decomposition rate was 50% in 60A and 65% in 180A.

Fig. 4 and Fig. 5 show the hydrogen content of gas

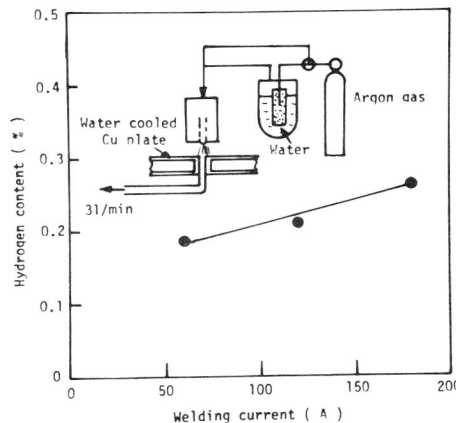

Fig. 3 Effect of current on decomposition of water vapour in argon gas

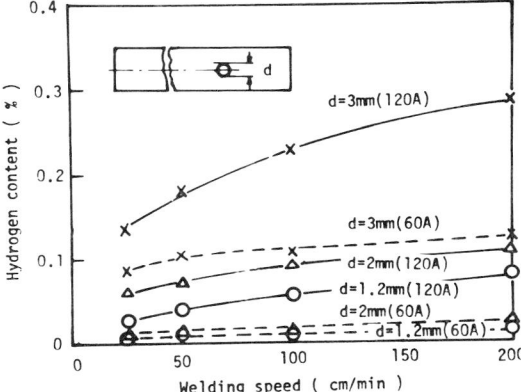

Fig. 4 Effects of current and diameter of gas sampling hole on hydrogen content in arc atmosphere

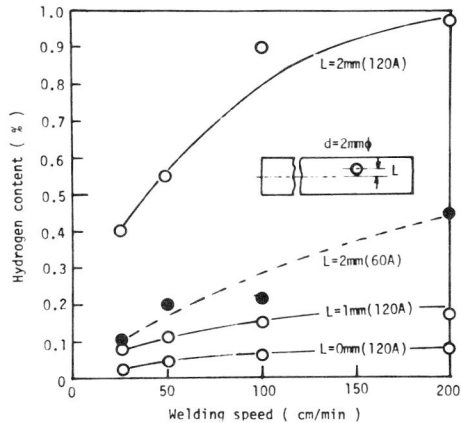

Fig. 5 Effect of position of gas sampling hole on hydrogen content in arc atmosphere

collected with the gas sampling apparatus shown in Fig. 1. Fig. 4 shows the effects of diameter of gas sampling hole and curren and Fig. 5 shows the effect of position of sampling hole and current. The copper block moves together with the torch in order to maintain the position of the electrode and sampling hole. The water from the proceeding direction of welding is therefore rather little but it comes from the side through the gap between the two copper blocks, which is different from the actual welding condition. The effect of welding speed was comparatively small due to the above mentioned reason. As the position of the water source did not change, the increase of the current resulted in increase of the evaporation of the water. The effect of the current was therefore larger than that of constant vapour content shown in Fig. 3. The most effective factor was the position of sampling hole and the size of it, that is to say, the distance between arc and water source. The shorter the distance is, the more the hydrogen generates.

The hydrogen content is only about 1% even in case of 120A and L = 2 mm in Fig. 5. The distance between the arc spot and the end of the water film is 5 mm even in the high welding speed of 100 cm/min as shown in Fig. 2. The hydrogen content in an actual welding is therefore supposed to be a condition of 120A, L = 0 mm and d = 1.2 mm in Fig. 5, that indicates 0.08% H_2.

3.3 Critical hydrogen content for generation of blowholes

The generation of blowholes in underwater welding must be caused mainly by the hydrogen gas[7]. Fig. 6 shows the results of X-ray inspections which show the relation between the amount of H_2 in shielding argon gas and blowholes, that is represented in Fig. 7 based on the JIS Z 3104.

The blowholes generated a little till 15% H_2 except for the starting point and the crater. (If the conventional small nozzle was used, the blowholes generated from 5% H_2,

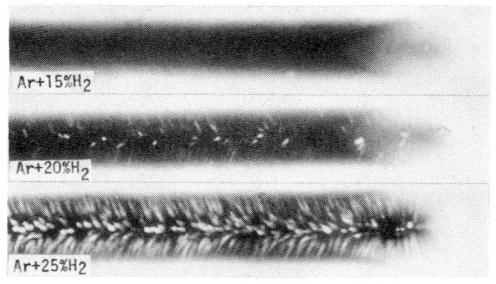

Fig. 6 X-ray inspection results

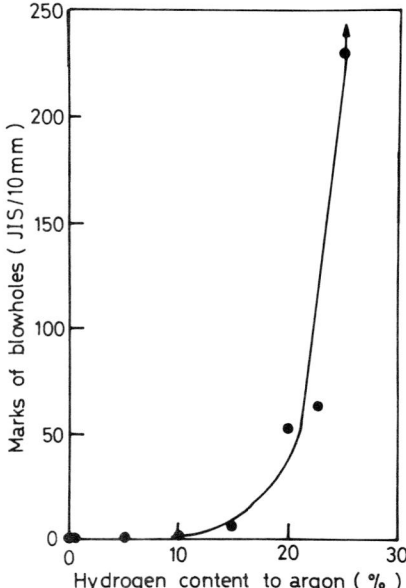

Fig. 7 Effect of hydrogen content in argon on blowholes

which was caused by the air contamination.) On the other hand wormholes coexisted at 20% H_2 and the large amount of wormholes existed at 25% H_2. The abrupt increase of the blowholes from 20% to 25% H_2 may be related to saturated solubility.

The solubility of the hydrogen to an iron in the equilibrium is shown as follows.[8]

$$S = \sqrt{P_{H_2}(1 - P_m)} \exp\left(-\frac{\Delta H}{2RT}\right) \quad (1)$$

where S   : equilibrium solubility of hydrogen

P_{H_2} : hydrogen partial pressure in atmosphere

P_m : metal vapour pressure in atmosphere

ΔH : heat of dissolution

R   : gas constant

T   : absolute temperature

The solubility S_f at a fusion temperature of the iron T_f is expressed as follows as the metal vapour pressure at T_f is considered to be zero.

$$S_f = \left[P_{H_2 f} \exp\left(-\frac{\Delta H}{RT_f}\right)\right]^{\frac{1}{2}} \quad (2)$$

At the temperature T_{max} corresponds to the maximum solubility of hydrogen, where the metal vapour pressure must be taken into consideration, then write S_{max} for the solubility at T_{max}, the next equation is given.

$$S_{max} = [\,P_{H_2 max}(1 - P_m) \exp(-\frac{\Delta H}{RT_{max}})\,]^{\frac{1}{2}} \quad (3)$$

In order not to generate the blowholes with a supersaturated hydrogen, the solubility of hydrogen at T_{max} must be less than that at the maximum solubility at T_f. So that the critical partial pressure of hydrogen $P_{H_2 cr}$ is given when S_f at $P_{H_2 f} = 1$atm and S_{max} are equall. Therefore putting $P_{H_2 f} = 1$atm, $P_{H_2 max} = P_{H_2 cr}$ and $S_f = S_{max}$ in eq. (2) and (3).

$$P_{H_2 cr} = \frac{1}{(1 - P_m)} \exp[\frac{\Delta H}{R}(\frac{1}{T_{max}} - \frac{1}{T_f})] \quad (4)$$

The metal vapour pressure at a high temperature and the temperature for maximum solubility of hydrogen under consideration of the metal vapour pressure are given by D.G.Howden et al, that is, $T_{max} = 2400°C$ and $P_m = 0.01$atm[9]. If $\Delta H = 15.6$kcal/mol[8], $R = 1,986$cal/mol°K and $T_f = 1,537°C$, then the critical hydrogen pressure is 0.24atm. This calculated value is agreed fairly with the result shown in Fig. 7 under consideration of false equilibrium caused by a rapid cooling rate in welding.

It was true that a cooling rate in the water curtain type underwater welding was larger than that in surface welding, but it was not affected largely until a solidification temperature because the rapid cooling began from 800-1000°C at which the curtain water struck the bead[10]. If the heat input is the same both in underwater and surface welding, the critical hydrogen content for generating the blowholes must not be much difference. In the water curtain type underwater welding, therefore, the blowholes based on the supersaturated hydrogen can not be generated.

3.4 Numbers of transfered droplets

Fig. 8 shows the relation between welding current and numbers of transfered droplets in underwater and surface welding. In a current range of globular transfer, the numbers of droplets in underwater is less than those in surface. But in a spray transfer condition, the difference is not recognized in both cases. The critical current for the spray transfer in underwater becomes a little higher than that of surface (If the critical current is defined as a diameter of the droplet is equal to that of wire (1.6 mmϕ only in this case), it is 265A in underwater and 260A in surface respectively).

Fig. 9 shows the relation between hydrogen content in shielding gas of argon and numbers of droplets which is

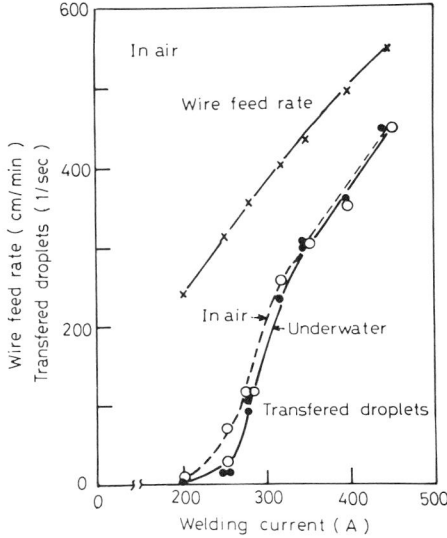

Fig. 8　Relation between welding current and transfered droplets in air and under water welding

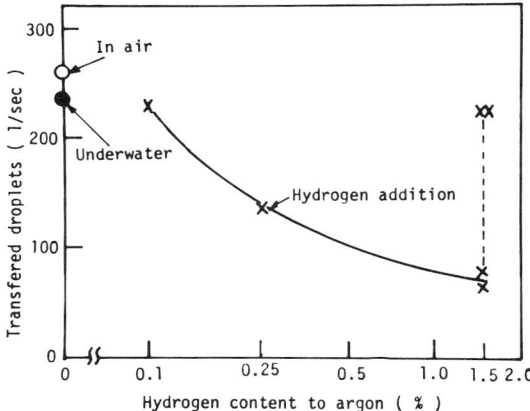

Fig. 9　Relation between hydrogen content in argon and transfered droplets

measured by a oscillogram in a constant welding condition of 29V-320A. The numbers of droplets decreased clearly in 0.25%H_2, and the spray or the globular transfer occurred alternately in 1.5%H_2.

The hydrogen content in the arc atmosphere in underwater MIG welding can be estimated about 0.1% by comparing the numbers of droplets in underwater and surface.

3.5 Diffusible hydrogen in underwater MIG welding

Fig. 10 shows the diffusible hydrogen in the MIG arc welding in surface in which the hydrogen is added to argon as a shielding gas. The abscissa axis is expressed as a

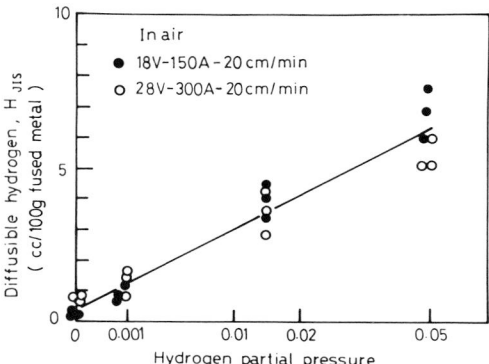

Fig. 10 Relation between hydrogen content in argon and diffusible hydrogen

square root of the hydrogen partial pressure, therefore the diffusible hydrogen is proportional to $\sqrt{P_{H_2}}$ and agrees with the Sieverts law. But even in case of no hydrogen addition, the diffusible hydrogen less than 1cc/100 g is measured, that is, 0.7cc/100 g for the spray arc and 0.3cc/100 g for the short circuiting arc. This difference based on the welding conditions may be caused by the amount of fused metal.

Taking no account of the difference, the relation between the diffusible hydrogen and the hydrogen partial pressure are described as follows.

$$H_{JIS} = 25.0 \sqrt{P_{H_2}} + 0.5 \text{ cc/100 g} \qquad (5)$$

On the other hand, Fig. 11 shows the diffusible hydrogen in underwater welding both in spray and short circuiting arc conditions. In both cases, the data were scattered

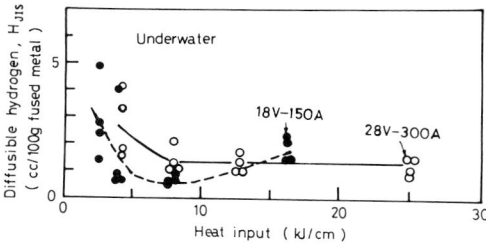

Fig. 11 Diffusible hydrogen in underwater welding

and the hydrogen content increased in the range of a high welding speed and a low heat input. But if the welding speed is too low, as shown in short circuiting arc condition, it had an inclination to increase, which was caused by the formation of a rectangular bead. Therefore it was less than 2cc/100 g in the practical welding speed of 20-40 cm/min. It was 1.1cc/100 g for the short circuiting arc and 1.3cc/100 g for spray arc condition as an average value. Appling these average values to eq. (5), the partial pressure of hydrogen in underwater welding atmosphere was calculated as P_{H_2} (short) = 0.00058 - 0.058% H_2 and P_{H_2} (spray) = 0.10% H_2 respectively. The results above mentioned are considered to be overestimated for spray arc and underestimated for short circuiting arc.

The hydrogen absorbed in a fused metal diffuses to a non-fused area or disperses to the exterior from the solidified metal[11]. The behaver of the hydrogen until the beginning of measuring the diffusible hydrogen is changed by the period of the arc generation (i.e. welding speed) and cooling rate. As the weld metal is cooled rapidly by the water just after the solidification in underwater welding, the hydrogen in the weld metal is apt to be larger than that in surface welding.

From these considerations above mentioned, the hydrogen content in arc atmosphere in underwater welding can be estimated less than 0.1% H_2.

4. Conclusion

Some experiments were performed related to hydrogen content in arc atmosphere. The results are as follows.

1) As a source of hydrogen in underwater welding, the thin water film on a base metal was considered and it was shown that the preheat effect by an arc acted effectivelly upon the evaporation of water in front of the arc.

2) The hydrogen content in arc atmosphere measured by a direct and an indirect method was estimated about 0.1% in a range of practical heat input.

3) The diffusible hydrogen in the water curtain type underwater MIG arc welding was less than 2cc/100 g (H_{JIS}) in practical heat input and this result was better than the other wet welding methods.

References

1) H. Ozaki, T. Naiman and K. Masubuchi, "A Study of Hydrogen Cracking in Underwater Steel Welds", Welding J., 56-8, P231s, 1977.

2) H.C. Cotton and D.B.J. Thomas, "Application of Underwater Welding to Offshore Structure", Underwater Const. Tech. Conf., Cardiff, April 1975.

3) Y. Arata, M. Hamasaki and J. Sakakibara, "Water Curtain Type Underwater MIG Arc Welding (The 1st Report)", J. of J.W.S., 46-9, P648, 1977.

4) Y. Arata, M. Hamasaki and J. Sakakibara, "Water Curtain Type Underwater MIG Arc Welding (The 2nd Report)", J. of J.W.S., 46-10, P278, 1977.

5) J. Sakakibara, M. Hamasaki and Y. Arata, "Water Curtain Type Underwater MIG Arc Welding at Water Depth of 200 m", J. of High Temp. Society, 6-6, P244, 1980.

6) J. Sakakibara, "Water Curtain Type Underwater CO_2 Arc Welding in Horizontal and Overhead Position", Report of the Govt. Industrial Research Inst., Shikoku, 8-1, P16, 1976.

7) I. Masumoto, A. Kondo, Y. Nakashima and K. Matsuda, "Study on the Underwater Welding (Report 2)", J. of J.W.S., 40-8, P748, 1971.

8) M. Mizuno, a private message.

9) D.G. Howden and D.R. Milner, "Hydrogen Absorption in Arc Welding", British Welding J., 10-6, P304, 1963.

10) M. Hamasaki and J. Sakakibara, "Underwater Welding of High Tensile Strength Steel", J. of J.W.S., 48-2, P115, 1979.

11) J. Tsuboi, S. Nakano and K. Sato, "The Behavior of Hydrogen in Arc Welding", J. of J.W.S., 42-3, P189, 1973.

Corrosion behavior of plasma-sprayed ceramic coated stainless steel at high humidity

1. Introduction

The environment for using metallic materials
is becoming stricter and severer these days.
Metal alone has a limit to its use, so the
surface of the metal is often coated with
ceramics and composite materials are used.
Spraying is one way of surface treatment.
When ceramics (oxides such as Al_2O_3, TiO_2,
ZrO_2 and carbides such as WC) are sprayed on
metal , the resistance against abrasive wear,
heat and corrosion increases, so that this
treatment is effectively used in many
industrial fields.
However, when oxide ceramic coating such as
Al_2O_3 is plasma-sprayed on stainless steel(SUS
304), and kept under humid conditions, the
Al_2O_3 sprayed surface forms brown spots and
some ceramics become swollen on the coated
surface. It is a big problem to clarify how
this occurs and to find a preventive measure.

In this study the corrosion behavior of
plasma-sprayed ceramic coating was observed
and the formation mechanism was studied.

2 Materials and Experimental Procedure

The oxide ceramic powders used for spraying in
this experiment are all marketed, and their
chemical compositions are shown in Table 1.
The particle size is varied from 5μ to 106μ .
The material used was SUS 304 steel
(60x60x1.2mm), and the plasma-spraying
instrument used was mainly the Metco 7M type.

Conditions: $Ar(150\sim80SCFH)$ and $H_2(15SCFH)$
gases, spraying voltage(60$\sim$70V), electric
current(450$\sim$500A), and fused Al_2O_3 was used
as blast material. The porosity of the
coatings was determined by the ferro-oxyl
test. Corrosion behavior of the ceramic
sprayed coating was clarify by the following
methods.

1) the accelerated atmospheric humidity test
2) the drop corrosion test
3) the electro-chemical measurment method

3 Experimental Results and Discussion

3.1 The Corrosion behavior in the accelerated atmospheric humidity test

In order to clarify the effect of each factor
in ceramic sprayed coating on corrosion
behavior under humid conditions, an
accelerated atmospheric humidity test was
conducted to promote corrosion in a corrosion
chamber under certain conditions. Running
water was sprayed on each ceramic sprayed coat
placed at a fixed position in the corrosion
chamber for 10 minutes once a day. Thus the
coating became equally wet with water and left
as it was. Then the formation of corrosion
(spots) on the surface was visually observed.
One of the results of the experiment is shown
in Table 2. The result of the SUS 304 steel
which was blasted only is also shown for
comparison. Corrosion was not observed on the
surface of the SUS 304 steel which was blasted
only, but brown spots formed on the surface of
the Al_2O_3 ceramic sprayed coating, as well as
on the ZrO_2 and the TiO_2 coatings. In this
experiment (30 days) many spraying conditions
were used, such as different thicknesses of
sprayed coating, various diameters of sprayed
particles and various corrosion atmospheres.
The results are summarized as follows;
1) Corrosion decreases as the coating becomes
   thicker, as shown in Table 2.
2) As shown in Table 2, corrosion did not
   occur easily when fine particles were used
   for coating at the same thickness.
3) Corrosion occurred on all the ceramics,
   though the amount of corrosion differed to
   some extent according to the kind of
   ceramics.

Table 1.
Chemical compositions of ceramic powders.

Material Number	Chemical Compositions (wt%)						
	Al_2O_3	ZrO_2	TiO_2	MgO	SiO_2	Fe_2O_3	CaO
A-1	98.95	--	--	--	0.27	0.06	0.39
A-2	94.85	--	2.69	0.11	1.36	0.47	0.25
A-3	88.70	--	11.03	--	--	0.05	0.22
A-4	64.23	--	35.22	0.21	--	--	trace
A-5	2.36	--	97.16	--	0.18	0.04	0.23
A-6	0.04	93.53	--	0.84	0.03	--	5.16

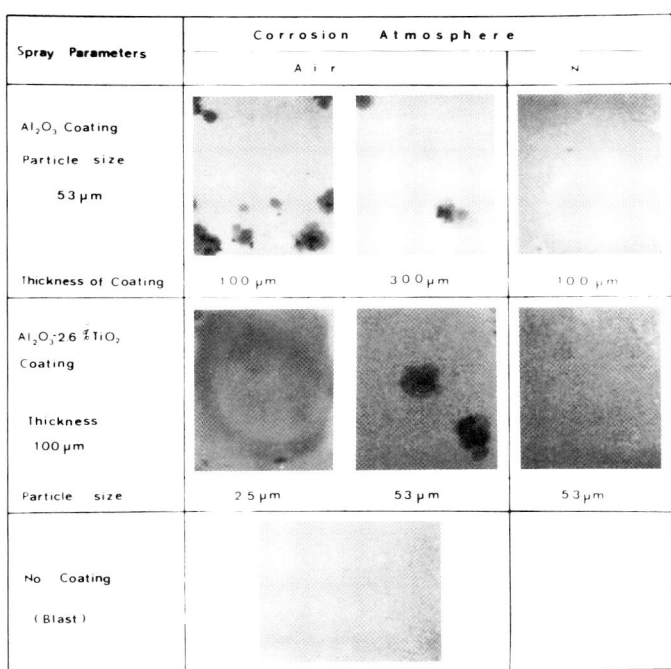

Table 2.
Corrosion results of the accelerated atmospheric humidity test.

4) Corrosion is influenced by O_2 in air. As shown in Table 2, corrosion is suppressed by N_2.
5) Corrosion is suppressed in a dry atmosphere.
6) Corrosion is suppressed with 80Ni-20Cr under coating, but the effect is not completely.

Next, to clarify these corrosion phenomena,

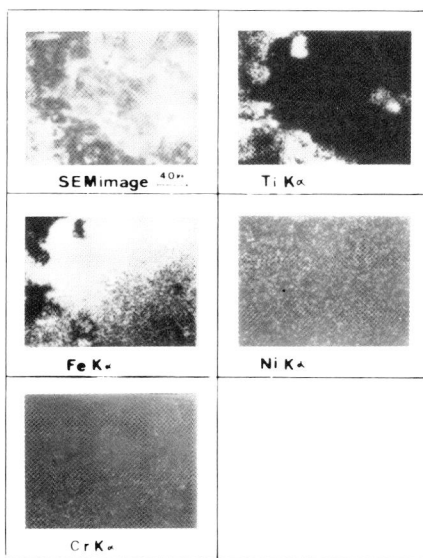

Figure 1. SEM micrograph and characteristic X-ray image results of coating surface after corrosion test.

the brown spots on each ceramic sprayed coat were analyzed using EPMA. The results for the TiO_2 coating are shown in Fig.1. It was proven that the spots were mainly iron compounds.

3.2 The influence of spraying factors on corrosion

It was observed that corrosion greatly differed according to the characteristics of the sprayed coating and the spraying conditions, therefore a study was conducted to know the influence of these factors. Table 3 shows in detail how spraying distance and thickness of the sprayed coating influenced the corrosion of Al_2O_3-2.3% TiO_2 in the accelerated atmospheric humidity test (10 days). Table 3, also shows the pores of the coatings made in the ferro-oxyl test.

Table 3. Effect of spray parameters on ferro-oxyl test and corrosion result.

Spray Distance (mm)	Thickness of Coating (µm)	Ferro-oxyl test	Corrosion Test in humid environment in air
80	200		
80	150		
140	150		

590

As shown in the table, when the thickness of the coatings was the same and the parameter of spraying distance was changed, corrosion occurred easily at farther distance and the number of pores penetrating to the material increased corresponding to the spraying distance. On the other hand, when the spraying distance was kept the same and the thickness of the coating was changed, the corrosion occurred easily for thinner coating and the number of pores increased as the coating became thinner, as shown in results of ferro-oxyl test. Fig.2 shows the formation of spots in the accelerated atmospheric humidity test with the corrosion test time when the diameter of the sprayed particles is changed and the thickness(100μm) is constant. It also shows the results of the ferro-oxyl test of both coatings(particles size, 25 and 106μm). It is seen that the corrosion is slow and suppressed when the diameter of the sprayed particles is small, therefore the number of pores in the coating is small.

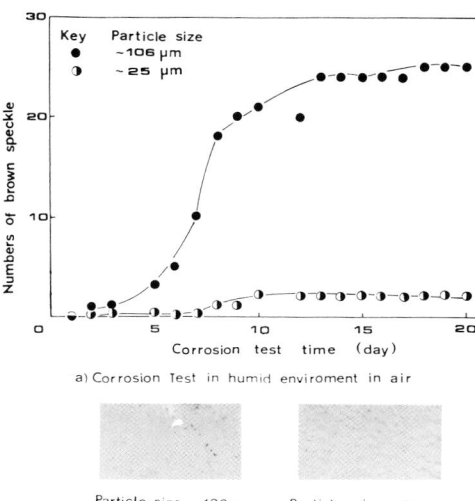

a) Corrosion Test in humid enviroment in air

Particle size ~106μm   Particle size ~25μm

b) Ferro-oxyl test

Figure 2. Corrosion test and ferro-oxyl test results for 100 μm Al$_2$O$_3$-2.650 % TiO$_2$ coatings (Particle size: ~106 μm and ~25 μm).

As stated above, when the influence of the spraying factors(spraying distance, diameter of sprayed particles, thickness of coating, kinds of ceramics) on corrosion in air under humid conditions was studied, it was proven that the number of pores that penetrate the coating changes with the change in the spraying factors, and that corrosion is influenced by the number of these pores. The more pores there are that reach the material through the sprayed coating, the more easily corrosion occurs, and it is assumed that the relation between the spraying factors and corrosion is related to the pores in the coating.

3.3 Corrosion mechanism of ceramic spray coating in air under humid conditions

3.3.1 Behavior of drop corrosion test

Based on the results of observation of the corrosion behavior of ceramic-sprayed coating

in an accelerated atmospheric humidity test as well as the correlation between spraying factors and the existence of pores in the coating, atmospheric corrosion, object of this report, is presumably caused as follows: generally the iron inside the stainless steel material is dissolved by the joint action of moisture(H$_2$O) and oxygen in the air entering through the pores in the ceramic-spray coating, reaching the stainless steel material and the iron compound thus formed emerging to the coat surface as rust (spots) through the pore.

In order to further clarify this corrosion phenomenon, the corrosion mechanism was studied by means of an accelerated testing method based on the drop corrosion test. The following test was carried out in order to study the degree of the influence of oxygen using the same testing method. A spray-coated test piece was placed in a dessicator; either oxygen or nitrogen gas was introduced after vacuum de-airing; under this ambient NaCl solution drops were placed on the coat surface and formation as well as growth of spots was visually observed. Fig.3 shows an example of the results thus obtained. Under oxygen ambient, spots were seen to form as soon as 3h later, but under nitrogen ambient, no spot was seen even after 72 h; thus suggesting the role played by oxygen in the case of this corrosion.

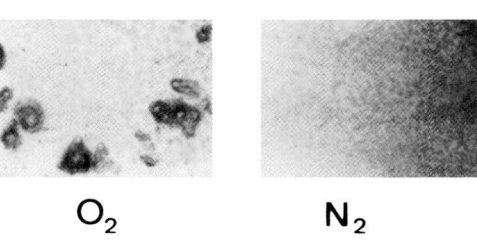

O$_2$ N$_2$

Figure 3. Results of drop corrosion test in O$_2$ or N$_2$ atmosphere.

The above suggests that corrosion produced under a humid atmospheric condition is differential aeration (crevice) forming an oxygen concentration cell produced by means of the pores in the ceramic coating. In other words, the surface(A) as shown in Fig.4 of the material at the pore in the ceramic-sprayed coating is exposed to the ambient air, consequently the oxygen density is high there, causing cathode reduction($H_2O + \frac{1}{2}O_2 + 2e^- \rightarrow 2OH^-$) and the formation of OH ion from water and oxygen. On the other hand, at part B where the ceramic has adhered to the stainless material, there is reduced oxygen density,an anode zone is formed, mainly producing dissolving reaction of iron within stainless steel($Fe \rightarrow Fe^{2+} + 2e^-$).
Then the Fe^2 thus formed reacts($Fe^{2+} + 2OH^- \rightarrow Fe(OH)_2$) on contact with OH$^-$. Through the reaction of

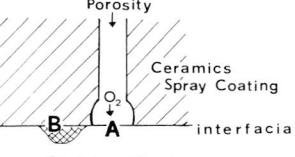

Figure 4. Schematic diagram of corrosion mechanism.

591

Fe(OH)$_2$ thus formed on contact with oxygen introduced through the pores, Fe$_2$O$_3$·nH$_2$O (red rust) is formed which emerges as brown spots through the pores on the ceramic-coated surface.

This corrosion also seems to be accelerated by active agents including Cl ion.

The above suggests that this corrosion is produced according to the amount of pores existing in the coating regardless of the kind of ceramics and that the more pores there are, the more specks appear on the coat surface.

Fe is promptly dissolved and OH$^-$ ion is quickly formed, but they require some time before appearing on the surface as red rust, thus suggesting some latent period until spots are produced.

3.3.2 Electro-chemical approach

An electro-chemical analysis was made in order to clarify, by measuring the electro-chemical polarization curve, the difference between blasted material and a material with ceramic-sprayed coating. Fig.5 shows an

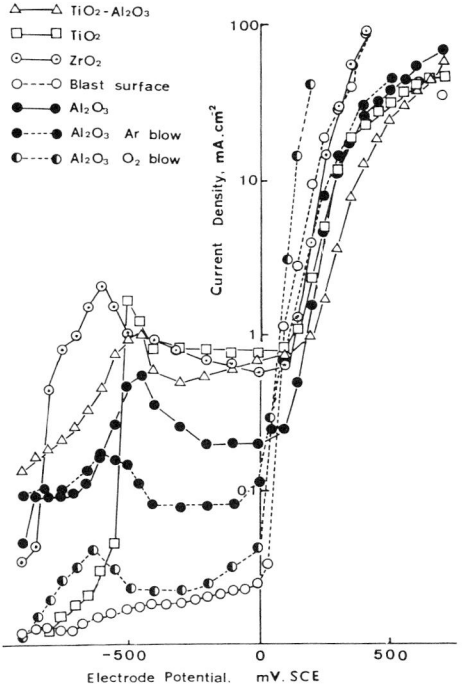

Figure 5. Electro-chemical polarization curves of ceramic coating.

example of the results thus obtained for ceramic-sprayed coating, blasted stainless steel(SUS 304) material, O$_2$ blowing, etc. In the drawing, the ceramic-sprayed coating presents a passive state area in the range between-500mv and 0mv, which, as compared with the passive state area of the blasted stainless steel surface, shows current density as high as 0.1 - 1mA/cm^2. This testifies to the fact that a ceramic-sprayed coating barely passes to a passive state area, besides the passive state layer is promptly destroyed, facilitating the progress of corrosion. By blowing oxygen into the solution, the passive state current inside the ceramic-sprayed coating is greatly reduced, thus suggesting that the increase of oxygen in the system accelerates the passage to the passive state during polarization on the coat-material surface.

The above suggests that compared to stainless steel material, the ceramic-sprayed coat barely passes to a passive state, thus easily accelerating corrosion, probably due to the formation of an oxygen concetration cell (differential aeration) in the surface mentioned above.

4. Conclusion

The following results were obtained through the observation of corrosion behaviour during the accelerated atmospheric humidity test and drop corrosion test concerning the corrosion of ceramic-sprayed coats under humid atomospheric ambient and the electro-chemical approach: This corrosion is believed to be induced by the formation of red rust appearing on the coating surface through pores after Fe ion dissolves out of the stainless steel material due to an oxygen concentration cell (differential aeration cell); this oxygen concentration cell is formed by the joint action of moisture and oxygen contained in the atmosphere reaching the material surface via the pores existing in the ceramic-sprayed coating.

5.Acknowledgement

The author would like to take this opportunity to thank the spray group staff at this institute as well as Prof. Yamaguchi of the Kinki University for their appropriate suggestions and earnest opinions on the subject. The author also owes thanks to Mr. Mochizuki for his active part in carrying out the present test.

The Present State of Ceramic Spraying in Japan

INTRODUCTION

In line with the development of industries today, the environment is getting stricter and severer for using metallic materials. As a consequence, the applications for ceramics(alumina, zirconia, etc.), which are highly heat resistant, abrasive wear resistant and hot oxidation resistant and gas corrosion resistant, have remarkably expanded. They are particularly looked to as structural materials. However,due to the poor machinability of ceramics per se, the shortcomings of ceramic material (weakness against heat shock, etc.) and the delay in the development of new metal materials, the development of composite materials, using more than two materials as a composite, has become an urgent matter.

Under such circumstances, attention has been paid to surface treatment technologies to bring entirely different material functions to the surface of the material by merely providing treatment on the surface. Flame spraying, one surface treatment technique, has been increasingly applied in various industrial fields owing to its features (universality for materials, efficiency etc.) over other surface treatment method. In particular, the flame spraying of ceramics has recently entered the limelight as a composite technique for effectively utilizing the above-mentioned characteristics of ceramics.

Hereinafter under the title of "The Present State of Ceramic Spraying in Japan", we discuss the following points:

   i) Ceramic Spraying Technique and the Process of its Development (including Applications)

  ii) Problems of Ceramic Spraying

 iii) Future Prospects of Ceramic Spraying

CERAMIC SPRAYING TECHNIQUE AND THE PROCESS OF ITS DEVELOPMENT

Ceramic spraying technique

Flame spraying is one surface treatment technique to form a film by spraying fine molten liquid particles at high speeds onto the surface of the material. Then, the heat is applied to a solid in the form of powder, wire or rod.

In ceramic spraying, various methods have been contrived, such as the use of plasma, to effectively melt the ceramics which are materials with higher melting points compared to metals and alloys. Table 1 shows the physical properties of sprayed alumina coating as an example of a sprayed ceramic coating.

Table 1 Physical properties of sprayed Al_2O_3-TiO_2 coating

Kinds of coating	$90\%Al_2O_3$ $-10\%TiO_2$	$60\%Al_2O_3$ $-40\%TiO_2$	$95\%Al_2O_3$ $-2.6\%TiO_2$	$94\%Al_2O_3$ $-2.5TiO_2$
Particle size range(μm)	$-53 \sim +15$	$-45 \sim +5$	$-25 \sim +5$	$-53 \sim +15$
Hardness (Cross section) DpH_{300}	850	850	$950 \sim 1000$	760
Density (g/cm^3)	3.5	3.5	3.4	3.3
Surface roughness(μm) (as sprayed)	---	$2.5 \sim 4.5Ra$	$2.5 \sim 3.8Ra$	7.5Ra
Dielectric constant (Volt/mil)	400	---	350	90
Melting point (°C)	~ 2000	1840	2040	2010
Coefficient of thermal expansion ($\times 10^6$/°C)	5.3 $(30 \sim 930$ C)	---	7.4 $(21 \sim 1480$°C)	7.4 $(21 \sim 1480$°C)

In gas flame spraying utilizing the combustion of acetylene-oxygen and others, it is hard to completely melt the ceramics with powdered materials for spraying, and the rod system (using a ceramic rod) is employed to melt the ceramics effectively. However, making the rod is itself not free from problems, and the kinds of ceramics available are limited (Al_2O_3, Cr_2O_3, ZrO_2, $ZrSiO_4$, 60% Al_2O_3-40% TiO_2, etc).

To effectively melt the ceramics, the use of a high energy density heat source is essential. Although flame spraying using a plasma jet is currently attracting attension, that system is greatly affected by the parameters of the spraying conditions, etc., so it is necessary that the parameters on the performance of coating film is made   clear. However, the development of a coating performance testing method remains unsolved as a future problem.

In addition the detonation system of acetylene-oxygen is utilized, but it is accompanied by problems of material size, universality, etc. in relation to the equipment.

The process of development in ceramic spraying(application examples)

Ceramic spraying was developed in line with the rapid progress in space engineering which requires materials which are highly resistant to heat, hot gas corrosion, abrasive wear, etc. to withstand severe conditions, and its applications have expanded to general industries utilizing its features.

In Japan ceramic spraying has been practiced since the late 1950's. Typical applications of ceramic spraying are as follows:

1) Solid fuel jet nozzle for rocket:

Flame spraying of ZrO_2 and Al_2O_3 has been conducted in the rod system, and applied to rockets ranging from the 'kappa' developed by the Institute of Space and Astronautical Science to the present 'Mu' rocket. Plasma spraying is also employed cuncurrently. Fig.1 shows Al_2O_3-ZrO_2 duplex coating of fuel jet nozzle for rockets.

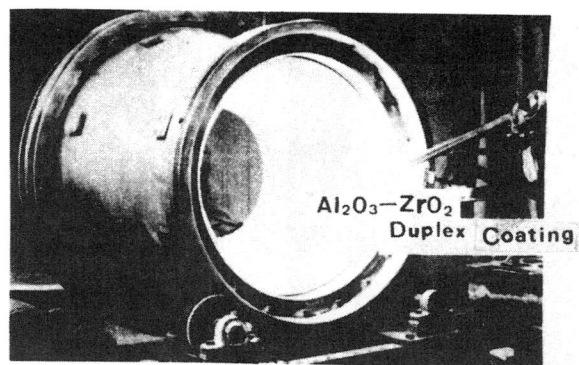

Fig.1 Al_2O_3-ZrO_2 duplex coating of solid fuel jet
nozzle for rockets

2) Insulation of generating channel for MHD generation

Flame spraying of Al_2O_3 was carried out in the rod system, taken up as a large scale energy-saving development project(Moonlight Project, 1966) by MITI and adopted in the Mark 7 generator. Currently development is in progress of the Mark 8 generator. Future developments are expected to be made in the direction of plasma spraying. Fig.2 shows Al_2O_3 coating of insulation of generating channel for MHD generation.

3) Jet engine

Flame spraying of ceramics was carried out in the parts of jet engine which require heat resistance, the resistance of hot gas corrosion and abrasion wear, etc. Fig.3 shows ceramic graded-coating of combustion chamber.

4) Gas turbine for generation

Plasma jet spraying of ZrO_2 is adopted in the combustion cylinder. In addition Cr_3C_2 spraying is also employed.

5) Tanker's propeller shaft and shaft sealing
The gas spraying of Cr_2O_3 in the rod system is conducted, but practically there are problems of how to treat seal holes, etc., so whether it will be adopted or not is presently under consideration.

6) Ocean excavating ship's propeller shaft sealing
Gas spraying of Cr_2O_3 in the rod system was adopted in 1976 in the ALSKAN STAR, a semisubmergible self-propelled ocean excavating ship.

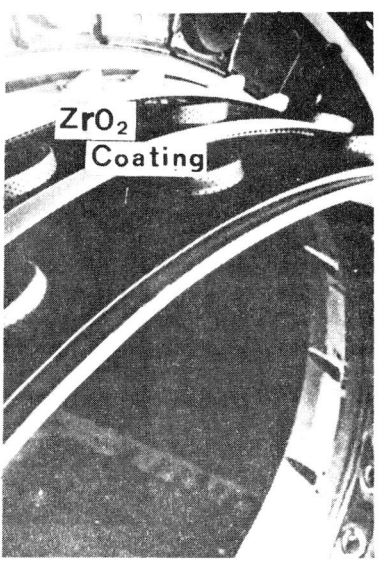

Fig.2 Al_2O_3 coating of insulation of genatrating channel for MHD generation

Fig.3 Ceramic graded-coating of combustion chamber of jet engine

7) Longitudinal expansion roll in film manufacture
Plasma jet spraying and gas spraying in the rod system of Al_2O_3 is adopted in the production line of ultra-thin polyester film(4-38μm) for video and audio condensers, as shown in Fig.4. The Al_2O_3 film made by flame spraying has solved the problem encountered with chrome galvanized rolls which is that, when the surface temperature rises to over 80°C the friction coefficient increases to produce speckled adhesive portions on the film and make it defective.

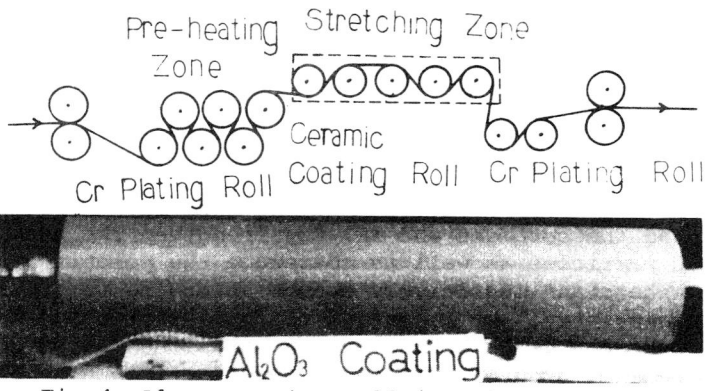

Fig.4 Al$_2$O$_3$ coating roll in polyester
film manufacture

8) Yarn guide of ultra-high speed extension false twister

Plasma spraying of Al$_2$O$_3$ is employed in the areas where wear is conspicuously caused by yarns at the high speed of 15m/sec.

As described above, ceramic spraying has been adopted in various fields to utilize the characteristics of ceramics, such as heat resistance, wear resistance, corrosion resistance, electrical insulation at high temperatures, etc.

PROBLEMS OF CERAMIC SPRAYING

In order to fully utilize ceramic spraying under severe conditions in any strict environment and make the best use of the characteristics of ceramics, such as heat resistance, hot gas corrosion resistance, wear resistance, etc., many problems have still to be overcome. For example, the quality of the sprayed ceramic coating should be improved (airtightness, etc.) and the adhesion between substrate-ceramics should be strengthened.

To solve these problems using existing methods and equipment for flame spraying, the quality of the sprayed film has to be improved by establishing the optimum spraying conditions and how to reproduce the film performance.

For that purpose, the establishment of a quantitative testing method, in which the performance of the film per se (porosity, wear resistance, adhesion between ceramics and materials, etc.) is sensitively reflected, is the prerequisite. Once the above problems are solved, then the optimum spraying conditions for maximum film performance can be expected to be set easily.

Relation between performance of sprayed ceramic coating and method of testing sprayed coating.

When ceramic spraying is applied in a field of industry requiring high reliability such as aircraft manufacture, a quality control system is established on the basis of practical simulation tests to meet the working environment. However, it is nearly impossible to conduct such tests in general industries with a wide range of applications and variety of environments. That is why a simple and practical method of testing sprayed coating which sensitively reflects the coating performance is required. The establishment of such a testing method will facilitate the checking of coating performance, enables higher-grade coatings to be developed and higher coating quality to be controlled.

As urgent matters in ceramic spraying technology at present, the following two points are considered: i) to reinforce the adhesion between the substrate and the ceramics and ii) to strengthen the bond between sprayed ceramic particles as well as to reduce the porosity.

These two phenomena sensitively change according to the spraying method, conditions, etc., so a coating performance testing method corresponding to the change is necessary, but existing method are inadequate. Table 2 shows some methods of testing sprayed coating currently in practice. These testing methods have respective characteristics.

To test the adhesive bond strength A) B) C) D), etc. are used, and to test the performance of coating per se E), F), G), etc. are used.

Table 2. Methods of testing sprayed coating

A)	Adhesion bond strength test using adhesive.
B)	Scratch hardness test.
C)	Indentation test.
D)	Hardness test.
E)	Microscopic observation.
F)	Wear abration test.
G)	Others

A) Conditions

   (1) Blast nozzle diameter; 5.2mmϕ
   (2) Blast distance; 100mm
   (3) Blast air pressure; 5kg/cm^2
      (induced suction type)
   (4) Blast test specimen size;
      50x60x3.2t mm
      Base material; Mild steel
      (SS-41)
   (5) Blast material; Molten
      alumina(-30~+60mesh)
   (6) Blast time; 10sec

B) Method

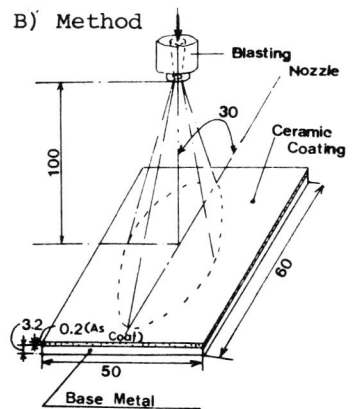

Fig.5 Conditions and method of blast erosion test

Other testing items include heat resistance, corrosion resistance, etc. Althouth E) is recommended as the method to best represent the coating performance, errors of sample surface finish, individual reading, etc. are likely to appear. To cope with such shortcomings a specially trained operator is required. The method (A) is greatly affected by the adhesive agent and is hard for displaying changes in the coating performance. Care should be taken in applying the indentation test to porous samples, so it seems hard to sensitively reflect the coating performance, which delicately varies according to the spraying conditions, in the test results.

As one attempt, we propose what we call the 'blast erosion test' and show a part of its results. In this method, Al_2O_3 ceramic particles controlled to a certain particle distribution are blown onto the sprayed coating surface by compressed air and the amount of reduction of the coating before and after the blast is measured. Fig.5 shows our testing conditions, testing methods and typical drawings respectively. Fig.6 shows an example of the results concerning a 95% Al_2O_3-2.6% TiO_2 coating according this testing method. Also, optical microscopic observation results of coating films are shown.

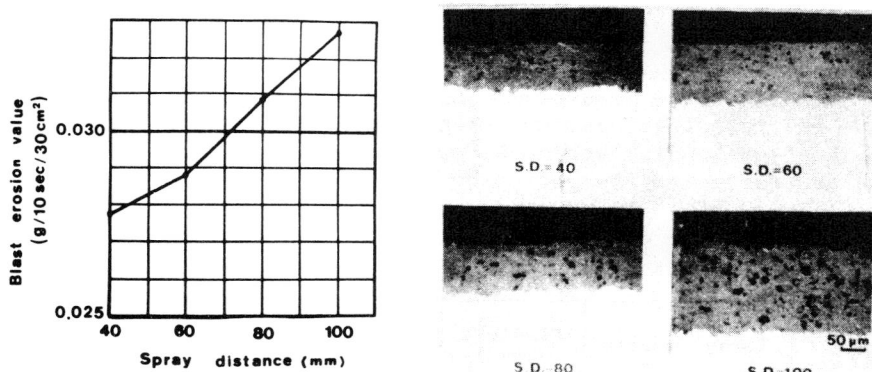

Fig.6 Results of blast erosion test and micro-structures of 95% Al_2O_3-2.6% TiO_2 coating($-25 \sim +5\mu m$)

From this figure, such condition change corresponds to the blast erosion test result of each coating, when the parameter of the spraying distance alone is varied as ceramic spraying condition. This also corresponds to the porosity change of the coating cross section. Fig.7 shows the results of the blast erosion test and hardness test depending on Al_2O_3 plasma spraying parameters. In Fig.7, we can see that changes in the spraying distance, the secondary gas (H_2) flow and the cross point distance of the air jet for cooling are sensitively reflected in both sets of test results. Also, the result of the blast erosion test well corresponds to the wear resistance of a sprayed ceramic coating.

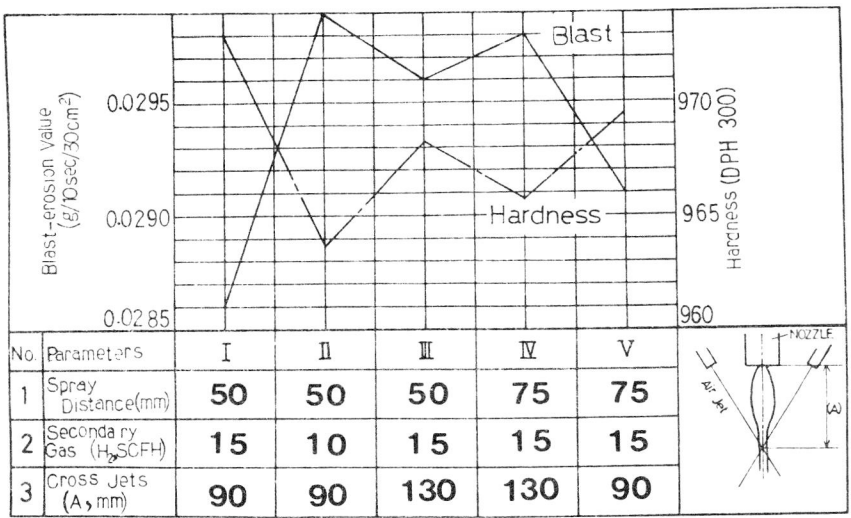

Fig.7 Results of blast erosion test and hardness test
depending on Al$_2$O$_3$ plasma spraying parameters

Table 3 shows the results of the blast erosion test on ceramics of different kinds. These results are shown as the amount of decrease in volume
with the bulk specific gravity taken into account. The contents of these
results (how the bonding strength, pore condition, etc. of the ceramic
coating are reflected) are yet to be studied in the future.

Table 3 Results of blast erosion test of various ceramic
coatings

Material	Spray system	Particle size range (μm)	Bulk specific gravity	Values of blast erosion test (cm^3/10sec/30cm^2)
Cr$_2$O$_3$	Rod spray		4.6	0.01130
	Plasma spray	$-53 \sim +15$	4.9	0.03654
Al$_2$O$_3$ (11%TiO$_2$)	Plasma spray	$-53 \sim +15$	3.5	0.01005
Al$_2$O$_3$ (2.5%TiO$_2$)	Plasma spray	$-53 \sim +15$	3.3	0.01910
Al$_2$O$_3$ (40%TiO$_2$)	Plasma spray	$-45 \sim + 5$	3.5	0.01260
Al$_2$O$_3$ (11%TiO$_2$)	Plasma spray	$-30 \sim + 5$	3.6	0.00664

Pores of sprayed ceramic coating and corrosion

When ceramics such as Al_2O_3, ZrO_2, etc. are sprayed onto the surface of metal materials (e.g. SUS 304 steel) highly resistant to corrosion in the atmosphere, as shown in Fig.8, in the humid environment of the atmosphere, brown speckles are noticed, and sometimes blisters are produced which cause pore corrosion of the stainless steel substance.

Fig.8 Brown speckles produced on Alumina
coating sprayed on SUS 304 steel

Oxygen and moisture in the air have a great influence on corrosion. The mechanism of corrosion depends on the dissolution of Fe in stainless steel due to the formation of an oxygen concentration cell on the interface. Through pores penetrating to the stainless steel substrate, and is accelerated by the presence of activators such as Cl^- ion. This corrosion also is accelerated and reproduced in the drop corrosion test of NaCl aqueous solution in the air, and the dependence of corrosivity on the permeating speed of NaCl aqueous solution into the ceramic coating was noticed.

However, regardless of the type of ceramics, the method of spraying, etc., corrosion is caused by pores(Permeation)found in the sprayed ceramic coating and the pores are unremovable by existing spraying methods. This is considered to become a major problem in the future in relation to hot gas corrosion and sea water corrosion in deep sea.

As countermeasures against such corrosion, some methods have been considered, such as eliminating the pores, impregnating the pores with resin or the like to seal them, and so on. Pore sealing is presently practiced, but even impregnation under reduced pressure enables the resin to permeate only about 0.1mm at best.

Applying resin with a brush can not get the resin to permeate more than about 0.02mm. So the development of a new impregnation method and the manufacture of new resin is urgently required.

Although efforts are being made to eliminate pores and develop a non-permeable coating, the existing sprayed ceramic coating is substantially accompanied by 0.5-1.2% pores according to the ferro-oxyl test.

Against the corrosion on a sprayed ceramic coating in the above-mentioned humid atmospheric environment, the corrosion are hard to be produced in the 13% TiO_2-Al_2O_3 coating, so we presume a partially non-permeable coating is formed. This phenomenon is most marked if the coating temperature right after spraying is kept at 200-250°C. Our object is expected to be attained in the future by utilizing composite materials for spraying, such as dissimilar ceramics, ceramic-metal.

PROSPECTS FOR CERAMIC SPRAYING

In the preceding section, we discussed some problems of ceramic spraying. As the working environment becomes harder and severer in the future, the characteristics of ceramics are expected to be more and more effectively utilized. We presume that there are many problems to be solved in the future; for instance; 1) the application for a gas turbine at higher temperatures; 2) the matter of the reaction of ceramics with coal slag as well as with gas at high temperatures in MHD power generation; 3) sea water corrosion in oceanic environments, particularly caused by the use of sea water pressure in place of oil pressure for submergible working machines, such as a deep-sea bed mineral resources explorer, sea bed cable layer, etc., 4) the matter of corrosion inside a living body in the field of bioceramics, and so on.

Regarding these matters, how to cope with the pores in a sprayed ceramic coating will remain a great obstacle. Some of the problems considered detrimental to the future use of ceramic spraying are listed below.
1) Pores in a sprayed ceramic coating
2) A high quality inspection method for sprayed ceramic coating has not yet been established.
3) Research has been scarecely done on changes in the physical properties of ceramics after spraying   in relation to nitrides and carbides
4) Insufficient adhesion of ceramics with the substrate.
5) Others
By solving the above problems in the future, the application of ceramic spraying is expected to be expanded to all possible fields including the above-mentioned areas.
For that purpose the sprayed ceramic coating must be formed in an almost non-permeable and poreless state(like sintered ceramics), and positive techniques should be applied, such as the effective use of heat by adopting a ultra high density heat source, high speed spraying of particles, etc.
In addition efforts should be made to improve the performance of the sprayed coating with existing spraying equipment and methods, e.g. to develop spraying materials using composites, to carry out better post-treatment of the sprayed coating, to study the optimum spraying conditions, etc.
Thus, the establishment of a coating performance testing method to sensitively reflect any change in the spraying condition is urgently desired.

List of Other Important Papers Concerning Ultra High Energy Density Beams

(note: This list doesn't include the papers written in Japanese.)

A. Plasma Beam

1. Studies of Plasma in Heavy Current Linear Pulse Discharges
   K. Nishiguchi, Y. Arata and M. Okada; Tech. Rept. of the Osaka Univ. 10-403 (1960) 423.

2. The Behavior of Plasma in Heavy Current Linear Pulse Discharge
   K. Nishiguchi, Y. Arata and M. Okada; Tech. Rept. of the Osaka Univ. 10-418 (1960) 593.

3. Dynamic Behaviors of Plasma in a Linear Pulse Discharge
   T. Ariyasu, Y. Arata and M. Okada; Tech. Rept. of the Osaka Univ. 11-449 (1961) 53.

4. Interaction of a Hot Gas on a Solid Wall
   Y. Arata and K. Inoue; Tech. Rept. of the Osaka Univ. 17-751 (1967) 23.

5. Automatic Control of Plasma Arc Welding
   Y. Arata and K. Inoue; IIW Doc. IV-54-71 (1971), XIIk-7-71 (1971).

6. New Perfect Automatic Control System for Arc Welding
   Y. Arata and K. Inoue; IIW Doc. XIIk-6-71 (1971).

7. Magnetic Control of Plasma Arc Welding
   Y. Arata and H. Maruo; Trans. of JWRI 1-1 (1972) 1.

8. Magnetic Control of Arc Plasma and Its Application for Welding (2nd Report)
   Y. Arata and H. Maruo; IIW Doc. IV-85-72 (1972).

9. Present Status of Research and Development concerning with Automatic Control of Welding in Japan
   Y. Arata; IIW Doc. IV-92-72 (1972).

10. Some Fundamental Characteristics of Microwave Plasma Beam as a Heat Source (1)
    Y. Arata, S. Miyake and S. Takeuchi; Trans. of JWRI 1-1 (1972) 115.

11. Automatic Control of Arc Welding by Monitoring Molten Pool
    Y. Arata and K. Inoue; Trans. of JWRI 1-1 (1972) 99.

12. Some Properties of Magnetically Controlled Plasma Arc
    Y. Arata, H. Maruo and K. Yasuda; Trans. of JWRI 2-1 (1973) 21.

13. Automatic Control of Arc Welding (Report II)
    Y. Arata and K. Inoue; Trans. of JWRI 2-1 (1973) 87, IIW Doc. XIIk-41-73 (1973).

14. High Power Microwave Plasma Beam as a Heat Source (Report 1)
    Y. Arata, S. Miyaka and S. Takeuchi; IIW Doc. VI-119-73 (1973), Trans. of JWRI 2-1 (1973) 27.

15. Automatic Control of Arc Welding (Report III)
    Y. Arata and K. Inoue; Trans. of JWRI 2-2 (1973) 240, IIW Doc. XIIk-50-74 (1974).

16. High Power Microwave Plasma Beam as a Heat Source (Report II)
    Y. Arata, S. Miyake, S. Takeuchi and A. Kobayashi; Trans. of JWRI 3-1 (1974) 21.

17. Automatisierung beim Lichtbogenschweißen — Beobachtung für Kalte Überlappungen —
    Y. Arata, K. Inoue, G. Kawasaki and Y. Horio; Trans. of JWRI 3-2 (1974) 229.

18. Experimental Investigation of Coaxial Microwave Plasmatron in Nitrogen Gas
    S. Miyake, S. Takeuchi and Y. Arata; J. Appl. Phys. Japan 13 (1974) 296.

19. High Power Microwave Plasma Beam as a Heat Source (Report III)
    Y. Arata, S. Miyake, A. Kobayashi and S. Takeuchi; Trans of JWRI 3-2 (1974) 137.

20. Plasma Welding of Centrifugal Cast Steel Pipe (HK-40) (Report I)
    Y. Arata, S. Murakami, H. Nishihara, S. Sone and T. Mihara; Trans of JWRI 4-1 (1975) 61.

21. High Power Microwave Plasma Beam as a Heat Source (Report IV)
    Y. Arata, S. Miyake, T. Innami and Y. Yoshioka; Trans. of JWRI 4-2 (1975) 105.
22. Automatic Control of Arc Welding (Report IV)
    Y. Arata and K. Inoue; Trans. of JWRI 4-2 (1975) 205, IIW Doc. XIIk-64-76 (1976).
23. Automatic Control of Arc Welding
    Y. Arata, K. Inoue, M. Morita, G. Kawasaki, Y. Horio and K. Numata; Proc. of the 2nd Inter. Symp. of the JWS No. 3-(6) (1976).
24. Automatic Control of Arc Welding (Report V)
    Y. Arata, K. Inoue, M. Morita and G. Kawasaki; Trans. of JWRI 5-1 (1976) 77, IIW Doc. XIIk-65-76 (1976).
25. Research of Stationary High Power Microwave Plasma at Atmospheric Pressure
    Y. Arata, S. Miyake, A. Kobayashi and S. Takeuchi; J. Phys. Soc. Japan 40 (1976) 1456.
26. Automatic Control of Arc Welding (Report VI)
    Y. Arata, K. Inoue, Y. Shibata, M. Tamaoki and H. Akashi; Trans. of JWRI 6-1 (1977) 7.
27. Pulsed High Current Heating of a Microwave Plasma at Atmospheric Pressure
    Y. Arata, S. Miyake, M. Ushio and Y. Yoshioka; J. Phys. Soc. Japan 44 (1978) 1483.
28. Automatic Control of Horizontal Narrow Gap Welding (Report 1)
    Y. Arata, K. Inoue, M. Tamaoki and H. Akashi; Trans. of JWRI 8-1 (1979) 1.
29. Investigation on Welding Arc Sound (Report I)
    Y. Arata, K. Inoue, M. Futamata and T. Toh; Trans. of JWRI 8-1 (1979) 25, IIW Doc. SG 212-451-79 (1979).
30. Investigation on Welding Arc Sound (Report II)
    Y. Arata, K. Inoue, M. Futamata and T. Toh; Trans. of JWRI 8-2 (1979) 193.
31. Stationary High Temperature Plasma in a Vortex Gas Tunnel at Atmospheric Pressure
    A. Kobayashi and Y. Arata; Proc. Inter. Conf. on Plasma Phys. 1 (1980) 183.
32. Automatic Control of Horizontal Narrow Gap Welding (Report II)
    K. Inoue, Y. Shibata, H. Tamaoki, H. Akashi and Y. Arata; Trans. of JWRI 9-1 (1980) 31.
33. Transient Arc Diagnosis with OMA
    Y. Arata, S. Miyake, Y. Yoshioka and H. Matsuoka; Trans. of JWRI 9-1 (1980) 47.
34. Investigation on Welding Arc Sound (Report III)
    Y. Arata, K. Inoue, M. Futamata and T. Toh; Trans. of JWRI 9-2 (1980) 167.
35. Development of Welding Robots and Their Practical Application
    T. Nakamura, H. Sakurai, E. Maeda, M. Imakita, S. Maruyama and Y. Arata; Proc. Inter. Conf. on Welding Research in the 1980's (1980) 245.
36. Automatic Control of Horizontal Narrow Gap Welding
    M. Tamaoki, H. Akashi, Y. Shibata, K. Inoue and Y. Arata; Proc. Inter. Conf. on Welding Research in the 1980's (1980) 239.
37. Automatic Control of Horizontal Narrow Gap Welding (Report III)
    K. Inoue, Y. Shibata, M. Tamaoki, H. Akashi and Y. Arata; Trans. of JWRI 9-2 (1980) 151.
38. Two-Dimensional Multichannel Spectroscopy of a Pulsed Arc
    Y. Arata, S. Miyake, H. Matsuoka and H. Kishimoto; Trans. of JWRI 10-1 (1981) 33.
39. Investigation on Welding Arc Sound (Report IV)
    Y. Arata, K. Inoue, M. Futamata and T. Toh; Trans. of JWRI 10-1 (1981) 39.
40. Forschung in Bezug auf das Schweißlichtbogengeräusch (Bericht 5)
    Y. Arata, K. Inoue, M. Futamata and T. Toh; Trans. of JWRI 11-1 (1982) 31.
41. Researches on the Improvement of Sprayed Coatings
    A. Ohmori and Y. Arata; Proc. of 3rd Asian-Pacific Corrosion Control Conf. (1983) 487.

42. Corrosion behavior of Plasma-sprayed ceramic coated stainless steel at high humidity
    Y. Arata, A. Ohmori, J. Morimoto, T. Kudoh and M. Kishida; Proc. of the 10th Inter. Thermal Spraying Conf. (1983) 197.

43. Multichannel Measurement of Soft X-ray with MCP and OMA System
    Y. Arata, S. Miyake, H. Kishimoto and N. Abe; Trans. of JWRI 12-1 (1983) 51.

44. Thermal Plasma Diagnostics Using Tunable Dye Laser (Report I)
    Y. Arata, S. Miyake and H. Matsuoka; Trans. of JWRI 12-1 (1983) 43.

45. Measurement of an obliquely incident soft X-ray with a microchannel plate
    Y. Arata, S. Miyake, H. Kishimoto and N. Abe; Review of Scientific Instruments 55-6 (1984) 991.

46. Thermal Plasma Diagnostics Using Tunable Dye Laser (Report II)
    Y. Arata, S. Miyake and H. Matsuoka; Trans. of JWRI 13-1 (1984) 13.

47. ECR Plasma in a High Power Millimeter-Wave Beam (Report 1)
    Y. Arata, S. Miyake, N. Abe, H. Kishimoto, Y. Agawa and Y. Kawai; Trans. of JWRI 13-2 (1984) 181.

48. Measurements of Stark broadening of He (II) 4686 Å
    Y. Arata, S. Miyake and H. Matsuoka; J. Quant. Spectrosc. Radiat. Transfer 32-4 (1984) 343.

49. Thermal Plasma Diagnostics Using Tunable Dye Laser (Report III)
    Y. Arata, S. Miyake and H. Matsuoka; Trans. of JWRI 14-1 (1985) 1.

B. Electron Beam

1. The Transtron Accelerator S-1 (Report 1)
   M. Okada and Y. Arata; Tech. Rept. of the Osaka Univ. 13-567 (1963) 289.

2. Nonvacuum Electron Beam (1)
   Y. Arata and M. Tomie; Tech. Rept. of the Osaka Univ. 17-777 (1967) 303.

3. A New Method for Ultra-High Power Electron-beam Welds
   Y. Arata; Tech. Rept. of the Osaka Univ. 17-936 (1967).

4. Some Metallugical Investigation on Electron-beam Welds
   F. Matsuda, T. Hashimoto and Y. Arata; IIW Doc. IV-22-70, Trans. of JWS 1 (1970) 176.

5. Electron Beam Welding Process by Introduction of Insert Metal
   Y. Arata, K. Terai, Y. Nagai, I. Futami, S. Shimizu and T. Aota; IIW Doc. IV-89-72 (1972).

6. The Present Stetus of Pervasion of Electron Beam in Japan
   Y. Arata; IIW Doc. IV-91-72 (1972).

7. Some Properties of 30kW Class Electron Beam for Welding
   Y. Arata, M. Tomie and Y. Kato; Trans. of JWRI 2-1 (1973) 7, IIW Doc. IV-111-73 (1973).

8. Weld Distortion of Disk Shaped Joint by Electron Beam Welding
   Y. Arata, F. Matsuda and N. Yokoshima; Trans. of JWRI 2-2 (1973) 174, IIW Doc. IV-115-73 (1973).

9. Additional Note on the Report Quench Hardening and Cracking in Electron Beam Weld Metal of Carbon and Low Alloy Hardenable Steels
   Y. Arata, F. Matsuda and K. Nakata; Trans. of JWRI 2-1 (1973) 123.

10. Some Dynamic Aspects of Weld Molten Metal in Electron Beam Welding
    Y. Arata, F. Matsuda and T. Murakami; Trans. of JWRI 2-2 (1973) 152, IIW Doc. IV-149-74 (1974).

11. Merkmale Des Elektronenstrahlsschweißverfahrens MIT Wechselstromsablenkung
    Y. Arata, K. Terai and S. Matsuda; Trans. of JWRI 3-2 (1974) 233.

12. Study on Characteristics of Weld Defect and its Prevention in Electron Beam Welding (Report II)
    Y. Arata, K. Terai and S. Matsuda; Trans. of JWRI 3-1 (1974) 69, IIW Doc. IV-147-74 (1974).

13. Energy Density in Cavity during Electron Beam Welding
    Y. Arata, H. Inagaki, T. Hashimoto and H. Irie; IIW Doc. IV-148-74 (1974).

14. Insert-type Electron Beam Welding Technology (Report 2)
    Y. Arata, K. Terai, H. Nagai, I. Futami, S. Shimizu and K. Satoh; IIW Doc. IV-146-74 (1974), Trans. of JWS 4-1 (1973) 223.

15. Study on Characteristics of Weld Defect and its Prevention in Electron Beam Welding (Report III)
    Y. Arata, K. Terai and F. Matsuda; Trans. of JWRI 3-2 (1974) 207.

16. 100kW Klasse-Elektronenstrahlen-Schweißtechnologie (Berichit II)
    Y. Arata, M. Tomie and Y. Kato; Trans. of JWRI 4-1 (1975) 1.

17. Energy Density in Cavity during Electron Beam Welding
    Y. Arata, M. Inagaki, Y. Hashimoto and H. Irie; Trans. of JWRI 4-1 (1975) 7.

18. Mechanical Properties on Electron Beam Welds of Constructional High Tension Steels (Report II)
    Y. Arata, F. Matsuda, Y. Shibata, S. Hozumi, Y. Ono and S. Fujihira, Trans. of JWRI 4-1 (1975) 65.

19. Characteristics of Insert-type Electron Beam Welding Technology
    Y. Arata, K. Terai, I. Futami, H. Nagai, S. Shimizu, T. Aota and K. Sato; Proc. 2nd Inter. Symp. of JWS (1975) 87.

20. Electron Beam Welding of Carbon Steel and Titanium Sheets Using Ag Insert Metal
    Y. Arata, F. Matsuda and S. Harada; Trans. of JWRI 4-2 (1975) 171,
    Y. Arata and H. Kanayama; Proc. 2nd Inter. Symp. of JWS (1975).

21. Study on Characteristics of Weld Defect and its Prevention in Electron Beam Welding (Report IV)
    Y. Arata, K. Terai and F. Matsuda; Trans. of JWRI 4-2 (1975) 189.

22. Effect of A.C. Deflected Beam on Characteristics of Weld Defects in Electron Beam Welding
    Y. Arata, K. Terai, K. Kita, S. Matsuda and T. Nakamura; Proc. 2nd Inter. Symp. of JWS (1975) 39.

23. Mechanical Properties on Electron Beam Welds of Constructional High Tension Steels (Report IV)
    Y. Arata, F. Matsuda, Y. Shibata, Y. Ono, M. Tamaoki and S. Fujihira; Trans. of JWRI 5-1 (1976) 27, IIW Doc. IV-195-76 (1976).

24. Fundamental Features of 100kW Class Electron Beam Welding Technology
    Y. Arata and M. Tomie; 7th Inter. Conf. of Electron and Ion Beam Science and Tech. (at Washington D.C., U.S.A.) (1976).

25. Mechanical Properties on Electron Beam Welds of Constructional High Tension Steels (Report V)
    Y. Arata, F. Matsuda, Y. Shibata, Y. Ono, M. Tamaoki and S. Fujihira; Trans. of JWRI 5-2 (1976) 199.

26. Elektronenstrahlschweißen von Magnesiumgußlegierung AZ91C
    Y. Arata, M. Ohsumi and Y. Hayakawa; Trans. of JWRI 5-2 (1976) 211.

27. Dynamic Behavior of Electron Beam Welding
    Y. Arata, E. Abe and M. Fujisawa; IIW Doc. IV-220-77 (1977).

28. Untersuchungen über die Elektronenstrahlschweißbarkeit des hochfesten Stahls für Konstruction

   Y. Arata, F. Matsuda, Y. Shibata, Y. Ono, M. Tamaoki and S. Fujihira; Trans. of JWRI 6-1 (1977) 63.

29. Fundamental Studies on Electron Beam Welding of Heat-resistant Superalloys for Nuclear Plants (Report II)

   Y. Arata, K. Terai, H. Nagai, S. Shimizu and T. Aota; Trans. of JWRI 6-1 (1977) 69.

30. Fundamentale Studien zum Elektronenstrahlschweißen von Hitzebeständigen Legierungen für Kernkraftanlagen (Bericht 3)

   Y. Arata, K. Terai, H. Nagai, S. Shimizu and T. Aota; Trans. of JWRI 6-2 (1977) 235.

31. Fundamental Studies on Electron Beam Welding of Heat-resistant Superalloys for Nulear Plants (Report 4)

   Y. Arata, K. Terai, H. Nagai, S. Shimizu, T. Aota and Y. Ikemoto; Trans. of JWRI 7-1 (1978) 41.

32. Tandem Electron Beam Welding — Prevention of internal defects —

   Y. Arata and E. Nabegata; 2nd Inter. Colloquium for Electron Beam Welding and Melting (1978).

33. Untersuchung zum Metallschmeltzprozeß im Elektronenstrahlschweißen

   Y. Arata, M. Inagaki, T. Hashimoto and H. Irie; Trans. of JWRI 7-1 (1978) 143.

34. Fundamental Studies of Electron Beam Welding of Heat-resistant Superalloys for Nuclear Plants (Report 5)

   Y. Arata, K. Terai, H. Nagai, S. Shimizu, T. Aota, K. Satoh and Y. Ikemoto; Trans. of JWRI 7-2 (1978) 221.

35. Tandem Electron Beam Welding (Report II)

   Y. Arata, E. Nabegata and N. Iwamoto; Trans. of JWRI 7-2 (1978) 233.

36. Fundamental Studies on Electron Beam Welding of Heat-resistant Superalloys for Nuclear Plants (Report VI)

   Y. Arata, K. Terai, H. Nagai, S. Shimizu, T. Aota and Y. Murakami; Trans. of JWRI 8-1 (1979) 33.

37. Preliminary Evaluation of Root Defects Elimination Methods in Partial Penetration EB Welding

   K.P. Friedel and Y. Arata; Proc. of Inter. Conf. on Welding Research in the 1980's (1980) 7.

38. On Electron Beam Welding of Heavy Section Steel Plates

   M. Nakanishi, J. Furusawa, S. Yasunaga, M. Tomie and Y. Arata; Proc. of Inter. Conf. on Welding Research in the 1980's (1980) 85.

39. Fundamental Research on Horizontal Electron Beam Welding

   Y. Arata and M. Tomie; Proc. of Inter. Conf. on Welding Research in the 1980's (1980) P1.

40. Fundamental Research on Horizontal Electron Beam Welding

   Y. Arata; Keynoto of the 60th anniversary of Harbin Institute of Technology, China (1980), IIW Doc. IV-308-81 (1981).

41. Transmission X-ray Observation of Beam Hole during Tandem Electron Beam Welding

   Y. Arata, N. Abe, K.P. Friedel, S. Yamamoto and E. Abe; Proc. of Inter. Conf. on Welding Research in the 1980's (1980) P2.

42. Study on Ultra High Power Heat Sources of Electron Beam and Application for Welding

   Y. Arata and M. Tomie; IIW Doc. IV-309-81 (1981).

43. Electron Beam Weldability of Heavy Section Steel Plates

   M. Nakanishi, J. Furusawa, S. Yasunaga, M. Tomie and Y. Arata; IIW Doc. IX-1237-82 (1982).

44. Tandem Electron Beam Welding (Report IV)
   Y. Arata, N. Abe and E. Abe; Trans. of JWRI 11-1 (1982) 1.

45. Entwicklung des Elektronenstrahlschweißens im Ortlichen Vakuum
   Y. Arata, S. Sato, T. Shimoyama and G. Takano; Trans. of JWRI 11-1 (1982) 25.

46. Tandem Electron Beam Welding (Report V)
   Y. Arata, N. Abe and F. Wang; Trans. of JWRI 11-2 (1982) 1.

47. Some Problems of the Weldability of High Strength Thick Plate Steels Subjected of Electron Beam Welding
   S. Piwowar, Y. Arata, H. Nakagawa and M. Tomie; Trans. of JWRI 11-2 (1982) 29.

48. Study on Fundamental Tandem Electron Beam Welding
   Y. Arata and N. Abe; Keynoto of the 30th anniversary of Division of Welding Engineering, Harbin Institute of Technology China (1982).

49. Dynamic Observation of Beam Hole during Electron Beam Welding in Carbon Steel
   Y. Arata, N. Abe, F. Wang, M. Tomie and E. Abe; Trans. of JWRI 12-1 (1983) 1.

50. Dynamic Observation of Beam Hole during Electron Beam Welding
   Y. Arata, N. Abe, F. Wang, M. Tomie and E. Abe; IIW Doc. IV-338-83 (1983).

51. Electron Beam Welding of Titanium and Ti-6Aℓ-4V Thick Plates
   A. Kohyama, Y. Arata, M. Tomie and N. Igata; The Third Topical Meeting on Fusion Reactor Materials (1983), Albuquerque, New Mexico, U.S.A. to be published in Journal of Nuclear Materials.

52. Electron Beam Welding of Heavy Section Steel Plates
   N. Nakanishi, J. Furusawa, S. Yasunaga, M. Tomie and Y. Arata; Trans of the Iron and Steel Institute of Japan, 23 (1983) 71.

53. Study on Electron Beam Welding of Dissimilar Materials for Nuclear Plant (Report 1)
   Y. Arata, S. Shimizu and T. Murakami; Trans. of JWRI 12-2 (1983) 183.

54. Mechanical Properties of Electron Beam Welds of Heavy Section Steel Plates
   Y. Arata, M. Tomie, M. Nakanishi and J. Furusawa; Trans. of JWRI 12-2 (1983) 263.

55. Beam Hole Behaviour in Carbon Steel during Electron Beam Welding
   Y. Arata, H. Wang, N. Abe, M. Tomie and E. Abe; Inter. Conf. on Quality and Reliability in Welding (1984) A-22.

56. Fundamental Research on Electron Beam Welding for Centrifugal Casting Steel Pipe
   Y. Arata, M. Tomie, H. L. Han, H. Nishihara and T. Mihara; The Inter. Conf. on Quality and Reliability in Welding (1984) A-23.

57. Wady spoin wykonywanych wiazka elektronów
   Y. Arata, K. Friedel and A. Halas; Przeglad Spawalnictwa 5-6 (1984) 15.

58. Weldability of 18%Ni Steel (HT210) by Ultra High Voltage Electron Beam
   Y. Arata, M. Tomie and S. Katayama; Trans. of JWRI 14-1 (1985) 97.

59. Electron Beam Welding of Tianium and Ti-6Aℓ-4V Thick Plates
   A. Kohyama, Y. Arata, M. Tomie and N. Igata; J. of Nuclear Materials 122 (1984) 772.

C. Laser Beam

1. Generation and Application of CW High Power CO_2 Laser
   Y. Arata and I. Miyamoto; Tech. Rept. of the Osaka Univ. 17-775 (1967) 285.

2. Gain Characteristics of CO_2 Laser Amplifier
   Y. Arata and I. Miyamoto; Tech. Rept. of the Osaka Univ. 19-886 (1969) 371.

3. Some Fundamental Properties of High Power CW Laser Beam as a Heat Source
   Y. Arata, I. Miyamoto and M. Kubota; IIW Doc. IV-4-69 (1969).

4. Analysis of Temperature Distribution by Non-Uniform Surface Heat Source and Its Appri-cation to Laser Heating
   Y. Arata, I. Miyamoto and T. Yamada; Tech. Rept. of the Osaka Univ. 20-936 (1970) 391.
5. Metal Heating by Laser Beam
   Y. Arata and I. Miyamoto; IIW Doc. IV-51-71 (1971).
6. Processing Mechanism of High Energy Density Beam (Report 1)
   Y. Arata and I. Miyamoto; Trans. of JWRI 2-2 (1973) 148.
7. Development of Light Beam Welding Process
   A. Arata, T. Oku, Y. Matsumoto and S. Yoshizumi; IIW Doc. IV-150-74 (1974).
8. Some Fundamental Properties of High Power CO_2 Laser Beam as a Heat Source
   Y. Arata and I. Miyamoto; Trans. of JWRI 3-1 (1974) 1.
9. Development of Light Beam Welding Process
   Y. Arata, T. Oku, Y. Matsumoto and S. Yoshizumi; Trans. of JWRI 4-1 (1975) 15, Proc. 2nd Inter. Symp. of JWS.
10. Dynamic Behavior of Laser Welding and Cutting
    Y. Arata, H. Maruo, I. Miyamoto and S. Takeuchi; IIW Doc. IV-194-76 (1976).
11. Laser Welding
    Y. Arata and I. Miyamoto; Technocrat 11-5 (1978) 33.
12. Experimental Study of Cutting Different Materials with a 1.5kW CO_2 Laser
    V.S. Kovalenko, Y. Arata, H. Maruo and İ. Miyamoto; Trans of JWRI 7-2 (1978) 249.
13. The Approach to Understand Hardening Mechanism at Laser Beam Irradiation of Material
    V.S. Kovalenko, Y. Arata and S. Nenno; J. of High Temp. Society 4 (1978) 280.
14. Characteristics of High Power CO_2 Laser Welding
    Y. Arata, K. Inoue, H. Maruo and I. Miyamoto; Inter. Beam Tech. Conf. in Dusseldorf (1980) 181.
15. К Вопросу механизма упрочнения материала при воэдействии непрерывного лазер-ного излучения
    В.С. Коваленко, К. Энами, Е. Арата, С. Ненно; ЭЛЕКТРОННАЯ ОБРАБОТКА МАТЕРИА-ЛОВ 1 (1985) 35.
16. Investigation on Parameters in CO_2 Laser Surface Hardening by Utilizing Data Base System
    K. Inoue, S. Matsumura and Y. Arata; Proc. Inter. Conf. on Welding Research in the 1980's (1980) P3.
17. Improvement of Cut Quality in Laser-Gas-Cutting Stainless Steels
    Y. Arata, H. Maruo, I. Miyamoto and S. Takeuchi; Proc. Inter. Conf. on Welding Research in the 1980's (1980) 95.
18. Laser Welding of 304 Stainless Steels
    Y. Arata, J. Miyata, S. Sasano and K. Inoue; Proc. Inter. Conf. on Welding Research in the 1980's (1980) 95.
19. Fundamental Properties of 5kW CO_2 Laser Welding
    Y. Arata, K. Inoue, H. Maruo and I. Miyamoto; IIW Doc. IV-306-81 (1981).
20. Welding Characteristics of 5kW Class CO_2 Laser
    K. Inoue and Y. Arata; The First Inter. Laser Proc. Conf. (1981).
21. Quality in Laser-Gas-Cutting Stainless Steel and its Improvement
    Y. Arata, H. Maruo, I. Miyamoto and S. Takeuchi; Trans. of JWRI 10-2 (1981) 129.
22. Welding Characteristics of 5kW Class CO_2 Laser
    K. Inoue, J. Miyata and Y. Arata; Trans. of JWRI 10-2 (1981) 141.
23. Condition Setting Method Utilizing Data Base System in CO_2 Laser Surface Hardening (Report II)
    K. Inoue, S. Matsumura and Y. Arata; Trans. of JWRI 11-1 (1982) 37.

24. Condition Setting Method Utilizing Data Base system in CO_2 Laser Surface Hardening (Report III)

    K. Inoue, S. Matsumura, and Y. Arata; Trans. of JWRI 12-1 (1983) 35.

25. Measurement of Intensity Profile of Focussed Laser Beam Using Acrylic Plastic

    I. Miyamoto, H. Maruo and Y. Arata, 3 éme CIS FFEL (1983).

26. Laser Nitriding of Titanium and Its Alloys

    S. Katayama, A. Matsunawa, A. Morimoto, S. Ishimoto and Y. Arata; 3rd Inter. Colloquium on Welding and Melting by Electrons and Laser Beam (1983) 219.

27. Surface Hardening of Titanium by Laser Nitriding

    S. Katayama, A. Matsunawa, A. Morimoto, S. Ishimoto and Y. Arata; ICALEO '83 (1983) 127.

28. Condition Setting Method Utilizing Data Base System in CO_2 Laser Surface Hardening

    Y. Arata, K. Inoue and S. Matsumura; ICALEO '83 (1983) 100.

29. Measurement Method of Laser Beam Energy density distributor

    H. Maruo, I. Miyamoto and Y. Arata; 3rd Inter. Colloquium on Welding and Melting by Electrons and Laser Beam (1983).

30. Effect of Assist Gas on Bead Formation in High Power Laser Welding

    Y. Arata, T. Oda and R. Nishio; Trans. of JWRI 12-2 (1983) 161.

31. Intensity Profile Measurement of Focused CO_2 Laser Beam Using PMMA

    I. Miyamoto, H. Maruo and Y. Arata; ICALEO '84 (1984) 313.

32. Fundamental Research on Laser Welding of Structural Steel

    Y. Arata and T. Oda; Trans. of JWRI 13-2 (1984) 227.

33. Fundamental Phenomena in Laser Welding

    Y. Arata, N. Abe and T. Oda; Gas Flow and Chemical Lasers (1984) 61.

34. Fundamental Phenomena in High Power CO_2 Laser Welding

    Y. Arata, N. Abe and T. Oda; Trans. of JWRI 14-1 (1985) 5.

35. Fundamental Phenomena during Vacuum Laser Welding

    Y. Arata, N. Abe and T. Oda; IIW Doc. IV-396-85 (1985).

D. Miscellaneous matters

1. High Energy Density Beam Welding

    Y. Arata; Invited Rept. Proc. of The Fourth Inter. Conf. on Vacuum Metallurgy (1974).

2. Analysis of Temperature Distribution Considering Latent Heat and Molten Pool in High Energy Density Beam Welding

    Y. Arata and H. Kanayama; Proc. 2nd Inter. Symp. of JWS (1975) 21.

3. Terms and Definitions for Electron Beam Welding, Laser Welding and Laser Cutting Used in Japan.

    Y. Arata; IIW Doc. IV-229-77 (1977).

4. New Weldability Concept for EB-Welds and Others

    Y. Arata, N. Ohji, N. Khosai and K. Nishiguchi; 2nd Inter. Colloquium for Electron Beam Welding and Melting (1978).

5. Basic Phenomena in High Energy Density Beam Welding and Cutting

    Y. Arata; First European Conf. on Cineradiography with Photons or Particles (1981).

6. Basic Characteristics of Large Output High Energy Density Heat Sources

    Y. Arata; The First Inter. Laser Proc. Conf. (1981).

7. Development of Ultra-High Power Energy Density Heat Source and Weldability of Thick Plate Materials for Fusion Reactor

    Y. Arata; Keynoto, AWS (1981).

8. Caratteristiche fondamentali di sorgenti termiche ad elevata densitá di energia di grande potenza
   Y. Arata; Rivista Italiana Della Saldatura 6 (1982) 343.

9. Some Topics of the Development in New Welding Technology in Japan
   Y. Arata, K. Horikawa; Invited Report in 30th Annual Convention of the Australian Welding Institute Welding Technology 82 (1982).

10. Dynamic Behaviour of Beam Hole during High Energy Density Beam Welding
    Y. Arata, N. Abe and T. Oda; 3rd Inter. Colloquium on Welding and Melting by Electrons and Laser Beam (1983).

11. Some Fundamental Beam Hole Behaviour during Ultra High Energy Density Beam Welding
    Y. Arata, N. Abe and T. Oda; IIW Doc. IV-374-84 (1984).

Yoshiaki Arata

1925   Born May 22 in Kyoto Prefecture, Japan.

1949   Graduation from Osaka University.
Kusumoto Award of Osaka University.

1956   Associate Professor of Osaka University.
Dr. of Engineering.

1964   Professor of Osaka University.

1965   Visiting Professor of Ohio State University, U.S.A.

1972   Professor of Welding Research Institute and Faculty of Engineering, Osaka University.
Member of the Council of Osaka University.

1975   First President of the High Temperature Society of Japan.
Tanigawa/Harris Prize in the Japan Institute of Metals.

1977   Director General of Welding Research Institute, Osaka University.
Japan Welding Society Award for Outstanding Thesis.

1980   Consulting Professor of Harbin Technical University, China.
Goldschmit-Clermont Prize of the International Institute of Welding (Gold Medal).
General Director's Prize of Science and Technology Agency at Scientific and Technical Film Prize (Japan).
Prize of the 18th Japan Industrial Film Festival (Highest Prize for Scientific and Technical Films).

1981   Director of Research Center for Ultra High Energy Density Heat Source, Osaka University.
Gold Prize of the Polish Mechanical Engineering Society (SIMP).
Letters of Appreciation from the International Laser Processing Conference (ILPC).

1982   Gold Badge of the Order of Merit of the Polish People's Republic.
Honorable Member of Welding Alumni, Ohio State University, U.S.A.

1984   Award of Merits for Distinguished Educational Contribution from Minister of Education.
Honorary Professor of Shanghai Jiao Tong University, China.
Honorary Professor of X'ian Jiao Tong University, China.

1985   Director of Advanced Material Processing International (AMPI, JSPW).
Japan Academy Prize.
Golden Honorary Badge of Governing Council of General Technical Organization ("NOT"), Poland.

- Specialty: Welding Engineering and High Temperature Engineering

- Research Papers and Reviews: About 550

- Books: "Plasma Engineering," Nikkan Kogyo Shinbun Sha, 1965
  "Welding Engineering," Asakura Shoten, 1980
  "Fundamental of Welding Method," Sanpo, 1979
  in "Welding Handbook," Maruzen, 1977
  in "Electron and Ion Beam Handbook," Sanpo Shuppan, 1979
  "Development of Ultra High Energy Density Heat Source and Its Application to Heat Processing," Okada Memorial Japan Society for the Promotion of Welding (JSPW), 1985

- Patents: 54 (19 under application)

- Others: Special Volume of the Journal issued from ISTITUTO ITALIANO DELLA SALDATURA (Welding Society of Italy), No. 6, 1982

Index

A

618

620

622

625

X